PROBABILITY
AND
STATISTICS

PROBABILITY
AND
STATISTICS

KEVIN J. HASTINGS
Knox College

ADDISON-WESLEY
ADVANCED SERIES
IN STATISTICS

519.2
H

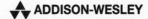 **ADDISON-WESLEY**

An imprint of Addison Wesley Longman, Inc.

Reading, Massachusetts • Menlo Park, California • New York • Harlow, England
Don Mills, Ontario • Sydney • Mexico City • Madrid • Amsterdam

Senior Editor: Julia Berrisford

Associate Editor: Jennifer Albanese

Production Services: Diane Freed

Production Supervisor: Peggy McMahon

Senior Manufacturing Manager: Roy Logan

Associate Marketing Manager: Benjamin Rivera

Cover Design Supervisor: Meredith Nightingale

Cover Designer: Darci Meehall

Compositor: Intercontinental Photocomposition, Ltd.

Technical Art Illustration: George Nichols

Library of Congress Cataloging-in-Publication Data

Hastings, Kevin J.
    Probability and statistics / by Kevin J. Hastings. -- 1st ed.
      p. cm.
    Includes bibliographical references (p   -) and index.
    ISBN 0-201-59278-9
    1. Probabilities.   2. Mathematical statistics.   I. Title.
QA273.H375 1997   96-11393
519.2–dc20   CIP

2 3 4 5 6 7 8 9 10 MA 999897

To my loving wife and child, who bring joy and laughter to my life

PREFACE

It is not an easy decision to write a textbook. You can be motivated by the negative point of view that there is not a book out there on your subject that is worth using. But this is not true in the case of probability and statistics. Or you can believe, enthralled by your own ego, that you have the definitive approach to which the world simply must be exposed. I hope that I had no such motivations. For me the reason to write this book had a lot to do with the joy of writing itself, and with the challenge. I did go so far as to think that I had several productive new ideas of dealing with an established topic that might interest and enlighten people and could make a worthwhile addition to the great library of probability and statistics.

The most important audience for a textbook is the students, and so I have at all times attempted to write with the student reader in mind. This does not mean talking down to the reader; quite the opposite, a proper style acknowledges the reader's intelligence and willingness to work. It does mean writing in a lively style, involving the reader in the process, avoiding obscurity and ambiguity, and not fearing the occasional use of humor. I have a little sermon to preach here on the subject of mathematical writing style. An early reviewer complained rather bitterly about what he or she considered my excessive use of the pronoun *we*, as in "we can now see that ...," but I disagree. This style makes us, the reader and I who are studying the objects of probability and statistics together, the subjects of sentences. The indirect passive style makes the objects of study into subjects and pushes the reader into the background, as in "Random variables are said to be independent if ..." This style is traditional in mathematics, but it is distinctly dry and uninvolving. It also makes sentences weak and unnatural through the use of passive verbs like *are*. Since I am at the mercy of my own imperfections and years of countertraining, I have not nearly attained my ideal of avoiding the passive voice completely throughout this long book. But I hope that the more active style that I have attempted, and other aspects of the book, will appeal to both students and instructors.

Let's turn now to issues of content and pedagogy. When I first started to think about this project, I had several main ideas in mind, the totality of which would make this book a little different from others in the area: (1) a full exploitation of matrix algebra and vector calculus; (2) an interactive, problem-solving approach; (3) a fuller coverage of graphical methods in data analysis; (4) computer laboratory investigations to augment the usual pencil-and-paper exercises and straightforward data analysis exercises using statistical packages; and (5) suggestions for student research projects. I want to elaborate on these features one by one.

It has always seemed to me that instructors do themselves and their students a disservice by attempting to package courses to be as independent of one another and prerequisite-free as possible. This gives students an unrealistic view of subject areas as being separate, rather than interdependent, and it gives them no incentive to carry knowledge, concepts, and techniques from one course to another. In our particular case, there are many ways to use matrix algebra and multivariate calculus to simplify, clarify, and extend results in probability and statistics, but most books do not fully exploit them, fearing perhaps that to do so would restrict the audience too much. I have used matrix analysis and multivariable calculus freely in this book, and so I assume that the student has had previous courses in these subjects. When it seemed appropriate, I reviewed the results I needed in context. I have included a short appendix (Appendix E) that summarizes briefly some of the most useful results in matrix algebra.

The second main theme of this book is harder to define. My philosophy is that the students need to be active participants in their education, not passive recipients. I tried every device I could think of to encourage this, including the stylistic one already discussed, and the springboard appendix to independent study. For example, I included numerous data sets, both in examples and exercises, which present the problem solver with a mass of numbers and an ill-stated question. The challenge is to make sense of the question, to get a feeling for what the numbers are saying, to construct a statistical procedure, and interpret the results in terms of the original problem. Every section includes several self-check questions for the reader to answer, as a device to monitor understanding. I left some of the theoretical material to the exercises, instead of laying out everything in a nice, neat package. I also let the students discover some of the cases of hypothesis tests and confidence intervals by solving problems instead of plodding mulelike through textual exposition of all of them. And probably the most important example of the implementation of the philosophy is the problem sets, which contain quite a few exercises that my own students have found challenging and thought-provoking.

The third main idea involves graphics. Statistics does not begin and end with analytical methods. Statistical consultants are coming around to the view that one of their most important roles is to present information to their clients that is understandable and convincing, and that using graphics is a good way to accomplish this. Moreover, it is now generally understood that the statistician must be concerned not only with graphical *methods*, but also with *principles* of good graph construction. Probability and statistics books have been successful at illustrating the methods, and this text does not attempt to break much new ground in that area. But the construction and design ideas are understressed in other books (perhaps because they don't seem mathematical enough), and I have given these more than the usual amount of attention. Appropriate graphs are regularly included in the more analytical chapters to solidify intuition.

The fourth of the distinctive qualities that I have given to the book is the set of computer investigations. I have included ten of them, in Appendix C, and have ideas for quite a few more. You the instructor may already have some of your own, or your imagination may suggest others by looking at the ones I have created. Now that the technology is widely available, it is possible to look at questions that we could not easily look at before. Parameter sensitivity, robustness against outliers, empirical power comparisons, and asymptotic properties are areas in which technology can assist understanding, enable the user to branch off in new directions, or go beyond the barrier of computational intensiveness. Simulation is a necessary means to study some of these problems, but I have found in my teaching experience that more can be said for simulation: Doing simulations helps the student better understand the fundamental concepts of randomness, random variables, and distributions. Hence simulation is not just a means but an end. In general, when used appropriately, technology is an indispensable means of enhancing education. These lab projects assume some familiarity with *Mathematica*, but they also give detailed assistance on the necessary commands.

And the fifth main idea is a lengthy appendix (Appendix D) to suggest and support projects for student investigation. Many books of this genre use the theory of probability mostly to set up statistical inference and do not take the opportunity to illustrate the beautiful applications of probability itself. I have chosen several applications (Markov chains, reliability, portfolios, and queueing) that I like and have worked on in the past, which the instructor could assign as small research projects culminating in an oral presentation and a written report. With a reluctant sigh, I decided to group these applications in Appendix D rather than integrating them, mainly because the book was getting rather long. There are suggestions for research in other areas of statistics such as Bayesian statistics and time series analysis, with references.

Apart from the differences discussed here, the organization and content of the book are fairly standard, and the instructor should not have much trouble adapting this text to an existing course. Here are some other less consequential ways in which this book differs from some others. There is a bit less coverage here than usual on the theoretical properties of point estimators and hypothesis tests. I preferred to devote more space to matrix methods and graphical data analysis and to extend discussion of the design (as opposed to just the analysis) of experiments. Nonparametric methods are integrated so as to present them along with the problem for which they are meant, rather than marginally.

Following is a quick run-through of the ordering of topics and other items worthy of note when preparing to use this book for a course. One important general observation is that I frequently use the device of giving sneak previews of later topics, and returning to earlier topics. I think that the reinforcement resulting from multiple exposure helps students to learn better and retain longer. A drawback of this construction, which is more like a network than a line, is that it is dangerous to skip things because there is so much interdependence. Here are some specifics about Chapters 1–6 on Probability.

Chapter 1 is meant to throw the reader into the deep water of the basic concepts of probability quickly: sample spaces, random variables, distributions, conditioning, and independence. The Law of Total Probability is highlighted in the text and in the exercises, and a rather general definition of the concept of distribution is used. Especially with regard to random variables and their distributions, the strategy is to expose students to these concepts and let them be puzzled for a while as they do exercises. Then they will come back to them in other contexts later in Chapters 2, 3, and 4 on discrete distributions, continuous distributions, and independence, respectively. For instance, my students have found that they have had trouble early with the idea of a continuous probability distribution, but by the end of Section 3.2 something had clicked, and they came away with a deep understanding that they put to use later in the book.

Chapter 2 brings the student to the point of familiarity with basic combinatorial probability (focusing on sampling), discrete distributions and applications related to binomial experiments, phenomena modeled by the Poisson distribution, and discrete expectation. The treatment is fairly standard here, with the possible exception of the extra thoroughness of the material on Poisson processes.

The third chapter explores continuous probability distributions and expectation, highlighting the gamma family and the normal distribution. There are vector expectation results given in the second section which are used occasionally later. The multivariate normal distribution is also introduced there, which becomes a recurring theme as the exposition progresses. We return to it in Sections 4.4 and 5.4 as we learn more about conditional distributions and transformations.

Chapter 4 is a straightforward discussion of conditional distributions and independence of random variables, including a couple of nice examples of the application of the Law of Total Probability for expectation. The third section on covariance and correlation includes a discussion of covariance matrices, and a result about the covariance of a constant matrix times a random vector, an unusual topic for books at this level.

The material in Chapter 5 on transformations is fairly mainstream, including the c.d.f. technique with an emphasis on the simulation theorem, multivariate change of variables, moment-generating functions, and applications to transformations of normal random variables. But there is consistent attention given to multivariate results and methods, especially as regards quadratic forms of normal vectors. These ideas are used heavily later in Chapters 10 and 11 on regression and the analysis of variance.

In a short sixth chapter on large sample theory, the Weak and Strong Laws of Large Numbers and the Central Limit Theorem are discussed. I have carefully tried to clarify the distinction between weak and almost everywhere convergence. The weak law depends on Chebyshev's inequality. The Central Limit Theorem is treated in the usual way. Here, and earlier in the section of Chapter 5 on normal transformations, a sneak preview of statistical analysis using the sample mean is given.

As for Chapters 7–12 on Statistics, Chapter 7 leads off the study with a treatment of random sampling and the elementary summary statistics. We then proceed to graphical methods, covering the usual types of graphs, including histograms, dot and box plots, and normal scores plots. But the chapter ends with a rather unique discussion of good graph-making techniques, done from a problem-solving point of view: If a graph is to tell a story, how should it be constructed, and how should the fine details be implemented so as to enhance, or at least not to obscure, the story to be told?

Chapter 8 deals with parameter estimation, highlighting maximum likelihood and the notions of unbiasedness and precision of estimators. Sufficiency is included, as is a short treatment of confidence intervals, which should dovetail well with the idea of precision of an estimator.

In the ninth chapter I have tried to cover as efficiently as possible the classical normal theory tests for means, variances, and proportions. But the chapter takes a conceptual approach, rather than a smorgasbord approach. The reader is encouraged to go at data analysis problems by asking questions: What am I trying to find out about a certain population? and What common-sense summary measures would be appropriate to look at? The first section introduces the basic terms and ideas and includes a discussion of both the large sample test for the proportion and the sign test. This is an attempt to get the reader to think of a hypothesis test as a problem to be solved creatively, rather than a standard technique to be applied. The rest of the chapter passes through one- and two-sample location tests, including integrated subsections on the Wilcoxon and Mann–Whitney tests, and then the chapter proceeds to dispersion tests. In Section 9.3 where dispersion tests are covered there is also some material on Mood's nonparametric test. Beyond just giving the nonparametric procedures their fair due, I think that the integration of classical normal theory with some distribution-free theory stresses to the student the fact that assumptions are important and that testing hypotheses is a creative, problem-solving enterprise. The chapter ends with a brief introduction to the likelihood ratio testing criterion.

Chapter 10 on regression and correlation takes more than the customary advantage of multivariate theory from the first part of the book. It considers the regression model as a matrix linear equation in the unknown parameters, and in so doing previews the following chapter. The first two sections set up the problem and cover least squares parameter estimation in roughly the usual way. Section 10.3 contains very powerful theorems, derived with matrix techniques, on the distribution of estimators and sums of squares. With these in hand, the student is ready to apply and understand the statistical inference procedures. There is some discussion of residuals and diagnostic checking in the fourth section of this chapter, using an example-oriented approach. Correlation analysis completes the chapter.

On a recent sabbatical I became rather interested in the area of experimental design, especially in the fact that despite the appearance of being a disjointed amalgamation of problems and techniques, it is thematically unified by the basic structure of the linear model. So, I decided to stress this idea in Chapter 11, and as well, to include a longer discussion of data gathering in designed experiments than is normally done (Section 11.1). We study one-factor problems, including random effects models and blocking, and two-factor problems including random effects. The topic of Latin Square designs is left as one of the research project possibilities in Appendix D. There is heavy usage throughout the chapter of the results on quadratic forms from Chapter 5.

Chi-square goodness-of-fit testing occupies most of our attention in Chapter 12. We begin by looking at the simple multinomial fit problem with all category probabilities known, then move on to testing the fit of distributions, and problems in which parameters are estimated. The Kolmogorov–Smirnov procedure is also discussed here. The chapter closes with a rather standard presentation of independence testing on contingency tables using the classical chi-square statistic. The nonparametric runs test for problems involving trends is also covered.

The supporting material in the appendices includes statistical tables, a short guide to the *Minitab* statistical system, computer exercises with hints on using *Mathematica*, research project ideas, and a linear algebra review. Answers to many of the exercises follow. A complete solutions manual is available.

I would like to thank the people at Addison-Wesley, particularly Julia Berrisford, for their help and their understanding when I was just a tad past our agreed upon deadlines. Much of whatever credit the book earns is really due to the great teachers and colleagues of probability and statistics that I have known and worked with: Rothwell Stephens at Knox College, Stan Pliska and my dissertation adviser, Erhan Cinlar, at Northwestern, Marcel Neuts and others at the University of Delaware, and Russ Lenth at the University of Iowa. The many students who I have had, including those whose projects are mentioned in the text and those who helped me with the solution manual, may have made as great a contribution to my education as I have to theirs, and for that I thank them. I am very grateful to the reviewers of the manuscript: Bernice Auslander, University of Massachusetts, Boston; James Conklin, Ithaca College; Joseph Glaz, University of Connecticut; H. Allan Knappenberger, Wayne State University; Karen Messer, California State University, Fullerton; Don Ridgeway, North Carolina State University; Lyndon Weberg, University of Wisconsin, River Falls; and Douglas Wolfe, The Ohio State University. Most importantly, I thank my wife, Gay Lynn, whose love and support have helped carry me through the second half of this project and who also helped with a good deal of the technical word processing. And to my precious baby, Emily Marie, who gives my life so much joy, thanks for going to sleep at 8:00 so that Mommy and Daddy could work.

I sincerely hope that this text will be a positive contribution to the field. In some ways it is on the demanding side, but not unreasonably so if the student has honestly attempted to commit calculus and linear algebra to long-term memory. More than anything, I hope that the approach that I decided to take will spark the same kind of excitement about probability and statistics that I felt when I was first learning. The subjects deserve no less.

KJH
Galesburg, IL

CONTENTS

10 Regression and Correlation — 435

11 Experimental Design and Analysis of Variance — 489

12 Goodness-of-Fit — 535

A Statistical Tables — 571

CHAPTER 1

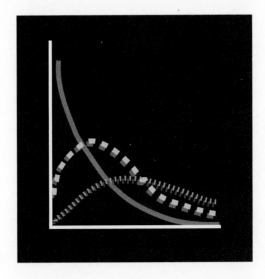

SAMPLE SPACES AND
RANDOM VARIABLES

1.1 | Introduction and Examples

Our investigation of probability and statistics must begin with an understanding of the word *random*. The Second College Edition of *Webster's New World Dictionary* makes several attempts at a definition, including: "without careful choice," "haphazardly," and the rather contradictory pair "not uniform" and "pertaining to a set of items, every member of which has an equal chance of occurring." The last definition comes closest to what we need, although it is far too restrictive. The subjects of probability and statistics exist because of a need to understand events whose occurrence we cannot predict with certainty. However, we may know how likely these events are. So for us, the word *random* will pertain to an experiment or phenomenon whose result (or *outcome*, in the language of probability) remains uncertain until the experiment is performed or the phenomenon is observed. The most familiar and elementary random phenomenon is probably the flip of a coin. It is unknown prior to the flip which of the two possible outcomes, "head" or "tail," will occur, but the outcome is observable after the flip. Note that we needn't assume that each possible outcome of a random experiment is as likely as each other outcome, just as we needn't assume that our coin is fair.

Here are several examples of random phenomena and their possible outcomes.

1. Will I or won't I be interrupted by a phone call while I am writing today? Until such a phone call comes or I complete my writing for the day, I will not know the answer to this question. Thus the event of being interrupted becomes a random one, with possible outcomes "yes" or "no." It turned out that I was interrupted (an outcome that seems to have high probability).

2. How much will IBM common stock go up today, January 13, 1992, on the New York Stock Exchange? We will not know what will happen until the close of trading, so this situation fits our idea of a random phenomenon. There are many possible outcomes, which are numbers usually reported as multiples of eighths. The observed outcome turned out to be $-5/8$. Past observation of the performance of the stock might give information on the likelihoods of outcomes like $-5/8$.

3. What will be the winning three-digit number today, January 13, 1992, on Michigan's Daily-3 Lottery game? Again, we cannot know until the number is selected by the lottery authorities; after the number is drawn, the outcome is known and it is of the form abc, where a, b, and c are digits from the set of digits 0 through 9. The winning outcome happened to be 517. Theoretically, each three-digit number should be as likely as any other if the mechanism for drawing numbers is fair.

4. What will be the official high temperature in Atlantic City today? Here, outcomes are numerical temperatures. (The observed outcome was 57°F.) At least ideally, if we could measure temperatures with perfect accuracy on a continuous scale, the possible outcomes would form some (uncountably infinite) interval on the real line. One would guess that temperatures nearer to the normal temperature for Atlantic City at this time of year would be more likely than extremely warm or cold temperatures.

—— **?Question 1.1.1** Think of at least three other random phenomena. What are their possible outcomes?

The purpose of this first section is to introduce you to the main concepts of probability, such as random phenomena, very informally, so that you begin the process of acquiring intuition. Then as the next few chapters unfold, we will look at the concepts and their applications more rigorously and in more detail.

Problems involving random phenomena or experiments entail collections of possible *outcomes*. These outcomes are indivisible, meaning that they cannot be broken down into more primitive components. We will refer to the set of all possible outcomes of the experiment as the *sample space* of the experiment. For instance, the sample space associated with the phone interruption phenomenon cited earlier is {yes, no}. As another example, a lottery game in which four different two-digit numbers are selected includes outcomes such as {02, 65, 73, 22}, {87, 01, 94, 45}, and so forth. A formal way to write this sample space is

$$\{\{x_1, x_2, x_3, x_4\} \mid x_i \in \{0, 1, \ldots, 99\}, x_i \neq x_j, \; i, j = 1, 2, 3, 4\}.$$

In Chapter 2 we will learn how to count the number of possible outcomes of an experiment like this.

—— **?Question 1.1.2** Write a formal set-theoretic description of the sample space for the Michigan Daily-3 Lottery game described earlier.

Notice from the examples that outcomes may sometimes be numbers, other times collections or sequences of numbers, and still other times nonnumerical character strings. Because outcomes can take on many different structures depending on the random phenomenon being considered, we will construct a framework for probability in which outcomes are left abstract. The idea of a universal sample space of possible outcomes, however, will remain the consistent unifying theme.

Indivisible outcomes are the building blocks of *events*, which are subsets of the sample space. In reference to some of the examples, the event that IBM rises in value by at least 2/8, the event that 23 is one of the winning Michigan lottery numbers, and the event that the high temperature in Atlantic City is at least 50° all consist of more than one outcome. We will also allow events to contain no outcomes or one outcome.

—— **?Question 1.1.3** For each of your phenomena from Question 1.1.1, give an example of an event.

Probability is a measure of the likelihood of events. Since events are sets of outcomes, it should be enough to define probability for outcomes. The probability associated with an event should then be the total of the probabilities of the outcomes in that event. This last observation is essentially true, although in the kind of idealized experiments mentioned here where the set of outcomes can be uncountably infinite, we need to think more carefully about what the total of outcome probabilities is and to consider whether it even makes sense to give outcomes nonzero probability.

To see the difference between finite (or countable) sample space models and uncountably infinite models, consider this example. A runner in the 100-meter dash will not always complete the run in the same amount of time. One instance may result in a time of 10.3 seconds, another 10.5, another 10.1, and so on. We could propose two probabilistic models.

1. If the stopwatch that we use to time the run is accurate only to the nearest 1/10 of a second, and our runner cannot run a faster time than 10 seconds or a slower time than 12 seconds, then the outcomes that form the sample space are {10.0, 10.1, 10.2, ..., 11.9, 12.0}. An event is a subset of this set, for instance, those times greater than 10.5. The sample space has finitely many elements, and therefore events also have finite cardinality. A simple model would assume that all 21 outcomes are equally likely, so we could measure the probability of an event by the ratio of the number of outcomes in the event to 21. We could arrive at the same probability by giving all outcomes probability 1/21, then summing the constant 1/21 over the outcomes in the event. Using either method, you can check that the probability of the event that the time is greater than 10.5 is 15/21. Of course, the assumption that probability is distributed uniformly among outcomes is a very severe one and is probably not valid for this random phenomenon. We will be studying nonuniform distributions of probability shortly.

2. On the other hand, if we had a perfect stopwatch, we could measure the runner's time to infinite accuracy, and the sample space would then consist of the uncountably infinitely many points in the interval [10, 12]. In an analogous simple uniform model, no point is to be preferred over any other. If we gave any outcome a nonzero probability p, then all of them would have to be given probability p. Events with infinitely many points would then have infinite probability, which is an undesirable situation.

—— **?Question 1.1.4**   Think carefully about the next question before moving on. How would you sensibly assign probabilities to intervals in the second model?

You should have decided in the last question that it is sensible under the assumptions to say that the probability of a region is the ratio of its length to the total length of the interval, which is 2. Then, for instance, the event [11, 12], which has half the length of the sample space [10, 12], would be given probability 1/2. Analogously, for sample spaces that are regions in the plane, a uniform model would define the probability of a set to be the ratio of its area to the area of the sample space.

Probabilities can come from several sources. In a lottery drawing, steps are usually taken to ensure that numbers or sequences of numbers drawn are equally likely. So, *theoretical considerations* provide one way to define and compute probabilities of events. Another way is to assign probabilities to events *subjectively* based on judgment and past experience. You hear such probability assignments frequently. If someone says to you, "I'll lay you 2 to 1 odds that we get a tax increase this year," the speaker is using whatever current information is available about legislative action, his memories of what has happened in past years, and probably a dose of pessimism to proclaim that the probability of a tax increase is 2/3. But he has no firm foundation other than a belief for these assignments of probabilities to events. Others might choose different probability values. We prefer to deal with situations

Table 1.1

Distribution of Student Time Evaluations

	Much More	**Somewhat More**	**About the Same**	**Somewhat Less**	**Much Less**
No. Responding	5	9	8	1	0
% of Class	21.7 %	39.1 %	34.8 %	4.3 %	0 %

where, with the same information in hand, rational individuals would assign probability in identical ways. For this reason, we will not discuss subjective probability in this text.

A third way of producing an assignment of probability to outcomes appears when past data suggest *empirical probabilities*. A survey I recently gave to one of my classes asked the students to rate the class on the basis of time spent compared to other classes, by selecting one of several responses. The frequency of student responses and the associated percentage of all respondents for each category are given in Table 1.1.

A bar chart, or a *histogram,* showing the distribution of results is given in Fig. 1.1. The heights of the bars are the proportions of the students who responded in the five possible ways. If we consider the students in this class as a representative sample of all students at my college who could have taken the class (a questionable assumption, I admit), we could estimate the proportion of all students who would spend much more time by the observed proportion .217. This number is the empirical probability of that outcome. Similarly, we could estimate the proportion of students who would spend at least somewhat more time by the observed proportion in the class who did, namely .391 + .217 = .608. In this way events also can be given estimated probabilities from the empirical data.

The last fundamental concept of probability that we will discuss is the idea of a *random variable*. A *random variable* is a function that gives a numerical value to each outcome of a random experiment. In the phone interruption example, we could let a random variable X encode the outcomes yes and

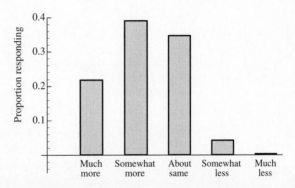

Figure 1.1 Histogram of student responses

no by 1 and 0, respectively. The formal definition of this random variable would be $X(\text{yes}) = 1$ and $X(\text{no}) = 0$. For the IBM price change example, we could define a random variable X to be the observed price change. The distinction between the random variable and the experimental outcome itself can become blurry in this case, because if ω denotes an outcome, then $X(\omega) = \omega$. Remember, however, that the random variable is the function and the outcome is its input. In the Michigan lottery game example, a random variable might be defined as the smallest digit among the winning three. As in our phone example, this random variable produces an output value with a structure entirely different from the outcome on which it operates. The common thread among all examples is that the domain of the random variable is the sample space and that its range is a set of numbers.

—— **?Question 1.1.5**   Consider the random phenomena that you constructed in Question 1.1.1. Define at least one random variable relating to each example.

The purpose of this section has been to induce you to think as a probabilist and to introduce informally the terms *experiment, outcome, sample space, event, probability,* and *random variable*. The next few sections formalize some of the ideas discussed here, as well as other ideas that will lay the foundation for the rest of this book.

1.2 | Axioms of Probability and Their Consequences

Like the terms *point* and *line* in geometry, the terms *random experiment* (or *phenomenon*) and *outcome* will be left with no formal definition, but with only our intuitive understanding of what they are. Our definitions begin at this point.

DEFINITION 1.2.1

The *sample space* Ω of a random experiment is the collection of all possible outcomes. An *event A* is a subset of the sample space, that is, a set of outcomes.

In the first section we discussed probability informally. Three main points follow from that discussion. First, the probability of an event is computed according to its share of the sample space; hence the likelihood of the whole sample space should be 1. Second, probability gives nonnegative values to likelihoods of events in sample spaces. Finally, like cardinality, length, and area, probability should be additive on disjoint sets. We will use these three observations in the formal definition of a probability measure.

DEFINITION 1.2.2

A *probability measure* on a sample space Ω of a random experiment is a function $P[\cdot]$ that maps events in Ω to real numbers such that

(a) $P[\Omega] = 1$,
(b) $P[F] \geq 0$ for all events F,
(c) $P[E_1 \cup E_2 \cup E_3 \cup \cdots] = P[E_1] + P[E_2] + P[E_3] + \cdots$,

where in (c), $E_1, E_2, E_3, \ldots$ is a finite or countably infinite sequence of events such that any pair of the events is disjoint.

To get a firmer grip on the definition, let us consider a few examples that typify later work. The first is the simplest possible case, where the sample space is finite and the outcomes have the same probability. In the second case, the sample space is infinite but countable and outcomes do not have the same probability. For these first two examples, it is very easy to create a probability measure by assigning probability to outcomes in such a way that the total probability over all outcomes is 1. The third example is a radical departure from the previous two, in the sense that the sample space is uncountably infinite and individual outcomes cannot be given nonzero probability. What unifies the examples is the fact that the axioms of probability (a)–(c) are always satisfied. By the way, we will occasionally breach strict notational correctness in favor of simplicity in the following way: Instead of using the correct notation $P[\{\omega\}]$ for the probability of the event consisting of the single outcome ω, we will simply write $P[\omega]$.

Example 1.2.1

Finitely Many, Equally Likely Outcomes   Suppose that two U.S. Senators among four—Bradley, Glenn, Moynihan, and Simon—are to be appointed as a committee by the Democratic leadership to lobby their Republican colleagues for some bill. Then the sample space of all possible committees has six outcomes, which we can label by the first letters of the senators' names:

$$\Omega = \{\{B, G\}\{B, M\}\{B, S\}\{G, M\}\{G, S\}\{M, S\}\}.$$

For example, the event E, which can be described in words as "Bradley is on the committee," is the subset of Ω consisting of the outcomes $\{B, G\}$, $\{B, M\}$, and $\{B, S\}$.

A valid probability measure can be defined on this space by supposing that no particular committee selection is favored over another, so that all six selections are equally likely. Define the probability of each outcome to be $1/6$, and the probability of an event to be the total of the probabilities of the outcomes in the event, that is, $P[E] = \sum_{\omega \in E} P[\omega]$. For instance, the probability of the event E that Bradley is selected is $3/6$. Because outcomes are equally likely here, the probability of an event is the number of outcomes in the event divided by the number of outcomes in the sample space (6). The probability of the whole sample space is $6/6 = 1$, as required by the first axiom; clearly event probabilities are nonnegative, so that the second axiom is also satisfied. It is easy to show that the third axiom holds, because if events $E_1, E_2, \ldots, E_n$ are pairwise disjoint—that is, $E_i \cap E_j = \emptyset \; \forall \; i \neq j$—then

$$P[E_1 \cup E_2 \cup \cdots \cup E_n] = \sum_{\omega \in E_1 \cup E_2 \cup \cdots \cup E_n} P[\omega]$$

$$= \sum_{\omega \in E_1} P[\omega] + \sum_{\omega \in E_2} P[\omega] + \cdots + \sum_{\omega \in E_n} P[\omega]$$

$$= P[E_1] + P[E_2] + \cdots + P[E_n]. \tag{1.1}$$

Nothing is inherently correct about our assumption of equally likely outcomes. The committee selection process could be stacked in some way in favor of Bradley, for example; the first three outcomes could be given total probability $2/3$ and the last three could be given total probability $1/3$. This requirement could be achieved by assigning probability $2/9$ to each of the first three outcomes, and $1/9$ to each of the last three. Physically, one could implement such a mechanism by putting in a hat two copies of each of the first three outcomes, together with one copy of each of the last three outcomes, and then drawing an outcome at random.

?Question 1.2.1   Why need we concern ourselves only with pairwise disjoint unions of *finitely* many events in the computation (1.1) in the last example?

Example 1.2.2   **Countably Infinite Sample Space, Unequally Likely Outcomes**   Flip a coin indefinitely, until the first tail comes up. Then the sample space can be written

$$\Omega = \{T, HT, HHT, HHHT, HHHHT, \ldots\}.$$

Because it is possible to flip 100, 1000, even 1 million times, there is no upper limit on the number of required flips; hence this sample space is countably infinite. To define a probability measure, we can again assign probabilities to outcomes and let the probability of an event be the sum of its outcome probabilities. Because each outcome requires a sequence of events, a multiplication rule for sequences, which we will study later, makes it reasonable to assign probabilities to outcomes as follows:

$$P[T] = \frac{1}{2}, \qquad P[HT] = \frac{1}{2} \cdot \frac{1}{2} = \frac{1}{4}, \qquad P[HHT] = \frac{1}{2} \cdot \frac{1}{2} \cdot \frac{1}{2} = \frac{1}{8}, \ldots.$$

Since the infinite geometric series $\frac{1}{2} + \frac{1}{2}^2 + \frac{1}{2}^3 + \cdots$ adds to 1, the probability of the sample space is one. You should justify to yourself that the other axioms of probability hold once again.

Example 1.2.3

Uncountable Sample Space   A local pizza parlor promises that it will deliver orders in a half hour and, if the delivery is more than 5 minutes late, the pizza will be free. Various unexpected factors affect the delivery time, so that one cannot perfectly predict how long a particular pizza will take. Thus we seem to have a random experiment suitable for a probabilistic model. Let us measure time in units of 10 minutes and suppose that an experimental outcome corresponds to the amount of time by which the pizza is late. If, for example, past data indicate that all deliveries are between 10 minutes early and 10 minutes late, then we can define the sample space to be $\Omega = [-1, 1]$, where negative numbers indicate early delivery and positive numbers indicate late delivery.

We would like to define a probability measure on the sample space in a sensible way. We will use the idea of area under a curve, generalizing from a histogram of data.

Suppose that we have observed many such deliveries and have recorded the number of them that landed in, say, 16 evenly spaced subintervals of $[-1, 1]$, each of width $1/8$. A bar chart with bar heights equal to the absolute numbers of observations in each subinterval would give information about the distribution of the data, but we can do something a little better. That is, we can make the area of a bar equal to the estimated probability that an observation falls into the associated subinterval. We proceed by making the bar heights equal to the relative proportions of observations that occur in the subintervals, divided by the subinterval width $1/8$:

$$\text{Subinterval bar height} = \frac{\text{observations in subinterval}}{(\text{total observations}) \cdot (\text{subinterval width})}. \quad (1.2)$$

Then the area of a bar is its height times the subinterval width, which is the observed proportion of observations in the subinterval, an empirical estimate of the probability that an observation falls into the subinterval.

So, playing on the idea of area as a measure of probability, we find some continuous curve that approximates the bar graph reasonably well. Figure 1.2 illustrates a typical bar chart, with a very simple continuous curve, namely an inverted parabola, superimposed. The function that gives rise to the parabola

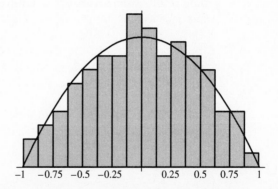

Figure 1.2   Pizza Lateness Distribution

has the form $f(x) = -ax^2 + b$ and its x-intercepts would be located at -1 and 1. Just as the total area of the bars is the total proportion falling in all subintervals, which must equal 1, the total area under the graph of f between -1 and 1 should be 1. You are asked to show in Exercise 16 at the end of this section that this implies that $a = b = 3/4$.

Now define the probability of an event to be the definite integral of f over that event, or over that subset of $[-1, 1]$. For instance,

$$P[\Omega] = \int_{-1}^{1} \frac{3}{4} - \frac{3}{4}x^2 \, dx = 1, \qquad P[\,[.5, 1]] = \int_{.5}^{1} \frac{3}{4} - \frac{3}{4}x^2 \, dx = \frac{5}{32},$$

$$P[[-1, -.5] \cup [0, 1]] = \int_{-1}^{-.5} \frac{3}{4} - \frac{3}{4}x^2 \, dx + \int_{0}^{1} \frac{3}{4} - \frac{3}{4}x^2 \, dx = \frac{21}{32},$$

$$P[\{.2\}] = \int_{.2}^{.2} \frac{3}{4} - \frac{3}{4}x^2 \, dx = 0.$$

Notice that the second probability in the first line is the probability that the pizza is at least 5 minutes late. (How could you describe in words the event in the second line?)

This construction does satisfy the axioms of probability. We chose the coefficients so as to make axiom (a) of probability hold. Axiom (b) holds because the integrand $f(x)$ is nonnegative over the set $[-1, 1]$. Axiom (c) follows from the additivity of the integral over disjoint regions.

—— **?Question 1.2.2**   Check the probabilities calculated in the last example. What integral represents the probability that the pizza is at least 5 minutes early? no more than 2 minutes late?

One of the advantages of an axiomatic development in mathematics is that the properties of the class of objects under study can be derived in general, then applied to particular examples. We now give a succession of important, basic results about probability, whose proofs rely only on the axioms and some results from set theory. Therefore these theorems are applicable to any examples we encounter throughout this book.

PROPOSITION 1.2.1

$P[\emptyset] = 0$.

■ **Proof**   Observe first that $\emptyset \cup \Omega = \Omega$, and $\emptyset \cap \Omega = \emptyset$. Therefore, by axioms (a) and (c) of probability,

$$1 = P[\Omega] = P[\Omega \cup \emptyset] = P[\Omega] + P[\emptyset] = 1 + P[\emptyset]$$

$$\Rightarrow \quad 1 = 1 + P[\emptyset].$$

Thus the empty set has zero probability.

PROPOSITION 1.2.2

For any event E, denote the complement of E by E^c. Then

$$P[E] + P[E^c] = 1. \tag{1.3}$$

Consequently, $P[E] = 1 - P[E^c]$ and $P[E^c] = 1 - P[E]$.

■ **Proof** The events E and E^c are disjoint, and their union makes up the whole sample space. Therefore, by axioms (a) and (c),

$$1 = P[\Omega] = P[E \cup E^c] = P[E] + P[E^c].$$

PROPOSITION 1.2.3

For any event E, $P[E] \le 1$.

■ **Proof** By axiom (b), $P[E^c] \ge 0$. Thus, by Proposition 1.2.2,

$$P[E] = 1 - P[E^c] \le 1.$$

PROPOSITION 1.2.4

For any events A and B,

$$P[A] = P[A \cap B] + P[A \cap B^c]. \tag{1.4}$$

■ **Proof** We leave this to the reader as Exercise 19.

PROPOSITION 1.2.5

If $E \subseteq F$ are events, then $P[E] \le P[F]$.

■ **Proof** As Fig. 1.3 indicates, we can decompose F into two disjoint subsets as

$$F = E \cup (F \cap E^c).$$

Then by axioms (c) and (b),

$$P[F] = P[E] + P[F \cap E^c] \ge P[E].$$

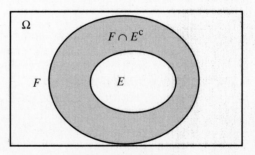

Figure 1.3 Subset Rule

PROPOSITION 1.2.6

If A and B are events, then

$$P[A \cup B] = P[A] + P[B] - P[A \cap B]. \qquad (1.5)$$

■ **Proof**   The event $A \cup B$ can be decomposed into disjoint subsets as $A \cup B = B \cup (A \cap B^c)$. (Draw a picture to justify this.) Hence, by axiom (c),

$$P[A \cup B] = P[B] + P[A \cap B^c]. \qquad (1.6)$$

But it is also clear that the two sets $A \cap B$ and $A \cap B^c$ are disjoint and that their union is A; hence by Axiom (c) again,

$$P[A \cap B^c] = P[A] - P[A \cap B]. \qquad (1.7)$$

Combining (1.6) and (1.7) finishes the proof.

This section has glossed over some technicalities, especially with regard to which subsets of the sample space qualify as events. When the sample space is finite, there is no difficulty in calling all subsets events. It is usually when the sample space is uncountably infinite that problems can occur. We will leave these issues for a more advanced course, except to say that we must assume that the complement of an event is an event and that unions (hence intersections) of countably many events are also events.

 **EXERCISES 1.2**

1. Consider the roll of two distinguishable six-sided dice. Describe the sample space that records the outcomes of both dice, and define a probability measure on it consistent with the assumption that the dice are fair. What is the probability that a sum of 10 or more is observed?

2. Referring to Exercise 1, describe the sample space that retains only the the sum of the two faces of the dice, and define a probability measure on that. Should all outcomes be equally likely? Find the probability that an even sum is observed.

3. Many results in probability have analogues in counting. For example, denoting the number of elements of a set F by $n(F)$, the analogue of Proposition 1.2.6 is $n(A \cup B) = n(A) + n(B) - n(A \cap B)$, assuming, of course, that A and B have finitely many elements. Give an argument that this formula is true.

4. In a lottery drawing of 1000 numbers, players pay for as many numbers as they want, and they win a prize if their number is drawn. Numbers can be selected by more than one player. If I pay for 100 numbers, you pay for 100 numbers, and between us we have 150 different numbers, construct an appropriate probability model to compute the probability that both of us win, and the probability that neither of us wins.

5. Prove that if A, B, and C are arbitrary events, then

$$\begin{aligned} P[A \cup B \cup C] = {} & P[A] + P[B] + P[C] \\ & - P[A \cap B] - P[A \cap C] - P[B \cap C] \\ & + P[A \cap B \cap C]. \end{aligned}$$

(*Hint:* First treat $A \cup B$ as a single set, then union with C.)

6. On a given broadcasting week, 80 television programs were surveyed. Forty were situation comedies. Thirty were about families. Thirty were reruns. Fifteen programs were sitcoms about families. There were five reruns about sitcom families. Four of the reruns about families were not sitcoms. Three of the sitcom reruns were not about families. If a television program is picked at random from this group, what is the probability that it is
   a. neither a rerun, nor a sitcom, nor about families?
   b. a first-run sitcom not about families?
   c. a rerun of a sitcom?
   (*Hint:* Use a Venn diagram to help count the number of elements in the sets of sitcoms, family programs, reruns, and all possible intersections.)

7. Two cards are dealt without replacement from a standard 52-card deck. Assuming that the deck is well shuffled, describe a sample space and a probability measure for this random phenomenon. Use your model to find the probability that a blackjack is dealt (i.e., an ace and a face card).

8. If $P[A] = .4$, $P[B] = .7$, and $P[A \cap B] = .3$, find $P[A \cup B]$, $P[(A \cup B)^c]$, $P[B \cap A^c]$, and $P[A \cap B^c]$.

9. What, if anything, is wrong with the following information: $P[A] = .3$, $P[B] = .7$, $P[A \cap B] = .4$?

10. Find $P[A^c \cup (A \cap B)]$, if $P[A] = .2$, $P[B] = .6$, and $P[A \cup B] = .7$.

11. A building in which your next class is located is 200 yards away. You walk toward it at some constant but randomly selected rate R, which we will assume is between 2 and 5 miles per hour, and no rate is to have any preference over any other rate. Find the probability that you will complete the walk in less than $1/25$ hour.

12. Define the *symmetric difference* operation on sets by

$$E \triangle F = (E \cap F^c) \cup (E^c \cap F).$$

   Show that $P[E \triangle F] = P[E] + P[F] - 2P[E \cap F]$.

13. True or false? $P[B \cap A^c] = P[B] - P[A]$. Explain. Are there any conditions under which the statement is true?

14. Show that
   a. if $E \subseteq F$, and E has probability 1, then F has probability 1.
   b. if $A \subseteq B$ and B has probability 0, then A has probability 0.

15. Three children are born into a family. Assuming that child genders are equally likely, construct a sample space and a probability measure that will model this phenomenon. Find the probability that at least one of the children is a girl.

16. In Example 1.2.3, verify the claim that the coefficients a and b of the quadratic function f are both $3/4$.

17. In the game of odds and evens, each player hides one hand behind his or her back, extending 0, 1, or 2 fingers, and then simultaneously the two players show their hands. Say that player A wins if the total number of fingers exposed is odd, otherwise player B wins. Construct a sample space and a probability measure to model this game, assuming that both players pick their number of fingers randomly and each choice of number

of fingers is equally likely. Find the probability that player A wins. Find the probability that player B wins by holding out an even number of fingers.

18. If the piecewise linear function shown in the figure is to be used to define a probability measure on the sample space $\Omega = [-2, 2]$ along the lines of Example 1.2.3, what value must b have? Find $P[[1, 2]]$.

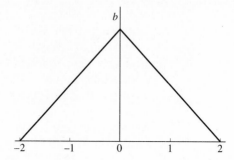

19. Prove Proposition 1.2.4.

1.3 | Random Variables and Distributions

1.3.1 Random Variables

In many probabilistic problems, we are interested in numerical-valued outcomes, or outcomes that can be coded into numerical values. Perhaps we observe among a group of ulcer patients the proportion whose condition improves under a dietary regimen. Or, we may observe a waiting line at a bank over a period of time and record its maximum length. In many cases, moreover, a single real number is not enough to describe the experiment fully. For example, an automobile tested in the city and on the highway would have two real numbers giving its gas mileage in the two situations. In a problem involving the game of bridge, four numbers giving the prevalence of each suit in a thirteen-card hand would all be of interest. In general, the numerical outcome may lie in some subset of n-dimensional Euclidean space $\mathbb{R}^n$. The notion of *random variable* allows us to study such problems. Figure 1.4 illustrates the meaning of the following definition.

DEFINITION 1.3.1 ─────────────────────────────────

Let Ω be a sample space, and let E be a subset of $\mathbb{R}^n$. A *random variable* X is a function from Ω into E. We call E the *state space* of the random variable X.

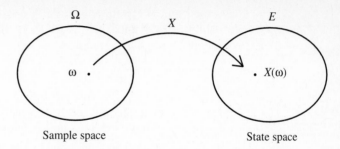

Figure 1.4 Action of a Random Variable

So, a random variable encodes an experimental outcome as a number, or a vector of real numbers in the multidimensional case. When a random variable has a multidimensional state space, we emphasize that fact by calling it a *random vector* and by using boldface style to denote it.

Often, we will be trying to compute the probability that the value $X(\omega)$ of a random variable lies in a specified subset A of the space of possible states. There is a set of outcomes B in the sample space that leads to this event, as shown in Fig. 1.5. We may write this set of outcomes as $B = \{\omega \mid X(\omega) \in A\}$, or $\{X \in A\}$ for short. Since B is a subset of the sample space, it makes sense to write $P[\{X \in A\}]$ (or $P[X \in A]$ to simplify notation). In higher-level courses one worries about whether this set B is really an event for a particular set A, but we will not do so here.

Example 1.3.1

In the example of the bank waiting line, the sample space might consist of outcomes of the form

$$\omega = (s_1, s_2, s_3, s_4, \ldots, s_n),$$

where s_k is the length of the line after the kth arrival or departure. The outcome $(1, 2, 3, 2, 3, 4)$ would mean that one customer entered the line, then another and another to make the line size 1, 2, and then 3. A departure occurred, decreasing the line size to 2, and then two more arrivals occurred. The maximum line size random variable $X(\omega) = \max\{s_1, s_2, s_3, s_4, \ldots, s_n\}$ would

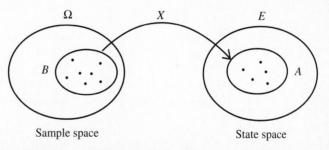

Figure 1.5 Random Variable Taking Values in Set A

have the value 4 for this outcome. We would be interested in knowing the probabilities of events such as $\{X < 3\}$, that is, the event that the line size was never as large as 3 people.

Example 1.3.2

In the bridge example, a typical outcome is $\omega = \{2, 3, 4, \text{ and } 7 \text{ of spades}; 5, \text{ jack, and ace of hearts}; 6, 8, 9, \text{ and king of clubs}; 3 \text{ and } 7 \text{ of diamonds}\}$. The random vector **Y**, which counts the numbers of each suit in the hand, has four component random variables $Y_1, Y_2, Y_3,$ and Y_4 for the numbers of spades, hearts, clubs, and diamonds, respectively. The value of **Y** for the outcome is $\mathbf{Y}(\omega) = (4, 3, 4, 2)$. Events related to this random variable that could be of interest include $\{Y_1 \geq 5\}$, that is, the event that there are at least 5 spades in the hand. We will be able to compute the probability of this event in the next chapter.

?Question 1.3.1

Give another example of a random variable, paying attention to the sample space on which it is defined.

1.3.2   Probability Distributions

The state space E of a random variable X inherits a probability measure from the measure P on the sample space Ω on which X is defined. To see how, look again at Fig. 1.5. The idea is simply to give the set A in E the same probability as the set B in Ω.

DEFINITION 1.3.2

The *probability distribution Q* of a random variable X is the probability measure on E defined by

$$Q(A) = P[X \in A], \qquad \text{for } A \subseteq E. \tag{1.8}$$

Since the concepts of random variable and distribution are rather abstract, a summary is in order at this point. Concentrate on Fig. 1.5. A random variable is a function that assigns values to outcomes of a random experiment. The values are points in a subset of $\mathbb{R}^n$ called the state space. Because there is a probability defined on the space of outcomes, you can find the probability of the set of outcomes for which the value of the random variable falls into a set A in the state space. This probability, looked at as a function of the set A, is what we call the probability distribution of the random variable.

?Question 1.3.2

Argue that Q as defined in (1.8) satisfies axioms (a) and (b) of probability—that is, $Q(E) = 1$—and if $A \subseteq E$, then $Q(A) \geq 0$.

In light of the last question, to show that Q qualifies as a probability measure, we must now check axiom (c) in Definition 1.2.2. If $A_1, A_2, A_3, \ldots$ is a sequence of pairwise disjoint sets in E, then

$$B \equiv \{\omega \mid X(\omega) \in (A_1 \cup A_2 \cup A_3 \cup \cdots)\}$$
$$= \{\omega \mid X(\omega) \in A_1\} \cup \{\omega \mid X(\omega) \in A_2\} \cup \{\omega \mid X(\omega) \in A_3\} \cup \cdots$$
$$\equiv B_1 \cup B_2 \cup B_3 \cup \cdots. \tag{1.9}$$

Also the B_i's are pairwise disjoint. (Why?) Thus

$$Q(A_1 \cup A_2 \cup A_3 \cup \cdots) = P[X \in (A_1 \cup A_2 \cup A_3 \cup \cdots)]$$
$$= P[B] = P[B_1 \cup B_2 \cup B_3 \cup \cdots]$$
$$= P[B_1] + P[B_2] + P[B_3] + \cdots$$
$$= P[X \in A_1] + P[X \in A_2] + P[X \in A_3] + \cdots$$
$$= Q(A_1) + Q(A_2) + Q(A_3) + \cdots. \tag{1.10}$$

The main idea of this argument is that Q satisfies the disjoint union property because P does.

In future work, we will concentrate on two classes of probability distributions that will allow us to characterize the distribution easily. These classes are illustrated in the next two examples.

Example 1.3.3

Let two six-sided dice be rolled in succession. The sample space is

$$\Omega = \{(1,1), \ldots, (1,6), \quad (2,1), \ldots, (2,6), \quad \ldots, \quad (6,1), \ldots, (6,6)\},$$

where each outcome is an ordered pair indicating the results of the first and second rolls, respectively. We can put a probability measure on the sample space by assigning probability $1/36$ to each of the 36 outcomes in Ω. Define a random variable X to be the total of the two upturned faces (more formally, $X(\omega) = X(\omega_1, \omega_2) = \omega_1 + \omega_2$). This random variable maps the sample space Ω to the set $E = \{2, 3, \ldots, 12\}$ as shown in Table 1.2.

If Q is the probability distribution of X, then for instance,

$$Q(\{3,4\}) = P[X \in \{3,4\}]$$
$$= P[\{(1,2)\ (2,1)\ (1,3)\ (2,2)\ (3,1)\}]$$
$$= 5/36.$$

Table 1.2

Distribution of the Sum of Two Dice

Outcomes in Ω	$\ldots$ Mapped by X to $\ldots$	Points $x \in E$	$Q(\{x\})$
(1, 1)	$\longmapsto$	2	1/36
(1, 2), (2, 1)	$\longmapsto$	3	2/36
(1, 3), (2, 2), (3, 1)	$\longmapsto$	4	3/36
$\vdots$	$\vdots$	$\vdots$	$\vdots$
(5, 6), (6, 5)	$\longmapsto$	11	2/36
(6, 6)	$\longmapsto$	12	1/36

Because the state space is discrete, to completely describe Q it is not necessary to specify the probability that it attaches to every set in E. It is enough to list the probabilities that Q gives to each point of E. These probabilities are in the last column of Table 1.2.

?Question 1.3.3   Fill in the rest of Table 1.2.

Example 1.3.4   Consider again Example 1.2.3 on pizza delivery. Define X to be the amount of time, in units of 10 minutes, that the pizza is late. By the way that we defined the sample space, the value that X gives to an outcome is just the outcome itself—$X(\omega) = \omega$—which is the lateness of delivery. We can then let $E = \Omega = [-1, 1]$. The probability distribution of X is

$$
\begin{aligned}
Q(A) &= P[\{\omega \mid X(\omega) \in A\}] \\
&= P[\{\omega \mid \omega \in A\}] \\
&= P[A] \\
&= \int_A \frac{3}{4} - \frac{3}{4}x^2 \, dx.
\end{aligned}
\tag{1.11}
$$

For instance,

$$
Q([0, 1]) = \int_0^1 \frac{3}{4} - \frac{3}{4}x^2 \, dx = \frac{1}{2}.
$$

We see that the distribution Q of X is identical to the probability measure P. In this example, the distribution is characterized not by a discrete collection of probabilities on states, but rather by the function in the integrand.

1.3.3   Cumulative Distribution Functions, Mass Functions, and Density Functions

The previous two examples let us glimpse ahead to the subjects of Chapters 2 and 3, discrete and continuous distributions, respectively. First we will define the *probability mass function*, which characterizes discrete distributions in an intuitive way. For example, the rightmost column in Table 1.2 gives the probability mass function for the random variable that returns the sum of the two dice.

DEFINITION 1.3.3

Suppose that a random variable X has a discrete (i.e., finite or countable) state space. The function $q : E \mapsto [0, 1]$ is called the *probability mass function (p.m.f.)* of X if

$$
q(x) = Q(\{x\}) = P[X = x].
\tag{1.12}
$$

To define a valid probability distribution, the p.m.f. must satisfy

$$q(x) \geq 0 \quad \forall x \in E \quad \text{and} \quad \sum_{x \in E} q(x) = 1. \tag{1.13}$$

In the discrete case, the distribution Q of the random variable X is fully characterized by its value on singleton states, that is, by the probability mass function. The conditions in Eq. (1.13) are necessary for the distribution to satisfy the axioms of probability.

Example 1.3.5

From the coin flipping example (Example 1.2.2), if X is the number of flips required to achieve the first tail, then the p.m.f. of X is

$$q(x) = P[X = x] = \left(\frac{1}{2}\right)^x, \qquad x = 1, 2, 3, \ldots. \tag{1.14}$$

We have previously shown that this function sums to 1 over all states x, and certainly its values are nonnegative, so it satisfies the criteria for a valid p.m.f. All relevant probabilities can be calculated using q; for example, the probability that three or more flips are required is

$$\sum_{x=3}^{\infty} \left(\frac{1}{2}\right)^x = 1 - \sum_{x=1}^{2} \left(\frac{1}{2}\right)^x = 1 - \frac{1}{2} - \frac{1}{4} = \frac{1}{4}.$$

Formula (1.11) of Example 1.3.4 points the way toward the characterization of the probability distribution for random variables with uncountable state space: The distribution Q of the random variable X is characterized by a function f called the *probability density function*. In that example the density of the number of 10-minute periods by which the pizza is late is

$$f(x) = \begin{cases} \frac{3}{4} - \frac{3}{4}x^2, & \text{if } x \in [-1, 1], \\ 0 & \text{otherwise.} \end{cases}$$

DEFINITION 1.3.4

A random variable X is said to have *probability density function* *(p.d.f.)* f if, for all subsets A of the state space,

$$Q(A) = P[X \in A] = \int_A f(x)\, dx. \tag{1.15}$$

When the dimension of the state space is higher than one, the integral is a multiple integral. To be a valid p.d.f., the function f must satisfy

$$f(x) \geq 0 \quad \forall x \in E \quad \text{and} \quad \int_E f(x)\, dx = 1. \tag{1.16}$$

Again, conditions (1.16) are required if the distribution Q is to satisfy the axioms of probability. The idea, explored more fully in Chapter 3, is that the function f measures how fast probability accumulates at each point; that is, it is not the mass but the density of probability. A related concept for one-dimensional random variables is the *cumulative distribution function*, which measures how much total probability is to the left of a point x.

DEFINITION 1.3.5

The *cumulative distribution function (c.d.f.)* is defined as

$$F(x) = P[X \le x]. \tag{1.17}$$

When the state space is the real line or some subinterval thereof, and the distribution is characterized by a probability density function f, the following relationships exist between the p.d.f. and its c.d.f. F:

$$F(x) = \int_{-\infty}^{x} f(t)\, dt \quad \text{and} \quad f(x) = F'(x). \tag{1.18}$$

?Question 1.3.4   Why are the equations displayed in (1.18) true?

Most of the rest of Chapters 1–6 is devoted to exploring the properties and applications of mass functions, densities, and c.d.f.'s. We close with a brief example.

Example 1.3.6

A plane has an estimated time of arrival of noon, plus or minus 15 minutes, and within this interval no time should be given preference over any other time. By the latter we mean that the density of probability is the same for each possible time. If we label noon as time 0 and measure time in minutes, then the state space is $[-15, 15]$. The probability density function $f(x)$ is a constant c, where c is chosen so that conditions (1.16) are satisfied. Then

$$1 = \int_{-15}^{15} c\, dx = 30c,$$

hence $c = 1/30$. The probability of the event that the plane arrives after 12:05 is then

$$P[X > 5] = \int_{5}^{15} c\, dx = \frac{1}{30}(15 - 5) = \frac{1}{3},$$

which is intuitively reasonable because the interval $(5, 15]$ is one-third as long as the whole state space $[-15, 15]$, and we are assuming for this example that the density of the arrival time is constant. The cumulative distribution function can be computed as follows:

$$F(x) = P[X \le x] = \int_{-15}^{x} c\, dt = \frac{1}{30}(x + 15), \qquad x \in [-15, 15].$$

This function yields immediately that the probability of an arrival time before 11:50 is $F(-10) = 5/30$.

?Question 1.3.5   Use the c.d.f. in the last example to find easily the probability that the arrival time is between 11:55 and 12:05.

EXERCISES 1.3

1. Six candidates, three men and three women, are interviewing for a job. The candidates are to be divided randomly between two personnel officers, who will interview three candidates apiece. Let W be the number of women assigned to the first personnel officer. Construct a sample space and state space for this experiment, and display the probability distribution of W in a table along the lines of Table 1.2.

2. The actual weight X of a box of detergent listed at 110 ounces is a random variable whose state space is the interval $[108, 112]$. The distribution of X has the property that the probability that the actual weight is in a subset of $[108, 112]$ is the length of the subset, divided by the total length of $[108, 112]$. Find the cumulative distribution function and the probability density function of X.

3. Let $\mathbf{Z} = (X, Y)$ be a point in the plane selected randomly within the triangle whose corners are $(0, 0)$, $(1, 0)$, and $(0, 1)$. By this we mean that there should be constant probability density in the region. Find the cumulative distribution function of X.

4. A dart is thrown randomly at a dartboard of radius 1, in the sense that the probability that the dart hits a subset of the board is the area of the subset divided by the area of the board. What is the state space of the random variable $R =$ distance of selected point from the center of the board? Characterize the distribution Q of R by computing the cumulative distribution function. Use the c.d.f. to find $Q((s, t])$ for each pair $s < t$.

5. Let a sample space Ω have four outcomes, $\omega_1, \omega_2, \omega_3, \omega_4$, with probabilities $1/4$, $1/2$, $1/8$, and $1/8$, respectively. If a random variable X with state space $E = \{x_1, x_2\}$ is defined by $X(\omega_1) = x_2, X(\omega_2) = x_2, X(\omega_3) = x_1, X(\omega_4) = x_1$, find the probability distribution of X.

6. A computer can simulate the selection of outcomes ω of equal probability density from the interval $[0, 1]$ on the real line. Define a random variable X by $X(\omega) = 2\omega + 1$. Find the state space, cumulative distribution function, and probability density function of X. Comment on how this result improves upon the simulation capabilities of the computer.

7. A roulette wheel has 38 slots around its edge, numbered 00, 0, 1, 2, 3, ..., 36. Half the integers from 1–36 are red, and half are black (0 and 00 are a special color). When you place a bet, the money you risked is lost, but if you win you get some multiple of what you risked in return. Suppose that each dollar bet on red pays two dollars if a red number comes up when the wheel is spun, and each dollar bet on a particular number pays 35 dollars if that number comes up. If X denotes the net

winnings for a dollar bet on red, and Y denotes the net winnings for a dollar bet on a single number, find the probability mass functions of X and Y.

8. a. Show that $Q(A) = \int_A e^{-x}\, dx$ is a valid probability distribution on $E = [0, \infty)$.

   b. Show that $Q(A) = \sum_{k \in A} e^{-\lambda}\lambda^k / k!$ is a valid probability distribution on $E = \{0, 1, 2, \ldots\}$, where λ is a positive constant.

   c. Find a constant C such that $Q(A) = \sum_{k \in A} C(1/3)^k$ is a valid probability distribution on $E = \{0, 1, 2, \ldots\}$.

9. Let X be a random variable with state space $E = [0, 1]$ and probability distribution

$$Q_X(A) = \int_A 2x\, dx.$$

   a. Find $P[X \le 1/2]$.
   b. Find $P[X > 1/4]$.
   c. If $Y = 4X$, find the cumulative distribution function of Y.
   d. With Y as defined in part (c), find $P[1 \le Y \le 2]$.

10. A *queueing network* is a waiting line system consisting of customers moving from service area to service area, with possible restrictions on their movements. One example might be a small airport, in which the "customers" are all planes that use the airport, and the "service areas" that the customers may enter are "refueling," "repair," "gate," and so forth. Suppose there is a queueing network with three service stations and four customers cycling through the system in the clockwise direction, as shown in the accompanying figure. An example configuration of customers is displayed here: two at station 1, one at station 2, and one at station 3. Assume that all stations are large enough to hold four customers. Construct a random variable describing a configuration and its sample space Ω and state space E, its domain and range, respectively.

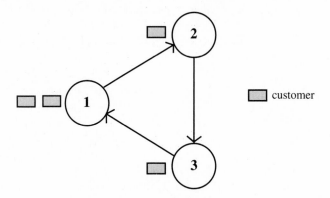

11. In the queueing network of Exercise 10, suppose that at some time instant the customers are in the configuration shown in the figure. Suppose also that the passage of one unit of time allows customers either to stay where they are or to move to the next station with equal probabilities. What are

the possible states for the random variable Y, defined as the configuration of customers after one unit of time? What is the probability mass function of Y?

1.4 | Conditional Probability

Sometimes we can make use of partial information to determine probabilities of events related to an experiment. A game such as blackjack allows us to make predictions about new cards based on the cards already visible on the table. A device subject to breakdown has a past history that may be of value in predicting the likelihood of future failure. Results of medical tests can tell the likelihood that a patient has a certain medical condition. In these kinds of examples, the problem is to find the probability that one event will occur given that another has occurred.

Look at the sample space shown in Fig. 1.6, and suppose that the event B is known to have occurred. Then B can be viewed as a *reduced sample space*, and the *conditional probability* that A occurs given B should be the relative weight of points in $A \cap B$ to points in B. If the outcomes in the figure are equally likely, then the conditional probability would be $3/9$.

—— **?Question 1.4.1** What is the conditional probability of B given A in Fig. 1.6?

This discussion suggests the following definition.

DEFINITION 1.4.1 ————————————————————————

If B is an event such that $P[B] > 0$, then the *conditional probability* of an event A given B is defined as

$$P[A \mid B] = \frac{P[A \cap B]}{P[B]}. \tag{1.19}$$

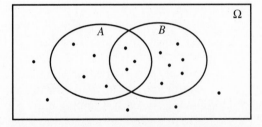

Figure 1.6 Reduced Sample Space in Event B

Grade

		A	B	C	D	F	Other
	W	26.5	36.6	17.5	3.6	1.6	14.2
University	X	23.9	44.9	18.9	4.3	1.2	6.8
	Y	19.9	45.8	23.1	4.1	.9	6.2
	Z	19.4	45.2	25.2	4.8	1.1	4.3

Figure 1.7 Grade Distribution

Example 1.4.1

I recently received information on the undergraduate grade distribution at a number of colleges and universities, a subset of which is in Fig. 1.7. (I have replaced the institutions' names by letters W, X, Y, and Z for confidentiality.) To simplify, suppose that all the institutions listed have 2000 undergraduate students. If a student is selected at random from among the 8000 students included in this table, the probability that the student received a C is

$$P[C] = \frac{2000(.175 + .189 + .231 + .252)}{8000} \approx .21,$$

that is, the total number of C's divided by the total number of students. But if the student is known to be at university Y, the conditional probability of receiving C is $P[C \mid Y] = .231$, directly from the table. Notice that $P[C \mid Y]$ and $P[C]$ are not the same; this is a property called *dependence* of the events C and Y, which we will explore in the next section. We can also compute

$$\begin{aligned}
P[Y \mid C] &= \frac{P(Y \cap C)}{P(C)} \\
&= \frac{n(Y \cap C)/n(\Omega)}{n(C)/n(\Omega)} \\
&= \frac{n(Y \cap C)}{n(C)} \\
&= \frac{2000 \cdot (.231)}{2000(.175 + .189 + .231 + .252)} \approx .27.
\end{aligned}$$

Notice also that $P[C \mid Y]$ and $P[Y \mid C]$ are not the same.

Example 1.4.2

Blackjack is a card game in which several players are each dealt one card face down and another card face up from an ordinary 52-card deck. The point values of cards of rank 2–10 are their ranks, face cards have point value 10, and aces can have either value 1 or 11 at the option of the player. Each player is offered the chance of drawing more cards, one at a time, with the object of coming the closest among players to a total of 21 points without going over 21. At a casino blackjack table the object is simply to beat the dealer, who plays for the house.

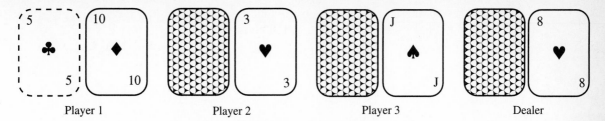

Figure 1.8 Blackjack Hands

Suppose you are player 1 among four players at a blackjack table. To make a point about independence of events, consider the first card that you receive. The probability that it is a king is $4/52 = 1/13$. The probability that it is a heart is $13/52$, and the probability that it is both a king and a heart is $1/52$. Therefore the conditional probability that the card is a king given that it is a heart is

$$P[\text{king} \mid \text{heart}] = \frac{P[\text{king and heart}]}{P[\text{heart}]} = \frac{1/52}{13/52} = \frac{1}{13}.$$

This shows that knowing that the card is a heart does not change the probability that the card is a king. This is the property that we will be calling *independence* in the next section.

Next, suppose that cards have been dealt by player 4 as in Fig. 1.8. Although you have a relatively high total of 15, the fact that the dealer could well have an 18 might persuade you to ask for another card. If you were naively to compute your chance of beating 18 with your next draw without going over 21 based only on what you know about your own cards, then you would count a total of eleven cards that are 4s, 5s, and 6s among the 50 cards other than your own. Hence the conditional probability of beating 18, given the knowledge of only your own hand, is $11/50$. You do know, of course, that three other up cards, none of which is a 4, 5, or 6, are also no longer available, so the conditional probability of beating 18 given the full information is $11/47$. Exercise 17 takes this example a bit further by asking for the probability that you beat the dealer's current total (which may not be 18) by drawing a card.

?Question 1.4.2 Given the configuration of cards in Fig. 1.8, what is the probability that the dealer has a blackjack (a hand consisting of an ace and a face card)?

Example 1.4.3 Suppose that the lifetime X of a lightbulb (in thousands of hours) is a random variable whose distribution is characterized by the probability density function

$$f(x) = xe^{-x}, \qquad x > 0.$$

Then the probability that the bulb lasts for more than 2000 hours given that it has lasted for more than 1500 hours is

$$P[X > 2 \mid X > 1.5] = \frac{P[X > 2 \cap X > 1.5]}{P[X > 1.5]}$$

$$= \frac{P[X > 2]}{P[X > 1.5]}$$

$$= \frac{\int_2^\infty xe^{-x}dx}{\int_{1.5}^\infty xe^{-x}dx}$$

$$= \frac{-xe^{-x} - e^{-x}|_2^\infty}{-xe^{-x} - e^{-x}|_{1.5}^\infty}$$

$$= \frac{3e^{-2}}{2.5e^{-1.5}} \approx .728.$$

The definition of conditional probability (1.19) can be rewritten as

$$P[A \cap B] = P[B] \cdot P[A \mid B]. \tag{1.20}$$

In this form it is usually referred to as the *multiplication rule* for conditional probability. The idea is that an experiment might be of the kind that can be broken up into stages, and knowledge of the probability of a first stage event B and the conditional probability of a second stage event A given the first stage event B yields the probability of the sequence of two events. The multiplication rule generalizes to many stages as follows.

PROPOSITION 1.4.1

Generalized Multiplication Rule   If $A_1, A_2, A_3, \ldots, A_n$ are events such that the following conditional probabilities are defined, then

$$P[A_1 \cap A_2 \cap A_3 \cap \cdots \cap A_n] = P[A_1] \cdot P[A_2 \mid A_1] \cdot P[A_3 \mid A_1 \cap A_2]$$
$$\cdots P[A_n \mid A_1 \cap A_2 \cap A_3 \cap \cdots \cap A_{n-1}].$$

$$\tag{1.21}$$

■ **Proof**   Formula (1.20) is the case $n = 2$. You may care to do a general inductive argument using this as the anchoring step. Here we will only do the case $n = 3$ as an illustration. Applying formula (1.20) to the two events $(A_1 \cap A_2)$ and A_3, we obtain

$$P[A_1 \cap A_2 \cap A_3] = P[(A_1 \cap A_2) \cap A_3] = P[A_1 \cap A_2] \cdot P[A_3 \mid A_1 \cap A_2].$$

$$\tag{1.22}$$

By formula (1.20) again, $P[A_1 \cap A_2] = P[A_1] \cdot P[A_2 \mid A_1]$. Substituting this into the right side of formula (1.22) finishes the proof for $n = 3$.

Example 1.4.4

Shakespeare wrote 10 tragedies, 17 comedies, and 10 histories. If Professor Brady of the English department puts together a collection of readings from this group of works by randomly picking four of them in sequence without repetition, what is the probability that the first two readings are comedies and the next two are tragedies?

The generalized multiplication rule applies. Let

$$A_1 = \text{"comedy on first"}$$
$$A_2 = \text{"comedy on second"}$$
$$A_3 = \text{"tragedy on third"}$$
$$A_4 = \text{"tragedy on fourth."}$$

Each factor in the product must acknowledge the fact that works previously selected may not be selected again. Using the given numbers, we obtain

$$P[A_1 \cap A_2 \cap A_3 \cap A_4] = P[A_1] \cdot P[A_2 \,|\, A_1] \cdot P[A_3 \,|\, A_1 \cap A_2]$$
$$\cdot P[A_4 \,|\, A_1 \cap A_2 \cap A_3]$$
$$= \frac{17}{37} \cdot \frac{16}{36} \cdot \frac{10}{35} \cdot \frac{9}{34} = .015.$$

The next proposition is a powerful computational tool for finding the probability of an event when we know the conditional probability that the event will occur given the occurrence of other events. Probabilists sometimes refer to the method as *conditioning and unconditioning*.

The idea is to break up the event A of interest into disjoint subevents, which are the intersections of A, with a collection of mutually exclusive sets $B_1, B_2, \ldots, B_n$, which exhaust the sample space, as shown in Fig. 1.9(a). Another way to depict the situation is by using a tree diagram as in Fig. 1.9(b). Think of the B_i events as a collection of alternatives for the first stage of a two-stage procedure; A and its complement are alternatives for the second stage. To achieve A on the second stage, exactly one of the exclusive paths B_i followed by A must be the experimental result. Of course, $P[B_i \cap A] = P[B_i] \cdot P[A \,|\, B_i]$, by formula (1.20). We *condition* by finding the

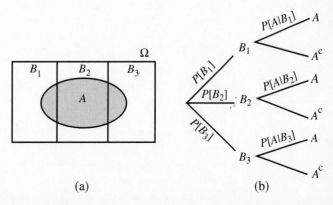

(a) (b)

Figure 1.9 Conditioning and Unconditioning

probability of A as if B_i had occurred, then *uncondition* by weighting the conditional probability by $P[B_i]$ and adding over the possible B_i's. The formal statement of the result, called the *Law of Total Probability*, follows.

PROPOSITION 1.4.2

Law of Total Probability   Let A be an event, and let $B_1, B_2, \ldots, B_n$ be mutually exclusive events of nonzero probability whose union is the sample space Ω. Then

$$P[A] = P[B_1] \cdot P[A \mid B_1] + P[B_2] \cdot P[A \mid B_2] + \cdots + P[B_n] \cdot P[A \mid B_n]. \quad (1.23)$$

■ **Proof**   By the assumed conditions on the events B_i, the events $A \cap B_i$ are pairwise disjoint, and their union is the event A. Therefore

$$\begin{aligned} P[A] &= P[(A \cap B_1) \cup (A \cap B_2) \cup \cdots \cup (A \cap B_n)] \\ &= P[A \cap B_1] + P[A \cap B_2] + \cdots + P[A \cap B_n] \\ &= P[B_1] \cdot P[A \mid B_1] + P[B_2] \cdot P[A \mid B_2] + \cdots + P[B_n] \cdot P[A \mid B_n]. \end{aligned}$$

The last line comes from the multiplication rule.

One of the most powerful ways to use the Law of Total Probability is in phenomena elapsing over time, when information can be gathered as time passes and used to make updated predictions of what the future will hold. The following example, which is of that kind, comes from the domain of study called *queueing theory*. (In Great Britain and elsewhere, a waiting line is called a *queue*.)

Example 1.4.5

Consider the busy hair styling salon displayed schematically in Fig. 1.10(a). A single stylist is working, and customers arrive at random times seeking service. If the stylist is currently busy, the customer waits in a first-come, first-served line. The stylist requires a random amount of time to serve each customer, and then the customer departs, leaving the stylist free to work on the next customer, if one is waiting.

Mark the passage of a unit of time by an event in the queueing system, that is, either an arrival or a departure of a customer. As time passes, we observe the values of a sequence $X_0, X_1, X_2, \ldots$ of random variables that count the number of customers in the system at the instants of time $0, 1, 2, \ldots$. Suppose that at the initial time 0, there are four customers in the system, and suppose also that the probabilistic arrival and service mechanism is such that whenever there is at least one customer, the next event will be an arrival of a new customer with probability p and a departure with probability $1 - p$. If no one is in the system, the next event must be an arrival. We will use the Law of Total Probability to obtain the probability distribution of X_2, given $X_0 = 4$.

The tree diagram in Fig. 1.10(b) points the way. Since the initial system size was 4, the system size X_1 at time 1 will be either 5 or 3, depending

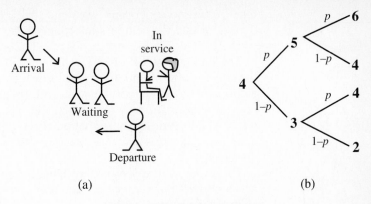

Figure 1.10 Queueing System

on whether the next event is an arrival or a departure. These two events happen with probability p and $1 - p$, respectively. If we knew the result at time 1, we could easily predict the result at time 2 in a similar way. Given $X_1 = j$, where $j \geq 1$, X_2 will be $j + 1$ with probability p (i.e., an arrival is next) or it will be $j - 1$ with probability $1 - p$ (a departure is next). So we can condition and uncondition on the system size at time 1 in order to get the system size distribution at time 2. The formulas that follow use the Law of Total Probability directly to do this; it is useful to note that on the tree, the total probability that $X_2 = k$ is being computed as the total of all path probabilities among paths ending in k, where a path probability is the product of the probabilities of the branches in the path.

$$P[X_2 = 6] = P[X_2 = 6 \mid X_1 = 5] \cdot P[X_1 = 5]$$
$$+ P[X_2 = 6 \mid X_1 = 3] \cdot P[X_1 = 3]$$
$$= p \cdot p + 0 \cdot (1 - p)$$
$$= p^2$$

$$P[X_2 = 4] = P[X_2 = 4 \mid X_1 = 5] \cdot P[X_1 = 5]$$
$$+ P[X_2 = 4 \mid X_1 = 3] \cdot P[X_1 = 3]$$
$$= (1 - p) \cdot p + p \cdot (1 - p)$$
$$= 2p(1 - p)$$

$$P[X_2 = 2] = P[X_2 = 2 \mid X_1 = 5] \cdot P[X_1 = 5]$$
$$+ P[X_2 = 2 \mid X_1 = 3] \cdot P[X_1 = 3]$$
$$= 0 \cdot p + (1 - p) \cdot (1 - p)$$
$$= (1 - p)^2$$

—— **?Question 1.4.3** The computation of the probability distribution of X_2 was done as if the probabilities involved were not dependent on the initial state X_0 of the system. To be more rigorous, we should write conditional probabilities given $X_0 = 4$. Would the computation change? (See Exercise 6 of this section.)

The Law of Total Probability also comes up when we are trying to judge the effectiveness of imperfect diagnostic techniques. A pilot study of subjects with known characteristics gives us information about how likely the diagnosis is to be right or wrong when the type of subject is known. Now we would like to turn that information around; if the diagnosis says that a new unknown subject has a characteristic, what is the likelihood that the subject does have it?

To answer such questions, we have the result known as *Bayes' theorem*, which permits the reversal of the order of conditioning in conditional probability expressions.

PROPOSITION 1.4.3

Bayes' Theorem Let the sets A and B_i, $i = 1, 2, \ldots, n$ satisfy the hypothesis of Proposition 1.4.2. Then for each $i = 1, 2, \ldots, n$,

$$P[B_i \mid A] = \frac{P[A \mid B_i] \cdot P[B_i]}{P[A \mid B_1] \cdot P[B_1] + P[A \mid B_2] \cdot P[B_2] + \cdots + P[A \mid B_n] \cdot P[B_n]}.$$

(1.24)

■ **Proof** By the definition of conditional probability,

$$P[B_i \mid A] = \frac{P[A \cap B_i]}{P[A]}.$$

Also,

$$P[A \cap B_i] = P[A \mid B_i] \cdot P[B_i],$$

by the multiplication rule. Substituting this result into the numerator of the first equation gives

$$P[B_i \mid A] = \frac{P[A \mid B_i] \cdot P[B_i]}{P[A]}.$$

The proof is completed by substituting the law of total probability formula (1.22) into the denominator for $P[A]$.

Example 1.4.6

The intradermal skin test for tuberculosis is subject to error. If the injection is too deep, the welt that indicates a positive test may not appear even in a patient with the disease. On the other hand, a healed tuberculosis or an infection by a different type of bacteria may cause a positive result on a patient without an active case of the disease. Estimates of the chance that a tuberculin test shows a positive result on people who have the virus vary, but 90% is a typical number. Suppose that the test also shows negative on 90% of subjects who do not have tuberculosis. Suppose that 1% of the population in a densely populated area has tuberculosis. If a randomly selected subject tests positive, what is the probability that the subject has the disease?

The magnitude of the number that we will compute will probably run counter to your intuition, in light of the apparently high success rates of the test. The events of interest are

$$B_1 = \text{the subject has tuberculosis,}$$
$$B_2 = B_1^c = \text{the subject does not have tuberculosis,}$$
$$A = \text{the skin test is positive.}$$

The problem statement tells us that

$$P[A \mid B_1] = .90, \qquad P[A \mid B_2] = 1 - .90 = .10, \qquad P[B_1] = .01, \qquad P[B_2] = .99.$$

We are asked to compute $P[B_1 \mid A]$. A direct application of Bayes' formula (1.23) yields

$$
\begin{aligned}
P[B_1 \mid A] &= \frac{P[A \mid B_1] \cdot P[B_1]}{P[A \mid B_1] \cdot P[B_1] + P[A \mid B_2] \cdot P[B_2]} \\
&= \frac{(.90)(.01)}{(.90)(.01) + (.10)(.99)} \\
&= \frac{.009}{.108} = .083.
\end{aligned}
$$

The positive test result has changed the probability that the subject has the disease from .01 (unconditionally) to .083 (conditionally, given the positive test). The probability has grown more than eight times, yet it is far smaller than you would probably expect. In fact, the numerator of the last expression, .009, is actually $P[A \cap B_1]$, and the denominator, .108, is $P[A]$. What has happened is that because B_1 has such small probability, its share of A is relatively small. Results like this give rationale to the practice of retesting, or doing other procedures like X rays in cases of positive test results.

As long as your mind is focused on conditional probability, a sneak preview seems to be in order. Later we will frequently be using conditional probability in the context of random variables. In Chapter 4 we will explore the idea of the conditional probability distribution of one random variable given the value of another. It makes sense to write

$$P[X_2 \in A_2 \mid X_1 \in A_1] = \frac{P[X_1 \in A_1 \cap X_2 \in A_2]}{P[X_1 \in A_1]}, \tag{1.25}$$

by the definition of conditional probability, as long as the event $\{X_1 \in A_1\}$ has nonzero probability. For discrete distributions therefore, it already is meaningful to condition on $\{X_1 = x_1\}$. However, we would like to give meaning to conditional probabilities like the preceding one for singleton sets $A_1 = \{x_1\}$ in the case of continuous random variables as well. You will see how this is done in Section 4.2.

EXERCISES 1.4

1. Two dice are rolled. Find (a) $P[\text{2nd} = 6 \mid \text{1st} = 1]$; (b) $P[\text{sum} > 6 \mid \text{1st} = 2]$.
2. Suppose that the actual numbers of students at the colleges in Example 1.4.1 are, respectively, 1450, 1690, 1320, and 1870. Find the probability that a randomly selected student receives a B, the conditional probability

that a student receives a B given that the student is selected from university X, and the conditional probability that a student is from university X given that the student received a B.

3. To study complex random phenomena by simulation, researchers frequently rely on transformations of numbers simulated repeatedly by a computer's random number generator. Each of these random numbers appears to have the characteristic of being selected randomly from the interval $[0, 1]$. That is, if X is one of the numbers and A is a subset of the interval $[0, 1]$, then $P[X \in A] = \int_A 1 \, dx = $ length of A. Find each of the following, and comment on how your answers to parts (a) and (b) can be compared to a stream of random numbers to give information about how well the random number generator is working.
   a. $P[X \in [1/2, 3/4] \mid X \in [1/4, 5/8]]$
   b. $P[X \in [1/2, 1] \mid X \in [1/4, 1]]$
   c. $P[X \in [1/4, 1] \mid X \in [1/2, 1]]$

4. a. If A and B are disjoint events, what is $P[A \mid B]$?
   b. For any event A, what is $P[A \mid \Omega]$?

5. A family has three children. Write out the sample space of possible genders of the oldest, middle, and youngest child, and define a reasonable probability measure on it. Then find
   a. $P[$youngest is a boy $\mid$ oldest is a boy$]$.
   b. $P[$youngest is a boy $\mid$ at least 2 children are boys$]$.

6. Show that for a fixed event B of nonzero probability, $Q(A) = P[A \mid B]$ defines a probability measure.

7. A simple model for diffusion of molecules follows. The box has two compartments, connected by a small opening. At each instant, either one molecule passes from the left to the right or one molecule passes from the right to the left. The molecule that moves is randomly selected. Suppose, as in the figure, that initially there are two type-1 molecules and three type-2 molecules on the left, and three type-1 molecules and two type-2 molecules on the right.
   a. Find the conditional probability that after the next move there will be at least two type-1 molecules on the left given that the move was left to right.
   b. Find the unconditional probability that after the next move there will be at least two type-1 molecules on the left.

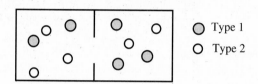

8. Let P be a probability measure on $\Omega = \{1, 2, 3, \ldots\}$ defined by
$$P[A] = \sum_{j \in A} (1/2)^j.$$
Find $P[\{5, 6, 7, \ldots\} \mid \{3, 4, 5, \ldots\}]$.

9. Let $\mathbf{X} = (X_1, X_2)$ be a random vector representing the attendance (in 1000s) at concerts given by two performers, where X_1 is the smaller and X_2 is the larger attendance. The arena in which they perform can hold 2000 people. For reasons that we will see later, a legitimate form for the probability density function of $\mathbf{X}$ is $f(\mathbf{x}) = 1/2$ for values of $\mathbf{x}$ in the interior of the triangle whose corners are $(0,0)$, $(2,2)$, and $(0,2)$. Find the probability that the more popular performer draws at least 1500 people given that the less popular performer draws at least 1000 people.

10. A notion of distance between two probability measures on a finite sample space $\Omega = \{\omega_1, \ldots, \omega_n\}$ has been proposed, which leads to an interesting interpretation of conditional probability (Ramer, 1990, pp. 336–337). Suppose that probability measure Q puts weight q_i on ω_i, and similarly probability measure P puts weight p_i on ω_i for $i = 1, \ldots, n$. Define the distance between Q and P by

$$d(Q, P) = \sum_{i=1}^{n} q_i \cdot \log(q_i/p_i).$$

Now let $B = \{\omega_1, \ldots, \omega_m\}$ be the subset of the sample space consisting of the first m outcomes. Let P be a fixed probability measure on Ω. Show that the conditional probability measure $Q_0(A) = P[A \mid B]$ has the smallest distance from P of all probability measures Q, subject to the constraint that $Q(B) = 1$. (Adopt the convention that $q_i \cdot \log(q_i/p_i) = 0$ if Q gives no weight to ω_i.)

11. The following table classifies a group of individuals according to hair color H and eye color E. Each individual has exactly one of three possible values of each characteristic: blond, red, or brown for hair, and blue, hazel, or brown for eyes. The cells contain the numbers of individuals in the group falling into each of the nine categories, but the count that should have gone in the brown hair–blue eyes cell has been lost and labeled as x. An individual is selected randomly from the group. Is there a possible value for x such that $P[\text{brown hair}] = P[\text{brown hair} \mid \text{blue eyes}]$? If so, find it.

		Eye		
		Blue	Hazel	Brown
	Blond	10	5	2
Hair	Red	10	10	5
	Brown	x	5	3

12. Consider again the diffusion model of Exercise 7. Find the probability that after two moves, the molecules return to their beginning configuration.

13. From Example 1.4.5, find (a) the probability distribution of X_2 given $X_0 = 1$; (b) the probability distribution of X_3 given $X_0 = 4$.

14. A car rental agency has two locations in a city, from either of which customers can rent cars. Cars must be returned to one of the locations, not necessarily the same one from which they were rented, one week later. The probability is .7 that a car rented at location 1 this week will be returned there next week, and the probability is .5 that a car rented at location 2 will be returned there. Suppose that initially half of the available cars are at each location. One car is chosen at random. What is the probability that the car will be in location 1 in two weeks?

15. A lie detector test correctly detects liars 80% of the time, and nonliars 90% of the time. If half the people who take lie detector tests lie and the test on one particular subject indicates a lie, what is the probability that the subject actually is lying?

16. Table 1.3 gives information about a population of potential bus riders in a community: the proportion of the community falling into each of four income classes and the probability that an individual from an income group rides the bus during a given week. If a person on a bus is sampled, find the probabilities that the person belongs to each of the four income categories. What conclusion can you draw about how well a sample from a bus represents the community?

Table 1.3

Income Distribution and Bus Ride Probabilities

Income Level	Proportion of Population	Probability of Bus Ride
$0–$20,000	10%	.2
$20,001–$30,000	35%	.3
$30,001–$40,000	25%	.2
Above $40,000	30%	.1

17. Find the probability that a single draw by player 1 in the blackjack scenario of Fig. 1.8 will beat the dealer's current total. (*Hint:* Condition and uncondition on the point value of the dealer's down card.)

18. An article in the *Scientific American* (Redfield and Burke, 1988) reported the results of a study of 906 patients afflicted with the HIV virus. At that time, scientists at Walter Reed Army Institute of Research had developed a six-stage classification system for the progression of the disease, with stage 1 being the earliest and stage 6 the latest stage. The patients were classified and then reclassified at a follow-up exam around 14 months later. The percentages of each of groups 1–5 who made the transition to each of groups 1–6 after 14 months are shown in the following table. Assuming the trends remain constant, what is the probability that a stage-1 individual is in each of the other possible stages after 28 months?

Stage in 14 Months

	1	2	3	4	5	6
1	45%	36%	9%	5%	4%	1%
2		65%	21%	7%	6%	1%
3			67%	11%	15%	7%
4				43%	30%	27%
5					31%	69%

Initial Stage (rows 1–5)

1.5 | Independent Events

The last section had a table of grades broken down by university. A student at one university could be more or less likely to get a C than a student at another, and so forth. That is, the event of getting a C and the event of being at a certain university are what we call *dependent events*. In this section we want to talk about *independent events*. Two events are independent if the conditional probability that the first event occurs given the second is the same as the unconditional probability of the first event.

DEFINITION 1.5.1

Events A and B are said to be *independent* of one another if

$$P[A \mid B] = P[A], \tag{1.26}$$

provided $P[B] > 0$.

The definition in formula (1.26) seems asymmetrical in the sense that the events A and B have different roles. To see that independence is a mutual property, let A and B be independent events of positive probability, and use the definition of conditional probability and the multiplication rule to write

$$P[B \mid A] = \frac{P[A \cap B]}{P[A]} = \frac{P[A \mid B] \cdot P[B]}{P[A]} = \frac{P[A] \cdot P[B]}{P[A]} = P[B]. \tag{1.27}$$

Therefore two events of positive probability are independent if the occurrence of one does not change the probability of the other.

—— **?Question 1.5.1** Are disjoint events independent? Give an answer, then read on.

Do not confuse independence with disjointness. If A and B are disjoint, then $P[A \cap B] = 0$; hence $P[B \mid A] = 0$. As long as B is not an event of zero probability, $P[B \mid A]$ cannot be equal to $P[B]$; that is, disjoint events of non-zero probability *cannot* be independent.

Example 1.5.1

For convenience, we reproduce Fig. 1.6 below as Fig. 1.11. If the outcomes in the sample space shown are equally likely, are the events A and B independent?

By counting outcomes, $P[A] = 7/18$. $P[A \mid B]$ is the ratio between the number of outcomes in $A \cap B$ and the number of outcomes in B, which is $3/9$, however. Knowledge that B occurs does change the probability of A; hence the events are not independent but dependent. You can also check that $P[B]$ and $P[B \mid A]$ are not equal to one another.

Example 1.5.2

Figure 1.12 displays the state space $E = [0, 2] \times [0, 2]$ of a random vector $\mathbf{X} = (X_1, X_2)$ that represents the times required for an oil change and a filter change at a quick lubrication station. Let the probability distribution of $\mathbf{X}$ be characterized by $P[\mathbf{X} \in A] = (\text{area of } A)/(\text{area of } E)$. Show that if A_1 and A_2 are time intervals that are subsets of $[0, 2]$, the event that the oil change time belongs to A_1 is independent of the event that the filter change time belongs to A_2.

The event $\{X_1 \in A_1\}$ corresponds to the rectangular strip whose base is A_1 and that extends vertically up to the top of the square, in mathematical notation, $A_1 \times [0, 2]$. Its probability is its area divided by the area of the square. The event that both $\{X_1 \in A_1\}$ and $\{X_2 \in A_2\}$ corresponds to the smaller rectangle labeled A in the figure. Hence

$$P[X_2 \in A_2 \mid X_1 \in A_1] = \frac{P[\{X_2 \in A_2\} \cap \{X_1 \in A_1\}]}{P[X_1 \in A_1]}$$

$$= \frac{(\text{length } A_1) \cdot (\text{length } A_2)/4}{(\text{length } A_1) \cdot 2/4} = \frac{\text{length } A_2}{2}.$$

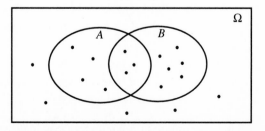

Figure 1.11 Non-independent sets A and B

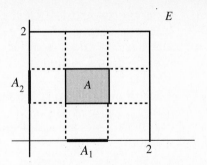

Figure 1.12 State Space of **X**

But the event $\{X_2 \in A_2\}$ corresponds to the rectangular strip whose left side is A_2 and that extends to the right as far as the right edge of the square. Thus

$$P[X_2 \in A_2] = \frac{(\text{length } A_2) \cdot 2}{4} = \frac{\text{length } A_2}{2},$$

which equals the previous conditional probability. Therefore the two events $\{X_1 \in A_1\}$ and $\{X_2 \in A_2\}$ are independent. This is an example in which two random variables satisfy the independence condition $P[X_2 \in A_2 \mid X_1 \in A_1] = P[X_2 \in A_2]$, for all intervals A_1 and A_2. We will explore independence of random variables more fully in Chapter 4.

—— **?Question 1.5.2** In the last example, what is the probability density function of **X**?

An important probability factorization rule involving independent sets is frequently used as the defining condition for independence. Its advantage is that, since it makes no direct reference to conditional probability, it can be used even when one or both sets have zero probability. Its disadvantage is that it is less intuitive than the definition we have used.

PROPOSITION 1.5.1

Let A and B be events of positive probability. Then A and B are independent of each other if and only if

$$P[A \cap B] = P[A] \cdot P[B]. \tag{1.28}$$

■ **Proof**   The proof is left to you as Exercise 3. [*Hint:* Look at the equations in (1.27).]

Note that if either A or B has zero probability, then so does $A \cap B$ (why?), and hence (1.28) also holds. It follows that if we extend the definition of independence using (1.28) to events of probability zero, such events would always be independent of any other event.

Example 1.5.3

A carton contains five GI Joe action figures. Mischievous pacifists have replaced the combat boots on two of the figures with fuzzy pink bunny slippers. One figure is drawn from the carton at random, replaced, and then another figure is drawn at random. Let A be the event that the first figure drawn is wearing fuzzy pink bunny slippers, and let B be the event that the second figure drawn is wearing the slippers. Show that the two events are independent.

If the GI Joes are numbered from 1 to 5, then we can describe the sample space as

$$\Omega = \{(i, j) \mid i, j \in \{1, 2, 3, 4, 5\}\}.$$

Since repetition is allowed, the first draw i can equal the second draw j, and therefore the size of the sample space is 25. For concreteness suppose that Joes 1 and 2 are the ones with the fuzzy pink bunny slippers. The randomness of the drawing suggests that outcomes should be equally likely. Then,

$$P[A] = P[\{(1, 1), (1, 2), \ldots, (1, 5), (2, 1), (2, 2), \ldots, (2, 5)\}] = 10/25,$$
$$P[B] = P[\{(1, 1), (1, 2), (2, 1), (2, 2), \ldots, (5, 1), (5, 2)\}] = 10/25,$$
$$P[A \cap B] = P[\{(1, 1)\,(1, 2)\,(2, 1)\,(2, 2)\}] = 4/25.$$

Since $10/25 = 2/5$ and $2/5 \cdot 2/5 = 4/25$, the factorization condition (1.28) holds for these events, and they are independent. This example illustrates a principle of random sampling: If a random sample is drawn with replacement, then the results of successive samples are independent; but if sampling is done without replacement, then successive samples are not independent. We will be able to show this more clearly in Chapter 2.

Example 1.5.4

In Fig. 1.13 is a 2×2 table of frequencies for 20 subjects classified into one of two categories for each of the two characteristics handedness and quality of language skills. Half the subjects are left-handed and half are right-handed. Twelve of the subjects are determined to have high language skills, and eight to have low skills. What must the table frequencies be in order that the event that a randomly selected subject is left-handed is independent of the event that the subject has high language skills? With those counts, is it also true that the event that a random subject is right-handed is independent of the event that the subject has low language skills?

Let m be the number of individuals in the upper left corner cell, left-handed and highly skilled. Then, because the rows must add to the row marginal totals and the columns must add to the column marginal totals, the counts in the other cells are uniquely determined as displayed in the table. For convenience let A_1 and A_2 denote the events of left- and right-handedness, and let B_1 and B_2 denote the events of high and low skill, respectively. By the independence condition,

$$\frac{m}{20} = P[A_1 \cap B_1] = P[A_1] \cdot P[B_1] = \frac{10}{20} \cdot \frac{12}{20} = \frac{6}{20}.$$

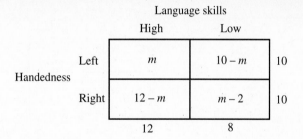

Figure 1.13   Classification Table for Independent Characteristics

We can see that $m = 6$. Using the preceding relationships, we see that each of the table rows is 6 on the left and 4 on the right. Therefore,

$$P[A_2 \cap B_2] = \frac{4}{20} = \frac{10}{20} \cdot \frac{8}{20} = P[A_2] \cdot P[B_2];$$

A_2 and B_2 are independent. Actual data will probably not conform exactly to theoretical predictions, even if the independence condition that we imposed is true. But the extent to which actual data differ from the predicted numbers gives insight into how much we believe the hypothesis of independence of the traits of handedness and language skill.

Now we turn to the case of more than two events. Although a definition of mutual independence of many events can be given in terms of conditional probabilities, it is easier to use the factorization criterion. Exercise 15 looks at some implications for conditional probabilities.

DEFINITION 1.5.2

Events $A_1, A_2, \ldots, A_n$ are said to be *mutually independent* if for any subcollection $A_{i_1}, A_{i_2}, \ldots, A_{i_k}$ of the events,

$$P[A_{i_1} \cap A_{i_2} \cap \cdots \cap A_{i_k}] = P[A_{i_1}] \cdot P[A_{i_2}] \cdot \, \cdots \, \cdot P[A_{i_k}]. \tag{1.29}$$

Thus many events are independent if any pair, triple, quadruple, and so on of the events satisfies the probability factorization criterion (1.29), which generalizes (1.28). This definition allows us to justify assignments of probability like those in Example 1.2.2, the repeated coin flip problem. If we assume independence of flips and that on each flip heads has probability p and tails has probability $(1 - p)$, then the probability of observing three heads and then a tail is

$$P[\{HHHT\}] = p \cdot p \cdot p \cdot (1 - p) = p^3 \cdot (1 - p).$$

Example 1.5.5

This example is in the area called *structural reliability*. A machine has two components, the first of which is held stable by three bolts, the second by two. The bolts are redundant safety features, in the sense that the component stays

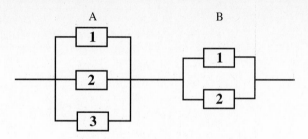

Figure 1.14   Reliability System

stable if at least one of its supporting bolts stays tight. Suppose that each bolt fails (that is, it loosens during a certain fixed time of use) with probability p, and bolts fail independently of one another. We will compute the probability that the machine suffers instability.

A diagram like that in Fig. 1.14 is very useful. The machine will be unstable if all of the component A bolts fail, or if both of the component B bolts fail. Let A_1, A_2, A_3, B_1, B_2 be the events that the five bolts fail, where the A bolts are on the left and the B bolts are on the right. Then,

$$P[\text{machine unstable}] = P[(A_1 \cap A_2 \cap A_3) \cup (B_1 \cap B_2)]$$
$$= P[A_1 \cap A_2 \cap A_3] + P[B_1 \cap B_2]$$
$$- P[(A_1 \cap A_2 \cap A_3) \cap (B_1 \cap B_2)]$$
$$= p^3 + p^2 - p^3 \cdot p^2.$$

The second line is from the general formula for unions (Proposition 1.2.6), and the third line results by the mutual independence of the bolts, and the fact that each bolt has the same probability p of failure.

?Question 1.5.3   How would the preceding formula for the probability of an unstable machine change if there were only two bolts for component A? What if component A had two bolts and component B had three bolts? What changes result to the original problem if the component A bolt failure probabilities are all p_1, and the component B failure probabilities are both p_2?

EXERCISES 1.5   1. A sample of 100 individuals classifies respondents according to gender and according to whether the individual is color-blind. The data are given in the following table. If a person is sampled at random from this group, are the events "the person is male" and "the person is color-blind" independent?

	Color-Blind	Not Color-Blind
Male	12	45
Female	2	41

2. One battery in a package of six does not work. Two batteries are sampled randomly from the package; the first is replaced before the second is chosen. Show that the events "bad battery on first" and "good battery on second" are independent. Also show that if the first battery is not replaced before the second is chosen, then these two events are not independent.

3. Prove Proposition 1.5.1.

4. A student guesses randomly on an eight-question multiple-choice test with four alternative answers per question. The student must answer at least seven questions correctly to get an A. What is the probability that this event happens? If you are making any extra assumptions to solve this problem, state them explicitly.

5. An indecisive diner decides to choose randomly from among eight luncheon specials. Among the entrées, two are chicken, two are fish, three are red meat, and one is vegetarian. The diner also has to pick a soup from among chicken noodle and New England clam chowder, and a salad dressing from among French, Russian, bleu cheese, and thousand island. If the diner's choices are made independently, what is the probability that the meal will not contain both a red meat entrée and clam chowder?

6. Show that if A and B are independent, then A^c and B^c are independent.

7. Show that if A and B are independent, then A and B^c are independent.

8. Let A_1, A_2, A_3 be mutually exclusive events whose union is the entire sample space, and let B_1, B_2, B_3 have the same property. Show that if each of A_1 and A_2 is independent of each of B_1 and B_2, then A_3 is independent of B_3.

9. The following 3×3 table with marginal totals is similar to the one in Fig. 1.13. Suppose that we require each of A_1 and A_2 to be independent of each of B_1 and B_2. Then by Exercise 8, A_3 is also independent of B_3. Show that in this case the rows of the table must be identical.

	B_1	B_2	B_3	
A_1				10
A_2				10
A_3				10
	15	12	3	

10. If A, B, and C are mutually independent events, is $A \cup B$ independent of C? If you only have that A is independent of C and that B is independent of C, is your answer the same?

11. Construct a probability measure on the sample space of the experiment of rolling a loaded die twice, with the following requirements: On a single roll 6 is twice as likely as 1, the rest of the faces are as likely as one another, and their probabilities total to 1/2. Are you making any additional assumptions? Find the probability distribution of the sum of the upturned faces.

12. Suppose that a reliability system consists of three modules connected in

parallel, each module having components in series as shown. A series system fails if and only if at least one of its components fails; a parallel system fails if and only if all of its components fail. For $i = 1, 2, 3$, the components in module i all have failure probability p_i. Write an expression for the probability that this system fails.

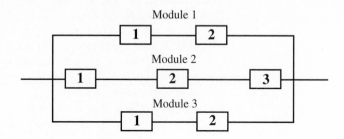

13. Let random variables X_1 and X_2 represent the times required respectively for seating a diner at a restaurant and taking that diner's order. Assume that the pair (X_1, X_2) has probability distribution $Q(A) = \int \int_A 2e^{-x_1 - 2x_2} \, dx_1 \, dx_2$ for $A \subseteq [0, \infty) \times [0, \infty)$. Show that for any subintervals $[a_1, b_1]$ and $[a_2, b_2]$ of $[0, \infty)$, the event that the seating time is in $[a_1, b_1]$ is independent of the event that the ordering time is in $[a_2, b_2]$.

14. A very badly preoccupied professor starts walking from his office at street intersection 3 in the figure. He makes a random decision to go a block in one of the possible directions initially, pauses at the next intersection, then makes another random decision to go a block in an available direction, independent of the first decision. Find the probability distribution of his final position after walking two blocks.

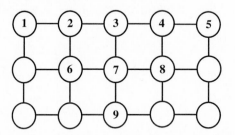

15. If A_1, A_2, A_3, A_4 are mutually independent events, show that (a) $P[A_1 \cap A_2 \mid A_3 \cap A_4] = P[A_1 \cap A_2]$; (b) $P[A_4 \mid A_1 \cap A_2 \cap A_3] = P[A_4]$.

CHAPTER 2

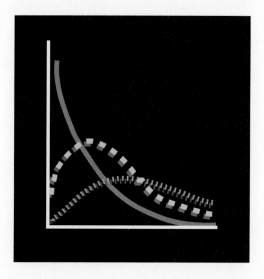

DISCRETE PROBABILITY

2.1 | Combinatorial Probability

2.1.1 Fundamental Counting Principle

Many random phenomena take place in stages, where each stage can result in finitely many possible outcomes. We would like to describe and measure the cardinality of sample spaces and events for such problems. In many staged phenomena, outcomes are equally likely; hence the probability of an event A would be calculated as

$$P[A] = \frac{n(A)}{n(\Omega)}, \tag{2.1}$$

where Ω is the sample space of the experiment. An important class of examples is that in which objects are sampled successively, with or without replacement, from a finite population. In this case the stages are the individual samples, and an outcome is the complete sequence or subset of objects sampled.

—— **?Question 2.1.1** If an experiment has two stages, the first has seven possible outcomes, and the second has three, how many outcomes does the combination of both stages have?

The preceding question points to a simple yet powerful counting method, which we now state. Part (a) of the following theorem is the most basic form of the counting principle, and parts (b) and (c) generalize this principle in two directions.

PROPOSITION 2.1.1

Fundamental Counting Principle (a) Suppose that an experiment has two stages. For the first stage, there are m possible outcomes, and for each of these, the second stage has n possible outcomes. Then the two-stage experiment has $m \cdot n$ outcomes.

(b) For a more general two-stage experiment, let the first-stage outcomes be labeled $i = 1, 2, \ldots, m$. Assume that if the first-stage outcome is i, then there are n_i possible outcomes for stage 2. Then the two-stage experiment has

$$\sum_{i=1}^{m} n_i \tag{2.2}$$

possible outcomes.

(c) Suppose that an experiment consists of k stages such that the first stage has m_1 possible outcomes, for each outcome of stage 1 there are m_2 possible outcomes of stage 2, for each combined outcome of the first two stages there are m_3 possible outcomes of stage 3, and so on. Then there are $m_1 \cdot m_2 \cdot m_3 \cdots m_k$ outcomes of the entire experiment.

■ **Proof** Since part (a) follows as a special case of part (b), let us prove the latter first. It will be helpful to refer to the tree diagram in Fig. 2.1, in which a two-stage procedure is depicted that has two first-stage outcomes, one of which has three possible second-stage outcomes and the other has two. An outcome for the two-stage procedure is a path of two edges from the root to a leaf.

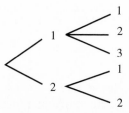

Figure 2.1 Fundamental Counting Principle

In general, the two-stage experiment has a sample space of pairs (i, j), where i is the first-stage outcome and j is one of its n_i possible second-stage outcomes. Let A_i be the event that outcome i occurs on the first stage, for $i = 1, 2, \ldots, m$. Then the events A_i are disjoint and their union is all of Ω. Therefore

$$n(\Omega) = \sum_{i=1}^{m} n(A_i) = \sum_{i=1}^{m} n_i.$$

Part (a) follows immediately by setting all $n_i = n$. We leave the proof of part (c) as Exercise 11.

Example 2.1.1

Recall that in blackjack, a player is dealt one card face down and then a second card face up from an ordinary deck. What is the probability that a player is dealt a blackjack (an ace and a face card)?

The outcomes are pairs (i, j) of cards, where no repetition is allowed. We suppose that the cards are sampled fairly from a well-shuffled deck, which justifies an assumption of equal likelihood of outcomes. For each of the 52 possible values of i, there are 51 values of j, hence by part (a) of the Fundamental Counting Principle,

$$n(\Omega) = 52 \cdot 51.$$

A regular deck contains four aces and twelve face cards. The cards are dealt in order, so that a blackjack may be obtained by having the ace down or the ace up. So, by the Counting Principle again, together with axiom (c) of probability,

$$
\begin{aligned}
P[\text{blackjack}] &= P[\text{blackjack with ace down}] \\
&\quad + P[\text{blackjack with ace up}] \\
&= \frac{4 \cdot 12}{52 \cdot 51} + \frac{12 \cdot 4}{52 \cdot 51} \\
&= \frac{8}{221} \approx .036.
\end{aligned}
$$

Example 2.1.2

A list of names of potential jurors contains 20 names beginning with each of the 6 letters A–F, 25 names beginning with each of the 14 letters G–T, and 10 names beginning with each of the 6 letters U–Z. A prosecuting attorney samples a name by first selecting a letter randomly, then sampling a name within that letter. How many outcomes are in the sample space? Is each person on the list equally likely to be chosen?

The sample space consists of pairs (i, j) where i = A, B, ..., Z. For i = A–F, j takes on $n_i = 20$ possible name values, for i = G–T, j takes on

$n_i = 25$ possible name values, and for $i = $ U–Z, j takes on $n_i = 10$ possible name values. By part (b) of the Fundamental Counting Principle,

$$n(\Omega) = \sum_{i=A}^{Z} n_i = \sum_{i=A}^{F} n_i + \sum_{i=G}^{T} n_i + \sum_{i=U}^{Z} n_i$$

$$= 6 \cdot 20 + 14 \cdot 25 + 6 \cdot 10 = 530.$$

To determine whether each person is equally likely to be chosen, look at the implications of assuming that names are equally likely. Each name corresponds uniquely to an outcome; hence if names were equally likely, each outcome would get probability $1/530$, from the preceding computation. On the surface there seems to be nothing wrong with that; however, it follows that

$P[\text{person selected has last name beginning with A}] = 20/530,$

$P[\text{person selected has last name beginning with G}] = 25/530,$

$P[\text{person selected has last name beginning with U}] = 10/530.$

This is contrary to the experimental procedure, in which first the letter is selected randomly, so that letters should be equally likely. Therefore people are not equally likely to be chosen.

Example 2.1.3

Writing a 16-character string of 0s and 1s can be viewed as an experiment of 16 stages, in which each stage can result in either 0 or 1, unaffected by other stages. Part (c) of the Fundamental Counting Principle implies that there are $2^{16} = 65{,}536$ possible character strings of this type. (This number is equal to $2^6 \cdot 2^{10} = 64 \cdot 2^{10}$, often referred to in computer jargon as $64K$, where $K = 2^{10}$ is around 1000.) This example is important due to the fact that computer data are encoded as strings of 0s and 1s, often in blocks called *words* of length 16. Older computers encoded integers with one word, so that fewer than 33,000 positive and 33,000 negative integers could be represented. Similarly, there are $2^7 = 128$ different 0-1 strings of length 7. Text characters are often encoded using 7 positions among an available 8 positions in a half-word (or *byte*). Thus 128 characters can be encoded this way, which is enough to include the upper- and lowercase alphabets, digits, punctuation marks, and numerous special symbols.

?Question 2.1.2

In a 32-bit computer, integers are represented as strings of 32 0s and 1s. In view of the fact that one bit will be reserved to characterize the sign of the encoded integer, what range of integers can be so represented?

Example 2.1.4

A confidence man wants to perpetrate a fraudulent activity in n cities in some order, with no city being visited more than once, for obvious reasons. How many possible routes among cities are there?

With successive cities as stages, there are n possibilities for the first city visited, $n-1$ for the second, $n-2$ for the third, and so on, culminating in 1 possibility for the nth city. By version (c) of the Counting Principle, the total number of routes is

$$n! \equiv n(n-1)(n-2)\cdots 3 \cdot 2 \cdot 1. \qquad (2.3)$$

The symbol $n!$ is read "n factorial."

This is a little piece of the famous traveling salesman problem, in which distances are given between cities and the salesman is looking for the shortest tour that hits each city exactly once. The sheer magnitude of the number of possible routes makes it infeasible for even a computer to brute force its way through all of them, without helpful problem-solving heuristics to produce good paths. The reason is that $n!$ grows very rapidly as the number of cities grows, resulting in what algorithm analysts sometimes call *combinatorial explosion*. For example, for three cities 3! is only 6, but 6! is 120, 10! is 3,628,800, and 13! is 6,227,020,800.

2.1.2 Permutations and Combinations

Example 2.1.4, the con man's tour, gave us an instance of a general counting situation, in which successive samples are taken from a group of objects without replacing the objects as sampling progresses. The following definition is relevant to this situation.

DEFINITION 2.1.1

A *permutation* of n objects $\{y_1, y_2, \ldots, y_n\}$, taken r at a time is an ordered list $(x_1, x_2, \ldots, x_r)$ selected from the original n objects, such that $x_i \neq x_j \ \forall \ i \neq j$. We denote the number of such permutations by $P_{n,r}$.

It is easy to find $P_{n,r}$. Using the Counting Principle, let r stages be associated with the sampled objects $x_1, \ldots, x_r$. The first object x_1 can be any of the n members of the population. Once x_1 is specified, exactly $n-1$ members of the population remain as possible values of x_2. Similarly, once x_1 and x_2 are both specified, x_3 has $n-2$ possible values, and so on. Then,

$$P_{n,r} = n \cdot (n-1) \cdot (n-2) \cdots (n-r+1) = \frac{n!}{(n-r)!}. \qquad (2.4)$$

Example 2.1.5

If you were at a small party and, at an unfortunate lull, you began to wonder whether there were two people at the party with the same birthday, how many partygoers would you expect to be necessary in order to have a good chance to have matching birthdays? (This is a famous problem whose answer will probably surprise you.)

It will be helpful to work with the complementary event of having no pair

Table 2.1

Birthday Matching Probabilities

Number of People	P[at least one match]
20	0.411
21	0.444
22	0.476
23	0.507
24	0.538
25	0.569
26	0.598
27	0.627
28	0.654
29	0.681
30	0.706

of matching birthdays at all among n people. The probability that we want can then be found by subtracting the probability of this complement from 1.

Consider asking all persons at the party in sequence to tell their birthday. By the Counting Principle, 365^n possible sequences of birthdays could result. The event of having no matches demands that we sample n days sequentially and without replacement from a collection of 365 birthdays. By formula (2.4) we can then write

$$P[\text{at least 1 pair of matching birthdays}] = 1 - P[\text{no matches}]$$
$$= 1 - \frac{365 \cdot 364 \cdots (365 - n + 1)}{365^n}.$$

Some of these probabilities are listed in Table 2.1. The party need have only 23 or more guests to have more than a 50% chance of finding two people with the same birthday; if there are 30 or more guests, the probability of at least one match rises to over 70%. The reason why this happens is that the fractions that are multiplied together to find the probability of no matches, although themselves near 1, combine to produce a small product if there are enough of them.

We have just studied permutations, the ordered lists of r objects selected without replacement from n objects. Now we turn to unordered subsets.

DEFINITION 2.1.2

A *combination* of n items $\{y_1, y_2, \ldots, y_n\}$, r at a time, is a subset $\{x_1, x_2, \ldots, x_r\}$ selected from the original n items, such that $x_i \neq x_j \ \forall \ i \neq j$. We denote the number of such combinations by $C_{n,r}$, or $\binom{n}{r}$. The latter is read "n choose r."

—— **?Question 2.1.3**   What is the difference between a permutation and a combination?

As with permutations, $C_{n,r}$ is easy to compute, and again the Counting Principle underlies the result. We need to be a little craftier than before, however. Consider the experiment of obtaining a permutation of r objects from n. We can break the experiment into a two-stage procedure by first drawing an unordered subset of r of the objects—that is, a combination— and then forming an ordered list by drawing objects one at a time without replacement out of the subset. The total number of permutations is therefore the product of the number of ways of doing the two stages:

$$P_{n,r} = C_{n,r} \cdot P_{r,r} = C_{n,r} \cdot r!.$$

Thus

$$C_{n,r} = \binom{n}{r} = \frac{P_{n,r}}{r!} = \frac{n!}{(n-r)! \cdot r!}. \tag{2.5}$$

You may have seen the quantity $\binom{n}{r}$ before in the Binomial Theorem as the coefficient of the rth term in the expansion of $(x+y)^n$. Equation (2.5) should agree with the form you saw in that context. For this reason, $\binom{n}{r}$ is often referred to as a *binomial coefficient*.

Example 2.1.6

If a 13-card bridge hand is drawn, what is the probability that there are 7 spades and 6 hearts in the hand?

The size of the sample space Ω is the number of ways of drawing 13 cards, without order or replacement, from the deck of 52, with no other restrictions on the content of the hand. This is $C_{52,13}$, by (2.5). Outcomes should be equally likely; hence it remains to compute the number of ways that the described hand can occur. Let A be the event of drawing 7 spades and 6 hearts. An outcome of the event A can be associated uniquely with a pair of sets,

$$(\{s_1, s_2, \dots, s_7\}, \{h_1, h_2, \dots, h_6\}),$$

where the s_i's are the particular seven spades and the h_i's are the particular six hearts in the hand. The cards of each set are sampled without order or replacement from a universe of 13 cards of that suit. Thus the cardinality of A is

$$n(A) = \binom{13}{7} \cdot \binom{13}{6},$$

and the desired probability is

$$P[A] = \frac{\binom{13}{7}\binom{13}{6}}{\binom{52}{13}} = .0000046.$$

Example 2.1.7

A job discrimination study was performed of a company that had filled four vice-presidential positions by promoting from within. All of those eventually promoted were men. Investigators determined that the pool under consideration consisted of six men and five women, all of comparable qualifications. Would the fact that none of the promotions went to women be unusual if indeed promotions were not gender-biased?

Under assumptions of equal qualifications and gender neutrality, it would be reasonable to suppose that outcomes of 4 selections from among the 11 candidates are equally likely. To evaluate the likelihood of discrimination, let us find the probability of the event A that all promoted people are men. The outcomes in A are in 1–1 correspondence with samples of 4 people from among the 6 men, without replacement or order. The total number of ways of filling the 4 positions—that is, the size of the sample space—is the number of ways of sampling 4 people from among 11, without replacement or order. Therefore

$$P[A] = \frac{n(A)}{n(\Omega)} = \frac{C_{6,4}}{C_{11,4}} = \frac{\frac{6!}{4!2!}}{\frac{11!}{4!7!}} = \frac{15}{330} = .045.$$

This event is less than 5% likely, which makes one suspect that the promotion process was not gender-blind. The probability of such a hiring configuration is not zero, however. Since it is possible, even if unlikely, that all of the lucky individuals were men, probability will never let us know for certain whether discrimination was present, but it does help quantify our degree of suspicion.

It would also be interesting here to check whether the assumption that the order of selection of candidates does not matter has any effect on the probability calculation. Suppose instead that there were first, second, third, and fourth vice-presidential positions, distinguishable from one another. Outcomes are now ordered lists instead of subsets of candidates, and the only change in the preceding computation is that permutations are counted instead of combinations. In this case,

$$P[A] = \frac{n(A)}{n(\Omega)} = \frac{P_{6,4}}{P_{11,4}} = \frac{\frac{6!}{2!}}{\frac{11!}{7!}} = \frac{15}{330} = .045.$$

There were factors of 4! in both numerator and denominator in the earlier computation, corresponding to the number of ways of rearranging four names into an ordered list. They canceled, making the two results agree. (See Exercise 18 at the end of the section for a similar result.)

Let us summarize three sampling situations that arise frequently in applications. In all three cases, r samples are taken from a population $\{y_1, y_2, \ldots, y_n\}$ of n objects.

1. If sampling is done in order and with replacement (i.e., items sampled may be selected again), then the total number of samples is $n \cdot n \cdots n = n^r$.
2. If sampling is done in order and without replacement (i.e., items sampled may not be selected again), then the total number of samples is

$$P_{n,r} = n(n-1) \cdots (n-r+1) = \frac{n!}{(n-r)!}.$$

3. If sampling is done without regard to order and without replacement, then the total number of samples is

$$C_{n,r} = \binom{n}{r} = \frac{n!}{r! \cdot (n-r)!}.$$

2.1.3 Other Combinatorial Problems

In this subsection we will touch on other counting problems that sometimes come up. In particular, we will study the following:

1. Sampling from indistinguishable objects
2. Partitioning a group of objects into subsets
3. Distributing indistinguishable objects into a group of cells

We will proceed by means of three examples, from which you should try to extract general principles.

Example 2.1.8

Suppose that you are trying to decide whether a sequence of data that is being gathered as time passes is showing a downward trend. The following data, reported in Fisher and Lorie (1977, p. 85), are annual rates of return between the years 1966 and 1976 on a common stock portfolio purchased in 1953:

10.7, 12.3, 12.6, 10.4, 9.7, 10.1, 10.0, 8.3, 6.5, 7.7, 8.4.

The larger values do tend to be earlier in the sequence, but could such a sequence occur randomly with reasonably high probability?

One way to recast the question is as follows. There are eleven data points. The median value of the sample is defined as the sixth smallest value in the list, here 10.0. Now, in place of the actual numerical values, write an A or a B, according to whether the data point is above or below the median, respectively, omitting the median itself. Then the sequence is recoded as

A, A, A, A, B, A, B, B, B, B.

If there was no downward trend (assuming other kinds of anomalies are also not present), then the observed data should be a random arrangement of 5 A symbols and 5 B symbols. Put another way, all arrangements of this many symbols are equally likely, but since only a few will result in most of the A's occurring before most of the B's, if this sequence occurs we tend to disbelieve the assumption of randomness. So, we can evaluate the likelihood that the sequence has no trend by checking how likely it is that a sequence with most of the B's to the right of most of the A's will occur. To make the problem more precise, we find the probability of the set of sequences of 5 A's and 5 B's in which at most one A lies to the right of a B.

To find the size of the sample space, we have to solve a new combinatorial problem: How many ordered samples of five indistinguishable A's and five indistinguishable B's are there? To answer the question, we can reason indirectly, much as we did when we derived the formula for the number of combinations of n objects r at a time. First suppose that the A and B symbols were each subscripted 1–5 so as to be distinguishable. Then we have ten different symbols, and the number of rearrangements of them is easily seen to be 10!. But we could obtain such an arrangement by a three-stage procedure: (1) Determine the locations for A's and B's without their subscripts; (2) arrange the subscripted A's within their possible locations; and (3) arrange the subscripted B's within their possible locations. By the Counting Principle,

10! = number of arrangements of indistinguishable A's and B's $\cdot$ 5! $\cdot$ 5!.

Hence the number of arrangements of indistinguishable A's and B's—that is, the size of the sample space that we are looking for—is

$$n(\Omega) = \frac{10!}{5! \cdot 5!} = \binom{10}{5} = 252.$$

To finish the example, it is easy to find that there are just six outcomes such that at most one A lies to the right of a B (where a single B is in one of the first six places, and the other four B's are in the rightmost four places in the sequence of ten). The probability of such a sequence is then $6/252 \approx .024$. Since this is unlikely to happen under the assumption of randomly ordered sequences, we have strong evidence of a decreasing trend.

?Question 2.1.4 Explain why we could have obtained the result for the size of the sample space by looking at the problem as one of selecting 5 positions in the list in which to put A symbols.

Example 2.1.9 To analyze the effectiveness of four stain-resistance treatments on carpet, an experiment is to be performed in which 20 pieces of carpet are randomly assigned to the four stain resistors, with each resistor to be applied to 5 carpet pieces. The levels of staining when the carpet samples are subjected to certain stain-producing actions will then be measured. In how many ways can the carpet samples be divided?

This is an example of a *partitioning problem*, in which we have a group of n distinguishable objects, partitioned into some number k of unordered subsets, perhaps with different sizes $r_1, r_2, \ldots, r_k$. Here there are 20 carpet pieces, to be separated into 4 groups, each of size 5. The possible partitions are obtained as the results of a four-stage procedure: (1) choose 5 carpet pieces from the original 20 without order or replacement for stain resistor 1; (2) choose 5 pieces from the remaining 15 without order or replacement for stain resistor 2; (3) choose 5 pieces from the remaining 10 without order or replacement for stain resistor 3; (4) choose 5 pieces from the remaining 5 without order or replacement for stain resistor 4. By the Counting Principle and the results on combinations, the number of partitions is

$$\binom{20}{5} \cdot \binom{15}{5} \cdot \binom{10}{5} \cdot \binom{5}{5} = 11504 \cdot 3003 \cdot 252 = 8,705,721,024.$$

?Question 2.1.5   Try to simplify the product of binomial coefficients on the left in the last equation. The result that you get might be familiar to you from other contexts as a *multinomial coefficient*. See also Exercise 23 of this section.

Example 2.1.10   Our final example illustrates a class of problems known as *occupancy problems*. In how many ways can r indistinguishable balls be distributed into n cells, with no restrictions on how many balls can be in each cell? Physicists have applied these ideas to distributions of photons among energy levels (the *Bose–Einstein statistics*) and other situations.

Outcomes correspond to specifications of the number of balls in each cell. A case where $n = 5$ cells and $r = 5$ balls is shown in Fig. 2.2, where cell 1 has two balls, cells 2, 4, and 5 have one each, and cell 3 is empty. The picture gives a clue as to how to think of counting the sample space. Distribute the balls into cells by writing a sequence of $n + 1$ bars and r balls in a row, forcing the two outermost symbols to be bars. The bars mark off the boundaries of the cells. Between those two outer bars, there are $n - 1$ indistinguishable bars and r indistinguishable balls that we may arrange as we please. Analyzing as in Example 2.1.8, we deduce that there are

$$\frac{(n+r-1)!}{(n-1)!r!} = \binom{n+r-1}{n-1}$$

ways of distributing r balls into n cells.

Figure 2.2 Partitioning

EXERCISES 2.1   1. A palindrome is a word that reads the same backward as forward. How many six-letter palindromes are there? How many seven-letter palindromes are there? (Don't be concerned whether the word makes sense; *ytqaaqty* is a perfectly good palindrome.) If a six-letter word is formed at random, what is the probability that it will be a palindrome?

2. A new demonstration event in the Winter Olympics is the 28-man bob-sled. The coach has 30 men available. How many teams can he form, assuming:
   a. The order in which they are seated in the sled matters.
   b. The order doesn't matter.
   c. The front man, who steers, and the rear man, who does most of the pushing, are distinct from the rest, whose order does not matter.

3. Sequences of numbers can be easier to remember when there are re-peated numbers in the sequence. If you are assigned a four-digit phone extension randomly, what is the probability that it will have at least two consecutive digits (e.g., 1222 or 3306)?

4. In Example 2.1.2, what is the probability that the person selected has a name beginning with a letter after R in the alphabet?

5. There are four departments in the science division of a college. Depart-ments 1 and 2 have 4 members, and departments 3 and 4 have 5 members. The dean of the college picks three departments at random, then selects one person at random from each of those departments to serve on a special committee.
   a. How many committees are possible?
   b. Are all committees equally likely? If not, describe probabilities on outcomes. Does each person have an equal chance of being selected?

6. Suppose that three cards are drawn in succession and without replace-ment from an ordinary 52-card deck. Justify using combinatorics that $P[\diamondsuit$ on 2nd $|\diamondsuit$ on 1st$] = 12/51$. Furthermore, justify a probability of $11/50$ that a diamond is drawn on the third given that the first two draws are also diamonds.

7. In a slot machine, you deposit a token, pull the machine's arm, and three ring-shaped spinners containing pictures begin to spin. Each spinner eventually comes to rest to expose some random location on the spinner through a window. Certain combinations of the three exposed pictures are winners, but there is also blank space (of about the same size as a picture) between each picture, on which the spinner can stop. The pictures on each spinner are one copy of the digit 7, one copy of a picture called Double Diamond, two copies of the BAR picture, two copies of the double BAR picture, two copies of the triple BAR picture, and two copies of a picture of cherries.
   a. What is the probability that all of the spinners come to rest on cher-ries?
   b. What is the probability that all spinners come to rest on some BAR symbol (single, double, or triple)?
   c. Double Diamond is used as a wild card, meaning that it can match any of the other pictures. In light of this, what is the probability that all spinners match?

8. In the transportation system depicted in the following diagram, the num-bered bubbles are stations and the arrows represent possible ways to travel between stations. How many routes from station 1 to station 10 are there?

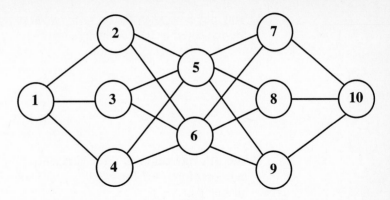

9. A family has four children. Assume all gender distributions are equally likely. Find
   a. P[all are boys].
   b. P[all are boys | at least 1 is a boy].
   c. P[all are boys | at least 2 are boys].
   d. P[all are boys | at least 3 are boys].

10. A group of people at a mall consists of 2 construction workers, 3 doctors, 2 interior designers, 2 college professors, and 3 electricians. If an interviewer selects three people at random from this group for the purpose of asking a question about a political issue, what is the probability that at least two will have the same profession?

11. Prove Proposition 2.1.1(c).

12. Many daily newspapers carry a popular word game called Jumble. (Many is the time that I have been in my office engaged in this educational activity when, to my embarrassment, the dean walked in.) The game lists four words, two of 5 letters and two of 6 letters, that are scrambled. The game player's task is to unscramble the letters, then to use the circled letters shown in the diagram as scrambled letters of a final puzzle, to which there is a clue in cartoon form which generally contains a horrible pun.
    a. In how many ways can the puzzle designer produce a puzzle with the letters shown?
    b. If, additionally, the clue to the final puzzle is, "What is the best thing to buy in a lousy shop" to which the answer is obviously "No lice," and the circled letters are in the positions in the diagram, how many possible scramblings do you now have to comb through? (Feel free to solve the puzzle.)

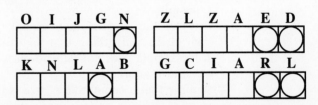

13. You may have seen *Pascal's triangle* in other contexts. The first few rows are indicated in the following diagram.

$$
\begin{array}{ccccccccccc}
 & & & & & 1 & & & & & \\
 & & & & 1 & & 1 & & & & \\
 & & & 1 & & 2 & & 1 & & & \\
 & & 1 & & 3 & & 3 & & 1 & & \\
 & 1 & & 4 & & 6 & & 4 & & 1 & \\
1 & & 5 & & 10 & & 10 & & 5 & & 1
\end{array}
$$

The nth row contains the coefficients $\binom{n}{k}$ of $x^k y^{n-k}$ in the binomial expansion of $(x+y)^n$ for $k = 0, \ldots, n$. Show that for all integers $n > 0$ and all $k = 1, \ldots, n$,

$$
\binom{n}{k} + \binom{n}{k-1} = \binom{n+1}{k}.
$$

Interpret what this implies about Pascal's triangle.

14. Give an intuitive argument based on counting numbers of subsets of a set of size n that

$$
\binom{n}{0} + \binom{n}{1} + \cdots + \binom{n}{n} = 2^n.
$$

15. In straight poker, a 5-card hand is dealt from an ordinary 52-card deck. A *flush* is a hand of all the same suit; a *full house* is a hand of three cards of one denomination and two of another. Find the following probabilities.
    a. A flush
    b. A full house
    c. Two pairs, but not a full house (nor another more valuable hand)

16. Suppose that in a particular sheet of 100 postage stamps, 3 have flaws. The inspection policy is to look at 5 randomly chosen stamps on a sheet and to release the sheet into circulation if none of those 5 have flaws. What is the probability that the sheet described here will be allowed to go into circulation?

17. A field has a population of M mice. A total of $N \leq M$ mice are trapped, tagged, then set free to mix with the others. A second group of n mice is trapped.
    a. What is the probability that there will be k tagged mice in the second sample?
    b. Now assume that M is no longer a known constant, but a random variable with possible values $M_1, M_2, \ldots, M_r$, occurring with probabilities $q_1, q_2, \ldots, q_r$. Write an expression for the probability that the population is M_i given that k tagged mice were caught in the second sample.

18. Consider the experiment of sampling k integers without replacement from $\{1, 2, \ldots, n\}$. Show that the probability that a particular integer is in the sample does not depend on whether we assume that the sample was drawn in order.

19. Suppose that a committee of five is to be selected at random from 100 senators. If there are 53 Democrats and 47 Republicans in the Senate,

what is the probability that the Republicans will have (at least) a majority on the committee?

20. In how many distinguishable ways can 6 red balls, 5 white balls, and 4 green balls be arranged in a row?

21. In the game of bridge there are four players, usually called North, South, East, and West due to their positions around a square table. All 52 cards in an ordinary deck are dealt, 13 to a player. What is the probability that North is dealt a hand of 6 hearts, 4 spades, 2 diamonds, and 1 club? Remember that the sample space for the complete experiment must acknowledge that four hands are being dealt, not just one. However, does this turn out to be important?

22. In a list of symbols, a *run* is a sequence of identical symbols. Consider a randomly arranged list of 8 letters composed of 4 X's and 4 Y's. Show that the probability distribution of the random variable R, which is the total number of runs in the sequence, is

$$P[R = 2k] = \frac{2\binom{3}{k-1}\binom{3}{k-1}}{\binom{8}{4}}, \qquad k = 1, 2, 3, 4,$$

$$P[R = 2k + 1] = \frac{2\binom{3}{k}\binom{3}{k-1}}{\binom{8}{4}}, \qquad k = 1, 2, 3.$$

(*Hint:* Note that X and Y runs alternate. Divide X's into runs by using markers, similarly for Y's, then insert the Y runs as the markers for the X runs to form the complete list.)

23. If n is a positive integer, and $r_1, r_2, \ldots, r_k$ are nonnegative integers summing to n, then the *multinomial coefficient* for these numbers is defined by

$$\binom{n}{r_1 \ r_2 \ \ldots \ r_k} = \frac{n!}{r_1! r_2! \cdots r_k!}.$$

Argue that this is the number of ways to partition a set of n elements into k subsets of sizes $r_1, r_2, \ldots, r_k$.

24. Wottsamatta University has 66 scholarship football players, 33 on offense and 33 on defense. The offensive players, being all of sound body but limited ability, are equally inept at all positions, and similarly for the defensive players. The coach vows to try all combinations of players on first, second, and third string on each of offense and defense until he finally wins a game. At worst, how many games might he have to play? (Assume that once a player is assigned to a string, it does not matter which position that player plays.) At a game a day, how many years could it take?

25. In a soft-drink taste test, experimenters are interested in whether subjects can distinguish among three brands. The skeptical hypothesis is that they cannot, but there is some prior evidence that brand 1 tastes better. Fifteen subjects are tested, and twelve of them prefer brand 1, one prefers brand 2, and two prefer brand 3. Comment on whether this seems like strong evidence that subjects are in fact not choosing their preferences randomly. State clearly any assumptions that you are using.

26. An arrangement of objects in a circle is called a *circular permutation*. In a circular permutation, there is no beginning or end; only the right and left neighbor relationships are important. An example is shown in the following figure. Suppose that six representatives to a peace conference are seated randomly around a circular table. What is the probability that representatives A and B will be seated next to each other?

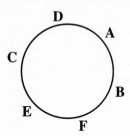

2.2 | Discrete Distributions

2.2.1 Uniform, Empirical, and Hypergeometric Distributions

Recall that we introduced random variables and their distributions in Section 1.3. When the state space is of finite or countable cardinality, the probability law of random variable X can be characterized by its *probability mass function (p.m.f.)*:

$$q(x) = P[X = x] \quad \text{for } x \in E. \tag{2.6}$$

For the distribution determined by q to satisfy the axioms of probability, q must satisfy $q(x) \geq 0$, $\forall x \in E$, and $\sum_{x \in E} q(x) = 1$. We understand that the mass function has value 0 for $x \notin E$. If the state space E is a subset of the real line, the *cumulative distribution function (c.d.f.)* of X can be defined as the function

$$F(x) = P[X \leq x] = \sum_{t \leq x} q(t). \tag{2.7}$$

In this section we will begin to list some distributions that frequently arise in applications. We will continue this task in Sections 2.3 and 2.4. We will also explore the multivariate case and introduce the notion of *marginal distributions*.

Before we proceed, here are a few elementary properties of the distribution function:

$$F(x) - F(x^-) = q(x) \qquad \forall x \in E \tag{2.8}$$

$$F(x) \text{ is a nondecreasing, nonnegative function} \tag{2.9}$$

$$\lim_{x \to \infty} F(x) = 1 \tag{2.10}$$

$$P[a < X \leq b] = F(b) - F(a) \tag{2.11}$$

In (2.8) the notation $F(x^-)$ means the left-hand limit of the function F at the point x. It is not too hard to show (see Exercise 4b) that $F(x^-) = P[X < x]$; hence (2.8) follows from the equation

$$P[X \leq x] - P[X < x] = P[X = x]. \tag{2.12}$$

Result (2.9) is true because $F(x)$ is a probability, and $P[X \leq x] \leq P[X \leq y]$ if $x \leq y$, since the former event is contained in the latter. Equation (2.10) must be true, since $P[X < \infty] = 1$ (see also Exercise 4c). Finally,

$$P[a < X \leq b] = P[X \leq b] - P[X \leq a], \tag{2.13}$$

from which (2.11) is immediate.

A sketch of a typical cumulative distribution function $F(x)$ for a random variable with four possible states r, s, t, and u is in Fig. 2.3. Note that it begins at zero, is flat between successive states, and jumps up at the states by an amount equal to the probability $q(x)$ associated to the state, by (2.8).

A special discrete distribution that arises frequently is the *discrete uniform distribution*, which puts equal mass on all states in a finite state space. Specifically, if the random variable X has states $\{x_1, x_2, \ldots, x_n\}$, then the uniform mass function is

$$q(x) = \frac{1}{n}, \qquad x \in \{x_1, x_2, \ldots, x_n\}. \tag{2.14}$$

The digit selected on the first draw of the Michigan Daily 3 Lottery is one example of a uniformly distributed random variable, with mass $1/10$ given to each of the ten states $\{0, 1, 2, \ldots, 9\}$. The slot among $0, 00, 1, \ldots, 36$ into which a roulette ball falls when the roulette wheel is spun is another uniform random variable, each of whose states has probability $1/38$. In general, any experiment in which one object is sampled at random from a set of finite cardinality leads to the discrete uniform distribution.

—— **?Question 2.2.1**   What would the cumulative distribution function of the discrete uniform mass function on $\{1, 2, \ldots, n\}$ look like?

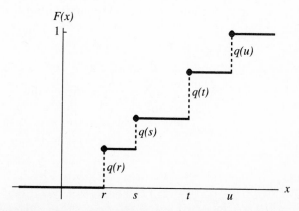

Figure 2.3   Discrete Cumulative Distribution Function

The uniform distribution also relates to the following statistical idea. Although there are sometimes theoretical reasons for assuming that a random variable X has a certain distribution, frequently there are no such reasons, and we must resort to data gathering in order to estimate the distribution by a so-called *empirical distribution*. Suppose that we are able to replicate the random experiment that gives rise to X under independent conditions, so that a finite sequence of independent random variables $X_1, X_2, \ldots, X_n$ all of which have the distribution of X is available. (By *independent*, we mean that all joint events involving the X_i's are mutually independent in the sense of Chapter 1. A formal definition will come later.) The proportion of the X_i's in this sample that take on the value x would be a good estimate of $q(x) = P[X = x]$. For example, if you were to roll a die 1000 times and face six came up on 150 of those rolls, then $150/1000 = .15$ would be a sensible estimate of the probability that the observed face on a single roll is six. We will have more to say in Chapter 6 about the proximity of empirical probabilities to theoretical probabilities. If all of our random variables X_i happened to take on different values x_i, then each x_i would have empirical probability $1/n$; that is, the distribution of X would be estimated to be uniform on the observed x_i's.

—— **?Question 2.2.2**   If you had n independent observations of a discrete random variable, what would be a sensible way to estimate the c.d.f. F? See the following definition for the answer.

DEFINITION 2.2.1 ——————————————————————————

Let $X_1, X_2, \ldots, X_n$ be independently sampled random variables all having the same probability distribution. Let $\{w_1, w_2, \ldots, w_m\}$ be the collection of states taken on by at least one of the X_i's. Then the *empirical probability mass function (emf)* of the sample is

$$\widehat{q}(w_j) = \frac{(\text{number of } X_i = w_j)}{n}. \tag{2.15}$$

The *empirical cumulative distribution function (edf)* of the sample is then the c.d.f. associated with $\hat{q}$ through equation (2.7), alternatively,

$$\hat{F}(w_j) = \frac{(\text{number of } X_i \leq w_j)}{n}. \tag{2.16}$$

Example 2.2.1

A recent survey conducted on the Internet studied grade point averages in calculus courses at 45 responding universities. For calculus I, the GPAs that were reported, rounded to the nearest tenth, were as follows:

1.9, 1.6, 2.5, 2.3, 2.0, 2.5, 2.5, 1.4, 1.8, 3.1,
3.0, 2.3, 2.2, 1.8, 2.5, 1.9, 2.3, 2.1, 2.2, 2.2,
1.5, 2.7, 2.0, 1.8, 2.2, 2.1, 2.0, 2.5, 2.3, 2.2,
1.7, 1.9, 2.4, 2.2, 2.3, 2.2, 2.1, 2.7, 2.2, 2.1,
2.0, 2.1, 2.7, 2.5, 2.6

Then the estimated probability that a university calculus I GPA is equal to 2.2 is 8/45, which is the number of instances of 2.2 in this list divided by the total number, $n = 45$, of universities. Similarly, the empirical mass function value at $x = 1.9$ is 3/45. A histogram showing the empirical mass function is given in Fig. 2.4. The associated empirical c.d.f. has the value 4/45 at $x = 1.7$; because of these 45 observations, four of them were less than or equal to 1.7. The complete empirical c.d.f. is sketched in Fig. 2.5.

The experiments involving sampling without replacement discussed in the last section give rise to a frequently observed distribution called the *hypergeometric distribution*. Let there be M objects in a population, N of which have a certain characteristic, and the remaining $M - N$ of which do not. The objects could be hearts and nonhearts in a deck of cards, city and rural residents polled on a tax referendum, type O and other blood types, and so on. Let n objects be sampled from the M without regard to order and without replacement. Define the random variable X to be the number in the sample of n that do have the specified characteristic. Notice that the nonnegative value k that X can take on cannot be more than the sample size n or more than the available N objects with the characteristic. More subtly, k also must

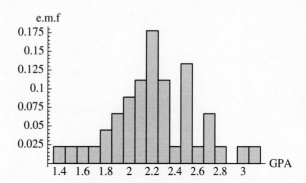

Figure 2.4   Empirical Mass Function

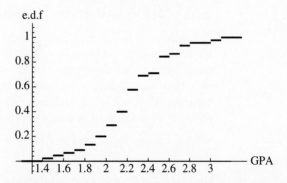

Figure 2.5   Empirical Distribution Function

be large enough that the remaining number of objects, $n - k$, in the sample is no more than the number of population objects, $M - N$, which do not have the special characteristic. In light of these restrictions, the probability mass function of X is

$$q(k) = P[X = k] = \frac{\binom{N}{k}\binom{M-N}{n-k}}{\binom{M}{n}} \quad \text{for } k = \max\{0, n - (M - N)\}, \ldots, \min\{N, n\}.$$

(2.17)

—— **?Question 2.2.3**   Justify (2.17) using the results from Section 2.1.

Example 2.2.2   Exercise 17 of Section 2.1 dealt with a capture–recapture model. Among a group of M mice, N are captured, tagged, and set free to mix with the others. If a second sample of n is subsequently caught, then the number of tagged mice in the second sample has the hypergeometric distribution (2.17). If the overall population has 100 mice, and the first capture tagged 20 of them, then the probability that there will be 2 or fewer tagged mice in a new sample of 5 is

$$P[X \leq 2] = P[X = 0] + P[X = 1] + P[X = 2]$$

$$= \frac{\binom{20}{0}\binom{80}{5}}{\binom{100}{5}} + \frac{\binom{20}{1}\binom{80}{4}}{\binom{100}{5}} + \frac{\binom{20}{2}\binom{80}{3}}{\binom{100}{5}}$$

$$= \frac{24,040,016 + 31,631,600 + 15,610,400}{75,287,520} \approx .947.$$

In view of this high probability, we should be suspicious of our assumption about the population size if we do observe more than two tagged mice. In that case, it would be more likely that the true population was less than 100. These considerations allow ecologists to gain information indirectly about populations that they may not be able to determine directly.

2.2.2   Multivariate Discrete Distributions

When $\mathbf{X} = (X_1, X_2, \ldots, X_n)$ is a discrete vector random variable of dimension more than one, we usually call $q(\mathbf{x}) = P[\mathbf{X} = \mathbf{x}]$ the *joint probability mass function* of the components of $\mathbf{X}$. We add values of the mass function just as in the one-dimensional case to calculate probabilities. The idea of the cumulative distribution function can also be extended to random vectors of dimension more than one as follows. If $\mathbf{X} = (X_1, X_2, \ldots, X_n)$, then let

$$F(\mathbf{x}) = F(x_1, x_2, \ldots, x_n) = P[\mathbf{X} \leq \mathbf{x}] = P[X_1 \leq x_1, X_2 \leq x_2, \ldots, X_n \leq x_n],$$

(2.18)

that is, let it be the joint probability that all of the random variables X_i do not exceed the corresponding real numbers x_i.

For example, if the mass function q for a bivariate random variable $\mathbf{X} = (X_1, X_2)$ gives equal probability weight $1/6$ to each of the integer grid points $\{(0,0), (1,0), (2,0), (1,1), (2,1), (2,2)\}$, then it follows that

$$P[X_1 = 2] = P[(2,0), (2,1), (2,2)] = 3/6,$$
$$F(1,1) = P[X_1 \leq 1, \ X_2 \leq 1] = P[(0,0), (1,0), (1,1)] = 3/6.$$

—— **?Question 2.2.4** Draw a picture of the state space of the random vector just described. What is $P[X_2 = 0]$? $P[X_2 = 1]$? $P[X_2 = 2]$? Does anything strike you about this collection of numbers?

This question suggests that when a random vector $\mathbf{X} = (X_1, \ldots, X_n)$ has a probability distribution $q(\mathbf{x})$, a probability distribution is also induced on each individual component X_i by summing the mass function over all possible states of the other random variables X_j for $j \neq i$. In fact, a probability distribution is induced on any subcollection of the X_i's, as indicated in the next definition.

DEFINITION 2.2.2 ————————————————————————

Let $\mathbf{X} = (X_1, \ldots, X_n)$ be a random vector with probability mass function $q(\mathbf{x})$. The *marginal mass function* of X_i is

$$q_i(x_i) = P[X_i = x_i] = \sum_{x_1} \sum_{x_2} \cdots \sum_{x_{i-1}} \sum_{x_{i+1}} \cdots \sum_{x_n} q(x_1, \ldots, x_n). \quad (2.19)$$

Similarly, the *joint marginal mass function* of a subcollection $X_{i_1}, \ldots, X_{i_k}$ of the random variables is

$$q(x_{i_1}, \ldots, x_{i_k}) = P[X_{i_1} = x_{i_1}, \ldots, X_{i_k} = x_{i_k}] = \sum \cdots \sum q(x_1, \ldots, x_n), \quad (2.20)$$

where the sum in (2.20) is taken over all indices i not in $\{i_1, \ldots, i_k\}$.

Note that we are actually applying the law of total probability in (2.19), which yields that the probability that $X_i = x_i$ is the sum of the intersection probabilities of this event with all possible configurations of the other random variables. The marginal of X_i is in fact the simple one variable distribution of the random variable X_i. Similarly, the joint marginal of a subcollection, found by adding q over all variables not in the list, is the joint mass function of the random variables in the list. The word *marginal* is used not so much to convey some different kind of mass function, but as a reminder that the random variables involved are a part of a larger group of random variables.

Example 2.2.3 An oil company is drilling at three sites. At each site, holes will be drilled at a sequence of locations until oil is struck. Let $\mathbf{X} = (X_1, X_2, X_3)$ represent the numbers of holes drilled at sites 1, 2, and 3, and suppose that $\mathbf{X}$ has joint probability mass function

$$q(x_1, x_2, x_3) = \frac{1}{12} \cdot \left(\frac{1}{2}\right)^{x_1-1} \cdot \left(\frac{3}{4}\right)^{x_2-1} \cdot \left(\frac{1}{3}\right)^{x_3-1},$$

$$x_1 = 1, 2, 3, \ldots, \quad x_2 = 1, 2, 3, \ldots, \quad x_3 = 1, 2, 3, \ldots.$$

We will compute the joint marginal p.m.f. of X_1 and X_2, and the marginal p.m.f. of X_1. Those results will shed light on the meaning of this joint probability mass function and on what is really being assumed about the experiment.

First, by (2.20), for $x_1, x_2 \geq 1$,

$$q_{12}(x_1, x_2) = P[X_1 = x_1, X_2 = x_2]$$

$$= \sum_{x_3=1}^{\infty} \frac{1}{12} \cdot \left(\frac{1}{2}\right)^{x_1-1} \cdot \left(\frac{3}{4}\right)^{x_2-1} \cdot \left(\frac{1}{3}\right)^{x_3-1}$$

$$= \frac{1}{12} \cdot \left(\frac{1}{2}\right)^{x_1-1} \cdot \left(\frac{3}{4}\right)^{x_2-1} \cdot \sum_{x_3=1}^{\infty} \left(\frac{1}{3}\right)^{x_3-1}$$

$$= \frac{1}{12} \cdot \left(\frac{1}{2}\right)^{x_1-1} \cdot \left(\frac{3}{4}\right)^{x_2-1} \cdot \frac{1}{1-1/3}$$

$$= \frac{1}{8} \cdot \left(\frac{1}{2}\right)^{x_1-1} \cdot \left(\frac{3}{4}\right)^{x_2-1}.$$

We can use this formula to find the marginal probability mass function of X_1 by adding $q_{12}(x_1, x_2)$ over all possible values of x_2:

$$q_1(x_1) = \frac{1}{8} \cdot \left(\frac{1}{2}\right)^{x_1-1} \cdot \sum_{x_2=1}^{\infty} \left(\frac{3}{4}\right)^{x_2-1}$$

$$= \frac{1}{8} \cdot \left(\frac{1}{2}\right)^{x_1-1} \cdot \frac{1}{1-3/4}$$

$$= \frac{1}{2} \cdot \left(\frac{1}{2}\right)^{x_1-1}, \quad x_1 = 1, 2, \ldots.$$

You are asked to show in Exercise 20 that the marginal p.m.f. of X_2 is $q_2(x_2) = \frac{1}{4}\left(\frac{3}{4}\right)^{x_2-1}$, $x_2 = 1, 2, \ldots$, and that the marginal p.m.f. of X_3 is $q_3(x_3) = \frac{2}{3}\left(\frac{1}{3}\right)^{x_3-1}$, $x_3 = 1, 2, \ldots$. A quick comparison of these three marginals against the joint p.m.f. shows that q is the product of the marginals q_1, q_2, and q_3—that is, the probability that it requires x_1 holes on site 1 and x_2 holes on site 2 and x_3 holes on site 3 factors into the product of the probabilities of the individual events. Consequently, one assumption in postulating the joint mass function q is that drilling successes at the three sites are mutually independent. Another implicit assumption is unearthed by looking at the structure of the marginals themselves. Each is of the form $p(1-p)^{x-1}$. This is just the mass function that would result by flipping a coin successively and independently until the first head comes up, where p is the probability of heads on a single flip. So the assumed joint mass function also says that successive drills at a single site are independent and that the chance of hitting oil on a single try is the same from one try to another (1/2 for site 1, 1/4 for site 2, and 2/3 for site 3).

—— **?Question 2.2.5** Consider the uniform distribution on the triangular grid of integers $\{(0,0),$ $(1,0), (2,0), (1,1), (2,1), (2,2)\}$ with which we started the subsection. Are the marginals of the first coordinate X_1 and the second coordinate X_2 uniform?

EXERCISES 2.2

1. Find and sketch the cumulative distribution function for the probability mass function q whose values at the states $-1, 0, 2, 4,$ and 5 are $1/4, 1/8,$ $1/4, 1/8,$ and $1/4$, respectively.

2. Could the following function be the cumulative distribution function of a discrete random variable? If so, find its mass function, and also find $P[3 < X < 6]$.

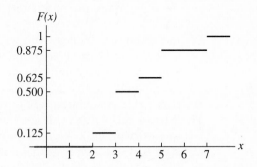

3. If X is a discrete random variable with c.d.f. F, express (a) $P[a < X < b]$; (b) $P[a \le X \le b]$ in terms of F.

4. Let $A_1 \subseteq A_2 \subseteq A_3 \subseteq \cdots$ be events. Show that

   a.

$$P\left[\bigcup_{n=1}^{\infty} A_n\right] = \lim_{n \to \infty} P[A_n].$$

   [*Hint:* Express A_n as $\cup_{k=1}^{n}(A_k \setminus A_{k-1})$.]

   b. Use the result of part (a) to show that $P[X < x] = F(x^-)$, where F is the cumulative distribution function of X.

   c. Use the result of part (a) to show that $\lim_{x \to \infty} F(x) = 1$.

5. Let $A_1 \supseteq A_2 \supseteq A_3 \supseteq \cdots$ be events.

   a. Show that

$$P\left[\bigcap_{n=1}^{\infty} A_n\right] = \lim_{n \to \infty} P[A_n].$$

   Together, the results in Exercises 4 and 5 are called the *monotone continuity of probability*. (*Hint:* Try complementation.)

   b. Use part (a) to show that $\lim_{x \to -\infty} F(x) = 0$, where F is the c.d.f. of a random variable.

6. If X is a random variable with the discrete uniform distribution on $\{a, a+ 1, \ldots, b\}$, and $a \le c < d \le b$, write an expression for $P[c < X \le d]$.

7. Suppose that the number of diners in a restaurant during the lunch hour is discretely uniformly distributed between 21 and 40. The restaurant owner wants to schedule a photographer to take some publicity photos

in which the restaurant seems to be relatively crowded. For how many days should he ask the photographer to come in order that the probability that at least one day has more than 30 diners is at least .9?

8. A roulette wheel has slots numbered $00, 0, 1, 2, \ldots, 36$. When the wheel is spun, the slot into which the roulette ball falls is uniformly distributed on these states. The slots 00 and 0 are always losers, but for every dollar bet on a number from 1–36, the bettor earns back the dollar, plus 35 more. Formulate a notion of "average winnings" on a bet of $100 on a single number, and compute the average winnings.

9. Consider a discrete distribution on the state space $\{1, 2, \ldots, n\}$ such that the probability given to every state is proportional to the number of the state. Write the probability mass function explicitly, and find $P[X \geq n - 2]$.

10. In certain dice games (with which this author is entirely unacquainted) a losing roll of two dice is one with a total of 2, 3, or 12. If two dice are rolled successively until the first losing roll, find the p.m.f. and c.d.f. of the random variable X, which counts the number of rolls required. Make reasonable assumptions about the underlying probability space.

11. While I was writing this chapter, a student of mine, Christopher Najim, was working on a research project on optimization of stock portfolios. As a part of his work he gathered information on weekly rates of return for several real stocks. (The weekly rate of return is the ratio of the change in a stock's price to the price at the start of the week, expressed as a percentage.) Thirty consecutive observations of the rate of return on AT&T, rounded to the nearest integer, were

$$0,\ 0,\ -3,\ 0,\ -1,\ -2,\ 0,\ 5,\ -3,\ 2,\ -1,\ -3,\ -2,\ 0,\ 2,$$
$$2,\ 2,\ 0,\ 2,\ 3,\ -1,\ -6,\ -3,\ -1,\ 5,\ -3,\ -1,\ 2,\ 1,\ 6.$$

Sketch the graphs of the empirical probability mass function and c.d.f. for these data.

12. Given is a list of numbers simulated by computer from the uniform distribution on $\{0, 1, 2, \ldots, 9\}$. Plot the empirical mass function and, on the same graph but in a different color, the underlying mass function being sampled from. Do the same thing for the empirical and underlying cumulative distribution functions. Comment on what you see.

$$8,\ 4,\ 2,\ 9,\ 6,\ 7,\ 3,\ 9,\ 9,\ 4,\ 7,\ 8,\ 8,$$
$$4,\ 8,\ 2,\ 5,\ 9,\ 5,\ 8,\ 9,\ 8,\ 2,\ 8,\ 7,\ 6,$$
$$3,\ 8,\ 6,\ 2,\ 1,\ 2,\ 3,\ 2,\ 6,\ 6,\ 1,\ 4,\ 8,\ 0.$$

13. Suppose that among the people in a small country town, 20 people are on file at the local hospital as having type O^- blood. About half of all of the people can be reached by phone and are willing to come in to donate at any one time. If an accident occurs and there is a need for at least 3 type O^- blood donors, what is the chance that the need cannot be filled by calling 8 of the potential donors?

14. Four players, North, South, East, and West, sit down to play bridge. Recall that in bridge, the entire 52-card deck is dealt, 13 cards to a player.

Bearing in mind that the sample space takes into account all four hands, write down and justify carefully the probability mass function of the number of spades in North's hand.

15. Suppose that 10% of a certain population is left-handed. A sample of 5 people is taken.

    a. What is the distribution of the number of left-handed people in the sample if the population has size 200?

    b. Suppose that the population is essentially infinite, so that sampling with and without replacement are nearly the same thing. Under a plan of sampling 5 randomly with replacement, what is the distribution of the number of left-handed people in the sample? (*Hint:* How many configurations of states $(x_1, \ldots, x_5)$ are there with k L's and $(5 - k)$ R's?)

16. Discuss the reasons why the hypergeometric distribution might pertain to problems of quality control performed by drawing samples for inspection.

17. Find the marginal mass functions and distribution functions of the random variables X_1 and X_2 whose joint p.m.f. is displayed in the figure. Find also: $P[X_1 \le 2,\ X_2 \le 2]$, $P[X_1 \le 2,\ X_2 = 1]$, and $P[X_1 \le 2 \mid X_2 = 1]$.

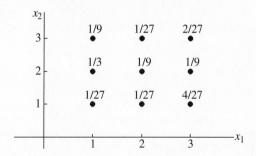

18. Find the marginal mass functions of the individual random variables X_1, X_2, X_3 jointly taking values on the corners of the box with the probabilities indicated.

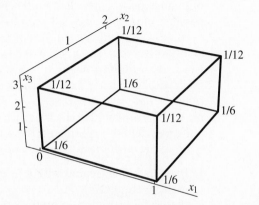

19. Let $\mathbf{X} = (X_1, X_2)$ be a random vector with the probability mass function illustrated in the figure. Compute
    a. $P[X_1 = 3, X_2 \geq 3]$.
    b. $F(2,2) = P[X_1 \leq 2, X_2 \leq 2]$.
    c. $P[X_1 \leq 3]$.

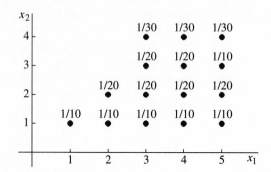

20. Show for the oil-drilling problem, Example 2.2.3, that the marginal p.m.f. of X_2 is $q_2(x_2) = \frac{1}{4} \left(\frac{3}{4}\right)^{x_2 - 1}$, $x_2 = 1, 2, \ldots$, and that the marginal p.m.f. of X_3 is $q_3(x_3) = \frac{2}{3} \left(\frac{1}{3}\right)^{x_3 - 1}$, $x_3 = 1, 2, \ldots$.

2.3 | Binomial Experiments

In this section we study a group of discrete probability distributions related to the following simple experimental scenario, which can be thought of as a generalization of the experiment of flipping a coin sequentially. A sequence of small experiments called *trials* is performed. Each trial results in one of two possible outcomes, labeled arbitrarily as "success" or "failure." The events of success and failure on different trials are mutually independent. The probability of success, say p, on a single trial does not change from trial to trial; hence the failure probability $q = 1 - p$ is also constant. If these conditions hold, we say that the experiment is a sequence of *Bernoulli trials*, sometimes called a *binomial experiment*. At the end of the section, we will generalize to experiments in which each trial can result in more than two possible outcomes. These experiments are called *multinomial experiments*.

Here are just a few examples of what trials could model:

1. Patients being treated with a new procedure, a success being an improvement in condition
2. People being polled about their opinions on a controversial issue, a success being a pro as opposed to a con opinion

3. Airline passengers scheduled for a flight; a success being a passenger who does show up
4. Shots taken by a basketball player during a game, a success being a made shot

—— **?Question 2.3.1** Think of at least two other examples of random phenomena that might be modeled as binomial experiments. Review the list of conditions that such an experiment is supposed to satisfy. Are there assumptions that must be made in your examples?

2.3.1 Bernoulli Trials

To begin a more careful mathematical treatment of a binomial experiment, consider first a single trial. A natural random variable can be written which codes success as 1 and failure as 0:

$$I(\omega) = \begin{cases} 1 & \text{if } \omega = S, \\ 0 & \text{if } \omega = F. \end{cases} \tag{2.21}$$

A random variable with state space $\{0, 1\}$ is called an *indicator variable*. Since p is the probability of success, the distribution of I, called the *Bernoulli distribution*, is clearly:

$$q(x) = P[I = x] = p^x(1 - p)^{1-x}, \qquad x = 0, 1. \tag{2.22}$$

Now consider a sequence of n independent trials. Let I_j be the indicator as in (2.21) for the jth trial; that is, $I_j = 1$ if the jth trial is a success and $I_j = 0$ otherwise. The joint p.m.f. of the indicators is easy to determine. For instance, in four Bernoulli trials,

$$\begin{aligned} q(0, 0, 1, 1) &= P[I_1 = 0, I_2 = 0, I_3 = 1, I_4 = 1] \\ &= P[I_1 = 0] \cdot P[I_2 = 0] \cdot P[I_3 = 1] \cdot P[I_4 = 1] \\ &= (1 - p) \cdot (1 - p) \cdot p \cdot p \\ &= p^2 \cdot (1 - p)^2 = p^2 \cdot q^2. \end{aligned}$$

In like manner, the joint probability mass function of the indicators in general is

$$\begin{aligned} q(x_1, x_2, \ldots, x_n) &= P[I_1 = x_1, I_2 = x_2, \ldots, I_n = x_n] \\ &= P[I_1 = x_1] \cdot P[I_2 = x_2] \cdots P[I_n = x_n] \\ &= p^{x_1}(1 - p)^{1-x_1} \cdot p^{x_2}(1 - p)^{1-x_2} \cdots p^{x_n}(1 - p)^{1-x_n} \\ &= p^k \cdot (1 - p)^{n-k} = p^k \cdot q^{n-k}, \end{aligned} \tag{2.23}$$

where $k = \sum_{i=1}^{n} x_i$ is the number of x_i's in the list that are equal to 1; hence $n - k = n - \sum x_i$ of the x_i's are 0.

The fact that the joint distribution of the indicators is not determined by the order of appearance of the 1s in the list but only by how many there are is helpful in the ensuing discussion.

We will mostly be concerned with the random variable X, which counts the total number of successes in binomial experiments. Its distribution is called the *binomial distribution*. It is easy to derive the form of the binomial probability mass function. Each sequence of trials resulting in exactly k successes has probability $p^k(1 - p)^{n-k}$ as in (2.23). The total number of such sequences is $\binom{n}{k}$, since to specify a sequence, we must pick k positions out of n in which to place the successes. Thus we have shown the following result.

PROPOSITION 2.3.1

Let X be the total number of successes in n Bernoulli trials. Then X has the *binomial probability mass function*:

$$q(k) = P[X = k] = \binom{n}{k} p^k (1 - p)^{n-k}. \tag{2.24}$$

(A shorthand notation used to refer to this distribution is $b(n, p)$.)

Figure 2.6 plots the binomial p.m.f. for $n = 5$ in each of the cases $p = .5$, which gives a symmetric distribution, and $p = .2$, which does not. Table 1 of Appendix A contains probabilities for the binomial p.m.f. for values of n from 2 to 12, and some common values of p. For instance, if X is $b(6, 1/5)$ distributed, then

$$P[X \le 1] = P[X = 0] + P[X = 1] = .2621 + .3932 = .6553.$$

Example 2.3.1

Airlines frequently overbook flights, expecting that some of the passengers who have reservations will not show up. Often the airline will offer a free flight or other compensation to passengers with reservations who are bumped from a flight due to overbooking. Suppose that 5% of reserved ticket holders tend to renege. If, on a flight of 200, the airline actually books 210 passengers, what is the probability that the airline will have to give away at least one free ticket?

Consider each of the 210 potential passengers as a trial. Let a success for a particular passenger be the event that the passenger reneges. Assume that passengers make their decisions independently and each has probability .05 of not showing up for the flight. (What may be wrong with these assumptions?) The event that the airline must give away at least one free ticket is the event

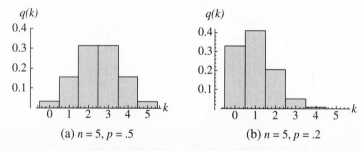

(a) $n = 5, p = .5$ (b) $n = 5, p = .2$

Figure 2.6 Binomial Mass Functions

that no more than 9 passengers renege, since the airline has overbooked by exactly 10. In the parlance of binomial experiments, this is the event that the number of successes X is less than or equal to 9, in an experiment with $n = 210$ trials and success probability $p = .05$ per trial. By (2.24), this probability is

$$\sum_{k=0}^{9} P[X = k] = \sum_{k=0}^{9} \binom{210}{k} (.05)^k (.95)^{210-k}.$$

This is a rather awkward list of numbers to sum. The PDF and CDF commands in *Mathematica* were used to compute the values of the mass function and cumulative distribution function of $b(210, .05)$ displayed in Table 2.2. Since $F(9) = P[X \le 9] = .39258$, the probability is about 39% that the airline will have to give away at least one free ticket.

?Question 2.3.2　What is the probability that the airline will give away exactly 3 free tickets? Try to draw some other conclusions about the overbooking problem using Table 2.2.

Example 2.3.2　Suppose that the shots taken by a basketball player form a sequence of Bernoulli trials, with a success defined as a made shot. There is something interesting that we can derive about the pattern of made shots. Let $I_1, I_2, \ldots, I_n$ be the Bernoulli trial indicator variables as before, and let X be the (binomially distributed) number of successfully made shots. We will compute the conditional probability

$$P[I_1 = x_1, \ldots, I_n = x_n \mid X = k].$$

Table 2.2

Binomial Probabilities, $n = 210$, $p = .05$

k	q(k)	F(k)
0	.00002	.00002
1	.00023	.00025
2	.00128	.00153
3	.00466	.00618
4	.01268	.01886
5	.02750	.04636
6	.04944	.09581
7	.07584	.17165
8	.10129	.27293
9	.11965	.39258

By the definition of conditional probability,

$$P[I_1 = x_1, \ldots, I_n = x_n \mid X = k] = \frac{P[I_1 = x_1, \ldots, I_n = x_n, \ X = k]}{P[X = k]}.$$

The event in the numerator that a particular sequence of k 1s and $n - k$ 0s occurs has the probability given in (2.23). The probability in the denominator is (2.24). Thus

$$P[I_1 = x_1, \ldots, I_n = x_n \mid X = k] = \frac{p^k(1 - p)^{n-k}}{\binom{n}{k} p^k (1 - p)^{n-k}} = \frac{1}{\binom{n}{k}}. \qquad (2.25)$$

There are $\binom{n}{k}$ arrangements of 0–1 sequences with k 1s, each of which has this probability. Thus the intuitive interpretation of this result is that given the number of made shots, every possible pattern of made shots is as likely as any other pattern. This is a by-product of the independence assumption and the consistency of the success probability from one trial to another.

Before leaving this subsection I would like to point at one of the suggested computer laboratories in the appendices. We now have two scenarios and two ways of computing in experiments involving sampling from a population that is divided into two subgroups, one whose members possess a certain characteristic and one whose members do not. Sampling can be done without replacement, in which case the number of individuals in the sample who have the characteristic has the hypergeometric distribution. Or, sampling can be done with replacement, so that the number in the sample who have the characteristic has the binomial distribution (Why?). You might well think that if the population was large, it would not matter much whether replacement was done, and so the binomial and hypergeometric probabilities should not differ by much. The laboratory exercise on the relation between binomial and hypergeometric probabilities investigates this conjecture.

2.3.2   Geometric and Negative Binomial Distributions

We now study an open-ended sequence of Bernoulli trials, performed until such time as the rth success is observed. Consider first the case $r = 1$.

PROPOSITION 2.3.2

Let T_1 be the random variable that returns the trial on which the first success occurs in a sequence of Bernoulli trials. Then T_1 has the *geometric distribution*

$$P[T_1 = n] = (1 - p)^{n-1} p, \qquad n = 1, 2, 3, \ldots. \qquad (2.26)$$

■ **Proof**   The event that $T_1 = n$ consists of the single outcome $(F, F, F, \ldots, S)$, where $n - 1$ failures precede a success. The formula for the p.m.f. of T_1 follows immediately.

—— **?Question 2.3.3**   Verify that the geometric probability mass function in (2.26) does satisfy the appropriate conditions for a p.m.f.

Example 2.3.3

Suppose that a court hears civil suit cases and that only 10% of the cases are decided in favor of the plaintiff. What is the probability that the first successful plaintiff in a given week is the seventh case? Conditioned on the next nine cases being ruled against the plaintiffs, what is the probability that the next twelve cases (including the nine) are ruled against?

In this example the Bernoulli trials, fittingly, are trials. A success is a trial that a plaintiff wins, which occurs with probability 1/10. We assume that independent decisions are made by the court. Then the probability that the first successful ruling is on the seventh case is

$$P[T_1 = 7] = \left(\frac{9}{10}\right)^6 \cdot \frac{1}{10} \approx .053,$$

since six failures and then one success are required to fulfill the condition. To answer the second question, note that the next k cases are lost by the plaintiffs if and only if $T_1 > k$. Then, using the definition of conditional probability,

$$\begin{aligned}
P[T_1 > 12 \mid T_1 > 9] &= \frac{P[T_1 > 12,\ T_1 > 9]}{P[T_1 > 9]} \\
&= \frac{P[T_1 > 12]}{P[T_1 > 9]} \\
&= \frac{(9/10)^{12}}{(9/10)^9} = \left(\frac{9}{10}\right)^3.
\end{aligned} \tag{2.27}$$

This is the same as the unconditional probability that the next three cases (after the ninth) are refused, without regard to the event that nine cases have been refused so far. Under our assumptions, the court is memoryless; it is not "due" to rule for a plaintiff just because nine straight plaintiffs have lost.

Now we move to the general case of the rth successful trial.

PROPOSITION 2.3.3

Let T_r be the trial on which the rth success occurs in a sequence of Bernoulli trials. Then T_r has the *negative binomial distribution*:

$$P[T_r = n] = \binom{n-1}{r-1} p^r (1-p)^{n-r}, \qquad n = r, r+1, r+2, r+3, \dots . \tag{2.28}$$

■ **Proof**   For values of n at least r, the event that $T_r = n$ is the event that there were $r - 1$ successes on the first $n - 1$ trials, and then a success on the nth trial. Applying the binomial distribution and the independence of trials gives us

$$P[T_r = n] = \binom{n-1}{r-1} p^{r-1} (1-p)^{n-r} \cdot p.$$

Formula (2.28) results.

Example 2.3.4

Rangers at a mountain resort set charges at various locations in order to induce controlled avalanches after snowfalls. An avalanche will start after two effective charges detonate, and the chance that a charge will be effective is about .3. Then the probability that the avalanche will occur between the detonation of charges 10 and 20 inclusive is

$$P[10 \leq T_2 \leq 20] = \sum_{n=10}^{20} \binom{n-1}{1} \left(\frac{3}{10}\right)^2 \left(\frac{7}{10}\right)^{n-2}$$

$$= \left(\frac{3}{10}\right)^2 \sum_{n=10}^{20} (n-1) \left(\frac{7}{10}\right)^{n-2}.$$

The last partial series is of the form $\sum (n-1)x^{n-2}$, which is the derivative of the partial series $\sum x^{n-1}$. Since it is well known that the closed form for the partial geometric series $\sum_{n=0}^{m} x^n$ is $(1 - x^{m+1})/(1 - x)$, we can finish the computation as follows:

$$P[10 \leq T_2 \leq 20] = \frac{9}{100} \frac{d}{dx} \left[\sum_{n=10}^{20} x^{n-1} \right]_{x=7/10}$$

$$= \frac{9}{100} \frac{d}{dx} \left[\sum_{n=0}^{19} x^n - \sum_{n=0}^{8} x^n \right]_{x=7/10}$$

$$= \frac{9}{100} \frac{d}{dx} \left[\frac{1 - x^{20}}{1 - x} - \frac{1 - x^9}{1 - x} \right]_{x=7/10}$$

$$= \frac{9}{100} \frac{d}{dx} \left[\frac{x^9 - x^{20}}{1 - x} \right]_{x=7/10}$$

$$= \frac{9}{100} \left[\frac{9x^8 - 8x^9 - 20x^{19} + 19x^{20}}{(1 - x)^2} \right]_{x=7/10} \approx .188.$$

2.3.3 Multinomial Distribution

Let us look at an extension of the binomial distribution to the multivariate case. Let the assumptions of the binomial experiment hold once again, with one exception: Instead of two categories, "success" and "failure," for trial outcomes, let there be k mutually exclusive and exhaustive categories, arbitrarily labeled $1, 2, \ldots, k$. The probability that a trial results in category i is denoted by p_i. It follows that $\sum_{i=1}^{k} p_i = 1$. Let X_i denote the number of trials among the total of n resulting in category i. Since there are exactly n trials, it must be that $\sum_{i=1}^{k} X_i = n$. The vector random variable $\mathbf{X} = (X_1, X_2, \ldots, X_k)$ has the *multinomial distribution*

$$q(x_1, x_2, \ldots, x_k) = P[X_1 = x_1, X_2 = x_2, \ldots, X_k = x_k]$$

$$= \binom{n}{x_1 \, x_2 \, \cdots \, x_k} p_1^{x_1} p_2^{x_2} \cdots p_k^{x_k}, \tag{2.29}$$

for values of x_i ranging through $0, 1, \ldots, n$ such that the total of the x_i's is identically n. A shorthand notation used to refer to the multinomial distribution is $m(n, p_1, p_2, \ldots, p_k)$. The multinomial coefficient in the formula is a generalization of the binomial coefficient, defined as

$$\binom{n}{x_1 \, x_2 \, \cdots \, x_k} = \frac{n!}{x_1! x_2! \cdots x_k!}. \tag{2.30}$$

When computing probabilities, you may have an easier time computing the multinomial coefficients in the equivalent form

$$\binom{n}{x_1 \, x_2 \, \cdots \, x_k} = \binom{n}{x_1}\binom{n - x_1}{x_2}\binom{n - x_1 - x_2}{x_3} \cdots \binom{n - x_1 - x_2 - \cdots - x_{k-1}}{x_k}. \tag{2.31}$$

You should verify that (2.31) is true.

For $k = 3$ categories, $n = 4$ trials, and category probabilities of $1/3$ each, the multinomial probability mass function (2.29) is displayed in Fig. 2.7 as a function of x_1 and x_2 only, using the fact that x_3 is forced to be $4 - x_1 - x_2$.

—— **?Question 2.3.4**   Look at slices of the figure perpendicular to the coordinate axes. Do you notice anything?

The form of the multinomial probability mass function follows from a generalization of the argument used to derive the binomial mass function. An outcome of the experiment can be thought of as a list, such as $(1, 2, 1, 3, 6, \ldots, 5, 3)$, of categories observed on the n trials. Any particular list with x_1 1s, x_2 2s, $\ldots, x_k$ ks, will have probability $p_1^{x_1} p_2^{x_2} \cdots p_k^{x_k}$, by the product rule for independent events. Also, the number of lists with this specified number of 1s, 2s, and so on, is the number of distinguishable ways of forming an n letter word from these digits. We can count this by counting the number of ways of partitioning the set of n positions in the list into subsets of (1) x_1 positions for the 1 symbols; (2) x_2 positions for the 2 symbols; $\ldots$; (k) x_k positions for the k symbols. Reasoning as in Section 2.1, the number of such partitions is the

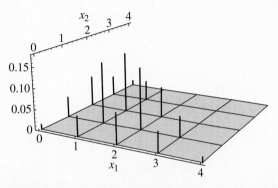

Figure 2.7   Multinomial p.m.f., $n = 4, p_1 = p_2 = 1/3$

multinomial coefficient in (2.31). The total probability in (2.29) is then the number of cases that lead to the event $X_1 = x_1, X_2 = x_2, \ldots, X_k = x_k$ times the probability per case.

Example 2.3.5

A politician who is basing his campaign on a certain controversial issue claims that among his constituents, 60% are in favor of the issue, 30% are against, and 10% are undecided. In an attempt to discredit his claim, his rival commissions a survey of a random sample of 20 of the constituents. Among these, just 7 are in favor, 6 are against, and 7 are undecided. Does this survey give strong evidence against the first politician?

To fit the problem conditions to the multinomial format, we assume that the sampling process is done with replacement, or that the population being sampled from is very large with respect to the sample size, in order to justify the assumption of independence of trials. Each individual surveyed is a multinomial trial, and there are $n = 20$ trials. A trial falls into one of $k = 3$ categories: 1, in favor; 2, against; and 3, undecided. If the first politician was right, then the probability that fewer than or equal to 7 people in the sample are for the issue and fewer than or equal to 6 are against is

$$P[X_1 \leq 7, X_2 \leq 6] = \sum_{x_1=0}^{7} \sum_{x_2=0}^{6} \binom{20}{x_1 \ x_2 \ 20 - x_2} (.6)^{x_1}(.3)^{x_2}(.1)^{20-x_1-x_2}$$

$$\approx .0004.$$

The rival politician can say that a polling result in which so many people are still undecided and so few are in favor would be extremely unlikely if the claim is true.

We close with a result about marginal distributions that should be very intuitive. If a trial has possible outcomes $1, 2, \ldots, k$ and we are interested in only the distribution of X_1, we can call category 1 a success and lump together the cases whose trial outcome is one of $2, 3, \ldots, k$ as failure events. Then we have reduced the problem to the binomial case, and X_1 should have the $b(n, p_1)$ distribution. An interesting calculation that makes this argument more rigorous is presented in the proof of the next proposition.

PROPOSITION 2.3.4

If $X_1, X_2, \ldots, X_k$ have the $m(n, p_1, p_2, \ldots, p_k)$ distribution, then for each i, X_i has the $b(n, p_i)$ distribution.

■ **Proof** We will prove the theorem only for $k = 3$, for random variable X_1. If X_1, X_2, X_3 have the $m(n, p_1, p_2, p_3)$ distribution, we find the marginal distribution of X_1. A general proof follows along the same lines.

Recall first that X_3 is required to take on the value $x_3 = n - x_1 - x_2$, where x_1 and x_2 are the values taken on by X_1 and X_2, respectively. Thus we can look at the joint p.m.f. as a function of x_1 and x_2 only:

$$q(x_1, x_2) = \binom{n}{x_1 \; x_2 \; (n - x_1 - x_2)} p_1^{x_1} p_2^{x_2} (1 - p_1 - p_2)^{n - x_1 - x_2}. \qquad (2.32)$$

Summing (2.32) over all possible values of x_2 for a given value of x_1, we obtain

$$q_1(x_1) = P[X_1 = x_1]$$

$$= \sum_{x_2=0}^{n-x_1} \frac{n!}{x_1! x_2! (n - x_1 - x_2)!} p_1^{x_1} p_2^{x_2} (1 - p_1 - p_2)^{n - x_1 - x_2}$$

$$= \frac{n!}{x_1! (n - x_1)!} p_1^{x_1} \sum_{x_2=0}^{n-x_1} \frac{(n - x_1)!}{x_2! (n - x_1 - x_2)!} p_2^{x_2} (1 - p_1 - p_2)^{n - x_1 - x_2}$$

$$= \binom{n}{x_1} p_1^{x_1} (p_2 + (1 - p_1 - p_2))^{n - x_1}$$

$$= \binom{n}{x_1} p_1^{x_1} (1 - p_1)^{n - x_1}. \qquad (2.33)$$

We have used the Binomial Theorem to go from the fourth line to the fifth line. The last line establishes that X_1 has a marginal $b(n, p_1)$ distribution.

—— **?Question 2.3.5** What do you think can be said about the joint marginal distribution of X_1 and X_2 if $X_1, X_2, \ldots, X_k$ have the $m(n, p_1, p_2, \ldots, p_k)$ distribution? Try to generalize.

EXERCISES 2.3

1. Suppose that there are five traffic lights between one end of town and another. Each red light lasts for 1 minute, yellow for 10 seconds, and green for 1 minute. What is your probability of being stopped by at least three of the lights as you drive through town? What assumptions are you making, and do they seem realistic?

2. A .250 hitter in baseball is one who hits safely on the average 25% of the time, or 1 hit in 4 official times at bat. Suppose that such a player bats 12 times during a particular series. What is the probability that the player's batting average for the series will be at least .200 (i.e., 20% safe hits)? Do not worry about special at-bats, such as walks and hit batsmen, for which the hitter is not officially charged.

3. Suppose that at each of the time instants 0, 1, 2, ... a ferocious hunting snipe occupies a region represented by an integer position on a line (see the diagram). At each instant, the snipe makes a decision to move to the neighboring region to the right or to the left, to hunt in that region the next instant of time. Being not as intelligent as it is ferocious, it makes its decisions with no memory of past decisions, that is, independently of them. Because the left side of its walnut-sized brain is heavier than the right, it always chooses to go right with probability 2/3, and left with

probability $1/3$. If the snipe starts in region 0 at time 0, what is the probability that it is located strictly to the left of region 5 at time 8?

4. TW Mangrove, a student at Smallville College, needs some books currently unavailable at the college library in order to write a paper. The librarian tells him that each book is 75% likely to arrive via interlibrary loan by the time TW needs them. He decides that he can get by with as few as two books but wishes to order more than two to be safe. How many books should he order on interlibrary loan in order to be at least 90% sure that he will have the resources he needs on time?

5. By direct enumeration of outcomes, show that, if three Bernoulli indicator random variables have joint distribution as in (2.23), then their sum has the $b(3, p)$ distribution.

6. For three Bernoulli random variables X_1, X_2, X_3 with the following joint mass function, find the mass function of the sum $X = X_1 + X_2 + X_3$ by direct enumeration of outcomes. Is it a binomial distribution?

$$q(x_1, x_2, x_3) = \begin{cases} 1/8 & \text{if } x_1 = 0, x_2 = 0, x_3 = 0 \text{ or } 1 \\ 1/16 & \text{if } x_1 = 1, x_2 = 0, x_3 = 0 \text{ or } 1 \\ 1/8 & \text{if } x_1 = 0, x_2 = 1, x_3 = 0 \text{ or } 1 \\ 1/8 & \text{if } x_1 = 1, x_2 = 1, x_3 = 0 \\ 1/4 & \text{if } x_1 = 1, x_2 = 1, x_3 = 1 \end{cases}$$

7. How long must a string of random digits be so that the probability that 0 appears in it is at least
   a. .5?
   b. .75?
   c. .9?

8. Show that if $q(x)$ is the $b(n, .5)$ mass function, then $q(n - x) = q(x)$ for $x = 0, 1, \ldots, n$. Conclude that this mass function is symmetric about $x = n/2$.

9. A system is made up of n smaller components. Assume that it is a member of the class of systems called the *k-out-of-n systems*; that is, it functions if and only if at least k of its components function. (We might envision a roof supported by several beams, which remains intact as long as at least some number of the beams are of a threshold level of strength.) In what way are systems whose components are connected in parallel members of this class? In what way are systems whose components are connected in series members of this class? If the components of a k-out-of-n system fail independently, each with probability p, over some fixed time period, then what is the reliability of the system, that is, the probability that it still functions at the end of this time? Apply your result to write expressions for the reliability of parallel and series systems.

10. Write an expression for the probability that n consecutive 5-card poker hands are dealt (replacing cards and shuffling after each hand), none of

which have any pair of cards of the same denomination (the denominations are ace, 2, 3, 4, and so on.).

11. For the purposes of this problem, denote by $b(k; m, p)$ the value of the binomial probability mass function at state k for a binomial experiment with m trials and success probability p. Show that

$$b(x; n+1, p) = p \cdot b(x-1; n, p) + (1-p) \cdot b(x; n, p)$$

for $1 \leq x \leq n$. Interpret the result intuitively.

12. For the sequence of traffic lights described in Exercise 1, what is the probability that a car can successfully pass through at least the first three of them before being stopped by a red light?

13. Derive the cumulative distribution function F of the geometric distribution with success parameter p. Show also that $1 - F(n+m) = (1 - F(n))(1 - F(m))$ for all $n, m \geq 0$. The geometric distribution is often called a "memoryless" discrete distribution. Interpret the last result in terms of a lack of memory of how many trials have gone by.

14. The game of Odd Man Out is played by N players, often to see who picks up the tab in various entertainment establishments. Each person flips a coin simultaneously, and if there is one person whose result, be it head or tail, matches none of the others, then that person is the Odd Man. If there is no such person, the entire group flips again, and so forth, until the Odd Man is found. Assume all coins used have head probability p and that all flips are mutually independent.
    a. Find the probability that there will be an Odd Man on the first round of flipping. Simplify the formula in the case $p = 1/2$.
    b. Find the probability that, if 4 people play and $p = 1/2$, it will require at least 4 rounds of flipping to determine the Odd Man.

15. What is the probability mass function of the number of at-bats required for the .250 batter of Exercise 2 to achieve the third base hit? Verify that this is a good p.m.f. by showing that it sums to 1 over all possible states. (*Hint:* Take two derivatives of a geometric series.)

16. Suppose that an automatic cake icing machine is starting to malfunction, in the sense that it can no longer form the letters *i* and *r*, but instead writes *a* for each. So, *Happy Birthday, Irma!* becomes *Happy Baathday, Aama!*, which is unfortunate. A quality control plan is in place whereby one cake in four is randomly selected and inspected for errors; the machine is stopped when the second error is found. Suppose that the letters *i* and *r* appear on half of the cakes. What is the probability that at least 10 cakes go by before the machine is stopped?

17. For $n = 5$, $k = 3$, $p_1 = 1/2$, $p_2 = 1/4$, and $p_3 = 1/4$, find and sketch as a function of x_1 and x_2 the values of the $m(n, p_1, p_2, p_3)$ probability mass function.

18. In genetics, some characteristics are passed on from two parents to an offspring by what is called *random Mendelian mating*. Each individual possesses a pair of alleles of a genetic trait, an allele being denoted by recessive *a* or dominant *A*. Thus the possible genotypes of parents and offspring are *AA*, doubly dominant; *aA*, hybrid; and *aa*, doubly recessive.

Each parent contributes one of its alleles, chosen at random, to the off-spring. Consider 10 matings of type aA parents. Find the probability of x_1 AA type, x_2 aA type, and x_3 aa type offspring among the ten. Find the probability of no more than 2 AA's among the ten offspring.

19. In a certain state, presidential candidate B has 30% of the voters in his camp, candidate C has 25%, and candidate P has 25%. The remaining voters are undecided. Find the probability that, if 25 voters are selected at random,
    a. At least 5 favor candidate B.
    b. Exactly 6 favor B, 7 favor C, and 5 favor P.
    c. Exactly 10 favor B and 6 favor C.

20. If X_1, X_2, X_3, and X_4 have the $m(n, p_1, p_2, p_3, p_4)$ distribution, find the joint marginal probability mass function of X_3 and X_4.

2.4 | Poisson Random Phenomena

2.4.1 Poisson Distribution

The *Poisson distribution*, whose properties we will study in this section, is a discrete distribution with state space $n = 0, 1, 2, \ldots$, that arises as a model of random phenomena in a way that differs from any of the special distributions that we have studied thus far. In the cases of the discrete uniform, hyperge-ometric, Bernoulli, binomial, geometric, and negative binomial probability distributions, the form of the probability mass function could be derived eas-ily from very primitive randomness assumptions using basic combinatorics. The Poisson distribution comes from limiting considerations. Perhaps more than the other distributions, it is important to use observed data to check how well the Poisson distribution fits reality (that is, to check whether the empirical distribution of a random sample fits the theoretical Poisson distri-bution reasonably well). We will investigate so-called "goodness-of-fit" tests in a later chapter.

The following example will help to motivate the limiting process from which the Poisson distribution is obtained.

Example 2.4.1

A student researcher in ecology by the name of Cathy Clover was interested in the geographical distribution of a certain species of jumping mouse in a prairie region. The region was divided into n small, equally sized zones, as in Fig. 2.8 where, on inspection, one could either find or not find a trapped mouse. The dots on the diagram are meant to indicate zones where mice were found. (Incidentally, the trapping process was done in such a way as not to harm the animals.)

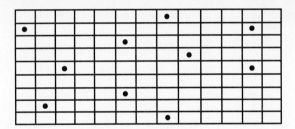

Figure 2.8 Trapping Zones

If the mice were *randomly distributed* around the region, but there were relatively few of them in total, it would be reasonable to model each zone as a Bernoulli trial, with some small success probability p of finding a trapped mouse. If we envision increasing the number n of zones while decreasing zone size, it stands to reason that p should decrease. Let us suppose that $p = p_n$ is proportional to $1/n$. If we call the proportionality constant λ, then

$$p_n = \frac{\lambda}{n} \implies \lambda = np_n. \tag{2.34}$$

Under the assumptions that the success probability is consistent from zone to zone, and that the events of finding trapped mice in different zones are mutually independent, the random variable X = number of trapped mice throughout the region would have the $b(n, p_n)$ distribution. For large n, on the other hand, we can approximate the values of the $b(n, p_n)$ probability mass function as indicated in the next proposition.

PROPOSITION 2.4.1

Under condition (2.34), for each fixed $k \geq 0$,

$$\binom{n}{k}(p_n)^k(1 - p_n)^{n-k} \longrightarrow \frac{e^{-\lambda}\lambda^k}{k!} \quad \text{as } n \to \infty. \tag{2.35}$$

■ **Proof** In the case that $k > 0$, we have that

$$\binom{n}{k}(p_n)^k(1 - p_n)^{n-k}$$

$$= \frac{n!}{(n-k)!k!}\left(\frac{\lambda}{n}\right)^k\left(1 - \frac{\lambda}{n}\right)^{n-k}$$

$$= \frac{\lambda^k}{k!} \cdot \left(1 - \frac{\lambda}{n}\right)^n \cdot \frac{n \cdot (n-1)\cdots(n-k+1)}{n \cdot n \cdots n}\left(1 - \frac{\lambda}{n}\right)^{-k}.$$

By the well-known limit result from calculus, $(1 - \lambda/n)^n$ approaches $e^{-\lambda}$ as $n \to \infty$. Since k is fixed, the third factor is the product of k ratios n/n, $(n-1)/n, \ldots, (n-k+1)/n$, all of which approach 1; hence the product also converges to 1. Also, since k and λ are fixed, the fourth factor approaches 1 as $n \to \infty$. This establishes (2.35).

—— **?Question 2.4.1** What happens to the limit-taking process in the proof in the case $k = 0$?

The preceding discussion leads us to the following definition of a new discrete distribution, whose c.d.f. appears in Table 2 of Appendix A.

DEFINITION 2.4.1 —————————————————————

The *Poisson probability mass function* with parameter λ is

$$q(k) = P[X = k] = \frac{e^{-\lambda}\lambda^k}{k!}, \qquad k = 0, 1, 2, \ldots. \qquad (2.36)$$

Table 2 of Appendix A contains probabilities for the Poisson p.m.f. for several values of λ between .5 and 10. For example, if X is Poisson(2.5) distributed, then

$$P[1 \leq X \leq 3] = .2052 + .2565 + .2138 = .6755.$$

In the case of the jumping mice and in other cases, the approximability of $b(n, p)$ probabilities by Poisson probabilities with $\lambda = np$ for large n and small p helps to justify using the Poisson distribution as a model for the distribution of the number of occurrences of some event in a fixed region of space or time, since the region can be broken into many small subsets containing at most one occurrence. The probability of observing an occurrence in a subset should converge to 0 as the subset size converges to 0. Also, the numbers of occurrences in different subsets should be independent in order for the Poisson distribution to be a reasonable model.

The Poisson mass function (2.36) does satisfy the properties that it should. It is clear that $q(k) \geq 0$ for all $k \in E$. Also,

$$\sum_{k=0}^{\infty} q(k) = \sum_{k=0}^{\infty} \frac{e^{-\lambda}\lambda^k}{k!} = e^{-\lambda} \sum_{k=0}^{\infty} \frac{\lambda^k}{k!} = e^{-\lambda} \cdot e^{\lambda} = 1.$$

The third equality in the string uses the Taylor expansion of the exponential function about 0.

A few examples of the Poisson probability mass function for several values of λ and the weightiest states are sketched in Fig. 2.9. Note that the distribution is not symmetric but becomes more so as the parameter λ increases. Also, the distribution flattens out and the weight moves out to the right as λ increases. There is a reason for this, as we will see when we study mean and variance later.

To get a feel for the accuracy of the Poisson approximation to the binomial distribution, look at Table 2.3, in which values of each distribution are displayed, in the case that the parameters are $n = 20$, $p = .1$, $\lambda = 2$. Exercise 7 asks you to create another such table. The approximation turns out to be fairly close even for such a small value of n.

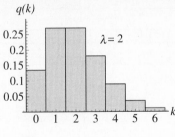

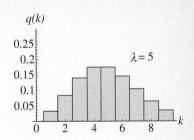

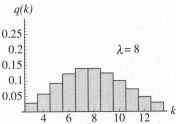

Figure 2.9 Poisson p.m.f.'s

Table 2.3

Poisson vs. Binomial Probabilities

k	Binomial	Poisson
0	.122	.135
1	.264	.271
2	.285	.271
3	.190	.180
4	.090	.090
5	.032	.036
6	.009	.012
7	.002	.004
8	.0003	.001

Example 2.4.2

Suppose that there are a large number of users of a computer lab, but during a particular short time interval each has a very small probability of wanting access to the network server. The number of users X who do want access might be modeled as a Poisson random variable, since it is the sum of many Bernoulli random variables, one for each potential user (with value 1 if the user wants access and 0 otherwise), and the success probability per user is small. Figure 2.10 illustrates the situation.

Suppose also that on the average 25% of the time, the server is not busy with any jobs from user machines. This gives us a way of estimating the parameter λ of the Poisson distribution, because

$$q(0) = P[X = 0] = \frac{e^{-\lambda}\lambda^0}{0!} = e^{-\lambda};$$

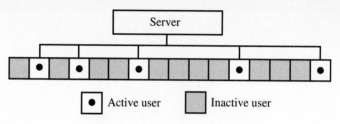

Figure 2.10 Computer Users

hence

$$\lambda = -\log q(0) = -\log(P[X=0]) = -\log(.25) \approx 1.39.$$

Note also that $e^{-\lambda} \approx .25$. We can use this information to compute, for example,

$$P[\text{exactly 3 users}] = \frac{e^{-\lambda}\lambda^3}{3!} = \frac{.25\,(1.39)^3}{6} \approx .11,$$

$$\begin{aligned}
P[\text{no more than 2 users}] &= \frac{e^{-\lambda}\lambda^0}{0!} + \frac{e^{-\lambda}\lambda^1}{1!} + \frac{e^{-\lambda}\lambda^2}{2!} \\
&= \frac{.25\,(1.39)^0}{1} + \frac{.25\,(1.39)^1}{1} + \frac{.25\,(1.39)^2}{2} \\
&\approx .84,
\end{aligned}$$

$$P[\text{at least 3 users}] = 1 - P[\text{no more than 2 users}] \approx .16.$$

?Question 2.4.2 Try to think of other examples of experimental situations where a Poisson model is appropriate, using the primitive assumption of many Bernoulli trials of small success probability.

2.4.2 Poisson Processes

Another area in which the Poisson distribution finds broad application is in random phenomena evolving over time. Suppose that for each time t, we count the number N_t of occurrences of an event up to and including t. For example, say that we are counting the arrivals of customers to a coffee shop. Then an experimental outcome ω consists of the arrival instants $(T_1(\omega), T_2(\omega), T_3(\omega), \ldots)$. For each time $t \in [0, \infty)$ we have a random variable $N_t(\omega)$, which is the total number of customers who have arrived by time t. For a fixed experimental outcome ω, we could plot this cumulative count as a function of t. (See Fig. 2.11.) The graph begins at state 0 at time 0, when no customers have arrived yet. It then jumps by exactly one unit at each arrival time of a customer.

By the device used earlier of breaking up the time interval $[0, t]$ into many small intervals, in each of which there is a very small probability of an arrival,

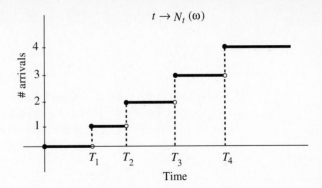

Figure 2.11 Poisson Process

it should be true that under suitable assumptions, the total number of arrivals N_t by time t has a Poisson distribution. We now detail those assumptions and use them to verify that N_t has a Poisson distribution with a parameter of a special form.

DEFINITION 2.4.2

A *Poisson process* is a family $(N_t)_{t \geq 0}$ of random variables whose paths are step functions beginning at state 0 at time 0, which jump by 1 at a sequence of random times $T_1, T_2, T_3, \ldots$. Additionally, the following conditions are assumed:

(i) The numbers of jumps within disjoint time intervals are independent.

(ii) The probability distribution of the number of jumps in a time interval depends only on the length of the interval, not on the initial time.

(iii) The probability of exactly one jump on a short time interval of length h is $\lambda h + o(h)$, the probability of two or more jumps is $o(h)$, and hence the probability of no jumps is $1 - \lambda h + o(h)$.

[Here the notation $o(h)$ indicates a generic function with the property that $o(h)/h \to 0$ as $h \to 0$.] We call λ the *rate parameter* of the Poisson process.

Condition (i) says intuitively that the Bernoulli trials formed by breaking up the time axis are independent, and similarly condition (ii) implies that the success probability will be consistent as long as the subintervals are equally sized. The last condition says that for very short time intervals $[0, h]$ only two events can happen with nonnegligible probability: There is one arrival in $[0, h]$, with approximate probability λh, or no arrivals in $[0, h]$, with approximate probability $1 - \lambda h$. This is the reason for using the term *rate* to describe λ. Under condition (iii), trials can have only two possible outcomes, no arrival or arrival, at least approximately.

We shall derive a formula for the function:

$$P_n(t) = P[N_t = n],$$

that is, the probability of exactly n arrivals by time t. This is, of course, just the p.m.f. $q(n)$ of the random variable N_t, but it proves to be useful to look at it as a family of functions of t, one for each n. The very clever approach is to try to set up a difference quotient in t for the function and a related first-order differential equation. To do this, we need an expression for $P_n(t + h) = P[N_{t+h} = n]$. The proof of the following proposition details the argument.

PROPOSITION 2.4.2

If $(N_t)_{t \geq 0}$ is a Poisson process with rate parameter λ, then

$$P[N_t = n] = \frac{e^{-\lambda t}(\lambda t)^n}{n!}; \tag{2.37}$$

that is, the number of arrivals up through time t has the Poisson(λt) distribution.

■ **Proof**   Consider the case $n \geq 1$ first. Use the Law of Total Probability to condition and uncondition the event $\{N_{t+h} = n\}$ on the value of N_t, which could be n, $n - 1$, $n - 2, \ldots, 0$. Since the event of two or more arrivals in $(t, t + h]$ has probability $o(h)$, we can write

$$
\begin{aligned}
P[N_{t+h} = n] &= P[N_t = n, N_{t+h} - N_t = 0] \\
&\quad + P[N_t = n - 1, N_{t+h} - N_t = 1] \\
&\quad + \sum_{k=0}^{n-2} P[N_t = k, N_{t+h} - N_t = n - k] \\
&= P[N_t = n]P[N_{t+h} - N_t = 0 \mid N_t = n] \\
&\quad + P[N_t = n - 1]P[N_{t+h} - N_t = 1 \mid N_t = n - 1] \\
&\quad + o(h).
\end{aligned}
$$

By Poisson process axiom (i), the number of arrivals $N_{t+h} - N_t$ during $(t, t + h]$ is independent of the number of arrivals N_t in $[0, t]$; hence the conditional probabilities on the right side can be changed to ordinary probabilities. Also, by axiom (ii), $N_{t+h} - N_t$ has the same distribution as the number of arrivals in $[0, h]$, so that these differences can be replaced by N_h. Lastly, axiom (iii) lets us express the probability of no arrivals as $1 - \lambda h$ and the probability of one arrival as λh, lumping the $o(h)$ terms into the one already present. In summary, the last expression above reduces to

$$P[N_{t+h} = n] = P_n(t + h) = P_n(t)(1 - \lambda h) + P_{n-1}(t)\lambda h + o(h). \tag{2.38}$$

Expanding out the product on the right, moving the $P_n(t)$ term to the left side, and dividing through by h gives

$$\frac{P_n(t+h) - P_n(t)}{h} = -\lambda P_n(t) + \lambda P_{n-1}(t) + \frac{o(h)}{h}.$$

On the limit as $h \to 0$, we obtain the system of differential equations:

$$P_n'(t) = -\lambda P_n(t) + \lambda P_{n-1}(t), \qquad n \geq 1. \tag{2.39}$$

In the case $n = 0$, similar reasoning (see Exercise 18) results in the following differential equation for $P_0(t) = P[N_t = 0]$:

$$P_0'(t) = -\lambda P_0(t), \qquad P_0(0) = 1. \tag{2.40}$$

The solution of (2.40) is

$$P_0(t) = P[N_t = 0] = e^{-\lambda t}, \tag{2.41}$$

which is formula (2.37) for the case $n = 0$. Straightforward substitution of formula (2.37) into the differential equation (2.39) verifies that it is a solution.

?Question 2.4.3 Why must the initial condition $P_0(0) = 1$ in (2.40) be true? What do you think might be the proper initial condition in (2.39) for $P_n(0)$?

?Question 2.4.4 Verify by substitution that $P_1(t) = e^{-\lambda t}(\lambda t)^1/1!$ satisfies (2.39).

Example 2.4.3

Let's return to our coffee shop example. Suppose that customers arrive according to a Poisson process with a rate of $\lambda = 10$ per hour. Recall that the meaning of the latter is that for short time intervals of length h hours, the probability of an arrival during the interval is about $10h$. Now the counter attendant, Nigel, is on the verge of a nervous breakdown for reasons that we would not like to get into here, and he will collapse if he has to serve more than four customers per hour. By Proposition 2.4.2, the probability of four or fewer customers during the next hour is (using Table 2 of Appendix A)

$$P[N_t \leq 4] = \sum_{n=0}^{4} P[N_t = n]$$

$$= \sum_{n=0}^{4} \frac{e^{-10 \cdot 1}(10 \cdot 1)^n}{n!}$$

$$= 0 + .0005 + .0023 + .0076 + .0189 \approx .0293.$$

Since $P[N_t > 4]$ is therefore about .97, Nigel is living on the edge.

Let us continue the example by finding the c.d.f. of the time of arrival of the second customer, and the probability that the second customer arrives between times 10 and 20 minutes, or $1/6$ and $1/3$ hour.

To compute the c.d.f. of T_2, note that the event that the second customer arrives by time t is the same as the event that the number of customers who have arrived by time t is at least 2. Therefore we can write

$$P[T_2 \le t] = P[N_t \ge 2]$$
$$= 1 - P[N_t \le 1]$$
$$= 1 - \sum_{n=0}^{1} \frac{e^{-10 \cdot t}(10 \cdot t)^n}{n!}$$
$$= 1 - e^{-10t}[1 + 10t].$$

Now we can use this result to answer the second question, because

$$P[T_2 \in (1/6, 1/3]] = P[T_2 \le 1/3] - P[T_2 \le 1/6]$$
$$= 1 - e^{-10/3}[1 + 10(1/3)] - (1 - e^{-10/6}[1 + 10(1/6)])$$
$$= \frac{8}{3}e^{-5/3} - \frac{13}{3}e^{-10/3} \approx .35.$$

EXERCISES 2.4

1. Give an argument that the number of knots in a 10-foot 1×6 board could reasonably be expected to have a Poisson distribution. What assumptions are necessary to provide for this?

2. (From Larson and Marx, 1986, p. 198) It is estimated that there could be 100 billion stars with planetary systems in our Milky Way galaxy. If p is the probability that a planetary system has intelligent life, then about how large must p be so that the probability that there is at least 1 system with intelligent life exceeds .5?

3. Suppose that the number of viruses in a small volume of blood has a Poisson distribution and that the proportion of such blood samples that have no viruses is .01. Find $P[3$ or fewer viruses$]$.

4. A Geiger counter is used to count radioactive emissions from a particular rock sample. Assume that the number of emissions in a fixed time interval has a Poisson distribution. Past observations of 16 such time intervals have yielded the following observed values of number of emissions:

$$0, \ 2, \ 4, \ 1, \ 2, \ 1, \ 3, \ 3, \ 2, \ 0, \ 4, \ 2, \ 1, \ 1, \ 3, \ 2.$$

Estimate $P[X = 5]$, $P[X \le 2]$. Do the preceding data seem to support the assumption of a Poisson distribution?

5. Suppose that gas stations tend to be distributed according to a Poissonian probability model at about one per square mile. For a 2×4 mile area, approximate the likelihood of the event that there are at least eight gas stations. (*Hint:* To arrive at an estimate of the Poisson parameter λ, break up the region into a large number n of subregions and make commonsense assumptions about what the probability of a gas station in a subregion should be.)

6. A *recurrence relation* is a formula giving the nth member of a sequence in terms of the $(n-1)$st member, and possibly preceding members. Find a recurrence relation for the values $q(n)$ of the Poisson probability mass function.

7. Make a table comparing the values of the Poisson mass function with parameter $\lambda = 1.5$ to the binomial mass function with parameters $n = 30$, $p = .05$, for states $k = 0, \ldots, 5$.

8. Plot a histogram of the Poisson probability mass function for each of the cases $\lambda = 1$, 6, 12. Include only those states that have probability of at least .02.

9. On parallel graphs, sketch the p.m.f. and c.d.f. of the Poisson(4) distribution.

10. Let X be a Poisson random variable such that $P[X = 1] = .149$ and $P[X = 2] = .224$. Find $P[X = 0]$.

11. If X is a random variable with the Poisson(1) distribution, compute the probability that X is an odd number.

12. a. This exercise is meant to set up the next section. Let X have the discrete uniform distribution on 1, 2, 3, 4, 5. Compute and try to interpret $\sum_{i=1}^{5} i P[X = i]$.

    b. Similarly, let X have the $b(3, 1/2)$ distribution. Compute and interpret $\sum_{i=0}^{3} i P[X = i]$.

    c. Now let X have the Poisson (λ) distribution. Show that

$$\sum_{i=0}^{\infty} i P[X = i] = \lambda.$$

    What is the intuitive meaning of the result?

13. Let X be Poisson(2) and let Y be Poisson(3), and suppose that the joint mass function of X and Y is the product of the marginals. Find $P[X + Y = m]$. What distribution does $X + Y$ have?

14. A city has three fire stations. During an especially hard economic period for the restaurant business, restaurant owners set fire to their establishments at the times of a Poisson process with rate 1 per hour. Assume that once a fire company goes out on an alarm, it remains occupied for at least four hours. Given that a particular day starts out with no fires, what is the probability that two hours later there will be at least one fire raging unchecked? What is the cumulative distribution function of the time of the third fire? What is the probability that the third fire occurs after the second hour but not after the fourth hour?

15. Let T_1 be the time of the first arrival of a Poisson process with rate λ. Show that the probability density function of T_1 is $\lambda e^{-\lambda t}$ for $t \geq 0$. (*Hint:* The event $\{T_1 \leq t\}$ is equivalent to what event in terms of N_t?)

16. Cars arrive at an intersection according to a Poisson process with a rate of 4 per minute. Let N_t denote the total number of cars that have arrived by time t. Find

    a. $P[N_5 = 8]$.

    b. $P[N_5 - N_2 = 6]$.

    c. $P[N_5 = 8 \,|\, N_2 = 2]$.

    d. $P[N_2 = 2 \,|\, N_5 = 8]$.

    e. $P[N_2 = k \,|\, N_5 = n]$.

17. During an 8-hour blood drive, Red Cross volunteers will serve a random stream of donors, which we will take to be a Poisson process with rate

4 per hour. Suppose that the probability that a donor is type A⁺ is .2. What is the distribution of the number of pints of type A⁺ blood that the Red Cross will receive, assuming that donations are always a pint? (*Hint:* Condition and uncondition on the total number of donors of all types.)

18. Derive the differential equation (2.40) for $P_0(t) = P[N_t = 0]$, where (N_t) is a Poisson process.

2.5 | Expected Value and Variance

2.5.1 Properties of Expectation

So far in our study of discrete probability we have made direct use of the entire probability distribution of the random variable under consideration. Now we would like to focus on certain summarizing aspects of a distribution, particularly its center of mass and degree of spread. We will find that, for many distributions, these summary quantities depend in a direct and simple way on the constant parameters that characterize the distribution.

Let us begin with a simple intuitive example. Suppose that a random variable X has two possible states, 1 and 3, as displayed in Fig. 2.12. The set of outcomes mapped by X to 1 has some probability p, and consequently the set of outcomes mapped to 3 has probability $1 - p$. What might be meant by the *average value* taken on by X? If both p and $1 - p$ are equal to 1/2, so that X is equally likely to take on each of the two values 1 and 3, then the obvious way to define the average value of X is the arithmetical average $\frac{1}{2} \cdot 1 + \frac{1}{2} \cdot 3 = 2$. (Note that the average value 2 is not itself a possible state of X.) As one state or the other receives more probability weight, the average value should be pulled toward that state. This suggests that we should take as our definition the following weighted average of the possible states:

$$\text{Average value of } X = p \cdot 1 + (1 - p) \cdot 3. \tag{2.42}$$

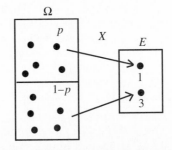

Figure 2.12 A Random Variable with Two States

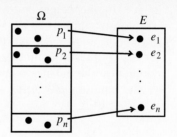

Figure 2.13   A Random Variable with Many States

This expression coincides with the computation in the $p = 1/2$ case, and in the extreme cases $p = 0$ and $p = 1$, the expression gives 3 and 1, respectively, which is in accordance with intuition.

More generally, in the situation displayed in Fig. 2.13 in which there are n possible states, the average value of X should be defined as the weighted average of the possible states e_i of X, where the weight of the ith state is the probability that X assumes that state. This is the reasoning behind the following definition.

DEFINITION 2.5.1

Let X be a discrete, real-valued random variable with state space $E = \{e_1, e_2, \ldots\}$ and probability mass function q. Then the *expected value* or *expectation* of X is

$$E[X] = \sum_i e_i P[X = e_i] = \sum_i e_i q(e_i), \qquad (2.43)$$

provided the series converges.

Defining the expected value in this way makes it a kind of balance point for the distribution. This is easiest to see in the case of two states, as shown in Fig. 2.14. Think of the probabilities as weights sitting at positions 1 and 3 on a board (like a seesaw) that pivots about a point. In Fig. 2.14(a), where the probabilities are equal, the board will balance if the pivot point is located at the average value of 2. Those of you who know a little physics will recall that the torque, or turning force, applied to the board by a weight is the weight times the distance of the weight from the pivot. In (a) the clockwise torque

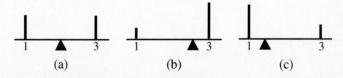

Figure 2.14   Expected Value as Balance Point

applied by the weight on the right exactly counterbalances the counterclock-wise torque applied by the weight on the left. In Fig. 2.14(b) where most of the weight is on the right, the balance point must also move right in order for the torques to counterbalance, and in Fig. 2.14(c) where the left weight dominates, the balance point moves left. (See Exercise 2 for more detail.)

The notation $\mu = E[X]$ is used very frequently in probability. We call μ the *mean* of X.

?Question 2.5.1   Must a nonnegative valued random variable have a nonnegative mean? Why?

Example 2.5.1

Fast Eddie likes to play the horses. In an attempt to diversify his investments, he decides to bet $8 on horse A, which is rated as a 20:1 shot to win; $4 on horse B, which is rated as a 5:1 shot; and $10 on horse C, rated as a 2:1 shot to win. If horse A wins, Eddie receives $18 on the dollar, similarly horse B pays $4, and horse C pays $1.50. Find Fast Eddie's expected winnings.

Let X be the random variable representing Eddie's net winnings, after deducting the $22 he has bet. Then X takes on these values: (a) $18 \cdot 8 - 22 = 122$ if horse A wins; (b) $4 \cdot 4 - 22 = -6$ if horse B wins; (c) $10 \cdot 1.5 - 22 = -7$ if horse C wins; and (d) $0 - 22 = -22$ if none of the horses A, B, or C win. The odds ratios may be translated into probabilities as follows. If the odds against an event are $m:n$, then the probability that the event occurs is $n/(n + m)$. Thus horse A wins with probability 1/21, horse B wins with probability 1/6, horse C wins with probability 1/3, and by complementation, the probability that none of these horses win is $1 - 1/21 - 1/6 - 1/3 = 19/42$. We have now found the state space of X and the probabilities on the states, listed in the following table:

Horse	A	B	C	None
Probability	1/21	1/6	1/3	19/42
Winnings	122	−6	−7	−22

Hence formula (2.43) yields

$$E[X] = \frac{1}{21} \cdot 122 + \frac{1}{6} \cdot (-6) + \frac{1}{3} \cdot (-7) + \frac{19}{42} \cdot (-22) = -\frac{314}{42} \approx -7.48.$$

It appears as though playing the horses is a losing proposition. The next question elaborates on this example.

?Question 2.5.2   What would Fast Eddie's expected net winnings be if he had placed all of his money on the long shot? What do you think of his diversification system? Supposing (very hypothetically) that you were inclined to bet on one of the three horses described, which would you choose and why?

Example 2.5.2

This example illustrates a case where a random variable does not have finite expectation. Suppose that a fair coin is flipped independently until the first head occurs. Then a typical outcome ω is a string of 0 or more tails, followed by a head. Let $X(\omega)$ have the value 2^n if ω is such that the first head is on the nth flip. Then $P[X = 2^n] = (1/2)^n$; consequently,

$$E[X] = \sum_{n=1}^{\infty} 2^n \cdot \left(\frac{1}{2}\right)^n = \sum_{n=1}^{\infty} 1 = \infty.$$

Example 2.5.3

The television game show "The Price Is Right" contains a game called Plinko, which is an amazingly rich illustration of probabilistic ideas. A contestant who has successfully priced some items earns one or more chips and has the opportunity to stand at the top of the Plinko board, sketched in Fig. 2.15, and release a chip from any of the topmost slots. The chip travels down the

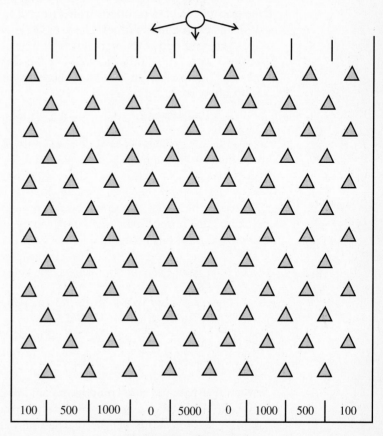

Figure 2.15   Plinko Board

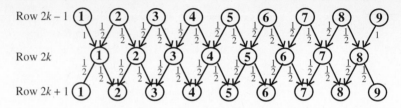

Figure 2.16 Successive Plinko Rows

board, bouncing either right or left by one column (presumably with equal probability) when it encounters a peg. When it hits a wall it just bounces off in the only available direction. Finally the chip lands in a bin at the bottom of the Plinko board, earning the contestant the amount shown. Contestants' behaviors vary as to where they let go of their chips, but there is a definite tendency to choose slots in the middle. The question is, Which slot should contestants choose to maximize the expected value of their winnings?

The winnings in each bin are given, so we are left with the problem of finding the probabilities of landing in each bin. You can count 13 rows of the board, including the top slots and the bottom bins. Odd-numbered rows have nine possible column positions from left wall to right (look at the tops of the pegs), and even-numbered rows have eight. Figure 2.16 illustrates consecutive odd, even, and odd rows schematically. Except for the fact that odd 1 goes to even 1 with probability one because the chip bounces off of the left wall, and odd 9 goes to even 8 with probability one, all other transitions occur with probability 1/2. Let

$$p_m(i) = \text{probability that chip is in column } i \text{ when it hits row } m.$$

Of course this probability also depends on which slot the chip was dropped from, but we will suppress that in the notation. We are trying to find the set of numbers $p_{13}(i), i = 1, 2, \ldots, 9$. In row 1, when the contestant chooses slot i from which to drop the chip, it means that $p_1(i) = 1$ and all other $p_1(j) = 0$.

Consider how the chip could have arrived at column i of an even row numbered $2k$. It must have been either in column i of the previous odd row and moved right or in column $i + 1$ of the previous odd row and moved left. Columns 1 and 8 are somewhat special because of the walls of the board. By the Law of Total Probability, we have the set of recurrence relations:

$$p_{2k}(1) = 1 \cdot p_{2k-1}(1) + \frac{1}{2} \cdot p_{2k-1}(2),$$

$$p_{2k}(i) = \frac{1}{2} \cdot p_{2k-1}(i) + \frac{1}{2} \cdot p_{2k-1}(i + 1), \ i = 2, \ldots, 7, \qquad (2.44)$$

$$p_{2k}(8) = \frac{1}{2} \cdot p_{2k-1}(8) + 1 \cdot p_{2k-1}(9).$$

Now look at how the chip could have come to column i of an odd row numbered $2k + 1$. Column 1 is accessible only through column 1 of the previous even row, and column 9 can be reached only through column 8 of the previous even row. For columns i in between, the previous column could have

Table 2.4

Plinko Expected Winnings

Bin	1	2	3	4	5	6	7	8	9	
Prize	$100	500	1000	0	5000	0	1000	500	100	*E[win]*
Slot 1	.23	.39	.24	.11	.03	.01	0	0	0	608
Slot 2	.19	.35	.25	.14	.06	.02	0	0	0	744
Slot 3	.12	.25	.24	.20	.12	.05	.02	0	0	997
Slot 4	.05	.14	.20	.23	.19	.12	.05	.02	0	1285
Slot 5	.02	.06	.12	.19	.23	.19	.12	.06	.02	1454
Slot 6	0	.02	.05	.12	.19	.23	.20	.14	.05	1285
Slot 7	0	0	.02	.05	.12	.20	.24	.25	.12	997
Slot 8	0	0	0	.02	.06	.14	.25	.35	.19	744
Slot 9	0	0	0	.01	.03	.11	.24	.39	.23	608

been either $i - 1$ or i in the even row. Again, by the Law of Total Probability, we have the recurrence relations:

$$p_{2k+1}(1) = \frac{1}{2} \cdot p_{2k}(1),$$

$$p_{2k+1}(i) = \frac{1}{2} \cdot p_{2k}(i - 1) + \frac{1}{2} \cdot p_{2k}(i), \qquad i = 2, \ldots, 8, \qquad (2.45)$$

$$p_{2k+1}(9) = \frac{1}{2} \cdot p_{2k}(8).$$

These difference equations can be solved analytically (think about using matrix multiplication to express one vector of probabilities p_m in terms of the preceding vector p_{m-1}), but numerical calculuation is good enough to answer the question. Indeed, the problem is ideally suited for computation on a spreadsheet. We set into a row the initial values $p_1(i)$ as 1 or 0 as already described, then put formulas in successive rows $2, 3, \ldots, 13$ according to (2.44) or (2.45). The desired probabilities come out in row 13. They are shown in Table 2.4, rounded to the nearest hundredth, with the resulting expected rewards for each initial slot in which the contestant could drop the chip. The table shows that it is the center slot that produces the highest expected reward, as you might have guessed.

Next is a formula for the expected value of a function of a random variable. If X is a discrete random variable with p.m.f. $q(x)$, and f is a real-valued function whose domain includes the state space of X, then the *expected value of $f(X)$* is

$$E[f(X)] = \sum_{x \in E} f(x) \cdot q(x), \qquad (2.46)$$

provided the series converges. Notice that it is not necessary to explicitly find the p.m.f. of $Y = f(X)$ to compute the expectation $E[Y] = E[f(X)]$, although you can. To illustrate, let X have state space $E = \{-2, -1, 0, 1, 2\}$, and p.m.f.

$$q_X(-2) = 1/10, \quad q_X(-1) = 1/5, \quad q_X(0) = 2/5, \quad q_X(1) = 1/5, \quad q_X(2) = 1/10.$$

Consider the transformed random variable $Y = f(X) = X^2$. Then Y has state space $E_Y = \{0, 1, 4\}$, and $Y = 0$ iff $X = 0$; $Y = 1$ iff either $X = -1$ or $X = 1$; and $Y = 4$ iff either $X = -2$ or $X = 2$. This means that Y has the p.m.f.

$$q_Y(0) = 2/5, \qquad q_Y(1) = 2/5, \qquad q_Y(4) = 1/5.$$

Therefore the expectation of Y using the definition of expectation is

$$E[Y] = \frac{2}{5} \cdot 0 + \frac{2}{5} \cdot 1 + \frac{1}{5} \cdot 4 = \frac{6}{5}.$$

But the sum on the right side of the last formula can also be written as

$$E[Y] = \frac{2}{5} \cdot 0^2 + \frac{1}{5} \cdot (-1)^2 + \frac{1}{5} \cdot 1^2 + \frac{1}{10} \cdot (-2)^2 + \frac{1}{10} \cdot 2^2$$

$$= \sum_{x=-2}^{2} x^2 \cdot q_X(x)$$

$$= \sum_{x=-2}^{2} f(x) \cdot q_X(x).$$

This means that in order to find the expected value of $f(X)$, it suffices to compute the weighted average of all $f(x)$ values for x in the state space of X, using the p.m.f. of X as the weighting system. The preceding argument, presented in the context of a special example, can be adapted so as to prove formula (2.46) in general.

Here are two very important computational results for expectation.

PROPOSITION 2.5.1

If c is a real constant, then

$$E[c] = c. \tag{2.47}$$

■ **Proof** There are two ways in which we could look at the meaning of $E[c]$. One of these is to let X be any discrete random variable with p.m.f. q, and to consider the function $f(X) = c$. By (2.46) the expectation is

$$E[c] = \sum_{x \in E} c \cdot q(x) = c \cdot \sum_{x \in E} q(x) = c \cdot 1 = c.$$

For the second approach, see the next question.

—— **?Question 2.5.3** Give an alternative proof of Proposition 2.5.1 using a random variable with only one state c.

PROPOSITION 2.5.2

If X and Y are random variables with finite expectation, then

$$E[aX + bY] = aE[X] + bE[Y]. \tag{2.48}$$

■ **Proof** Let $q(x, y)$ be the joint p.m.f. of X and Y. Then,

$$E[aX + bY] = \sum_x \sum_y (ax + by) \cdot q(x, y)$$

$$= a \sum_x \sum_y x \cdot q(x, y) + b \sum_x \sum_y y \cdot q(x, y)$$

$$= a \sum_x x \sum_y q(x, y) + b \sum_y y \sum_x q(x, y)$$

$$= a \sum_x x q_x(x) + b \sum_y y \cdot q_y(y)$$

$$= aE[X] + bE[Y]. \tag{2.49}$$

It should be easy to see from the proof that formula (2.48), called the *linearity of expectation*, generalizes to linear combinations of more than two random variables.

Example 2.5.4

To gain information on the expected time in the recovery room of patients who have undergone a certain surgical procedure, a sample of 14 patients was taken. Their times, recorded to the nearest minute, follow:

58, 75, 53, 60, 62, 80, 65, 54, 78, 63, 58, 54, 60, 63.

These times are observed values of random variables $X_1, X_2, \ldots, X_{14}$, which we will suppose have the same probability distribution and hence the same expectation. We again denote the mean of the distribution by $\mu = E[X_i]$. Intuitively, the simple arithmetical average of the observed values $\sum_{i=1}^{14} x_i / 14 = 63.07$ should be a good estimate of the underlying mean μ. One reason is that, by Proposition 2.5.2,

$$E\left[\frac{1}{14} \sum_{i=1}^{14} X_i\right] = \frac{1}{14} \sum_{i=1}^{14} E[X_i] = \frac{1}{14} \cdot 14\mu = \mu. \tag{2.50}$$

That is, the sample average has the same expectation as the identically distributed random variables that contribute to the average. But what makes the sample average a better estimator than, say, the first observation 58, or the average of the first two observations $(58 + 75)/2 = 66.5$? It would seem that averaging a large number of observations smoothes out variation that can severely affect one or two observations, thus providing more precision. Later, when we discuss the variance of a distribution, we will be able to say more.

2.5.2 Variance and Moments

The mean μ is a useful measure of the central tendency of the probability distribution, but the general nature of a distribution depends on more than just its center. Another important feature of a distribution is its spread, as measured by the expected square difference between the random variable and its mean.

DEFINITION 2.5.2

The *variance* of a real random variable X is

$$\sigma^2 = \text{Var}(X) = E[(X - \mu)^2], \tag{2.51}$$

provided the expectation is finite. The square root of the variance, σ, is referred to as the *standard deviation* of the random variable.

We also refer to the *variance* and *standard deviation* of a *distribution*, which are the same quantities as in the last definition for a random variable possessing that distribution.

Example 2.5.5

To gain an intuitive appreciation for what the variance measures, consider the three probability mass functions sketched in Fig. 2.17.

All three have mean equal to 2 (check this). The distribution q_x in part (a) is our baseline. The distribution q_y in part (b) differs from it in the sense that the states have been moved away from center, although the probabilities have not changed. The distribution q_z in (c) differs from the baseline in the sense that the probabilities have been redistributed away from the center, although the states are the same. In both cases we have a dispersion of probability away from center as compared to the baseline distribution. The variances are

$$\sigma_x^2 = E[(X - \mu_x)^2] = \frac{1}{4}(1 - 2)^2 + \frac{1}{2}(2 - 2)^2 + \frac{1}{4}(3 - 2)^2 = \frac{1}{2},$$

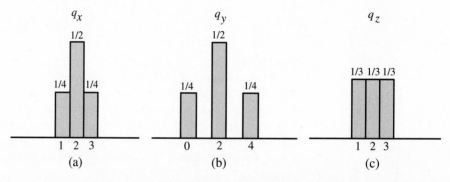

Figure 2.17 Three p.m.f.'s with Different Variances

$$\sigma_y^2 = E[(Y - \mu_y)^2] = \frac{1}{4}(0-2)^2 + \frac{1}{2}(2-2)^2 + \frac{1}{4}(4-2)^2 = 2,$$

$$\sigma_z^2 = E[(Z - \mu_z)^2] = \frac{1}{3}(1-2)^2 + \frac{1}{3}(2-2)^2 + \frac{1}{3}(3-2)^2 = \frac{2}{3}.$$

It should be no surprise that the variances of Y and Z came out to be larger than that of X. You should make term-by-term comparisons in the computation preceding to solidify your understanding of exactly why.

?Question 2.5.4 Referring to the last example, compute the variance for a distribution that disperses probability in both senses; that is, the states are 0, 2, and 4, and the probabilities are each $1/3$. How does it compare to the three preceding variances?

The following formula can frequently simplify computation of the variance:

$$\begin{aligned} \sigma^2 = E[(X - \mu)^2] &= E[X^2 - 2\mu X + \mu^2] \\ &= E[X^2] - 2\mu E[X] + \mu^2 \\ &= E[X^2] - 2\mu^2 + \mu^2 \\ &= E[X^2] - \mu^2. \end{aligned} \tag{2.52}$$

Both the linearity property of expectation and the result on constants were used in this derivation.

Example 2.5.6 Let X be a random variable with the uniform distribution on $\{1,2,3,4,5,6\}$. The mean, variance, and standard deviation of X are

$$\mu = E[X] = \frac{1}{6}(1+2+3+4+5+6) = \frac{21}{6} = \frac{7}{2}$$

$$\sigma^2 = E[X^2] - \mu^2 = \frac{1}{6}(1^2 + 2^2 + 3^2 + 4^2 + 5^2 + 6^2) - \left(\frac{7}{2}\right)^2 = \frac{35}{12}$$

$$\sigma = \sqrt{\sigma^2} = \sqrt{35/12} \approx 1.71.$$

Had we used the defining formula (2.51) rather than the computational formula (2.52) to find σ^2, we would have added terms like $(1 - 7/2)^2, (2 - 7/2)^2$, and so on, necessitating much more arithmetic of fractions.

The variance has an interesting property that will give us more information about the estimation question raised in Example 2.5.4.

PROPOSITION 2.5.3

If X is a real random variable with finite variance and a and b are real constants, then

$$\text{Var}(aX + b) = a^2 \text{Var}(X). \tag{2.53}$$

■ **Proof**   By linearity of expectation, the mean μ_y of the random variable $Y = aX + b$ is $\mu_y = E[aX + b] = a\mu_x + b$, where μ_x is the expected value of X. Then,

$$\begin{aligned}
\text{Var}(aX + b) = E[(Y - \mu_y)^2] &= E[((aX + b) - (a\mu_x + b))^2] \\
&= E[(a(X - \mu_x))^2] \\
&= a^2 E[(X - \mu_x)^2] \\
&= a^2 \text{Var}(X).
\end{aligned}$$

—— **?Question 2.5.5**   What is the variance of a constant random variable $X(\omega) = b$, $\forall \omega \in \Omega$?

Remember that in Example 2.5.4 we were concerned with estimating the mean recovery time of patients by the sample average of 14 observed times. Generalizing a bit, for a collection of n random observations $X_1, X_2, \ldots, X_n$ from the same probability distribution, define the *sample mean* to be

$$\overline{X} = \frac{1}{n} \sum_{i=1}^{n} X_i. \tag{2.54}$$

As in the example it is easy to check that $\overline{X}$ has expected value equal to the common expectation μ of the X_i's. The variance of $\overline{X}$ is

$$\text{Var}(\overline{X}) = \text{Var}\left(\frac{1}{n} \sum_{i=1}^{n} X_i\right) = \frac{1}{n^2} \text{Var}\left(\sum_{i=1}^{n} X_i\right),$$

by the previous proposition.

Here we hit a roadblock, because we do not know yet how to find the variance of a sum of random variables in terms of their individual variances. But in Chapter 4 we will show that under an independence condition on the X_i's, the variance of the sum is the sum of the variances. Accepting this result for the time being, we then obtain

$$\text{Var}(\overline{X}) = \frac{1}{n^2} \sum_{i=1}^{n} \text{Var}(X_i) = \frac{1}{n^2} \sum_{i=1}^{n} \sigma^2 = \frac{1}{n} \sigma^2, \tag{2.55}$$

where σ^2 is the common variance of the X_i's. This means that the sample mean $\overline{X}$ has average value μ, and its variance approaches 0 at the rate of $1/n$ as the sample size $n \to \infty$. The distribution of probability weight for $\overline{X}$ will therefore be tightly packed around μ for large n, which will make $\overline{X}$ a very precise estimator of μ.

The mean and variance are special cases of expectations called the *moments* of a probability distribution, defined as follows.

DEFINITION 2.5.3 ———————————————————————————

The *r*th *moment* of the distribution of a real random variable X is $E[X^r]$, provided the expectation exists. Similarly the *r*th *central moment* or *moment about the mean* is $E[(X - \mu)^r]$.

The first moment of the distribution of X is the mean; the second moment about the mean is the variance of the distribution. Apart from these, the moment that probably commands the most attention among practitioners is the third moment about the mean $E[(X - \mu)^3]$, called the *skewness*. The next example illustrates the information that this expectation supplies about a distribution.

Example 2.5.7

Consider the three probability distributions displayed in Fig. 2.18.

You can verify easily that the means are $E[X] = 3$, $E[Y] = 37/15$, $E[Z] = 53/15$. The first distribution is symmetric about its mean, the second is asymmetric in the sense that it has a long right tail, and the third is asymmetric in the sense that it has a long left tail. The distribution in part (b) would be called *skewed to the right* or *positively skewed*, and the distribution in part (c), *skewed to the left* or *negatively skewed*. We can compute the skewness of X as

$$E[(X - \mu_x)^3] = \frac{1}{10} \cdot (1-3)^3 + \frac{2}{10} \cdot (2-3)^3 + \frac{4}{10} \cdot (3-3)^3$$
$$+ \frac{2}{10} \cdot (4-3)^3 + \frac{1}{10} \cdot (5-3)^3 = 0.$$

Notice that because of the symmetry of the distribution of X, the negative terms in the computation exactly subtract away the positive terms. It is true in general for symmetric distributions that the skewness, if finite, is zero. The skewness of Y is

$$E[(Y - \mu_y)^3] = \frac{1}{15} \cdot \left(1 - \frac{37}{15}\right)^3 + \frac{9}{15} \cdot \left(2 - \frac{37}{15}\right)^3 + \frac{3}{15} \cdot \left(3 - \frac{37}{15}\right)^3$$
$$+ \frac{1}{15} \cdot \left(4 - \frac{37}{15}\right)^3 + \frac{1}{15} \cdot \left(5 - \frac{37}{15}\right)^3 = 1.08.$$

So, the right-skewed distribution has a positive skewness. Finally, the skewness of Z is

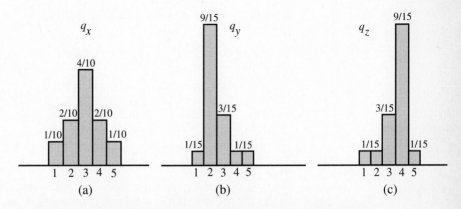

Figure 2.18 Three p.m.f.'s with Different Skewnesses

$$E[(Z - \mu_z)^3] = \frac{1}{15} \cdot \left(1 - \frac{53}{15}\right)^3 + \frac{1}{15} \cdot \left(2 - \frac{53}{15}\right)^3 + \frac{3}{15} \cdot \left(3 - \frac{53}{15}\right)^3$$

$$+ \frac{9}{15} \cdot \left(4 - \frac{53}{15}\right)^3 + \frac{1}{15} \cdot \left(5 - \frac{53}{15}\right)^3 = -1.08.$$

Thus the left-skewed distribution has a negative skewness, of the same magnitude as the mirror image right-skewed distribution in part (b).

We note in passing that most authors prefer a standardized measure of skewness, which divides the third central moment by the cube of the standard deviation, thereby correcting for spread.

2.5.3 Special Distributions

In this subsection, we list for future reference the means and variances of some of the important discrete distributions encountered earlier in the chapter.

PROPOSITION 2.5.4

If X has the Bernoulli distribution with success parameter p, then

$$E[X] = p; \quad \text{Var}(X) = p(1 - p). \tag{2.56}$$

■ **Proof** The proof is left as Exercise 16.

Although we can compute expectations of Bernoulli random variables from scratch, it also happens that the result is the special case $n = 1$ of the following result.

PROPOSITION 2.5.5

If X has the $b(n, p)$ distribution, then

$$E[X] = np; \quad \text{Var}(X) = np(1 - p). \tag{2.57}$$

■ **Proof** By the definition of expected value,

$$E[X] = \sum_{k=0}^{n} k \cdot \binom{n}{k} p^k (1 - p)^{n-k}$$

$$= n \cdot p \cdot \sum_{k=1}^{n} \frac{(n-1)!}{(k-1)!(n-k)!} p^{k-1} (1 - p)^{n-k}$$

$$= n \cdot p \cdot \sum_{l=0}^{n-1} \frac{(n-1)!}{l!(n-1-l)!} p^l (1 - p)^{n-1-l}$$

$$= n \cdot p \cdot 1.$$

In the second line the $k = 0$ term is dropped, n and p are factored out of the sum, and k is canceled from the $k!$ factor in the denominator. The third line results from substituting $l = k - 1$. The sum compresses to 1 in the fourth line because it is the sum of all values of the $b(n-1, p)$ probability mass function.

To find the variance, we take the circuitous, but surprisingly helpful route of finding $E[X(X-1)]$. This can be done very similarly to the mean computation:

$$E[X(X-1)] = \sum_{k=0}^{n} k(k-1) \cdot \binom{n}{k} p^k (1-p)^{n-k}$$

$$= n(n-1)p^2 \cdot \sum_{k=2}^{n} \frac{(n-2)!}{(k-2)!(n-k)!} p^{k-2} (1-p)^{n-k}$$

$$= n(n-1)p^2 \cdot \sum_{l=0}^{n-2} \frac{(n-2)!}{l!(n-2-l)!} p^l (1-p)^{n-2-l}$$

$$= n(n-1)p^2.$$

(Make sure that you understand each line of this derivation.) Now we can compute

$$\mathrm{Var}(X) = E[X^2] - (E[X])^2 = E[X(X-1)] + E[X] - (E[X])^2$$
$$= n(n-1)p^2 + np - (np)^2$$
$$= n^2 p^2 - np^2 + np - n^2 p^2$$
$$= np - np^2 = np(1-p).$$

Note that for a multinomial random variable **X** with parameters $n, p_1, p_2, \ldots, p_k$, the marginal distribution of X_i is $b(n, p_i)$; hence the expected number of objects of category i in a multinomial experiment is np_i. This fact will become important later when we discuss goodness-of-fit tests.

?Question 2.5.6 Give an alternative proof of the mean and variance formulas for the binomial distribution using the results for the Bernoulli distribution, linearity, and independence.

Example 2.5.8 A contractor submits bids on ten building projects, on each of which four other contractors will also bid. How many bids would the contractor expect to win? State assumptions carefully.

In the experiment described in the question, there is a sequence of ten trials, one for each project. A success corresponds to the event that the contractor in question gets a bid. Under the assumptions that awards on different projects are independent of one another (what could be wrong with this?), and that each bidder is as likely as any other to win the contract, we have a binomial experiment with ten trials and success probability per trial of $1/5$. Multiplying the number of contracts $n = 10$ by the success probability per contract $p = 1/5$ gives the expected value of 2.

Although in this example the expected value of X = number of contracts won is indeed a possible state for X, this needn't be the case, as we see by simply considering a similar scenario with only three other bidders. The average number of contracts won is $10 \cdot (1/4) = 5/2$, which is not in the state space of X. One way of interpreting this number might be that if a large number of groups of ten projects were observed, and the numbers of contracts won on each were averaged, the average number won would be about $5/2$.

PROPOSITION 2.5.6

If X has the Poisson(λ) distribution, then

$$E[X] = \lambda; \quad \text{Var}(X) = \lambda. \tag{2.58}$$

■ **Proof**   The Poisson distribution is another distribution amenable to the sort of computational tricks we used on the binomial distribution. First, the expected value is

$$E[X] = \sum_{n=0}^{\infty} n \cdot \frac{e^{-\lambda} \lambda^n}{n!}$$

$$= \lambda \cdot \sum_{n=1}^{\infty} \frac{e^{-\lambda} \lambda^{n-1}}{(n-1)!}$$

$$= \lambda \cdot \sum_{m=0}^{\infty} \frac{e^{-\lambda} \lambda^m}{m!} = \lambda.$$

In the second line a common factor of λ is pulled out, the $n = 0$ term, which is 0, is dropped, and an n is canceled from $n!$. The third line changes the variable of summation to $m = n - 1$. Then the resulting series sums to 1, since it is the sum of all values of the Poisson(λ) p.m.f.

A similar trick works for the variance; first,

$$E[X(X-1)] = \sum_{n=0}^{\infty} n(n-1) \cdot \frac{e^{-\lambda} \lambda^n}{n!}$$

$$= \lambda^2 \cdot \sum_{n=2}^{\infty} \frac{e^{-\lambda} \lambda^{n-2}}{(n-2)!}$$

$$= \lambda^2 \cdot \sum_{m=0}^{\infty} \frac{e^{-\lambda} \lambda^m}{m!} = \lambda^2.$$

Then as in the derivation for the binomial distribution,

$$\text{Var}(X) = E[X(X-1)] + E[X] - (E[X])^2 = \lambda^2 + \lambda - \lambda^2 = \lambda.$$

Referring back to Section 2.4, since λ is the mean of the Poisson distribution, it is reasonable to estimate it by the sample mean of a collection of n

observations. First, the larger n is, the more precise the estimate is, that is, the smaller is its variance. Second, since λ is both the mean and the variance of the distribution, as λ grows the probability weight not only shifts to the right but also spreads out, as we noticed in an earlier figure. Third, recall that the number N_t of arrivals in a Poisson process by time t has the Poisson distribution with parameter λt. We now know that λt is the expected number of arrivals by time t; hence λ itself is the expected number of arrivals per unit time. This is another reason for calling λ the arrival rate of the process.

PROPOSITION 2.5.7

If X has the geometric distribution with parameter p, then

$$E[X] = \frac{1}{p}; \quad \text{Var}(X) = \frac{1-p}{p^2}. \tag{2.59}$$

If X has the negative binomial distribution with parameters r and p, then

$$E[X] = \frac{r}{p}; \quad \text{Var}(X) = r \cdot \frac{1-p}{p^2}. \tag{2.60}$$

■ **Proof** The proof is left for Exercise 17.

From this proposition we obtain the intuitive result that the expected number of trials until the first success in a sequence of Bernoulli trials with, for instance, success probability $p = 1/3$ is 3. The expected number of trials until the second success would be $2 \cdot 3 = 6$.

2.5.4 Multivariate Expectation

To this point, we have worked with one-dimensional random variables, that is, those whose state spaces are discrete subsets of the real line. It is easy to extend the idea to the case of multidimensional random vectors.

DEFINITION 2.5.4

If $\mathbf{X} = [X_1 \ X_2 \ \ldots \ X_n]'$ is a random vector, then the *expected value* of $\mathbf{X}$ is the vector

$$E[\mathbf{X}] = \begin{bmatrix} E[X_1] \\ E[X_2] \\ \vdots \\ E[X_n] \end{bmatrix}, \tag{2.61}$$

provided the individual component expectations exist.

There is also an extension of the concept of variance to multidimensional random vectors, but we will postpone consideration of that until Chapter 4. There is no ambiguity about what the component expectations mean, because if we computed using the joint mass function $q(x_1, x_2, \ldots, x_n)$, we would find:

$$E[X_1] = \sum_{x_1}\sum_{x_2}\cdots\sum_{x_n} x_1 \cdot q(x_1, x_2, \ldots, x_n)$$

$$= \sum_{x_1} x_1 \sum_{x_2}\cdots\sum_{x_n} q(x_1, x_2, \ldots, x_n)$$

$$= \sum_{x_1} x_1 \cdot q_1(x_1). \tag{2.62}$$

The last formula is the mean computed using the marginal p.m.f. of X_1. Of course, the same argument can be applied to the other component expectations.

Example 2.5.9

If X_1 and X_2 have the discrete uniform distribution on the triangular integer grid shown in Fig. 2.19, find $E[X_2]$ and $E[X_1 \cdot X_2]$.

There are ten equally likely points (i, j) in the state space of $\mathbf{X} = (X_1, X_2)$, each of which have probability $q(i, j) = 1/10$. Then

$$E[X_2] = \sum_i\sum_j j \cdot \frac{1}{10}$$

$$= \frac{1}{10}(1 + 1 + 1 + 1 + 2 + 2 + 2 + 3 + 3 + 4) = 2.$$

For the second question, we have

$$E[X_1 \cdot X_2] = \sum_i\sum_j i \cdot j \cdot \frac{1}{10}$$

$$= \frac{1}{10}(1 \cdot 1 + 2 \cdot 1 + 3 \cdot 1 + 4 \cdot 1 + 1 \cdot 2$$
$$+ 2 \cdot 2 + 3 \cdot 2 + 1 \cdot 3 + 2 \cdot 3 + 1 \cdot 4)$$

$$= \frac{35}{10}.$$

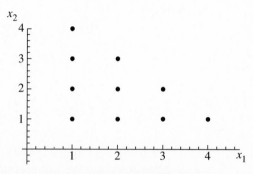

Figure 2.19 A Two-Variable Discrete Uniform Distribution

—— **?Question 2.5.7**   In Example 2.5.9, what is $E[\mathbf{X}]$?

The linearity property of expectation carries over to the multivariate case, as the next proposition shows.

PROPOSITION 2.5.8

(a) Let $\mathbf{X}$ and $\mathbf{Y}$ be random vectors of the same dimension n, and let a and b be scalar constants. Then,

$$E[a\mathbf{X} + b\mathbf{Y}] = aE[\mathbf{X}] + bE[\mathbf{Y}], \tag{2.63}$$

provided the expectations of $\mathbf{X}$ and $\mathbf{Y}$ exist.

(b) Let $\mathbf{X}$ be an n-dimensional random vector and let A be an $m \times n$ matrix of constants. Then,

$$E[A\mathbf{X}] = AE[\mathbf{X}], \tag{2.64}$$

provided the expectation of $\mathbf{X}$ exists.

■ **Proof**   (a) See Exercise 25.

(b) It will be helpful to think of matrix multiplication in the following way. Let A be a matrix as hypothesized in part (b) and let $\mathbf{b}$ be an n-dimensional column vector. Denote the rows of A as $\mathbf{a}_1, \mathbf{a}_2, \ldots, \mathbf{a}_m$. Then the product $A\mathbf{b}$ has m rows $\mathbf{a}_1 \cdot \mathbf{b}, \mathbf{a}_2 \cdot \mathbf{b}, \ldots, \mathbf{a}_m \cdot \mathbf{b}$. The symbol "$\cdot$" indicates vector dot product here. Therefore

$$E[A\mathbf{X}] = E\left[\begin{bmatrix} \mathbf{a}_1 \cdot \mathbf{X} \\ \mathbf{a}_2 \cdot \mathbf{X} \\ \vdots \\ \mathbf{a}_m \cdot \mathbf{X} \end{bmatrix}\right] = \begin{bmatrix} E[\mathbf{a}_1 \cdot \mathbf{X}] \\ E[\mathbf{a}_2 \cdot \mathbf{X}] \\ \vdots \\ E[\mathbf{a}_m \cdot \mathbf{X}] \end{bmatrix}.$$

It is easy to prove (see Exercise 24) that

$$E[\mathbf{a}_i \cdot \mathbf{X}] = \mathbf{a}_i \cdot E[\mathbf{X}]; \tag{2.65}$$

hence

$$E[A\mathbf{X}] = \begin{bmatrix} \mathbf{a}_1 \cdot E[\mathbf{X}] \\ \mathbf{a}_2 \cdot E[\mathbf{X}] \\ \vdots \\ \mathbf{a}_m \cdot E[\mathbf{X}] \end{bmatrix} = A \cdot E[\mathbf{X}].$$

We will find the matrix version of linearity to be particularly useful later in the book when we study regression and experimental design. As a brief note on the use of (2.64), we can find in one computation the expected values of several linear combinations of the components of a vector random variable. Suppose $\mathbf{X} = [X_1 \ X_2 \ X_3]'$ has expected value $[0 \ 2 \ -1]'$. Then the expected values of $X_1 - X_2$, $X_2 - X_3$, and $2X_1 + X_2 +$

X_3 can be found by computing as follows:

$$E\left[\begin{bmatrix} X_1 - X_2 \\ X_2 - X_3 \\ 2X_1 + X_2 + X_3 \end{bmatrix}\right] = E\left[\begin{bmatrix} 1 & -1 & 0 \\ 0 & 1 & -1 \\ 2 & 1 & 1 \end{bmatrix}\begin{bmatrix} X_1 \\ X_2 \\ X_3 \end{bmatrix}\right]$$

$$= \begin{bmatrix} 1 & -1 & 0 \\ 0 & 1 & -1 \\ 2 & 1 & 1 \end{bmatrix}\begin{bmatrix} 0 \\ 2 \\ -1 \end{bmatrix}$$

$$= \begin{bmatrix} -2 \\ 3 \\ 1 \end{bmatrix}.$$

EXERCISES 2.5

1. Find the expected sum of the upturned faces of two fairly rolled fair dice.

2. Let the torque applied by a state x relative to a fixed point c be defined as the product of the probability weight given to x and the difference $c - x$. Consider a distribution with mass confined to two states a and b on the real line. Show that the mean of the distribution is the unique fixed point relative to which the total torque applied by both states is 0. State and prove a similar result for a three-point distribution on the real line.

3. As a variation on Example 2.5.2, suppose that you are playing a game that pays $2 if the first head occurs on the first flip of a fair coin, $4 if the first head occurs on the second flip, $8 if it occurs on the third flip, and so on, up to a maximum of $1024 if the first head occurs on the tenth flip or thereafter. What entry fee would make the expected net winnings equal to 0?

4. A $30,000 term insurance policy is written so that it pays only if its owner dies within the next 25 years. Until death or until the 25 years is up, the owner pays premiums of p dollars each year. Suppose that the owner is certain to die within the next 50 years, and the probability that he will die in the nth year is proportional to n. What value of p marks the break-even point for the insurance company? (You needn't account for inflation of money.)

5. An oil company submits a bid of $1 million on an offshore area that the government is releasing for drilling. The company will win the bid and be awarded exclusive rights to the area with probability .4. If awarded the bid, the company will drill and will find oil worth $6 million with probability 1/3, otherwise the hole will be dry and yield nothing. The company's drilling costs are $1 million. Find the expected value of the deal to the company.

6. In Example 2.5.3, on the Plinko game, express for each column i and each odd numbered row $2k + 1$ the relationship between the probabilities $p_{2k+1}(i)$ and the set of probabilities for the previous odd-numbered row $\{p_{2k-1}(1), p_{2k-1}(2), \ldots, p_{2k-1}(9)\}$.

7. A 5-card poker hand is dealt from an ordinary deck without regard to order and without replacement. Compute the expected value and variance of the number of kings in the hand.

8. Let μ be the mean of a random variable X. Show that $b = \mu$ is the point at which the minimum value of the expected square difference function $f(b) = E[(X - b)^2]$ is achieved.

9. Suppose that an observation of temperature has an expected value of 70° Fahrenheit. What is the expected temperature in degrees Celsius? Be sure to justify your answer. Discuss the effect on mean and variance in general when a change in units of measurement of a random quantity is made.

10. If X is a random variable with finite expectation and state space $\{0, 1, 2, \ldots\}$, show that

$$E[X] = \sum_{n=0}^{\infty} P[X > n].$$

(*Hint:* Write the expectation as a double sum.)

11. Suppose that a street hustler offers you a game in which you get to pick a ball from an opaque jar 4 times, replacing and mixing each time. Each orange ball drawn pays $100, but you must pay $10 to play. The hustler is willing to tell you that there is one orange ball among the remaining white balls in the jar. Given that this person is probably not in the business of giving away money, at least how many balls must be in the jar? If there are 50 balls in the jar, what are your expected net winnings?

12. An investor is considering two stocks to combine in a $1000 portfolio. Her investment counselor estimates that the first has a random rate of return over the next six months with a mean of .05 and a variance of .0001. The second has a mean rate of return of .03 and a variance of .00005. Upon analyzing her feelings toward risk, she decides that she will balance her portfolio so as to maximize the mean dollar return minus twice the variance of the return. How should she split the money between the two stocks? (Assume for this example that the stocks behave independently, which means that the variance of a sum is the sum of the variances, as mentioned in the section.)

13. Suppose that X is a random variable with state space $E = \{x_1, x_2, \ldots, x_n\}$, a finite subset of the real line. Let f be a 1–1 function from E into another subset F of the real line. Show that $E[f(X)]$ computed by formula (2.46) is the same as $E[Y]$, computed using the probability mass function of $Y = f(X)$.

14. a. Suppose that a device fails at the first time T in time set $\{1, 2, 3, \ldots\}$ that it receives a shock. A shock occurs at time k with probability $1/10$, independently of other times. Find the mean and variance of the time to failure.

    b. Consider a similar experiment, except the device fails at the time of the third shock. Find the mean time to failure.

15. Let X have the discrete uniform distribution on $\{1, 2, \ldots, n\}$. Find $g(t) = E[t^X]$. Find and interpret $g'(1)$. (Treat the point $t = 1$ with care.)

16. Prove Proposition 2.5.4 without using Proposition 2.5.5.

17. Prove Proposition 2.5.7.

18. Suppose that 210 passengers are booked on a 200-seat flight and that the probability that a passenger does not show up for the flight is .05,

independent of other passengers. All seats for this flight cost $270. The policy of the airline is to offer a cash rebate of $300 to each person who is bumped from the flight due to overbooking and to refund only half of the ticket price to no-show passengers. What is the expected amount that the airline will have to pay out? What is the expected net profit for the airline for the flight?

19. Suppose that customers arrive to the gift wrapping station of a department store according to a Poisson process with rate $\lambda = 3$ arrivals per hour. Find the expected value and the standard deviation of the number of customers who arrive between 2:00 P.M. and 5:00 P.M.

20. Find the skewness of the following distributions. Sketch their probability mass functions.
    a. $b(3, .5)$
    b. $b(3, .25)$
    c. $b(3, .75)$

21. a. Prove: If X and Y are discrete real random variables with finite expectation defined on the same sample space Ω, and if $X(\omega) \leq Y(\omega)$ for all $\omega \in \Omega$, then

$$E[X] \leq E[Y].$$

(Expectation is therefore a *monotone operator* on random variables.)
    b. Show that if a random variable is *bounded*—that is, there exists a constant C such that $|X(\omega)| \leq C$ for all $\omega \in \Omega$—then the expectation of X is finite.
    c. Show that if $E[|X|]$ exists, then $|E[X]| \leq E[|X|]$.

22. Let X and Y have the joint probability mass function

$$q((1,1)) = \frac{1}{12}, \qquad q((2,1)) = \frac{1}{6}, \qquad q((3,1)) = \frac{1}{12},$$

$$q((1,2)) = \frac{1}{6}, \qquad q((2,2)) = \frac{1}{3}, \qquad q((3,2)) = \frac{1}{6}.$$

Find the following
    a. $E[X]$
    b. $E[Y]$
    c. $E[XY]$
    d. $E[X + Y]$

23. Suppose that $\mathbf{X} = [X_1 \ X_2 \ \ldots \ X_n]'$ is a random vector whose components X_i all have the same expectation μ. If $\mathbf{c}$ is a constant row vector of n entries such that $E[\mathbf{c} \cdot \mathbf{X}] = \mu$, then what can be said about $\mathbf{c}$?

24. Prove formula (2.65).

25. Prove part (a) of Proposition 2.5.8.

CHAPTER 3

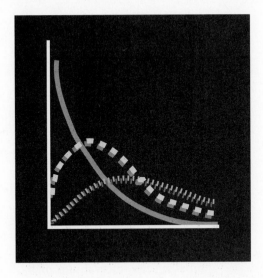

CONTINUOUS
PROBABILITY

3.1 | Densities and Distribution Functions

3.1.1 Motivation and Review

This chapter will parallel the development of Chapter 2 for the family of *continuous distributions*, those whose probabilistic law is characterized by a density function. Actually the word *continuous* might more properly be replaced by *differentiable*, since the common feature of members of this class is that their c.d.f.'s have derivatives.

For continuous distributions, the way in which probabilities are calculated is very analogous to the calculation of the physical mass of objects that are not necessarily uniformly dense. Consider a cylindrical bar of cross-sectional area one, which has a continuous density $f(x)$ at the point at a distance of x from its left end. (See Fig. 3.1.) Over a short interval $[x, x + \Delta x]$ the bar is approximately uniformly dense with density $f(x)$. The mass of the bar in this

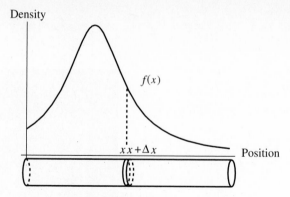

Density

$f(x)$

$x\ x+\Delta x$

Position

Figure 3.1   Cross-Sectional Area of Bar $= 1$

interval is approximately its density times its volume,

$$f(x) \cdot \Delta x \cdot 1,$$

since the slice is a cylinder with height Δx and cross-sectional area one. The mass of the bar over an interval $[a, b]$ can be approximated closely by breaking the interval into many small intervals $[x_i, x_{i+1}]$, $i = 0, 1, \ldots, n-1$, where $x_{i+1} = x_i + \Delta x$. The total mass of the bar on $[a, b]$ is the sum of the masses on all the subintervals, which is approximately

$$\sum_{i=0}^{n-1} f(x_i)\Delta x.$$

It is known from calculus that, because f is continuous, the preceding sum converges to

$$\int_a^b f(x)\,dx,$$

as the number of subintervals approaches infinity. Thus the mass of the bar in $[a, b]$ is the definite integral of its density function over $[a, b]$. Similarly, for the probability models considered in this chapter, the probability that a random variable takes a value in a given interval in its state space will be the definite integral of a probability density function over the interval. Probabilities therefore will be areas under density curves.

There is no question about the usefulness of analogies in understanding mathematical ideas, but what makes this particular analogy valid? What right have we to think of the probability that a random variable falls into an interval as an area under a density curve?

We can better understand continuous distributions by considering them as limiting cases of discrete distributions as the state space becomes denser. Let $E = \{x_1, x_2, \ldots, x_n\}$ be a state space of equally spaced points on the real line, with spacing Δx. Let probability masses $p_1, p_2, \ldots, p_n$ be placed on these points. Construct a histogram whose bars have height $p_i/\Delta x$ and whose bases, of length Δx, are centered about the points x_i, as in Fig. 3.2. Note that

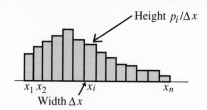

Figure 3.2   Approximating a Continuous Distribution

the heights are then probabilities per unit length, which brings to mind the density analogy.

The area of the ith rectangle is

$$\frac{p_i}{\Delta x} \cdot \Delta x = p_i = P[X = x_i].$$

Hence, even in the discrete case it is possible to think of probability as area. For very large n and very small spacing Δx, the state space resembles a continuous interval and, if the assignment of probability is smooth enough, the step function formed by the tops of the rectangles resembles a continuous function. The area under that function in an interval is approximately the total area of the thin rectangles within that interval, which is the probability that the random variable associated with the distribution falls into the interval.

—— **?Question 3.1.1**   What would have happened to the foregoing argument if the histogram rectangles were given a height of p_i instead of $p_i/\Delta x$? What would happen to the histogram as the number of states increases?

Recall from Chapter 1 that a random variable X has *probability density function* (p.d.f.) f if its probability distribution is

$$Q(A) = P[X \in A] = \int_A f(x)\, dx. \tag{3.1}$$

The requirements on f necessary to make $Q(A)$ in (3.1) a probability measure are $f(x) \geq 0$, $\forall x \in E$, and $\int_E f(x)dx = 1$. We always understand that $f(x) = 0$ outside of the state space E. Note that for this kind of distribution,

$$P[X = c] = \int_c^c f(x)\, dx = 0; \tag{3.2}$$

that is, single points receive no probability.

If X is a continuous random variable, then the *cumulative distribution function* (c.d.f.) of X is defined as in the discrete case: $F(x) = P[X \leq x]$. Thus (3.1) implies that

$$F(x) = \int_{-\infty}^x f(t)\, dt, \tag{3.3}$$

and consequently the Fundamental Theorem of Calculus gives the following relation between the density and c.d.f.:

$$F'(x) = f(x). \tag{3.4}$$

In addition, these properties are easy to check:

$$F \text{ is a nondecreasing, nonnegative function,} \tag{3.5}$$

$$\lim_{x \to -\infty} F(x) = 0, \tag{3.6}$$

$$\lim_{x \to \infty} F(x) = 1, \tag{3.7}$$

$$P[a < X \le b] = P[a \le X \le b] = P[a \le X < b]$$
$$= P[a < X < b] = F(b) - F(a). \tag{3.8}$$

Example 3.1.1

Some probability distributions are hybrids of the two classes that we have considered, discrete and continuous. Consider this simple reliability replacement scenario. The failure time T (in months) of a streetlight bulb has the probability density

$$f(t) = \frac{1}{5}e^{-\frac{1}{5}t}, \qquad t \in (0, \infty).$$

The local city manager has a plan whereby streetlight bulbs are replaced either in 6 months or when they fail, whichever comes first. Find the c.d.f. of the random variable S, the replacement time.

The extreme cases are simple: $F(s) = 0$ if $s < 0$, and $F(s) = 1$ if $s \ge 6$, since the bulb cannot be replaced before time 0 and must be replaced by time 6. Now let $s \in [0, 6)$. According to the plan, S is the smaller of T and 6. Thus, for $s \in [0, 6)$, the event that the replacement time S is less than or equal to s is the same as the event that the bulb fails by time s. Therefore

$$F(s) = P[S \le s] = P[T \le s] = \int_0^s \frac{1}{5}e^{-\frac{1}{5}t}\, dt = 1 - e^{-\frac{1}{5}s},$$

for $s \in [0, 6)$. Notice that the limit of $F(s)$ as $s \to 6$ from the left is $1 - e^{-6/5}$, but $F(6) = 1$. The point 6 in the state space of S has positive probability weight $e^{-6/5}$, but in $(0, 6)$ the c.d.f. is differentiable with the same density f as the failure time T. The distribution of S is therefore a hybrid of the two main classes that we have studied, discrete and continuous. The graph of the c.d.f. of S is shown in Fig. 3.3.

3.1.2   Uniform and Empirical Distributions

Example 3.1.2

This example illustrates the *continuous uniform distribution*. Suppose that a friend tells you that he will meet you for lunch at a restaurant at noon, plus or minus ten minutes. What is the probability that your companion will arrive between 12:05 and 12:08?

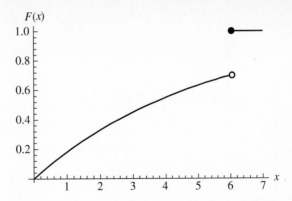

Figure 3.3 Hybrid Discrete, Continuous Distribution

We need to make some assumptions about the probability distribution of the arrival time. Start by labeling noon as time 0 minutes. Then the state space E is the time interval $[-10, 10]$. If no time is more likely than any other, then probability ought to be distributed in a uniform fashion across the states. To distribute probability uniformly means to make the density of probability constant in this interval; that is, $f(t) = c$ for $t \in [-10, 10]$ and f is zero otherwise. For f to qualify as a p.d.f., we must have

$$1 = \int_{-10}^{10} f(t) \, dt = \int_{-10}^{10} c \, dt = 20c.$$

Hence $c = 1/20$, which of course is the reciprocal of the length of the state space $[-10, 10]$. Thus, denoting the arrival time by T,

$$P[T \in [5, 8]] = \int_{5}^{8} \frac{1}{20} \, dt = \frac{3}{20}.$$

It should come as no surprise that in this uniform model, the probability that the random variable takes a value in an interval is the ratio of the length of the interval to the length of the state space.

Also, note that the c.d.f. F of this distribution is

$$F(t) = P[T \le t] = \int_{-10}^{t} \frac{1}{20} \, du = \frac{1}{20}(t + 10), \quad t \in [-10, 10].$$

It is easy to see that $F(t) = 0$ for $t < -10$, and $F(t) = 1$ for $t > 10$. By (3.8), we could equivalently have computed $P[T \in [5, 8]]$ by $F(8) - F(5)$. The density and c.d.f. are sketched in Fig. 3.4.

In general, the *continuous uniform density* on an interval $[a, b]$ has the form

$$f(x) = \begin{cases} \dfrac{1}{b - a} & \text{if } x \in [a, b], \\ 0 & \text{otherwise.} \end{cases} \tag{3.9}$$

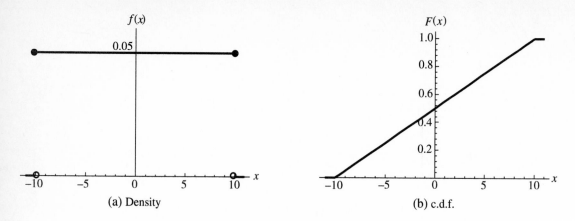

Figure 3.4 A Continuous Uniform Distribution

The associated cumulative distribution function is (see Exercise 8)

$$F(x) = \begin{cases} 0 & \text{if } x < a, \\ \dfrac{x-a}{b-a} & \text{if } x \in [a,b], \\ 1 & \text{if } x > b. \end{cases} \tag{3.10}$$

One class of distributional parameters is used occasionally to characterize both the center and the spread of the distribution. We will study them briefly and calculate them in the context of the uniform distribution. Exercises 10 and 11 ask you to look at some non-uniform distributions.

The *median* of a continuous distribution is a constant m such that

$$P[X \le m] = F(m) = \frac{1}{2}. \tag{3.11}$$

That is, exactly half the probability weight of the distribution lies to the left of m. More generally, the *p×100th percentile* of a continuous distribution is a number x_p such that

$$P[X \le x_p] = F(x_p) = p, \tag{3.12}$$

meaning that $p \times 100\%$ of the probability weight is to the left of x_p. Therefore the median is the 50th percentile of the distribution. The situation is depicted in Fig. 3.5.

For the uniform distribution, the percentiles are particularly easy to find and to grasp intuitively. Consider for instance the uniform densities of the form $f(x) = 1/c$, $x \in [0, c]$. To find the median, we set

$$1/2 = \int_0^m \frac{1}{c}\, dx$$
$$= m \cdot \frac{1}{c}$$
$$\Rightarrow m = \frac{c}{2}.$$

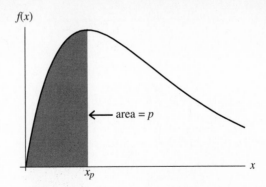

Figure 3.5 Percentiles of a Distribution

The median is therefore exactly halfway through the state space $[0, c]$. Computing in the same way for the pth percentile,

$$p = \int_0^{x_p} \frac{1}{c}\, dx$$

$$= x_p \cdot \frac{1}{c}$$

$$\Rightarrow x_p = cp.$$

The pth percentile is therefore a proportion of the way p through the state space. Of course, these results depend heavily on the fact that probability is uniformly distributed.

The idea of the *empirical distribution function* based on a sample of observations $X_1, X_2, \ldots, X_n$ from a given distribution is the same in the continuous case as in the discrete case. That is, the empirical distribution function $\widehat{F}_n(x)$ is

$$\widehat{F}_n(x) = \frac{\text{number of } X_i\text{'s} \leq x}{n}. \tag{3.13}$$

We will see later that there are theoretical reasons to believe that for large sample size n, $\widehat{F}_n(x)$ will be close to the true c.d.f. value $F(x)$ for all x.

The *empirical mass function* of a random sample from a continuous distribution, defined as in the discrete case to give weight $1/n$ to each of the sample values, is not the most useful object for displaying the shape of the density. Usually the practitioner chooses how to divide up the state space into categories, then constructs an empirical probability mass function by counting the number of observations in each category and dividing by both n and the category size to scale the vertical distances for better comparison with the density, as in the next example.

Example 3.1.3

Following are 100 simulated observations from a distribution with p.d.f. $f(x) = 3x^2$, $x \in [0, 1]$. Such figures could come, for example, from a simple model of a death age distribution in which there is high probability density at the end

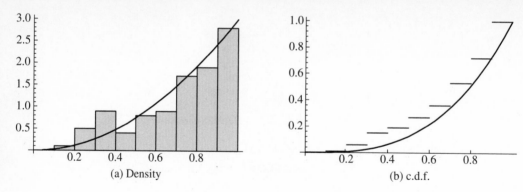

(a) Density   (b) c.d.f.

Figure 3.6 Empirical Distribution

and a low density at the start. For brevity, the exact observations have not been recorded, only the left endpoints of the intervals $[0, .1)$, $[.1, .2)$, $[.2, .3)$, and so on, to which the observations belonged.

.3, .8, .8, .7, .8, .6, .6, .4, .3, .9, .9, .8, .5, .4, .7, .6, .3, .8, .9, .5,
.5, .4, .8, .7, .7, .3, .5, .9, .9, .9, .7, .4, .9, .7, .9, .9, .7, .9, .8, .9,
.8, .3, .8, .8, .9, .9, .8, .9, .3, .6, .6, .9, .5, .7, .8, .9, .9, .8, .3, .9,
.7, .6, .9, .2, .7, .7, .9, .1, .5, .7, .9, .9, .8, .6, .2, .3, .9, .6, .7, .6,
.8, .5, .2, .3, .9, .9, .2, .9, .8, .7, .7, .5, .2, .9, .8, .9, .8, .7, .7, .8

A histogram of these values is plotted in Fig. 3.6(a), where the bar heights are the sample proportions of 0s, .1s, .2s, and so on, divided by the subinterval size 0.1. Superimposed on this graph is a graph of the probability density function $f(x)$ listed here. Part (b) of the figure plots the empirical distribution function together with the theoretical c.d.f. $F(x) = x^3$ for $x \in [0, 1]$. There is fairly close agreement between the theoretical and empirical distributions.

3.1.3 Multivariate Continuous Distributions

We now discuss the extension of the idea of continuous distribution of probability to more than one random variable. When $\mathbf{X}$ is a multivariate random vector, the density $f(\mathbf{x})$ in (3.1) is a real-valued function of a vector variable and the integral in (3.1) is a multiple integral. It is still true that the probability that $\mathbf{X}$ will fall into a set A is the integral of the density over A. If, for example, the state space of $\mathbf{X} = (X_1, X_2, X_3)$ is $[0, \infty) \times [0, \infty) \times [0, \infty)$, and the density is $f(x_1, x_2, x_3)$, then the probability that $\mathbf{X}$ belongs to a box-shaped region $[0, 2] \times [1, 3] \times [1, 2]$ could be written as

$$P[X \in [0, 2] \times [1, 3] \times [1, 2]] = \int_0^2 \int_1^3 \int_1^2 f(x_1, x_2, x_3) \, dx_3 \, dx_2 \, dx_1.$$

For bivariate random vectors $\mathbf{X} = (X_1, X_2)$, we can grasp geometrically the meaning of the double integral $\int \int_A f(x_1, x_2) \, dx_2 \, dx_1$, which represents the chance that $\mathbf{X}$ falls into A. This is just the volume of the region bounded

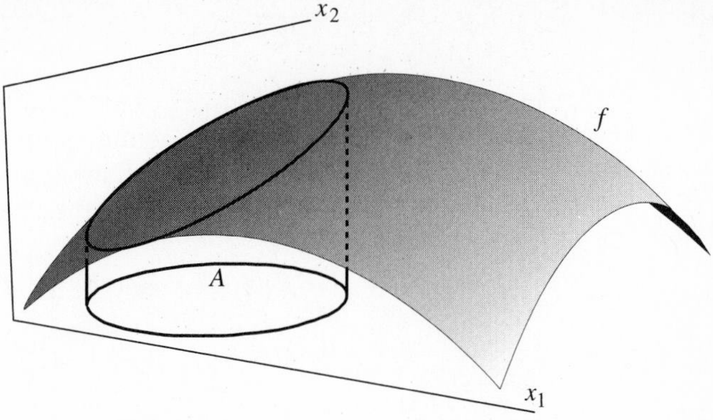

Figure 3.7   Joint Probability as Volume

above by the surface $z = f(x_1, x_2)$ in $\mathbb{R}^3$ and below by the part of the $x_1 - x_2$ plane determined by the intersection of A and the state space, as shown in Fig. 3.7.

The concept of cumulative distribution function can also make sense in the multivariate case if the inequality $\mathbf{X} \le \mathbf{x}$ is interpreted as a joint inequality in the components of $\mathbf{X} = (X_1, X_2, \ldots, X_n)$:

$$F(\mathbf{x}) = P[\mathbf{X} \le \mathbf{x}] = P[X_1 \le x_1, X_2 \le x_2, \ldots, X_n \le x_n]. \qquad (3.14)$$

The next examples show how to set up integrals to compute multivariate probabilities.

Example 3.1.4

Suppose that a random vector $\mathbf{X} = (X_1, X_2)$ describes, respectively, the longer and shorter utilization times of a mainframe computer by two jobs. The operating system is constructed so as to give jobs a time slice of at most 2 time units. For reasons that we will see in Chapter 5 when we study order statistics, it is reasonable to assume that $\mathbf{X}$ has the following bivariate density function:

$$f(x_1, x_2) = \begin{cases} c & \text{if } x_1 - x_2 \ge 0, \ x_1 \le 2, \ x_1, x_2 \ge 0, \\ 0 & \text{otherwise.} \end{cases}$$

Find c, find the probability that the longer exceeds the shorter by at least 1, and find $P[X_1 > x]$ for all real x.

The region of the $x_1 - x_2$ plane over which f is nonzero—that is, the state space E of $\mathbf{X}$—is the triangle in Fig. 3.8. For f to be a density function, c must be nonnegative and

$$1 = \int_E f(x_1, x_2) \, dx_2 \, dx_1$$

$$= \int_0^2 \int_0^{x_1} c \, dx_2 \, dx_1$$

$$= \int_0^2 c x_1 \, dx_1$$

$$= c \cdot \left. \frac{x_1^2}{2} \right|_0^2 = 2c \implies c = 1/2.$$

The event that the larger exceeds the smaller by at least 1 can be written $\{X_1 \geq X_2 + 1\}$. The set of points in the state space $A = \{x_1 \geq x_2 + 1\}$ into which the pair (X_1, X_2) must fall in order for the event to occur is the shaded region in Fig. 3.8. The integral of the density over this region is

$$P[X_1 \geq X_2 + 1] = \int_1^2 \int_0^{x_1 - 1} \frac{1}{2} \, dx_2 \, dx_1$$

$$= \frac{1}{2} \int_1^2 x_1 - 1 \, dx_1$$

$$= \frac{1}{2} \left(x_1^2/2 - x_1 \Big|_1^2 \right) = \frac{1}{4}.$$

(How would you have done this calculation if the inner integral were taken over x_1 instead of x_2?)

To find $P[X_1 > x]$, first eliminate the trivial cases: When $x < 0$, $P[X_1 > x] = 1$, and when $x > 2$, $P[X_1 > x] = 0$. The most interesting case is shown in Fig. 3.9, where $x \in [0, 2]$. Then,

$$P[X_1 > x] = \int_x^2 \int_0^{x_1} \frac{1}{2} \, dx_2 \, dx_1$$

$$= \int_x^2 \frac{1}{2} x_1 \, dx_1$$

$$= \frac{1}{4} x_1^2 \Big|_x^2 = 1 - \frac{1}{4} x^2.$$

Notice that we can immediately write a formula for the c.d.f. of X_1, which

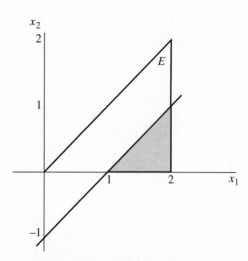

Figure 3.8   Event $X_1 \geq X_2 + 1$

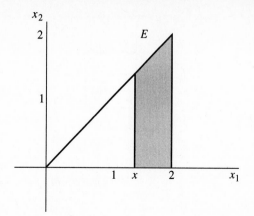

Figure 3.9 Event $X_1 > x$ for $0 \le x \le 2$

is the complementary probability to this one: $F_1(x) = \frac{1}{4}x^2$ if $x \in [0, 2]$. The density of X_1 is $f_1(x) = F_1'(x) = \frac{1}{2}x$ if $x \in (0, 2)$.

?Question 3.1.2 Find the density of X_2 in the last example.

As Example 3.1.4 indicated, a joint density of many random variables gives rise to *marginal densities* of the component random variables. We leave the following proposition without proof, because its full proof requires techniques beyond our scope.

PROPOSITION 3.1.1

Let $\mathbf{X} = (X_1, X_2, \ldots, X_n)$ be a random vector with probability density function $f(\mathbf{x})$ and state space E. The marginal density of X_i is

$$f_i(x_i) = \int \int_{E(x_i)} \cdots \int f(x_1, x_2, \ldots, x_n) \, dx_n \cdots dx_{i+1} \, dx_{i-1} \cdots dx_1, \quad (3.15)$$

where $E(x_i) = \{(x_1, \ldots, x_{i-1}, x_{i+1}, \ldots x_n) \mid (x_1, \ldots, x_n) \in E\}$. In other words, the marginal density of X_i is found by integrating the joint density over all combinations of possible states for the other component random variables given a fixed state x_i of X_i.

?Question 3.1.3 How would the joint marginal density of a pair X_i and X_j be computed? a triple X_i, X_j, X_k? For the answer, see the next example.

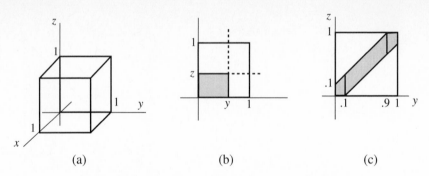

Figure 3.10   Joint Distributions and Probabilities

Example 3.1.5

Let $\mathbf{X} = (X, Y, Z)$ represent three delivery times of supplies to a college cafeteria (one unit of time is 10 hours). Suppose that the component random variables have the joint p.d.f.

$$f(x, y, z) = \begin{cases} 8xyz & \text{if } x, y, z \in [0, 1], \\ 0 & \text{otherwise.} \end{cases}$$

Find the joint marginal density and c.d.f. of Y and Z, and find the probability that deliveries Y and Z are within an hour of each other.

The state space is the unit cube shown in Fig. 3.10(a). To find the joint marginal density of Y and Z, for fixed y, z integrate the given joint density over all possible values of x, in this case the interval $[0, 1]$:

$$f_{23}(y, z) = \int_0^1 8xyz \; dx = 8yz \cdot \frac{x^2}{2}\Big|_0^1 = 4yz,$$

which is nonzero for (y, z) such that both y and z are in the interval $[0, 1]$. The joint marginal c.d.f. can be computed in more than one way. One alternative is to use the marginal density f_{23} that was just computed to find (see Fig. 3.10(b))

$$F_{23}(y, z) = P[Y \leq y, Z \leq z] = \int_0^y \int_0^z 4uv \; dv \; du$$

$$= 4 \int_0^y u \; du \cdot \int_0^z v \; dv$$

$$= 4 \cdot \frac{y^2}{2} \cdot \frac{z^2}{2} = y^2 z^2, \qquad y, z \in [0, 1].$$

(You should reason out what F_{23} is when either y or z is outside of the interval $[0, 1]$.) Notice that $f_{23} = \partial^2 F_{23}/\partial y \partial z$, which is a result that is true in general and which extends the analogous result $F' = f$ in the one-dimensional case. Another method of computing F_{23} is to note that $\{Y \leq y, Z \leq z\} = \{X \in [0, 1], Y \leq y, Z \leq z\}$. The probability of this event is then found using the original joint density:

$$\int_0^1 \int_0^y \int_0^z 8t \, uv \; dv \; du \; dt.$$

You can check that this formula reduces to the first integral in the preceding computation.

Finally, the event described in the problem statement translates to $|Y - Z| \le .1$. This occurs if and only if the pair (Y, Z) falls into the shaded region in Fig. 3.10(c). To integrate over that region requires us to divide the region of integration into the three areas shown in Fig. 3.10, since the limits of integration for the z variable depend on whether the y variable is in the interval $[0, .1)$, $[.1, .9)$, or $[.9, 1]$. We can compute, after some calculation,

$$P[|Y - Z| \le .1] = \int_0^{.1} \int_0^{y+.1} 4yz \; dz \; dy + \int_{.1}^{.9} \int_{y-.1}^{y+.1} 4yz \; dz \; dy$$

$$+ \int_{.9}^1 \int_{y-.1}^1 4yz \; dz \; dy$$

$$= \int_0^{.1} 2y(y + .1)^2 \, dy + \int_{.1}^{.9} 2y(.4y) \, dy + \int_{.9}^1 2y(1 - (y - .1)^2) \, dy$$

$$\approx .247.$$

EXERCISES 3.1

1. Suppose that the amount of time T required to complete a transaction at an automated teller machine is a continuous random variable with p.d.f. $f(t) = 4te^{-2t}$ for $t > 0$. Find the cumulative distribution function, and $P[2 < T < 3]$.

2. Suppose that the time interval S between successive beeps of an automobile radar detector is a continuous random variable with cumulative distribution function $F(s) = 1 - e^{-3s}$ for $s > 0$ and 0 otherwise. Find the density function and $P[S \in (1, 5]]$.

3. a. If the graph of a density function $f(x)$ of a random variable is as in the figure, draw a rough sketch of the associated c.d.f.
   b. If the graph of a cumulative distribution function $F(x)$ of a random variable is as in the figure, draw a rough sketch of the associated density.

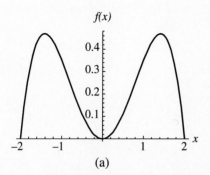

(a)

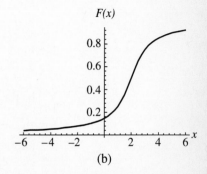

(b)

4. In Example 3.1.3 on death ages, find (a) the probability of dying after age 50; (b) the probability of dying between ages 60 and 80. Compare these

theoretical probabilities to estimated probabilities using the simulated sample.

5. A high-intensity projector bulb lasts for a random number of hours T with the density $f(t) = .02e^{-.02t}$, $t \in [0, \infty)$. Find
   a. the probability that a new bulb lasts for more than 50 hours.
   b. the conditional probability that a new bulb lasts for more than 70 hours given that it has lasted for more than 20 hours.
   Comment on what the results of (a) and (b) say about the aging properties of the bulb.

6. For their morning grooming rituals, two people share a single bathroom that can accommodate only one person at a time. Suppose that the proportion of the available time used by the first person is given by a random variable X with the p.d.f. $f(x) = 6x(1-x)$, $x \in [0, 1]$. Compute the c.d.f. of the distribution, and find the probability that the first person uses at least half of the time.

7. Suppose that the random number generator of a computer is able to supply an observation X from the continuous uniform distribution on $[0, 1]$. Devise a way of obtaining from X an observation Y from the uniform distribution on $[2, 4]$, and show that your method works.

8. Verify the form of the c.d.f. (3.10) of the uniform distribution. Sketch its graph.

9. Following are 40 observations of a random variable X that measures the reaction times of drivers under the influence of alcohol to a coordination test. The reaction time is some nonzero delay time c common to all drivers, followed by a nonnegative, random residual time. We will suppose that X has a density of the form $f(x) = 2e^{-2(x-c)}$ for $x > c$, and 0 otherwise. How would you use the data to obtain a rough estimate of the unknown parameter c? Using your estimate, sketch the empirical and theoretical c.d.f.'s on the same graph.

   3.447, 3.968, 4.702, 3.929, 3.331, 3.361, 3.181, 4.381, 3.056, 3.132,
   3.012, 4.097, 3.027, 3.530, 3.635, 4.286, 3.298, 3.091, 3.618, 3.525,
   3.136, 3.666, 5.528, 3.114, 3.377, 3.079, 3.198, 4.293, 3.449, 3.015,
   3.775, 3.303, 3.850, 3.087, 4.009, 3.243, 3.025, 3.030, 3.111, 3.905

10. Find the medians of the distributions with the following densities.
    a. $f(x) = 2x$, $x \in [0, 1]$
    b. $f(x) = \frac{1}{2}e^{-|x|}$, $-\infty < x < \infty$

11. Find
    a. the 25th and 75th percentiles of the distribution with density $f(x) = 3x^2$, $x \in [0, 1]$.
    b. the $p \times 100$th percentile of the distribution with density $f(x) = c/x^2$, $x \in [1, \infty]$. (Determine the value of c first.)

12. Let $f(x_1, x_2) = \frac{1}{2}e^{-x_1}$ for $x_1 \in [0, \infty)$, $x_2 \in [0, 2]$ and 0 otherwise. Find the joint cumulative distribution function. Find the marginal densities and cumulative distribution functions. Do you notice anything special?

13. Let $f(x, y) = c(x + y)$ for $x, y \in [0, 4]$ and 0 otherwise. Find c such that this function is a valid p.d.f., find the marginal density functions of X and Y, and find $P[X + Y > 1]$.

14. Two people agree to meet at a restaurant for dinner at 6:00 P.M. There is 10 minutes uncertainty in either direction in each of their arrival times. What is the probability that neither person will have to wait more than 5 minutes for the other? (*Hint:* Formulate an appropriate bivariate uniform model.)

15. A projector has a system of two high-intensity bulbs as in Exercise 5. When the first burns out, the operator flips a switch and the second comes into use. Only when both bulbs are burned out must the projector be serviced. Assume that the joint density of the lifetimes T_1 and T_2 of the two bulbs is the product of the marginal densities. Find the probability that the time S until the projector must be serviced exceeds t for fixed $t > 0$.

16. Let $\mathbf{X} = (X, Y, Z)$ have constant density on the prism $\{(x, y, z) \mid x, y, z > 0 \text{ and } x + y + z \leq 1\}$. Find $P[X - Y > 0]$.

17. The diamond-shaped dartboard in the figure is such that each of its diagonals is 12 inches long and the inner circle is 2 inches in diameter. Point values of 100, 50, and 0 are earned if a thrown dart hits the inner circle, the ring, or the region outside the outer circle, respectively. A dart thrown at random lands at some point on the board. Find the expected value of the number of points earned.

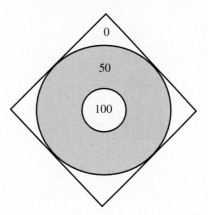

3.2 | Expectation of Continuous Random Variables

In Chapter 2 on discrete distributions we moved immediately from the general information to a catalog of common distributions, and we went on to discuss expectation thereafter. In this chapter, because the parameters of some of the continuous distributions depend so intimately on mean and variance, we will first study continuous expectation and then look at the common distributions in the next section.

Recall that we defined the expectation of a discrete random variable X with state space E to be

$$E[X] = \sum_{x \in E} x P[X = x], \tag{3.16}$$

that is, the weighted average of the states using their probabilities as weights. To see how to carry this notion to the setting of continuous random variables, let X be a real-valued continuous random variable with continuous probability density function f. To simplify, suppose that the state space of X is some bounded interval $[a, b]$. (This is not a crucial assumption.) The area under the density curve in an interval $[x, x + \Delta x]$ is approximately the area of the rectangle whose height is $f(x)$ and whose base has length Δx when Δx is very small. In symbols,

$$P[X \in [x, x + \Delta x]] = \int_x^{x+\Delta x} f(t)\, dt \approx f(x)\Delta x. \tag{3.17}$$

Approximate the state space $[a, b]$ by finitely many equally spaced points $x_1, x_2, \ldots, x_n$, with very small spacing Δx. Let the left endpoint x_i of each interval $[x_i, x_i + \Delta x]$ inherit the whole probability of the interval, which is about $f(x_i)\Delta x_i$, by formula (3.17). But then the sum that defines the mean of the approximate discrete distribution in (3.16) would be

$$\sum_{i=1}^n x_i P[X = x_i] \approx \sum_{i=1}^n x_i f(x_i)\Delta x.$$

In the limit as the number of states $n \to \infty$, and $\Delta x \to 0$, the sum on the right may or may not converge, but if it does, it converges to the definite integral $\int_E x f(x)\, dx$.

Here is the formal definition of continuous expectation.

DEFINITION 3.2.1

The *expected value* of a real-valued, continuous random variable X with p.d.f. f and state space E is

$$E[X] = \int_E x \cdot f(x)\, dx, \tag{3.18}$$

provided the integral exists.

It is possible to prove that the expected value of a function $g(X)$ is

$$E[g(X)] = \int_E g(x) \cdot f(x)\, dx. \tag{3.19}$$

As with discrete real random variables, we adopt terminology and notation as follows. The *mean* of X is

$$\mu = E[X]; \tag{3.20}$$

the *variance* of X is

$$\text{Var}(X) = \sigma^2 = E[(X - \mu)^2] = \int_E (x - \mu)^2 f(x) \, dx; \qquad (3.21)$$

the *standard deviation* of X is $\sigma = \sqrt{\sigma^2}$; the *rth moment* of X is

$$\mu_r = E[X^r] = \int_E x^r f(x) \, dx; \qquad (3.22)$$

and the *rth moment about the mean* of X is

$$\mu'_r = E[(X - \mu)^r] = \int_E (x - \mu)^r f(x) \, dx. \qquad (3.23)$$

The mean is a measure of the center of the probability distribution, and the variance is a measure of the distribution's spread, since σ^2 will be large if large values of $(x - \mu)^2$ have high probability density. The *skewness* is the third moment about μ, and it is a rough measure of asymmetry of the distribution.

—— **?Question 3.2.1** Look back at the results of Section 2.5 on discrete expectation. Which ones seem to carry through to the continuous case? Which of the proofs are independent of the exact definition of expectation, depending only on basic properties such as linearity? (We will summarize the results shortly.)

Example 3.2.1 Let X have the continuous uniform distribution on $[a, b]$. Find the mean, variance, and skewness of X.

We are given that the density of X is

$$f(x) = \begin{cases} 1/(b - a) & \text{if } x \in [a, b], \\ 0 & \text{otherwise.} \end{cases}$$

By the definition of expectation (3.18),

$$E[X] = \int_a^b x \cdot \frac{1}{b - a} \, dx = \frac{1}{b - a} \cdot \frac{x^2}{2} \bigg|_a^b = \frac{1}{2} \cdot \frac{b^2 - a^2}{b - a} = \frac{1}{2}(a + b). \qquad (3.24)$$

As intuition would suggest, the average value of a uniform random variable is just the midpoint of the state space. By (3.21),

$$\begin{aligned}
\text{Var}(X) &= \int_a^b (x - \mu)^2 \cdot \frac{1}{b - a} \, dx \\
&= \frac{1}{b - a} \cdot \frac{(x - \mu)^3}{3} \bigg|_a^b \\
&= \frac{1}{3(b - a)} [(b - \mu)^3 - (a - \mu)^3] \\
&= \frac{1}{3(b - a)} \left[\left(b - \frac{a + b}{2} \right)^3 - \left(a - \frac{a + b}{2} \right)^3 \right] \\
&= \frac{1}{3(b - a)} \left[\left(\frac{b - a}{2} \right)^3 - \left(\frac{a - b}{2} \right)^3 \right] \\
&= \frac{(b - a)^2}{12}. \qquad (3.25)
\end{aligned}$$

The skewness is

$$E[(X - \mu)^3] = \int_a^b (x - \mu)^3 \cdot \frac{1}{b - a}\, dx$$

$$= \frac{1}{4(b - a)} \cdot (x - \mu)^4 \Big|_a^b$$

$$= \frac{1}{4(b - a)} [(b - \mu)^4 - (a - \mu)^4]$$

$$= \frac{1}{4(b - a)} \left[\left(\frac{b - a}{2} \right)^4 - \left(\frac{a - b}{2} \right)^4 \right]$$

$$= 0. \tag{3.26}$$

The fourth line uses the same algebra as in the variance calculation to simplify $b - \mu$ and $a - \mu$. This time, however, the quantities $(b - a)^4$ and $(a - b)^4$ are identical, and this forces the skewness of the uniform distribution to be 0. This is to be expected, since the uniform density is symmetric about its mean.

?Question 3.2.2   What can you say about all odd-order moments about the mean of the uniform distribution?

Example 3.2.2   Consider a density $f(x) = c/x^m$ for $x \in [1, \infty)$, where c is chosen in order to make the density integrate to 1 over the state space $[1, \infty)$. Up to what order r does the rth moment of the distribution exist?

The rth moment, by (3.22), is

$$E[X^r] = \int_1^\infty x^r \cdot \frac{c}{x^m}\, dx$$

$$= c \cdot \int_1^\infty x^{r-m}\, dx$$

$$= c \cdot \lim_{B \to \infty} \int_1^B x^{r-m}\, dx$$

$$= c \cdot \begin{cases} \lim_{B \to \infty} \ln x \big|_1^B & \text{if } r = m - 1, \\ \lim_{B \to \infty} \dfrac{x^{r-m+1}}{r - m + 1} \Big|_1^B & \text{if } r \neq m - 1. \end{cases}$$

The top case clearly gives an infinite limit, as does the bottom case when $r - m + 1 > 0$, that is, $r > m - 1$. Thus the rth moment does not exist when $r \geq m - 1$. The limit does exist when $r < m - 1$; in fact, it equals $-c/(r - m + 1)$. Thus, for instance, the density c/x^2 on $[1, \infty)$ has neither a mean ($r = 1 = m - 1$) nor a second moment; the density c/x^3 has a mean ($r = 1 = m - 2$) but not a second moment ($r = 2 = m - 1$); and the density c/x^4 has both a mean ($r = 1 = m - 3$) and a second moment ($r = 2 = m - 2$).

—— **?Question 3.2.3**   What must c equal for $f(x) = c/x^m$, $x \in [1, \infty)$ to be a density?

We will now list without proof the main results about expectation and variance, which are directly parallel to those for discrete expectation. The proofs, analogous to the discrete proofs, take advantage of the fact that integrals possess the same linearity properties as sums.

If c is a real constant, then

$$E[c] = c. \tag{3.27}$$

Let X and Y be random variables of continuous type, and let a and b be constants. Then,

$$E[aX + bY] = aE[X] + bE[Y], \tag{3.28}$$

provided the expectations of X and Y exist.

If X is a real random variable of continuous type with finite variance and a and b are real constants, then

$$\text{Var}(aX + b) = a^2 \text{Var}(X). \tag{3.29}$$

In particular the variance of a constant is 0.

If X is a real random variable of continuous type with finite variance and mean μ, then

$$\text{Var}(X) = E[X^2] - \mu^2. \tag{3.30}$$

If $X_1, X_2, \ldots, X_n$ is a random sample from a continuous distribution with mean μ and variance σ^2, and $\overline{X} = (\sum_{i=1}^n X_i)/n$, then

$$E[\overline{X}] = \mu, \qquad \text{Var}(\overline{X}) = \frac{\sigma^2}{n}. \tag{3.31}$$

Expected value is frequently used in decision-making settings where there is underlying uncertainty, as in the following example.

Example 3.2.3

The latest blockbuster super-hero movie, *Tick Boy*, has come out, and Fast Eddie is looking to make a few dollars marketing T-shirts. He needs to set a price p per shirt. Eddie is relatively sure that the cost per shirt for materials and labor would be \$5. He is uncertain about two areas, however: (1) the number N of shirts he would sell if the price is p, and (2) the setup cost B for licensing and acquisition of equipment to make the shirts. Suppose that N has the Poisson distribution with parameter $\mu = 600 - 25p$, where p is the price in dollars, and B has the continuous uniform density on [\$100, \$200]. What should Fast Eddie do?

You may well have set up optimization problems of this kind in calculus in a nonrandom setting. We will use expectation to convert our problem to one of these more familiar problems.

Eddie's profit is his total revenue (number of shirts sold times price per shirt) minus his total cost (cost per shirt times number of shirts plus setup cost). In our notation, his profit is

$$P = Np - (5N + B).$$

This result is random, however, since N and B are random. It is reasonable to try to find p to maximize the *expected value* of his profit. We can proceed as follows:

$$h(p) = E[P] = E[Np - (5N + B)]$$
$$= pE[N] - 5E[N] - E[B].$$

From the results of Chapter 2 for Poisson random variables, $E[N] = \mu = 600 - 25p$. As shown earlier in this section, $E[B] = (100 + 200)/2 = 150$. Substituting these quantities into the last equation gives

$$h(p) = p(600 - 25p) - 5(600 - 25p) - 150 = -25p^2 + 725p - 3150.$$

This expected profit function is quadratic and concave down; hence it takes its maximum value at its lone critical point, which satisfies

$$h'(p) = -50p + 725 = 0 \Rightarrow p = 14.5.$$

Therefore Eddie's *Tick Boy* T-shirts should be priced at $14.50 apiece, which is about triple his expected cost to make each shirt.

Expectation in the multivariate continuous case is also analogous to the discrete case. The expected value $E[\mathbf{X}]$ of a random vector $\mathbf{X} = (X_1, X_2, \ldots, X_n)'$ is once again the vector of component expectations $(E[X_1], E[X_2], \ldots, E[X_n])'$. For a real-valued function g of a vector variable, the definition of $E[g(\mathbf{X})]$ is the same as in (3.19); in this case the integral is a multiple integral over all components. The linearity theorem Proposition 2.5.8, whose proof only depends on the linearity of one-dimensional expectation, holds once again.

Here is an example computation.

Example 3.2.4

In the city of Evanston, just north of Chicago, express and nonexpress commuter trains to the Chicago Loop run on weekday mornings. Suppose that at time 0, we are sitting at the Davis Street stop, and the joint density of the times T_e and T_n (in minutes) of arrival of the next express and nonexpress train is

$$f(t_e, t_n) = \frac{1}{50} e^{-\frac{1}{10} t_e - \frac{1}{5} t_n}, \qquad t_e, t_n > 0.$$

Compute the expected time interval between arrivals of the two trains.

The expected amount of time between the two arrivals is $E[|T_n - T_e|]$. The formula for defining the function $g(t_e, t_n) = |t_n - t_e|$ on the region $t_e \leq t_n$ differs from that for $t_e > t_n$; hence it makes sense to break the integral into the sum of two, as follows:

$$E[|T_n - T_e|] = \int_0^\infty \int_0^\infty |t_n - t_e| \cdot f(t_e, t_n) \, dt_e \, dt_n$$
$$= \int\int_{t_e \leq t_n} |t_n - t_e| \cdot f(t_e, t_n) \, dt_e \, dt_n$$

$$+ \int \int_{t_e > t_n} |t_n - t_e| \cdot f(t_e, t_n) \, dt_e \, dt_n$$

$$= \int_0^\infty \int_0^{t_n} (t_n - t_e) \cdot \frac{1}{10} e^{-1/10 t_e} e^{-1/5 t_n} \, dt_e \, dt_n$$

$$+ \int_0^\infty \int_{t_n}^\infty (t_e - t_n) \cdot \frac{1}{50} e^{-1/10 t_e} e^{-1/5 t_n} \, dt_e \, dt_n$$

$$= \int_0^\infty t_n \left(\int_0^{t_n} \frac{1}{10} e^{-1/10 t_e} \, dt_e \right) \frac{1}{5} e^{-1/5 t_n} \, dt_n$$

$$- \int_0^\infty \left(\int_0^{t_n} t_e \cdot \frac{1}{10} e^{-1/10 t_e} \, dt_e \right) \frac{1}{5} e^{-1/5 t_n} \, dt_n$$

$$+ \int_0^\infty \left(\int_{t_n}^\infty t_e \cdot \frac{1}{10} e^{-1/10 t_e} \, dt_e \right) \frac{1}{5} e^{-1/5 t_n} \, dt_n$$

$$- \int_0^\infty t_n \left(\int_{t_n}^\infty \frac{1}{10} e^{-1/10 t_e} \, dt_e \right) \frac{1}{5} e^{-1/5 t_n} \, dt_n$$

$$= \int_0^\infty t_n (1 - e^{-1/10 t_n}) \frac{1}{5} e^{-1/5 t_n} \, dt_n$$

$$- \int_0^\infty (10 - (t_n + 10) e^{-1/10 t_n}) \frac{1}{5} e^{-1/5 t_n} \, dt_n$$

$$+ \int_0^\infty ((t_n + 10) e^{-1/10 t_n}) \frac{1}{5} e^{-1/5 t_n} \, dt_n$$

$$- \int_0^\infty t_n e^{-1/10 t_n} \frac{1}{5} e^{-1/5 t_n} \, dt_n$$

$$= \int_0^\infty \frac{1}{5} t_n e^{-1/5 t_n} \, dt_n - 2 \int_0^\infty e^{-1/5 t_n} \, dt_n + 4 \int_0^\infty e^{-3/10 t_n} \, dt_n.$$

You should look most carefully at the limits of integration in the first few lines; the rest of it is lengthy but straightforward computation in which we use integration by parts several times to do integrals of the form $\int t e^{-ct} \, dt$. The integrals in the last line come out to be 5, 5, and 10/3, respectively. Therefore the ultimate result is

$$E[|T_n - T_e|] = 5 - 2 \cdot 5 + 4 \cdot \frac{10}{3} = \frac{25}{3}.$$

EXERCISES 3.2

1. Show that the mean of the probability distribution with p.d.f. $f(x) = c e^{-cx}$ for $x > 0$ is $1/c$ and the variance is $1/c^2$.

2. Find the mean, variance, and standard deviation of the amount of time T required to complete a transaction at an automated teller machine, assuming as in Exercise 1 of Section 3.1 that T is a continuous random variable with p.d.f. $f(t) = 4t e^{-2t}$ for $t > 0$.

3. Does the density function $f(x) = 1/[\pi(1 + x^2)]$, $x \in \mathbb{R}$ have a mean? a variance?

4. Suppose that the lifetime T of a life insurance policy holder has probability density function $f(t) = c \cdot t$, $t \in [0, 50]$. Time is measured in years from the initial date of the policy. The terms of the policy are that the holder is to pay the insurance company a premium of p dollars per year while he is living, in exchange for which the company agrees to pay the sum of \$30,000 to the holder's heirs if the holder dies within 30 years. In this case, the premium is prorated if the holder dies in midyear. If the holder dies after 30 years have passed, the policy expires and no money is paid. At least what premium must the company charge in order to break even in expected value? (Ignore the issue of monetary inflation.)

5. Find the expected value and variance of the proportion of the available bathroom time used by the first person in Exercise 6 of Section 3.1, which had the p.d.f. $f(x) = 6x(1 - x)$, $x \in [0, 1]$.

6. Find the mean time until the projector of Exercise 15, Section 3.1, needs service.

7. Let X be a real-valued random variable with state space $[0, \infty)$. Show that if $E[X^2]$ is finite, then

$$E[X] = \int_0^\infty P[X > x] \, dx.$$

(*Hint:* Integrate by parts.)

8. Suppose that you have \$1000 to invest in either a risky stock or a non-risky bond. At the end of the year, the bond is certain to yield 4 cents for every dollar invested in it. The stock will yield a random number R of cents per dollar invested, where R has the p.d.f. sketched in the figure. Write an expression for the total yield Y in dollars. How should you allocate your money so as to maximize the difference between the mean of your total yield and half of the variance of the yield?

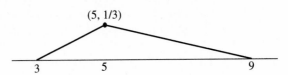

9. Prove Eq. (3.27).
10. Prove Eq. (3.28).
11. Prove Eq. (3.29).
12. Compute the expected total time $E[X_1 + X_2]$ for the two computer jobs of Example 3.1.4, in which the density of $\mathbf{X} = (X_1, X_2)$ is

$$f(x_1, x_2) = \begin{cases} c & \text{if } x_1 - x_2 \geq 0, \ x_1 \leq 2, \ x_1, x_2 \geq 0, \\ 0 & \text{otherwise.} \end{cases}$$

13. In Example 3.2.4 on commuter trains, what is the expected (nonabsolute) difference between the arrival times of the express and nonexpress trains? Does your answer strike you as inconsistent with the number $25/3$ obtained in the example? Why or why not?

14. Find $E[X_1^2 \cdot X_2]$ if $\mathbf{X} = (X_1, X_2)$ has the density $f(x_1, x_2) = \frac{1}{2}e^{-x_1}$ for $x_1 \in [0, \infty)$, $x_2 \in [0, 2]$ and 0 otherwise. Do you notice anything special about the way the computation proceeds?

15. Suppose that X and Y have a joint density of the form $f(x, y) = f_1(x)f_2(y)$, $x \in [a, b]$, $y \in [c, d]$. Show that if g_1 is a real-valued function whose domain contains the state space of X and g_2 is a real-valued function whose domain contains the state space of Y, then

$$E[g_1(X)g_2(Y)] = E[g_1(X)] \cdot E[g_2(Y)],$$

provided the expectations exist.

16. Suppose that X_1, X_2, and X_3 have means 2, -1, and 3, respectively. Find $E[A\mathbf{X}]$, where $\mathbf{X} = (X_1, \ X_2, \ X_3)'$ and

$$A = \begin{bmatrix} 1 & 1 & 0 \\ 2 & 2 & -1 \\ 6 & 0 & 3 \end{bmatrix}.$$

You obtain three expectation results in this way. Write them out explicitly.

3.3 | Examples of Continuous Distributions

A number of single-variable continuous distributions come up frequently in applications. Our task in this section is to list them and to discuss some of their properties. One such distribution is the continuous uniform density $f(x) = 1/(b - a)$, $x \in [a, b]$, which was described earlier in this chapter. Because this density is constant on the state space $[a, b]$, however, it is appropriate only for problems that do not favor one state over another. To treat a broader range of applications, we must also look at some nonconstant density functions.

3.3.1 Gamma Family

We will begin with several members of a class of densities called the *gamma family*. We will see that these densities are intimately connected with the Poisson process.

DEFINITION 3.3.1 ────────────────────────────

The *exponential density* is

$$f(t) = \lambda e^{-\lambda t}, \qquad t > 0. \tag{3.32}$$

This distribution, referred to briefly as exp(λ), is characterized by one parameter $\lambda > 0$. Sketches of the exp(λ) density for several values of λ are given in Fig. 3.11. You can see from part (a) of the figure that the probability

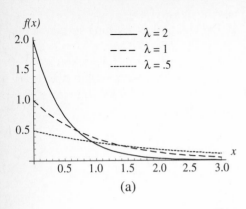

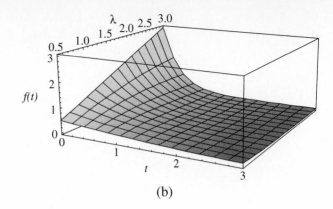

Figure 3.11 Exponential Densities

weight shifts toward 0 as the parameter λ increases. A three-dimensional look at the same phenomenon appears in part (b), where we see that the cross sections of the surface perpendicular to the λ-axis are the density curves for fixed λ. As λ moves from .5 to 3, probability gathers more tightly around 0.

The exponential distribution is often used to model the time required for an event to occur, such as the arrival of a plane to an airport or a request for processing to a computer.

Example 3.3.1

This example is an interesting application of the exponential distribution that I ran across recently (Badalamente et al., 1994). A group of literary researchers concerned with poetry writing styles analyzed several poems, including Frost's "The Road Not Taken." They found the frequency distribution of the number of words required to reach the next word not already used before in the poem. For instance, in the sentence "There is gold in there but it is too deep," the first four words all have a delay of 0 until a new word is reached, but there is a delay of 1 at the second *there*, which is not new in the sentence. There is a similar delay of 1 at the second *is*. So the sequence of delays for this sentence is 0, 0, 0, 0, 1, 0, 1, 0. For "The Road Not Taken," a histogram of the delays is shown in Fig. 3.12. An exponential distribution with parameter $\lambda = 1.86$ is superimposed (this is the reciprocal of the mean delay; see Proposition 3.3.1, page 136), and the fit to the histogram is good. Chapter 8 takes up the question of how to determine parameter estimates in order to fit density curves to data.

?Question 3.3.1 Show that the c.d.f. of the exponential distribution is

$$F(t) = 1 - e^{-\lambda t}, \quad \text{if } t \geq 0. \tag{3.33}$$

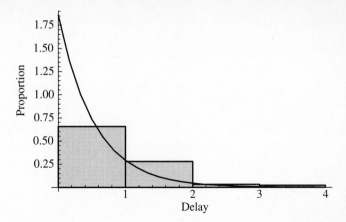

Figure 3.12 Distribution of Delays Until New Word in "The Road Not Taken"

The exponential distribution has an interesting property that can allow you to accept it or reject it on logical grounds as a model for a random time: It is *memoryless*, in the sense illustrated in the following example.

Example 3.3.2

Suppose that the time interval between arrivals of buses to a particular stop is a random variable T (in hours) with the exp(2) distribution. You arrive, panting, to the stop just after a bus leaves. What is the probability that you will have to wait at least 15 minutes for the next bus? Suppose that the next bus has not arrived 15 minutes later when a second person comes along. What is the probability that the newcomer will have to wait at least 15 more minutes?

Since 15 minutes is one quarter of an hour, the answer to the first question is

$$P\left[T > \frac{1}{4}\right] = \int_{1/4}^{\infty} 2e^{-2t}\, dt = -e^{-2t}\Big|_{1/4}^{\infty} = e^{-1/2} \approx .607.$$

In the second question, we are asked for the conditional probability that the bus does not arrive for 30 minutes (that is $\frac{1}{2}$ hour), given that it does not arrive for 15 minutes. This probability is

$$P\left[T > \frac{1}{2} \,\Big|\, T > \frac{1}{4}\right] = \frac{P[T > \frac{1}{2}, T > \frac{1}{4}]}{P[T > \frac{1}{4}]}$$

$$= \frac{P[T > \frac{1}{2}]}{P[T > \frac{1}{4}]}$$

$$= \frac{e^{-2(1/2)}}{e^{-2(1/4)}} = \frac{e^{-1}}{e^{-1/2}} = e^{-1/2}. \tag{3.34}$$

The third line follows from a computation similar to the preceding one, or by complementation of (3.33). But this conditional probability is the same as the unconditional probability that the newcomer waits 15 minutes, computed as if a bus has just left prior to his arrival. Having already waited 15 minutes gains you nothing, probabilistically, because the next bus does not seem to remember that you were waiting.

The next proposition gives the first two moments of the exponential distribution.

PROPOSITION 3.3.1

If X has the exponential density (3.32), then

$$E[X] = \frac{1}{\lambda}, \quad \text{Var}(X) = \frac{1}{\lambda^2}. \tag{3.35}$$

■ **Proof** The proof is left as Exercise 2 at the end of the section.

As we already noted, the probability weight packs more tightly around 0 as the parameter λ increases. It is therefore not a surprise that the mean, $1/\lambda$, approaches 0 as λ approaches ∞. Because the variance $1/\lambda^2$ also decreases as λ increases, the probability is more concentrated when λ is large.

The exponential distribution is related to the Poisson process. Recall that Poisson arrivals occur at a sequence of random times $T_1, T_2, T_3, \ldots$, and the number N_t of arrivals by a fixed time t has the Poisson distribution with parameter λt, where λ is the constant that is called the rate of the process. Now consider the first arrival time T_1. The event $T_1 > t$ is the same as the event that the number of arrivals by time t is 0. Thus

$$P[T_1 > t] = P[N_t = 0] = \frac{e^{-\lambda t}(\lambda t)^0}{0!} = e^{-\lambda t}. \tag{3.36}$$

Since the right side is one minus the c.d.f. of the $\exp(\lambda)$ distribution, we have shown that the first arrival time in a Poisson process has the $\exp(\lambda)$ distribution. A more general result is true, whose proof we omit. (Don't worry about the formal definition of independence; we'll get to it later.)

PROPOSITION 3.3.2

Let $T_1, T_2, T_3, \ldots$ be the arrival times of a Poisson process with rate λ. Then the interarrival time random variables $T_1, T_2 - T_1, T_3 - T_2, \ldots$ each have the $\exp(\lambda)$ distribution. Furthermore, the interarrival times are independent.

A generalization of the exponential density is the following.

DEFINITION 3.3.2

The *gamma density* is

$$f(t) = \frac{\lambda^\alpha}{\Gamma(\alpha)} t^{\alpha-1} e^{-\lambda t}, \qquad t > 0, \tag{3.37}$$

where the *gamma function* is defined by

$$\Gamma(\alpha) = \int_0^\infty x^{\alpha-1} e^{-x} \, dx. \tag{3.38}$$

Note in (3.38) that the gamma function is a function of α, not x, and that the improper integral will converge for all $\alpha > 0$.

We often use the shorthand notation $\Gamma(\alpha, \lambda)$ when referring to this distribution. In the case that $\alpha = 1$, formula (3.37) reduces to the formula for the $\exp(\lambda)$ density, because

$$\Gamma(1) = \int_0^\infty e^{-x} \, dx = 1. \tag{3.39}$$

The gamma density in (3.37) does satisfy the property necessary for it to be a valid probability density, because

$$\int_0^\infty f(t) \, dt = \int_0^\infty \frac{\lambda^\alpha}{\Gamma(\alpha)} t^{\alpha-1} e^{-\lambda t} \, dt \qquad (x = \lambda t, \; dx = \lambda \, dt)$$

$$= \frac{1}{\Gamma(\alpha)} \int_0^\infty \lambda^\alpha \left(\frac{x}{\lambda}\right)^{\alpha-1} e^{-x} \frac{1}{\lambda} \, dx$$

$$= \frac{1}{\Gamma(\alpha)} \int_0^\infty x^{\alpha-1} e^{-x} \, dx = \frac{1}{\Gamma(\alpha)} \cdot \Gamma(\alpha) = 1.$$

Gamma densities for several parameter values are sketched in Fig. 3.13. Part (a) of the figure shows the density for $\lambda = 1$ and values of $\alpha = .5, 1, 2$. These three cases illustrate the three basic shapes of the gamma density: (1) When $\alpha \in (0, 1)$, the graph is asymptotic to the y-axis; (2) when $\alpha = 1$ (the exponential case), the graph has a y-intercept and is strictly decreasing; (3) when $\alpha > 1$ the graph has a humpback shape. Part (b) of the figure is a three-dimensional graph showing more completely the dependence of the density on the α parameter (in the case $\alpha > 1$) for fixed $\lambda = 1$. Similarly, the surface of part (c) of Fig. 3.13 shows how the density depends on the λ parameter for fixed $\alpha = 2$. We see clearly how the mass moves toward the origin as λ increases and away from it as α increases. (See also Exercise 9.)

The gamma function in formula (3.38) is defined in a strange way, but it is not hard to work with because of three convenient properties, namely (3.39) and

$$\Gamma(\alpha + 1) = \alpha \cdot \Gamma(\alpha), \tag{3.40}$$

$$\Gamma(1/2) = \sqrt{\pi}. \tag{3.41}$$

You are asked for proofs of these results in Exercise 11. Meanwhile, to see how helpful they are, answer the following question.

----- **?Question 3.3.2**   What is $\Gamma(2)$? $\Gamma(3)$? $\Gamma(4)$? Show that $\Gamma(n) = (n-1)!$ when n is a positive integer. Now find $\Gamma(3/2)$ and $\Gamma(5/2)$, and devise a similar result for $\Gamma(m/2)$, where m is an odd positive integer.

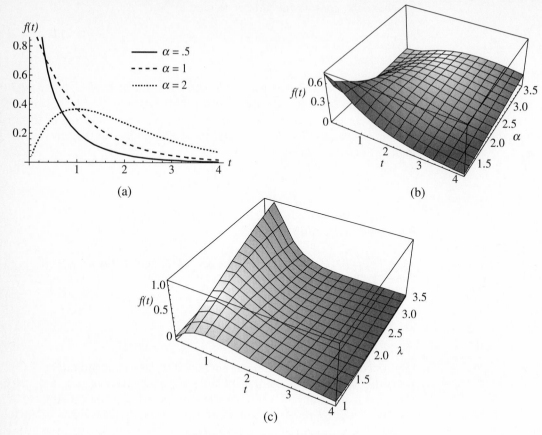

Figure 3.13 Gamma Densities

When the parameter α is a positive integer n, the gamma density is often called the *Erlang density* in honor of an applied mathematician who did early work on queueing theory in which this distribution played an important role. Like the exponential distribution, the gamma distribution has a connection to Poisson processes. The reason, as we will show in Chapter 5, is that the $\Gamma(n, \lambda)$ distribution is the distribution of the sum of n independent $\exp(\lambda)$ distributed random variables. An arrival time in a Poisson process can be decomposed as $T_n = T_1 + (T_2 - T_1) + (T_3 - T_2) + \cdots + (T_n - T_{n-1})$, that is, as the sum of n $\exp(\lambda)$ interarrival times, by Proposition 3.3.2. Therefore the following proposition is true. (Try to argue for the second statement of this proposition yourself.)

PROPOSITION 3.3.3

Let $T_1, T_2, T_3, \ldots$ be the arrival times of a Poisson process with rate λ. Then T_n has the $\Gamma(n, \lambda)$ distribution. Furthermore, the time between the mth and $(m + n)$th arrivals $T_{m+n} - T_m$ has the $\Gamma(n, \lambda)$ distribution for $m, n > 0$.

Example 3.3.3

Suppose that cars pass an expressway exit according to a Poisson process with a rate of 10 per minute. Find the probability that the second car passes the exit between times 5 and 8 seconds.

Since 5 seconds is $\frac{1}{12}$ minute, 8 seconds is $\frac{2}{15}$ minute, and the time T_2 of passage of the second car has the $\Gamma(2, 10)$ distribution,

$$P\left[\frac{1}{12} \leq T_2 \leq \frac{2}{15}\right] = \int_{1/12}^{2/15} \frac{10^2}{1!} t^1 e^{-10t}\, dt$$

$$= \left. (-10te^{-10t} - e^{-10t})\right|_{1/12}^{2/15}$$

$$= \frac{11}{6} e^{-5/6} - \frac{7}{3} e^{-4/3} \approx .18.$$

The first two moments of the gamma distribution are computed in the next proposition. Notice that the formulas therein reduce to the proper ones in the exponential case ($\alpha = 1$).

PROPOSITION 3.3.4

If T has the $\Gamma(\alpha, \lambda)$ distribution, then

$$E[T] = \frac{\alpha}{\lambda}, \quad \text{Var}(T) = \frac{\alpha}{\lambda^2}. \tag{3.42}$$

■ **Proof** We can calculate the mean by a simple substitution and an appeal to properties of the gamma function as follows:

$$E[T] = \int_0^\infty t \cdot \frac{\lambda^\alpha}{\Gamma(\alpha)} t^{\alpha-1} e^{-\lambda t}\, dt$$

$$= \frac{1}{\Gamma(\alpha)} \int_0^\infty (\lambda t)^\alpha e^{-\lambda t}\, dt \qquad (x = \lambda t, \quad dx = \lambda\, dt)$$

$$= \frac{1}{\lambda \Gamma(\alpha)} \int_0^\infty x^\alpha e^{-x}\, dx$$

$$= \frac{1}{\lambda \Gamma(\alpha)} \cdot \Gamma(\alpha + 1)$$

$$= \frac{1}{\lambda \Gamma(\alpha)} \cdot \alpha \Gamma(\alpha) = \frac{\alpha}{\lambda}.$$

Recall that $\text{Var}(T) = E[T^2] - (E[T])^2$. The second moment of T is

$$E[T^2] = \int_0^\infty t^2 \cdot \frac{\lambda^\alpha}{\Gamma(\alpha)} t^{\alpha-1} e^{-\lambda t}\, dt$$

$$= \frac{1}{\lambda \Gamma(\alpha)} \int_0^\infty (\lambda t)^{\alpha+1} e^{-\lambda t}\, dt \quad (x = \lambda t, \quad dx = \lambda\, dt)$$

$$= \frac{1}{\lambda^2 \Gamma(\alpha)} \int_0^\infty x^{\alpha+1} e^{-x}\, dx$$

$$= \frac{1}{\lambda^2 \Gamma(\alpha)} \cdot \Gamma(\alpha + 2)$$

$$= \frac{1}{\lambda^2 \Gamma(\alpha)} \cdot (\alpha + 1)\Gamma(\alpha + 1)$$

$$= \frac{1}{\lambda^2 \Gamma(\alpha)} \cdot (\alpha + 1)\alpha\Gamma(\alpha) = \frac{\alpha(\alpha + 1)}{\lambda^2}.$$

Subtraction of $(E[T])^2 = (\alpha/\lambda)^2$ from $E[T^2]$ yields the formula in (3.42) for the variance of T.

Then, for example, the mean and variance of the time at which the third car passes the expressway exit of Example 3.3.3 are $E[T_3] = 3/10$ and $\mathrm{Var}(T_3) = 3/100$.

Another important subcase of the gamma family has little to do with times of occurrence of random phenomena.

DEFINITION 3.3.3

The *chi-square density* is

$$f(x) = \frac{1}{2^{n/2}\Gamma(n/2)} x^{n/2-1} e^{-x/2}, \qquad x > 0, \tag{3.43}$$

where n is a positive integer parameter that is called the *degrees of freedom* of the distribution.

We use the shorthand $\chi^2(n)$ when referring to this distribution. By comparing formula (3.43) with formula (3.37), you can see that $\chi^2(n) = \Gamma(n/2, 1/2)$. The mean and variance, from Proposition 3.3.4, are therefore $\mu = n$ and $\sigma^2 = 2n$.

Some instances of the density are sketched in Fig. 3.14. You can see that the probability weight shifts to the right and spreads out as the parameter n increases.

We will deal with the chi-square distribution in Chapter 5, then frequently in the second half of this book, where it plays a role nearly as important as that of the normal distribution in the statistical analysis of variability of data.

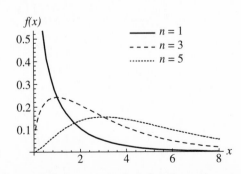

Figure 3.14 Chi-Square Densities

Table 4 in Appendix A gives points x such that $P[X \leq x] = \alpha$, for a number of specified values of the degrees of freedom r, and specified probability values α, where X is a $\chi^2(r)$ random variable. For example, when $n = 14$, the appropriate line of the table indicates that

$$P[X \leq 4.66] \approx .01, \qquad P[X \leq 6.57] \approx .05, \qquad P[X \leq 29.14] \approx .99.$$

Therefore, for instance, we can calculate such probabilities as

$$P[X > 4.66] \approx .99, \qquad P[X \in (6.57, 29.14]] \approx .99 - .05 = .94.$$

—— **?Question 3.3.3** If X is $\chi^2(10)$, for what x is $P[X \leq x] = .05$? What is $P[X > 4.87]$? If X is $\chi^2(18)$, what is $P[X \in [9.39, 28.87]]$?

3.3.2 Normal Distribution

Figure 3.15 displays a histogram of some actual data on natural birthrates in 33 countries that have been subject to forced mass migration recently (Wood, 1994). The data are grouped into intervals of length .3, starting at birthrate 0.7. The histogram of frequencies takes on a rough bell shape that is characteristic of histograms of many data samples, including physical quantities like heights and weights, measurement errors in scientific experiments, and test scores and grades (see, for example, Fig. 2.4 on the calculus GPAs).

The *normal distribution* that we are about to study is important not only as a good empirical fit, but for theoretical reasons as well. A major thrust of Chapter 6 will be to show in general that the distribution of any sum of n independent and identically distributed random variables converges as $n \to \infty$ to the normal distribution. For this reason, the distribution with the following density is the most important distribution in probability and statistics.

DEFINITION 3.3.4 ————————————————————————

The *normal density* is

$$f(x) = \frac{1}{\sqrt{2\pi\sigma^2}} \exp\left(-\frac{(x-\mu)^2}{2\sigma^2}\right), \qquad -\infty < x < \infty. \qquad (3.44)$$

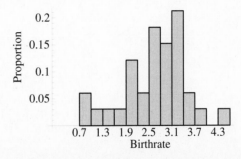

Figure 3.15 Histogram of Birthrates

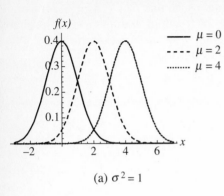

(a) $\sigma^2 = 1$

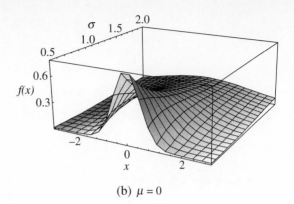

(b) $\mu = 0$

Figure 3.16 Normal Densities

It is characterized by two parameters: $\mu \in \mathbb{R}$ and $\sigma^2 > 0$, and abbreviated $N(\mu, \sigma^2)$. We will show very shortly that the notation is well chosen; μ is in fact the mean and σ^2 is the variance of the distribution.

Sketches for fixed $\sigma^2 = 1$ and several values of μ are given in Fig. 3.16(a). As μ increases, the density curve translates to the right. A three-dimensional look at the dependence of the density on σ^2 for fixed $\mu = 0$ is in part (b) of the figure. As the variance increases, the curve spreads out.

—— **?Question 3.3.4**   Show that the normal density is symmetric about μ. (For an interesting geometric interpretation of σ^2, read Exercise 15.)

Before proceeding, we should check that the function in (3.44) does define a valid p.d.f. Clearly $f(x) \geq 0 \; \forall \; x \in \mathbb{R}$. It remains to show that the integral of f over $(-\infty, +\infty)$ equals 1, for which it suffices to show that the *square* of the integral is 1. Substituting $u = (x - \mu)/\sigma$, $du = dx/\sigma$, we obtain

$$\left(\int_{-\infty}^{+\infty} \frac{1}{\sqrt{2\pi\sigma^2}} \exp\left(-\frac{(x - \mu)^2}{2\sigma^2} \right) dx \right)^2$$

$$= \left(\int_{-\infty}^{+\infty} \frac{1}{\sqrt{2\pi}} e^{-u^2/2} \, du \right)^2$$

$$= \left(\int_{-\infty}^{+\infty} \frac{1}{\sqrt{2\pi}} e^{-u^2/2} \, du \right) \left(\int_{-\infty}^{+\infty} \frac{1}{\sqrt{2\pi}} e^{-v^2/2} \, dv \right)$$

$$= \int_{-\infty}^{+\infty} \int_{-\infty}^{+\infty} \frac{1}{2\pi} e^{-u^2/2 - v^2/2} \, dv \, du.$$

In the double integral, change to polar coordinates with the substitution $u = r\cos\theta$, $v = r\sin\theta$, $dv\,du = r\,dr\,d\theta$. Then,

$$\left(\int_{-\infty}^{+\infty} \frac{1}{\sqrt{2\pi\sigma^2}} \exp\left(-\frac{(x - \mu)^2}{2\sigma^2} \right) dx \right)^2 = \int_0^\infty \int_0^{2\pi} \frac{1}{2\pi} r e^{-r^2/2} \, d\theta \, dr$$

$$= \left(\int_0^\infty r e^{-r^2/2} \, dr \right) \left(\int_0^{2\pi} \frac{1}{2\pi} \, d\theta \right)$$

$$= \left(-e^{-r^2/2} \Big|_0^\infty \right) \left(\frac{1}{2\pi} \theta \Big|_0^{2\pi} \right)$$

$$= 1 \cdot 1 = 1. \tag{3.45}$$

Thus the normal density does integrate to 1 over the state space $(-\infty, +\infty)$. We had to use the trick of squaring and computing a double integral because the single integral of the normal density is not expressible in closed form.

Next we will verify the claim made earlier about the parameters μ and σ^2.

PROPOSITION 3.3.5

If X has the $N(\mu, \sigma^2)$ distribution, then

$$E[X] = \mu, \quad \text{Var}(X) = \sigma^2. \tag{3.46}$$

■ **Proof**   To prove that $E[X] = \mu$, we write

$$E[X - \mu] = \int_{-\infty}^{+\infty} (x - \mu) \cdot \frac{1}{\sqrt{2\pi\sigma^2}} \exp\left(-\frac{(x-\mu)^2}{2\sigma^2} \right) dx$$

$$\left(u = \frac{x - \mu}{\sigma}, \; du = dx/\sigma \right)$$

$$= \sigma \cdot \int_{-\infty}^{+\infty} u \cdot \frac{1}{\sqrt{2\pi}} e^{-u^2/2} \, du$$

$$= \frac{\sigma}{\sqrt{2\pi}} \left(-e^{-u^2/2} \right) \Big|_{-\infty}^{+\infty} = 0.$$

By linearity of expectation, $E[X] = \mu$.

By using the general properties of variance, we can recast the desired result into something computationally easier as follows:

$$\text{Var}(X) = \sigma^2 \iff \text{Var}(X - \mu) = \sigma^2$$

$$\iff \frac{\text{Var}(X - \mu)}{\sigma^2} = 1$$

$$\iff \text{Var}\left(\frac{X - \mu}{\sigma} \right) = 1.$$

$E[\frac{X-\mu}{\sigma}] = 0$, moreover, so it suffices to show that $E\left[((X - \mu)/\sigma)^2 \right] = 1$. (Justify this statement.)

$$E\left[\left(\frac{X - \mu}{\sigma} \right)^2 \right] = \int_{-\infty}^{+\infty} \left(\frac{x - \mu}{\sigma} \right)^2 \cdot \frac{1}{\sqrt{2\pi\sigma^2}} \exp\left(-\frac{(x-\mu)^2}{2\sigma^2} \right) dx$$

$$\left(y = \frac{x - \mu}{\sigma}, \; dy = dx/\sigma \right)$$

$$= \frac{1}{\sqrt{2\pi}} \cdot \int_{-\infty}^{+\infty} y^2 \cdot e^{-y^2/2} \, dy$$

$$\text{(parts, } u = y, \; dv = ye^{-y^2/2} \, dy)$$

$$= \frac{1}{\sqrt{2\pi}} \left(-ye^{-y^2/2} \Big|_{-\infty}^{+\infty} + \int_{-\infty}^{+\infty} e^{-y^2/2} \, dy \right)$$

$$= 0 + 1 = 1.$$

The last integral on the right is 1, since, after bringing the $1/\sqrt{2\pi}$ factor back into the integral, the integrand is the normal density with $\mu = 0$ and $\sigma^2 = 1$, which integrates to 1 over the state space $(-\infty, +\infty)$. This completes the proof.

The special case $\mu = 0$, $\sigma^2 = 1$ is called the *standard normal density* $N(0,1)$, specifically:

$$f(x) = \frac{1}{\sqrt{2\pi}} e^{-x^2/2}, \qquad x \in (-\infty, +\infty). \tag{3.47}$$

Although empirical distributions rarely fit this particular member of the normal family of distributions, it is still a very important density because probabilistic calculations involving other normal random variables can be reduced to calculations involving the standard normal distribution. This point is clarified by the next proposition.

PROPOSITION 3.3.6

If a random variable X has the $N(\mu, \sigma^2)$ distribution, then the random variable Z defined by

$$Z = \frac{X - \mu}{\sigma} \tag{3.48}$$

has the standard normal distribution. Therefore

$$P[X \le b] = P\left[Z = \frac{X - \mu}{\sigma} \le \frac{b - \mu}{\sigma} \right] = \int_{-\infty}^{(b-\mu)/\sigma} \frac{1}{\sqrt{2\pi}} e^{-z^2/2} \, dz. \tag{3.49}$$

(The algebraic operation of subtracting μ and dividing by σ is known as *standardizing*.)

■ **Proof**   We can compute the c.d.f. of Z as follows:

$$F(z) = P[Z \le z] = P\left[\frac{X - \mu}{\sigma} \le z \right] = P[X \le \mu + \sigma z].$$

Since X was assumed to be normally distributed,

$$F(z) = \int_{-\infty}^{\mu + \sigma z} \frac{1}{\sqrt{2\pi\sigma^2}} \exp\left(-\frac{(x - \mu)^2}{2\sigma^2} \right) \, dx.$$

By the Fundamental Theorem of Calculus and the chain rule, the density of Z is

$$f(z) = F'(z) = \sigma \cdot \frac{1}{\sqrt{2\pi\sigma^2}} \exp\left(-\frac{(\mu + \sigma z - \mu)^2}{2\sigma^2} \right) = \frac{1}{\sqrt{2\pi}} e^{-z^2/2},$$

that is, Z is standard normal. Formula (3.49) is self-evident.

The standard normal cumulative probabilities $F(z) = P[Z \leq z]$ needed to do calculations by the standardization process illustrated in (3.49) are contained in Table 3 of Appendix A. Recall that the normal density does not have a closed form integral, so these tabulated probabilities have been found by numerical approximation. By using them, by cleverly employing the symmetry of the $N(0,1)$ density about 0, and by using complementation when necessary, we can compute any normal probabilities that we desire. The following example will show how this is done. Note that in the standard normal table, the units and tenths place of the number z are located in the left margin and the hundredths place of z is found in the top margin. If z is known to the thousandths place, linear interpolation between adjacent table values provides sufficiently accurate probabilities.

Example 3.3.4

Suppose that the load X required to break a 1×10 board is normally distributed with mean 2.50 and standard deviation .24. Then the probability that the piece breaks at a load of 2.61 or less is

$$P[X \leq 2.61] = P\left[Z = \frac{X - \mu}{\sigma} \leq \frac{2.61 - 2.50}{.24}\right] = P[Z \leq 0.46] \approx .6772.$$

This is the area under the standard normal curve to the left of .46 shown in Fig. 3.17(a). The result of .6772 is found by referring to the standard normal table (Table 3 in Appendix A). Look for 0.4 along the left margin, and 6 along the top margin.

The probability that the board breaks at a load of more than 2.39 is

$$P[X > 2.39] = P\left[Z = \frac{X - \mu}{\sigma} > \frac{2.39 - 2.50}{.24}\right] = P[Z > -0.46].$$

This probability is the area shown in Fig. 3.17(b) to the right of -0.46. By the symmetry of the $N(0,1)$ density about 0, however, this result is the same as the probability in part (a), hence $P[X > 2.39] \approx .6772$.

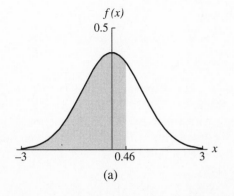

(a)

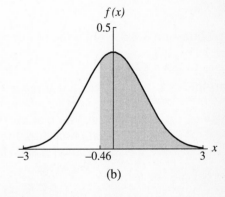

(b)

Figure 3.17 Normal Probabilities

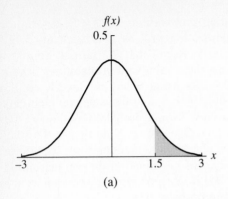
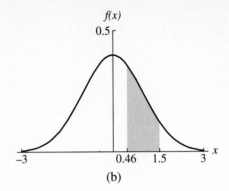

Figure 3.18   More Normal Probabilities

The probability that the breaking load is at least 2.86 is

$$P[X \geq 2.86] = P\left[Z \geq \frac{2.86 - 2.50}{.24}\right] = P[Z \geq 1.50],$$

which is the area to the right of 1.50 shown in Fig. 3.18(a). By complementation,

$$P[Z \geq 1.50] = 1 - P[Z < 1.50] \approx 1 - .9332 = .0668,$$

from the standard normal table.

Finally, the probability that the breaking load is in the interval $[2.61, 2.86]$ is

$$P[2.61 \leq X \leq 2.86] = P\left[\frac{2.61 - 2.50}{.24} \leq Z \leq \frac{2.86 - 2.50}{.24}\right]$$
$$= P[.46 \leq Z \leq 1.50].$$

This area is between .46 and 1.50, as shown in Fig. 3.18(b). The desired area equals the area to the left of 1.50, minus the area to the left of .46, which we know from the computations above to be $.9332 - .6772 = .2560$.

?**Question 3.3.5**   From the last example, what is $P[Z \leq -1.50]$? $P[-1.50 \leq Z \leq -.46]$?

Example 3.3.5

Recall that the $p \times 100$th percentile x_p of a distribution satisfies $P[X \leq x_p] = p$, in other words, the area under the p.d.f. of X to the left of x_p is exactly p. Suppose that for the birthrate data with which we began the section, observations are approximately normally distributed with mean 2.58 and standard deviation .836. What are the 90th and 80th percentiles?

We would like to find $x_{.9}$ such that

$$.9 = P[X \leq x_{.9}] = P\left[Z \leq \frac{x_{.9} - 2.58}{.836}\right].$$

From Table 3 in Appendix A, the value $z_{.9} = 1.28$ is such that $P[Z \leq z_{.9}] \approx .9$. (To see this, locate .8997 in the body of the table, and then read the margins to see the value of z corresponding to that probability.) In light of the preceding equation, we have

$$\frac{x_{.9} - 2.58}{.836} = z_{.9} \approx 1.28 \Longrightarrow x_{.9} \approx 2.58 + 1.28 \cdot (.836) \approx 3.65.$$

In the data set, 31 out of 33 or about 94% of the data were less than or equal to 3.65.

To do a similar calculation for the 80th percentile, note that, from Table 3, $P[Z \leq .84] \approx .80$. Paralleling the derivation for the 90th percentile,

$$\frac{x_{.8} - 2.58}{.836} \approx .84 \Longrightarrow x_{.8} \approx 2.58 + .84 \cdot (.836) \approx 3.28.$$

In the data set 28 out of 33, or about 85% of the data, were less than or equal to 3.28. The fact that the distribution of the data was a little heavy on the left side accounts for these discrepancies between our theoretical predictions and the actual percentiles.

—— **?Question 3.3.6**   Justify that $x_p = \mu + \sigma z_p$ is a general formula relating the percentiles of the $N(\mu, \sigma^2)$ distribution to the standard normal percentiles.

3.3.3   Other Distributions

Let us finish the hard work that we have accomplished in this section by simply listing a few other density functions that arise occasionally. It is not worthwhile to try to memorize these functions; you can look up their functional forms as needed. You should use a graphing device to check the shapes of the graphs of these densities, as well as their dependence on parameters.

Perhaps the most important is the *Weibull distribution*, which is widely used to model time to failure of components or systems in reliability problems. The form of the *Weibull*(λ, β) *density* is

$$f(t) = \beta \lambda^\beta t^{\beta-1} e^{-(\lambda t)^\beta}, \qquad t > 0, \tag{3.50}$$

where λ and β are positive real parameters called the *scale* and *shape* parameters, respectively. Note that in the $\beta = 1$ case, the Weibull density becomes the $\exp(\lambda)$ density.

By varying the β parameter, a variety of aging properties can be modeled by members of the Weibull family of densities. To see this, let T be a failure time random variable with some density $f(t)$ and c.d.f. $F(t)$. Then the probability of failure in the time interval $(t, t + \Delta t]$, given survival until time t, is

$$P[T \in (t, t + \Delta t] \mid T > t] = \frac{P[T \in (t, t + \Delta t]]}{P[T > t]} \approx \frac{f(t)\Delta t}{1 - F(t)}, \tag{3.51}$$

for small Δt, by (3.17). If we define the *failure rate function* by

$$h(t) = \frac{f(t)}{1 - F(t)}, \qquad (3.52)$$

then

$$P[T \in (t, t + \Delta t) \mid T > t] \approx h(t)\Delta t, \qquad (3.53)$$

for small Δt. Thus the failure rate $h(t)$ measures the density of failure at time t given survival until that time. If T has the Weibull distribution, you are asked to show in Exercise 22 that $F(t) = 1 - e^{-(\lambda t)^\beta}$ for $t > 0$; hence the failure rate is

$$h(t) = \frac{f(t)}{1 - F(t)} = \frac{\beta \lambda^\beta t^{\beta-1} e^{-(\lambda t)^\beta}}{e^{-(\lambda t)^\beta}} = \beta \lambda^\beta t^{\beta-1}. \qquad (3.54)$$

This failure rate is decreasing for $0 < \beta < 1$, constant when $\beta = 1$ (the exponential case), and increasing for $\beta > 1$. One notion of aging is increasing failure rate, a condition that we now see that we can model with a Weibull density but not with an exponential density.

The *lognormal density* with parameters $\mu \in \mathbb{R}$ and $\sigma^2 > 0$ is

$$f(y) = \frac{1}{y} \cdot \frac{1}{\sqrt{2\pi\sigma^2}} \cdot \exp\left[-\frac{(\ln(y) - \mu)^2}{2\sigma^2}\right], \qquad y > 0. \qquad (3.55)$$

This distribution is sometimes used to model price behavior of financial instruments. It has a relationship to the normal density that we will see when we study transformations of random variables.

The *Pareto density* with parameter $\theta > 0$ is

$$f(x) = \frac{\theta}{(1 + x)^{\theta+1}}, \qquad x > 0. \qquad (3.56)$$

This distribution has been found to be a reasonable empirical fit to incomes above a threshold value and other economic random phenomena.

The *Cauchy density* with parameter μ is

$$f(x) = \frac{1}{\pi[1 + (x - \mu)^2]}, \qquad -\infty < x < +\infty. \qquad (3.57)$$

Its shape is similar to that of the normal density except for its heavier tails. In Exercise 3 of Section 3.2 you showed that its mean and variance are undefined.

The *beta density* with parameters $\alpha,\ \beta > 0$ is

$$f(x) = \frac{\Gamma(\alpha + \beta)}{\Gamma(\alpha)\Gamma(\beta)} x^{\alpha-1}(1 - x)^{\beta-1}, \qquad 0 < x < 1. \qquad (3.58)$$

By results on transformations of random variables in Chapter 5, it can be shown that the beta distribution is the distribution of the proportion that one gamma random variable has within the total of two independent gamma random variables. For instance, in a Poisson process, since the time until the third arrival T_3 and the time between the third and fifth arrivals $T_5 - T_3$ are independent and have gamma distributions, the ratio $T_3/(T_3 + T_5 - T_3) =$

T_3/T_5 would have a beta distribution. In words, the ratio is the proportion of the time required for the first five arrivals that is taken by the first three arrivals.

Two other distributions, called the *t- and F-distributions*, are extremely important in statistical analysis, but since they too require transformation methods in order to understand their relationships to the normal distribution, we will postpone discussion of them until a special section in Chapter 5.

EXERCISES 3.3

1. Suppose that patients arrive to an emergency room according to a Poisson process with rate 1 per minute. Find
   a. $P[T_1 > 3]$.
   b. $P[T_2 \in (1,3]]$.
   c. $E[T_5 - T_1]$.

2. Prove Proposition 3.3.1.

3. The authors cited in Example 3.3.1 also analyzed Lord Byron's poem "She Walks in Beauty." Reenact their steps by looking up the poem, counting the frequencies of waiting times until new words as we did, and plotting the histogram. Compute the average waiting time, and use its reciprocal as an estimate of the exponential parameter in order to superimpose an exponential density graph on the histogram.

4. Find the skewness $E[(X - \mu)^3]$ for the exp(λ) distribution. What happens as λ grows?

5. Find a general formula for the $p \times 100$th percentile of the distribution of T_1, the time until the first arrival occurs in a Poisson process of rate λ.

6. Show directly that the gamma density with parameters $\alpha = 3$ and $\lambda = 6$ is a valid p.d.f.

7. Show that if a random variable T of continuous type with state space $(0, \infty)$ is such that
$$P[T > t + s \mid T > t] = P[T > s],$$
then T must have the exponential density. (*Hint:* Investigate the derivative of the function $g(s) = P[T > s]$.)

8. Find an equation for the median of the $\Gamma(2, 3.5)$ distribution. Solve the equation numerically. (Recall that the median of a distribution is its 50th percentile.)

9. For $\alpha > 1$, find the point at which the $\Gamma(\alpha, \lambda)$ density achieves its maximum. What happens to this point as the parameters α and λ grow?

10. Let X have the $\Gamma(\alpha, \lambda)$ distribution. Show that the random variable $Y = cX$, where c is a positive constant, also has a gamma distribution, with the same α parameter.

11. Prove properties (a) (3.40) and (b) (3.41) of the gamma function. [*Hint on (b):* Substitute $x = u^2$.]

12. If X is a $\chi^2(8)$ random variable, find
    a. a value of x such that $P[X > x] = .05$.
    b. $P[3.49 < X \le 13.36]$.
    c. constants c and d such that $P[c < X < d] = .90$. Are c and d unique? Why or why not?

13. Cockroaches found in the penthouse apartment of a certain New York real estate tycoon have lengths X that are normally distributed with mean 1.5 inches and standard deviation .4 inch. Find
    a. $P[X > 1.9]$.
    b. $P[X \leq 1]$.
    c. $P[X \in [1.3, 1.7]]$.
    d. the expected length of a necklace made of 12 such cockroaches laid end to end (which he has given to his ex-wife for their anniversary).

14. Comrie (1994), studying the connection between ozone in a forested area of Pennsylvania and the location of pollutant sources, obtained daily ozone readings (in parts per billion) for the growing seasons of 1988–1990. Based on a part of his data set (estimated from a graph) from August 1988, the ozone level X in this region is approximately normally distributed with mean about 65.9 and standard deviation about 23.3. Find
    a. $P[X > 80]$.
    b. $P[40 < X < 90]$.

15. Show that the inflection points of the $N(\mu, \sigma^2)$ density curve occur at $\mu \pm \sigma$.

16. After a recent blood test, I was given a report on various blood constituents that indicated that a reference range of sodium concentration was 135–146 milliequivalents per liter (meq/l). My own reading was 143 meq/l. Assuming a normal distribution for sodium concentration so that the reference range interval corresponds to two standard deviations above and below the mean, in what percentile is my reading?

17. If X has the $N(\mu, \sigma^2)$ distribution, find
    a. $P[|X - \mu| \leq 2\sigma]$.
    b. $P[|X - \mu| > 3\sigma]$.
    c. $P[X > \mu + \sigma]$. (It is interesting that probabilities such as these do not depend on the particular values of the mean and variance.)

18. Suppose that the average calculus GPA at university X is normally distributed with mean 2.2, and 95% of the time, the GPA is between 1.6 and 2.8. Find the standard deviation of X.

19. Show that the skewness $E[(X - \mu)^3]$ of the normal distribution is 0.

20. Suppose that 60% of cans of paint filled at a certain factory come from fill line 1, and the rest from fill line 2. Cans from line 1 are filled to an average of 1.1 gallons, with a standard deviation of .05 gallon, and cans from line 2 are filled to an average of .9 gallon, with a standard deviation of .08 gallon.
    a. If a can is tested, and its paint volume is more than 1.05 gallons, what is the probability that the can came from line 1?
    b. Consider a strategy that guesses that a can must have come from line 1 if its fill volume is more than c. What value should c have in order that the probability of misclassifying a line 1 can is .05?
    c. Continuing part (b), for this value of c, what is the probability of misclassifying a line 2 can?

21. Suppose that the lifetime T of a microcomputer hard disk drive has the

Weibull(2, 1.5) distribution. Find the probability that the drive fails at some time in the interval $[1, 3]$.

22. Show that the c.d.f. of the Weibull(λ, β) distribution is $F(t) = 1 - e^{-(\lambda t)^\beta}$.

23. Show that the expected value of the Weibull distribution with parameters $\lambda = 1$ and $\beta = 2$ is $\Gamma(3/2) = \sqrt{\pi}/2$.

24. a. Show that the Pareto density in (3.56) is a valid p.d.f.
    b. Show that the lognormal density in (3.55) is a valid p.d.f.

25. Use calculus techniques to graph accurately the Cauchy density with parameter $\mu = 0$. Compare several of its functional values to those of the standard normal density, to get a feeling for the relative heaviness of the tails of the two distributions.

3.4 | Multivariate Normal Distribution

This section begins the study of a multivariate analog of the normal distribution. We will return to it in Chapters 4 and 5. The multivariate normal distribution is used as a model in multidimensional measurements of the same kinds of random numerical phenomena for which the single-variable normal distribution is appropriate (e.g., heights and weights, SAT and ACT scores, temperature, humidity, and barometric pressure, etc.), as well as examples like the following.

Example 3.4.1

Van Buren (1995), interested in the economic exploitation of workers in Mexico, analyzed average prices and worker salaries in the United States and Mexico and used the results to come up with estimates of the numbers of minutes that U.S. and Mexican workers must work to earn enough to buy certain standard quantities of grocery staples. The data follow; the first coordinate is the time required by an average U.S. worker, and the second is the time for a Mexican worker. A plot of these points (called a *scattergram*) is given in Fig. 3.19, together with some superimposed curves.

(4.3, 32.3), (22.1, 250.0), (10.7, 60.0), (4.1, 45.1), (7.3, 315.8), (1.5, 11.9), (12.6, 214.3), (4.5, 87.0), (8.4, 117.6), (9.3, 111.1), (5.9, 11.9), (11.9, 176.5), (7.0, 84.5), (3.1, 136.4), (7.3, 127.7), (4.2, 32.3), (12.2, 142.9), (3.0, 51.7), (8.4, 166.7), (8.2, 150.0), (13.5, 69.0), (1.8, 20.4), (8.4, 96.8), (5.7, 20.4)

There does seem to be an increasing relationship between the two variables, roughly linear, but with high variability. There is at least one exceptional observation [(7.3, 315.8), which corresponds to a pound of butter] far away from the pattern set by the other points. The data seem to spread out in a sort of elliptical cloud, denser toward the center, with major axis having a slope of about 10. It follows that for every extra minute that the U.S. worker

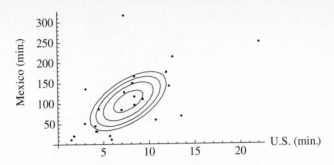

Figure 3.19 Time to Work for Food Staples, U.S. and Mexico

puts in for added consumer benefit, the Mexican worker must put in around ten, depending on some random wage, price, availability, and general market factors.

The multivariate normal distribution in two variables provides a theoretical model for the behavior of pairs of random quantities like the U.S.–Mexican work hours in the foregoing example. To illustrate a simple special case of the distribution, consider the joint density

$$f(x_1, x_2) = \frac{1}{2\pi\sigma_1\sigma_2} \exp\left(-\frac{1}{2}\left[\frac{x_1^2}{\sigma_1^2} + \frac{x_2^2}{\sigma_2^2}\right]\right), \qquad x_1, x_2 \in \mathbb{R}. \qquad (3.59)$$

Figure 3.20(a) shows the graph of this function for $\sigma_1 = \sigma_2 = 1$, and part (b) shows the case $\sigma_1 = 2$, $\sigma_2 = 1$. Slices of the surface by planes perpendicular to the x_1-axis have a bell shape, as do slices perpendicular to the x_2-axis. In case (b), when $\sigma_1 > \sigma_2$, the slices perpendicular to x_2, looked at as functions of x_1, have wider spread than slices perpendicular to x_1, looked at as functions of x_2.

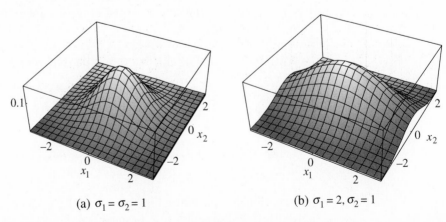

(a) $\sigma_1 = \sigma_2 = 1$ (b) $\sigma_1 = 2, \sigma_2 = 1$

Figure 3.20 Bivariate Normal Density, Case $\rho = 0$

—— **?Question 3.4.1**   What does the graph of the density look like in the case $\sigma_1 = 1$, $\sigma_2 = 2$? What shape is taken on by slices of the surface parallel to the $x_1 - x_2$ plane? How would the graph be affected if, in the exponent of (3.59), the x_i^2 terms were replaced by $(x_i - \mu_i)^2$, where μ_1 and μ_2 are constants?

It is easy to see that f factors:

$$f(x_1, x_2) = \frac{1}{\sqrt{2\pi\sigma_1^2}} e^{-x_1^2/2\sigma_1^2} \cdot \frac{1}{\sqrt{2\pi\sigma_2^2}} e^{-x_2^2/2\sigma_2^2}, \qquad x_1, x_2 \in \mathbb{R}.$$

Therefore the marginal density of X_1 is

$$f_1(x_1) = \int_{-\infty}^{+\infty} f(x_1, x_2)\, dx_2$$

$$= \frac{1}{\sqrt{2\pi\sigma_1^2}} e^{-x_1^2/2\sigma_1^2} \cdot \int_{-\infty}^{+\infty} \frac{1}{\sqrt{2\pi\sigma_2^2}} e^{-x_2^2/2\sigma_2^2}\, dx_2$$

$$= \frac{1}{\sqrt{2\pi\sigma_1^2}} e^{-x_1^2/2\sigma_1^2}, \qquad x_1 \in \mathbb{R}. \tag{3.60}$$

Similarly, the marginal density of X_2 is

$$f_2(x_2) = \frac{1}{\sqrt{2\pi\sigma_2^2}} e^{-x_2^2/2\sigma_2^2}. \tag{3.61}$$

Thus the joint density f factors into the product of normal marginal densities with means equal to zero.

The joint density (3.59) can be written in matrix form as follows:

$$f(\mathbf{x}) = \frac{1}{2\pi\sqrt{\det(\Sigma)}} e^{-(1/2)\mathbf{x}'\Sigma^{-1}\mathbf{x}}, \tag{3.62}$$

where

$$\mathbf{x} = \begin{bmatrix} x_1 \\ x_2 \end{bmatrix}, \quad \Sigma = \begin{bmatrix} \sigma_1^2 & 0 \\ 0 & \sigma_2^2 \end{bmatrix}, \quad \det(\Sigma) = \sigma_1^2\sigma_2^2, \quad \Sigma^{-1} = \begin{bmatrix} 1/\sigma_1^2 & 0 \\ 0 & 1/\sigma_2^2 \end{bmatrix}. \tag{3.63}$$

—— **?Question 3.4.2**   Check the representation in (3.62).

A slight generalization of formula (3.62) to nonzero means leads to the following definition.

DEFINITION 3.4.1 ——————————————————————————

A random vector $\mathbf{X} = (X_1, X_2, \ldots, X_n)'$ is said to have the *multivariate normal distribution* with parameters μ and Σ if its density is

$$f(\mathbf{x}) = \frac{1}{(2\pi)^{n/2}\sqrt{\det(\Sigma)}} e^{-1/2(\mathbf{X}-\mu)'\Sigma^{-1}(\mathbf{X}-\mu)}, \qquad \mathbf{x} \in \mathbb{R}^n, \tag{3.64}$$

where Σ is an $n \times n$ symmetric matrix, assumed to be positive definite (i.e., $\mathbf{y}'\Sigma\mathbf{y} > 0 \ \forall \mathbf{y} \neq \mathbf{0}$), and $\boldsymbol{\mu}$ is an n-vector. The parameter $\boldsymbol{\mu}$ is called the *mean vector* and Σ is called the *covariance matrix* of the distribution.

Example 3.4.2

Suppose that $\mathbf{X} = (X_1, X_2, X_3)'$ has a trivariate normal distribution with mean $\boldsymbol{\mu} = [1, 2, 0]'$ and diagonal covariance matrix

$$\Sigma = \begin{bmatrix} 1 & 0 & 0 \\ 0 & 4 & 0 \\ 0 & 0 & 9 \end{bmatrix}.$$

Then,

$$\Sigma^{-1} = \begin{bmatrix} 1 & 0 & 0 \\ 0 & 1/4 & 0 \\ 0 & 0 & 1/9 \end{bmatrix}, \quad \det(\Sigma) = 1 \cdot 4 \cdot 9.$$

Also, the quadratic form in the exponent in (3.64) can be expanded as

$$(\mathbf{x} - \boldsymbol{\mu})'\Sigma^{-1}(\mathbf{x} - \boldsymbol{\mu}) = \begin{bmatrix} x_1 - 1 & x_2 - 2 & x_3 \end{bmatrix} \begin{bmatrix} 1 & 0 & 0 \\ 0 & 1/4 & 0 \\ 0 & 0 & 1/9 \end{bmatrix} \begin{bmatrix} x_1 - 1 \\ x_2 - 2 \\ x_3 \end{bmatrix}$$

$$= (x_1 - 1)^2 + \frac{1}{4}(x_2 - 2)^2 + \frac{1}{9}x_3^2.$$

Consequently, the joint density in expanded form is

$$f(x_1, x_2, x_3) = \frac{1}{(2\pi)^{3/2}\sqrt{1 \cdot 4 \cdot 9}} e^{-(1/2)[(x_1-1)^2 + 1/4(x_2-2)^2 + (1/9)x_3^2]}$$

$$= \frac{1}{\sqrt{2\pi \cdot 1}} e^{-(1/2)(x_1-1)^2/1} \cdot \frac{1}{\sqrt{2\pi \cdot 4}} e^{-(1/2)(x_2-2)^2/4} \cdot \frac{1}{\sqrt{2\pi \cdot 9}} e^{-(1/2)x_3^2/9}.$$

$$(3.65)$$

The absence of terms off of the diagonal in Σ^{-1} led to a sum of perfect squares in the variables when the quadratic form was expanded, and this in turn led to a factorization of the joint density into the product of normal marginal densities with means 1, 2, and 0 and variances 1, 4, and 9, respectively, for $X_1, X_2,$ and X_3. When Σ is diagonal, this always happens, as Proposition 3.4.1 states. Also, we see that the components of the mean vector $\boldsymbol{\mu}$ give the means μ_1, μ_2, μ_3 of the component random variables, and the diagonal components of the covariance matrix Σ give the component variances $\sigma_1^2, \sigma_2^2, \sigma_3^2$. This much is true even when Σ is not diagonal, as we shall show shortly in the two-variable case.

PROPOSITION 3.4.1

Let $\mathbf{X} = (X_1, X_2, \ldots, X_n)'$ have the multivariate normal distribution with mean $\boldsymbol{\mu} = (\mu_1, \mu_2, \ldots, \mu_n)'$ and covariance Σ. If Σ is a diagonal matrix with entries $\sigma_1^2, \ldots, \sigma_n^2$ on its diagonal, then the joint density is

$$f(\mathbf{x}) = f_1(x_1) \cdot f_2(x_2) \cdots f_n(x_n),$$

where f_i is the $N(\mu_i, \sigma_i^2)$ density. Consequently, under these assumptions the marginal distribution of each X_i is $N(\mu_i, \sigma_i^2)$.

■ **Proof** The proof is left as Exercise 4.

We have not said anything yet about the off-diagonal elements of the covariance matrix Σ. To do this analysis, from now until the end of the section we concentrate on the $n = 2$ case, called the *bivariate normal distribution*. The results that we will discover have analogues for $n \geq 3$ (see, e.g., Anderson, 1984). To simplify notation, we call the two random variables X and Y rather than X_1 and X_2 and we make corresponding changes to the notation for the means and variances.

The covariance matrix has the form

$$\Sigma = \begin{bmatrix} \sigma_{11} & \sigma_{12} \\ \sigma_{21} & \sigma_{22} \end{bmatrix}, \quad \text{where } \sigma_{12} = \sigma_{21}. \tag{3.66}$$

Recall that Σ is positive definite. It is a fact that the diagonal entries of a positive definite matrix must be positive (see Exercise 8), so we will write them as squares σ_x^2 and σ_y^2. Also, we parameterize σ_{12} as a multiple ρ of the product of standard deviations. Then the covariance matrix can be rewritten

$$\Sigma = \begin{bmatrix} \sigma_x^2 & \rho\sigma_x\sigma_y \\ \rho\sigma_x\sigma_y & \sigma_y^2 \end{bmatrix}. \tag{3.67}$$

The bivariate normal distribution is now characterized by five parameters: μ_x, μ_y, σ_x^2, σ_y^2, and ρ.

The condition that Σ is positive definite implies that its determinant is positive; hence

$$\sigma_x^2\sigma_y^2 - \rho^2\sigma_x^2\sigma_y^2 = \sigma_x^2\sigma_y^2(1 - \rho^2) > 0 \implies -1 < \rho < 1. \tag{3.68}$$

The parameter ρ will be called the *correlation coefficient* of X and Y. It will be a focus of our study of dependence in Chapter 4. For now, we would simply like to see its role in determining the geometry of the bivariate normal density surface.

First let us expand the bivariate normal density. Recall that the inverse of a 2×2 matrix is the reciprocal of its determinant times the matrix obtained by interchanging its diagonal elements and changing the signs of its off-diagonal elements. Referring to (3.67), we get

$$\det(\Sigma) = \sigma_x^2\sigma_y^2(1 - \rho^2), \quad \Sigma^{-1} = \frac{1}{\sigma_x^2\sigma_y^2(1 - \rho^2)} \begin{bmatrix} \sigma_y^2 & -\rho\sigma_x\sigma_y \\ -\rho\sigma_x\sigma_y & \sigma_x^2 \end{bmatrix}.$$

By inserting these expressions into the matrix version of the formula for the density (3.64), it is not difficult to obtain

$$f(x, y) = \frac{1}{2\pi\sigma_x\sigma_y\sqrt{1 - \rho^2}} \exp[-Q], \tag{3.69}$$

where

$$Q = \frac{1}{2(1 - \rho^2)} \left[\frac{(x - \mu_x)^2}{\sigma_x^2} - 2\rho\frac{(x - \mu_x)(y - \mu_y)}{\sigma_x\sigma_y} + \frac{(y - \mu_y)^2}{\sigma_y^2} \right]. \tag{3.70}$$

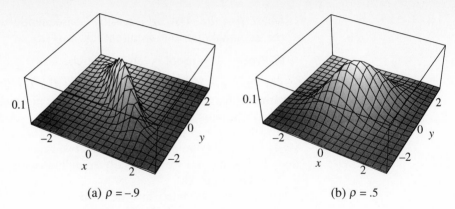

(a) $\rho = -.9$ (b) $\rho = .5$

Figure 3.21   Bivariate Normal Densities, $\sigma_x^2 = \sigma_y^2 = 1$

To investigate the dependence of the density on the correlation coefficient ρ, we sketch $f(x, y)$ for fixed $\mu_x = \mu_y = 0$, fixed $\sigma_x^2 = \sigma_y^2 = 1$ and $\rho = -.9, .5$ in Figs. 3.21 (a) and (b). The contrast to the $\rho = 0$ case displayed in Fig. 3.20 is striking. When $\rho = -.9$ in Fig. 3.21(a), the probability mass is concentrated heavily around the line $y = -x$. It would seem that in this case the value that X takes on has a great deal of bearing on what Y is likely to be. Similarly, when $\rho = .5$ in Fig. 3.21(b), most of the weight is near the line $y = x$, except that the probability is not as concentrated around the line. At least for the μ and σ^2 values we are using, it seems that the magnitude of ρ dictates the extremity of accumulation of probability weight around the line, and the sign of ρ dictates which of the two lines $y = -x$ or $y = x$ is the focus of the probability weight.

—— **?Question 3.4.3**   Use a computer grapher to see what happens to the graph when the variances are unequal. Check that the equal variance assumption is directly responsible for the fact that the lines around which probability weight is concentrated have slopes -1 and 1.

To explain the role that ρ plays in shaping the bivariate density surface, note that by (3.64), the slices of the surface parallel to the x–y plane, determined by the equation $f(\mathbf{x}) = f(x, y) = k$, are the solution sets of

$$(\mathbf{x} - \boldsymbol{\mu})'\Sigma^{-1}(\mathbf{x} - \boldsymbol{\mu}) = c_1. \tag{3.71}$$

Using (3.70), this equation can be rewritten as

$$\sigma_y^2(x - \mu_x)^2 - 2\rho\sigma_x\sigma_y(x - \mu_x)(y - \mu_y) + \sigma_x^2(y - \mu_y)^2 = c_2\sigma_x^2\sigma_y^2. \tag{3.72}$$

You probably recognize equation (3.72) as the equation of an ellipse with center at (μ_x, μ_y). The major and minor axes will not be parallel to the coordinate axes unless $\rho = 0$, in which case the cross-product term vanishes. In calculus you may have learned a direct formula for finding the angle of rotation of a rotated conic section corresponding to the equation $A(x - h)^2 + B(x - h)(y - k) + C(y - k)^2 = D$. The angle of rotation α satisfies

$$\cot 2\alpha = \frac{A - C}{B} = \frac{\sigma_y^2 - \sigma_x^2}{-2\rho\sigma_x\sigma_y}. \tag{3.73}$$

In the cases we graphed in Fig. 3.21, $\sigma_y^2 = \sigma_x^2$, and so $\cot 2\alpha = 0$, which yields immediately that the angle of rotation is $45°$. Upon performing a change in coordinate system to translate the origin to the point $(h, k) = (\mu_x, \mu_y)$ and rotate through angle α, the new equation is $A'x'^2 + C'y'^2 = D$, where

$$\begin{aligned} A' &= A \cos^2\alpha + B \cos\alpha \sin\alpha + C \sin^2\alpha, \\ C' &= A \sin^2\alpha - B \cos\alpha \sin\alpha + C \cos^2\alpha. \end{aligned} \tag{3.74}$$

In the case $\sigma_y^2 = \sigma_x^2 = \sigma^2$, both the cosine and sine factors are $\sqrt{2}/2$, which gives

$$\begin{aligned} A' &= \frac{1}{2}(A + B + C) = \frac{1}{2}(\sigma_y^2 - 2\rho\sigma_x\sigma_y + \sigma_x^2) = \sigma^2(1 - \rho), \\ C' &= \frac{1}{2}(A - B + C) = \frac{1}{2}(\sigma_y^2 + 2\rho\sigma_x\sigma_y + \sigma_x^2) = \sigma^2(1 + \rho), \end{aligned} \tag{3.75}$$

and our contour equation (3.72) reduces to

$$\sigma^2(1 - \rho)x'^2 + \sigma^2(1 + \rho)y'^2 = D = c_2\sigma_x^2\sigma_y^2 = c_2\sigma^4$$
$$\Rightarrow (1 - \rho)x'^2 + (1 + \rho)y'^2 = c_2\sigma^2. \tag{3.76}$$

Suppose that $\rho > 0$. Then $(1 + \rho)$ is the larger coefficient and the longer elliptical axis will be in the x' direction. This will mean that the contours will appear to focus around the line $y = x$. If, on the other hand, $\rho < 0$, then $(1 - \rho)$ is the larger coefficient and the longer axis will be in the y' direction. The contours will then focus around the line $y = -x$. Pictures of the contours for the surfaces graphed in Fig. 3.21 are in Fig. 3.22; they bear out our conclusions. Remember that (3.76) is appropriate only when $\sigma_y^2 = \sigma_x^2 = \sigma^2$. In the general case, the angle of rotation in (3.73) and the coefficients of the ellipse equation in (3.74) depend in a complicated way on ρ, σ_1, and σ_2. But the principle that sets of constant probability density are ellipses centered at (μ_x, μ_y) and the highest density is at the center remains true in general. This principle also explains the fact that the bivariate normal distribution is a reasonable model for paired data like the U.S.–Mexico work times.

If you look back to Fig. 3.21 [part (b) is easier to see], it seems that cross sections of the bivariate normal surface perpendicular to the coordinate axes have the bell shape. This distribution suggests that the individual random variables X and Y may have marginal normal densities. In the next proposition, we show this. The proof contains two identities, (3.79) and (3.82), which will also be of use to us when we study conditional distributions in Chapter 4.

PROPOSITION 3.4.2

Let $\mathbf{X} = (X, Y)$ have the bivariate normal density (3.69). Then X has the $N(\mu_x, \sigma_x^2)$ distribution, and Y has the $N(\mu_y, \sigma_y^2)$ distribution.

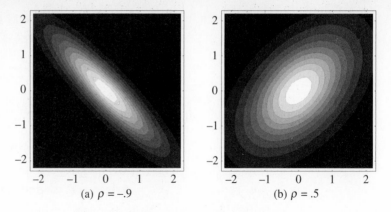

(a) $\rho = -.9$ (b) $\rho = .5$

Figure 3.22   Contours for Surfaces in Fig. 3.21

■ **Proof**   In Exercise 9 you are to show that completing the square on y in the exponent Q in (3.70) yields

$$Q = \frac{(y - \mu_{y|x})^2}{2\sigma_y^2(1 - \rho^2)} + \frac{(x - \mu_x)^2}{2\sigma_x^2}, \tag{3.77}$$

where

$$\mu_{y|x} \equiv \mu_y + \rho\frac{\sigma_y}{\sigma_x}(x - \mu_x). \tag{3.78}$$

(The reason for the odd notation $\mu_{y|x}$ will be explained in the next chapter.) Then f can be rewritten as

$$f(x, y) = \frac{1}{\sqrt{2\pi\sigma_x^2}}\exp\left[-\frac{(x - \mu_x)^2}{2\sigma_x^2}\right] \cdot \frac{1}{\sqrt{2\pi\sigma_y^2(1 - \rho^2)}}\exp\left[-\frac{(y - \mu_{y|x})^2}{2\sigma_y^2(1 - \rho^2)}\right]. \tag{3.79}$$

The marginal density of X is the integral between $-\infty$ and $+\infty$ of (3.79) with respect to y. But the leading terms in x, which make up the $N(\mu_x, \sigma_x^2)$ density, can be drawn outside the integral, and the remaining integral is the integral of a normal density, which equals 1. This shows that X has the desired density.

Working similarly for Y, we can also write the exponent Q as

$$Q = \frac{(x - \mu_{x|y})^2}{2\sigma_x^2(1 - \rho^2)} + \frac{(y - \mu_y)^2}{2\sigma_y^2}, \tag{3.80}$$

where

$$\mu_{x|y} = \mu_x + \rho\frac{\sigma_x}{\sigma_y}(y - \mu_y). \tag{3.81}$$

Then the bivariate density f has the form

$$f(x, y) = \frac{1}{\sqrt{2\pi\sigma_y^2}}\exp\left[-\frac{(y - \mu_y)^2}{2\sigma_y^2}\right] \cdot \frac{1}{\sqrt{2\pi\sigma_x^2(1 - \rho^2)}}\exp\left[-\frac{(x - \mu_{x|y})^2}{2\sigma_x^2(1 - \rho^2)}\right]. \tag{3.82}$$

The marginal density of Y is the integral of (3.82) with respect to x, which, by the foregoing reasoning, reduces to the $N(\mu_y, \sigma_y^2)$ density. This completes the proof.

EXERCISES 3.4

1. In Section 3.3 we made reference to data on birthrates in countries with mass forced migration, such as Bosnia, Afghanistan, Guatemala, and others. The Population Crisis Committee made estimates of the Human Suffering Index for 28 of those countries. Given the data following, in which the first component is the birthrate and the second is the H.S.I., do you think that a bivariate normal distribution with a diagonal covariance matrix would make a good model? Why or why not?

$$(0.8, 44), \ (3.2, 76), \ (3.2, 75), \ (2.7, 93), \ (2.8, 85), \ (2.8, 89),$$
$$(2.6, 84), \ (3.1, 86), \ (3.1, 92), \ (1.0, 44), \ (2.3, 76), \ (3.1, 89),$$
$$(2.3, 73), \ (3.6, 71), \ (2.1, 61), \ (2.6, 61), \ (3.7, 65), \ (2.8, 77),$$
$$(1.4, 58), \ (2.5, 84), \ (2.0, 63), \ (2.8, 85), \ (3.3, 88), \ (3.1, 69),$$
$$(2.6, 64), \ (2.5, 82), \ (1.9, 81), \ (3.0, 70)$$

2. Write out the formula for the trivariate normal density with mean $[0\ 0\ 1]'$ and covariance matrix

$$\Sigma = \begin{bmatrix} 4 & -1 & 0 \\ -1 & 4 & 0 \\ 0 & 0 & 1 \end{bmatrix}.$$

3. Are the cross sections of the density surface in Fig. 3.20(a) perpendicular to the x_2-axis densities in x_1? What normalizing 'constant' (actually, function of x_2) must you multiply by to change the cross section into a density? What density do you get?

4. Prove Proposition 3.4.1.

5. Suppose that the random vector (T_1, T_2) of high temperatures in Cleveland and Cincinnati on July 17 has the bivariate normal distribution with mean vector $\mu = [84\ 87]'$ and covariance matrix

$$\Sigma = \begin{bmatrix} 49.2 & 40.8 \\ 40.8 & 56.3 \end{bmatrix}.$$

a. Compute $P[T_1 \leq 78]$.
b. Compute $P[T_2 > 90]$.
c. Find the correlation coefficient of T_1 and T_2.

6. If $\mathbf{X} = (X_1, X_2, X_3)$ is multivariate normal with mean $\mu = [0\ 1\ 2]'$ and the following covariance matrix, find the indicated quantities.

$$\Sigma = \begin{bmatrix} 16 & 0 & 0 \\ 0 & 9 & 0 \\ 0 & 0 & 4 \end{bmatrix}.$$

a. $P[X_1 \geq 0]$
b. $P[X_2 \leq 0]$
c. $P[X_3 \geq 4]$
d. $P[X_1 \geq 0, \ X_2 \leq 0, \ X_3 \geq 4]$

  e. $E[X_1^2]$

  f. $E[X_1^2 X_2^2]$.

7. Arterial blood pH can increase under conditions of oxygen deprivation. Suppose that blood pH X and blood oxygen concentration Y (% by volume) have the bivariate normal distribution with mean $\mu = [7.4 \; 18.5]'$ and covariance matrix

$$\Sigma = \begin{bmatrix} .000625 & -.02844 \\ -.02844 & 3.0625 \end{bmatrix}.$$

  a. Find the correlation ρ.

  b. Find $P[X > 7.45]$.

  c. Find $P[Y \le 15]$.

8. Show that if Σ is the covariance matrix of a multivariate normal distribution, then each of its diagonal entries σ_{ii} is strictly positive. (*Hint:* Apply the positive definite condition to a special vector **y**.)

9. Verify formula (3.77).

10. Suppose that the covariance matrix of a four variable normal density with mean $[0\;0\;0\;0]$ is, in block form,

$$\Sigma = \begin{bmatrix} \Sigma_{11} & 0 \\ 0 & \Sigma_{22} \end{bmatrix},$$

where Σ_{11} and Σ_{22} are 2×2 symmetric positive definite matrices. Find a simplified form of the joint density. (*Hint:* The determinant of Σ is $\det(\Sigma_{11}) \cdot \det(\Sigma_{22})$, and you can guess a form for the inverse of Σ.)

11. Find the angle of rotation and the rotated equation for the elliptical contours of the bivariate normal density with variances $\sigma_x^2 = 16$, $\sigma_y^2 = (\sqrt{17}-1)^2$, and correlation $\rho = \sqrt{3}/2$.

12. Use Eq. (3.79) to compute $E[XY]$ for the bivariate normal distribution. Find also

$$\frac{E[XY] - E[X]E[Y]}{\sigma_x \sigma_y}.$$

13. (From Hogg and Craig, 1978, p. 121) The following joint density is clearly not of bivariate normal form, yet it has normal marginals. Show this. (So having normal marginals is a necessary, but not sufficient condition for a bivariate distribution to be normal.)

$$f(x, y) = \frac{1}{2\pi} e^{-1/2(x^2+y^2)}(1 + xy e^{-1/2(x^2+y^2-2)}).$$

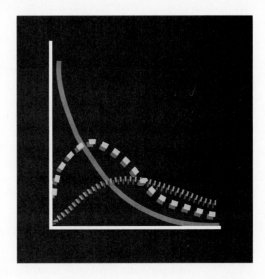

CHAPTER 4

CONDITIONAL
DISTRIBUTIONS AND
INDEPENDENCE

4.1 | Independence of Random Variables

The preceding chapters have hinted at some ideas pertaining to the dependence of random variables on other random variables. The central question is this: Does the observed value of one random variable influence the probability distribution of others, and if so, how? The goal of this chapter is to crystallize the intuition that you have already started to build about this issue.

Remember that in Chapter 1 we called events E and F independent if

$$P[E \mid F] = \frac{P[E \cap F]}{P[F]} = P[E]. \qquad (4.1)$$

To move toward a well-defined notion of independence of random variables, let us start with (4.1), applied to events involving two random variables

X and Y. Specifically, it would be natural to define X and Y to be independent if

$$P[X \in A \mid Y \in B] = \frac{P[X \in A \, , \, Y \in B]}{P[Y \in B]} = P[X \in A], \qquad (4.2)$$

for all subsets A and B of the state spaces of X and Y, respectively, such that $P[Y \in B] \neq 0$. Condition (4.2) says that the knowledge that Y takes a value in B does not alter the probability that X will fall into A, if X and Y are independent.

The rightmost equality in (4.2) implies a factorization that we will take as our definition of independence of random variables.

DEFINITION 4.1.1

Random variables X and Y are said to be *independent* of one another if

$$P[X \in A, \, Y \in B] = P[X \in A] \cdot P[Y \in B], \qquad (4.3)$$

for all sets A, B. Random variables $X_1, X_2, \ldots, X_n$ are called *mutually independent* if for any subcollection of them $X_{i_1}, X_{i_2}, \ldots, X_{i_k}$, $k \leq n$, and corresponding subsets $B_{i_1}, B_{i_2}, \ldots, B_{i_k}$ of their state spaces,

$$P[X_{i_1} \in B_{i_1}, \, X_{i_2} \in B_{i_2}, \, \ldots \, , \, X_{i_k} \in B_{i_k}]$$
$$= P[X_{i_1} \in B_{i_1}] \cdot P[X_{i_2} \in B_{i_2}] \cdots P[X_{i_k} \in B_{i_k}]. \qquad (4.4)$$

Random variables that are not independent are called *dependent*.

Exercise 8 at the end of this section shows that in order to prove independence of many random variables, it is enough to prove the factorization (4.4) for the entire collection of random variables $X_1, X_2, \ldots, X_n$, rather than for every possible subcollection. Although we have defined independence via factorization, the intuition is best served by the conditional probability version (4.2); that is, random variables are independent of one another if the value taken on by one does not influence the probability law of another. The definition also formalizes some ideas we had foreshadowed earlier, particularly with regard to the Poisson process. We now know what it really means to say that the interarrival times $S_1 = T_1, S_2 = T_2 - T_1, S_3 = T_3 - T_2, \ldots$ are independent random variables.

—— **?Question 4.1.1** Show that Definition 4.1.1 implies that for any pair X_i and X_j of the collection of independent random variables, $P[X_i \in A \mid X_j \in B] = P[X_i \in A]$.

Example 4.1.1 A state lottery game called Pick 4 operates by putting Ping-Pong balls numbered 0–9 in each of four mixing machines, then drawing out a ball from each machine, in succession, to choose the digits of the winning number. If the mixing is good and the machines operate independently, it makes sense to assume that the random variables X_1, X_2, X_3, X_4, defined as the observed digits on machines 1–4, respectively, are independent and have discrete uniform distributions on $\{0, 1, \ldots, 9\}$. Then, for example,

$$P[X_1 > 5, \ X_2 < 4] = P[X_1 > 5] \cdot P[X_2 < 4] = \left(\sum_{i=6}^{9} \frac{1}{10}\right) \cdot \left(\sum_{j=0}^{3} \frac{1}{10}\right)$$

$$= \frac{4}{10} \cdot \frac{4}{10} = \frac{16}{100};$$

$$P[X_1 = 0, X_2 < 2, X_3 = 8, X_4 > 7]$$
$$= P[X_1 = 0] \cdot P[X_2 < 2] \cdot P[X_3 = 8] \cdot P[X_4 > 7]$$
$$= \frac{1}{10} \cdot \frac{2}{10} \cdot \frac{1}{10} \cdot \frac{2}{10} = \frac{4}{10,000},$$

and in general, for $i, j, k, l \in \{0, 1, \ldots, 9\}$,

$$P[X_1 = i, \ X_2 = j, \ X_3 = k, \ X_4 = l]$$
$$= P[X_1 = i] \cdot P[X_2 = j] \cdot P[X_3 = k] \cdot P[X_4 = l]$$
$$= \frac{1}{10} \cdot \frac{1}{10} \cdot \frac{1}{10} \cdot \frac{1}{10} = \frac{1}{10,000}.$$

The joint distribution of the X's, or the distribution of the random vector $\mathbf{X} = (X_1, X_2, X_3, X_4)$ equivalent to the winning number, is therefore uniform on the set of 10,000 possible Pick 4 winners 0–9999.

The next proposition characterizes independence in terms of the joint probability distribution of the random variables.

PROPOSITION 4.1.1

The following are equivalent:
  (a) $X_1, X_2, \ldots, X_n$ are independent random variables.
  (b) If $F(x_1, x_2, \ldots, x_n)$ is the joint c.d.f. of $X_1, X_2, \ldots, X_n$ and $F_1(x_1)$, $F_2(x_2), \ldots, F_n(x_n)$ are the marginal c.d.f.'s, then

$$F(x_1, x_2, \ldots, x_n) = F_1(x_1) \cdot F_2(x_2) \cdots F_n(x_n). \tag{4.5}$$

  (c) If $f(x_1, x_2, \ldots, x_n)$ is the joint probability density function (mass function in the discrete case) of $X_1, X_2, \ldots, X_n$ and $f_1(x_1), \ f_2(x_2), \ldots, f_n(x_n)$ are the marginal densities (or mass functions), then

$$f(x_1, x_2, \ldots, x_n) = f_1(x_1) \cdot f_2(x_2) \cdots f_n(x_n). \tag{4.6}$$

■ **Proof**   The strategy for showing this triple equivalence will be to show that (a) $\Rightarrow$ (b) $\Rightarrow$ (c) $\Rightarrow$ (a).

(a) $\Rightarrow$ (b): If the random variables are independent, then we can apply Eq. (4.4) to the sets $B_1 = (-\infty, x_1]$, $B_2 = (-\infty, x_2]$, ..., $B_n = (-\infty, x_n]$, to obtain

$$
\begin{aligned}
F(x_1, x_2, \ldots, x_n) &= P[X_1 \le x_1, X_2 \le x_2, \ldots, X_n \le x_n] \\
&= P[X_1 \in (-\infty, x_1], X_2 \in (-\infty, x_2], \ldots, X_n \in (-\infty, x_n]] \\
&= P[X_1 \in (-\infty, x_1]] \cdot P[X_2 \in (-\infty, x_2]] \cdots P[X_n \in (-\infty, x_n]] \\
&= F_1(x_1) \cdot F_2(x_2) \cdots F_n(x_n).
\end{aligned}
$$

(b) $\Rightarrow$ (c): We will prove this implication only for $n = 2$; the argument in the continuous case extends in a straightforward way, but the discrete argument is tedious. (You will be asked in Exercise 9 for a proof when $n = 3$, which should convince you that the methods can be adapted for any desired n.) So, let X_1 and X_2 be random variables whose joint distribution function factors into the product of the marginal distribution functions. If X_1 and X_2 are of the continuous class, then the Fundamental Theorem of Calculus applied twice yields that

$$
\frac{\partial^2 F}{\partial x_1 \partial x_2} = \frac{\partial^2}{\partial x_1 \partial x_2} \int_{-\infty}^{x_1} \int_{-\infty}^{x_2} f(t_1, t_2)\, dt_1\, dt_2 = f(x_1, x_2), \qquad (4.7)
$$

where f is the joint density of X_1 and X_2. But also,

$$
\frac{\partial^2 F}{\partial x_1 \partial x_2} = \frac{\partial^2}{\partial x_1 \partial x_2} (F_1(x_1) \cdot F_2(x_2)) = f_1(x_1) \cdot f_2(x_2),
$$

where f_1 and f_2 are the marginals. Equating the two representations of the second partial of F proves the result in the continuous case.

If X_1 and X_2 are of the discrete class, let $B_1 = (a_1, b_1]$ and $B_2 = (a_2, b_2]$. From Fig. 4.1, it is easy to see that the probability that $X_1 \in B_1$ and $X_2 \in B_2$ can be broken apart as follows:

$$
\begin{aligned}
P[X_1 \in B_1,\ X_2 \in B_2] &= P[a_1 < X_1 \le b_1,\ a_2 < X_2 \le b_2] \\
&= P[X_1 \le b_1,\ X_2 \le b_2] - P[X_1 \le a_1,\ X_2 \le b_2] \\
&\quad - P[X_1 \le b_1,\ X_2 \le a_2] + P[X_1 \le a_1,\ X_2 \le a_2] \\
&= F(b_1, b_2) - F(a_1, b_2) - F(b_1, a_2) + F(a_1, a_2) \\
&= F_1(b_1)F_2(b_2) - F_1(a_1)F_2(b_2) \\
&\quad - F_1(b_1)F_2(a_2) + F_1(a_1)F_2(a_2) \\
&= (F_1(b_1) - F_1(a_1))(F_2(b_2) - F_2(a_2)) \\
&= P[a_1 < X_1 \le b_1]P[a_2 < X_2 \le b_2] \\
&= P[X_1 \in B_1] \cdot P[X_2 \in B_2]. \qquad (4.8)
\end{aligned}
$$

If b_1 and b_2 are arbitrary points of positive probability for the discrete distribution, then in the limit as $a_1 \to b_1$ and $a_2 \to b_2$ from below, the foregoing computation shows that

$$
\begin{aligned}
P[X_1 = b_1,\ X_2 = b_2] &= P[X_1 = b_1] \cdot P[X_2 = b_2] \\
&\Rightarrow f(b_1, b_2) = f_1(b_1) \cdot f_2(b_2),
\end{aligned}
$$

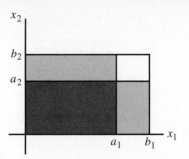

Figure 4.1 Light = Total − Two Grays (Including Dark) + Dark

where f is the joint mass function of X_1 and X_2 and f_1 and f_2 are the marginals. This finishes the proof of (b)⇒(c).

(c) ⇒ (a): Suppose that the joint density factors as in (c). Then we can write, in the continuous case,

$$P[X_1 \in B_1, \ldots, X_n \in B_n] = \int_{B_1} \cdots \int_{B_n} f(x_1, \ldots, x_n) \, dx_n \ldots \, dx_1$$

$$= \int_{B_1} \cdots \int_{B_n} f_1(x_1) \cdots f_n(x_n) \, dx_n \ldots \, dx_1$$

$$= \int_{B_1} f_1(x_1) \, dx_1 \cdots \int_{B_n} f_n(x_n) \, dx_n$$

$$= P[X_1 \in B_1] \cdots P[X_n \in B_n].$$

Thus (c) ⇒ (a) for continuous random variables. The discrete case is similar.

?Question 4.1.2 Show that if the joint density factors into a product of any functions of the individual variables, not just the marginals, then the associated random variables are independent.

Example 4.1.2

Suppose that an ecologist is interested in whether a species of prairie mouse and a species of vole tend to avoid each other. One way of checking that they do not is to see whether the mouse and vole populations in small regions are independent. Numerous traps are set in a field, and for each of 100 days the catch is recorded as a pair (M, V), where M is the (random) number of mice caught and V is the (random) number of voles caught. Suppose that the frequency of days on which each catch combination was observed is as appears in Fig. 4.2(a). Using an empirical estimate of the distribution of (M, V), does it seem as if the number of mice caught is independent of the number of voles caught?

It is reasonable to estimate the probability that m mice and v voles are caught by the proportion of days among the 100 on which that event happened. This is done in the table of Fig. 4.2(b). Now these numbers will only

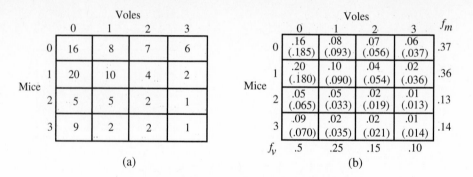

Figure 4.2   Mouse and Vole Trap Probabilities

be estimates of some unknown theoretical joint probabilities $f(m, v)$, and since independence requires the very restrictive factorization condition in (c) of Proposition 4.1.1 to hold, it would be remarkable indeed if the estimated probabilities worked out exactly right. We will judge whether the factorization of f into the product of marginal mass functions f_M and f_V holds approximately.

The marginal probability mass functions are found by totaling the joint probabilities across rows for f_M, and columns for f_V. These probabilities are also displayed in Fig. 4.2(b). In parentheses in this figure, we have computed for each (m, v) pair the product $f_M(m) \cdot f_V(v)$, for purposes of comparison with the joint probabilities $f(m, v)$. The most serious discrepancies are at the extremes when either 3 mice or 3 voles are caught, but otherwise the agreement is close. We have not yet developed an analytical method to decide how well the two tables fit each other, but we will study this question when we deal with *goodness-of-fit* tests in Chapter 12. For now, we have no striking evidence against the independence of M and V.

Example 4.1.3

Three components are connected in series to make up a device as in Fig. 4.3. The components fail at random times T_1, T_2, T_3, which are independent of each other and have the Weibull distribution with parameters $\lambda = 1$, $\beta = 2$. Therefore the components have an increasing failure rate. Find the c.d.f. and density function of the system failure time. Does the system have an increasing failure rate?

Let T be the system failure time. Because of the series structure, the system still survives at time t if and only if all of the components survive at

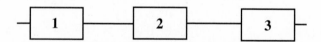

Figure 4.3   Series Reliability System

time t. This observation allows us to take advantage of the independence of the component failure times in the following computation:

$$
\begin{aligned}
P[T > t] &= P[T_1 > t,\ T_2 > t,\ T_3 > t] \\
&= P[T_1 > t] \cdot P[T_2 > t] \cdot P[T_3 > t] \\
&= (1 - P[T_1 \le t]) \cdot (1 - P[T_2 \le t]) \cdot (1 - P[T_3 \le t]) \\
&= e^{-t^2} \cdot e^{-t^2} \cdot e^{-t^2} = e^{-3t^2}.
\end{aligned}
$$

Independence is used in the second line, and the fourth line uses the formula in Exercise 21 of Section 3.3 for the c.d.f. of the Weibull distribution. By complementing the preceding probability, we obtain the system c.d.f.:

$$
F(t) = P[T \le t] = 1 - e^{-3t^2} = 1 - e^{-(\sqrt{3}t)^2}, \qquad t > 0.
$$

This is the c.d.f. of the Weibull distribution with $\lambda = \sqrt{3}$ and $\beta = 2$. Its derivative, the density function, is

$$
f(t) = 6te^{-3t^2}, \qquad t > 0.
$$

The failure rate function for the system is

$$
h(t) = \frac{f(t)}{1 - F(t)} = \frac{6te^{-3t^2}}{e^{-3t^2}} = 6t,
$$

which is an increasing function of t.

Example 4.1.4

Sugar Blasters cereal comes in boxes that are listed as 32 oz (by weight, before settling, of course). Past experience suggests that the actual fill weight is normally distributed with a standard deviation of 1 oz. If four such boxes were to be inspected, and all were found to have a weight of under 31 oz, would there be cause for suspicion of systematic underfilling?

Let X_1, X_2, X_3, and X_4 be the weights of the four boxes. To proceed, we will assume that the X_i's are independent and that all have the $N(32, 1^2)$ distribution. Then,

$$
\begin{aligned}
P[X_1 < 31, \ldots, X_4 < 31] &= P[X_1 < 31] \cdots P[X_4 < 31] \\
&= P\left[\frac{X_1 - 32}{1} < \frac{31 - 32}{1}\right] \cdots P\left[\frac{X_4 - 32}{1} < \frac{31 - 32}{1}\right] \\
&= (P[Z < -1])^4 \approx (.16)^4 \approx .00065.
\end{aligned}
$$

This computation shows that the event that four boxes are underfilled by at least 1 oz is highly unlikely to occur, if our assumptions are true. However, we cannot jump to conclusions too quickly about the implications of the analysis. One of our assumptions is likely to be false, but which? It needn't be the case that the manufacturer is purposely filling the boxes to an average weight that is less than 32 oz. The independence assumption may be violated, particularly if the sampled boxes came from the same batch, and this violation could give the event a probability higher than .00065. Or, the standard deviation could be more than 1, producing the same effect. So we really need tests for means, variances, and independence, which will occupy a good deal of our attention in Chapter 9.

?Question 4.1.3   In the computation of the last example, we used independence to factor in the first line, and then we standardized each inequality in the second line. Do you think that the order of these two operations could have been reversed? Give a preliminary answer now, then reconsider the question later after you have read Proposition 4.1.2.

The next proposition holds for a broad class of functions f_i, the only restriction being that $f_i(X_i)$ is a legitimate random variable. A more advanced course explores the conditions for belonging to that class.

PROPOSITION 4.1.2

Suppose that $X_1, X_2, \ldots, X_n$ are mutually independent random variables, and suppose that $f_1, f_2, \ldots, f_n$ are functions whose domains include the state spaces of the corresponding $X_1, X_2, \ldots, X_n$. Then $f_1(X_1), f_2(X_2), \ldots, f_n(X_n)$ are mutually independent random variables.

■ **Proof**   If A is a set in the range of a function f, define the *inverse image* of A under f as the following subset of the domain of f:

$$f^{-1}(A) = \{x \mid f(x) \in A\}.$$

Then note that $f_i(X_i) \in A_i$ if and only if $X_i \in f^{-1}(A_i)$. Consequently,

$$
\begin{aligned}
P[f_1(X_1) \in A_1, \ldots, f_n(X_n) \in A_n] &= P[X_1 \in f_1^{-1}(A_1), \ldots, X_n \in f_n^{-1}(A_n)] \\
&= P[X_1 \in f_1^{-1}(A_1)] \cdots P[X_n \in f_n^{-1}(A_n)] \\
&= P[f_1(X_1) \in A_1] \cdots P[f_n(X_n) \in A_n].
\end{aligned}
$$

By Exercise 8, this is sufficient to show the independence of the random variables $f_i(X_i)$.

Example 4.1.5   One of many useful consequences of Proposition 4.1.2 is that independence of random variables is a fundamental structural property that is not lost if the random variables are measured in a different system of units. If X and Y are independent random variables, then the rescaled random variables $X' = aX + b$ and $Y' = cY + d$ are also independent.

Yet another factorization formula follows from independence.

PROPOSITION 4.1.3

Suppose that $X_1, X_2, \ldots, X_n$ are independent random variables, and suppose that $h_1, h_2, \ldots, h_n$ are functions whose domains include the state spaces of the corresponding $X_1, X_2, \ldots, X_n$. Then

$$E[h_1(X_1) \cdot h_2(X_2) \cdots h_n(X_n)] = E[h_1(X_1)] \cdot E[h_2(X_2)] \cdots E[h_n(X_n)], \quad (4.9)$$

provided the expectations exist.

■ **Proof** Since $Y_1 = h_1(X_1), \; Y_2 = h_2(X_2), \; \ldots, \; Y_n = h_n(X_n)$ are independent, it is enough to show that for any family of independent random variables $Y_1, Y_2, \ldots, Y_n$,

$$E[Y_1 \cdot Y_2 \cdots Y_n] = E[Y_1] \cdot E[Y_2] \cdots E[Y_n]. \quad (4.10)$$

In the discrete case, by (4.6),

$$\begin{aligned}
E[Y_1 \cdot Y_2 \cdots Y_n] &= \sum \sum \cdots \sum y_1 \cdot y_2 \cdots y_n \, f(y_1, y_2, \ldots, y_n) \\
&= \sum \sum \cdots \sum y_1 \cdot y_2 \cdots y_n \, f_1(y_1) \cdot f_2(y_2) \cdots f_n(y_n) \\
&= \left(\sum y_1 \, f_1(y_1) \right) \cdot \left(\sum y_2 \, f_2(y_2) \right) \cdots \left(\sum y_n \, f_n(y_n) \right) \\
&= E[Y_1] \cdot E[Y_2] \cdots E[Y_n].
\end{aligned}$$

The continuous case is similar.

An important corollary of Proposition 4.1.3 is the next result on variances.

PROPOSITION 4.1.4

If $X_1, X_2, \ldots, X_n$ are independent random variables, then

$$\text{Var}\left(\sum_{i=1}^{n} c_i X_i \right) = \sum_{i=1}^{n} c_i^2 \text{Var}(X_i), \quad (4.11)$$

provided the variances exist.

■ **Proof** By Proposition 4.1.2, the random variables $Y_i = c_i X_i$ are independent, and by Proposition 2.5.3, $\text{Var}(Y_i) = \text{Var}(c_i X_i) = c_i^2 \text{Var}(X_i)$. Therefore it suffices to show that if $Y_1, Y_2, \ldots, Y_n$ is any collection of independent random variables, then

$$\text{Var}\left(\sum_{i=1}^{n} Y_i \right) = \sum_{i=1}^{n} \text{Var}(Y_i). \quad (4.12)$$

To do this, note that $E[\sum_{i=1}^{n} Y_i] = \sum_{i=1}^{n} \mu_i$, where μ_i is the mean of Y_i. Thus

$$\text{Var}\left(\sum_{i=1}^{n} Y_i \right) = E\left[\left(\left(\sum_{i=1}^{n} Y_i \right) - \left(\sum_{i=1}^{n} \mu_i \right) \right)^2 \right]$$

$$= E\left[\left(\sum_{i=1}^{n}(Y_i - \mu_i)\right)^2\right]$$

$$= E\left[\sum_{i=1}^{n}(Y_i - \mu_i)^2 + \sum_{1\leq j,k\leq n; j\neq k}(Y_j - \mu_j)(Y_k - \mu_k)\right]$$

$$= \sum_{i=1}^{n}E\left[(Y_i - \mu_i)^2\right] + \sum_{1\leq j,k\leq n; j\neq k}E\left[(Y_j - \mu_j)(Y_k - \mu_k)\right].$$

(4.13)

The first summation is the sum of the variances of the Y_i's, which is the desired quantity. The second summation is zero, since, by the independence of the Y_i's and Proposition 4.1.3,

$$E\left[(Y_j - \mu_j)(Y_k - \mu_k)\right] = E[Y_j - \mu_j] \cdot E[Y_k - \mu_k] = 0 \cdot 0 = 0.$$

—— **?Question 4.1.4** Write down formula (4.13) for just two random variables Y_1 and Y_2. Try to interpret the meaning of the mixed product term that is added to the sum of the variances.

Example 4.1.6

When investors decide how to spread their wealth among common stocks and other risky assets, they must consider the probabilistic behavior of the *rate of return* on the stocks, that is, the dollars that will be profited per dollar invested. If $R_1, R_2, \ldots, R_n$ denote the random rates of return on n stocks, and if for each $i = 1, 2, \ldots, n$ the investor devotes a fraction w_i of his wealth to stock i, then the rate of return on the combination or *portfolio* of assets is the random variable

$$R = w_1 R_1 + w_2 R_2 + \cdots + w_n R_n.$$

(4.14)

The expected value of the portfolio rate of return is just the linear combination of expected rates of return on the individual stocks:

$$E[R] = w_1 E[R_1] + w_2 E[R_2] + \cdots + w_n E[R_n].$$

(4.15)

If the stock rates of return are independent, then the portfolio rate of return variance can be represented in terms of the stock variances using Proposition 4.1.4:

$$\text{Var}(R) = \text{Var}(w_1 R_1 + w_2 R_2 + \cdots + w_n R_n)$$

$$= w_1^2 \text{Var}(R_1) + w_2^2 \text{Var}(R_2) + \cdots + w_n^2 \text{Var}(R_n).$$

(4.16)

These expressions become important (see, e.g., Exercise 10) in writing an objective function for a problem of maximizing expected return while minimizing risk. One could quantify risk by multiplying $\text{Var}(R)$ by a constant c whose magnitude depends on how averse to risk a particular investor is. The objective would then be to maximize $E[R] - c \cdot \text{Var}(R)$ subject to

$w_1 + \cdots + w_n = 1$. The relationship between the variance of the portfolio rate of return and the variances of the stock rates of return in the dependent case will be discussed later in this chapter.

⎿_____

We close this section with a note on a term that has been used a few times, but has not been fully explained yet. A *random sample* $X_1, X_2, \ldots, X_n$ is a collection of n independent and identically distributed (abbreviated i.i.d.) random variables. The intuition is that sequential observations of a phenomenon are made, or items are selected from a population, in such a way that the outcome of one observation does not affect others. In an actual experiment of sampling from a finite population, sampling must be done with replacement, and consistently from observation to observation in order for the independence and identical distribution assumptions to hold. In this case, the joint density (or mass function) of the sample variables is

$$f(x_1, x_2, \ldots, x_n) = g(x_1) \cdot g(x_2) \cdots g(x_n), \tag{4.17}$$

where g is the common density of each X_i.

EXERCISES 4.1

1. a. A recent study (Nelligan, 1994) looked for gender differences in opinions of school children about their physical education classes. The responses of tenth-grade boys and girls to the question "Have you been refused any options because of your gender?" are given in the following table. If a child is selected at random from this group, and the child's gender X_1 (male or female) and response X_2 (yes or no) are observed, do X_1 and X_2 seem to be independent random variables?

	Yes	No
Male	5	27
Female	3	41

   b. A further question was posed to more students—"Do you prefer mixed or single-gender classes?"—to which they could respond "mixed," "single," or "don't care." The results follow. Does the response seem to be independent of gender? Justify your answer.

	Mixed	Single	Don't Care
Male	33	6	6
Female	17	21	11

2. Suppose that X has the $N(0, 1)$ distribution and Y has the $N(1, 4)$ distribution. If X and Y are independent, find $P[X > 1 \cup Y > 1]$.

3. A sharpshooting forward in the National Basketball Association hits on about 35% of his 3-point shot attempts. Find the joint probability mass function of the random variables $T_1 = $ number of shots until the first successful attempt, $T_2 = $ number of shots strictly after the first success

until the second, and T_3 = number of shots strictly after the second success until the third. Make clear the assumptions you are using. Compute
$P[T_1 > 2, \ T_2 \leq 3, \ T_3 = 1]$.

4. Researchers (Reisinger et al., 1994) analyzing political value differences
in three former Soviet republics reported data on a 1992 survey in which
people from Russia, the Ukraine, and Lithuania responded to the statement "Party competition will make the political system stronger." Respondents could answer (1) fully agree, (2) agree, (3) neutral, (4) disagree, or (5) fully disagree. The results follow. If a person is randomly
selected from this group of interviewees, and the person's national origin
and response are recorded, would these two random variables appear to
be independent?

	1	2	3	4	5
Russia	14	40	18	25	4
Ukraine	16	40	23	18	3
Lithuania	13	34	25	23	6

5. Suppose that the numbers of workers $X_1, \ldots, X_5$ at five of a company's
plants, all in different states, who call in sick on a particular day are independent, Poisson (10) distributed random variables. If, during a period
of difficult labor–management negotiations, each factory had at least 13
workers call in sick, would you suspect a cause other than the flu?

6. a. Find the joint c.d.f. and p.d.f. of the first two interarrival times S_1 and
S_2 of a Poisson process with rate λ.
b. If $T_1 = S_1$ and $T_2 = S_1 + S_2$ are the first two arrival times of a Poisson
process with rate $\lambda = 2$, find $P[T_1 < 1, T_2 > 3]$.

7. If X_1, X_2, X_3, X_4 are mutually independent, show that

$$P[X_1 \in A_1 \mid X_3 \in A_3, \ X_4 \in A_4] = P[X_1 \in A_1].$$

8. Show that if the following factorization holds for all $B_1, \ldots, B_n$, then the
random variables $X_1, \ldots, X_n$ are independent.

$$P[X_1 \in B_1, \ldots, X_n \in B_n] = P[X_1 \in B_1] \cdots P[X_n \in B_n]$$

9. Prove the $n = 3$ discrete case of the implication (b) $\Rightarrow$ (c) in Proposition
4.1.1.

10. A portfolio of three risky assets is to be assembled. The expected rates
of return are 5%, 7.5%, and 10%, respectively, the variances of the rates
of return are 1%, 4%, and 16%, and the rates of return are independent. How should an investor with a risk aversion constant of $c = 2$ (see
Example 4.1.6) allocate her wealth?

11. Exercise 1 of Section 3.4 gave data on the birthrates and human suffering
indices of several countries. If these two variables X and Y are independent with, respectively, the $N(2.8, .36)$ and $N(79, 100)$ distributions, find
a rectangle in the $x - y$ plane that will contain (X, Y) with a probability
of at least .81. Is your rectangle the only one possible?

12. Suppose that a system is composed of four components connected in parallel, as shown in the figure. The system fails only when all of its components have failed. Each component failure time T_i has the Weibull distribution with parameters $\lambda = 1$, $\beta = 3$, and these times are independent. Find the c.d.f. and density function of the system failure time T.

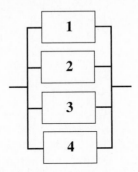

13. Let X_1, X_2, X_3 be a random sample from the exponential distribution with parameter λ. Find the mean and variance of (a) $\bar{X}$; (b) $(X_1 + 2X_2 + X_3)/4$.
14. If X_1, X_2, X_3 are i.i.d. gamma random variables with parameters $\alpha = 2$, $\lambda = 1/2$, find
$$E[2(X_1 - 1)^2 X_2(X_3 + 1)].$$
15. If $X_1, X_2, \ldots, X_n$ is a random sample from the $N(\mu, \sigma^2)$ distribution, show that the joint density is multivariate normal, and find the mean vector and covariance matrix.

4.2 | Conditional Distributions of Random Variables

Having dealt with the concept of independent random variables, we will now take a close look at dependence. Specifically, we will study the probability distribution of a random variable given the observed value of another. We will start with the discrete case, then we will pass on to continuous random variables, and finally we will investigate conditional expectation.

4.2.1   Discrete Conditional Distributions

The idea of a conditional probability distribution in the discrete case is not really new, as the following example should point out.

Example 4.2.1

As I write this example, my doctor is conducting a survey of patients who have recently had an appointment. The survey form contains many items or statements about service, to which respondents are to answer either (1)

Waiting time

		1	2	3	4	Total
	1	15	22	16	8	61
Doctor time	2	10	25	18	12	65
	3	5	18	20	17	60
	4	3	9	10	15	37
Total		33	74	64	52	223

Figure 4.4 Doctor Survey

very satisfied, (2) somewhat satisfied, (3) somewhat dissatisfied, or (4) very dissatisfied. One item involves the amount of time spent waiting in the reception area, and another asks the amount of time the doctor took to answer questions. Suppose that the number of patients responding in each of the possible combinations comes out as in Fig. 4.4. A look at the numbers seems to suggest that as the waiting time increases, patients are more likely to be dissatisfied with the doctor's performance.

If X is the response of a randomly selected patient to the waiting time item and Y is the response of the same patient to the doctor time item, then using the given marginal totals we can find, for example,

$$P[Y = 1 \mid X = 2] = 22/74 \approx .297,$$
$$P[Y = 2 \mid X = 2] = 25/74 \approx .338,$$
$$P[Y = 3 \mid X = 2] = 18/74 \approx .243,$$
$$P[Y = 4 \mid X = 2] = 9/74 \approx .122,$$

but

$$P[Y = 1 \mid X = 4] = 8/52 \approx .154,$$
$$P[Y = 2 \mid X = 4] = 12/52 \approx .231,$$
$$P[Y = 3 \mid X = 4] = 17/52 \approx .327,$$
$$P[Y = 4 \mid X = 4] = 15/52 \approx .288.$$

So the probability distribution of the patient's evaluation of the doctor changes according to the patient's experience in the waiting room. Notice that in each offset group, the total probability adds to 1 (taking rounding into account), so each list of numbers defines a valid probability mass function. Also notice that, since the total number of patients surveyed was 223, these probabilities could have been computed in the following way:

$$P[Y = 3 \mid X = 2] = 18/74 = \frac{18/223}{74/223} = \frac{P[Y = 3 \cap X = 2]}{P[X = 2]}.$$

We are now ready to state the following definition.

DEFINITION 4.2.1

If X and Y are discrete random variables with joint probability mass function $f(x, y)$, and f_X and f_Y are the marginal mass functions, then the *conditional probability mass function* of Y given $X = x$ is

$$f(y \mid x) = P[Y = y \mid X = x] = \frac{P[X = x, \ Y = y]}{P[X = x]} = \frac{f(x, y)}{f_X(x)}, \qquad (4.18)$$

provided $f_X(x) > 0$. Similarly, the *conditional probability mass function* of X given $Y = y$ is

$$f(x \mid y) = P[X = x \mid Y = y] = \frac{P[X = x, \ Y = y]}{P[Y = y]} = \frac{f(x, y)}{f_Y(y)}, \qquad (4.19)$$

provided $f_Y(y) > 0$.

—— **?Question 4.2.1** Use the table of Fig. 4.4 to find the conditional p.m.f. of the waiting time evaluation given that the doctor's evaluation is 1, and the conditional p.m.f. of the waiting time evaluation given that the doctor's evaluation is 3.

—— **?Question 4.2.2** Verify that if $f(y \mid x) = f_Y(y)$, then X and Y are independent. Similarly, argue that if $f(x \mid y) = f_X(x)$, then X and Y are independent.

It is an easy matter to extend the idea of the conditional distribution to the joint conditional distribution of several random variables given several others. Consider a random vector $\mathbf{X} = (X_1, X_2, \ldots, X_n)$ with joint p.m.f. $f(x_1, x_2, \ldots, x_n)$. The *joint conditional p.m.f.* of $X_{m+1}, \ldots, X_n$ given $X_1, \ldots, X_m$ is

$$f(x_{m+1}, \ldots, x_n \mid x_1, \ldots, x_m) = \frac{f(x_1, x_2, \ldots, x_n)}{f_{1,\ldots,m}(x_1, \ldots, x_m)}, \qquad (4.20)$$

where $f_{1,\ldots,m}$ is the joint marginal p.m.f. of $X_1, \ldots, X_m$. Conditional distributions given groups of X_i's other than just the first m are defined similarly, by dividing the overall joint probability mass function by the joint marginal mass function of those X_i's being conditioned on.

Example 4.2.2 Random variables X, Y, and Z representing, respectively, the observed face on the first of three rolled dice, the sum of the faces on the first two dice, and the sum of the faces on all three dice have the joint p.m.f.

$$f(x, y, z) = \left(\frac{1}{6}\right)^3, \qquad 1 \le x \le 6, \quad x + 1 \le y \le x + 6, \quad y + 1 \le z \le y + 6,$$

where x, y, and z are positive integers. (Justify this.) Find the conditional distribution of Z given X and Y.

By (4.20), the conditional mass function is

$$f(z \mid x, y) = \frac{f(x, y, z)}{f_{XY}(x, y)}. \qquad (4.21)$$

The joint marginal p.m.f. of X and Y in the denominator of (4.21) is found by summing the given joint p.m.f. of X, Y, and Z over the possible values of z:

$$f_{XY}(x, y) = \sum_{z=y+1}^{y+6} \left(\frac{1}{6}\right)^3 = 6 \cdot \left(\frac{1}{6}\right)^3 = \left(\frac{1}{6}\right)^2, \quad 1 \le x \le 6, \quad x+1 \le y \le x+6.$$

The quotient in (4.21) becomes

$$f(z \mid x, y) = \frac{(\frac{1}{6})^3}{(\frac{1}{6})^2} = \frac{1}{6}, \quad y+1 \le z \le y+6.$$

It turns out (see Exercise 7) that the conditional mass function $f(z \mid x, y)$ is the same as $f(z \mid y)$; that is, conditioned on Y, the additional knowledge of the value of X does not change the probability distribution of Z. In this situation we call Z and X *conditionally independent* of each other, given Y.

4.2.2 Continuous Conditional Distributions

Examine formula (4.18) again. In the discrete case it makes sense to write $P[Y = y \mid X = x]$, and we obtain the definition of conditional distribution directly from the definition of conditional probability of an event given another. Since continuous random variables satisfy $P[X = x] = 0$, however, we cannot quite do the same thing. Yet, it is tempting to define the conditional density just as in (4.18), as the quotient of the joint density and the marginal density of the random variable being conditioned on. To motivate that this is precisely the thing to do, we will take two different angles of attack.

The first is an algebraic approach based on a limiting argument. It is reasonable to try to define the *conditional c.d.f.* of Y given $X = x$ as the limit of the following as $h \to 0$, if the limit exists.

$$P[Y \le y \mid X \in [x, x+h]] = \frac{P[Y \le y, \, x \le X \le x+h]}{P[x \le X \le x+h]}$$

$$= \frac{\int_x^{x+h} (\int_{-\infty}^y f(t, u) \, du) \, dt}{\int_x^{x+h} f_X(t) \, dt}. \tag{4.22}$$

Since both the numerator and denominator approach 0 as $h \to 0$, l'Hôpital's rule applies to find the limit of the quotient. By the Fundamental Theorem of Calculus, the quotient of the derivative of the numerator with respect to h and the derivative of the denominator with respect to h is

$$\frac{\int_{-\infty}^y f(x+h, u) \, du}{f_X(x+h)}.$$

As h goes to zero the limit is

$$P[Y \le y \mid X = x] = \frac{\int_{-\infty}^y f(x, u) \, du}{f_X(x)} = \int_{-\infty}^y \frac{f(x, u)}{f_X(x)} \, du. \tag{4.23}$$

We obtain the formula $f(x,y)/f_X(x)$ for the conditional density of Y given $X = x$, as anticipated, by taking the derivative of both sides of (4.23) with respect to y.

A second approach to justifying that (4.18) is a good defining formula in the continuous case is geometrical. Consider the joint density function depicted in Fig. 4.5 for two random variables X and Y that have the bivariate normal distribution with both means 0, correlation 0, and variances 3 and 2, respectively.

Focus your attention on a cross section of the surface at a fixed x (such as $x = 1$, which stands out fairly well on the figure). For this fixed x, $f(x,y)$ gives relative probability densities for y values, but it is not itself a valid probability density function in y, since the total integral is not equal to 1:

$$\int_{-\infty}^{+\infty} f(x,y)\ dy = f_X(x).$$

However,

$$\int_{-\infty}^{+\infty} \frac{f(x,y)}{f_X(x)}\ dy = 1, \tag{4.24}$$

so that for fixed x, the function of y defined by $f(x,y)/f_X(x)$ is a valid density, which maintains relative densities for different y's. Geometrically, a vertical stretching of the cross-sectional curve at x, accomplished by multiplying the joint density by the constant $1/f_X(x)$, produces a curve of the same general shape that is a valid p.d.f. in y.

—— **?Question 4.2.3** In the example that we have just considered, find a simplified form for the conditional density of Y given $X = x$. Geometrically, does the curve obtained after stretching the cross section depend on x?

We now have ample evidence that the following is a good way to define conditional distributions in the continuous case.

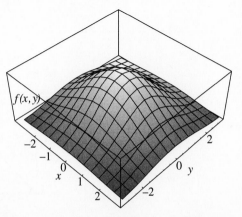

Figure 4.5 Bivariate Normal Density, $\sigma_x^2 = 3$, $\sigma_y^2 = 2$, $\rho = 0$

DEFINITION 4.2.2

If X and Y are continuous random variables with joint probability density function $f(x,y)$, and f_X and f_Y are the marginal density functions, then the *conditional probability density function* of Y given $X = x$ is

$$f(y \mid x) = \frac{f(x,y)}{f_X(x)}, \qquad (4.25)$$

provided $f_X(x) > 0$. Similarly, the *conditional probability density function* of X given $Y = y$ is

$$f(x \mid y) = \frac{f(x,y)}{f_Y(y)}, \qquad (4.26)$$

provided $f_Y(y) > 0$.

Joint conditional densities of continuous random variables given other continuous random variables are defined just as in (4.20) and the surrounding discussion.

Example 4.2.3

Consider the joint density

$$f(x,y) = \begin{cases} 1/2 & \text{if } x - y \geq 0, \ x \leq 2, \ x, y \geq 0, \\ 0 & \text{otherwise.} \end{cases}$$

The state space E is the triangle and its interior displayed in Fig. 4.6.

It is easy to compute that the marginal density of X is $f(x) = \frac{1}{2}x$, for $x \in [0,2)$. Then the conditional density of Y given $X = x$ is

$$f(y \mid x) = \frac{f(x,y)}{f_X(x)} = \frac{\frac{1}{2}}{\frac{1}{2}x} = \frac{1}{x}, \qquad 0 \leq y \leq x.$$

This is a continuous uniform density, but its weight is on an interval that depends on the observed value of X. Then, for example, if the event $X = 1$ is known to have occurred,

$$P[Y \in [0, 1/2] \mid X = 1] = \int_0^{1/2} f(y \mid 1) \, dy = \int_0^{1/2} 1/1 \, dy = 1/2.$$

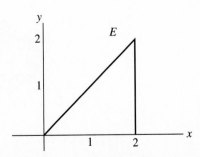

Figure 4.6 State Space of (X, Y)

Example 4.2.4

If an event has nonzero probability, the conditional distribution of a random variable given that the event has occurred is easy to characterize using the cumulative distribution function and our original definition of conditional probability. For instance, let us compute the conditional distribution of the first arrival time T_1 in a Poisson process with rate λ, given the event $N_t = 1$.

Given $N_t = 1$, the first arrival must occur somewhere in the time interval $[0, t]$. Thus the conditional c.d.f. of T_1 is the following conditional probability:

$$
\begin{aligned}
F(s \mid N_t = 1) &= P[T_1 \leq s \mid N_t = 1] \\
&= \frac{P[T_1 \leq s, \ N_t = 1]}{P[N_t = 1]} \\
&= \frac{P[N_s = 1, \ N_t - N_s = 0]}{P[N_t = 1]} \\
&= \frac{P[N_s = 1]P[N_t - N_s = 0]}{P[N_t = 1]} \\
&= \frac{(e^{-\lambda s}(\lambda s)^1/1!)(e^{-\lambda(t-s)}(\lambda(t - s))^0/0!)}{e^{-\lambda t}(\lambda t)^1/1!} \\
&= \frac{e^{-\lambda s}e^{-\lambda(t-s)}(\lambda s)}{e^{-\lambda t}(\lambda t)} = \frac{s}{t}, \qquad s \in [0, t].
\end{aligned}
$$

The third line says that the compound event $\{T_1 \leq s\}$ and $\{N_t = 1\}$ is the same as the event that one arrival occurs during $[0, s]$ and none occurs during $(s, t]$. We can factor in the fourth line because the numbers of arrivals during disjoint time intervals are independent random variables.

The derivative of this c.d.f. with respect to s—that is, the conditional density of T_1 given $N_t = 1$—is $f(s \mid N_t = 1) = 1/t$, $s \in [0, t]$. Thus, if we know that exactly one arrival occurs during the first t time units, the time of that arrival is uniformly distributed on $[0, t]$. Note that the rate of the process has no effect on the final answer.

4.2.3   Conditional Expectation

DEFINITION 4.2.3

The *conditional expectation* $E[g(Y) \mid X = x]$ of a function of a continuous random variable given the observed value of another continuous random variable is

$$
E[g(Y) \mid X = x] = \int_{-\infty}^{+\infty} g(y)f(y \mid x) \, dy. \tag{4.27}
$$

The integral is replaced by a sum in the discrete case.

It is easy to see from (4.27) that conditional expectation satisfies the same linearity properties as ordinary expectation.

Two special conditional expectations are the *conditional mean* of Y given

$X = x$,

$$\mu_{Y|x} = E[Y|X = x] = \int_{-\infty}^{+\infty} y \cdot f(y|x) \, dy, \qquad (4.28)$$

and the *conditional variance* of Y given $X = x$,

$$\sigma_{Y|x}^2 = E[(Y - \mu_{Y|x})^2 | X = x] = \int_{-\infty}^{+\infty} (y - \mu_{Y|x})^2 f(y|x) \, dy. \qquad (4.29)$$

You can show (Exercise 16) that the following computational formula holds:

$$\sigma_{Y|x}^2 = E[Y^2 | X = x] - (\mu_{Y|x})^2. \qquad (4.30)$$

—— **?Question 4.2.4**   What does $E[g(Y)|X = x]$ reduce to if X and Y are independent? (The answer will appear later.)

Example 4.2.5   Let random variables X and Y have joint density:

$$f(x, y) = \frac{1}{64}(x + y), \qquad x, y \in [0, 4].$$

The marginal density of X is easy to compute:

$$f_X(x) = \int_0^4 \frac{1}{64}(x + y) \, dy = \frac{1}{16}x + \frac{1}{8}, \qquad x \in [0, 4].$$

Therefore the conditional density of Y given $X = x$ is

$$f(y|x) = \frac{f(x, y)}{f_X(x)} = \frac{\frac{1}{64}(x + y)}{\frac{1}{16}(x + 2)} = \frac{x + y}{4(x + 2)}, \qquad y \in [0, 4].$$

With this conditional density in hand, we can compute the conditional mean as

$$\mu_{Y|x} = E[Y|X = x] = \int_0^4 y \cdot \frac{x + y}{4(x + 2)} \, dy$$

$$= \frac{1}{4(x + 2)} \int_0^4 xy + y^2 \, dy$$

$$= \frac{8x + 64/3}{4x + 8}, \qquad (4.31)$$

after simplification. The conditional expected square of Y is computed similarly:

$$E[Y^2 | X = x] = \int_0^4 y^2 \cdot \frac{x + y}{4(x + 2)} \, dy$$

$$= \frac{1}{4(x + 2)} \int_0^4 xy^2 + y^3 \, dy$$

$$= \frac{\frac{64}{3}x + 64}{4x + 8}. \qquad (4.32)$$

Therefore, by the computational formula (4.30), the conditional variance of Y given $X = x$ is

$$\sigma^2_{Y|x} = E[Y^2 | X = x] - (\mu_{Y|x})^2$$

$$= \frac{\frac{64}{3}x + 64}{4x + 8} - \left(\frac{8x + 64/3}{4x + 8}\right)^2$$

$$= \frac{64}{9} \cdot \frac{3x^2 + 12x + 8}{(4x + 8)^2}. \tag{4.33}$$

Except in the special case where X and Y are independent, in which $E[g(Y) | X = x]$ is just $E[g(Y)]$, it is typical that $E[g(Y) | X = x]$ is a function of x. We saw this in all three of the expectations (4.31)–(4.33) of the last example. Denote this function as

$$h(x) = E[g(Y) | X = x]. \tag{4.34}$$

We will now study a version of the law of total probability for expectation that involves the composition of this function h with the random variable X. We adopt the notation

$$E[g(Y) | X] = h(X), \tag{4.35}$$

where $h(x)$ is the function in (4.34). The next proposition says that the average value of $g(Y)$ is the average, over all possible values x of X, of the conditional expectation of $g(Y)$ given $X = x$. Intuitively we can compute $E[g(Y)]$ by conditioning on the observed value of X, then "unconditioning" by averaging over X.

PROPOSITION 4.2.1

If the expectations of $g(Y)$ and $h(X)$ described earlier exist, then

$$E[g(Y)] = E[E[g(Y) | X]]. \tag{4.36}$$

■ **Proof**   We will do the proof in the discrete case this time; the continuous case is very similar. As usual, let $f(x, y)$ be the joint p.m.f. of X and Y, let $f_X(x)$ be the marginal p.m.f. of X, and let $f(y | x)$ be the conditional p.m.f. of Y given $X = x$. Then, by (4.34),

$$h(x) = E[g(Y) | X = x] = \sum_y g(y)f(y | x).$$

The expected value of the function $h(X)$ is therefore

$$E[h(X)] = \sum_x h(x)f_X(x) = \sum_x \sum_y g(y)f(y | x)f_X(x). \tag{4.37}$$

From the definition of conditional mass functions, however, $f(y | x)f_X(x) = f(x, y)$; hence

$$E[E[g(Y) | X]] = E[h(X)] = \sum_x \sum_y g(y)f(x, y) = E[g(Y)].$$

Example 4.2.6

A duplicating machine malfunctions at the times T_i of a Poisson process with rate 1/8 per week. When a breakdown happens, it requires one of four levels of repair, with probabilities 1/3, 1/3, 1/6, and 1/6; respectively. The costs incurred in having the repair performed are $50, $100, $200, and $400 for the four levels. Find the expected total cost of repairs through the tth week. Would a service contract that cost $15 per week be a good deal?

The number of repairs by time t, call it N_t, has the Poisson distribution with parameter $\lambda t = \frac{1}{8}t$. Let $C_1, C_2, C_3, \ldots$ be the costs of the first, second, third, and so on, repairs. Each C_i has the p.m.f. $q(50) = 1/3$, $q(100) = 1/3$, $q(200) = 1/6$, $q(400) = 1/6$. It is easy to compute that $E[C_i] = \$150$. The total cost C up to time t depends on the number of repairs N_t that are necessary, specifically

$$C = C_1 + C_2 + \cdots + C_{N_t} = \sum_{i=1}^{N_t} C_i.$$

The crucial observation is that we could compute the expected value of C if we knew what N_t was. To solve the problem, we condition on the value of N_t, then average out the result over all possible values of N_t. We have

$$h(n) = E\left[\sum_{i=1}^{N_t} C_i \,|\, N_t = n\right] = E\left[\sum_{i=1}^{n} C_i \,|\, N_t = n\right]$$

$$= \sum_{i=1}^{n} E[C_i \,|\, N_t = n]$$

$$= \sum_{i=1}^{n} 150 = 150n.$$

Therefore,

$$E\left[\sum_{i=1}^{N_t} C_i\right] = E\left[E\left[\sum_{i=1}^{N_t} C_i \,|\, N_t\right]\right]$$

$$= E[h(N_t)]$$

$$= E[150 \cdot N_t]$$

$$= 150 \cdot \frac{1}{8}t = 18.75t.$$

We can expect to pay $18.75 per week for repairs. Therefore a service contract costing $15 a week would be cost effective.

EXERCISES 4.2

1. Palm and Hodgson (1992) conducted a survey about the geographical distribution of California homeowners who purchased earthquake insurance. Frequency counts from four counties and three insurance categories are given (computed from the authors' given purchase percentages and survey sample sizes). If a random respondent is selected, and X denotes

his or her insurance category and Y the county of residence, find the conditional probability mass function of X given $Y =$ Santa Clara, the conditional probability mass function of X given $Y =$ Los Angeles, and the conditional probability mass function of Y given $X =$ "have insurance."

	Contra Costa	Santa Clara	Los Angeles	San Bernardino
Have Insurance	117	222	133	109
Previously Insured	28	26	10	14
Never Insured	376	307	193	249

2. Four distinguishable coins are flipped. Let $X =$ number of heads on the first three, and let $Y =$ number of heads on the last three. Find the conditional p.m.f. of Y given $X = 1$.

3. Suppose that the joint density of the times T_1 and T_2 of first and second arrival of calls to a psychic help line is

$$f(t_1, t_2) = 4e^{-2t_2}, \qquad 0 \le t_1 \le t_2.$$

Find

a. the conditional density of T_2 given $T_1 = t$.
b. the probability that $T_2 > 1$ given that $T_1 = 1/2$.
c. the expected value of T_2 given that $T_1 = 1/2$.

4. Show that $f(y \mid x)$ defined in (4.18) is a valid p.m.f.

5. Exercise 4 from Section 4.1 gave data on the results of a survey of some residents of Russia, Ukraine, and Lithuania about their reaction to the statement: "Party competition will make the political system stronger." Responses, coded as 1 (fully agree) to 5 (fully disagree) were tabulated. Compute the conditional p.m.f. and conditional expectation of the response of a randomly selected person given that the person is from each of the ex-Soviet republics.

6. Let $f(x, y) = 3e^{-(x+3y)}$, $x, y > 0$ be the joint density of two random variables X and Y that represent, respectively, a commuter's drive time to a city area and time required to find a parking place (in hours). Find the conditional density of the parking time given that the drive time was x, and compare it to the marginal density of Y. What can you conclude?

7. Verify that $f(z \mid y) = f(z \mid x, y)$ in Example 4.2.2.

8. Find the joint conditional p.m.f. of Y and Z given $X = x$ in Example 4.2.2.

9. Let $f(x, y, z) = \frac{1}{2}(x + 2y + z)$, $x, y, z \in (0, 1)$ be the joint density of three random variables X, Y, and Z. Find
a. $f(x \mid Y = \frac{1}{2}, Z = \frac{1}{2})$.
b. $f(x, y \mid Z = \frac{1}{2})$.
c. $P[X < \frac{1}{4} \mid Y = \frac{1}{2}, Z = \frac{1}{2}]$.
d. $E[X^2 \mid Z = \frac{1}{2}]$.

10. Joe Shlabotnick is an undistinguished baseball player with a lifetime .200 batting average; that is, his chance of getting a hit on a single at bat is .2. (You couldn't even get a Bob Uecker card for three Joe Shlabotnicks.) Find the conditional distribution of the at-bat on which Joe gets his second hit of the season, given that his first hit occurs on his tth at-bat.

11. Suppose that each offspring in a generation of flies carries two alleles, A or a, of a certain genetic trait. Therefore the possible genotypes are aa, AA, and aA. Assume that genetic laws predict a distribution of offspring of 1/4 aa, 1/4 AA, and 1/2 aA. For a group of n offspring, find the conditional distribution of the number of type aa flies given that there are m type AA flies in the group, $m \leq n$. In addition to a rigorous derivation, explain the result intuitively.

12. Diners arrive to a fast-food restaurant according to a Poisson process with a rate of 2 per minute. Customers' bills have approximate $\Gamma(8,2)$ distributions and are mutually independent. Use the law of total probability for expectation to find the expected gross receipts for the restaurant over a one-hour period.

13. Let X_1, X_2 be two numbers simulated via computer from the uniform distribution on $(0, 1)$, and let $Y_1 = \min\{X_1, X_2\}$ and $Y_2 = \max\{X_1, X_2\}$. Under the assumption that the algorithm for simulating a sequence of random numbers mimics independence of the numbers well, find the conditional density of Y_2 given $Y_1 = y_1$. (*Hint:* Find the joint c.d.f. by sketching the region of points in the unit square such that $y_1 \leq u$, $y_2 \leq v$.)

14. Suppose that we are concerned with two characteristics of individuals: One is encoded by a random variable Y that equals 1 if the individual is schizophrenic and 0 otherwise, and the other characteristic X is the score of the individual on a psychological profile. Schizophrenics' scores can be approximated by a normal distribution with parameters $\mu = 30$ and $\sigma^2 = 9$, and nonschizophrenics' scores have an approximate normal distribution with $\mu = 20$, $\sigma^2 = 16$. Assume that 5% of those who are profiled are in fact schizophrenics. Find the conditional distribution of Y given that an individual has a score of 25. (*Hint:* You will have to make a reasonable definition of conditional distributions in the case that one random variable is discrete and the other is continuous, then take a Bayes's theorem approach to solving the problem.)

15. Suppose that cars arrive to a certain intersection according to a Poisson process with a rate of 10 per minute. Among these cars, 1/4 turn right, 1/8 turn left, and 5/8 go straight. Drivers make their decisions independently of other drivers. Use Proposition 4.2.1 to show that the expected number of cars X during a five-minute period who do not go straight is 18.75. (*Hint:* Consider the random variables defined by $X_i = 1$ if the ith car turns, and 0 otherwise.)

16. a. Derive formula (4.30) for the conditional variance.
    b. Is the conditional variance the same thing as $E[(Y - \mu_Y)^2 \,|\, X = x]$? Why or why not?

17. In *Bayesian statistics* one takes the point of view that the parameter of a probability distribution is unknown but has a known probability distri-

bution of its own, called a *prior distribution*. One or more observations X are taken from the distribution characterized by the parameter. From them, we can compute a *posterior distribution* of the parameter, which is simply a conditional distribution of the parameter given the observations. For example, suppose that the parameter Λ of an exponential distribution has a prior uniform distribution on $[2, 4]$. One sample X is observed, which has a conditional $\exp(\lambda)$ distribution given $\Lambda = \lambda$. If the observed value of X is 1, find the posterior distribution of Λ. (*Hint:* Take a Bayes's theorem approach adapted to continuous densities.)

18. In reliability theory, a quantity of interest is the *conditional survival distribution*, that is, the distribution of the time of failure of a device given that the device has survived until time t. If a device has a Weibull(λ, β) distributed lifetime, find the conditional density of the lifetime given that it has survived until time t. (*Hint:* Define and find the conditional c.d.f. first.)

19. In the Plinko example (Example 2.5.3), find the conditional p.m.f. and expectation of the winnings given that the column of the chip when it is in row 11 is 5.

4.3 | Covariance and Correlation

4.3.1   Main Ideas

In much the same way as it is useful to have summarizing parameters like the mean and the variance for the distribution of one random variable, it is useful to have summarizing parameters for the dependence between two or more random variables. Since there is only so much information that one constant can contain, we cannot expect a perfect description of dependence. However, the quantities that we study in this section, the *covariance* and *correlation*, are relatively good measures of the degree to which one random variable is a linear function of another. They also enable us to fill in a remaining gap in our knowledge about the variance of a sum of random variables, and they increase our understanding of the multivariate normal distribution whose study is continued in the next section.

Consider the bivariate normal joint density with parameters $\mu_X = 0$, $\mu_Y = 0$, $\sigma_X^2 = 1$, $\sigma_Y^2 = 2$, and $\rho = .8$ sketched in Fig. 4.7(a). The contour plot of curves of equal probability density is given in part (b) of the figure. It uses levels of shading to indicate the height of the density surface in part (a); the lighter the shading, the higher the density.

Notice that much of the probability weight is located around a line through the origin whose slope is about 2. In terms of the random variables X and Y associated with this density, given an X value, the Y value tends to be around $2X$ with high probability. In particular, Y tends to be large when X

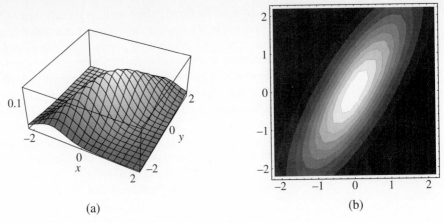

(a) (b)

Figure 4.7 Bivariate Normal Density: $\mu_X = 0$, $\mu_Y = 0$, $\sigma_X^2 = 1$, $\sigma_Y^2 = 2$, $\rho = .8$

is large, and small when X is small. This would imply that for this density the product

$$(X - \mu_X)(Y - \mu_Y) \tag{4.38}$$

would be large in magnitude and have positive sign with high probability; hence the product should have a large positive expected value. This discussion should help to motivate the following definition.

DEFINITION 4.3.1

The *covariance* of two real random variables X and Y is

$$\sigma_{XY} = \text{Cov}(X, Y) = E[(X - \mu_X)(Y - \mu_Y)], \tag{4.39}$$

provided the expectation exists. If the covariance and the marginal variances exist, then the *correlation* between X and Y is

$$\rho = \rho_{XY} = \text{Corr}(X, Y) = E\left[\left(\frac{X - \mu_X}{\sigma_X}\right)\left(\frac{Y - \mu_Y}{\sigma_Y}\right)\right] = \frac{\text{Cov}(X, Y)}{\sigma_X \sigma_Y}. \tag{4.40}$$

Notice that the covariance generalizes the variance, because

$$\text{Cov}(X, X) = E[(X - \mu_X)(X - \mu_X)] = E[(X - \mu_X)^2] = \text{Var}(X). \tag{4.41}$$

—— **?Question 4.3.1** Expand formula (4.39) to derive the computational formula:

$$\text{Cov}(X, Y) = E[XY] - E[X]E[Y] = E[XY] - \mu_X \mu_Y. \tag{4.42}$$

The correlation expresses the product of differences from the means in terms of standard deviation units: If $(X - \mu_X)/\sigma_X = c$, for example, then $(X - \mu_X) = c\sigma_X$; that is, X differs from its mean by c standard deviations.

Standardizing in this way will give us an absolute measure of dependence, as we will see shortly.

Example 4.3.1

To illustrate the computations, let us find the covariance and correlation of the random variables X and Y with the joint density in Example 4.2.3:

$$f(x,y) = \frac{1}{2}, \quad \text{if } x - y \geq 0, \quad x \leq 2, \quad x, y \geq 0.$$

The marginal density of X is $f_X(x) = \frac{1}{2}x$, $x \in [0, 2]$. The marginal of Y is computed easily as

$$f_Y(y) = \int_y^2 \frac{1}{2} \, dx = \frac{1}{2}(2 - y) = 1 - \frac{1}{2} \cdot y, \ y \in [0, 2].$$

The mean and variance of X are:

$$\mu_X = \int_0^2 x \cdot \frac{1}{2}x \, dx = \frac{4}{3},$$

$$\sigma_X^2 = E[X^2] - \mu_X^2 = \int_0^2 x^2 \cdot \frac{1}{2}x \, dx - \mu_X^2 = 2 - \frac{16}{9} = \frac{2}{9}.$$

You should check similarly that the mean and variance of Y are

$$\mu_Y = \frac{2}{3}, \qquad \sigma_Y^2 = \frac{2}{9}.$$

Also,

$$E[XY] = \int_0^2 \int_0^x \frac{1}{2}xy \, dy \, dx$$

$$= \int_0^2 \frac{1}{2}x \left(\frac{1}{2}y^2 \Big|_0^x \right) dx$$

$$= \int_0^2 \frac{1}{4}x^3 \, dx = \frac{1}{16} x^4 \Big|_0^2 = 1.$$

Using the computational formulas (4.42) and (4.40), we obtain

$$\text{Cov}(X, Y) = E[XY] - \mu_X\mu_Y = 1 - \frac{4}{3} \cdot \frac{2}{3} = \frac{1}{9},$$

$$\rho = \frac{\text{Cov}(X, Y)}{\sigma_X \sigma_Y} = \frac{1/9}{\sqrt{2/9}\sqrt{2/9}} = 1/2.$$

Since the covariance and correlation are positive, each random variable tends to be large when the other is, but as yet we have no good handle on the meaning of the magnitudes of these numbers.

Example 4.3.2

Let X and Y have the joint discrete mass function $f(x, y)$ that puts probability weight $1/4$ on the point $(1, 1)$, $1/2$ on the point $(2, 3/2)$, and $1/4$ on the point $(3, 2)$. Find the covariance and correlation of X and Y.

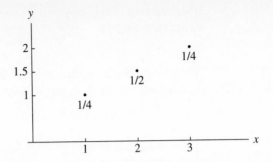

Figure 4.8 Perfectly Correlated Discrete Random Variables

From Fig. 4.8 you can see that the states lie along a line, which should cause you to anticipate a high correlation. We can derive the following results:

$$E[X] = \frac{1}{4} \cdot 1 + \frac{1}{2} \cdot 2 + \frac{1}{4} \cdot 3 = 2,$$

$$\text{Var}(X) = \frac{1}{4}(1-2)^2 + \frac{1}{2}(2-2)^2 + \frac{1}{4}(3-2)^2 = \frac{1}{2},$$

$$E[Y] = \frac{3}{2},$$

$$\text{Var}(Y) = \frac{1}{8},$$

$$E[XY] = \frac{1}{4}(1 \cdot 1) + \frac{1}{2}\left(2 \cdot \frac{3}{2}\right) + \frac{1}{4}(3 \cdot 2) = \frac{13}{4}.$$

Thus the covariance and correlation are

$$\text{Cov}(X,Y) = E[XY] - E[X]E[Y] = \frac{13}{4} - 2 \cdot \frac{3}{2} = \frac{1}{4},$$

$$\rho = \frac{\text{Cov}(X,Y)}{\sigma_X \sigma_Y} = \frac{1/4}{\sqrt{1/2}\sqrt{1/8}} = 1. \tag{4.43}$$

So in this case, in which Y is certain to be a linear function of X, we have obtained a positive value for the covariance, which means that Y tends to be large when X is. Again, the magnitude of the covariance is difficult to interpret, but the rather special value of 1 for the correlation is no coincidence, as we will show shortly.

?Question 4.3.2 Does the fact that $\rho = 1$ in this example depend in any way on the assignment of probabilities 1/4, 1/2, 1/4 to the three states? Try a few alternatives to see. Is the covariance affected by a change in these probabilities?

We will now prove a series of properties of covariance and correlation. The first proposition shows that the correlation is unchanged if X and Y are measured according to a different system of units, but the covariance does change.

PROPOSITION 4.3.1

(a) If X and Y are real random variables, and a, b, c, d are real constants with $b, d \neq 0$, then

$$\text{Cov}(a + bX, c + dY) = b \cdot d \cdot \text{Cov}(X, Y). \tag{4.44}$$

(b) Under the hypotheses of part (a),

$$\text{Corr}(a + bX, c + dY) = \begin{cases} \text{Corr}(X, Y) & \text{if } b, d \text{ have the same sign,} \\ -\text{Corr}(X, Y) & \text{if } b, d \text{ have opposite signs.} \end{cases} \tag{4.45}$$

■ **Proof** (a) By linearity of expectation,

$$E[a + bX] = a + b\mu_X, \qquad E[c + dY] = c + d\mu_Y.$$

Thus, by the definition of covariance,

$$\text{Cov}(a + bX, c + dY) = E\left[((a + bX) - (a + b\mu_X))((c + dY) - (c + d\mu_y))\right]$$
$$= E[b \cdot d \cdot (X - \mu_X)(Y - \mu_Y)] = b \cdot d \cdot \text{Cov}(X, Y).$$

(b) By Propostion 2.5.3,

$$\text{Var}(a + bX) = b^2 \text{Var}(X), \quad \text{Var}(c + dY) = d^2 \text{Var}(Y).$$

Thus

$$\text{Corr}(a + bX, c + dY) = \frac{\text{Cov}(a + bX, c + dY)}{\sqrt{\text{Var}(a + bX)}\sqrt{\text{Var}(c + dY)}}$$
$$= \frac{b \cdot d \cdot \text{Cov}(X, Y)}{|b| \cdot |d|\sigma_X\sigma_Y} = \frac{b \cdot d}{|b \cdot d|} \cdot \text{Corr}(X, Y).$$

The quantity $bd/|bd|$ is equal to $+1$ or -1 according to whether $bd > 0$ or $bd < 0$, or respectively, whether or not b and d have the same sign. This result establishes (4.45).

Thus, although covariance has a slightly simpler definition, correlation is a better measure of dependence than covariance from the standpoint that it is intrinsic to the random variables involved and is not dependent on their units. It is also better because it is bounded and its extreme values have a clear meaning, as the next proposition shows.

PROPOSITION 4.3.2

(a) If X and Y are real random variables with correlation ρ, then

$$|\rho| \leq 1.$$

(b) Moreover, $|\rho| = 1$ if and only if there are constants a, b with $b \neq 0$ such that $Y = a + bX$ with probability 1.

(c) If X and Y are independent, then both $\text{Cov}(X, Y) = 0$ and $\rho = 0$.

■ **Proof** (a) The following expectation is nonnegative for all real t:

$$E\left[((X - \mu_X) + t(Y - \mu_Y))^2\right] \geq 0.$$

Expanding out the square, it follows that

$$E[(X - \mu_X)^2] + 2tE[(X - \mu_X)(Y - \mu_Y)] + t^2 E[(Y - \mu_Y)^2] \geq 0,$$

that is,

$$\sigma_X^2 + 2t\sigma_{XY} + t^2\sigma_Y^2 \geq 0. \tag{4.46}$$

The quadratic function of t defined by the left side of (4.46) therefore has either exactly one or no real roots. This implies that the discriminant of the quadratic is less than or equal to 0, which means that

$$(2\sigma_{XY})^2 - 4\sigma_X^2\sigma_Y^2 \leq 0 \iff \sigma_{XY}^2 \leq \sigma_X^2\sigma_Y^2$$

$$\iff \left(\frac{\sigma_{XY}}{\sigma_X\sigma_Y}\right)^2 \leq 1$$

$$\iff \rho^2 \leq 1$$

$$\iff |\rho| \leq 1. \tag{4.47}$$

(b) If constants a, b, with $b \neq 0$ exist such that $Y(\omega) = a + bX(\omega)$ except for some outcomes $\omega \in \Omega$ of no total probability, then also the expected value of Y must be $a + b\mu_X$ and the expected value of the product $(X - \mu_X)(Y - \mu_Y)$ must be the same as the expected value of $(X - \mu_X) \cdot ((a + bX) - (a + b\mu_X))$. Thus, by (4.44),

$$\text{Cov}(X, Y) = \text{Cov}(X, a + bX) = b \cdot \text{Cov}(X, X) = b \cdot \sigma_X^2,$$

which implies

$$|\text{Corr}(X, Y)| = \left|\frac{\text{Cov}(X, Y)}{\sigma_X\sigma_Y}\right| = \left|\frac{b\sigma_X^2}{\sigma_X\sqrt{b^2\sigma_X^2}}\right| = \left|\frac{b}{|b|}\right| = 1.$$

Conversely, suppose that $|\rho| = 1$. Then the chain of inequalities in (4.47) are equalities, and the quadratic function of t in (4.46) has exactly one real root t_0. For this t_0, retracing the argument in part (a) in reverse gives

$$\sigma_X^2 + 2t_0\sigma_{XY} + t_0^2\sigma_Y^2 = 0 \implies E\left[((X - \mu_X) + t_0(Y - \mu_Y))^2\right] = 0.$$

The nonnegative random variable $((X - \mu_X) + t_0(Y - \mu_Y))^2$ has expected value equal to 0; hence it is impossible that it can take positive values on a set of positive probability. Therefore

$$((X - \mu_X) + t_0(Y - \mu_Y))^2 = 0 \implies (X - \mu_X) + t_0(Y - \mu_Y) = 0,$$

with probability 1. For all but perhaps some exceptional outcomes of total probability 0, Y is thus a linear function of X.

(c) Since X and Y are independent,

$$\sigma_{XY} = E[(X - \mu_X)(Y - \mu_Y)] = E[X - \mu_X]E[Y - \mu_Y] = 0;$$

hence $\rho = 0$.

It is worthwhile to review what Proposition 4.3.2 says. The correlation has extremes of -1 and 1, and a special value of 0. The closer that ρ is to -1, the more that Y tends to be a (negatively sloped) linear function of X. The value of X would then be very important in predicting the value of Y. When ρ is at the extreme of $+1$, Y is a (positively sloped) linear function of X. If X and Y are independent, then they have correlation 0. (We frequently use the word *uncorrelated* for random variables with $\rho = 0$.) Exercise 6 shows, however, that the converse is not true, that is, there are uncorrelated random variables that are not independent.

The next result generalizes Proposition 4.1.4.

PROPOSITION 4.3.3

If $X_1, X_2, \ldots, X_n$ are real random variables and $a_1, a_2, \ldots, a_n$ are real constants, then

$$\text{Var}\left(\sum_{i=1}^{n} a_i X_i\right) = \sum_{i=1}^{n} a_i^2 \cdot \text{Var}(X_i)$$

$$+ \sum_{\substack{j,k=1 \\ j \neq k}}^{n} a_j \cdot a_k \cdot \text{Cov}(X_j, X_k), \qquad (4.48)$$

provided the variances and covariances exist. Consequently, if each pair X_j, X_k is uncorrelated, then

$$\text{Var}\left(\sum_{i=1}^{n} a_i X_i\right) = \sum_{i=1}^{n} a_i^2 \cdot \text{Var}(X_i). \qquad (4.49)$$

■ **Proof**   We will do the proof in the case that all $a_i = 1$. Answer Question 4.3.3 following this proof for the general case. The proof comes directly from the computation (4.13) in Section 4.1, repeated here for your convenience:

$$\text{Var}\left(\sum_{i=1}^{n} X_i\right) = E\left[\left(\left(\sum_{i=1}^{n} X_i\right) - \left(\sum_{i=1}^{n} \mu_i\right)\right)^2\right]$$

$$= E\left[\left(\sum_{i=1}^{n}(X_i - \mu_i)\right)^2\right]$$

$$= E\left[\sum_{i=1}^{n}(X_i - \mu_i)^2 + \sum_{\substack{j,k=1 \\ j \neq k}}^{n}(X_j - \mu_j)(X_k - \mu_k)\right]$$

$$= \sum_{i=1}^{n} E\left[(X_i - \mu_i)^2\right] + \sum_{\substack{j,k=1 \\ j \neq k}}^{n} E\left[(X_j - \mu_j)(X_k - \mu_k)\right].$$

$$(4.50)$$

The first sum on the right side of the last line is the sum of the individual variances, and the second sum is the sum of the covariances, as desired.

?Question 4.3.3 Extend the proof of Proposition 4.3.3 to the case where the coefficients a_i are not necessarily equal to 1.

Example 4.3.3 From a list of M prospects, an insurance salesman selects n names at random, in order and without replacement. He plans to call them (during their dinner hour, of course) to sell them accidental dismemberment insurance. Suppose that among the M people, $N < M$ will be interested. Compute the mean and the variance of the number of interested people in the sample of n.

Let the random variable X be the number of interested people in the sample. Then $X = \sum_{i=1}^{n} X_i$, where

$$X_i = \begin{cases} 1 & \text{if sample person } i \text{ is interested,} \\ 0 & \text{otherwise.} \end{cases}$$

Each X_i has the Bernoulli distribution with success parameter N/M, but the X_i's are dependent, since, if one person in the sample is known to be interested, then there are only $N-1$ other interested people out of the remaining $M-1$ total names, so the success probabilities for the other X's change. Since X_i is Bernoulli distributed,

$$E[X_i] = 1 \cdot P[X_i = 1] + 0 \cdot P[X_i = 0] = 1 \cdot \frac{N}{M} = \frac{N}{M},$$

and

$$\text{Var}(X_i) = \frac{N}{M} \left(1 - \frac{N}{M}\right) = \frac{N(M-N)}{M^2}.$$

By linearity of expectation,

$$E[X] = \sum_{i=1}^{n} E[X_i] = \sum_{i=1}^{n} \frac{N}{M} = n \cdot \frac{N}{M}. \tag{4.51}$$

Also, Proposition 4.3.3 can be applied to find the variance of the sum. To use it, we must compute $\text{Cov}(X_j, X_k)$ for $j \neq k$. Since the product $X_j \cdot X_k$ can take on only the values 0 and 1, its expected value is

$$E[X_j \cdot X_k] = 1 \cdot P[X_j = 1, \ X_k = 1] + 0 = \frac{N(N-1)}{M(M-1)}.$$

(Make sure that you know why the Fundamental Counting Principle implies this result.) By (4.42),

$$\text{Cov}(X_j, X_k) = \frac{N(N-1)}{M(M-1)} - \left(\frac{N}{M}\right)^2 = \frac{N(N-M)}{M^2(M-1)}.$$

Thus

$$\mathrm{Var}(X) = \mathrm{Var}(X_1 + X_2 + \cdots + X_n)$$

$$= \sum_{i=1}^{n} \mathrm{Var}(X_i) + \sum_{\substack{j,k=1 \\ j \neq k}}^{n} \mathrm{Cov}(X_j, X_k)$$

$$= n \cdot \frac{N(M-N)}{M^2} + n(n-1) \cdot \frac{N(N-M)}{M^2(M-1)}$$

$$= \frac{nN(M-N)(M-n)}{M^2(M-1)}. \tag{4.52}$$

?Question 4.3.4   In the last example, for what sample size is the variance the smallest, and what is the expected number of successful calls for that sample size?

Example 4.3.4   Recall Example 4.1.6 on the rate of return $R = \sum_{i=1}^{n} w_i R_i$ of a portfolio of n assets with individual rates of return R_i and portfolio weights w_i. The expected return is $E[R] = \sum_{i=1}^{n} w_i E[R_i]$. The objective is to maximize $E[R] - c \cdot \mathrm{Var}(R)$ subject to $w_1 + \cdots + w_n = 1$. Previously, we were able to find the portfolio variance only for independent rates of return. We now know from (4.49) that the weaker assumption of uncorrelated rates of return would have sufficed. But even better, a direct application of formula (4.48) yields that in general the portfolio variance is

$$\mathrm{Var}(R) = \mathrm{Var}\left(\sum_{i=1}^{n} w_i R_i \right)$$

$$= \sum_{i=1}^{n} w_i^2 \cdot \mathrm{Var}(R_i) + \sum_{\substack{j,k=1 \\ j \neq k}}^{n} w_j \cdot w_k \cdot \mathrm{Cov}(R_j, R_k). \tag{4.53}$$

Notice the possibility that if some of the covariances are negative, then the portfolio variance could be smaller than it would be if asset rates of return were independent. This is the motivation for the strategy of hedging, that is, protecting oneself by including in the portfolio assets that tend to rise as others go down.

To solve a particular problem, suppose that two risky assets have mean rates of return 5.1% and 5%, their standard deviations are 2.5% and 3%, respectively, and their correlation is $-.8$. Suppose that the risk aversion constant c is 1. Let w be the proportion of money invested in the first asset, and hence $(1 - w)$ is invested in the second. The covariance of the rates of return is $\rho \cdot \sigma_1 \cdot \sigma_2$, and so the objective function reduces to

$$f(w) = E[R] - \text{Var}(R)$$
$$= .051w + .05(1 - w) - (.000625)w^2 - .0009(1 - w)^2$$
$$+ .8(.025)(.03)w(1 - w)$$
$$= -.0491 + .0034w - .002125w^2.$$

You can check that upon setting $f' = 0$, the maximum is achieved at $w = .8$; that is, 80% of the investor's wealth should be put into the first asset. The value of f at this optimal w is .05046. If you redo the problem (try it) with a correlation of 0, thereby eliminating the rightmost term in the second line above, you will get a maximum value for f of .05038 at about $w = .918$. If an investor ignores an existing correlation of $-.8$, then that investor is inclined to spend much more money on asset 1, which has the slightly higher mean return and lower variance, but if so the new portfolio earns about $1/100$ of a percent less in personal value. It is counter to intuition that the latter optimum point allows any money at all to be devoted to asset 2, since it has both lower mean and higher standard deviation, but that is what the mathematics indicates.

It may not be surprising that the ideas of covariance and correlation can be tied to conditional expectation. As you can see from formula (4.54) of the next proposition, when the value of X is known, the mean of Y changes by an amount proportional to the difference between X and its mean. The magnitude of the proportionality constant increases with ρ, so that the more extreme the correlation is, the more the expectation of Y changes when X deviates from μ_X.

PROPOSITION 4.3.4

Suppose that X and Y are random variables such that the conditional mean of Y given $X = x$ is a linear function of x. Then

$$\mu_{Y|x} = \mu_Y + \frac{\rho\sigma_Y}{\sigma_X}(x - \mu_X). \tag{4.54}$$

■ **Proof**   By assumption, $h(x) = \mu_{Y|x} = E[Y \mid X = x] = mx + b$ for some slope coefficient m and intercept b. Proposition 4.2.1 implies that

$$\mu_Y = E[Y] = E[E[Y \mid X]] = E[h(X)] = E[mX + b] = m\mu_X + b.$$

Thus $b = \mu_Y - m\mu_X$, and the linear function h has the form

$$h(x) = \mu_{Y|x} = mx + b = mx + \mu_Y - m\mu_X = \mu_Y + m(x - \mu_X). \tag{4.55}$$

It remains to show that the slope $m = \rho\sigma_Y/\sigma_X$.
   We can compute that

$$\rho\sigma_X\sigma_Y = \sigma_{XY} = E[(X - \mu_X)(Y - \mu_Y)]$$
$$= E[E[(X - \mu_X)(Y - \mu_Y) \mid X]]$$
$$= E[(X - \mu_X)\,E[(Y - \mu_Y) \mid X]]$$

$$= E[(X - \mu_X)(h(X) - \mu_Y)]$$
$$= E[(X - \mu_X) \cdot m(X - \mu_X)]$$
$$= m \cdot \sigma_X^2$$
$$\implies m = \frac{\rho \sigma_Y}{\sigma_X}. \tag{4.56}$$

The definition of ρ is used in the first line of (4.56). Proposition 4.2.1 yields the second line of the computation. In the third line, given X, $X - \mu_X$ is a constant that can be removed from the conditional expectation. (To justify this more rigorously, you can write $g(x) = E[(X - \mu_X)(Y - \mu_Y) \,|\, X = x]$, and find $g(X)$.) The fourth line follows from the linearity of conditional expectation, and the fifth line uses the result of (4.55) to substitute for $h(X)$.

Example 4.3.5

In Section 3.4 we saw an example data set in which X was the number of minutes that a U.S. worker had to work for a unit of food staple and Y was the number of minutes that a Mexican worker had to work for the same food. In Chapter 8 we will have ways of using data to produce good estimates of means, variances, and correlations. For now we will accept that the values of those estimates for this economic data are

$$\mu_X = 7.73, \qquad \mu_Y = 105.51, \qquad \sigma_X = 4.60, \qquad \sigma_Y = 79.20, \qquad \rho = .60.$$

Employing Proposition 4.3.4 necessitates that we assume that the conditional means are linear in the variables being conditioned on. In the next section we show that this assumption is correct if the joint distribution is bivariate normal. Then the conditional expectation of the number of minutes required by a Mexican worker for an item that requires a U.S. worker to work for 10 minutes is

$$\mu_{Y|x} = \mu_Y + \frac{\rho \sigma_Y}{\sigma_X}(x - \mu_X)$$
$$= 105.51 + .60 \cdot \frac{79.20}{4.60}(10 - 7.73) = 128.96.$$

Also, the conditional expectation of the number of minutes required by a U.S. worker for an item that requires a Mexican worker to work for 60 minutes is

$$\mu_{X|y} = \mu_X + \frac{\rho \sigma_X}{\sigma_Y}(y - \mu_Y)$$
$$= 7.73 + .60 \cdot \frac{4.60}{79.20}(60 - 105.51) = 6.14.$$

We will have much more to say about linear relationships between random variables in Chapter 10.

4.3.2   Multivariate Results

Matrix theory simplifies the characterization and study of dependence among many random variables. The key parameters are the matrices in the following definition.

DEFINITION 4.3.2

Let $\mathbf{X} = [X_1\ X_2\ \ldots\ X_n]'$ be a random vector. The *covariance matrix* of $\mathbf{X}$ is the $n \times n$ symmetric matrix

$$\Sigma = \mathrm{Cov}(\mathbf{X}) = \begin{bmatrix} \sigma_1^2 & \sigma_{12} & \cdots & \sigma_{1n} \\ \sigma_{21} & \sigma_2^2 & \cdots & \sigma_{2n} \\ \vdots & \vdots & \ddots & \vdots \\ \sigma_{n1} & \sigma_{n2} & \cdots & \sigma_n^2 \end{bmatrix}, \tag{4.57}$$

where $\sigma_i^2 = \mathrm{Var}(X_i)$ and $\sigma_{ij} = \sigma_{ji} = \mathrm{Cov}(X_i, X_j)$. The *correlation matrix* of $\mathbf{X}$ is the $n \times n$ symmetric matrix

$$\Upsilon = \mathrm{Corr}(\mathbf{X}) = \begin{bmatrix} 1 & \rho_{12} & \cdots & \rho_{1n} \\ \rho_{21} & 1 & \cdots & \rho_{2n} \\ \vdots & \vdots & \ddots & \vdots \\ \rho_{n1} & \rho_{n2} & \cdots & 1 \end{bmatrix}, \tag{4.58}$$

where $\rho_{ij} = \rho_{ji} = \mathrm{Corr}(X_i, X_j)$.

?Question 4.3.5   Why are the covariance and correlation matrices symmetric?

The covariance matrix contains the marginal variances on its main diagonal, and the pairwise covariance between the ith and jth random variable in its i, j off-diagonal component. Exercise 15 at the end of this section has direct algebraic relationships between the correlation and covariance matrices, so it is not necessary to go into detail on both matrices. Instead, we will spend most of our effort on the covariance matrix.

Another interesting characterization of the covariance matrix indicates the manner in which it generalizes the idea of variance to many dimensions. Let $\boldsymbol{\mu} = [\mu_1\ \mu_2 \ldots \mu_n]'$ be the mean vector of $\mathbf{X} = [X_1\ X_2 \ldots X_n]'$. Then,

$$(\mathbf{X} - \boldsymbol{\mu})(\mathbf{X} - \boldsymbol{\mu})'$$

$$= \begin{bmatrix} X_1 - \mu_1 \\ X_2 - \mu_2 \\ \vdots \\ X_n - \mu_n \end{bmatrix} \begin{bmatrix} X_1 - \mu_1 & X_2 - \mu_2 & \cdots & X_n - \mu_n \end{bmatrix}$$

$$= \begin{bmatrix} (X_1 - \mu_1)^2 & (X_1 - \mu_1)(X_2 - \mu_2) & \cdots & (X_1 - \mu_1)(X_n - \mu_n) \\ (X_2 - \mu_2)(X_1 - \mu_1) & (X_2 - \mu_2)^2 & \cdots & (X_2 - \mu_2)(X_n - \mu_n) \\ \vdots & \vdots & \ddots & \vdots \\ (X_n - \mu_n)(X_1 - \mu_1) & (X_n - \mu_n)(X_2 - \mu_2) & \cdots & (X_n - \mu_n)^2 \end{bmatrix}.$$

$$\tag{4.59}$$

Taking the expected value entry by entry, we obtain

$$E[(\mathbf{X} - \boldsymbol{\mu})(\mathbf{X} - \boldsymbol{\mu})'] = \text{Cov}(\mathbf{X}). \tag{4.60}$$

Using this result, you can show (see Exercise 20, this section) that the co-variance matrix is nonnegative definite.

A simple yet powerful result on covariance matrices is the following.

PROPOSITION 4.3.5

Let $\mathbf{X}$ be a vector random variable of n components whose covariance matrix Σ exists, and let A be a constant $m \times n$ matrix. Then,

$$\text{Cov}(A \cdot \mathbf{X}) = A \cdot \text{Cov}(\mathbf{X}) \cdot A' = A\Sigma A'. \tag{4.61}$$

■ **Proof**   By (4.60) we can write:

$$\begin{aligned}
\text{Cov}(A \cdot \mathbf{X}) &= E[(A\mathbf{X} - A\boldsymbol{\mu})(A\mathbf{X} - A\boldsymbol{\mu})'] \\
&= E[A(\mathbf{X} - \boldsymbol{\mu})(\mathbf{X} - \boldsymbol{\mu})'A'] \\
&= A \cdot E[(\mathbf{X} - \boldsymbol{\mu})(\mathbf{X} - \boldsymbol{\mu})'] \cdot A' = A\Sigma A'.
\end{aligned}$$

[The third line follows from the result in Exercise 20(a).]

Using Proposition 4.3.5, an alternative proof of Proposition 4.3.3 can be done with the proper choice of a coefficient matrix A (see Exercise 17).

Proposition 4.3.5 also yields a result that involves covariances between two different linear combinations of random variables. To set it up, note that one random vector $\mathbf{X} = [X_1 \ X_2 \ \ldots \ X_m]'$ can be stacked on another $\mathbf{Y} = [Y_1 \ Y_2 \ \ldots \ Y_n]'$ to produce a vector $\mathbf{Z}$ with $m + n$ components; the covariance matrix of $\mathbf{Z}$ would then have the following block structure:

$$\mathbf{Z} = \begin{bmatrix} X_1 \\ X_2 \\ \vdots \\ X_m \\ Y_1 \\ Y_2 \\ \vdots \\ Y_n \end{bmatrix} \Rightarrow \Sigma_Z = \text{Cov}(\mathbf{Z}) = \begin{bmatrix} \Sigma_X & \Sigma_{XY} \\ \Sigma_{YX} & \Sigma_Y \end{bmatrix}. \tag{4.62}$$

Here Σ_X is the $m \times m$ covariance matrix of $\mathbf{X}$, Σ_Y is the $n \times n$ covariance matrix of $\mathbf{Y}$, Σ_{XY} is the $m \times n$ matrix whose i, j component is $\text{Cov}(X_i, Y_j)$, and $\Sigma_{YX} = \Sigma'_{XY}$. (Make sure that you understand this before reading on.) Then the following is true.

PROPOSITION 4.3.6

Let $\mathbf{X}$ and $\mathbf{Y}$ be random vectors as described in the preceding discussion, and let $\mathbf{a} = [a_1 \ a_2 \ \ldots \ a_m]$ and $\mathbf{b} = [b_1 \ b_2 \ \ldots \ b_n]$ be constant row vectors.

Then

$$\text{Cov}\left(\sum_{i=1}^{m} a_i X_i, \sum_{j=1}^{n} b_j Y_j\right) = \sum_{i=1}^{m}\sum_{j=1}^{n} a_i b_j \text{Cov}(X_i, Y_j). \qquad (4.63)$$

■ **Proof** Form the random vector **Z** displayed in (4.62), and let

$$A = \begin{bmatrix} \mathbf{a} & \mathbf{0}_n \\ \mathbf{0}_m & \mathbf{b} \end{bmatrix},$$

where $\mathbf{0}_n$ and $\mathbf{0}_m$ are row vectors consisting entirely of zeros, of lengths n and m, respectively. Then

$$A \cdot \mathbf{Z} = \begin{bmatrix} \displaystyle\sum_{i=1}^{m} a_i X_i \\ \displaystyle\sum_{j=1}^{n} b_j Y_j \end{bmatrix},$$

and hence the covariance that we are looking for is the 1,2 component of the covariance matrix of $A \cdot \mathbf{Z}$. By Proposition 4.3.5,

$$\text{Cov}(A \cdot \mathbf{Z}) = A \cdot \text{Cov}(\mathbf{Z}) \cdot A'$$
$$= \begin{bmatrix} \mathbf{a} & \mathbf{0}_n \\ \mathbf{0}_m & \mathbf{b} \end{bmatrix} \begin{bmatrix} \Sigma_X & \Sigma_{XY} \\ \Sigma_{YX} & \Sigma_Y \end{bmatrix} \begin{bmatrix} \mathbf{a}' & \mathbf{0}'_m \\ \mathbf{0}'_n & \mathbf{b}' \end{bmatrix}$$
$$= \begin{bmatrix} \mathbf{a}\Sigma_X \mathbf{a}' & \mathbf{a}\Sigma_{XY}\mathbf{b}' \\ \mathbf{b}\Sigma_{YX}\mathbf{a}' & \mathbf{b}\Sigma_Y\mathbf{b}' \end{bmatrix}.$$

The desired covariance is therefore

$$\mathbf{a}\Sigma_{XY}\mathbf{b}' = \begin{bmatrix} a_1 & a_2 & \cdots & a_m \end{bmatrix} \begin{bmatrix} \text{Cov}(X_1, Y_1) & \cdots & \text{Cov}(X_1, Y_n) \\ \vdots & \ddots & \vdots \\ \text{Cov}(X_m, Y_1) & \cdots & \text{Cov}(X_m, Y_n) \end{bmatrix} \begin{bmatrix} b_1 \\ b_2 \\ \vdots \\ b_n \end{bmatrix}.$$

You can verify that when the matrix product is computed, formula (4.63) results.

Example 4.3.6 If R_1, R_2, R_3, and R_4 are random rates of return on four risky assets, then the portfolio rates of return on two portfolios that use only the first two and the second two assets, respectively, are $w_1 R_1 + w_2 R_2$, and $w_3 R_3 + w_4 R_4$. Proposition 4.3.6 allows us to write the covariance between these two portfolio returns as

$$\text{Cov}(w_1 R_1 + w_2 R_2, w_3 R_3 + w_4 R_4) = w_1 w_3 \text{Cov}(R_1, R_3) + w_1 w_4 \text{Cov}(R_1, R_4)$$
$$+ w_2 w_3 \text{Cov}(R_2, R_3) + w_2 w_4 \text{Cov}(R_2, R_4).$$
$$(4.64)$$

Seeing the proposition applied in this way highlights the bilinearity—that is, linearity in each argument—that the covariance operator possesses. It is even conceivable that if the weights w_i are chosen wisely, the two portfolios may be uncorrelated despite the existence of nonzero correlations among the assets. (See Exercise 22.)

EXERCISES 4.3

1. Suppose that X and Y have the joint density
$$f(x, y) = \frac{3}{10}(4 - (2x - y)^2), \qquad x, y \in [0, 1].$$
Produce a graph of the related surface and a contour plot. In light of the defining formula for f and these graphs, would you expect the correlation between X and Y to be high or to be low? Explain carefully.
2. Find the covariance and correlation of the random variables whose joint density is in Exercise 1.
3. A small convenience store has two checkout stations. Suppose that the joint probability mass function of the random variables X = number of customers at station 1 and Y = number of customers at station 2 is as tabulated next. Compute the covariance and correlation between X and Y.

			Y	
	No. Customers	***0***	***1***	***2***
	0	.3	.08	.02
X	***1***	.08	.2	.05
	2	.02	.05	.2

4. Argue that the derivation of Example 4.3.3 actually derives the mean and variance of the hypergeometric distribution, thereby filling a gap in our previous work. (*Hint:* Does the distribution of X, the number of successes in the sample, really depend on the assumption of selecting in order?)
5. For two random variables X and Y with joint density $f(x, y) = 3x$, $x \in [0, 1], y \in [0, x]$, predict the value of Y that is expected to occur if a value $x = 1/2$ is observed.
6. a. Show for the following joint probability mass function of two discrete random variables that the covariance is zero, but that the random variables are not independent:
$$f(x, y) = \begin{cases} 1/4 & \text{if } (x, y) \in \{(0, 0), (1, 1), (1, -1), (2, 0)\}, \\ 0 & \text{otherwise.} \end{cases}$$
   b. Show that if X has the continuous uniform distribution on $[-1, 1]$ and $Y = X^4$, then X and Y are uncorrelated (but clearly they are not independent).
7. A street intersection has gas stations on three corners. Let X be the proportion of total sales receipts among all three earned by station 1, and let Y be the proportion earned by station 2. (The rest is earned by station 3.) If the joint density of X and Y is
$$f(x, y) = 24xy, \quad \text{if } x, y \in [0, 1] \text{ and } x + y \leq 1,$$

then what are the covariance and correlation? Is the conditional mean of Y given $X = x$ linear as a function of x?

8. Most of the protein in human blood is in the form of albumin, which helps to regulate the distribution of water via the pressure of osmosis, or globulin, which has immunological properties and also has a secondary role in regulating osmotic pressure. Sometimes imbalances occur, which can indicate liver and other problems. An estimate of the average albumin concentration for an individual is 4.0 grams/deciliter, and the average globulin concentration is 3.1 grams/deciliter. Suppose that the standard deviations are .5 and .4, respectively. Suppose also that the concentrations have a correlation of $-.4$. What albumin concentration would be predicted for an individual whose globulin concentration is 3.3? What globulin concentration would be predicted for an individual whose albumin concentration is 5.0?

9. Let $(X_i, Y_i)_{i=1,\dots,n}$ be a random sample of pairs from a bivariate continuous distribution. The empirical p.m.f. would then take the value $1/n$ at each observed pair (x_i, y_i). Find formulas for the covariance and correlation for the empirical distribution.

10. Show that if the conditional mean of Y given $X = x$ is linear in x, then the conditional variance $g(x) = \sigma^2_{Y|x}$ satisfies

$$E[g(X)] = \sigma^2_Y(1 - \rho^2).$$

11. Suppose that the mean rates of return on two risky assets are 6% and 5%, their standard deviations are 4% and 2%, respectively, and their correlation is $-.6$. For an investor whose risk aversion constant c is 2, find the optimal proportion w of money to invest in the first asset.

12. Suppose that $Y = mX + b + \epsilon$, where m and b are constants, X has the $N(\mu_x, \sigma^2_x)$ distribution, ϵ has the $N(0, \sigma^2)$ distribution, and X and ϵ are independent. Find $\text{Cov}(X, Y)$.

13. Each of n people brings a grab bag gift to a holiday party. Everyone picks a gift from the bag at random. Find the mean and variance of the number of people who pick the same gift that they brought.

14. Show that $m = \rho \sigma_Y / \sigma_X$ minimizes the following function, which is the expected square difference between a centered Y and a multiple of a centered X. In this sense, m is the slope of the line that best predicts Y as a linear function of X.

$$f(m) = E\left[\left((Y - \mu_Y) - m(X - \mu_X)\right)^2\right]$$

15. The square root $V^{1/2}$ of a matrix V is a matrix such that $V^{1/2}V^{1/2} = V$. The matrix $V^{-1/2}$ is the square root of V^{-1}, provided the inverse exists. If Y is the correlation matrix of a random vector, Σ is the covariance matrix, and V is the diagonal matrix

$$V = \begin{bmatrix} \sigma^2_1 & 0 & \cdots & 0 \\ 0 & \sigma^2_2 & \cdots & 0 \\ \vdots & \vdots & \ddots & \vdots \\ 0 & 0 & \cdots & \sigma^2_n \end{bmatrix},$$

then show that

    a. $\Sigma = V^{1/2}\Upsilon V^{1/2}$.
    b. $\Upsilon = V^{-1/2}\Sigma V^{-1/2}$.

16. Suppose that X_1, X_2, and X_3 have the following covariance matrix. Find $\text{Cov}(2X_1 + X_3, X_1 - X_2 - X_3)$.

$$\Sigma = \begin{bmatrix} 2 & -.5 & .2 \\ -.5 & 6 & -.4 \\ .2 & -.4 & 1 \end{bmatrix}$$

17. Prove Proposition 4.3.3 using matrix techniques. (*Hint:* Let A be the $1 \times n$ matrix $[a_1\ a_2\ \ldots\ a_n]$ and use Proposition 4.3.5.)

18. Show that if a pair of random variables X and Y have equal variances, then the random variables $Z_1 = X + Y$ and $Z_2 = X - Y$ are uncorrelated.

19. Show that the correlation matrix of a random vector $\mathbf{X} = (X_1, \ldots, X_n)'$ is the covariance matrix of the standardized random vector whose ith component is $Z_i = (X_i - \mu_i)/\sigma_i$.

20. a. Show that if $\mathbf{X}$ is a random matrix and A and B are constant matrices of the proper dimension, then

$$E[A \cdot \mathbf{X} \cdot B] = A \cdot E[\mathbf{X}] \cdot B.$$

    b. Show that the covariance matrix of a random vector is nonnegative definite; that is, $\mathbf{y}'\Sigma\mathbf{y} \geq 0$ for all n-vectors $\mathbf{y}$. [*Hint:* Use Eq. (4.60).]

21. Use Proposition 4.3.6 to write a formula for

$$\text{Cov}\left(\sum_{i=1}^{m} a_i X_i, \sum_{i=1}^{m} b_i X_i\right).$$

22. For Example 4.3.6, devise a collection of portfolio weights and covariances such that the resulting portfolios are uncorrelated. (*Hint:* Try letting $w_1 = w_2$.)

4.4 | More on the Multivariate Normal Distribution

An appropriate end to this chapter is to apply the ideas studied here to gain more insight into the multivariate normal distribution. Recall that its density, written in matrix form, is

$$f(\mathbf{x}) = \frac{1}{(2\pi)^{n/2}\sqrt{\det(\Sigma)}} e^{-1/2(\mathbf{X}-\mu)'\Sigma^{-1}(\mathbf{X}-\mu)}, \qquad \mathbf{x} \in \mathbb{R}^n, \qquad (4.65)$$

where μ is the constant $n \times 1$ vector parameter called the *mean vector* and Σ is the constant, symmetric $n \times n$ matrix called the *covariance matrix*. The special case where $n = 2$, called the *bivariate normal density*, can be expanded as

$$f(x, y) = \frac{1}{2\pi\sigma_x\sigma_y\sqrt{1-\rho^2}}$$

$$\times \exp\left[-\frac{1}{2(1-\rho^2)}\left[\frac{(x-\mu_x)^2}{\sigma_x^2} - 2\rho\frac{(x-\mu_x)(y-\mu_y)}{\sigma_x\sigma_y} + \frac{(y-\mu_y)^2}{\sigma_y^2}\right]\right].$$

$$(4.66)$$

Graphs of the bivariate normal density (see Fig. 3.21) indicate that the magnitude of ρ determines the degree of concentration of probability density around a line in the $x - y$ plane. The sign of ρ determines whether that line is positively or negatively sloped. Cross sections of the density surface perpendicular to the coordinate axes have a normal bell shape, and contour curves are elliptical, with centers at (μ_x, μ_y) and tilt angles and axis radii dependent on σ_x^2, σ_y^2, and ρ.

With the ideas of independence, conditional distributions and expectation, and covariance formally defined and explored, we are now in a position to clarify and extend some of our earlier observations about this important multivariate distribution. First, Proposition 3.4.1 said that if the covariance matrix is diagonal, then the joint density $f(\mathbf{x})$ factors into the product $f_1(x_1)f_2(x_2)\cdots f_n(x_n)$ of single-variable normal densities. It is easy to see that the converse holds as well. But then Proposition 4.1.1, the factorization theorem for independent random variables, immediately yields the following result.

PROPOSITION 4.4.1

Let the random vector $\mathbf{X} = [X_1\ X_2\ \ldots\ X_n]'$ have the multivariate normal distribution with mean vector $\boldsymbol{\mu} = [\mu_1\ \mu_2\ \ldots\ \mu_n]'$ and covariance matrix Σ. Then Σ is a diagonal matrix with diagonal entries $\sigma_1^2, \sigma_2^2, \ldots, \sigma_n^2$ if and only if $X_1, X_2, \ldots, X_n$ are mutually independent and X_i has the $N(\mu_i, \sigma_i^2)$ distribution.

—— **?Question 4.4.1**   If a bivariate normal density has a diagonal covariance matrix, what will the cross sections of the density surface by planes parallel to the $x - y$ coordinate plane look like?

For the remainder of this section we will concentrate on the dependent case, beginning with the bivariate normal distribution. In Proposition 3.4.2 we have already shown that if $\mathbf{X} = [X\ Y]'$ has the bivariate normal density, then the marginal distributions of X and Y are $N(\mu_x, \sigma_x^2)$ and $N(\mu_y, \sigma_y^2)$, respectively. Remember that, although we used the term *covariance matrix* for Σ, we have not completely shown that this is indeed what Σ is. We have the marginal variances on the main diagonal as needed, but it remains to show that $\text{Cov}(X, Y) = \rho\sigma_x\sigma_y$, or equivalently, $\text{Corr}(X, Y) = \rho$. This fact will come as a by-product of another important result, for which we have already set up the necessary machinery.

PROPOSITION 4.4.2

If $\mathbf{X} = [X\ Y]'$ has the bivariate normal density as described earlier, then the conditional density of Y given $X = x$ is $N(\mu_{y|x}, \sigma^2_{y|x})$, where

$$\mu_{y|x} = \mu_y + \rho\frac{\sigma_y}{\sigma_x}(x - \mu_x); \qquad \sigma^2_{y|x} = \sigma^2_y(1 - \rho^2). \tag{4.67}$$

Similarly, the conditional density of X given $Y = y$ is $N(\mu_{x|y}, \sigma^2_{x|y})$, where

$$\mu_{x|y} = \mu_x + \rho\frac{\sigma_x}{\sigma_y}(y - \mu_y); \qquad \sigma^2_{x|y} = \sigma^2_x(1 - \rho^2). \tag{4.68}$$

(Notice that the conditional variance does not depend on the value of the variable being conditioned on.)

■ **Proof**   Recall the factorization (3.79) from the proof of Proposition 3.4.2:

$$f(x, y) = \frac{1}{\sqrt{2\pi\sigma_x^2}} \exp\left(-\frac{(x - \mu_x)^2}{2\sigma_x^2}\right)$$

$$\cdot \frac{1}{\sqrt{2\pi\sigma_y^2(1 - \rho^2)}} \exp\left(-\frac{(y - \mu_{y|x})^2}{2\sigma_y^2(1 - \rho^2)}\right).$$

The first two factors make up the marginal density f_x of the random variable X. Hence the conditional density of Y given $X = x$ is

$$f(y\,|\,x) = \frac{f(x, y)}{f_x(x)} = \frac{1}{\sqrt{2\pi\sigma_y^2(1 - \rho^2)}} \exp\left(-\frac{(y - \mu_{y|x})^2}{2\sigma_y^2(1 - \rho^2)}\right), \tag{4.69}$$

which is the desired normal density. You can easily do a similar argument to show that the conditional distribution of X given $Y = y$ is $N(\mu_{x|y}, \sigma^2_{x|y})$ using factorization (3.82).

This proposition allows us to tie up another loose thread. The conditional mean of Y given $X = x$ is now known to be linear in x with slope coefficient $\rho\sigma_y/\sigma_x$. Referring to Proposition 4.3.4, in the case where the conditional mean of Y is linear in x, the slope coefficient is the correlation between X and Y times the ratio of standard deviations. Equating this to the slope coefficient in the bivariate normal conditional mean, it follows that ρ is indeed the correlation between X and Y.

Example 4.4.1

The following information comes from a medical text (Fischbach, 1980). Fibrinogen is a protein in the blood that aids in coagulation. Its normal range of concentrations is 200–400 milligrams per 100 milliliters. A simple test for fibrinogen deficiency, the whole blood clotting test, measures the clotting time of a small volume of blood placed into a glass tube. The normal range of clotting times is 5–10 minutes. Increasing the expected clotting time to 12 minutes requires a decrease of fibrinogen concentration to about 50. If an individual has a clotting time of 15 minutes, what is the chance that his fibrinogen concentration has decreased to a level below 100?

Let us make some reasonable interpretations and assumptions about what the given information means. We will assume that among the general population of tested individuals, the fibrinogen concentration X and the clotting time Y have the bivariate normal distribution. The phrase *normal range* will mean an interval of radius two standard deviations centered about the mean. Therefore, for fibrinogen, the interval $[200, 400]$ gives us a mean $\mu_x = 300$ and a standard deviation $\sigma_x = 50$. For clotting time, the interval $[5, 10]$ yields a mean $\mu_y = 7.5$ and a standard deviation $\sigma_y = 1.25$. We will assume that the other information provided to us means that the conditional expectation of clotting time for a value of fibrinogen concentration of 50 is $\mu_{y|50} = 12$. Our problem is to compute $P[X < 100 \mid Y = 15]$.

First, we can compute the correlation coefficient from the given information, because

$$\mu_{y|50} = 12 = \mu_y + \rho \frac{\sigma_y}{\sigma_x}(x - \mu_x) = 7.5 + \rho \frac{1.25}{50}(50 - 300)$$

$$\Rightarrow 4.5 = \rho \cdot (-6.25) \Rightarrow \rho = -.72.$$

Next, given $Y = 15$, X is normal with mean and variance

$$\mu_{x|y} = \mu_x + \rho \frac{\sigma_x}{\sigma_y}(y - \mu_y) = 300 - .72 \cdot \frac{50}{1.25}(15 - 7.5) = 84,$$

$$\sigma_{x|y}^2 = \sigma_x^2(1 - \rho^2) = 2500(1 - (-.72)^2) = 1204.$$

Standardizing in the usual way gives

$$P[X < 100 \mid Y = 15] = P\left[Z < \frac{100 - 84}{\sqrt{1204}}\right]$$

$$= P[Z < .46] = .6772.$$

Clever use of some results from matrix algebra will allow us to generalize what we have done from the bivariate to the multivariate case. You should now review the results on block matrices, their determinants, and their inverses, presented in Appendix E, on linear algebra.

Let $\mathbf{X} = [X_1 \ X_2 \ \ldots \ X_n]'$ have the multivariate normal distribution with mean vector $\boldsymbol{\mu} = [\mu_1 \ \mu_2 \ \ldots \ \mu_n]'$ and positive definite covariance matrix Σ. Partition the components of $\mathbf{X}$ into two groups, the first m and the remaining $(n - m)$ random variables. Partition $\boldsymbol{\mu}$ and Σ accordingly:

$$\mathbf{X} = \begin{bmatrix} \mathbf{X}_1 \\ \mathbf{X}_2 \end{bmatrix}, \qquad \boldsymbol{\mu} = \begin{bmatrix} \mu_1 \\ \mu_2 \end{bmatrix}, \qquad \Sigma = \begin{bmatrix} \Sigma_{11} & \Sigma_{12} \\ \Sigma_{21} & \Sigma_{22} \end{bmatrix}, \tag{4.70}$$

where $\mathbf{X}_1$ and μ_1 are $m \times 1$ column vectors, $\mathbf{X}_2$ and μ_2 are $(n - m) \times 1$ column vectors, Σ_{11} is an $m \times m$ symmetric matrix, Σ_{22} is an $(n - m) \times (n - m)$

symmetric matrix, Σ_{12} is an $m \times (n - m)$ matrix, and $\Sigma_{21} = \Sigma'_{12}$. Because Σ is positive definite, Σ, Σ_{11}, and Σ_{22} are all invertible. The following theorem generalizes the earlier bivariate results on marginal and conditional distributions.

PROPOSITION 4.4.3

Let $\mathbf{X}$, $\boldsymbol{\mu}$, and Σ be as already described. Then,

(a) $\mathbf{X}_1$ has the multivariate normal distribution with mean vector $\boldsymbol{\mu}_1$ and covariance matrix Σ_{11}.

(b) $\mathbf{X}_2$ has the multivariate normal distribution with mean vector $\boldsymbol{\mu}_2$ and covariance matrix Σ_{22}.

(c) Conditioned on $\mathbf{X}_1 = \mathbf{x}_1$, $\mathbf{X}_2$ has the multivariate normal distribution with mean vector and covariance matrix

$$\boldsymbol{\mu}_{2|1} = \boldsymbol{\mu}_2 + \Sigma_{21}\Sigma_{11}^{-1}(\mathbf{x}_1 - \boldsymbol{\mu}_1); \ \Sigma_{2|1} = \Sigma_{22} - \Sigma_{21}\Sigma_{11}^{-1}\Sigma_{12}. \tag{4.71}$$

(d) Conditioned on $\mathbf{X}_2 = \mathbf{x}_2$, $\mathbf{X}_1$ has the multivariate normal distribution with mean vector and covariance matrix

$$\boldsymbol{\mu}_{1|2} = \boldsymbol{\mu}_1 + \Sigma_{12}\Sigma_{22}^{-1}(\mathbf{x}_2 - \boldsymbol{\mu}_2); \ \Sigma_{1|2} = \Sigma_{11} - \Sigma_{12}\Sigma_{22}^{-1}\Sigma_{21}. \tag{4.72}$$

■ **Proof** We will only prove parts (b) and (d) [parts (a) and (c) are analogous]. Proofs of both will fall out simultaneously once we obtain a factorization of the joint density $f(\mathbf{x})$ in (4.65). To that end, it can be verified (check the matrix products yourself in forthcoming Question 4.4.2) that

$$\begin{bmatrix} I & -\Sigma_{12}\Sigma_{22}^{-1} \\ 0 & I \end{bmatrix} \cdot \Sigma \cdot \begin{bmatrix} I & 0 \\ -\Sigma_{22}^{-1}\Sigma_{21} & I \end{bmatrix} = \begin{bmatrix} \Sigma_{11} - \Sigma_{12}\Sigma_{22}^{-1}\Sigma_{21} & 0 \\ 0 & \Sigma_{22} \end{bmatrix}, \tag{4.73}$$

where the I's are identity matrices of the appropriate size. Since both matrices surrounding Σ have determinant equal to 1, equating the determinants of both sides yields

$$\det(\Sigma) = \det(\Sigma_{11} - \Sigma_{12}\Sigma_{22}^{-1}\Sigma_{21}) \cdot \det(\Sigma_{22}) = \det(\Sigma_{1|2}) \cdot \det(\Sigma_{22}). \tag{4.74}$$

The relation (4.73) also allows us to factor Σ^{-1}. Multiply on the left and right by the inverses of the two matrices that surround Σ to obtain

$$\Rightarrow \Sigma = \begin{bmatrix} I & -\Sigma_{12}\Sigma_{22}^{-1} \\ 0 & I \end{bmatrix}^{-1} \begin{bmatrix} \Sigma_{11} - \Sigma_{12}\Sigma_{22}^{-1}\Sigma_{21} & 0 \\ 0 & \Sigma_{22} \end{bmatrix} \begin{bmatrix} I & 0 \\ -\Sigma_{22}^{-1}\Sigma_{21} & I \end{bmatrix}^{-1}$$

$$\Rightarrow \Sigma^{-1} = \begin{bmatrix} I & 0 \\ -\Sigma_{22}^{-1}\Sigma_{21} & I \end{bmatrix} \begin{bmatrix} (\Sigma_{11} - \Sigma_{12}\Sigma_{22}^{-1}\Sigma_{21})^{-1} & 0 \\ 0 & \Sigma_{22}^{-1} \end{bmatrix} \begin{bmatrix} I & -\Sigma_{12}\Sigma_{22}^{-1} \\ 0 & I \end{bmatrix}. \tag{4.75}$$

This factorization enables us to write

$$(\mathbf{x} - \boldsymbol{\mu})'\Sigma^{-1}(\mathbf{x} - \boldsymbol{\mu}) = (\mathbf{x}_1 - \boldsymbol{\mu}_{1|2})' \cdot \Sigma_{1|2}^{-1} \cdot (\mathbf{x}_1 - \boldsymbol{\mu}_{1|2})$$

$$+ (\mathbf{x}_2 - \boldsymbol{\mu}_2)'\Sigma_{22}^{-1}(\mathbf{x}_2 - \boldsymbol{\mu}_2). \tag{4.76}$$

(Prove this result in Exercise 11 at the end of this section.)

By Eqs. (4.74)–(4.76), the multivariate normal density (4.65) can therefore be factored as

$$f(\mathbf{x}) = \frac{1}{(2\pi)^{m/2}\sqrt{\det(\Sigma_{1|2})}} \cdot e^{-1/2(\mathbf{X}_1 - \boldsymbol{\mu}_{1|2})'\Sigma_{1|2}^{-1}(\mathbf{X}_1 - \boldsymbol{\mu}_{1|2})}$$

$$\cdot \frac{1}{(2\pi)^{(n-m)/2}\sqrt{\det(\Sigma_{22})}} \cdot e^{-1/2(\mathbf{X}_2 - \boldsymbol{\mu}_2)'\Sigma_{22}^{-1}(\mathbf{X}_2 - \boldsymbol{\mu}_2)}. \tag{4.77}$$

Since the first pair of factors in (4.77) forms a multivariate normal density in $\mathbf{x}_1$, the integral of $f(\mathbf{x})$ over $\mathbf{x}_1$ gives the second pair of factors. In other words, the marginal density of $\mathbf{X}_2$ is multivariate normal with mean vector $\boldsymbol{\mu}_2$ and covariance matrix Σ_{22}. The first pair of factors is therefore the quotient of the joint density $f(\mathbf{x})$ and the marginal density $f_2(\mathbf{x}_2)$, that is, the conditional density of $\mathbf{X}_1$ given $\mathbf{X}_2 = \mathbf{x}_2$. The parameters of this conditional density match those in the statement of part (d); hence both (b) and (d) are proved.

?Question 4.4.2   Check Eq. (4.73).

The matrix Σ that we have been calling the covariance matrix actually is the covariance matrix as defined earlier in this chapter, that is, its i, j entry is $\text{Cov}(X_i, X_j)$ for $i \neq j$. By renumbering the X's if necessary, it suffices to show that the $1, 2$ entry of Σ is $\text{Cov}(X_1, X_2)$. By part (a) of the last proposition, the vector $[X_1 \ X_2]'$ has a (marginal) bivariate normal distribution, and the off-diagonal entry of its 2×2 covariance matrix Σ_{11} is the same as the $1, 2$ entry of the original Σ. But we have already shown in the bivariate case that this entry is the covariance between X_1 and X_2.

Example 4.4.2   In Exercise 14 of Section 3.3 we mentioned a data set on ozone levels in Pennsylvania forests. It is reasonable to think that from one day to the next, and possibly to the third day, there is a correlation between ozone levels, but as more time elapses, levels are nearly independent. On this basis, I looked at some of the data and produced a sample of triples (X_1, X_2, X_3) on successive days, where one triple was separated by a few days from another. From these I estimated means, variances, and the covariance matrix using methods we will discuss in Chapters 7 and 8. Using a trivariate normal model with the following mean and covariance, let us compute the expected value of the ozone level on the third day and find an interval centered about that mean in which X_3 should lie with probability 95%, if on the first two days levels of 70 and 75 were observed:

$$\boldsymbol{\mu} = \begin{bmatrix} 52.8 \\ 52.4 \\ 55.3 \end{bmatrix}, \qquad \Sigma = \begin{bmatrix} 472.75 & 177.83 & 156.77 \\ 177.83 & 311.74 & 157.94 \\ 156.77 & 157.94 & 247.21 \end{bmatrix}.$$

From Proposition 4.4.3, the conditional distribution of X_3 is normal. To compute its conditional mean and variance, use part (c) of the proposition, with $\mathbf{X}_1 = [X_1 \ X_2]'$ and $\mathbf{X}_2 = [X_3]'$. Then

$$\boldsymbol{\mu}_1 = \begin{bmatrix} 52.8 \\ 52.4 \end{bmatrix}, \qquad \boldsymbol{\mu}_2 = [55.3], \qquad \Sigma_{11} = \begin{bmatrix} 472.75 & 177.83 \\ 177.83 & 311.74 \end{bmatrix},$$

$$\Sigma_{12} = \begin{bmatrix} 156.77 \\ 157.94 \end{bmatrix}, \qquad \Sigma_{22} = [247.21].$$

To answer the first question, we must find $E[X_3 \mid X_1 = 70, X_2 = 75]$. By (4.71) we have

$$\boldsymbol{\mu}_{2|1} = \boldsymbol{\mu}_2 + \Sigma_{21}\Sigma_{11}^{-1}(\mathbf{x}_1 - \boldsymbol{\mu}_1)$$

$$= [55.3] + \begin{bmatrix} 156.77 & 157.94 \end{bmatrix} \begin{bmatrix} .00269 & -.00154 \\ -.00154 & .00408 \end{bmatrix} \begin{bmatrix} 17.2 \\ 22.6 \end{bmatrix}$$

$$= 67.5.$$

For the second question we will need the conditional variance of X_3, which by (4.71) is

$$\Sigma_{2|1} = \Sigma_{22} - \Sigma_{21}\Sigma_{11}^{-1}\Sigma_{12}$$

$$= [247.21] - \begin{bmatrix} 156.77 & 157.94 \end{bmatrix} \begin{bmatrix} .00269 & -.00154 \\ -.00154 & .00408 \end{bmatrix} \begin{bmatrix} 156.77 \\ 157.94 \end{bmatrix}$$

$$= 155.2.$$

From the standard normal table, the point z such that 95% of the area under the density is between $-z$ and z is $z = 1.96$. (Check this.) Thus

$$.95 = P[-1.96 \le Z \le 1.96]$$

$$= P\left[-1.96 \le \frac{X_3 - 67.5}{\sqrt{155.2}} \le 1.96 \,\middle|\, X_1 = 70, X_2 = 75 \right]$$

$$= P[67.5 - 1.96\sqrt{155.2} \le X_3 \le 67.5 + 1.96\sqrt{155.2} \mid X_1 = 70, X_2 = 75]$$

$$= P[43.1 \le X_3 \le 91.9 \mid X_1 = 70, X_2 = 75].$$

The interval [43.1, 91.9] is very wide; hence the information about the first two days does not give a very precise prediction of what the ozone level on the third day will be.

—— **?Question 4.4.3**   In the previous example, find the 80% prediction interval centered about the mean. Is it substantially shorter, and if so, why would you have expected it to be?

EXERCISES 4.4

1. Explain the appearance of the cross sections of the bivariate normal density surface perpendicular to the x_1 and x_2 coordinate axes in Fig. 3.21, in light of the new developments of this section.
2. Suppose that the random vector $[Y \ D]'$, where Y is the crop yield per unit area and D is the distance upslope from the bottom of a hill, has

the bivariate normal distribution with parameters

$$\mu = \begin{bmatrix} 440 \\ 100 \end{bmatrix}, \qquad \Sigma = \begin{bmatrix} 1600 & -600 \\ -600 & 900 \end{bmatrix}.$$

Find the mean and variance of yield for values of (a) $D = 50$; (b) $D = 150$.

3. The applicants to a certain small college who take the SAT exam receive verbal score V and mathematics score M, which are jointly normal with parameters $\mu_V = 540$, $\mu_M = 560$, $\sigma_V^2 = 200$, $\sigma_M^2 = 180$, $\rho = .6$. What is the probability that all of a group of five students with math scores of 570 have verbal scores under 540?

4. Suppose that for a certain group of people, the joint distribution of systolic blood pressure P and cholesterol level L is approximately bivariate normal with parameters $\mu_P = 120$, $\sigma_P^2 = 25$, $\mu_L = 160$, $\sigma_L^2 = 36$, and $\rho = .25$.

   a. Find the predicted blood pressure of an individual whose cholesterol level is 190.

   b. Find the probability that an individual whose blood pressure is 130 has a cholesterol level exceeding 175.

   c. Find the 95th percentile of cholesterol level for people in this group whose blood pressure is 130.

5. My house has a gas heat and hot water system and runs the air conditioning on electricity. Consequently, I would hypothesize that the gas usage X (therms) and electrical usage Y (kilowatt-hours) tend to have a negative relationship. Using the following estimated mean vector and covariance matrix, what is the correlation between X and Y? What change is a unit increase in gas usage expected to make in electrical usage? Compute the probability that on a month when the electrical usage is 600 kwh, the gas usage is 40 therms or less.

$$\mu = \begin{bmatrix} 96.6 \\ 422.1 \end{bmatrix}, \qquad \Sigma = \begin{bmatrix} 5257 & -3547 \\ -3547 & 34360 \end{bmatrix}$$

6. Suppose that in the context of Example 4.4.2 some pairs of ozone measurements on successive days yielded estimates of the mean vector and covariance matrix as follows. If an ozone reading on a certain day is 35, find the conditional mean and variance of the reading on the next day, and find an interval centered about the conditional mean in which the next day's reading is 90% likely to lie.

$$\mu = \begin{bmatrix} 51.3 \\ 54.7 \end{bmatrix}, \qquad \Sigma = \begin{bmatrix} 421.8 & 204.3 \\ 204.3 & 365.9 \end{bmatrix}$$

7. Let $\mathbf{X} = [X_1\ X_2\ X_3\ X_4]'$ have the multivariate normal distribution with parameters

$$\mu = \begin{bmatrix} 1 \\ 0 \\ -1 \\ 0 \end{bmatrix}, \qquad \Sigma = \begin{bmatrix} 1 & .1 & .4 & .6 \\ .1 & 1 & 1 & 1 \\ .4 & 1 & 4 & 2.4 \\ .6 & 1 & 2.4 & 4 \end{bmatrix}.$$

Find the conditional distribution of $\mathbf{X}_1 = [X_1\ X_2]'$ given $\mathbf{X}_2 = [X_3\ X_4]' = [0\ 1]'$.

8. A gymnast's scores X_1 on the uneven bars, X_2 on the balance beam, and X_3 on floor exercise are jointly normal with mean vector $\mu = [9.8\ 9.5\ 9.6]'$ and covariance matrix

$$\Sigma = \begin{bmatrix} .08 & .0016 & .0096 \\ .0016 & .1 & .0036 \\ .0096 & .0036 & .12 \end{bmatrix}.$$

Find

a. the joint density of the scores on balance beam and floor exercise.

b. the marginal density of the score on uneven bars.

c. the expected score on floor exercise given scores of 9.9 on uneven bars and 9.6 on balance beam.

d. the conditional joint distribution of the scores on uneven bars and balance beam given a score of 9.5 on floor exercise.

9. Prove Proposition 4.4.2 as a special case of Proposition 4.4.3.

10. A study of the Peoria, Illinois, area elementary schools included many variables for each school, including expenditures per pupil (X_1), average years of experience of the teaching staff (X_2), and average student performance on a state reading exam (X_3). If the estimated mean vector and covariance matrix of these three variables is as follows, which produces the bigger change in the conditional expectation of student reading score: an increase of $500 in expenditure per pupil coupled with an increase of 2 years' experience, or an increase of $1000 per pupil coupled with an increase of 1 year experience?

$$\mu = \begin{bmatrix} 3754 \\ 15 \\ 288 \end{bmatrix}, \qquad \Sigma = \begin{bmatrix} 1,147,851 & 994 & 10,696 \\ 994 & 7.5 & 39.7 \\ 10,696 & 39.7 & 893 \end{bmatrix}.$$

11. Verify the identity (4.76).

12. In Example 4.4.2, find the conditional joint distribution of the second and third days' ozone level given the first day.

13. Let $\mathbf{X} = [X_1\ X_2\ X_3\ X_4]'$ have a multivariate normal distribution. Find in two ways the mean vector and covariance matrix of the random vector $\mathbf{Y} = A\mathbf{X}$, where

$$A = \begin{bmatrix} 0 & 0 & 1 & 0 \\ 1 & 0 & 0 & 0 \end{bmatrix}.$$

CHAPTER 5

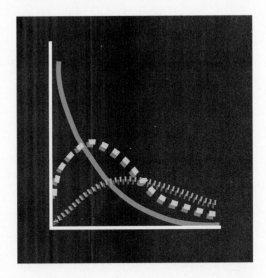

TRANSFORMATIONS OF
RANDOM VARIABLES

5.1 | Distribution Function Technique

We have already encountered several problems in which it is necessary to know about the probability distribution of a function of a random variable, or a combination of several random variables. For example, the binomial distribution, with all of its many useful applications, arises as the distribution of the sum of independent and identically distributed Bernoulli random variables. If a random variable X has the $N(\mu, \sigma^2)$ distribution, then the transformed random variable $Z = (X - \mu)/\sigma$ is standard normal, which permits us to rely on the standard normal table to find arbitrary normal probabilities. The time T_n of the nth arrival in a Poisson process is the sum $S_1 + S_2 + \cdots + S_n$ of exponentially distributed interarrival times, and we have used without proof the fact that the distribution of T_n is of the gamma family.

So the idea of transforming random variables is not new to us, but now we would like to study transformations more systematically and with more depth. This chapter, mostly theoretical in its content, is extremely important for many reasons. Perhaps the most important reason is that the results on transformations of normal random variables presented here are prerequisite for an understanding of the statistical techniques that we will develop later in this book. However, it should also be stressed that the same theoretical tools complete our development of the Poisson process and lay the groundwork for the pursuit of other topics in probability theory and its applications, which I hope that you will be interested in doing after you finish the book.

The subject of this first section is a method called the *c.d.f. technique* for finding the distribution of a real-valued, continuous function of a continuous random variable. Let X be a continuous random variable with a known density function $f_X(x)$, and suppose that we are interested in a new random variable $Y = g(X)$, where g is a real-valued function whose domain includes the state space of X. The c.d.f. of Y is

$$F_Y(y) = P[Y \le y] = P[g(X) \le y]. \tag{5.1}$$

The c.d.f. method is simply to rearrange algebraically the inequality $g(X) \le y$ into an equivalent form in which X is isolated. A set of outcomes of the form $\{X \in B_y\}$ will result, where B_y is some subset of the state space of X dependent on y. The probability of this set is found by writing the integral of the p.d.f. of X over this set. The technique is summarized by the following string of equations:

$$F_Y(y) = P[g(X) \le y] = P[X \in B_y] = \int_{B_y} f_X(x)\, dx. \tag{5.2}$$

Equation (5.2) gives an expression for the c.d.f. of the new random variable Y, but it is not in closed form. You can use the Fundamental Theorem of Calculus to differentiate $F_Y(y)$ directly, in order to produce the density of Y. It is usually easier to work this way than to evaluate the definite integral in (5.2) first, and then to differentiate. Thus

$$f_Y(y) = F_Y'(y) = \frac{d}{dy} \int_{B_y} f_X(x)\, dx. \tag{5.3}$$

If you look back at the proof of the normal standardization theorem, Proposition 3.3.6, you will find that we did exactly this.

Notice that nothing we have said so far precludes the possibility that the original random variable X is multidimensional. The only restriction that we are working under in this section is that the function $Y = g(X)$ is one-dimensional.

Example 5.1.1

In the financial analysis of prices of commodities of various kinds, it is sometimes assumed that from one time, labeled 0, to the next, labeled 1, the logarithm of the ratio P_1/P_0 of prices at the two times is normally distributed with some mean μ and variance σ^2. Denote by A the normal random variable $\ln(P_1/P_0) = \ln(P_1) - \ln(P_0)$. It is easy to invert this relationship between A and P_1 to find that $P_1 = P_0 \cdot e^A$. Considering P_0 to be fixed, what is the distribution of P_1?

The c.d.f. of P_1 is

$$F(p) = P[P_1 \leq p] = P[P_0 \cdot e^A \leq p]$$
$$= P[e^A \leq p/P_0]$$
$$= P[A \leq \ln(p/P_0) = \ln(p) - \ln(P_0)]$$
$$= \int_{-\infty}^{\ln(p)-\ln(P_0)} \frac{1}{\sqrt{2\pi\sigma^2}} \exp\left(-\frac{(a-\mu)^2}{2\sigma^2}\right) da, \qquad p > 0.$$

Not only is it easier to differentiate the integral than to evaluate it first, we have no other choice because the integrand has no closed form indefinite integral. Differentiating with respect to p using the Fundamental Theorem and the chain rule yields the following formula for the density of P_1:

$$f(p) = \frac{1}{p} \cdot \frac{1}{\sqrt{2\pi\sigma^2}} \exp\left[-\frac{(\ln(p) - \ln(P_0) - \mu)^2}{2\sigma^2}\right]. \tag{5.4}$$

We can choose our monetary unit so that the initial price P_0 is equal to 1. Then $\ln(P_0) = 0$, and the density in (5.4) reduces to the *lognormal density* with parameters μ and σ^2, which was mentioned in Section 3.3. This language comes from the fact that the lognormal density is the density of a random variable P_1 whose logarithm $\ln(P_1) = A$ has a normal density.

The cumulative distribution function technique can also be used to show a general result that has important implications to simulation. There are widely available algorithms, used by many computers, that can simulate an observation U which has the continuous uniform distribution on $[0, 1]$. We would like to transform U into another random variable that has the distribution we would like to simulate.

To set up the result, consider the task of generating an observation from a discrete distribution with a state space of three points $x_1 < x_2 < x_3$, and respective probability weights p_1, p_2, p_3. Define

$$X(\omega) = \begin{cases} x_1 & \text{if } 0 \leq U(\omega) \leq p_1, \\ x_2 & \text{if } p_1 < U(\omega) \leq p_1 + p_2, \\ x_3 & \text{otherwise.} \end{cases} \tag{5.5}$$

In that case, the event $\{X = x_1\}$ is equivalent to the event $\{0 \leq U \leq p_1\}$; hence the two events have the same probability, namely p_1. Also, $P[X = x_2] = P[p_1 < U(\omega) \leq p_1 + p_2] = p_2$, and $P[X = x_3] = P[p_1 + p_2 < U(\omega) \leq p_1 + p_2 + p_3] = p_3$. Thus the random variable X defined by (5.5) has the desired distribution.

?Question 5.1.1   Try to generalize (5.5) to a formula suitable for an n-point discrete distribution.

A graphical implementation of a procedure for simulating an observation from this distribution is illustrated in Fig. 5.1. A uniform random number U

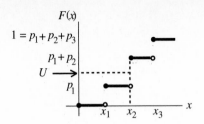

Figure 5.1 Simulating from a Discrete Distribution

is sampled (from the random number generator of a computer) and placed on the y-axis of a graph of the c.d.f. F of the distribution that is our target. The output random observation X is found on the x-axis as the smallest x such that $F(x) \geq U$. In the case shown, where U is between p_1 and $p_1 + p_2$, x_2 is the smallest such x. Notice that we are finding X as a kind of inverse image of U under F.

The graphical method of thinking about the simulation problem motivates the transition to the continuous case. Consider a plot of a continuous, strictly increasing c.d.f. F of a continuous real random variable X that we would like to simulate, as in Fig. 5.2. Simulate a uniform random variable U, and plot it along the y-axis. This time, a unique inverse image in the usual sense exists. (Why?) The value $X = F^{-1}(U)$ is shown on the x-axis. We will now use the c.d.f. technique to show that X has the distribution F, and while we are at it, we will prove a converse.

PROPOSITION 5.1.1

Let F be a continuous, strictly increasing c.d.f. on a state space I that is a subinterval of the real line. Suppose that U has the uniform $(0, 1)$ distribution. Then, $X = F^{-1}(U)$ is a random variable with the c.d.f. F. Conversely, suppose that X is a random variable with a continuous, strictly increasing c.d.f. F. Then the random variable $U = F(X)$ has the uniform $(0, 1)$ distribution.

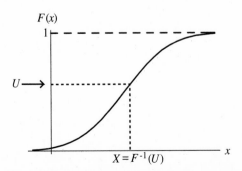

Figure 5.2 Simulating from a Continuous Distribution

■ **Proof** Beginning with the first implication, the cumulative distribution function of X is

$$P[X \leq x] = P[F^{-1}(U) \leq x] = P[U \leq F(x)] = F(x), \ x \in I.$$

In the third equation we take advantage of the fact that the c.d.f. of U is the identity function. Thus X has the c.d.f. F.

For the converse, assume that X has c.d.f. F. Since F^{-1} exists and is strictly increasing (you should verify this), the c.d.f. of $U = F(X)$ is

$$P[U \leq u] = P[F(X) \leq u] = P[X \leq F^{-1}(u)] = F(F^{-1}(u)) = u.$$

This computation is valid as long as u is in the domain of F^{-1}, which is the range of F, namely $(0, 1)$. (One or more of the endpoints 0 or 1 may also be in the range of F, depending on whether the state space I is bounded on one end or the other. This detail does not change the gist of the argument.) Since the c.d.f. of U has the proper form, U is uniformly distributed on $(0, 1)$.

Example 5.1.2

Many practitioners of probability must simulate large, complicated service systems that are not amenable to exact analysis, in order to reach conclusions about the efficiency of the system. This often entails simulation of the interarrival times of a Poisson process. As we have learned, a typical interarrival time S of a Poisson process has the exponential (λ) distribution, whose associated c.d.f. is

$$F(s) = 1 - e^{-\lambda s}, \qquad s > 0. \tag{5.6}$$

The inverse of F, which you will check in the next question, is

$$F^{-1}(u) = -\frac{1}{\lambda} \ln(1 - u), \qquad u \in [0, 1). \tag{5.7}$$

By the last proposition, if U has the uniform $(0, 1)$ distribution, then the random variable $S = F^{-1}(U) = -\ln(1 - U)/\lambda$ is exponential with parameter λ. A sequence of such uniform random variables then gives rise, by this transformation, to a sequence of interarrival times, which determines a particular outcome of a Poisson process.

?Question 5.1.2 Verify that the function in (5.7) is the inverse of the exponential c.d.f.

It is plain that the c.d.f. technique is very useful in deriving the distribution of a function of a single real random variable, which, by virtue of the application to simulation alone, is quite important. The next example illustrates how this method can also be used to find the distribution of a function of more than one random variable.

Example 5.1.3

Let X_1 and X_2 be independent and identically distributed random variables with density $f(x) = 2x$, $x \in (0,1)$, and let $Y = X_1 + X_2$. Find the density of Y.

Since X_1 and X_2 are independent, their joint density is the product of their marginals:

$$f(x_1, x_2) = 4x_1 x_2, \qquad x_1, x_2 \in (0,1).$$

By construction of Y, its c.d.f. is

$$F_Y(y) = P[Y \le y] = P[X_1 + X_2 \le y].$$

Now we must write out this expression more explicitly in terms of y. The integral that represents this probability depends on whether $y \in [0,1]$ or $y \in (1,2]$, as shown in Figs. 5.3(a) and (b).

In the case $y \in [0,1]$,

$$F_Y(y) = P[X_1 + X_2 \le y] = \int_0^y \int_0^{y-x_1} 4x_1 x_2 \, dx_2 \, dx_1$$

$$= \frac{1}{6} y^4.$$

The computation of the last integral is straightforward. For $y \in (1,2]$, you can check the result:

$$F_Y(y) = \int_0^{y-1} \int_0^1 4x_1 x_2 \, dx_2 \, dx_1 + \int_{y-1}^1 \int_0^{y-x_1} 4x_1 x_2 \, dx_2 \, dx_1$$

$$= 1 - \frac{8}{3} y + 2y^2 - \frac{1}{6} y^4.$$

The derivative of F_Y—that is, the density of Y—changes functional form at the breakpoint $y = 1$. Using the two preceding formulas easily results in the following:

$$f_Y(y) = \begin{cases} \frac{2}{3} y^3 & \text{if } y \in (0,1), \\ -\frac{8}{3} + 4y - \frac{2}{3} y^3 & \text{if } y \in (1,2). \end{cases}$$

The limiting value of the density at 1 is $2/3$ from both the left and the right side, so that nothing would be lost if we impose the condition $f_Y(1) = 2/3$ to produce a density function that is continuous on the interval $(0,2)$. [Is this density differentiable on $(0,2)$?]

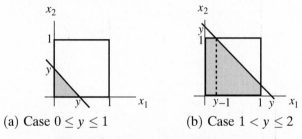

(a) Case $0 \le y \le 1$ (b) Case $1 < y \le 2$

Figure 5.3   Finding the Distribution of a Sum

So far, we have said little about transformations on discrete random variables. The reason is that when they do come up, it is often in the special setting of sums of i.i.d. random variables, which can be treated by the generating function techniques introduced later.

EXERCISES 5.1

1. Show that if Z is a standard normal random variable, then $X = \mu + \sigma Z$ has the $N(\mu, \sigma^2)$ distribution.
2. Suppose that X has the density function $f(x) = 2x, \; x \in (0,1)$. Find the density of $Z = 4X$.
3. a. Let X have the uniform distribution on the interval $[1,3]$. Find the density function of $Y = X^2$.
   b. Repeat part (a) given that X has the uniform distribution on $[-1,1]$.
4. Let $f(x)$ be the triangular density function shown here. Explain carefully how you would go about simulating an observation from this distribution.

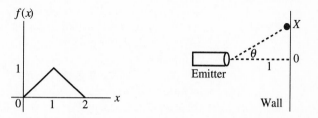

5. A particle emitter randomly fires particles toward a wall one unit away, such that the angle Θ made by the path of the particle relative to the perpendicular segment from the emitter to the wall is uniformly distributed on $(-\pi/2, \pi/2)$. If the wall is coordinatized as shown above, with 0 at the foot of the perpendicular, and X denoting the point where the particle hits the wall, what is the probability density of X? To what named family does it belong? (See the end of Section 3.3.)
6. Suppose that the lifetime of an electrical component has the Weibull(λ, β) distribution when measured in hours. Show that the lifetime still has a Weibull distribution when measured in days. What are the parameters of the latter distribution?
7. The simulation language GPSS has a statement, GENERATE a, b, that essentially creates a new object (called a *transaction*) at a random time of $a \pm b$ units, with no time in the range favored over any other time in the range. How do you think that GPSS goes about producing such a random transaction time?
8. Describe in detail how to simulate an observation from the distribution of Exercise 2.
9. Obtain from a table or a computer program 50 simulated uniform (0, 1) observations. Convert them to 50 simulated observations from the distribution of Exercise 2. Now compare a histogram of the converted observations to the theoretical density function, and comment on the fit.
10. Consider the problem of using uniform (0, 1) observations to simulate normal random variables. If the observed value of U is .5, what would be

the associated value of a simulated $n(0, 1)$ random variable? Repeat the problem, if the observed value of U is .95. For these two U values, what would be the associated values of a simulated $n(10, 4)$ random variable? (*Hint:* See Exercise 1.)

11. Use the c.d.f. technique to derive the density of $Y = X_1 + X_2$, where X_1 and X_2 are i.i.d. exponential(λ) random variables. To what family does this density belong?

12. In Exercise 6 of Section 3.1 we proposed a distribution for the proportion of bathroom time used by one of two people in the morning, whose density was $f(x) = 6x(1 - x)$, $x \in [0, 1]$. Then in Section 3.3 we made reference to the beta density, and if you look back you can now see that this f is the beta density with parameters $\alpha = \beta = 2$. Show that if T_1 and T_2 are the times taken by the two people, and they are i.i.d. $\Gamma(2, \lambda)$ random variables for some λ, then f is the density of the proportion $T_1/(T_1 + T_2)$.

13. One way to simulate an observation from the $b(4, .6)$ distribution is to list the states and probabilities as in the discussion prior to Proposition 5.1.1, and utilize a uniform $(0, 1)$ value in the analogous way. Describe a different way.

14. Let X_1 and X_2 be independent $N(0, 1)$ random variables. Use the c.d.f. technique to find the density of $Y = X_2 - X_1$.

15. Cars arrive to a toll station according to a Poisson process with a rate of 2 per minute. Each car pays a 30-cent toll. Find the probability mass function of the random variable Y defined as the total money paid over a 5-minute period.

5.2 | Multivariate Transformations

5.2.1 Distributions of Transformed Random Vectors

Consider a dart thrown randomly at a circular board such as the one shown in Fig. 5.4, which is centered at the origin and has radius A. One way of interpreting the word *randomly* is to say that the joint density of the coordinates (X, Y) of the landing point of the dart is uniform over the disk:

$$f(x, y) = \frac{1}{\pi A^2}, \quad \text{if } x^2 + y^2 \leq A^2. \tag{5.8}$$

The landing point can be described in more than one way, however. Instead of representing the point $\mathbf{X} = (X, Y)$ in rectangular coordinates, we could represent it in polar coordinates (R, Θ), where the radius R and the angle Θ are also random variables. They are related to X and Y by the usual joint transformation formulas $R^2 = X^2 + Y^2$, $\tan(\Theta) = Y/X$. Then the question is: What is the joint density of R and Θ? Is it also uniform?

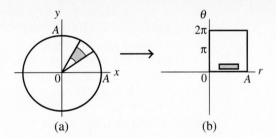

Figure 5.4 Polar Transformation

The state space of the vector (R, Θ) is easy to see; the disk in Fig. 5.4(a) is mapped by the joint transformation into the rectangle $[0, A] \times [0, 2\pi]$ in part (b) of the figure. A subset of the state space of (X, Y) like the shaded circular wedge bounded by two constant angles and two constant radii is taken into a rectangular region in the state space of (R, Θ). Shortly we will make use of this idea to find the joint density of R and Θ.

In this section we develop the machinery to find joint distributions of random variables obtained by transforming other random variables. The climax of this study will be the determination of the joint and marginal distributions of the *order statistics* $Y_1, Y_2, \ldots, Y_n$ of a random sample $X_1, X_2, \ldots, X_n$. The order statistics are simply the sample values themselves written in ascending order; they are very useful in certain data analysis problems. Later in the chapter, we will also be able to use the multivariate change of variable results to show that a linear transformation applied to a multivariate normal random vector produces another multivariate normal random vector. Therefore, the material in this section, although rather technical and complicated, yields worthwhile products once accomplished.

A fresh look at some results from single- and multivariable calculus on changes of variables in integrals is a good idea. Suppose that you are asked to evaluate the definite integral:

$$\int_0^\infty 3x^2 \cdot e^{-x^3} \, dx.$$

You quickly decide that the substitution $u = g(x) = -x^3$, $du = -3x^2 dx$ will simplify the integral. Let us do the substitution in a way that leads to the generalization to many dimensions. The function $u = g(x) = -x^3$ is a 1–1 function from the region of integration $[0, \infty)$ onto the interval $(-\infty, 0]$, and its inverse function is $x = g^{-1}(u) = -\sqrt[3]{u}$. The substitution could also be done as follows:

$$x = -\sqrt[3]{u}, \qquad dx = -\frac{1}{3} u^{-2/3} \, du$$

$$\implies \int_0^\infty 3x^2 \cdot e^{-x^3} \, dx = \int_0^{-\infty} 3(-\sqrt[3]{u})^2 e^{-(-\sqrt[3]{u})^3} \left(-\frac{1}{3} u^{-2/3} \right) \, du.$$

The last integral is of the form

$$\int_{-\infty}^0 3[g^{-1}(u)]^2 e^{-[g^{-1}(u)]^3} \left| \frac{d}{du} g^{-1}(u) \right| \, du.$$

Because $g^{-1}(u)$ happened to be a decreasing function here, the negative sign that is produced by differentiating it is used to restore the limits of integration to their usual order, leaving $|\frac{d}{du}g^{-1}(u)|$ in the integrand. In general, an integral $\int_A f(x)\,dx$ is transformed by an invertible transformation $u = g(x)$ by

$$\int_A f(x)\,dx = \int_B f(g^{-1}(u)) \left| \frac{d}{du}g^{-1}(u) \right| du, \tag{5.9}$$

where the set B is the image of set A under the transformation, that is, the set of all points $g(x)$ formed as x ranges through A.

When a transformation involves two variables in a double integral, a change of variables formula very similar to (5.9) holds. In the integral

$$\int_A \int f(x,y)\,dy\,dx, \tag{5.10}$$

let $u = g_1(x,y)$ and $v = g_2(x,y)$. For simplicity, we will assume that the transformation that takes (x,y) to (u,v) is 1–1; hence it is invertible, with inverse transformation

$$x = h_1(u,v), \qquad y = h_2(u,v). \tag{5.11}$$

Again let B be the image of A under the forward transformation, that is, the set of all points of the form $(u,v) = (g_1(x,y),g_2(x,y))$ as (x,y) ranges through A. Just as x was replaced by $g^{-1}(u)$ in the single variable case, the variables x and y in (5.10) are replaced by $h_1(u,v)$ and $h_2(u,v)$. The analogous factor to $|\frac{d}{du}g^{-1}(u)|$ in (5.9) is the absolute value $|J|$, where J is the determinant of the matrix of partial derivatives of the inverse transformation in (5.11):

$$J = J(u,v) = \det \begin{bmatrix} \partial h_1/\partial u & \partial h_1/\partial v \\ \partial h_2/\partial u & \partial h_2/\partial v \end{bmatrix}. \tag{5.12}$$

J is called the *Jacobian* of the transformation. The two-variable transformation formula is then

$$\int_A \int f(x,y)\,dy\,dx = \int_B \int f(h_1(u,v), h_2(u,v))\,|J(u,v)|\,du\,dv. \tag{5.13}$$

—— **?Question 5.2.1**   Using the two-variable transformation formula as a model, guess at the general n variable transformation formula. (We will return to it shortly.)

To solve problems like that of the dartboard, we must carry the multivariate calculus result (5.13) into the domain of probability. To begin, we let X and Y be continuous random variables with joint density $f(x,y)$. Consider a transformation of the vector $\mathbf{X} = (X,Y)$ into a new vector $\mathbf{U} = (U,V)$ by functions

$$U = g_1(X,Y), \qquad V = g_2(X,Y). \tag{5.14}$$

Assume that this transformation is 1–1, with inverse

$$X = h_1(U,V), \qquad Y = h_2(U,V). \tag{5.15}$$

As above, let $J = J(u, v)$ be the Jacobian of the transformation. Let B be a set in the state space of $\mathbf{U} = (U, V)$, and let A be its inverse image under (5.15). Then $(X, Y) \in A$ if and only if $(U, V) \in B$; hence the following two events in the sample space are identical:

$$\{\omega \in \Omega \,|\, (X(\omega), Y(\omega)) \in A\} = \{\omega \in \Omega \,|\, (U(\omega), V(\omega)) \in B\}. \qquad (5.16)$$

Since these events are the same, they have the same probability. Moreover, the probability of the event on the left is exactly the left side of (5.13). Hence the right side of (5.13) must be the probability of the event on the right of (5.16); that is,

$$P[(U, V) \in B] = \int_B \int f(h_1(u, v), h_2(u, v)) \, |J(u, v)| \, du \, dv. \qquad (5.17)$$

Therefore the integrand in (5.17) is the joint density of U and V. This argument establishes the following proposition.

PROPOSITION 5.2.1

Suppose that X and Y are continuous random variables with joint density $f(x, y)$. Let U and V be obtained from X and Y via the invertible transformation (5.14), whose inverse is in (5.15). Let J be the Jacobian of the transformation. Then the joint density of U and V is

$$f(h_1(u, v), h_2(u, v)) \, |J(u, v)|. \qquad (5.18)$$

Example 5.2.1

We can now find the joint density of R and Θ in the dartboard example. The usual inverse polar transformation is

$$x = r \cos \theta, \qquad y = r \sin \theta.$$

The Jacobian of the transformation is therefore

$$J = J(r, \theta) = \det \begin{bmatrix} \partial x/\partial r & \partial x/\partial \theta \\ \partial y/\partial r & \partial y/\partial \theta \end{bmatrix} = \det \begin{bmatrix} \cos \theta & -r \sin \theta \\ \sin \theta & r \cos \theta \end{bmatrix}.$$

This determinant simplifies easily to r; recall that we have restricted r to positive values (and θ to the interval $[0, 2\pi]$). Thus applying formula (5.18) to the joint density in (5.8) yields

$$f_{R,\Theta}(r, \theta) = f(r \cos \theta, r \sin \theta) \cdot |J(r, \theta)| = \frac{1}{\pi A^2} r, \qquad r \in [0, A], \qquad \theta \in [0, 2\pi].$$

Perhaps surprisingly, this density is not uniform, because larger radii r have higher probability density. So, the primitive assumption that equal areas have equal probability is not the same as saying that no radius is favored over another.

?Question 5.2.2 Try to explain the nonuniformity of the joint density in Example 5.2.1. What is the marginal density of Θ?

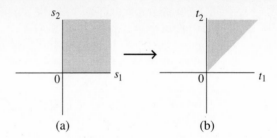

Figure 5.5   Transformation from Interarrival to Arrival Times

Example 5.2.2

Let S_1 and S_2 be the first two interarrival times of a Poisson process with rate λ, and let $T_1 = S_1$ and $T_2 = S_1 + S_2$ be the associated arrival times. Compute the joint density of T_1 and T_2.

Since S_1 and S_2 are i.i.d. exponential (λ) random variables, their joint density is

$$f_{S_1,S_2}(s_1, s_2) = \lambda^2 e^{-\lambda(s_1+s_2)}, \qquad s_1, s_2 > 0.$$

The transformation from the interarrival time variables to the arrival times is

$$t_1 = g_1(s_1, s_2) = s_1, \qquad t_2 = g_2(s_1, s_2) = s_1 + s_2,$$

which has inverse transformation

$$s_1 = h_1(t_1, t_2) = t_1, \qquad s_2 = h_2(t_1, t_2) = t_2 - t_1.$$

The Jacobian of this transformation is

$$J = \det \begin{bmatrix} \partial s_1/\partial t_1 & \partial s_1/\partial t_2 \\ \partial s_2/\partial t_1 & \partial s_2/\partial t_2 \end{bmatrix} = \det \begin{bmatrix} 1 & 0 \\ -1 & 1 \end{bmatrix} = 1.$$

Substitution of the inverse functions h_1 and h_2 into the joint density of S_1 and S_2 produces the following formula for the joint density of T_1 and T_2:

$$f_{T_1,T_2}(t_1, t_2) = \lambda^2 e^{-\lambda t_2}.$$

The state space of the original pair (S_1, S_2) is the first quadrant of the $s_1 - s_2$ plane, shown in part (a) of Fig. 5.5. Under the transformation $t_1 = s_1$, $t_2 = s_1 + s_2$, the possible values of t_1 are the same as those of s_1—namely $t_1 \in (0, \infty)$—but for a given $s_1 = t_1$ the possible values of $t_2 = t_1 + s_2$ are points in the interval (t_1, ∞). Hence the new state space on which f_{T_1,T_2} is nonzero is the shaded set in part (b) of Fig. 5.5, which is the solution set of the inequality system $0 < t_1$, $t_1 < t_2$.

?Question 5.2.3   Devise a general theorem analogous to Proposition 5.2.1 about the density of a one-dimensional continuous random variable $Y = g(X)$, obtained as an invertible function of another one-dimensional continuous random variable.

Proposition 5.2.1 extends in a natural way to joint transformations of more than two random variables. Let

$$U_1 = g_1(X_1, \ldots, X_n), \quad U_2 = g_2(X_1, \ldots, X_n), \quad \ldots, \quad U_n = g_n(X_1, \ldots, X_n) \tag{5.19}$$

be a 1–1 transformation with inverse

$$X_1 = h_1(U_1, \ldots, U_n), \quad X_2 = h_2(U_1, \ldots, U_n), \quad \ldots, \quad X_n = h_n(U_1, \ldots, U_n). \tag{5.20}$$

Define the Jacobian of the transformation, as in the two-variable case, to be the determinant of the matrix of partial derivatives of the inverse functions:

$$J = J(u_1, u_2, \ldots, u_n) = \det \begin{bmatrix} \partial h_1/\partial u_1 & \partial h_1/\partial u_2 & \cdots & \partial h_1/\partial u_n \\ \partial h_2/\partial u_1 & \partial h_2/\partial u_2 & \cdots & \partial h_2/\partial u_n \\ \vdots & \vdots & \ddots & \vdots \\ \partial h_n/\partial u_1 & \partial h_n/\partial u_2 & \cdots & \partial h_n/\partial u_n \end{bmatrix}. \tag{5.21}$$

Then

$$\int \int_A \cdots \int f(x_1, x_2, \ldots, x_n) \, dx_n \, \ldots \, dx_2 \, dx_1$$
$$= \int \int_B \cdots \int f(h_1, h_2, \ldots, h_n) \, |J| \, du_1 \, du_2 \, \ldots \, du_n, \tag{5.22}$$

where the set B is the image of the set A under the transformation (5.19), which implies that the joint density of $U_1, U_2, \ldots, U_n$ is

$$f(h_1, h_2, \ldots, h_n) \, |J|. \tag{5.23}$$

(To simplify the notation, we have omitted the arguments $u_1, \ldots, u_n$ from the functions $h_1, \ldots, h_n$ and J.)

Multivariable transformations can be useful even when there is just one univariate transformation $Y = Y_1 = g(X_1, \ldots, X_n)$ of most interest in an application. To find the distribution of such a function of many random variables, you can invent other, presumably simple functions $Y_2 = g_2(X_1, \ldots, X_n)$, $\ldots, Y_n = g_n(X_1, \ldots, X_n)$ in such a way that the entire multivariate transformation is 1–1. Then find the joint density of the random variables $Y_1, Y_2, \ldots, Y_n$, and finally compute the marginal density of $Y = Y_1$ as usual, by integrating out the other random variables Y_j. The following example illustrates the process.

Example 5.2.3

Suppose that you are writing a paper for a history class, which is due in five weeks. The project requires three phases: library research, the production of a first draft, and the revision of the first draft to the final draft. One phase must be completed before its successor can begin. If the time required for the research is distributed uniformly on the interval 0 week to 3 weeks, the time for the first draft is uniform on 0 week to 1 week, and the time for the final draft is uniform on 1 week to 2 weeks, what is the distribution of the completion time of the paper? What is the probability that it will be late?

Let X_1, X_2, and X_3 denote the times required for the research, first draft, and final draft, respectively. Since the tasks are done sequentially, the paper completion time in which we are interested is the random variable $Y = X_1 + X_2 + X_3$. To find its distribution, we shall also assume that the individual phase times are independent of one another. (What do you think about these assumptions?) Then the joint density of X_1, X_2, and X_3 is the product of the given marginal uniform densities:

$$f(x_1, x_2, x_3) = \frac{1}{3} \cdot 1 \cdot 1 = \frac{1}{3}, \qquad x_1 \in [0,3], \quad x_2 \in [0,1], \quad x_3 \in [1,2].$$

A joint transformation that is easy to invert and that includes the transformation of interest is

$$Y_1 = X_1, \qquad Y_2 = X_2, \qquad Y_3 = Y = X_1 + X_2 + X_3.$$

The inverse transformation is

$$X_1 = Y_1, \qquad X_2 = Y_2, \qquad X_3 = Y_3 - X_1 - X_2 = -Y_1 - Y_2 + Y_3.$$

Hence the Jacobian is

$$J = \det \begin{bmatrix} \partial x_1/\partial y_1 & \partial x_1/\partial y_2 & \partial x_1/\partial y_3 \\ \partial x_2/\partial y_1 & \partial x_2/\partial y_2 & \partial x_2/\partial y_3 \\ \partial x_3/\partial y_1 & \partial x_3/\partial y_2 & \partial x_3/\partial y_3 \end{bmatrix} = \det \begin{bmatrix} 1 & 0 & 0 \\ 0 & 1 & 0 \\ -1 & -1 & 1 \end{bmatrix} = 1.$$

Since the original density had constant value on the state space of the X's and the Jacobian is 1, the formula for the joint density of the Y's is particularly simple; by (5.23) it is the constant $1/3$. The complication lies in the state space of the Y's. Clearly, since Y_1 and Y_2 are identical to X_1 and X_2, respectively, we have the restrictions $y_1 \in [0,3]$ and $y_2 \in [0,1]$. But also, since $Y_3 = Y_1 + Y_2 + X_3$ and X_3 takes values in the interval $[1,2]$, we have the restriction that Y_3 must lie in $[Y_1 + Y_2 + 1, Y_1 + Y_2 + 2]$. Thus a full specification of the joint density of Y_1, Y_2, and Y_3 is

$$f_{Y_1, Y_2, Y_3}(y_1, y_2, y_3) = \frac{1}{3},$$

$$y_1 \in [0,3], \quad y_2 \in [0,1], \quad y_1 + y_2 + 1 \le y_3 \le y_1 + y_2 + 2. \qquad (5.24)$$

To find the marginal of the project completion time $Y = Y_3$, we must integrate f_{Y_1, Y_2, Y_3} with respect to both y_1 and y_2 for fixed y_3. This problem is made cumbersome by the fact that the shape of the region of integration in the $y_1 - y_2$ plane depends strongly on the value of y_3, because the third condition in (5.24) implies that

$$y_3 - 2 \le y_1 + y_2 \le y_3 - 1.$$

It is not hard to see that y_3 can be as small as 1 and as large as 6, and the shape of the integration region changes at each integer value in between, as shown in parts (a)–(e) of Fig. 5.6. To compute the marginal, integrate the constant function $1/3$ over the shaded set, case by case, which is the same as multiplying the area of the shaded set by $1/3$. (See Exercise 8 at the end of this section.) The results are

$$f_Y(y) = \begin{cases} (y^2 - 2y + 1)/6 & \text{if } y \in (1,2), \\ (-y^2 + 6y - 7)/6 & \text{if } y \in (2,3), \\ 1/3 & \text{if } y \in (3,4), \\ (-y^2 + 8y - 14)/6 & \text{if } y \in (4,5), \\ (y^2 - 12y + 36)/6 & \text{if } y \in (5,6). \end{cases} \qquad (5.25)$$

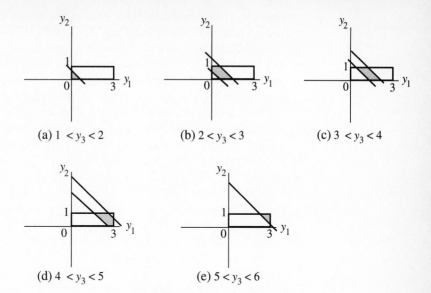

(a) $1 < y_3 < 2$ (b) $2 < y_3 < 3$ (c) $3 < y_3 < 4$

(d) $4 < y_3 < 5$ (e) $5 < y_3 < 6$

Figure 5.6   Cases for Marginal of Y_3

The paper will be late if and only if the completion time is between 5 and 6. Using the last case of f_Y in formula (5.25),

$$P[Y > 5] = \int_5^6 \frac{1}{6}(y^2 - 12y + 36) \, dy = \frac{1}{18}.$$

As in Section 5.1, we have been concentrating our attention on continuous random variables. In the discrete case, the joint probability mass function of the transformed random variables is rather easily obtainable by direct substitution of the inverse transformation into the original probability mass function. (See, for example, Exercise 23.)

5.2.2   Order Statistics

When a random sample $X_1, X_2, \ldots, X_n$ is taken from a distribution, it is possible to gain information about that distribution by examining the sample values in ascending order. We denote these ordered sample values by $Y_1 \leq Y_2 \leq \cdots \leq Y_n$ (some authors use the notation $X_{(1)}, \ldots, X_{(n)}$), and we refer to them as the *order statistics* of the sample. The extremes have particular importance. For example, if the sample values are the results of a poll in which the respondents are asked to rate the effectiveness of a political office-holder on a scale of 0–100, the smallest order statistic Y_1 and the largest Y_n give an indication of the range of public sentiment. The smallest and largest order statistics also appear in the context of reliability problems: if a system of n identical, independent components with failure times $X_1, X_2, \ldots, X_n$, respectively, is built in a series structure, then the system failure time is the smallest X_i. If the components are organized in a parallel structure, then the system failure time is the largest X_i.

More than the extreme values of a random sample are of interest here. In fact, in some samples the extreme values result from errors in data gathering or reporting. These errors can sometimes have drastic effects on the sample mean if the mean is used to measure the center point of the data. The middle value of the sample, however, is not affected by changes to the extremes. This statistic is called the *median* of the sample. In the case that the sample size n is odd, there is a unique middle value of the sample, namely $Y_{(n+1)/2}$, and if n is even, the median is defined as the average of the two adjacent middle values $(Y_{n/2} + Y_{n/2+1})/2$. For example, in the polling situation described, if a group of respondents gave ratings of 0, 54, 65, 70, 75 to the politician, the lone malcontent who contributed the 0 brings the mean down to 52.8, whereas the median of 65 seems to represent the data more closely.

So it is important to know how the probability distributions of the order statistics are found from the distribution from which the sample was taken. To solve this problem, we will need an extended change of variables formula for transformations that are not quite 1–1.

Let $U = g(X)$ be a transformation such that the state space E_X of X can be partitioned into several subsets $E_1, E_2, \ldots, E_k$, as in Fig. 5.7, where g is 1–1 with an inverse transformation h_i when restricted to E_i. Since these h_i are also vector transformations, they are composed of inverse mappings:

$$x_1 = h_{i1}(u_1, u_2, \ldots, u_n), \qquad x_2 = h_{i2}(u_1, u_2, \ldots, u_n), \qquad \ldots,$$
$$x_n = h_{in}(u_1, u_2, \ldots, u_n).$$

If J_i denotes the Jacobian associated with the ith inverse transformation h_i, then the generalized transformation theorem states that the joint density of $U = (U_1, U_2, \ldots, U_n)$ is the sum, over all of the transformations, of terms just like those in formula (5.23):

$$f_U(u_1, u_2, \ldots, u_n) = \sum_{i=1}^{k} f_X(h_{i1}, h_{i2}, \ldots, h_{in}) \cdot |J_i|. \tag{5.26}$$

You are asked to prove this result in Exercise 15 at the end of this section.

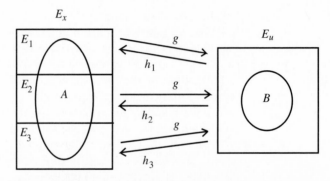

Figure 5.7   A Noninvertible Transformation

This result is precisely the one we need to derive the joint density of the order statistics.

PROPOSITION 5.2.2

Let $Y_1 \leq Y_2 \leq \cdots \leq Y_n$ be the order statistics of a random sample $X_1, X_2, \ldots, X_n$ taken from a continuous distribution with p.d.f. f. Then the joint density of the Y's is

$$f_Y(y_1, y_2, \ldots, y_n) = n! \cdot f(y_1) \cdot f(y_2) \cdots f(y_n), \qquad y_1 < y_2 < \cdots < y_n.$$
(5.27)

■ **Proof** Since the X's have a continuous distribution, we can safely ignore the zero probability events that sample values coincide with one another, and therefore we can break the state space of the X's into $n!$ subsets, on each of which a different ordering of the coordinates $(x_1, x_2, \ldots, x_n)$ of the state of the random vector is in effect. For example, here are a few such subsets, with the associated inverse transformations taking the observed x_i's back to the y_i's:

On $E_1 = \{x_1 < x_2 < x_3 < \cdots < x_n\}$,

$$x_1 = h_{11}(\mathbf{y}) = y_1, \qquad x_2 = h_{12}(\mathbf{y}) = y_2, \qquad x_3 = h_{13}(\mathbf{y}) = y_3, \qquad \ldots,$$
$$x_n = h_{1n}(\mathbf{y}) = y_n.$$

On $E_2 = \{x_2 < x_1 < x_3 < \cdots < x_n\}$,

$$x_1 = h_{21}(\mathbf{y}) = y_2, \qquad x_2 = h_{22}(\mathbf{y}) = y_1, \qquad x_3 = h_{23}(\mathbf{y}) = y_3, \qquad \ldots,$$
$$x_n = h_{2n}(\mathbf{y}) = y_n.$$

On $E_3 = \{x_1 < x_3 < x_2 < \cdots < x_n\}$,

$$x_1 = h_{31}(\mathbf{y}) = y_1, \qquad x_2 = h_{32}(\mathbf{y}) = y_3, \qquad x_3 = h_{33}(\mathbf{y}) = y_2, \qquad \ldots,$$
$$x_n = h_{3n}(\mathbf{y}) = y_n.$$

Because of the special form of each of the inverse transformations, in each of the Jacobians J_i the matrix of partials has exactly one 1 in every row and every column and the rest of the matrix entries are 0. It is not hard to show that the determinant of such a matrix is equal to 1 or to -1; hence the $|J_i|$ term in (5.26) is simply equal to 1. Also, since the X_i's are a random sample, the joint density of the X's has the form $f_X(\mathbf{x}) = f(x_1) \cdot f(x_2) \cdots f(x_n)$. Regardless of which inverse transformation h_i you look at, when it is substituted into f_X there will be exactly one $f(y_1)$ factor, exactly one $f(y_2)$ factor, and so forth, contained in the product. Therefore the joint density of the order statistics simplifies to

$$f_Y(y_1, y_2, \ldots, y_n) = \sum_{i=1}^{n!} f(y_1) f(y_2) \cdots f(y_n) \cdot 1$$
$$= n! \cdot f(y_1) \cdot f(y_2) \cdots f(y_n), \qquad y_1 < y_2 < \cdots < y_n.$$

The last theorem in this section gives a formula for the marginal distribution of the kth order statistic Y_k. The proof is a straightforward but lengthy computation in which the joint density (5.27) is integrated to produce the desired marginal density. But here is another way of looking at the marginal distribution that, although not rigorous, is highly instructive. The infinitesimal probability $f_k(y_k)\,dy_k$ that the kth smallest order statistic takes the value y_k is the probability that the n X_i values fall into three categories:

1. $k-1$ of the X_i's are less than y_k.
2. Exactly 1 X_i equals y_k.
3. The remaining $(n-k)$ X_i's are greater than y_k.

The quantity $F(y_k)$ is the probability that a single X falls into the first category, $f(y_k)\,dy_k$ is the infinitesimal probability that an X falls into the second category, and $1 - F(y_k)$ is the probability that an X falls into the third category. Formula (5.28) is then just the multinomial probability of $k-1$ successes of type 1, 1 success of type 2, and $n - k$ successes of type 3 among the n X_i's.

PROPOSITION 5.2.3

Let Y_k be the kth order statistic in a random sample $X_1, X_2, \ldots, X_n$ taken from a continuous distribution with density function $f(x)$ and c.d.f. $F(x)$. The density of Y_k is

$$f_k(y_k) = \frac{n!}{(k-1)!(n-k)!} \cdot f(y_k) \cdot [F(y_k)]^{k-1} \cdot [1 - F(y_k)]^{n-k}. \qquad (5.28)$$

■ **Proof**   In several of the following computations we make use of the fact that $F'(x) = f(x)$, and hence we can substitute $u = F(x)$, $du = f(x)\,dx$.

Suppose that the state space of each X_i is (a, b). (Either a or b may be infinite.) The marginal density of Y_k is the integral of the joint density in (5.27) with respect to all other Y_i's, and after removing factors this integral becomes

$$f_k(y_k) = n!f(y_k) \int_a^{y_k} f(y_1) \int_{y_1}^{y_k} f(y_2) \cdots \int_{y_{k-2}}^{y_k} f(y_{k-1}) \int_{y_k}^b f(y_{k+1}) \cdots$$

$$\cdots \int_{y_{n-2}}^b f(y_{n-1}) \int_{y_{n-1}}^b f(y_n) \, dy_n dy_{n-1} \cdots dy_{k+1} dy_{k-1} \cdots dy_2 dy_1$$

$$= n!f(y_k) \int_a^{y_k} f(y_1) \int_{y_1}^{y_k} f(y_2) \cdots \int_{y_{k-2}}^{y_k} f(y_{k-1}) \int_{y_k}^b f(y_{k+1}) \cdots$$

$$\cdots \int_{y_{n-2}}^b f(y_{n-1})(1 - F(y_{n-1})) \, dy_{n-1} \cdots dy_{k+1} dy_{k-1} \cdots dy_2 dy_1.$$

(Make sure you understand the limits of integration.) The second line evaluates the innermost integral in the first line, using the fact that $F(b) = 1$.

To carry on the computation, a substitution $u = 1 - F(y_{n-1})$ in the innermost integral in the last line above results in the following:

$$\int_{y_{n-2}}^{b} f(y_{n-1})(1 - F(y_{n-1})) \, dy_{n-1} = \frac{1}{2 \cdot 1}(1 - F(y_{n-2}))^2.$$

The next integral would then have the form

$$\int_{y_{n-3}}^{b} f(y_{n-2}) \cdot \frac{1}{2 \cdot 1}(1 - F(y_{n-2}))^2 \, dy_{n-2} = \frac{1}{3 \cdot 2 \cdot 1}(1 - F(y_{n-3}))^3,$$

by a similar substitution. This pattern continues throughout the computation of the innermost $n - k$ integrals, with respect to the variables $y_n, \ldots, y_{k+1}$. At the end of this phase of computation, we have reduced the marginal density of Y_k to the form

$$f_k(y_k) = n! f(y_k) \int_{a}^{y_k} f(y_1) \int_{y_1}^{y_k} f(y_2) \cdots$$

$$\cdots \int_{y_{k-2}}^{y_k} f(y_{k-1}) \cdot \frac{1}{(n-k)!}(1 - F(y_k))^{n-k} \, dy_{k-1} \cdots dy_2 dy_1$$

$$= \frac{n!}{(n-k)!} f(y_k)(1 - F(y_k))^{n-k} \int_{a}^{y_k} f(y_1) \int_{y_1}^{y_k} f(y_2) \cdots$$

$$\cdots \int_{y_{k-2}}^{y_k} f(y_{k-1}) \, dy_{k-1} \cdots dy_2 dy_1.$$

Now the innermost integral is

$$\int_{y_{k-2}}^{y_k} f(y_{k-1}) \, dy_{k-1} = F(y_k) - F(y_{k-2}).$$

Application of a substitution $u = F(y_k) - F(y_{k-2})$ to the next integral in line gives

$$\int_{y_{k-3}}^{y_k} (F(y_k) - F(y_{k-2})) f(y_{k-2}) \, dy_{k-2} = \frac{1}{2 \cdot 1} \cdot (F(y_k) - F(y_{k-3}))^2.$$

The $k - 2$ integrals with respect to the variables $y_{k-1}, \ldots, y_2$ follow this pattern, leaving us with one last integral:

$$f_k(y_k) = \frac{n!}{(n-k)!} f(y_k)(1 - F(y_k))^{n-k}$$

$$\cdot \int_{a}^{y_k} f(y_1) \cdot \frac{1}{(k-2)!} \cdot (F(y_k) - F(y_1))^{k-2} \, dy_1$$

$$= \frac{n!}{(n-k)!} f(y_k)(1 - F(y_k))^{n-k} \cdot \left[-\frac{(F(y_k) - F(y_1))^{k-1}}{(k-1)!} \right]_{a}^{y_k}$$

$$= \frac{n!}{(k-1)!(n-k)!} \cdot f(y_k) \cdot [F(y_k)]^{k-1} \cdot [1 - F(y_k)]^{n-k},$$

which completes the derivation.

—— **?Question 5.2.4**   Write down the formulas to which (5.28) reduces in the two cases $k = 1$ (the smallest order statistic) and $k = n$ (the largest order statistic). Interpret the expressions as multinomial probabilities.

Example 5.2.4

Consider the series structure in Fig. 5.8(a) and the parallel structure in Fig. 5.8(b). Suppose that system components 1, 2, and 3 each work properly for a length of time governed by the Weibull distribution with parameters $\lambda = 2$, $\beta = 2$, and assume that their failure times are independent. Find the distribution of the system lifetime in each case.

For the series structure in part (a), the system fails at the smallest of the three component failure times X_1, X_2, X_3, or the first order statistic Y_1. The density and c.d.f. of an individual component failure time are given to be

$$f(x) = 8xe^{-(2x)^2}, \quad x > 0; \qquad F(x) = 1 - e^{-(2x)^2}, \quad x > 0.$$

Thus from formula (5.28) we have the density of the series system:

$$\frac{3!}{0!(3-1)!} \cdot f(y_1) \cdot (F(y_1))^{1-1} \cdot (1 - F(y_1))^{3-1} = 24y_1 e^{-12y_1^2}, \qquad y_1 > 0.$$

You can recognize this as another Weibull density. (What are its parameters?)

The parallel structure in part (b) fails at the largest of the failure times X_i, which is the third-order statistic Y_3. Its density is

$$\frac{3!}{(3-1)!(3-3)!}f(y_3)(F(y_3))^{3-1}(1 - F(y_3))^{3-3} = 24y_3 e^{-4y_3^2}(1 - e^{-4y_3^2})^2,$$

$$y_3 > 0.$$

We can use these results to compare the performance of the two structures. For example, the likelihood that the series structure survives for at least .5 time unit is

$$P[Y_1 > .5] = \int_{.5}^{\infty} 24y_1 e^{-12y_1^2}\, dy_1 = -e^{-12y_1^2}\Big|_{.5}^{\infty} = e^{-3} \approx .05.$$

The likelihood that the parallel structure survives for at least the same length of time is

$$P[Y_3 > .5] = \int_{.5}^{\infty} 24y_3 e^{-4y_3^2} \cdot (1 - e^{-4y_3^2})^2\, dy_3$$

$$= (1 - e^{-4y_3^2})^3\Big|_{.5}^{\infty} = 1 - (1 - e^{-1})^3 \approx .75.$$

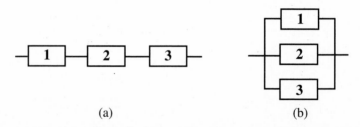

(a)   (b)

Figure 5.8   Series and Parallel Reliability Structures

The improvement in system reliability from the series to the parallel structure is very striking.

Example 5.2.5

A Pick 4 state lottery drawing claims to select four digits 0–9 randomly and independently in order to form a number ranging from 0 to 9999. We will approximate the discrete distribution of 10,000 numbers by a continuous uniform distribution on $[0, 1]$ (in units of 10,000). Under uniformity, you would expect the middle of the distribution to be around .5 (i.e., the actual number 5000). Suppose that among seven consecutive draws, the median number drawn was .7637. Would you have cause to doubt the uniformity of drawing?

The continuous uniform density is $f(x) = 1$ on $(0, 1)$, and its c.d.f. is $F(x) = x$. The sample median among seven sample values is the fourth-order statistic Y_4. We can recast the question as: Is it very likely under the uniformity assumption that Y_4 can be as big as .7637?

By formula (5.28) the density of Y_4 is

$$\frac{7!}{3!3!} \cdot f(y_4) \cdot (F(y_4))^3 \cdot (1 - F(y_4))^3 = 140 y_4^3 (1 - y_4)^3, \qquad y_4 \in [0, 1].$$

Then,

$$P[Y_4 \geq .7637] = \int_{.7637}^{1} 140 y_4^3 (1 - y_4)^3 \, dy_4 \approx .059.$$

So if the lottery drawing is as claimed, there is only about a $1/20$ chance that the sample median could be so large. This by itself is not conclusive evidence of some kind of conspiracy, but it is enough to warrant some further investigation.

EXERCISES 5.2

1. Let X_1 and X_2 be independent random variables with the $N(0, 1)$ distribution. Show that the random variable $Y = X_1/X_2$ has a Cauchy distribution.
2. Suppose that X_1 and X_2 are i.i.d. $\Gamma(n, \lambda)$ random variables. Show that the random variables $U_1 = X_1 + X_2$ and $U_2 = X_1/X_2$ are independent.
3. Let X_1 and X_2 be independent $N(0, 1)$ random variables. Show that the joint distribution of the random variables $W_1 = X_1$, $W_2 = X_1 + 2X_2$ is bivariate normal. What is the correlation coefficient associated with the joint distribution?
4. Find the probability distribution of $X_1/(X_1 + X_2)$, where X_1 has the $\Gamma(n_1, \lambda)$ distribution, X_2 has the $\Gamma(n_2, \lambda)$ distribution, and X_1 and X_2 are independent. What connection does this problem have to Poisson processes?
5. Let X and Y be independent and identically distributed random variables with density function $f(x) = 3x^2$, $x \in (0, 1)$. Find the joint density of $W = XY$ and $Z = Y$, and use it to find the marginal density of W.
6. Use the result of Example 5.2.2 on the first two times of arrival in a Poisson process to show that, conditioned on $T_2 = t_2$, the distribution of T_1 is uniform on the interval $[0, t_2]$.

7. For each of the following single-variable transformations, use your response to Question 5.2.3 to find the density of the transformed random variable.

   a. cX, where X is a continuous random variable with density $f(x)$ and c is a constant

   b. $X/(1 + X)$, where X is a positive-valued continuous random variable with density $f(x)$

8. Fill in the details of the computation of the density f_Y in formula (5.25) of Example 5.2.3.

9. A company is purchasing a new computer system. It will take 3 ± 2 weeks to receive and set up the equipment and 2 ± 1 week to train the employees to use the system once it is set up. Find the probability that the employees will be completely trained by the sixth week.

10. Let X_1 and X_2 be independent random variables with the Cauchy density

$$f(x) = \frac{1}{\pi(1 + x^2)}, \qquad -\infty < x < +\infty.$$

Find the joint density of $Y_1 = 2X_1 - X_2$ and $Y_2 = X_1 + X_2$.

11. Example 3.1.4 gave the joint density $f(x_1, x_2) = 1/2$ if $0 \le x_2 \le x_1 \le 2$ for the times of the longer and shorter of two computer jobs. Show that this density results from the simpler assumption that the two unordered times are i.i.d. random variables with the uniform $(0, 2)$ distribution.

12. The following transformation, called the *Box–Muller transformation*, is very useful in simulating normally distributed random variables. Let X_1 and X_2 be i.i.d. uniform random variables on the interval $(0, 1)$. Define $Z_1 = \sqrt{-2\ln(X_1)}\cos(2\pi X_2)$, $Z_2 = \sqrt{-2\ln(X_1)}\sin(2\pi X_2)$. Show that Z_1 and Z_2 are i.i.d. standard normal random variables.

13. A candidate for the U.S. Senate is interested in knowing about his potential support statewide. He commissions a poll among five randomly selected counties, and in them he earns $45\%, 57\%, 55\%, 51\%$, and 58% support. Can the senatorial candidate be reasonably sure that his median support among all counties is greater than 50%? (You may assume that his support in all counties is uniformly distributed and that his most popular and least popular counties differ by no more than 40%. Play the devil's advocate: What is implied if the median support were 50% or less?)

14. Let X and Y be continuous random variables with densities $f_X(x)$ and $f_Y(y)$, respectively, and suppose that X and Y are independent. Use the transformation $U = X + Y$, $V = X$, to show that the density of U satisfies the *convolution equation*:

$$f_U(u) = \int_{-\infty}^{+\infty} f_X(v) f_Y(u - v)\, dv.$$

15. Verify formula (5.26). [*Hint:* Let B be in the state space E_U. Split the event $\{U \in B\}$ into the disjoint subevents $X \in h_1(B), X \in h_2(B), \ldots$]

16. Write a general expression for the cumulative distribution function of the lifetime of a parallel system, in terms of the c.d.f.'s of the components' lifetimes.

17. Redo Example 5.2.4 under the assumption that the component lifetimes have the exponential distribution with parameter $\lambda = 2$.

18. A "k-out-of-n" reliability system consists of n components and works if and only if at least k of the components work. Find the probability density function of the system failure time of a 5-out-of-7 system in which each component has an exponentially distributed lifetime with parameter 1.

19. Give a plausibility argument similar to the one following Proposition 5.2.3 to justify that the joint density of two order statistics Y_i, Y_j for $i < j$ is

$$f_{ij}(y_i, y_j) = \frac{n!}{(i-1)!(j-i-1)!(n-j)!}[F(y_i)]^{i-1}[F(y_j) - F(y_i)]^{j-i-1}$$
$$\cdot [1 - F(y_j)]^{n-j}f(y_i)f(y_j),$$

for $a < y_i < y_j < b$.

20. Let $X_1, \ldots, X_6$ be a random sample from the uniform $(0, 1)$ distribution. Find the probability that the *range* $Y_6 - Y_1$—that is, the difference between the largest and smallest sample values—exceeds .5. (Use the result of Exercise 19.)

21. Show that the kth order statistic of a random sample $X_1, \ldots, X_n$ from the uniform $(0, 1)$ distribution has a beta distribution (see Section 3.3). Identify the parameters.

22. Three numbers are drawn at random and without replacement from the set $\{1, 2, 3, 4, 5\}$. By listing outcomes, find the joint distribution of the minimum, median, and maximum of the numbers. What is the probability that the minimum equals 1?

23. Let X_1 and X_2 be independent Poisson random variables with means μ_1 and μ_2, respectively. Find the joint distribution of the random variables $Y_1 = X_1$, $Y_2 = X_1 + X_2$ by direct substitution of the inverse transformation into the joint p.m.f. of the X's. Use the joint p.m.f. of the Y's to show that the marginal distribution of Y_2 is Poisson$(\mu_1 + \mu_2)$.

5.3 | Generating Functions

In the last section we saw how change-of-variable methods from calculus allow us to find the distribution of transformed random variables. Although this approach is very general, it can be difficult to apply, especially if more than two random variables are involved or if the transformation is complicated. We now introduce a much simpler method, one that works for the special problem of finding the distribution of the sum of independent random variables. This method uses a tool called the *moment-generating function*. Besides its usefulness in the transformation problem, the moment-generating function has a connection to the moments of the distribution with which it is associated that can produce some computational results.

DEFINITION 5.3.1

The *moment-generating function (m.g.f.)* of a real-valued random variable X is the function

$$M(t) = M_X(t) = E[e^{tX}], \tag{5.29}$$

which is defined for all real values of t such that the expectation is finite.

Notice that in the case where X is a continuous random variable with probability density function f on a state space E, the moment-generating function is given by

$$M(t) = E[e^{tX}] = \int_E e^{tx} f(x) \, dx. \tag{5.30}$$

For discrete random variables, the integral is replaced by a sum over all states. Look carefully at (5.30) to become familiar with the role of t, the argument of the m.g.f. It is a new, nonrandom, real variable that is not related to the variable of integration x or the random variable X. If you have studied differential equations and Laplace transforms, integrals like this will be familiar; in fact, the m.g.f. method for finding probability distributions of sums parallels the Laplace transform method for solving differential equations.

Let us first calculate a few moment-generating functions, then examine how we can make use of them.

Example 5.3.1

Consider the Poisson distribution with parameter μ. We can compute the m.g.f. as follows:

$$M(t) = E[e^{tX}] = \sum_{n=0}^{\infty} e^{tn} \frac{e^{-\mu} \mu^n}{n!} = e^{-\mu} \cdot \sum_{n=0}^{\infty} \frac{(\mu e^t)^n}{n!}. \tag{5.31}$$

Recall that the Taylor series expansion of the natural exponential function is $e^x = \sum_{n=0}^{\infty} x^n/n!$, which converges for all real x. Therefore the series in the last expression above converges to $e^{\mu e^t}$, which implies that the Poisson(μ) m.g.f. is

$$M(t) = e^{-\mu} e^{\mu e^t} = e^{\mu(e^t - 1)}, \qquad t \in \mathbb{R}. \tag{5.32}$$

?Question 5.3.1 Compute the derivative of the Poisson moment-generating function with respect to t, then evaluate it at $t = 0$. Try to explain what you get.

Example 5.3.2

The moment-generating function of the $\Gamma(\alpha, \lambda)$ distribution is the following integral:

$$M(t) = E[e^{tX}] = \int_0^{\infty} e^{tx} \cdot \frac{\lambda^\alpha}{\Gamma(\alpha)} x^{\alpha-1} e^{-\lambda x} \, dx$$

$$= \frac{\lambda^\alpha}{\Gamma(\alpha)} \cdot \int_0^{\infty} x^{\alpha-1} e^{-(\lambda-t)x} \, dx.$$

The substitution $u = (\lambda - t)x$, $du = (\lambda - t)\,dx$ produces the following integral, which converges only for $t < \lambda$:

$$
\begin{aligned}
M(t) &= \frac{\lambda^\alpha}{\Gamma(\alpha)} \cdot \int_0^\infty \left(\frac{u}{\lambda - t}\right)^{\alpha-1} e^{-u} \cdot \frac{1}{\lambda - t}\, du \\
&= \left(\frac{\lambda}{\lambda - t}\right)^\alpha \cdot \frac{1}{\Gamma(\alpha)} \cdot \int_0^\infty u^{\alpha-1} e^{-u}\, du \\
&= \left(\frac{\lambda}{\lambda - t}\right)^\alpha \cdot \frac{1}{\Gamma(\alpha)} \cdot \Gamma(\alpha) = \left(\frac{\lambda}{\lambda - t}\right)^\alpha, \qquad t < \lambda.
\end{aligned}
\tag{5.33}
$$

?Question 5.3.2 Use (5.33) to write expressions for the m.g.f.'s of the important subcases of the gamma family: the exponential and χ^2 distributions.

In Exercise 4 at the end of this section you will be asked to show that the m.g.f. of the $N(\mu, \sigma^2)$ distribution is

$$
M(t) = \exp\left(\mu t + \frac{1}{2}\sigma^2 t^2\right).
\tag{5.34}
$$

The main reason why we are concerned with moment-generating functions of distributions is that the m.g.f. of the sum $Y = X_1 + X_2 + \cdots + X_n$ of independent random variables can be expressed easily in terms of the m.g.f.'s of the X_i's. Why is this helpful? There is a result in probability theory (see, e.g., Parzen, 1960, p. 400) that states that a probability distribution is uniquely characterized by its moment-generating function; that is no two distributions share the same m.g.f. Thus, if the m.g.f. of the sum matches a moment-generating function of a distribution with which we are familiar, then Y must have that distribution. This gives us a simple and powerful method of finding the distribution of a sum of independent random variables.

The proofs of the following propositions illustrate the moment-generating function approach, so you should pay close attention to them.

PROPOSITION 5.3.1

Let $X_1, X_2, \ldots, X_n$ be independent random variables having the Poisson distribution with parameters $\mu_1, \mu_2, \ldots, \mu_n$, respectively. Then the random variable $Y = X_1 + X_2 + \cdots + X_n$ has the Poisson distribution with parameter $\mu_1 + \mu_2 + \cdots + \mu_n$.

■ **Proof** From formula (5.32), the moment-generating function of X_i is

$$
M_{X_i}(t) = E[e^{tX_i}] = e^{\mu_i(e^t - 1)}.
$$

Thus the m.g.f. of Y is

$$\begin{aligned}
M_Y(t) = E[e^{tY}] &= E[e^{t(X_1+X_2+\cdots+X_n)}] \\
&= E[e^{tX_1}]E[e^{tX_2}] \cdots E[e^{tX_n}] \\
&= M_{X_1}(t)M_{X_2}(t) \cdots M_{X_n}(t) \\
&= e^{\mu_1(e^t-1)}e^{\mu_2(e^t-1)} \cdots e^{\mu_n(e^t-1)} \\
&= e^{(\mu_1+\mu_2+\cdots+\mu_n)(e^t-1)}.
\end{aligned} \tag{5.35}$$

Since this is the moment-generating function of the Poisson $(\mu_1 + \mu_2 + \cdots + \mu_n)$ distribution, the proposition is proved.

PROPOSITION 5.3.2

Let $X_1, X_2, \ldots, X_n$ be independent random variables having gamma distributions with alpha parameters $\alpha_1, \alpha_2, \ldots, \alpha_n$, respectively, and common lambda parameter λ. Then the random variable $Y = X_1 + X_2 + \cdots + X_n$ has the gamma distribution with parameters $\alpha_1 + \alpha_2 + \cdots + \alpha_n$ and λ.

■ **Proof**   By formula (5.33), the m.g.f. of X_i is

$$M_{X_i}(t) = \left(\frac{\lambda}{\lambda - t} \right)^{\alpha_i}.$$

Then the m.g.f. of Y is

$$\begin{aligned}
M_Y(t) = E[e^{tY}] &= E[e^{t(X_1+X_2+\cdots+X_n)}] \\
&= E[e^{tX_1}]E[e^{tX_2}] \cdots E[e^{tX_n}] \\
&= M_{X_1}(t)M_{X_2}(t) \cdots M_{X_n}(t) \\
&= \left(\frac{\lambda}{\lambda - t} \right)^{\alpha_1} \left(\frac{\lambda}{\lambda - t} \right)^{\alpha_2} \cdots \left(\frac{\lambda}{\lambda - t} \right)^{\alpha_n} \\
&= \left(\frac{\lambda}{\lambda - t} \right)^{\alpha_1+\alpha_2+\cdots+\alpha_n}.
\end{aligned} \tag{5.36}$$

We recognize the last formula as the m.g.f. of the $\Gamma(\alpha_1 + \alpha_2 + \cdots + \alpha_n, \lambda)$ distribution, which completes the proof.

Recall that the $\chi^2(r)$ density is the special case of the gamma density for which $\alpha = r/2$ and $\lambda = 1/2$. Therefore our theorem about sums of gamma random variables specializes to a similar result about sums of χ^2 random variables. (Be sure that you see why the next result is true.)

PROPOSITION 5.3.3

Let $X_1, X_2, \ldots, X_n$ be independent random variables having χ^2 distributions with parameters $r_1, r_2, \ldots, r_n$, respectively. Then the random variable $Y = X_1 + X_2 + \cdots + X_n$ has the $\chi^2(r_1 + r_2 + \cdots + r_n)$ distribution.

The full impact of Proposition 5.3.3 is not apparent yet. But as we begin to see in the next section, this result is fundamental to an understanding of the variance of a random sample; hence it underlies much of the work that we will do in data analysis.

Proposition 5.3.2 also tells us that the distribution of the sum of n i.i.d. $\exp(\lambda)$ [i.e., $\Gamma(1, \lambda)$] random variables is $\Gamma(n, \lambda)$. Because of this result, the time of the nth arrival $T_n = S_1 + S_2 + \cdots + S_n$ in a Poisson process has the $\Gamma(n, \lambda)$ distribution.

The m.g.f. method produces some very powerful results just by writing the m.g.f. of the sum random variable Y as the product $M_{X_1}(t)M_{X_2}(t)\cdots M_{X_n}(t)$ of the m.g.f.'s of the X's in the sum, and then simplifying and recognizing that product. This approach also works for normal random variables, as the next proposition states. Prove it in this section's Exercise 8.

PROPOSITION 5.3.4

Suppose that $X_1, X_2, \ldots, X_n$ are independent normal random variables with means $\mu_1, \mu_2, \ldots, \mu_n$ and variances $\sigma_1^2, \sigma_2^2, \ldots, \sigma_n^2$, respectively. Let $a_1, a_2, \ldots, a_n$ be constants. Then $Y = \sum_{i=1}^{n} a_i X_i$ has the $N(\sum_{i=1}^{n} a_i \mu_i, \sum_{i=1}^{n} a_i^2 \sigma_i^2)$ distribution.

Example 5.3.3

Proposition 5.3.1 has implications for superpositions of Poisson processes. Suppose that calls to a computer software vendor for technical support, sales, and account questions arrive to a central access number according to three independent Poisson processes with rates λ_1, λ_2, and λ_3, respectively. Let N_t^1, N_t^2, and N_t^3 represent the total number of calls of the three types that have arrived by time t. Then N_t^1 has the Poisson distribution with parameter $\lambda_1 t$, N_t^2 has the Poisson distribution with parameter $\lambda_2 t$, and N_t^3 has the Poisson distribution with parameter $\lambda_3 t$. The sum $N_t = N_t^1 + N_t^2 + N_t^3$ represents the total number of requests for service of all of the three types that have occurred by time t. By Proposition 5.3.1 N_t has a Poisson distribution with parameter $\lambda_1 t + \lambda_2 t + \lambda_3 t = (\lambda_1 + \lambda_2 + \lambda_3)t$, and this is true for all times $t \geq 0$. It is possible to show from this that the axioms that define a Poisson process are satisfied by (N_t); hence we see that the superposition of independent Poisson processes is also a Poisson process with rate equal to the sum of the rates of the component processes.

Example 5.3.4

Suppose that scores on a mathematics placement exam are normally distributed with a mean of 18 and a variance of 4. Let $\overline{X} = (X_1 + \cdots + X_n)/100$ be the mean of a random sample of 100 of the scores. By Proposition 5.3.4 $\overline{X}$ is also normally distributed, with mean and variance

$$\mu_{\overline{X}} = \frac{1}{100}(18 + \cdots + 18) = 18,$$

$$\sigma_{\overline{X}}^2 = \frac{1}{100^2}(4 + \cdots + 4) = \frac{4}{100}.$$

Then the standard deviation of the sample mean is .2, which means that a very large percentage (about 95%) of the probability weight of the distribution of $\overline{X}$ is within two standard deviations or .4 units of the mean 18. The implication is that the sample mean is very likely to approximate the population mean closely.

Besides their valuable applications to sums of independent random variables, moment-generating functions are also connected to the moments of their distributions in a way that is sometimes helpful to exploit.

To see how this works, consider a generic m.g.f. $M(t) = E[e^{tX}]$, and suppose for a moment that it is proper to bring a derivative with respect to t inside the expectation. Then,

$$\frac{d}{dt}M(t) = E\left[\frac{d}{dt}e^{tX}\right] = E[Xe^{tX}] \tag{5.37}$$

$$\implies \frac{d}{dt}M(t)\Big|_{t=0} = E[Xe^{tX}]\Big|_{t=0} = E[X]. \tag{5.38}$$

We have shown that the first moment about 0 of the distribution is also equal to $M'(0)$. We can differentiate again to obtain

$$\frac{d^2}{dt^2}M(t) = E\left[\frac{d^2}{dt^2}e^{tX}\right] = E[X^2e^{tX}]$$

$$\implies \frac{d^2}{dt^2}M(t)\Big|_{t=0} = E[X^2e^{tX}]\Big|_{t=0} = E[X^2]. \tag{5.39}$$

It is clear that as long as it continues to be proper to interchange the derivative with the expectation, this pattern will continue. In general, the moments about 0 can be obtained from the formula

$$M^{(n)}(0) = E[X^n], \tag{5.40}$$

where $M^{(n)}(t)$ denotes the nth derivative of M with respect to t. This result explains the name *moment-generating function*.

?Question 5.3.3   How could you use the m.g.f. to compute the moments about the mean?

At the level of this book, we do not have the analytical tools to justify the exchange of the derivative and the expectation in all cases, though it is usually true. There is no difficulty justifying this in the simple case where X has finitely many states. Then the expectation $E[e^{tX}]$ is just a finite sum, and the exchange is equivalent to differentiation term by term, which is legal.

Example 5.3.5   Use the moment-generating function to calculate the mean and variance of the $\Gamma(\alpha, \lambda)$ distribution.

The gamma m.g.f. is $M(t) = \lambda^{\alpha}(\lambda - t)^{-\alpha}$ for $t < \lambda$, by (5.33). Its first derivative is

$$M'(t) = -\alpha\lambda^\alpha(\lambda - t)^{-\alpha-1}(-1) = \frac{\alpha\lambda^\alpha}{(\lambda - t)^{\alpha+1}}.$$

Evaluation at $t = 0$ gives

$$E[X] = M'(0) = \frac{\alpha\lambda^\alpha}{\lambda^{\alpha+1}} = \frac{\alpha}{\lambda},$$

which agrees with the previous result in Section 3.3.

Now we can find the variance by computing the second moment $E[X^2] = M''(0)$ and using the computational formula for variance. We have

$$M''(t) = \alpha(-\alpha - 1)\lambda^\alpha(\lambda - t)^{-\alpha-2}(-1) = \frac{\alpha(\alpha + 1)\lambda^\alpha}{(\lambda - t)^{\alpha+2}}$$

so that $E[X^2] = \alpha(\alpha + 1)/\lambda^2$. Thus

$$\mathrm{Var}(X) = E[X^2] - (E[X])^2 = \frac{\alpha(\alpha + 1)}{\lambda^2} - \frac{\alpha^2}{\lambda^2} = \frac{\alpha}{\lambda^2}.$$

This result also agrees with the formula derived in Section 3.3.

We can learn more about the multivariate normal distribution if we generalize the concept of moment-generating function to a random vector, which is done as follows.

DEFINITION 5.3.2

The *moment-generating function of a random vector* $\mathbf{X} = (X_1, \ldots, X_n)$ is the following real-valued function of a vector variable $\mathbf{t} = (t_1, \ldots, t_n)$:

$$M_{\mathbf{X}}(\mathbf{t}) = E[e^{\mathbf{t}'\cdot\mathbf{X}}] = E[\exp(\Sigma_{i=1}^n t_i X_i)], \qquad (5.41)$$

which is defined for all $\mathbf{t} \in \mathbb{R}^n$ such that the expectation is finite.

Example 5.3.6

Compute the moment-generating function of the multivariate normal distribution.

The desired m.g.f. is the multiple integral:

$$M(\mathbf{t}) = E[e^{\mathbf{t}'\cdot\mathbf{X}}]$$

$$= \int_{-\infty}^{+\infty} \cdots \int_{-\infty}^{+\infty} e^{\mathbf{t}'\cdot\mathbf{x}} \frac{1}{(2\pi)^{n/2}\sqrt{\det(\Sigma)}} e^{-1/2(\mathbf{x}-\boldsymbol{\mu})'\Sigma^{-1}(\mathbf{x}-\boldsymbol{\mu})} \, d\mathbf{x}.$$

When we regroup the integrand by combining the two exponential expressions, the new exponent is

$$-\frac{1}{2}[(\mathbf{x} - \boldsymbol{\mu})'\Sigma^{-1}(\mathbf{x} - \boldsymbol{\mu}) - 2\mathbf{t}'\mathbf{x}].$$

An old probabilist's trick (used frequently, but not exclusively by old probabilists) is to factor expressions out of the integrand in order to leave a valid density function inside, which then integrates to 1. In this particular case, a vector version of the technique of completing the square on the exponent yields the identity

$$-\frac{1}{2}[(\mathbf{x} - \boldsymbol{\mu})'\Sigma^{-1}(\mathbf{x} - \boldsymbol{\mu}) - 2\mathbf{t}'\mathbf{x}]$$

$$= -\frac{1}{2}[(\mathbf{x} - (\boldsymbol{\mu} + \Sigma\mathbf{t}))'\Sigma^{-1}(\mathbf{x} - (\boldsymbol{\mu} + \Sigma\mathbf{t}))] + \boldsymbol{\mu}'\mathbf{t} + \frac{1}{2}\mathbf{t}'\Sigma\mathbf{t}. \qquad (5.42)$$

You are asked to verify (5.42) in Exercise 16. [I did not obtain this identity by magical incantations. I actually looked at the one-dimensional case first, where the usual method of completing the square is straightforward. That gave me a clue to the proper form on the right side of (5.42).] Because of this identity, the expression $\exp[\boldsymbol{\mu}'\mathbf{t} + \mathbf{t}'\Sigma\mathbf{t}/2]$ factors out of the integral. The multivariate normal m.g.f. therefore simplifies to

$$M(\mathbf{t}) = e^{\boldsymbol{\mu}'\mathbf{t} + \frac{1}{2}\mathbf{t}'\Sigma\mathbf{t}}$$

$$\cdot \int_{-\infty}^{+\infty} \cdots \int_{-\infty}^{+\infty} \frac{1}{(2\pi)^{n/2}\sqrt{\det(\Sigma)}} e^{-1/2[(\mathbf{x}-(\boldsymbol{\mu}+\Sigma\mathbf{t}))'\Sigma^{-1}(\mathbf{x}-(\boldsymbol{\mu}+\Sigma\mathbf{t}))]} \, d\mathbf{x}.$$

Since the integrand is a multivariate normal density with mean vector $\boldsymbol{\mu} + \Sigma\mathbf{t}$ and covariance matrix Σ, the multiple integral equals 1. Hence

$$M(\mathbf{t}) = e^{\boldsymbol{\mu}'\mathbf{t} + \frac{1}{2}\mathbf{t}'\Sigma\mathbf{t}}, \qquad \mathbf{t} \in \mathbb{R}^n. \qquad (5.43)$$

Example 5.3.7

The mixed partials of the joint moment-generating function (5.41) of a random vector are useful in finding expected products of the components of the vector. To illustrate, let us use the m.g.f. to verify that the covariance of bivariate normal random variables X_1 and X_2 is $\rho\sigma_1\sigma_2$.

The form of the m.g.f. is $M(\mathbf{t}) = M(t_1, t_2) = E[e^{\mathbf{t}'\mathbf{X}}] = E[e^{t_1 X_1 + t_2 X_2}]$. The first-order partials of M with respect to t_1 and t_2, and the mixed second-order partial are as follows:

$$\frac{\partial M}{\partial t_1} = \frac{\partial}{\partial t_1} E[e^{t_1 X_1 + t_2 X_2}] = E[X_1 e^{t_1 X_1 + t_2 X_2}] \Rightarrow \left.\frac{\partial M}{\partial t_1}\right|_{(0,0)} = E[X_1], \qquad (5.44)$$

$$\frac{\partial M}{\partial t_2} = \frac{\partial}{\partial t_2} E[e^{t_1 X_1 + t_2 X_2}] = E[X_2 e^{t_1 X_1 + t_2 X_2}] \Rightarrow \left.\frac{\partial M}{\partial t_2}\right|_{(0,0)} = E[X_2], \qquad (5.45)$$

$$\frac{\partial^2 M}{\partial t_1 \partial t_2} = \frac{\partial}{\partial t_1}\frac{\partial}{\partial t_2} E[e^{t_1 X_1 + t_2 X_2}] = \frac{\partial}{\partial t_1} E[X_2 e^{t_1 X_1 + t_2 X_2}] = E[X_1 X_2 e^{t_1 X_1 + t_2 X_2}]$$

$$\Rightarrow \left.\frac{\partial^2 M}{\partial t_1 \partial t_2}\right|_{(0,0)} = E[X_1 X_2]. \qquad (5.46)$$

Equations (5.44)–(5.46) would apply to all bivariate random vectors, subject to the assumption that the exchange of derivative and expectation is valid. At this point we bring in formula (5.43) for the m.g.f. of the multivariate normal distribution. When $n = 2$, this formula reduces to

$$M(t_1, t_2) = \exp\left[\mu_1 t_1 + \mu_2 t_2 + \frac{1}{2}\sigma_1^2 t_1^2 + \rho\sigma_1\sigma_2 t_1 t_2 + \frac{1}{2}\sigma_2^2 t_2^2\right]. \qquad (5.47)$$

(Verify this result.) Then,

$$E[X_1] = \left.\frac{\partial M}{\partial t_1}\right|_{(0,0)} = (\mu_1 + \sigma_1^2 t_1 + \rho\sigma_1\sigma_2 t_2)\cdot M(t_1, t_2)\Big|_{(0,0)} = \mu_1,$$

$$E[X_2] = \left.\frac{\partial M}{\partial t_2}\right|_{(0,0)} = (\mu_2 + \sigma_2^2 t_2 + \rho\sigma_1\sigma_2 t_1)\cdot M(t_1, t_2)\Big|_{(0,0)} = \mu_2,$$

$$E[X_1 X_2] = \left.\frac{\partial^2 M}{\partial t_1 \partial t_2}\right|_{(0,0)}$$

$$= \begin{array}{c}(\rho\sigma_1\sigma_2 M(t_1, t_2) + (\mu_2 + \sigma_2^2 t_2 + \rho\sigma_1\sigma_2 t_1)) \\ \cdot(\mu_1 + \sigma_1^2 t_1 + \rho\sigma_1\sigma_2 t_2)\cdot M(t_1, t_2))\end{array}\Bigg|_{(0,0)}$$

$$= \rho\sigma_1\sigma_2 + \mu_1\mu_2.$$

(You should also check these computations.) Since $\text{Cov}(X_1, X_2) = E[X_1 X_2] - E[X_1]E[X_2]$, we obtain the desired formula $\rho\sigma_1\sigma_2$ for the covariance of the bivariate normal distribution.

—— **?Question 5.3.4**   What moments can be determined from the second-order partials $\partial^2 M/\partial t_1^2$, $\partial^2 M/\partial t_2^2$ for a general bivariate random vector? Can you generalize?

Before we close this section, a couple of remarks on other types of generating functions are appropriate. One such function is the *probability generating function*:

$$P(t) = E[t^X], \qquad (5.48)$$

which is particularly useful for discrete distributions on the nonnegative integers. In such a case it would take the form

$$P(t) = \sum_{k=0}^{\infty} t^k P[X = k].$$

Queueing theorists sometimes use the probability generating function to derive queue length probabilities and moments. Several of its properties are investigated in the exercises.

Another generating function is the *characteristic function*:

$$\phi(t) = E[e^{itX}], \qquad (5.49)$$

where i is the complex number $i = \sqrt{-1}$ and the complex exponential function is defined by

$$e^{iu} = \cos(u) + i\cdot\sin(u).$$

We will not dwell on the characteristic function here because we are not presupposing complex analysis. But you may know that the complex exponential function has similar algebraic and calculus properties to the real exponential function. In view of this fact, it is not surprising that the characteristic function can be used to do the same things as the moment-generating function. The characteristic function is actually better, however, in the sense that it exists for all real t and is bounded in magnitude by 1.

EXERCISES 5.3

1. Find the moment-generating function of the uniform distribution on (a, b).

2. Show that the m.g.f. of the $b(n, p)$ distribution is $M(t) = ((1 - p) + pe^t)^n$, $t \in \mathbb{R}$.

3. Find the m.g.f. of the distribution with density function $f(x) = 3x^2$, $x \in (0, 1)$.

4. Show by direct computation, rather than by appealing to (5.43), that the moment-generating function of the univariate normal distribution with parameters μ and σ^2 is $M(t) = \exp[\mu t + \sigma^2 t^2 / 2]$, $t \in \mathbb{R}$. [*Hint:* Start with the $N(0, 1)$ m.g.f., then express the general normal m.g.f. in terms of it.]

5. Show by direct computation that the moment-generating function of the exponential(λ) distribution is $M(t) = \lambda/(\lambda - t)$, $t < \lambda$.

6. Express the m.g.f. of the random variable $Y = cX + d$ in terms of the m.g.f. of X. Assume that c and d are nonzero constants. Use your result to show that $E[cX + d] = cE[X] + d$.

7. Use moment-generating functions to show that if X is $N(\mu, \sigma^2)$, then $cX + d$ has the $N(c\mu + d, c^2\sigma^2)$ distribution.

8. Prove Proposition 5.3.4.

9. a. Find the moment-generating function of the discrete distribution that gives probabilities 1/12, 1/6, 1/4, 1/4, 1/6, 1/12, respectively, to the states $-2, -1, 0, 1, 2, 3$.

   b. Similarly, find the moment-generating function of the discrete distribution that gives probabilities 1/6, 1/6, 1/6, 1/3, 1/12, 1/12, respectively, to the states 1, 2, 3, 4, 5, 6.

   c. Devise a general formula for the moment-generating function of a discrete distribution that gives probabilities $p_1, p_2, \ldots, p_n$, respectively, to the states $x_1, x_2, \ldots, x_n$.

   d. Use (a)–(c) to find the distribution of $X + Y$, where X has the distribution in (a), Y has the distribution in (b), and X and Y are independent.

10. a. Find the m.g.f. of the geometric distribution with parameter p.

    b. Find the m.g.f. of the negative binomial distribution with parameters 2 and p.

    c. Use (a) and (b) to show that the distribution of the number of Bernoulli trials required to reach the second success is the negative binomial distribution with parameters 2 and p.

11. Use moment-generating functions to show that if $X_1, X_2, \ldots, X_n$ are independent Bernoulli(p) random variables, then $X_1 + X_2 + \cdots + X_n$ has the $b(n, p)$ distribution. (See Exercise 2.)

12. Show that if X_1 has the $b(n_1, p)$ distribution, X_2 has the $b(n_2, p)$ distribution, ..., X_m has the $b(n_m, p)$ distribution, and $X_1, X_2, \ldots, X_m$ are mutually independent, then $X_1 + X_2 + \cdots + X_m$ has the $b(n_1 + n_2 + \cdots + n_m, p)$ distribution. Interpret this result in terms of Bernoulli trials experiments.

13. Use m.g.f.'s to verify the formulas for the mean and variance of the $b(n, p)$ distribution. (See Exercise 2.)

14. Use m.g.f.'s to verify the formulas for the mean and variance of the $N(\mu, \sigma^2)$ distribution. (See Exercise 4.)

15. Use m.g.f.'s to verify the formulas for the mean and variance of the Poisson(μ) distribution.

16. Verify formula (5.42).

17. Let the random variables X and Y have the bivariate normal distribution with parameters $\mu_X, \mu_Y, \sigma_X^2, \sigma_Y^2, \rho$. Use the joint moment-generating function to find $E[Y^3]$.

18. Express the m.g.f. of the random vector $A\mathbf{X} + \mathbf{b}$ in terms of the m.g.f. of $\mathbf{X}$, where $\mathbf{X}$ is an n-dimensional random vector, A is an $n \times n$ matrix, and $\mathbf{b}$ is an $n \times 1$ constant vector.

19. Let a function $\Psi(t)$ be defined by $\Psi(t) = \ln(M(t))$, where M is the moment-generating function of a distribution. What do $\Psi'(0)$ and $\Psi''(0)$ represent?

20. Let $Y = X_1 + X_2 + \cdots + X_n$ be a sum of independent random variables. Express the probability generating function of Y in terms of those of the X_i's.

21. a. Find the probability generating function of the Poisson(μ) distribution.

    b. Use the probability generating function to show that the sum of independent Poisson random variables is Poisson.

22. Let $P(t)$ be the probability generating function of a discrete random variable with state space $E = \{0, 1, 2, \ldots\}$. Find (a) $P^{(n)}(0)$; (b) $P^{(n)}(1)$.

5.4 Transformations of Normal Random Variables

5.4.1 Basic Results

Many data analysis problems are concerned with inference about random samples $X_1, X_2, \ldots, X_n$ from distributions that are either exactly or approximately normal with some mean μ and some variance σ^2. We have mentioned before that the central tendency of the distribution, measured by μ, can be estimated from the data by the *sample mean*

$$\bar{X} = \sum_{i=1}^{n} \frac{X_i}{n}. \tag{5.50}$$

Similarly the spread of the distribution, measured by σ^2, can be estimated by the average squared distance of data points from $\bar{X}$, called the *sample variance*:

$$S^2 = \sum_{i=1}^{n} \frac{(X_i - \bar{X})^2}{n-1}. \tag{5.51}$$

(The reason for the division by $n-1$ instead of n will be discussed in a later chapter.) How reliable are these estimates? How likely are they to be close to the underlying parameters μ and σ^2? To answer these questions, we must find the probability distributions of these particular transformations of normal random variables. The sample mean is a linear transformation of independent normal random variables, and the sample variance is a combination of squares and first powers of normals. Thus we will focus on linear and quadratic transformations in this section.

Let us review what is known already. The standardization theorem, Proposition 3.3.6, says that

$$X \sim N(\mu, \sigma^2) \Rightarrow Z = \frac{X - \mu}{\sigma} \sim N(0, 1). \tag{5.52}$$

We are using the symbol '$\sim$' to stand for the phrase *is distributed as*. This is one special kind of linear transformation of a single normal random variable, whose distribution was found using the c.d.f. technique. It is also easy (Exercise 1) to use the c.d.f. method to derive the following more general result.

PROPOSITION 5.4.1

If X has the $N(\mu, \sigma^2)$ distribution, then the random variable $Y = cX + d$ has the $N(c\mu + d, c^2\sigma^2)$ distribution.

We also had Proposition 5.3.4 on linear combinations of independent normal random variables:

$$X_1, \ldots, X_n \text{ independent } N(\mu_i, \sigma_i^2) \Rightarrow Y = \sum_{i=1}^{n} c_i X_i \sim N\left(\sum_{i=1}^{n} c_i \mu_i, \sum_{i=1}^{n} c_i^2 \sigma_i^2\right). \tag{5.53}$$

If $X_1, \ldots, X_n$ is a random sample, so that all of the means $\mu_i = \mu$ and all of the variances $\sigma_i^2 = \sigma^2$, we can set all of the coefficients $c_i = 1/n$ to obtain the following result about the distribution of the sample mean.

PROPOSITION 5.4.2

Let $X_1, \ldots, X_n$ be a random sample from the $N(\mu, \sigma^2)$ distribution. Then

$$\overline{X} \sim N\left(\mu, \frac{\sigma^2}{n}\right); \tag{5.54}$$

hence

$$Z = \frac{\overline{X} - \mu}{\sigma/\sqrt{n}} \sim N(0,1). \tag{5.55}$$

Some authors refer to the distribution in (5.54) as the *sampling distribution of the mean*, and they use the phrase *standard error of the mean* for its standard deviation $\sigma/\sqrt{n}$. The latter tells how much variability the sample mean has, which is important to know in order to use it with confidence as an estimator of the true mean.

Example 5.4.1

Suppose that past evidence suggests that the calorie count in a cup of a brand of low-fat yogurt is normally distributed with mean 150 calories and standard deviation 10 calories. An improved formulation of the yogurt has been concocted, for which the standard deviation has not changed but the mean may have been reduced. If 100 cups are measured and the sample average calorie content is 145, is there strong evidence to believe that the true mean calorie content has been reduced?

We can quantify the question by asking how likely is the event that the sample mean is 145 or less if the true mean is still 150. A small probability would indicate strong evidence that 150 is no longer the mean number of calories. It is easy to compute this probability by standardizing $\overline{X}$ as follows:

$$P[\overline{X} \leq 145] = P\left[Z = \frac{\overline{X} - \mu}{\sigma/\sqrt{n}} \leq \frac{145 - 150}{10/\sqrt{100}}\right] = P[Z \leq -5] \approx 0.$$

Note that since its standard error is $10/\sqrt{100} = 1$, $\overline{X}$ does not have too much variability. The observed deviation of five standard errors from the old mean of 150 is a very rare event, which is indeed strong evidence that the population mean has been reduced.

Let us now move to simple quadratic transformations of normal random variables. The next proposition will form our basis.

PROPOSITION 5.4.3

If Z has the $N(0,1)$ distribution, then the transformed random variable $Y = Z^2$ has the $\chi^2(1)$ distribution. Hence, if X is $N(\mu, \sigma^2)$, then

$$Y = \left(\frac{X - \mu}{\sigma}\right)^2 \sim \chi^2(1). \tag{5.56}$$

■ **Proof**   The c.d.f. of Y is

$$G_Y(y) = P[Y \leq y] = P[Z^2 \leq y]$$
$$= P[-\sqrt{y} \leq Z \leq \sqrt{y}]$$
$$= 2 \cdot P[0 \leq Z \leq \sqrt{y}], \qquad y > 0.$$

The final equation follows from the symmetry of the $N(0,1)$ density about 0. Therefore the density of Y is

$$g_Y(y) = \frac{d}{dy} G_Y(y) = \frac{d}{dy} \left(2 \cdot \int_0^{\sqrt{y}} \frac{1}{\sqrt{2\pi}} e^{-z^2/2} \, dz \right)$$

$$= 2 \cdot \frac{1}{2} y^{-1/2} \cdot \frac{1}{\sqrt{2\pi}} e^{-(\sqrt{y})^2/2}$$

$$= \frac{1}{2^{1/2}\Gamma(1/2)} y^{1/2-1} e^{-y/2}, \qquad y > 0.$$

In the last line, we have used the fact that $\Gamma(1/2) = \sqrt{\pi}$. The density of Y is of the $\Gamma(1/2, 1/2)$ form, which is the same as the $\chi^2(1)$ density; hence the first statement is proved. Statement (5.56) follows from the fact that the random variable in parentheses has the standard normal distribution.

Thus the χ^2 distribution has arisen in a significant way for the first time: It is the distribution of the square of a standard normal random variable. Next is a theorem about several squares of normal random variables.

PROPOSITION 5.4.4

Let $Z_1, Z_2, \ldots, Z_n$ be independent and identically distributed $N(0,1)$ random variables. Then

$$Y = Z_1^2 + Z_2^2 + \cdots + Z_n^2 \sim \chi^2(n). \tag{5.57}$$

In particular, if $X_1, \ldots, X_n$ are normal random variables with means $\mu_1, \ldots, \mu_n$ and variances $\sigma_1^2, \ldots, \sigma_n^2$, respectively, then

$$Y = \left(\frac{X_1 - \mu_1}{\sigma_1} \right)^2 + \cdots + \left(\frac{X_n - \mu_n}{\sigma_n} \right)^2 \sim \chi^2(n). \tag{5.58}$$

—— **?Question 5.4.1**   How does Proposition 5.4.4 follow from previous results?

The special case of formula (5.58) in which $X_1, X_2, \ldots, X_n$ is a random sample from the $N(\mu, \sigma^2)$ distribution states that the random variable

$$\sum_{i=1}^n \frac{(X_i - \mu)^2}{\sigma^2} \tag{5.59}$$

has the $\chi^2(n)$ distribution. Notice how closely the form of this random variable resembles the form of a simple function of the sample variance:

$$\frac{(n-1)S^2}{\sigma^2} = \frac{(n-1)}{\sigma^2} \cdot \frac{1}{(n-1)} \sum_{i=1}^n (X_i - \bar{X})^2 = \sum_{i=1}^n \frac{(X_i - \bar{X})^2}{\sigma^2}. \tag{5.60}$$

The only difference between the random variable in (5.59) and the one in (5.60) is that the true mean μ is estimated by the sample mean $\bar{X}$ in the latter formula. This observation gives us cause to hope that $(n-1)S^2/\sigma^2$ has a χ^2

distribution. It turns out that the estimation of μ by $\bar{X}$ results in a loss of one degree of freedom in the χ^2 parameter, reducing the parameter to $n - 1$. Quite surprisingly, the sample variance also turns out to be independent of the sample mean. We will show these results in the next subsection.

PROPOSITION 5.4.5

Let $X_1, X_2, \ldots, X_n$ be a random sample from the $N(\mu, \sigma^2)$ distribution. Then the random variable

$$\frac{(n-1)S^2}{\sigma^2} = \sum_{i=1}^{n} \frac{(X_i - \bar{X})^2}{\sigma^2} \tag{5.61}$$

has the $\chi^2(n-1)$ distribution and furthermore is independent of $\bar{X}$.

Example 5.4.2

A cutting machine saws two-by-four wood boards to a length X that is normally distributed with mean 8 feet. What is the probability that, among 16 such boards, the sample variance of the board lengths is at least twice the population variance?

By Proposition 5.4.5, the random variable $Y = (n-1)S^2/\sigma^2 = 15S^2/\sigma^2$ has the $\chi^2(15)$ distribution. The probability that is called for in the problem statement is

$$P[S^2 \geq 2\sigma^2] = P\left[\frac{S^2}{\sigma^2} \geq 2\right] = P\left[\frac{(n-1)S^2}{\sigma^2} \geq 2 \cdot 15\right] = P[Y \geq 30].$$

Using the χ^2 table in Appendix A or a statistical program, you can find that the probability of the last event is about .01.

5.4.2 Multivariate Transformations of Normal Random Vectors

Two reasons underline our need to carry the development further. First, we would like a proof of Proposition 5.4.5. Second, there is an important topic in statistics that relies heavily on stronger theorems than we have at present: The design and analysis of experiments, whose purpose is to gain information about the influence of one or more factors on the mean of a normally distributed response variable. For these reasons, we will now extend our study to linear and quadratic functions of multivariate normal data.

First we have a theorem that generalizes Proposition 5.4.1: Linear transformations of multivariate normal random vectors produce other multivariate normal random vectors. The proof given here uses the multivariate change of variable techniques of Section 5.2. In Exercise 12, you will be asked to give a proof using the joint moment-generating function of the multivariate normal distribution.

PROPOSITION 5.4.6

Let A be an $m \times n$ matrix of row rank m, where $m \leq n$, and let $\mathbf{b}$ be an $m \times 1$ column vector. If $\mathbf{X} = [X_1 \ X_2 \ \dots \ X_n]'$ is a multivariate normal random vector with mean vector $\boldsymbol{\mu}_X = [\mu_1 \ \mu_2 \ \dots \ \mu_n]'$ and covariance matrix Σ_X, then the random vector

$$\mathbf{Y} = A\mathbf{X} + \mathbf{b}$$

has the multivariate normal distribution with mean vector $\boldsymbol{\mu}_Y = A\boldsymbol{\mu}_X + \mathbf{b}$ and covariance matrix $\Sigma_Y = A\Sigma_X A'$.

■ **Proof** We will first reduce to the case where A is an $n \times n$ invertible matrix. If the number of rows m of A is strictly less than n, then since A has row rank m, its rows comprise m linearly independent vectors in $\mathbb{R}^n$. These rows may be supplemented by an additional $n - m$ vectors in $\mathbb{R}^n$ to form a basis. Let the new vectors be the rows of an $(n - m) \times n$ matrix B, which we can adjoin to the bottom of A to form an $n \times n$ matrix:

$$C = \begin{bmatrix} A \\ \cdots \\ B \end{bmatrix}.$$

Since the rows of C are linearly independent, C is an invertible matrix. Now adjoin $(n - m)$ 0's to the bottom of the constant vector $\mathbf{b}$ to obtain a new constant vector $\mathbf{c}$ and a new transformation:

$$\mathbf{W} = C\mathbf{X} + \mathbf{c} = \begin{bmatrix} A \\ \cdots \\ B \end{bmatrix} \begin{bmatrix} X_1 \\ \vdots \\ X_n \end{bmatrix} + \begin{bmatrix} \mathbf{b} \\ \cdots \\ \mathbf{0} \end{bmatrix} = \begin{bmatrix} A\mathbf{X} + \mathbf{b} \\ \cdots \\ B\mathbf{X} \end{bmatrix}.$$

Since the joint marginal distributions of multivariate normal vectors are also multivariate normal [see Proposition 4.4.3(a)], if we can show that the entire vector $\mathbf{W}$ is multivariate normal, then its first block $A\mathbf{X} + \mathbf{b}$ is also multivariate normal. Thus it suffices to prove the proposition in the case that the coefficient matrix is an invertible $n \times n$ matrix. From this point, we return to the original notation and abandon $\mathbf{W}$, C, and $\mathbf{c}$.

The inverse transformation of $\mathbf{Y} = A\mathbf{X} + \mathbf{b}$ is $\mathbf{X} = A^{-1}(\mathbf{Y} - \mathbf{b})$. You can check (see Question 5.4.2) that the Jacobian is just A^{-1}. The determinant of A^{-1} is the same as $(\det A)^{-1}$. By formula (5.23), the density of $\mathbf{Y} = A\mathbf{X} + \mathbf{b}$ is therefore

$$f_Y(\mathbf{y}) = f_X(A^{-1}(\mathbf{y} - \mathbf{b})) \cdot |\det A|^{-1},$$

and since $\mathbf{X}$ is multivariate normal, this density is

$$f_Y(\mathbf{y}) = \frac{1}{(2\pi)^{n/2}\sqrt{\det \Sigma}} \cdot \frac{1}{|\det A|}$$
$$\cdot \exp\left[-\frac{1}{2}(A^{-1}(\mathbf{y} - \mathbf{b}) - \boldsymbol{\mu}_X)'\Sigma^{-1}(A^{-1}(\mathbf{y} - \mathbf{b}) - \boldsymbol{\mu}_X)\right]$$

$$= \frac{1}{(2\pi)^{n/2}\sqrt{(\det \Sigma)(\det A)(\det A)}}$$

$$\cdot \exp\left[-\frac{1}{2}(A^{-1}(\mathbf{y} - (A\boldsymbol{\mu}_X + \mathbf{b})))'\Sigma^{-1}(A^{-1}(\mathbf{y} - (A\boldsymbol{\mu}_X + \mathbf{b})))\right]$$

$$= \frac{1}{(2\pi)^{n/2}\sqrt{(\det A)(\det \Sigma)(\det A')}}$$
$$\cdot \exp\left[-\frac{1}{2}(\mathbf{y} - (A\boldsymbol{\mu}_X + \mathbf{b}))'(A^{-1})'\Sigma^{-1}A^{-1}(\mathbf{y} - (A\boldsymbol{\mu}_X + \mathbf{b}))\right].$$

Since $(\det A)(\det \Sigma)(\det A') = \det(A\Sigma A')$ and $(A^{-1})'\Sigma^{-1}A^{-1} = (A\Sigma A')^{-1}$, we have reduced the density of $\mathbf{Y}$ to the bivariate normal form, and in the process we have obtained the correct formulas for the mean vector and covariance matrix.

?Question 5.4.2   Confirm that the Jacobian of $\mathbf{Y} = A\mathbf{X} + \mathbf{b}$, $\mathbf{X} = A^{-1}(\mathbf{Y} - \mathbf{b})$ is A^{-1}.

Example 5.4.3   Let X_1, X_2, X_3 be i.i.d. $N(\mu, \sigma^2)$ random variables, so that the vector $\mathbf{X} = [X_1\ X_2\ X_3]'$ is multivariate normal with mean vector $\boldsymbol{\mu}_X = [\mu\ \mu\ \mu]'$ and covariance matrix $\sigma^2 I$, where I is a 3×3 identity matrix. Let a constant matrix A be defined by

$$A = \begin{bmatrix} 1/3 & 1/3 & 1/3 \\ -1 & 0 & 1 \\ 0 & 1 & -1 \end{bmatrix}.$$

Then the transformed vector $\mathbf{Y} = A\mathbf{X}$ has the form

$$\mathbf{Y} = \begin{bmatrix} Y_1 \\ Y_2 \\ Y_3 \end{bmatrix} = A\mathbf{X} = \begin{bmatrix} 1/3 & 1/3 & 1/3 \\ -1 & 0 & 1 \\ 0 & 1 & -1 \end{bmatrix}\begin{bmatrix} X_1 \\ X_2 \\ X_3 \end{bmatrix} = \begin{bmatrix} (X_1 + X_2 + X_3)/3 \\ -X_1 + X_3 \\ X_2 - X_3 \end{bmatrix}.$$

By Proposition 5.4.6, $\mathbf{Y}$ is also multivariate normal, with mean vector

$$\boldsymbol{\mu}_Y = A\boldsymbol{\mu}_X = \begin{bmatrix} 1/3 & 1/3 & 1/3 \\ -1 & 0 & 1 \\ 0 & 1 & -1 \end{bmatrix}\begin{bmatrix} \mu \\ \mu \\ \mu \end{bmatrix} = \begin{bmatrix} \mu \\ 0 \\ 0 \end{bmatrix}$$

and covariance matrix

$$\Sigma_Y = A\Sigma_X A' = \begin{bmatrix} 1/3 & 1/3 & 1/3 \\ -1 & 0 & 1 \\ 0 & 1 & -1 \end{bmatrix}\begin{bmatrix} \sigma^2 & 0 & 0 \\ 0 & \sigma^2 & 0 \\ 0 & 0 & \sigma^2 \end{bmatrix}\begin{bmatrix} 1/3 & -1 & 0 \\ 1/3 & 0 & 1 \\ 1/3 & 1 & -1 \end{bmatrix}$$

$$= \begin{bmatrix} \sigma^2/3 & 0 & 0 \\ 0 & 2\sigma^2 & -\sigma^2 \\ 0 & -\sigma^2 & 2\sigma^2 \end{bmatrix}.$$

From this mean and covariance we can make the following observations. The random variable $Y_1 = (X_1 + X_2 + X_3)/3 = \bar{X}$ is marginally normal, with mean μ and variance $\sigma^2/3$, which confirms our earlier work. The random variables $Y_2 = -X_1 + X_3$ and $Y_3 = X_2 - X_3$ have the $N(0, 2\sigma^2)$ distribution, which should also come as no surprise (Why?). Since the 2–3 element of the covariance matrix of $\mathbf{Y}$ is negative, the random variables Y_2 and Y_3 are negatively correlated. Note, however, that the 1–2 and 1–3 elements of the new covariance matrix are 0, so that $Y_1 = \bar{X}$ is correlated neither with Y_2 nor with Y_3. Since $\mathbf{Y}$ is multivariate normal, the lack of correlation actually implies that Y_1 is independent of both Y_2 and Y_3, even though their functional definitions are based on the same X_i's.

?Question 5.4.3   What is the correlation between Y_2 and Y_3 in Example 5.4.3?

To generalize what we did in Section 5.4.1 with quadratic transformations, we now study the distribution of quadratic forms in multivariate normal random vectors, that is, transformed random vectors of the form

$$\mathbf{Y} = \mathbf{X}'Q\mathbf{X}, \tag{5.62}$$

where Q is a symmetric matrix and $\mathbf{X} = [X_1 \ X_2 \ \ldots \ X_n]'$ has the multivariate normal distribution with mean vector $\boldsymbol{\mu}$ and covariance matrix Σ.

To show why the random variable in (5.62) is useful for the the problem of finding the distribution of $(n-1)S^2/\sigma^2$, consider a random sample $X_1, X_2, \ldots, X_n$ from the $N(\mu, \sigma^2)$ distribution. You can easily check the identity

$$\sum_{i=1}^{n} X_i^2 = \sum_{i=1}^{n}(X_i - \bar{X})^2 + n\bar{X}^2 \Rightarrow \frac{\sum_{i=1}^{n} X_i^2}{\sigma^2} = \frac{(n-1)S^2}{\sigma^2} + \frac{n\bar{X}^2}{\sigma^2}. \tag{5.63}$$

This is how quadratic forms fit into the picture. Let $\mathbf{X} = [X_1 \ X_2 \ \ldots \ X_n]'$ be the column vector of sample variables. The three terms in (5.63) can be written as quadratic forms as follows:

$$\sum_{i=1}^{n} X_i^2 = \mathbf{X}' \begin{bmatrix} 1 & 0 & \ldots & 0 \\ 0 & 1 & \ldots & 0 \\ & & \ddots & \\ 0 & 0 & \ldots & 1 \end{bmatrix} \mathbf{X} = \mathbf{X}'I_n\mathbf{X},$$

$$n\bar{X}^2 = \mathbf{X}' \begin{bmatrix} 1/n & 1/n & \ldots & 1/n \\ 1/n & 1/n & \ldots & 1/n \\ & & \ddots & \\ 1/n & 1/n & \ldots & 1/n \end{bmatrix} \mathbf{X} = \mathbf{X}'Q_1\mathbf{X},$$

$$\sum_{i=1}^{n}(X_i - \bar{X})^2 = \mathbf{X}'Q_2\mathbf{X} \quad \text{where } Q_2 = I_n - Q_1. \tag{5.64}$$

You are asked to verify these expressions in this section's Exercise 17. Our task is therefore to prove that the quadratic form $\mathbf{X}'Q_2\mathbf{X}/\sigma^2 \sim \chi^2(n-1)$ and also that it is independent of the quadratic form $\mathbf{X}'Q_1\mathbf{X}$. The next propositions will show these results.

In Appendix E, on linear algebra, you can review the following facts. A symmetric matrix Q is called *idempotent* if $Q^2 = Q$. An idempotent matrix Q can have eigenvalues only equal to 1 or 0, and the number of eigenvalues equal to 1 is the rank r of the matrix. Such a matrix can then be converted to a diagonal matrix with the eigenvalues on the diagonal as follows:

$$N'QN = \begin{bmatrix} I_r & 0 \\ 0 & 0 \end{bmatrix} = A \Longrightarrow Q = NAN', \tag{5.65}$$

where N is an orthogonal matrix $(NN' = N'N = I_n)$ whose columns are the eigenvectors of Q.

PROPOSITION 5.4.7

Suppose that $\mathbf{X} = [X_1\ X_2\ \ldots\ X_n]'$ is a random vector that has the multivariate normal distribution with mean vector $\boldsymbol{\mu} = [\mu\ \mu \ldots \mu]'$ and covariance matrix $\sigma^2 \cdot I_n$. Let Q be an $n \times n$ symmetric, idempotent matrix of rank r. If either $\boldsymbol{\mu} = \mathbf{0}$ or all row sums of Q are 0, then the random variable $Y = \mathbf{X}'Q\mathbf{X}/\sigma^2$ has the $\chi^2(r)$ distribution.

■ **Proof** We will show that the moment-generating function of Y agrees with the moment-generating function of the $\chi^2(r)$ distribution. Because $\boldsymbol{\mu}$ has identical entries, in either the case that the row sums of Q are zero or $\boldsymbol{\mu} = \mathbf{0}$, we have

$$\mathbf{X}'Q\boldsymbol{\mu} = \boldsymbol{\mu}'Q\mathbf{X} = 0 \quad \text{and} \quad \boldsymbol{\mu}'Q\boldsymbol{\mu} = 0$$
$$\Rightarrow \mathbf{X}'Q\mathbf{X} = (\mathbf{X} - \boldsymbol{\mu})'Q(\mathbf{X} - \boldsymbol{\mu}).$$

Because the two quadratic forms that are equated in the last line are the same, their moment-generating functions are the same; hence we may as well proceed as if we had a random vector $\mathbf{Z} = \mathbf{X} - \boldsymbol{\mu}$ with mean vector $\mathbf{0}$. By the definition of the m.g.f.,

$$M_Y(t) = E[e^{t\mathbf{Z}'Q\mathbf{Z}/\sigma^2}]$$
$$= \int_{-\infty}^{+\infty} \int_{-\infty}^{+\infty} \cdots \int_{-\infty}^{+\infty} e^{t\mathbf{Z}'Q\mathbf{z}/\sigma^2} \frac{1}{(2\pi\sigma^2)^{n/2}} \exp\left(-\frac{1}{2\sigma^2}\mathbf{z}'I_n\mathbf{z}\right)\ d\mathbf{z}$$
$$= \int_{-\infty}^{+\infty} \int_{-\infty}^{+\infty} \cdots \int_{-\infty}^{+\infty} \frac{1}{(2\pi\sigma^2)^{n/2}} \exp\left[-\frac{1}{2\sigma^2}\mathbf{z}'(I_n - 2tQ)\mathbf{z}\right]\ d\mathbf{z}.$$
$$\tag{5.66}$$

In the last integral, multiply and divide by the quantity $\sqrt{|\det(I_n - 2tQ)^{-1}|}$ $= |\det(I_n - 2tQ)|^{-1/2}$. Then

$$M_Y(t) = |\det(I_n - 2tQ)|^{-1/2} \int_{-\infty}^{+\infty} \int_{-\infty}^{+\infty} \cdots$$

$$\cdots \int_{-\infty}^{+\infty} \frac{1}{(2\pi)^{n/2}\sqrt{(\sigma^2)^n|\det(I_n - 2tQ)^{-1}|}} \exp\left[-\frac{1}{2\sigma^2}\mathbf{z}'(I_n - 2tQ)\mathbf{z}\right]\ d\mathbf{z}.$$

The integrand is now a multivariate normal density with mean vector $\mathbf{0}$ and covariance matrix $\sigma^2(I_n - 2tQ)^{-1}$ (which exists if t is small enough).

Therefore the integral reduces to 1, and we have the following formula for the m.g.f. of Y (which is true even when Q is not idempotent):

$$M_Y(t) = E[e^{t\mathbf{Z}'Q\mathbf{Z}/\sigma^2}] = |\det(I_n - 2tQ)|^{-1/2}. \tag{5.67}$$

Now, since Q is idempotent, matrices N and A that satisfy the relations in (5.65) exist. Since $NN' = NI_nN' = I_n$, we can write the m.g.f. of Y as

$$
\begin{aligned}
M_Y(t) &= |\det(I_n - 2tQ)|^{-1/2} \\
&= |\det(NI_nN' - 2tNAN')|^{-1/2} \\
&= |\det(N(I_n - 2tA)N')|^{-1/2} \\
&= |\det(N)|^{-1/2}|\det(I_n - 2tA)|^{-1/2}|\det(N')|^{-1/2} \\
&= |\det(I_n - 2tA)|^{-1/2} = (1 - 2t)^{-r/2}.
\end{aligned}
\tag{5.68}
$$

In the fourth line of (5.68) we use the fact that the determinant of a product is the product of the determinants. In the fifth line we observe that since N is orthogonal, the determinant of its transpose is the reciprocal of its determinant. You should verify the last equation yourself by answering Question 5.4.4. Since we have now put the m.g.f. of Y into the form of the $\chi^2(r)$ m.g.f., the proof is complete.

—— **?Question 5.4.4**   Verify that $\det(I_n - 2tA) = (1 - 2t)^r$ in formula (5.68) of the proof of Proposition 5.4.7.

Look again at the quadratic forms in (5.64). The matrix Q_1 is idempotent, since each row $[1/n \ \cdots \ 1/n]$ times each column $[1/n \ \cdots \ 1/n]'$ gives n copies of $1/n^2$ added together, which is again $1/n$. But then Q_2 is also idempotent, because

$$Q_2^2 = (I_n - Q_1)(I_n - Q_1) = I_n - 2Q_1 + Q_1^2 = I_n - 2Q_1 + Q_1 = I_n - Q_1 = Q_2.$$

The rank of the matrix Q_2 is easily seen to be $n - 1$ (because if you add its rows you get 0, but there are no other dependencies). Also, in each row of Q_2 the sum of the entries is 1 minus n copies of $1/n$, which equals 0. By Proposition 5.4.7, we obtain the anticipated result:

$$\frac{(n-1)S^2}{\sigma^2} = \frac{\mathbf{X}'Q_2\mathbf{X}}{\sigma^2} \sim \chi^2(n-1). \tag{5.69}$$

Our final goal is to show that $\bar{X}$ and $(n-1)S^2/\sigma^2$ are independent. This fact rests on the following theorem.

PROPOSITION 5.4.8

Suppose that $\mathbf{X} = [X_1 \ X_2 \ \ldots \ X_n]'$ is a random vector that has the multivariate normal distribution with mean vector $\boldsymbol{\mu} = [\mu \cdots \mu]'$ and covariance matrix $\sigma^2 I_n$. Let Q_1 and Q_2 be symmetric $n \times n$ matrices such that $Q_1Q_2 = 0$. If either $\boldsymbol{\mu} = \mathbf{0}$ or the row sums of both Q_1 and Q_2 are all zero, then the random variables $Y_1 = \mathbf{X}'Q_1\mathbf{X}/\sigma^2$ and $Y_2 = \mathbf{X}'Q_2\mathbf{X}/\sigma^2$ are independent.

■ **Proof**   We will show that the joint moment-generating function of Y_1 and Y_2 factors into the product of the two marginal m.g.f.'s, which is sufficient for independence. By the same derivation as the one in Proposition 5.4.7 that led to formula (5.67), this joint m.g.f. is

$$
\begin{aligned}
M(t_1, t_2) = E[e^{t_1 Y_1 + t_2 Y_2}] &= E[\exp(t_1 \mathbf{X}'Q_1\mathbf{X}/\sigma^2 + t_2 \mathbf{X}'Q_2\mathbf{X}/\sigma^2)] \\
&= E[\exp(\mathbf{X}'(t_1 Q_1 + t_2 Q_2)\mathbf{X}/\sigma^2)] \\
&= |\det(I_n - 2(t_1 Q_1 + t_2 Q_2))|^{-1/2}.
\end{aligned}
\tag{5.70}
$$

The hypothesis that $Q_1 Q_2 = 0$ allows us to write

$$
\begin{aligned}
I_n - 2(t_1 Q_1 + t_2 Q_2) &= I_n - 2(t_1 Q_1 - 2t_1 t_2 Q_1 Q_2 + t_2 Q_2) \\
&= (I_n - 2t_1 Q_1)(I_n - 2t_2 Q_2).
\end{aligned}
$$

Thus the determinant in (5.70) factors:

$$
M(t_1, t_2) = |\det(I_n - 2t_1 Q_1)|^{-1/2} |\det(I_n - 2t_2 Q_2)|^{-1/2}.
$$

By (5.67) again, the two factors are the marginal m.g.f.'s $M_{Y_1}(t_1)$ and $M_{Y_2}(t_2)$, which finishes the proof. (You should note that we did not assume that the matrices were idempotent here. Independence of the forms follows strictly from the zero product hypothesis. The fact that each has a χ^2 distribution follows strictly from idempotency.)

Returning to the question of the independence of $\bar{X}$ and $(n-1)S^2/\sigma^2$, note that if we subtract the population mean μ from every sample variable to form new random variables $Z_i = X_i - \mu$, then the sample mean of the Z sample is $\bar{Z} = \bar{X} - \mu$, while the sample variance of the Z sample is the same as that of the X sample. To show the independence we want, it is enough therefore to show the independence of the sample mean $\bar{Z}$ from $(n-1)S^2/\sigma^2$. This means that we lose no generality in supposing that the X sample has population mean $\mu = 0$, as the Z sample does.

This means that we can apply Proposition 5.4.8 to the quadratic forms $\mathbf{X}'Q_1\mathbf{X}/\sigma^2 = n\bar{X}^2/\sigma^2$ and $\mathbf{X}'Q_2\mathbf{X}/\sigma^2 = (n-1)S^2/\sigma^2$. Since Q_1 is idempotent,

$$
Q_1 Q_2 = Q_1(I_n - Q_1) = Q_1 - Q_1^2 = Q_1 - Q_1 = 0.
$$

Therefore the two quadratic forms, hence $\bar{X}$ and $(n-1)S^2/\sigma^2$ as well, are independent.

EXERCISES 5.4

1. Use the c.d.f. technique to prove Proposition 5.4.1.
2. A point $\mathbf{X} = (X, Y)$ is selected in the plane in such a way that its coordinates X and Y are independent standard normal random variables. Find the probability that the radial vector from the origin to $\mathbf{X}$ has a length of at least 2.14.
3. Give a direct proof using moment-generating functions that the sample mean $\bar{X}$ of a random sample of size n from the $N(\mu, \sigma^2)$ distribution has the $N(\mu, \sigma^2/n)$ distribution.

4. For the previously cited set of data (see Exercise 10 in Section 4.4) on Peoria school district performance, comparative data were given on average performance by students in Elgin, Illinois, on the state reading assessment test for eighth-graders. The Elgin students averaged 256, and the average scores among students in several Peoria area districts follow. Assuming that the Peoria performance on the reading test follows a normal distribution with standard deviation 29, is there strong evidence that the Peoria students are doing better than the Elgin students?

$$276, \ 296, \ 247, \ 286, \ 239, \ 297, \ 346, \ 292, \ 289, \ 348, \ 307,$$
$$244, \ 285, \ 286, \ 257, \ 288, \ 285, \ 268, \ 348, \ 291, \ 285, \ 278$$

5. Let $X_1, X_2, \ldots, X_{36}$ be a random sample from the $N(\mu, 4)$ distribution. Find a constant $c > 0$ such that the probability that the interval $(\bar{X} - c, \bar{X} + c)$ contains μ is 95%.

6. Let $X_1, X_2, \ldots, X_{16}$ be a random sample from the $N(\mu, \sigma^2)$ distribution. Find constants c and d $(c < d)$ such that the probability that the interval (cS^2, dS^2) contains σ^2 is 95%.

7. Given are 20 observations that are purported to have been taken from a normally distributed population with variance 9. Does this data set seem to support the claim that the population mean μ is 0? Give probabilistic evidence for your answer.

$$2.51, \ 0.69, \ -2.26, \ 7.29, \ 0.52, \ -3.31, \ 3.74, \ 5.62, \ 1.51, \ -1.10,$$
$$10.70, \ -4.38, \ 0.70, \ 1.63, \ -0.38, \ 3.99, \ 1.13, \ 2.86, \ 2.32, \ 3.18$$

8. The following 20 observations are purported to have been taken from a normally distributed population. Does this data set seem to support the claim that the population variance σ^2 is 16? Give probabilistic evidence for your answer.

$$8.76, \ 16.49, \ 11.37, \ 11.35, \ 8.88, \ 8.93, \ 13.48, \ 11.31, \ 7.89, \ 24.76,$$
$$17.70, \ 15.27, \ 17.94, \ 13.43, \ 3.30, \ 5.31, \ 6.37, \ 11.67, \ 8.38, \ 5.96$$

9. Let $\bar{X}$ be the sample mean of a random sample of size 25 from the $N(\mu, \sigma^2)$ distribution. Compute $P[|\bar{X} - \mu| > k\sigma]$, for values of $k = .1, .2, .3, .4$.

10. Find the probability density function of the sample variance S^2 of a random sample of size n from the $N(\mu, \sigma^2)$ distribution. What parameters does it depend on? (*Hint:* S^2 may be obtained by a simple transformation of a random variable that is known to have a χ^2 distribution.)

11. A recent report on civilian and military employees at U.S. Army installations in the fifty states and Puerto Rico (*Green Book*, 1994) contains counts by base; a sample of numbers of civilians at 36 bases follows.
   a. Is there strong evidence that the standard deviation of the number of civilian employees differs from 2100?
   b. Assuming that the standard deviation is indeed 2100, is there strong evidence that the mean number of civilian employees at all bases is

greater than 3000?

$$7978, \ 1350, \ 4200, \ 9000, \ 5000, \ 2634, \ 350, \ 3730, \ 1533, \ 4652,$$
$$1700, \ 6051, \ 829, \ 526, \ 3573, \ 3300, \ 1700, \ 500, \ 2200, \ 1715,$$
$$1000, \ 2285, \ 5700, \ 981, \ 619, \ 4000, \ 1250, \ 1100, \ 7000, \ 4013,$$
$$6425, \ 1033, \ 3442, \ 2000, \ 1265, \ 4500$$

12. Prove using moment-generating functions that if $\mathbf{X} = (X_1, X_2, \ldots, X_n)'$ is multivariate normal, then $A\mathbf{X} + \mathbf{b}$ is also multivariate normal, where $\mathbf{b}$ is an $n \times 1$ vector of constants and A is an $n \times n$ matrix of constants.

13. Find the joint distribution of $Y_1 = X_1 - X_2 + X_3$ and $Y_2 = X_1 + X_3$ if $X_1 \sim N(0, 1)$, $X_2 \sim N(4, 4)$, $X_3 \sim N(-1, 2)$, and the X_i's are independent.

14. Let $\mathbf{X}$ have the bivariate normal distribution with means $\mu_1 = \mu_2 = 0$, variances $\sigma_1^2 = \sigma_2^2 = \sigma^2$, and correlation coefficient ρ. First, find the eigenvalues and eigenvectors of the covariance matrix Σ, and use them to write Σ as NDN^{-1}, where N is an orthogonal matrix. Now let the transformed random variable $\mathbf{Y}$ be defined by $\mathbf{Y} = N\mathbf{X}$. What distribution does $\mathbf{Y}$ have?

15. Prove the following: Let $X_1, X_2, \ldots, X_n$ be independent normal random variables with common variance σ^2. Let $\mathbf{a} = [a_1 \ a_2 \ \ldots \ a_n]'$ and $\mathbf{b} = [b_1 \ b_2 \ \ldots \ b_n]'$ be linearly independent vectors such that $\mathbf{a}'\mathbf{b} = \mathbf{b}'\mathbf{a} = \mathbf{0}$. Then the two random variables

$$Y_1 = \mathbf{a}'\mathbf{X} = a_1 X_1 + a_2 X_2 + \cdots + a_n X_n,$$
$$Y_2 = \mathbf{b}'\mathbf{X} = b_1 X_1 + b_2 X_2 + \cdots + b_n X_n$$

are independent and normally distributed.

16. Verify the decomposition

$$\sum_{i=1}^{n} X_i^2 = \sum_{i=1}^{n} (X_i - \overline{X})^2 + n\overline{X}^2$$

in formula (5.63).

17. Verify the identities

$$\sum_{i=1}^{n} X_i^2 = \mathbf{X}' I_n \mathbf{X}, \qquad n\overline{X}^2 = \mathbf{X}' Q_1 \mathbf{X}, \qquad \sum_{i=1}^{n} (X_i - \overline{X})^2 = \mathbf{X}' Q_2 \mathbf{X}$$

in formula (5.64).

18. Discuss how Proposition 5.4.4 and the linearity result in formula (5.53) follow as special cases of the multivariate results in Section 5.4.2.

19. Show that the following two quadratic forms in three standard normal random variables X_1, X_2, and X_3 are independent:

$$Y_1 = X_1^2 + X_2^2 + X_3^2 + 2X_1 X_2 + 2X_1 X_3 + 2X_2 X_3,$$
$$Y_2 = \frac{1}{2}X_2^2 - X_2 X_3 + \frac{1}{2}X_3^2.$$

Does either quadratic form have a χ^2-distribution?

5.5 | *t-* and *F-*Distributions

Recall Example 5.4.1, in which we were to decide whether the mean calorie content of a cup of yogurt could be 150 if the sample mean of 100 cups was 145 calories. We assumed a known population standard deviation $\sigma = 10$ in order to answer the question. But it is really not reasonable to assume that we know the standard deviation a priori. In place of the random variable that we used in that example,

$$Z = \frac{\bar{X} - \mu}{\sigma/\sqrt{n}}, \tag{5.71}$$

which has the $N(0,1)$ density, we would like to use the random variable

$$T = \frac{\bar{X} - \mu}{S/\sqrt{n}}. \tag{5.72}$$

In this random variable, the population standard deviation σ is estimated by the sample standard deviation S. This section discusses a continuous distribution, called the *t-distribution*, that applies to the random variable in (5.72). We will be using it frequently from Chapter 8 onward.

?Question 5.5.1   Recall the distributions of $\bar{X}$ and $(n-1)S^2/\sigma^2$ derived in the last section. Try to put these random variables together in such a way that the random variable in (5.72) results.

We will also discuss a very useful distribution for problems involving comparisons of variabilities of two random phenomena, and for the analysis of designed experiments. It will turn out that the so-called *F-distribution* gives the proper probabilistic model for the ratio of two sample variances S_1^2 and S_2^2.

DEFINITION 5.5.1 ——————————————————————————

The *t-*distribution with parameter r (called the *degrees of freedom* of the distribution) is the distribution of the random variable,

$$T = \frac{Z}{\sqrt{V/r}}, \tag{5.73}$$

where $Z \sim N(0,1)$, $V \sim \chi^2(r)$, and Z and V are independent. We use $t(r)$ as a shorthand notation for this distribution.

Sketches of the density function of the *t-*distribution for several parameter values r are contained in Fig. 5.9, along with a sketch of the standard normal density. Notice that the *t-*densities are symmetric about 0 and that the area under the $t(r)$ curve accumulates in the middle as the number of degrees

$f(x)$

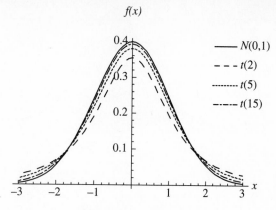

Figure 5.9 t-Densities

of freedom increases. As the graphs suggest, it can be shown that the $t(r)$ density converges to the $N(0,1)$ density as $r \to \infty$.

How is the formula for the t-density obtained? The multivariate transformation techniques of Section 5.2 can be applied to the joint transformation

$$T = \frac{Z}{\sqrt{V/r}}, \qquad U = V \tag{5.74}$$

to find the joint distribution of T and U. From this, the marginal density of T can be computed by integrating the joint density over the states of U. You will do this in Exercise 8. The end result of the derivation is the following formula for the p.d.f. of the $t(r)$ distribution:

$$f(t) = \frac{\Gamma(\frac{r+1}{2})}{\Gamma(\frac{r}{2})\sqrt{\pi r}} \left(1 + \frac{t^2}{r}\right)^{-(r+1)/2}, \qquad -\infty < t < +\infty. \tag{5.75}$$

The symmetry of the density about $t = 0$ is easy to see from formula (5.75).

It is not customary to compute probabilities by using the t-density formula directly, which is why we have treated it in a rather perfunctory way. Probabilities of events involving t-random variables may be found using a statistical program, or a table such as the one in Table 5 of Appendix A. The tabulated probabilities are not as complete as those for the $N(0,1)$ distribution because the probabilities change as the degrees of freedom change; hence a complete tabulation would require a separate table for each parameter value r. Instead, we list in a single table certain key percentiles of the distribution, which can of course be complemented to produce other probabilities. For example, when $r = 6$,

$$P[T \leq 1.44] = .90, \qquad P[T > 1.94] = .05.$$

These probabilities are the shaded areas under the $t(6)$ density curve displayed in Figs. 5.10(a) and (b), respectively.

?Question 5.5.2   For $r = 10$, what is $P[T > 2.76]$? For what value of t is $P[T \leq t] = .975$?

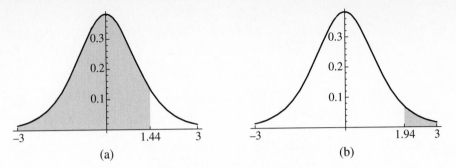

Figure 5.10 Some *t*-Percentiles

The *t*-distribution has direct bearing on the problem of estimating the normal mean μ by the sample mean $\bar{X}$ without knowing the variance σ^2.

PROPOSITION 5.5.1

Let $\overline{X}$ and S^2 be the sample mean and variance of a random sample of size n from the $N(\mu, \sigma^2)$ distribution. Then

$$T = \frac{\bar{X} - \mu}{S/\sqrt{n}} \sim t(n-1). \tag{5.76}$$

■ **Proof** We have from Propositions 5.4.2 and 5.4.5 that

$$Z = \frac{\bar{X} - \mu}{\sigma/\sqrt{n}} \sim N(0,1), \qquad V = \frac{(n-1)S^2}{\sigma^2} \sim \chi^2(n-1),$$

and Z and V are independent. By the definition of the *t*-distribution, it follows that

$$T = \frac{Z}{\sqrt{V/(n-1)}} = \frac{\dfrac{\bar{X} - \mu}{\sigma/\sqrt{n}}}{\sqrt{\dfrac{(n-1)S^2}{\sigma^2}/(n-1)}} = \frac{\bar{X} - \mu}{S/\sqrt{n}} \sim t(n-1).$$

Example 5.5.1

We recently opened a small computer lab in my building. A sign-in system was instituted that allowed the computer center to monitor the number of users in a particular day. Following are the results from 29 days in January 1995. Suppose that the center will decide to staff the lab with an assistant if the true average number of daily users is more than 30, but clear evidence is needed to justify the expense. Should the assistant be hired?

30, 32, 29, 9, 25, 25, 32, 41, 49, 44, 37, 12, 13, 44, 31,
45, 55, 52, 16, 26, 39, 40, 46, 58, 38, 9, 15, 39, 46

One way of making this judgment is to suppose that the population mean μ is 30, and to see whether the data yield an unreasonably high value of $\bar{X}$

under that assumption. To decide what is unreasonable, we also need to estimate the variability of the demand; that is, we must estimate the population standard deviation σ by the sample standard deviation S. For the given data the observed numerical values of the sample mean and standard deviation are

$$\bar{x} = 33.69, \qquad s = 13.98.$$

If we are willing to assume that the data form a random sample from the normal distribution with mean 30, then the random variable T in (5.76) has the $t(28)$ distribution, since the sample size is 29. Using the preceding sample statistics, we compute that the observed value of T is

$$t = \frac{\bar{x} - \mu}{s/\sqrt{n}} = \frac{33.69 - 30}{13.98/\sqrt{29}} = 1.42.$$

Is this unreasonably large? The t-table in Appendix A indicates that for a $t(28)$-distributed random variable,

$$P[T \geq 1.31] = .1 \Rightarrow P[T \geq 1.42] \leq .1.$$

The second inequality follows because the event $\{T \geq 1.42\}$ is contained in the event $\{T \geq 1.31\}$. The statistical program *Minitab* reported the value .0833 for $P[T \geq 1.42]$. Thus it is only about 8% likely to have observed a t-value of at least the magnitude that we did observe, assuming a mean of 30 users. This is fairly strong evidence that $\mu > 30$, although how strong is in the eye of the beholder. If the administration is particularly concerned about service, then they probably should hire the assistant.

Next we would like to define the F-distribution mentioned earlier.

DEFINITION 5.5.2

The *F-density* with parameters r_1, r_2 (both called *degrees of freedom*) is the p.d.f. of the random variable

$$F = \frac{U/r_1}{V/r_2}, \tag{5.77}$$

where $U \sim \chi^2(r_1)$, $V \sim \chi^2(r_2)$, and U and V are independent random variables. We use the shorthand notation $f(r_1, r_2)$ when referring to this distribution.

Like the t-density, the F-density is lengthy and awkward to use. For your reference, its form is

$$g(x) = \frac{\Gamma(\frac{r_1+r_2}{2})(r_1/r_2)^{r_1/2}}{\Gamma(r_1/2)\Gamma(r_2/2)} \frac{x^{r_1/2-1}}{(1+\frac{r_1}{r_2}x)^{(r_1+r_2)/2}}, \qquad 0 < x < \infty. \tag{5.78}$$

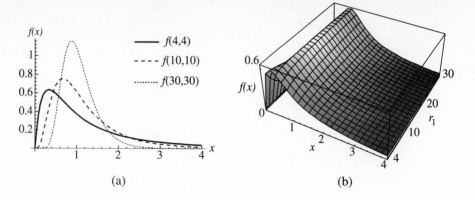

Figure 5.11   *F*-Densities

This can be derived using the joint transformation

$$F = \frac{U/r_1}{V/r_2}, \qquad Y = V.$$

Graphs of the F-density in formula (5.78) for parameter combinations (4,4), (10,10), and (30,30) appear in Fig. 5.11(a). As the common degrees of freedom increase, the probability weight becomes more tightly packed around the peak, moves out somewhat to the right, and becomes more symmetric. The surface in part (b) of the figure is the collection of F-density functions for a fixed value of $r_2 = 4$, and values of r_1 from 4 up to 30. There is a mild tendency for the peak to move to the right and for less of the probability weight to be given to large values of x as r_1 increases. A graphing program or calculator will allow you to check these patterns in more detail; you can see what happens when r_1 is held fixed and r_2 is allowed to grow.

Values of the F-percentiles for selected probability levels and various combinations of degrees of freedom can be found in Table 6 of Appendix A. The numerator degrees of freedom r_1 is found on the left margin and the denominator degrees of freedom r_2 is on the top margin. The body of the table contains constants f_α such that

$$P[F \leq f_P] = P, \tag{5.79}$$

for the given combination of degrees of freedom and the indicated percentiles $P = .95, .975, .99$. For example, for an $f(5,6)$ random variable F, $P[F \leq 4.39] = .95$ and $P[F > 5.99] = .025$. These percentiles f_P can also be obtained using statistical packages. By the way, in subsequent work we will abuse our notation a bit and follow the usual convention by writing f_α for the point to the right of which is small probability α, that is $P[F > f_\alpha] = \alpha$.

—— **?Question 5.5.3**   If $F \sim f(8, 10)$, find $P[F \leq 5.06]$, and find a constant c such that $P[F > c] = .05$.

The F-table is not quite as incomplete as it might seem. Complementation gives upper tail probabilities for small values of α, but certain lower tail probabilities can also be found using the following device. The random variable $F = (U/r_1)/(V/r_2)$ has the $f(r_1, r_2)$ distribution. If we form the reciprocal of F, we still have a ratio of independent χ^2 random variables divided by their degrees of freedom, which has an F-distribution with the roles of the two parameters reversed. Thus

$$P\left[F_{r_2,r_1} = \frac{1}{F_{r_1,r_2}} < \frac{1}{f_\alpha(r_1, r_2)}\right] = P[F_{r_1,r_2} > f_\alpha(r_1, r_2)]. \tag{5.80}$$

From the F-table in Appendix A, the area to its right of the point 4.53 under the $f(4, 6)$ density is .05. By what we have just found, the area to the left of its reciprocal $1/4.53 = .220$ under the $f(6, 4)$ density is also .05; that is, $P[F_{6,4} < .220] = .05$.

The next proposition shows how the F-distribution is involved in variance comparison.

PROPOSITION 5.5.2

Let $X_1, \ldots, X_m$ and $Y_1, \ldots, Y_n$ be independent random samples from normal populations with variances σ_X^2 and σ_Y^2, respectively. Denote the sample variances by S_X^2 and S_Y^2. Then

$$F = \frac{S_X^2/\sigma_X^2}{S_Y^2/\sigma_Y^2} \sim f(m-1, n-1). \tag{5.81}$$

■ **Proof** Since the samples are independent, by Proposition 5.4.5 we can form two independent χ^2 random variables:

$$U = \frac{(m-1)S_X^2}{\sigma_X^2} \sim \chi^2(m-1), \qquad V = \frac{(n-1)S_Y^2}{\sigma_Y^2} \sim \chi^2(n-1).$$

By (5.77),

$$F = \frac{(m-1)S_X^2/\sigma_X^2(m-1)}{(n-1)S_Y^2/\sigma_Y^2(n-1)} = \frac{S_X^2/\sigma_X^2}{S_Y^2/\sigma_Y^2} \sim f(m-1, n-1).$$

Note that if $\sigma_X^2 = \sigma_Y^2$, then the ratio S_X^2/S_Y^2 itself will have the $f(m-1, n-1)$ distribution. Extreme values of this ratio, which are either too close to zero or too large, furnish evidence against the equality of the population variances.

Example 5.5.2

A compilation of average faculty salaries by rank for 25 institutions in our consortium recently came across my desk. I began to wonder whether there was any difference in variability between salaries of assistant and full professors, hypothesizing that institutions might compete strongly to hire new

Ph.D.s by matching other offers, which would tend to level out the assistant professor salaries, but they may have different practices on merit raises, which would make salaries of senior people tend to vary more. If our consortium is typical of all colleges, we can extrapolate from it to the universe of all colleges (a risky proposition, admittedly). The data follow.

Assistant

42725, 41602, 40965, 40547, 39959, 38585, 38214, 38131, 37385, 37167, 37141, 36499, 36458, 36224, 36186, 35927, 35764, 35703, 34613, 34474, 34469, 33067, 32994, 31618, 31451

Full

69306, 68411, 65136, 65010, 63736, 63041, 62741, 59813, 59520, 58000, 57495, 57487, 56728, 56484, 55564, 55004, 54116, 53984, 53802, 53496, 52740, 50207, 48610, 45509, 42100

Presuming that the salaries are normally distributed and the samples are independent (though since the same schools are represented in each, the latter may also be doubtful), we would like to decide whether the normal variance σ_x^2 of assistant professor salaries is less than the normal variance σ_y^2 of full professor salaries.

If indeed there is no difference, then $\sigma_x^2 = \sigma_y^2$. In that case, from formula (5.81), the random variable

$$F = \frac{S_x^2}{S_y^2}$$

has the $f(24, 24)$ distribution. Our data yield the observed values $s_x = 2961$, $s_y = 6700$. Thus the observed f is $(2961)^2/(6700)^2 = .195$. How likely is it that an F-ratio as small as .195 could occur under the assumption of equal variance? From the F-table, since the reciprocal of an $f(24, 24)$ random variable is also $f(24, 24)$,

$$P\left[\frac{1}{F_{24,24}} > 2.66\right] = .01 \Rightarrow P\left[F_{24,24} < \frac{1}{2.66} = .375\right] = .01.$$

Since our observed f is even smaller than .375, it must be less likely than .01. *Minitab* reports that $P[F_{24,24} \le .195] = .0001$. This result gives very strong evidence against equality of population variances.

?**Question 5.5.4**   Plot histograms of the salary data for both ranks in the last example. Are there any striking departures from normality?

EXERCISES 5.5   1. If T is a random variable with the t-distribution with 15 degrees of freedom, find $P[1.34 \le T \le 2.13]$.

2. A weight-loss program was applied to 28 subjects over a two-week period. The average weight loss in the group was 2.1 pounds, and the sample standard deviation of weight loss was 3.2 pounds. Is there good reason to believe that this weight-loss program works? State any assumptions that you are making.

3. Exercise 4 in Section 5.4 presented data on reading scores on an eighth-grade assessment test. The mean Peoria performance was to be compared to the Elgin mean of 256. Redo that example without assuming that the standard deviation of the distribution of Peoria scores is known.

4. In Exercise 14 of Section 3.3 we made reference to a set of ozone observations taken in Pennsylvania forests. Following are 21 readings taken in August 1989. Is there strong evidence that the average ozone level exceeds 48?

$$69, \ 70, \ 40, \ 35, \ 59, \ 40, \ 82, \ 48, \ 49, \ 35, \ 60,$$
$$59, \ 55, \ 46, \ 78, \ 77, \ 49, \ 60, \ 37, \ 38, \ 56$$

5. a. Find a number c such that $P[-c < T < c] = .95$, where T has the $t(17)$ distribution.

   b. Suppose that you have obtained a random sample of size 18 from the $N(\mu, \sigma^2)$ distribution. Use part (a) to find an interval centered about the sample mean $\bar{X}$ in which the population mean μ lies with probability .95.

6. From Example 5.5.1 on the usage of the computer lab, there are also data on the number of users during 28 days of February, shown below.

   a. Use the t-table and the t-distributed random variable in (5.76) to find a constant c so that $P[-c < T < c] = .90$, for both 28 and 27 degrees of freedom.

   b. Use your results in part (a) to isolate μ so as to find an interval in which the January population mean lies with probability .90. Do the same for the February population mean. Does there seem to be a significant difference in mean usage from one month to another?

$$53, \ 62, \ 39, \ 14, \ 25, \ 45, \ 42, \ 70, \ 53, \ 47, \ 18, \ 25, \ 46, \ 38,$$
$$52, \ 57, \ 38, \ 17, \ 25, \ 57, \ 43, \ 63, \ 51, \ 43, \ 17, \ 27, \ 68, \ 57$$

7. Verify that if T has the $t(r)$ distribution, then $E[T] = 0, r > 1$.

8. Derive formula (5.75) for the t-density using the approach suggested in the text.

9. If $T \sim t(r)$, then what distribution does T^2 have? (*Hint:* Look at the definition of the t-distribution.)

10. If F is a random variable with the $f(6, 6)$ distribution, find $P[F > .172]$.

11. Let S_1^2 and S_2^2 be the sample variances of random samples of size 13 and 7, respectively, from normal distributions with variances $\sigma_1^2 = 10$ and $\sigma_2^2 = 12$. Find the probability that S_1^2 / S_2^2 exceeds 3.33.

12. Suppose that 10 bags of sugar filled by one machine had a sample variance of bag weight of .1 lb^2 and that 13 bags from a second machine had a sample variance of .4 lb^2. Is there cause to believe that the second machine does not work as well as the first? State any assumptions that you are making.

13. Using the data in Example 5.5.1 from January and the data in Exercise 6 from February, check for a significant difference in the variability of computer lab usage between months.

14. Referring to Exercise 4, following are 21 more ozone readings taken in August 1990. Is there evidence that there is any difference in the variability of ozone concentration between years, which could indicate increasing instability of pollution conditions?

$$27, \ 46, \ 47, \ 88, \ 44, \ 52, \ 46, \ 70, \ 69, \ 77, \ 41,$$
$$43, \ 75, \ 70, \ 37, \ 55, \ 56, \ 62, \ 28, \ 58, \ 69$$

15. Exercise 4 of Section 4.1 contained distributions of responses by Russians, Ukrainians, and Lithuanians to the question of whether party competition would make the political system stronger. There were just five possible responses, from 1 (fully agree) to 5 (fully disagree). We will learn later that even when the underlying distribution is discrete, when sample sizes are large, normal theory techniques of the kind we studied in this section give good approximate results. With this in mind, check for a difference in variability between the Russian and Lithuanian response distributions.

16. Let X and Y have joint density $f(x,y) = e^{-x-y}$, $x, y > 0$. Show that the random variable $F = X/Y$ has an F-distribution, assuming $r_2 > 2$. What are the degrees of freedom of the distribution?

17. By using the definition of the F-distribution (5.77) and the independence of U and V, find the mean of the $f(r_1, r_2)$ distribution. [*Hint:* First find $E[1/V]$, where $V \sim \chi^2(r_2)$.]

18. a. If $F \sim f(10, 8)$, find two numbers c and d such that $P[F > d] = .025$ and $P[F < c] = .025$.

    b. Let S_1^2 and S_2^2 be the sample variances of random samples of size 11 and 9, respectively, from $N(\mu_1, \sigma_1^2)$ and $N(\mu_2, \sigma_2^2)$ distributions. Use the result of part (a) to find an interval whose endpoints depend on the ratio S_1^2/S_2^2, which will contain the ratio σ_1^2/σ_2^2 with probability .95.

CHAPTER 6

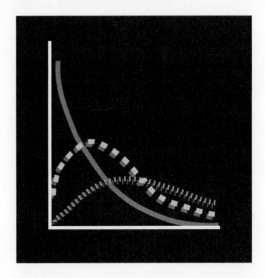

ASYMPTOTIC THEORY

6.1 | Laws of Large Numbers

The mission of this short chapter is to expose you to some ideas pertaining to the limiting behavior of sequences of random variables. Probably the best reason why the study of limiting behavior is important is that when we use random samples to estimate distributional parameters, we would like to know that as the sample size gets larger, the estimates are likely to be close to the parameters that they are estimating. The first section starts by deriving an inequality called *Chebyshev's inequality* on the closeness of a random variable to its mean. As you might suspect, the variance of the random variable plays a big role. We then use the inequality to prove the convergence of a sequence of sample means to the population mean, a theorem referred to as the *Law of Large Numbers*. Since this result justifies our use of empirical frequencies to estimate actual probabilities, it lies at the very heart of probability and statistics. The second section studies the Central Limit Theorem.

6.1.1 Chebyshev's Inequality

To begin, we will consider a random variable X that has a finite mean μ and a finite variance σ^2. We want to study the likelihood that X takes a value in the interval centered about μ with radius $h\sigma$ for $h > 0$. If μ and σ are fixed, then as h grows, the probability that $X \in (\mu - h\sigma, \mu + h\sigma)$ should increase. (Why?) It is not so clear what will happen to this probability as σ is allowed to increase; the interval widens, but on the other hand, as σ grows the probability weight spreads away from μ.

—— **?Question 6.1.1** If $X \sim N(\mu, \sigma^2)$, does $P[X \in (\mu - h\sigma, \mu + h\sigma)]$ depend on σ?

The next proposition gives a lower bound for the probability that X takes a value within h standard deviations of μ. This bound is valid for all distributions that have finite mean and variance.

PROPOSITION 6.1.1

Chebyshev's Inequality If a random variable X has a finite mean μ and variance σ^2, then for all $h > 0$,

$$P[|X - \mu| \geq h\sigma] \leq \frac{1}{h^2}, \tag{6.1}$$

and hence

$$P[|X - \mu| < h\sigma] \geq 1 - \frac{1}{h^2}. \tag{6.2}$$

■ **Proof** We do the proof under the assumption that X is a continuous random variable with p.d.f. f. The proof for the discrete case is similar. Consider the expected value of the function $g(X) = (X - \mu)^2 / \sigma^2 h^2$, which by linearity of expectation equals $1/h^2$. (Check this result.) The expectation can be broken into the sum of integrals over three disjoint regions as follows:

$$
\begin{aligned}
\frac{1}{h^2} = E[g(X)] &= \int_{-\infty}^{+\infty} \frac{(x - \mu)^2}{\sigma^2 h^2} f(x)\, dx \\
&= \int_{-\infty}^{\mu - h\sigma} \frac{(x - \mu)^2}{\sigma^2 h^2} f(x)\, dx + \int_{\mu - h\sigma}^{\mu + h\sigma} \frac{(x - \mu)^2}{\sigma^2 h^2} f(x)\, dx \\
&\quad + \int_{\mu + h\sigma}^{+\infty} \frac{(x - \mu)^2}{\sigma^2 h^2} f(x)\, dx.
\end{aligned}
\tag{6.3}
$$

The middle integral in (6.3) has a nonnegative integrand, and hence the integral itself is greater than or equal to zero. Therefore $1/h^2$ exceeds the sum of the two integrals on the left and right. In the integral on the left, for values of x in $(-\infty, \mu - h\sigma]$, we have that

$$x \leq \mu - h\sigma \Rightarrow \frac{x - \mu}{h\sigma} \leq -1 \Rightarrow \frac{(x - \mu)^2}{h^2 \sigma^2} \geq 1.$$

Similarly, in the integral on the right of (6.3), for $x \in [\mu + h\sigma, +\infty)$, you can show that $(x - \mu)^2 / h^2 \sigma^2 \geq 1$. So, in both of the remaining two integrals,

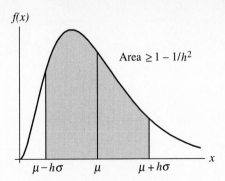

Figure 6.1 Chebyshev's Inequality

we can replace the expression for $g(x)$ by 1:

$$\frac{1}{h^2} \geq \int_{-\infty}^{\mu-h\sigma} \frac{(x-\mu)^2}{\sigma^2 h^2} f(x)\, dx + \int_{\mu+h\sigma}^{+\infty} \frac{(x-\mu)^2}{\sigma^2 h^2} f(x)\, dx$$

$$\geq \int_{-\infty}^{\mu-h\sigma} 1 \cdot f(x)\, dx + \int_{\mu+h\sigma}^{+\infty} 1 \cdot f(x)\, dx$$

$$= P[X \leq \mu - h\sigma] + P[X \geq \mu + h\sigma]$$

$$= P[|X - \mu| \geq h\sigma].$$

This proves formula (6.1); formula (6.2) follows easily by complementation.

Chebyshev's inequality is illustrated graphically in Fig. 6.1 for a continuous density. The area under the density curve between $\mu - h\sigma$ and $\mu + h\sigma$ is at least $1 - 1/h^2$, and the area outside of this interval is no more than $1/h^2$. When $h = 3$, for example, this inequality says that regardless of the assumed distribution of X, the probability is no more than $1/3^2 = 1/9$ that the random variable takes on a value more than three standard deviations away from its mean.

?Question 6.1.2 Find an upper bound for $P[|X - \mu| \geq \epsilon]$ and a lower bound for $P[|X - \mu| < \epsilon]$, where $\epsilon > 0$ is a constant. Does the role of the variance in your bounds make intuitive sense?

The sheer generality of the Chebyshev bound might lead you to believe that it sacrifices some precision in favor of breadth of applicability. In individual cases, the probability that $|X - \mu| < h\sigma$ may be greater than $1 - 1/h^2$, as the next example illustrates.

Example 6.1.1

Suppose that X is a random variable with the $N(\mu, \sigma^2)$ distribution. Then, denoting as usual the standardized X by Z, we can make the following comparisons between the actual probabilities and the Chebyshev bounds:

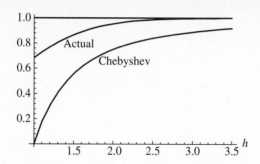

Figure 6.2 Comparison of Chebyshev Bound to Normal Probabilities

Normal	**Chebyshev**
$P[|X - \mu| < \sigma] = P[|Z| < 1] \approx .682$	$1 - \dfrac{1}{1^2} = 0$
$P[|X - \mu| < 2\sigma] = P[|Z| < 2] \approx .954$	$1 - \dfrac{1}{2^2} = .750$
$P[|X - \mu| < 3\sigma] = P[|Z| < 3] \approx .997$	$1 - \dfrac{1}{3^2} = .889$

If the distribution of X is known to be normal, then the actual probabilities that X takes a value within one, two, and three standard deviations, respectively, of its mean are substantially greater than the Chebyshev bounds. Plots of this actual normal probability and the Chebyshev lower bound as functions of h are shown in Fig. 6.2, and the looseness of the bound is apparent.

?Question 6.1.3 Refer to the proof of Chebyshev's inequality, particularly to the decomposition (6.3). Where does the sacrifice of accuracy really come from, and why would you expect a serious loss of accuracy for a distribution like the normal?

Example 6.1.2 A political candidate is interested in her current popularity. She would therefore like to estimate p, the proportion of voters who favor her, in other words, the probability that a randomly selected voter is a supporter. If she wants to obtain an estimate of p accurate to within $\epsilon = .05$ with probability at least 90%, how many people does Chebyshev's inequality say should be polled?

To form a probabilistic model, let $X_1, X_2, \ldots, X_n$ be i.i.d. Bernoulli random variables with success parameter p, which are 1 or 0, respectively, according to whether or not the person favors our candidate. Then the sample average of the X_i's, $\bar{X} = \sum X_i/n$, is also the proportion of successes $\widehat{p}$ among the n trial random variables, which is the sample proportion who favor the candidate. We would like to force the probability that $\bar{X}$ will be within a small distance $\epsilon = .05$ of p to be at least .90.

By answering Question 6.1.2 you should have produced the following variant of Chebyshev's inequality:

$$P[|X - \mu| < \epsilon] \geq 1 - \frac{\sigma^2}{\epsilon^2}. \tag{6.4}$$

[This follows from (6.2) upon setting $\epsilon = h\sigma$.] Apply this inequality to $X = \overline{X}$, whose mean is $\mu_{\overline{X}} = p$, and whose variance is $\sigma_{\overline{X}}^2 = \sigma^2/n = p(1-p)/n$:

$$P[|\overline{X} - p| < \epsilon] \geq 1 - \frac{p(1-p)}{n\epsilon^2}. \tag{6.5}$$

This lower bound on $P[|\overline{X} - p| < \epsilon]$ is as small as it can be when the quantity $p(1-p)$ is as large as it can be. This happens at $p = 1/2$ (check it), at which $p(1-p) = 1/4$. Thus, whatever the true value of p is,

$$P[|\overline{X} - p| < \epsilon] \geq 1 - \frac{1}{4n\epsilon^2}, \tag{6.6}$$

for any $\epsilon > 0$. To guarantee that this probability is at least .9 for $\epsilon = .05$, it suffices to have

$$1 - \frac{1}{4n(.05)^2} \geq .9 \Rightarrow .1 \geq \frac{1}{4n(.05)^2} \Rightarrow n \geq \frac{1}{4(.1)(.05)^2} = 1000.$$

This is a fairly large number of observations. Fortunately, this bound is an artifact of the original crudeness of the Chebyshev bound. We will obtain a much better lower bound later when we study the Central Limit Theorem.

Inequality (6.6) says something very important about estimating the probability of an event by its relative frequency among many replications of the experiment. Regardless of how small a distance $\epsilon > 0$ is chosen, the probability that $\overline{X} = \hat{p}$ is within ϵ units of p approaches 1 as the number of replications approaches ∞.

6.1.2   Weak Law of Large Numbers

The infamous Murphy's Law states that if something can go wrong, then it will. It turns out that Murphy is something of a probabilist, if his law is expressed in the following way. If there is a nonzero probability that an event will happen in an experiment, then if the experiment is repeated indefinitely the event will eventually happen. In fact, more is true. If p is the probability of the event, then if the experiment is repeated indefinitely, the relative frequency of occurrence of the event will approach p.

This statement is a case of a very important theorem in probability called the *Law of Large Numbers*. The general form of this theorem deals with the convergence of sample means to population means, but since the sample mean is the same as the relative frequency of occurrence in the Bernoulli trials case (see Example 6.1.2), this form of the theorem applies to relative frequencies as well.

We do have to be somewhat careful, because there is more than one sense in which the sequence of sample means can converge. The two forms of convergence that we will define, *convergence in probability* and *almost sure convergence*, give rise to two forms of the Law of Large Numbers, respectively, called the *weak* and the *strong* laws. The reason for this choice of adjectives, although we will not show it here, is that the almost sure convergence of a sequence of random variables implies the convergence in probability of the sequence, but not vice versa, in general. (See Exercise 19 at the end of this section.)

We begin with the weak law. In Example 6.1.2, we had a sequence $X_1, X_2,$ $X_3, \ldots$ of independent Bernoulli random variables with success parameter p. Let $\bar{X}_n$ denote the sample mean of the first n of them, which is the same as the sample proportion of successes among the first n trials. Thus we have a sequence of random variables:

$$\bar{X}_1 = X_1, \qquad \bar{X}_2 = \frac{X_1 + X_2}{2}, \qquad \bar{X}_3 = \frac{X_1 + X_2 + X_3}{3}, \ldots. \qquad (6.7)$$

By taking n large enough, inequality (6.6) says that we can make the probability of the event that $\bar{X}_n$ is within a small positive distance ϵ of its mean p as near to 1 as we please. Note that we pay *two* prices for randomness: first, $\bar{X}_n$ will not in general equal p exactly; second, we can only be very confident, not absolutely certain, that $\bar{X}_n$ will fall within a prescribed distance of p for large n.

The discussion of the preceding paragraph sets up the following definition.

DEFINITION 6.1.1

A sequence of random variables (Y_n) is said to *converge in probability* to a constant a if, for any $\epsilon > 0$,

$$P[|Y_n - a| < \epsilon] \to 1, \quad \text{as } n \to \infty, \qquad (6.8)$$

or equivalently,

$$P[|Y_n - a| \geq \epsilon] \to 0, \quad \text{as } n \to \infty. \qquad (6.9)$$

You can see now that because of inequality (6.6), the sample proportion of successes $\bar{X}_n$ in an open-ended sequence of Bernoulli trials converges in probability to the success parameter p. This means that we are doing a sensible thing when we estimate the probability of an event using its frequency of occurrence in many replications of an experiment. A more general result is the following.

PROPOSITION 6.1.2

Weak Law of Large Numbers   Let $X_1, X_2, X_3, \ldots$ be a sequence of i.i.d. random variables with finite mean μ and variance σ^2. Let $\bar{X}_n = \sum_{i=1}^{n} X_i / n$

be the sample mean of the first n X_i's. Then the sequence $(\bar{X}_n)$ converges in probability to μ.

■ **Proof**   The expected value of $\bar{X}_n$ is μ and the variance is σ^2/n. By Chebyshev's inequality, for arbitrary $\epsilon > 0$,

$$P[|\bar{X}_n - \mu| < \epsilon] \geq 1 - \frac{\sigma^2}{\epsilon^2 n}. \tag{6.10}$$

(You are asked to show this result in Exercise 15.) Since the right side converges to 1 as $n \to \infty$, the criterion for convergence in probability is satisfied.

Example 6.1.3

If an investor wants to gain information about the unknown mean weekly rate of return μ on a risky asset, for how many weeks should the investor observe its rates of return in order to produce an estimate that falls within a tolerance of .005 of the true μ, with a probability of at least 80%? Suppose that the variance of the asset rate of return is no more than .0004.

We estimate μ by the sample average $\bar{X}_n$ of n weekly rates of return. The weak law of large numbers says that for large n, $\bar{X}_n$ is likely to be close to μ. From (6.10),

$$P[|\bar{X}_n - \mu| < .005] \geq 1 - \frac{\sigma^2}{n(.005)^2} \geq 1 - \frac{.0004}{n(.005)^2}.$$

It suffices to choose n large enough that the right side is at least .80. Then,

$$1 - \frac{.0004}{n(.005)^2} \geq .80 \Rightarrow n \geq \frac{.0004}{(.005)^2(.2)} = 80.$$

?Question 6.1.4   How does the required number of observations in the last example change if the investor demands a tolerance of only .01?

6.1.3   Strong Law of Large Numbers

Convergence in probability, you may agree, is a rather complicated concept. We turn briefly to a simpler, more powerful mode of convergence. To illustrate, consider the experiment of flipping a fair coin indefinitely. A typical experimental outcome, such as $\omega_1 = (H, T, H, H, T, \ldots)$, is an infinite sequence of head and tail symbols. Many outcomes are logically possible, even including the case of all tails $\omega_2 = (T, T, T, \ldots)$, in which the sample proportion of heads among the first n flips is 0 for any n. Another extreme case is $\omega_3 = (H, T, T, T, \ldots)$, in which the only head is on the first flip. The sample proportion of heads among the first n flips for this outcome is $1/n$, which approaches 0 as $n \to \infty$.

Since we already know that there is one sense in which the sample proportion of heads will converge to $1/2$, we intuitively feel that outcomes like ω_2

and ω_3 are unlikely. A more refined expression of our intuition is that outcomes like ω_2 and ω_3 should belong to a set of outcomes of small, perhaps even no, total probability. In fact, what is true is that the set of outcomes for which the proportion of heads among the first n flips does not converge to $1/2$ as $n \to \infty$ has probability 0. Along with outcomes like ω_2 and ω_3, an outcome like $\omega_4 = (H, T, T, H, T, T, \ldots)$ in which the sample proportion of heads converges to $1/3$ also belongs to this set of probability zero.

The scenario described in the last paragraph is an instance of the kind of convergence defined next.

DEFINITION 6.1.2

A sequence of random variables (Y_n) is said to *converge almost surely* to a constant a if

$$\lim_{n \to \infty} Y_n(\omega) = a \qquad (6.11)$$

for all outcomes ω except possibly those in some event $N \subseteq \Omega$ of probability 0. [The limit in formula (6.11) is understood to be in the usual sense of convergence of real numbers to a limit.]

Almost sure convergence is close to the kind of convergence of sequences that you became familiar with in calculus. Anticipating the strong law for a moment, almost sure convergence merely says in the coin flip case that the sequence of numbers $\bar{X}_n(\omega)$, which are the proportions of heads among the first n flips, converges to the number $p = 1/2$ (for a fair coin) for those experimental outcomes ω that do not belong to an exceptional set whose probability is 0. Here is another example that may clarify the idea.

Example 6.1.4

Let the sample space Ω of an experiment consist of outcomes of the form

$$\omega = (\pm 1, \pm 1/2, \pm 1/3, \pm 1/4, \ldots).$$

As in the coin flip problem, let there be a probability measure on the sample space such that the probability that a single component is positive is $1/2$, and hence the probability that it is negative is also $1/2$. For each n, define the component random variable $Y_n(\omega) = \omega_n$. We claim that (Y_n) converges almost surely to 0. This model is far easier to work with than the coin flip problem, because Y_n takes the possible values $\{+1/n, -1/n\}$, each with probability $1/2$. Thus, for *every* outcome ω, the distance between $Y_n(\omega)$ and 0 is

$$|Y_n(\omega) - 0| = 1/n,$$

which obviously approaches 0 as $n \to \infty$. Here the exceptional set of outcomes is the empty set.

To prepare for the Strong Law of Large Numbers, consider Fig. 6.3. I simulated 400 observations from the $N(0, 1)$ distribution and progressively,

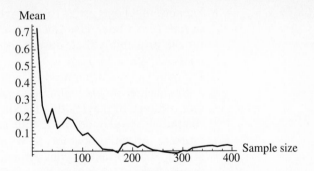

Figure 6.3   Strong Law of Large Numbers

as each ten new observations went by, recomputed the sample mean for all sample values to date. That is, I computed $\bar{X}_{10} = (X_1 + \cdots + X_{10})/10$, $\bar{X}_{20} = (X_1 + \cdots + X_{20})/20, \ldots, \bar{X}_{400} = (X_1 + \cdots + X_{400})/400$. The figure shows the running means wandering a bit but generally tending toward 0, which is the mean of the distribution being sampled from.

We can now state the Strong Law.

PROPOSITION 6.1.3

Strong Law of Large Numbers   Let $X_1, X_2, X_3, \ldots$ be a sequence of i.i.d. random variables with finite mean μ and variance σ^2. Let $\bar{X}_n = \sum_{i=1}^{n} X_i/n$ be the sample mean of the first n X_i's. Then the sequence $(\bar{X}_n)$ converges almost surely to μ.

Note that the strong law also implies that with probability 1, the sample proportion of successes in an infinite sequence of Bernoulli trials converges to the success probability p.

We do not have the analytical tools to prove this theorem here, nor does it really lead us to new ways of solving problems, yet its statement is well worth knowing. Both the weak and the strong laws are now so much a part of the philosophical milieu of probability that a serious student of the subject absolutely must know them and understand the difference between them. The strong law is actually an idealization, albeit a useful one. If it was possible to repeat an experiment countably infinitely many times, then it would be certain that the sample mean would converge to the theoretical mean. The weak law says only that for enough trials it is highly probable that the sample mean will be near μ.

The laws of large numbers have a long history. Chung (1979, pp. 228 ff) notes that although Jakob Bernoulli was able to prove the weak law for proportions in a laborious algebraic way as early as the 1690s, the "easier" proof did not exist until Chebyshev proved it for means in the late 1800s. Shortly after the turn of this century, Borel discovered the strong law for proportions, and in the 1930s Kolmogorov extended it to the case of means, showing in addition that the hypothesis of finite variance could be eliminated.

EXERCISES 6.1

1. Suppose that boats arrive to a large harbor according to a Poisson process of rate 25 per hour. Use Chebyshev's inequality to obtain a lower bound for the probability that, during a one-hour interval, between 15 and 35 boats arrive.

2. For purely theoretical purposes, you want to estimate the probability that a slot machine yields a winning combination on a single play. You decide on a strategy of playing the machine some large number n of times and computing the proportion of wins among the n trials. You would like to estimate the win probability to within .05, and you would also like to be at least 80% certain of your estimate. How many times should you play the machine? How would this number change if you knew that the win probability was around $1/4$?

3. Let X have the continuous uniform distribution on the interval $[0, b]$. Find an expression for $f(h) = P[|X - \mu| > h\sigma]$, and plot this function on the same graph with the Chebyshev upper bound on this probability.

4. Repeat Exercise 3 for the continuous density $g(x) = 3/x^4$, $x \in [1, \infty)$.

5. Repeat Exercise 3 for the discrete uniform distribution on $\{1, 2, \ldots, 9\}$.

6. A psychological test to determine inclination toward suicide is constructed. It is scored on a basis of 1–20. To develop norms, it is to be administered to a group of subjects with known suicidal tendencies. At least how many subjects should be used in the study in order to estimate the mean score for a suicidal person to within 1 unit with probability at least .85? Assume that the standard deviation of scores is no greater than that of the uniform distribution on $\{1, 2, \ldots, 20\}$.

7. A test that can be used to evaluate how severely a person's air passages are obstructed measures the amount of air that can be exhaled rapidly after a deep inhale. The value for a healthy person is about 4.8 liters. A physician obviously cannot have a sick patient repeat the test 100 times to obtain a very accurate reading. If at most five repetitions of the test are feasible (three is typical) and if the standard deviation of the amount of exhaled air is no greater than .6 liter, what is the highest precision guaranteed by Chebyshev's inequality for estimation of the true mean of exhaled air by the sample mean to within that precision with probability .9?

8. Use an approach similar to the one in the proof of Chebyshev's inequality to show that if Z_n and Z are random variables with finite second moments, then

$$P[|Z_n - Z| > \epsilon] \le \frac{E[(Z_n - Z)^2]}{\epsilon^2}.$$

If Z is a constant a, and a sequence of random variables $Z_1, Z_2, Z_3, \ldots$ is such that the second moment of each Z_n about a converges to 0 as $n \to \infty$, describe in words what can be concluded about the sequence (Z_n).

9. Let S^2 be the sample variance of a normal random sample of size n. Devise a lower bound for the probability that S^2 is within a distance ϵ of the population variance σ^2. [*Hint:* What distribution, mean, and variance

does the random variable $(n-1)S^2/\sigma^2$ have?] What can be concluded about S^2 as an estimator of σ^2 as the sample size grows?

10. Let $\hat{F}(y)$ be the empirical c.d.f. of a random sample from a distribution whose c.d.f. is F. Show that

$$P[|\hat{F}(y) - F(y)| < \epsilon] \geq 1 - \frac{1}{4n\epsilon^2}.$$

What does this say intuitively about the empirical distribution function?

11. The mean lifetime of a new battery is unknown. Based on past versions of the battery, it is reasonable to suppose that the standard deviation of the lifetime is no more than 50 hours. At least how many batteries should be tested to estimate the mean lifetime to within 20 hours with 80% certainty? with 90% certainty?

12. If a die is to be rolled repeatedly in order to gain information about the unknown probability p of rolling a 6, how many times must it be rolled in order to produce an estimate that falls within a tolerance of .01 of the true p, with a probability of at least 90%?

13. Let $\bar{X}_n$ be the sample mean of a random sample of size n from the $\Gamma(\alpha, 2)$ distribution. Show that the sequence of random variables defined by $Y_n = 2\bar{X}_n$ converges in probability and almost surely to α.

14. Let $\bar{X}_n$ be the sample mean of a random sample of size n from the uniform $[0, b]$ distribution. Show that the sequence of random variables defined by $Y_n = 2\bar{X}_n$ converges in probability and almost surely to b.

15. Verify inequality (6.10).

16. Let (Y_n) be the sequence of random variables defined in Example 6.1.4. Does the sequence approach 0 in probability? Why?

17. Let $X_1(\omega) = \omega$ be a random variable on the sample space $\Omega = [0, 1]$ on which the uniform probability measure is placed. Define $X_2 = X_1^2$, $X_3 = X_1^3$, $X_4 = X_1^4, \ldots$. Show that the sequence (X_n) approaches 0 both in probability and almost surely.

18. Consider the experiment of flipping a fair coin repeatedly. Show that the set of outcomes such that at most one head occurs after the sixth flip has probability 0. (Notice that for outcomes in this set, the sample proportion of heads would not converge to $1/2$.)

19. The following example illustrates that a sequence of random variables can converge in probability without converging almost surely. First, let $\Omega = [0, 1]$, and place the uniform probability measure P on Ω. Define random variables Y_{kj} to have the value 1 for outcomes ω in the sets listed below, and 0 otherwise:

$$Y_{21}: \left[0, \frac{1}{2}\right], \qquad Y_{22}: \left[\frac{1}{2}, 1\right],$$

$$Y_{31}: \left[0, \frac{1}{3}\right], \qquad Y_{32}: \left[\frac{1}{3}, \frac{2}{3}\right], \qquad Y_{33}: \left[\frac{2}{3}, 1\right],$$

$$Y_{41}: \left[0, \frac{1}{4}\right], \qquad Y_{42}: \left[\frac{1}{4}, \frac{2}{4}\right], \qquad Y_{43}: \left[\frac{2}{4}, \frac{3}{4}\right], \qquad Y_{44}: \left[\frac{3}{4}, 1\right],$$

$$\vdots$$

In general, $Y_{kj}(\omega) = 1$ if $\omega \in [(j-1)/k, j/k]$ and 0 otherwise, for $j =$

$1, 2, \ldots, k$. Form a sequence of random variables X_n by reading across a row of Y's, then moving to the next row, for example, $X_1 = Y_{21}$, $X_2 = Y_{22}$, $X_3 = Y_{31}$, $X_4 = Y_{32}$, $X_5 = Y_{33}$, $X_6 = Y_{41}, \ldots$. Explain intuitively why the sequence (X_n) converges to 0 in probability, but not almost surely.

6.2 | Central Limit Theorem

The laws of large numbers tell us that the sample mean $\bar{X}$ of a random sample of size n from a distribution with finite mean μ and variance σ^2 converges to μ as $n \to \infty$, regardless of the exact form of the distribution from which we are sampling. We would now like to investigate whether even more is true. For large n, does the *distribution* of $\bar{X}$ converge to some distribution, regardless of the underlying distribution of the sample?

From earlier work on expectation, $E[\bar{X}] = \mu$ and $\text{Var}(\bar{X}) = \sigma^2/n$. Thus the standardized random variable

$$Z = \frac{\bar{X} - \mu}{\sigma/\sqrt{n}} \tag{6.12}$$

has mean 0 and variance 1, regardless of the underlying distribution. Moreover, in the special case that the underlying sample values are themselves normal, the distribution of $\bar{X}$ is $N(\mu, \sigma^2/n)$, hence the random variable Z in (6.12) has the standard normal distribution. So if there is a common limiting distribution, then that distribution must be normal. This is the content of the major theorem of probability that we study in this section, the *Central Limit Theorem*.

Do we have any right to expect such a theorem to be true? One way to check is to simulate a large number of random samples, each of the same (large) size n, compute $\bar{X}$ for each sample, and plot a histogram of the values of $\bar{X}$. The results of two such experiments are displayed in Fig. 6.4. In part (a) of the figure, 200 random samples of size 40 from the uniform distribution on $[2, 4]$ were simulated. The histogram bears the classic normal bell shape. To see whether a distributional limit theorem could hold when the underlying distribution is discrete, I also tried a simulation of 100 random samples of size 36 from the Poisson distribution with parameter 4. The histogram of sample means for this experiment, shown in part (b) of Fig. 6.4, also has roughly the shape of the normal density.

—— **?Question 6.2.1** For each of the distributions in Fig. 6.4, find the theoretical standard deviation of $\bar{X}$. Are the standard deviations consistent with the spread of probability in the histograms? Are the center points of the histograms where you would expect them to be?

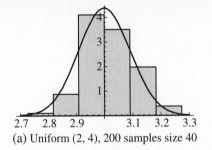
(a) Uniform (2, 4), 200 samples size 40

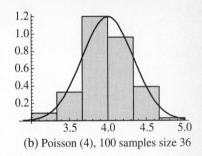
(b) Poisson (4), 100 samples size 36

Figure 6.4 Simulated Distributions of $\bar{X}$

What is responsible for this phenomenon? Recall that the defining formula for $\bar{X}$ is $(X_1 + X_2 + \cdots + X_n)/n$. Some of the X_i's will be smaller than μ and others larger. It would be rare, however, for excessively many X_i's to be smaller than μ or larger than μ. The small and large X_i's will "average out" most of the time, producing a value of $\bar{X}$ that is frequently near the center. Therefore an observed histogram of $\bar{X}$ values ought to have large frequencies near μ, and small frequencies as we move farther from μ. Shortly we will give a detailed motivation using moment-generating functions that an observed histogram of $\bar{X}$ values should have the shape of the normal density. To prepare for this, we will need the so-called *continuity theorem* for m.g.f.'s, which we must state without proof. (For a proof see Billingsley, 1979, p. 303.)

PROPOSITION 6.2.1

Let (F_n) be a sequence of cumulative distribution functions with associated moment-generating functions (M_n). Suppose that there is an interval $I = (-\epsilon, +\epsilon)$ about 0 such that for each t in I, $M_n(t) \to M(t)$ as $n \to \infty$, where M is a moment-generating function associated with a c.d.f. F. Then $F_n(x) \to F(x)$ as $n \to \infty$ at all points x at which F is continuous.

To paraphrase Proposition 6.2.1, if a sequence of moment-generating functions converges to another moment-generating function, then the associated sequence of c.d.f.'s converges to the c.d.f. of the limit moment-generating function. This theorem enables us to find the limiting distribution of a sequence of random variables by finding the limit moment-generating function. (See Exercises 1 and 2 at the end of this section.)

Let us apply the continuity theorem to our problem. We are given a random sample $X_1, X_2, \ldots, X_n$ from some distribution with mean μ and variance σ^2. Let $M_n(t)$ be the moment-generating function of Z in (6.12). Then, by independence of the X_is,

$$M_n(t) = E\left[\exp\left(t \cdot \frac{\bar{X} - \mu}{\sigma/\sqrt{n}}\right)\right]$$

$$= E\left[\exp\left(\frac{\sqrt{n}t}{n} \cdot \sum_{i=1}^{n} \frac{X_i - \mu}{\sigma}\right)\right]$$

$$= \prod_{i=1}^{n} E\left[\exp\left(\frac{t}{\sqrt{n}} \cdot \frac{X_i - \mu}{\sigma}\right)\right] = \left[M\left(\frac{t}{\sqrt{n}}\right)\right]^n, \qquad (6.13)$$

where M is the moment-generating function of the random variable $(X_i - \mu)/\sigma$. Recall that

$$M(0) = E[e^0] = 1, \qquad M'(0) = E\left[\frac{X_i - \mu}{\sigma}\right] = 0,$$

$$M''(0) = E\left[\left(\frac{X_i - \mu}{\sigma}\right)^2\right] = \frac{\sigma^2}{\sigma^2} = 1.$$

A third-order Taylor expansion of M about 0 yields

$$M(s) = M(0) + \frac{1}{1!}M'(0)s + \frac{1}{2!}M''(0)s^2 + \frac{1}{3!}M'''(r)s^3, \qquad r \in (0, s). \quad (6.14)$$

Evaluating this expression at $t/\sqrt{n}$ as required by (6.13) and using the preceding results for the derivatives of M, we obtain

$$M\left(\frac{t}{\sqrt{n}}\right) = 1 + \frac{1}{2} \cdot \frac{t^2}{n} + \frac{1}{6} \cdot M'''(r) \cdot \frac{t^3}{n^{3/2}}, \qquad r \in \left(0, \frac{t}{\sqrt{n}}\right). \quad (6.15)$$

Therefore, by formulas (6.13) and (6.15),

$$M_n(t) = \left(1 + \frac{t^2/2 + M'''(r)t^3/(6\sqrt{n})}{n}\right)^n. \qquad (6.16)$$

A standard limit result from calculus says that $\lim_{n\to\infty}(1 + a/n)^n = e^a$, where a is constant. This result can be extended to cases like ours in which, instead of a constant a, we have a sequence $a_n = t^2/2 + M'''(r)t^3/(6\sqrt{n})$ that approaches a finite limit $t^2/2$, as long as M''' is well behaved. Such an extension gives us the limit moment-generating function:

$$M_n(t) \to M_0(t) = e^{t^2/2} \quad \text{as } n \to \infty.$$

The latter m.g.f. is that of the $N(0, 1)$ distribution. By the continuity theorem, the c.d.f. of Z approaches the standard normal c.d.f. as the sample size approaches ∞. This argument can be made rigorous using the complex-valued characteristic function $\phi(t) = E[e^{itZ}]$ mentioned in Section 5.3, whose boundedness properties avoid certain technical problems. But we have at least given a very convincing argument for the following theorem.

PROPOSITION 6.2.2

Central Limit Theorem Let $X_1, X_2, X_3, \ldots$ be a sequence of independent and identically distributed random variables with finite mean μ and variance σ^2. Let $\bar{X}_n = \sum_{i=1}^{n} X_i/n$. Then the c.d.f. $F_n(x)$ of the random variable

$$Z_n = \frac{\bar{X}_n - \mu}{\sigma/\sqrt{n}} \qquad (6.17)$$

converges to the standard normal c.d.f. at all points x.

The identical distribution hypothesis of the Central Limit Theorem can be weakened, in which case the conclusion involves the standardized sums $S_n = \sum_{i=1}^{n} X_i$ (see e.g., Chung, 1974, p. 196). Even the hypothesis of independence of the X_i's can be slightly weakened (see Billingsley, 1979, p. 315). There are also error estimates (see Loeve, 1977, p. 300), which in the i.i.d. case say that the absolute difference between the true c.d.f. of the standardized sum and the standard normal c.d.f. is no larger than a constant times the third moment of the underlying distribution, divided by the square root of the sample size. The experience of many in the statistical community suggests that the normal approximation in the Central Limit Theorem is good for sample sizes as small as 25 or less. The more symmetric is the underlying distribution, the better.

?Question 6.2.2 Go to the library and investigate the history of the Central Limit Theorem. What special case concerned DeMoivre and Laplace? What did Lyapunov and Lindeberg do, and when?

Example 6.2.1 Recall Example 6.1.2, in which the political candidate wondered how many people to poll in order to estimate her support p to within .05, with confidence 90%. For the large sample size that will be required, it will be reasonable to approximate the distribution of $\hat{p} = \bar{X}$ by a normal distribution. The underlying sample random variables $X_1, X_2, X_3, \ldots$ are i.i.d. Bernoulli (p); hence their common mean is p and their variance is $p(1 - p)$. Then,

$$.9 = P[|\bar{X}_n - p| < .05]$$

$$= P\left[-\frac{.05}{\sqrt{p(1-p)}/\sqrt{n}} < \frac{\bar{X}_n - p}{\sqrt{p(1-p)}/\sqrt{n}} < \frac{.05}{\sqrt{p(1-p)}/\sqrt{n}}\right].$$

The distribution of the random variable Z in the middle of the string of inequalities is approximately $N(0, 1)$, by the Central Limit Theorem. The point z such that $P[-z < Z < z] = .9$ is found from the standard normal table to be $z = 1.645$. To achieve the desired precision and confidence, we must therefore ensure that

$$\frac{.05\sqrt{n}}{\sqrt{p(1-p)}} > 1.645 \Rightarrow \sqrt{n} > \frac{1.645\sqrt{p(1-p)}}{.05}.$$

The product $p(1 - p)$ can be no larger than $1/4$ (at $p = 1/2$), and so it suffices to ensure that

$$\sqrt{n} > \frac{1.645\sqrt{1/4}}{.05} \Rightarrow n > 270.6.$$

The new required sample size of 271 is a substantial improvement over the sample size of 1000 found using the Chebyshev inequality.

Example 6.2.2 Suppose that 30 students in a freshman chemistry class are doing an experiment to estimate the concentration of a certain solute in a solution given to them from a well-mixed common stock. All of the students perform independent experiments, and they obtain concentrations $X_1, X_2, \ldots, X_{30}$. Strangely enough (or not so strangely if you have ever done a chemistry lab) the readings are all different. Let us postulate a model in which the readings have some common distribution with mean μ equal to the true concentration and variance 6. If we estimate μ by the average concentration in the class $\bar{X}$, then the probability that the estimate differs from μ by more than 1 unit is

$$P[|\bar{X} - \mu| > 1] = P\left[\left|\frac{\bar{X} - \mu}{\sigma/\sqrt{n}}\right| > \frac{1}{\sqrt{6}/\sqrt{30}}\right] \approx P[|Z| > 2.236] \approx .02.$$

Although any one observation has a standard deviation of $\sqrt{6} \approx 2.45$, the sample mean is not very likely to be more than a unit from the true concentration. It appears that there is safety in numbers, even in chemistry lab.

EXERCISES 6.2

1. Let X_n $(n \geq 2)$ be a discrete random variable with probability mass function $q_n(0) = 1/2 + 1/n, q_n(1 + 1/n) = 1/2 - 1/n$, and $q_n(x) = 0$ otherwise. Find the moment-generating function $M_n(t)$ of X_n, and show that it converges to an m.g.f. $M(t)$ as $n \to \infty$. Then find the c.d.f. $F_n(x)$ of X_n and show directly that for all $x \neq 1$, $F_n(x) \to F(x)$ as $n \to \infty$, where F is the c.d.f. associated with M.

2. Let the random variable X_n have the probability mass function putting weights of $1/8 + 1/(2n)$, $3/4 - 1/n$, and $1/8 + 1/(2n)$ on the states $-1 - 1/n$, 0, and $1 + 1/n$, respectively. Find the limit as $n \to \infty$ of the moment-generating function of this distribution, and find the c.d.f. associated with that limit.

3. Consider a reliability system of n components connected in series. Assume that the components fail independently at times having the exponential distribution with parameter λ. Find the limit as $n \to \infty$ of the distribution of the random variable $n \cdot T$, where T is the failure time of the system.

4. When a bank computes the amount of interest it pays in a monthly cycle to its depositers, it rounds the exact value to the nearest cent. Suppose that a small bank has 150 savings accounts that draw interest in a month. Approximate the probability that the total interest paid to all accounts is within 10 cents of the true total interest. (*Hint:* Let X_i be the difference between the true interest and the rounded interest for the ith account.)

5. Estimate the probability that the sample mean $\bar{X}$ of 100 observations from some distribution with mean μ and variance σ^2 falls within a distance of $\sigma/10$ of μ.

6. Suppose that calls from each of 81 phone extensions to a college switchboard occur according to a Poisson process with rate 1 per hour and that the 81 processes are independent of one another. Estimate the probability that at least 160 calls occur during a two-hour period.

7. An airline runs a 200-seat flight by selling 215 seats. Those passengers who are closed out due to overcrowding must be offered a refund. Each passenger with a ticket will show up for the flight with probability .9,

independently of other passengers. Approximate the probability that the airline will have to offer at least 5 refunds.

8. A brother and sister raid their parents' penny jar, then pitch pennies against a wall on a rainy afternoon. In the game, each player throws one of his or her pennies, and the closest to the wall wins both. Suppose that the sister, who is a slightly better player, has a 55% chance of winning on a single play. Approximate the probability that the sister is at least 20 cents up on her brother after 200 plays. After how many plays is there a probability of at least .95 that she is at least 20 cents up on her brother?

9. The atomic mass of element number 66 Dysprosium is about 162.50. Suppose that experimental error causes empirical estimates of this mass to follow a triangular distribution that rises linearly from a value of 0 at 162 to a maximum at 162.50, then falls linearly to 0 at 163. Use the Central Limit Theorem to approximate how many experimental measurements must be made so that their average estimates the atomic weight to within .1 with probability 90%.

10. It is estimated that there are between 100 and 150 mg of caffeine in a cup of coffee. I drink about 5 cups per day, or 35 per week. Assuming that the mean amount of caffeine per cup is 125, and the variance is that of an appropriate uniform distribution, approximate the probability that I have a caffeine intake of more than 4550 mg in a week.

11. Redo Example 6.1.3 using the Central Limit Theorem.

12. The following scenario describes a model for the diffusion of particles. A particle, constrained to move along a line, begins at position 0. At each member of a sequence of discrete times, the particle moves either to the left by a distance of Δx or to the right by Δx, with equal probability. These steps are assumed to be independent of one another. The times are equally spaced: $t_1 = \Delta t$, $t_2 = 2\Delta t, \ldots, t_n = n\Delta t$. Let X_s be the position of the particle at time $s = n\Delta t$. Find the approximate probability distribution of X_s. Considering s to be fixed, what relationship between Δx and $\Delta t = s/n$ permits the limiting distribution of X_s to exist as $n \to \infty$?

13. Explain how the following *DeMoivre–Laplace* approximation theorem follows from the Central Limit Theorem: If X is a random variable with the $b(n, p)$ distribution, then

$$P\left[a < \frac{X - np}{\sqrt{np(1-p)}} < b\right] \approx \int_a^b \frac{1}{\sqrt{2\pi}} e^{-s^2/2}\, ds.$$

CHAPTER 7

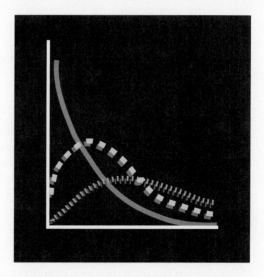

INTRODUCTORY
DATA ANALYSIS

7.1 | Random Samples and Summary Statistics

We now begin our study of statistics, which makes use of many of the results and ideas of probability contained in the previous chapters. In the broadest sense, a *statistician* is a gatherer and interpreter of information. The statistician is often prompted by an experimenter or researcher who has a problem or phenomenon to investigate. It is the responsibility of the statistician, in concert with the researcher–client, to develop a plan to gather and summarize relevant data, to construct pictures that facilitate understanding of the data, and to answer questions about the population from which the data were sampled, based on a probabilistic model. The statistician must also check the validity of this assumed model. We will be addressing all of these topics in the coming chapters. [I highly recommend that you read the first few chapters of Deming (1960), which elaborate on the general role of a statistician.]

This chapter will focus on some elementary ideas involved in gathering, summarizing, and graphically presenting data. This section contains a brief discussion of random sampling, and introduces some commonly used numerical summaries of data. Section 7.2 proceeds to various graphical methods for representing data. Whereas numerical summaries edit the data set down to the point at which important detail can be lost, well-designed graphs can retain most or all of the data set yet still provide explanatory power. The final section of the chapter discusses some relatively recent research on general principles of graph construction.

7.1.1 Random Samples

Recall the mathematical definition of a random sample. It starts with a population of individuals, some numerical characteristic of which we are interested in studying. A single individual drawn from this population gives a value of X for the numerical characteristic, and we suppose that X is random, with some probability distribution whose density (or mass function) is $f(x; \theta)$. Here θ is a constant parameter of the distribution; often, we want to use the sample to estimate it. A *random sample* is a sequence $X_1, X_2, \ldots, X_n$ of independent identically-distributed random variables with the distribution $f(x; \theta)$. We conceive of these random variables as observations sampled independently from the underlying population. Although this mathematical characterization seems to be precise and complete, there are some practical details relative to "randomness" of sampling that must be considered.

First, a sample can be random only relative to the universe of discourse for the problem. To illustrate, consider the following scenario.

Example 7.1.1

A cosmetics firm is developing a new antiperspirant for men. It would like to know how well the product functions under conditions of stress. The company therefore plans to test the protective power of the product on 30 men, each of whom will walk on a treadmill for 20 minutes.

These 30 men are certainly not a random sample of all people, since women are purposely excluded. Given that the intended segment of the market who will use this product is male, however, it makes sense to restrict the universe to male subjects. Before going on, respond to the following question.

?Question 7.1.1

What possible difficulties might the cosmetics firm face in gathering a random sample from the universe of men?

Example 7.1.1

Continued It should not have taken you long to realize that, for example, a sample of men drawn from the Phoenix city basketball league may not have the same physical characteristics as other potential consumers. That is, such factors as age, physical condition and home climate may lead to systematic overestimation or underestimation of distributional parameters if the sample is actually taken from a subpopulation of the universe. A little walking on a treadmill may not make a frequent basketball player sweat a lot, and hence the makeup of the sample may make the antiperspirant appear to be more effective than it really is.

In a truly random sample each member of the universe of discourse has an equal chance of appearing in the sample. This means that the universe is well described and that some randomization procedure is used to select individuals from the universe.

It becomes apparent that obtaining a random sample is a more difficult task than one would think. To clarify the universe, the cosmetics firm may decide that its market consists of all American males between the ages of 18 and 60. A selection procedure to produce a truly random sample would then require a list of all of these people (usually called a *frame* by sampling theorists), which could be numbered from 1 to N, where N is the size of the population. By selecting n numbers from a table of random numbers or a computer's random number generator, the subjects to be included in the sample can be determined. Note that for the independence assumption on the data values $X_1, X_2, \ldots, X_n$ to hold, the selection should be done *with* replacement as opposed to *without* replacement (recall the binomial and hypergeometric distributions), but if the size N of the frame is very large relative to the size n of the sample, then replacement should not make much difference.

Questions do remain. For example, how are we going to compile a complete list of American males aged 18–60 from which to select? And if we cannot compile it, or compilation is too expensive, how do we produce a sample that approximates the characteristics of the population? What happens if the individuals we select do not consent to participate in the study? These questions are better left to a thorough course on samples and surveys. The only purpose in this introduction is to alert you to the first principles of sampling: defining and framing the population of interest, then selecting randomly from it.

It is relatively simple to use a random number table to select the members of a sample. A table of random numbers contains a long sequence of digits 0–9, which to all intents and purposes form simulated values of an independent and identically distributed stream of random variables with state space $E = \{0, 1, \ldots, 9\}$. Moreover, these digits can be grouped into pairs, in which case they simulate values of a stream of random variables with state space $E = \{00, 01, \ldots, 99\}$. Triples, quadruples, and so on, can be used in the same way depending on the size of the numbers that describe the sampling frame.

An example sequence of random numbers follows. The blank spaces after every four digits are not significant other than to enhance readability.

$$4496 \ 6375 \ 0215 \ 5700 \ 5255 \ 5357 \ 9082$$

Suppose that from a frame of 65 individuals you were to sample five. The least power of ten greater that 65 is 100, so that the 65 members of the frame could be encoded as two-digit numbers 00–64, with the two-digit numbers 65–99 unused. From the table, pick two-digit numbers successively, skipping unused numbers, until there are five numbers. The sample then consists of the individuals associated with those numbers. Here, if we start on the far left of the sequence of random digits (anywhere would be equally good), we would pick individual 44, (skip 96), individual 63, (skip 75), individual 02, individual 15, and individual 57 to make up our sample.

Consult the manual for your particular software or computer to determine how to simulate random numbers via machine. Often only one command or statement is necessary, which allows you to specify the possible range of values for the random number, and the number of simulated values desired. In *Mathematica*, for example, there are predefined names for the most important distributions, such as *NormalDistribution[0, 1]*, and the command *Table[Random[distribution],{n}]* produces a list with *n* simulated observations from whatever distribution is named. Such routines depend upon the simulation methods that we discussed in Section 5.1 and the ability of computers to generate pseudo-random sequences of numbers that appear to be uniformly distributed on [0, 1]. [Incidentally, one of the most common ways of doing machine simulation is to let the uniform random number be the quantity *seed/modulus*, where *seed* is initialized arbitrarily and updated by the equation *seed* = (*m · seed* + *inc*) *mod modulus*, where *m*, *inc*, and *modulus* are suitable constants. See Exercise 7 at the end of this section.]

It is tempting to be lazy and avoid consulting a random number table or computer to produce the subset of the universe for the sample. Suppose that the subjects in the frame are numbered 1 to *N* and that you are to sample *n* of them. You might just start haphazardly picking numbers 2, 5, 8, 12, and so on, and stop when you have a total of *n* subjects. The problem is that you may not be a very good judge of randomness when selecting a subset of numbers, as the next question and example illustrate.

?Question 7.1.2 Try this experiment with some of your friends. Ask them to give you three random integers from 1 to 8, in ascending order. Do you notice them avoiding consecutive numbers? Now count the number of possible ordered samples without replacement that do not have any consecutive integers (a tree diagram may help). See whether the proportion of people who gave lists with no consecutive integers is larger than the theoretical probability of the event that there are no consecutive integers.

Example 7.1.2 A strange human perception is that odd numbers are more random than even numbers. I asked each person in a group of 18 people (who I must shamefacedly admit were a sample of convenience to me rather than a random sample) to give me five random integers from 1–20. For a random sample, the probability that there are *k* odd numbers in the sample is

$$P[k \text{ odd}] = \frac{\binom{10}{k}\binom{10}{5-k}}{\binom{20}{5}}, \qquad k = 0, 1, \dots, 5. \qquad (7.1)$$

The actual probability distribution of the number of odds in the sample, computed from this formula, is displayed in the first line of Table 7.1. The

Table 7.1

Distribution of Number of Odd Numbers

	0	1	2	3	4	5
Actual	.02	.13	.35	.35	.13	.02
Empirical	0	.17	0	.5	.17	.17

empirical distribution in my convenience sample is in the second line. It indicates a marked preference for odd numbers, in contradiction to true randomness. Later we will discuss a quantitative check for the goodness-of-fit of hypothesized probabilities to observed proportions that can be applied to this example.

You might be interested in performing a similar experiment with prime numbers, because I have noticed a tendency for people to consider primes as more random than nonprimes. The moral is to use a randomizing method rather than haphazard human judgment when choosing a sample from a frame.

One of the most problematic sampling issues is that of the self-selected sample. If responses to a survey are submitted voluntarily, those responses may be biased toward the segment of the frame that is most likely to return the survey, as illustrated by the next example.

Example 7.1.3

A team of researchers (Romana and Jose, 1991) studying Filipino images of Japan received 1200 responses to a 1988 survey that included the question "Do you think that the Japanese government can be trusted?" The results were yes, 17%; no, 52%; and undecided, 29%. A similar survey in 1989 yielded yes, 24%; no, 39%; and undecided, 36%. These numbers seem to show a distrust of the Japanese that is waning as time goes by. Let us suppose that the survey was conducted by mail, surveys were sent to 2500 Filipinos whose names were obtained from income tax data, and responses were mailed back voluntarily. Would you be very comfortable drawing conclusions from these data if these were the conditions under which they were gathered?

This survey entails several possible problems. The frame consists of people who presumably have high enough incomes to be on the tax rolls, which may exclude the poor, and the dependents of the people responsible for the taxes. Thus the frame may only include a special segment of all living Filipinos. It is possible that the frame itself may include a disproportionate number of people who feel themselves to be in competition with, or disadvantaged by, the Japanese. Moreover, those who return the surveys are likely to be people who care more about the issue than those who do not return the surveys. Human nature being what it is, we tend to be more outspoken about the things that we do not like than the things that we do like. The group of people who returned their surveys may include more people who have some complaint against the Japanese than it would if the responses were truly random.

The solutions to these problems are not easy. The random sampler must obtain as complete a census as is possible, and if the frame is inherently incomplete, the sampler must be aware of the exclusions and report them. Once the list of people to whom surveys are to be mailed is randomly selected, the investigator must also take special steps to obtain data from all of the designated sample members. This may involve follow-up mailings, phone calls, or visits. In summary, the main goals in random sampling are to have the frame correspond as closely as possible to the desired universe and to give all people in the frame an equal chance of appearing in the final data set.

7.1.2 Summary Statistics

Once a random sample $X_1, X_2, \ldots, X_n$ has been obtained, the information that it carries must be interpreted and presented. If the sample is large, it can be difficult to digest this information without some kind of simplification and organization. Special functions of the X_i's called *statistics* can summarize important features of the sample.

We have already seen several summary statistics. The sample mean and the sample variance

$$\overline{X} = \sum_{i=1}^{n} \frac{X_i}{n}, \qquad S^2 = \sum_{i=1}^{n} \frac{(X_i - \overline{X})^2}{n-1} \tag{7.2}$$

measure the center and the spread of the data, respectively. Exercise 9 contains an alternative computational formula for the sample variance, which you will find useful if you do not have a statistical calculator with a variance key.

Incidentally, we will be using the following notational distinction. Since a statistic is a function of sample random variables, it is itself a random variable. Hence a statistic is conceptually different from its observed numerical value, which is computable after the data are taken. We will normally reserve capital letters for statistics, and their corresponding lowercase letters will be used when the observed numerical data are substituted into the defining formula for the statistic. Thus $\overline{x}$ and s^2 are the observed values of the random variables $\overline{X}$ and S^2.

?Question 7.1.3 Some statisticians prefer the *coefficient of variation* $S/\overline{X}$ as a measure of spread. Can you think of reasons why this might be?

One family of summary statistics arises from the order statistics of the sample. Recall that the *order statistics* $Y_1, Y_2, \ldots, Y_n$ of a random sample $X_1, X_2, \ldots, X_n$ are the sample values written in ascending order. The notation

$X_{(i)}$ is also used for the ith smallest order statistic Y_i. The joint and marginal distributions of the order statistics were found in Section 5.2.

Three simple statistics that give information about the variation of the data are the *minimum* value in the sample $X_{(1)}$, the *maximum* value in the sample $X_{(n)}$, and the *range* $X_{(n)} - X_{(1)}$. These statistics are useful in summarizing the extremes of the data, but they give no information about how the data set may be distributed between the two extremes. For more detail, we can adapt the notion of the percentile to the context of samples.

—— **?Question 7.1.4** Watch your local weather forecast. Does the forecaster seem to compute the "average" daily temperature in a way that you did not expect, using the smallest and largest order statistics? Does this approach appear to be reasonable? Can you think of an alternative?

Roughly speaking, the $100 \times p$th percentile of a sample should be the order statistic that lies a proportion p of the way through the list of n order statistics. To be more precise, define the $100 \times p$th *percentile of a sample* to be the mth order statistic if $(n + 1) \times p$ is an integer m; otherwise, if $(n + 1) \times p = m + u$, where m is the integer part of $(n + 1) \times p$ and u is the remainder, define the $100 \times p$th percentile to be the weighted average

$$(1 - u) \times X_{(m)} + u \times X_{(m+1)} \tag{7.3}$$

of the mth and $(m + 1)$st order statistics. By doing this, we essentially interpolate between the two nearest order statistics to the percentile we want.

Example 7.1.4 In Example 5.5.1 we looked at a data set of numbers of users of a computer lab during a month of 29 days. The order statistics follow.

$$9, \ 9, \ 12, \ 13, \ 15, \ 16, \ 25, \ 25, \ 26, \ 29, \ 30, \ 31, \ 32, \ 32, \ 37,$$
$$38, \ 39, \ 39, \ 40, \ 41, \ 44, \ 44, \ 45, \ 46, \ 46, \ 49, \ 52, \ 55, \ 58$$

To find, for example, the 25th percentile of the data, first compute $(n + 1) \cdot p = (29 + 1) \cdot .25 = 7.5$. The integer part of this number is 7, and the remainder is .5. Then the desired percentile is

$$.5 \cdot x_{(7)} + .5 \cdot x_{(8)} = .5 \cdot (25) + .5 \cdot (25) = 25.$$

Similarly, the 75th percentile of the sample can be found by first computing $(29 + 1) \cdot .75 = 22.5$, then finding the weighted average:

$$.5 \cdot x_{(22)} + .5 \cdot x_{(23)} = .5 \cdot (44) + .5 \cdot (45) = 44.5.$$

—— **?Question 7.1.5** Find the 90th percentile of the data in Example 7.1.4.

The *median* of a sample, which we will denote by M, is the 50th percentile of the sample. It is not hard to see from the definition of percentiles that the median is the middle data value in the ordered list if the number of data points is odd, and the median is the simple average of the two middle values if the number of data points is even. (Check this.) For the computer users data set in Example 7.1.4, we see that the median number of users is the $30(.5) = 15$th order statistic, which has the value 37.

The other most commonly used summary statistic based on the percentiles is the *interquartile range*, defined to be the difference between the 75th and 25th percentiles of the sample, that is, the region in which the middle half of the data lies. The interquartile range is an alternative to the sample variance for describing the spread of the data. Since we found in the last example that the 75th percentile is 44.5 users, and the 25th percentile is 25 users, the value of the interquartile range is $44.5 - 25 = 19.5$.

Both the sample mean and median measure the central tendency of the data. How do they relate to each other? The answer, surprisingly, is that not only is there no direct formula that gives one of these statistics in terms of the other, but they do not even have to be close to each other (see Exercise 13). In fact, the difference between them can yield important information about the data set. To see how, consider these four sets of five numbers.

$$
\begin{array}{lll}
1, 2, 3, 4, 5; & \bar{x} = 3, & m = 3. \\
1, 2, 3, 4, 20; & \bar{x} = 6, & m = 3. \\
1, 2, 3, 12, 16; & \bar{x} = 6.8, & m = 3. \\
1, 5, 10, 11, 12; & \bar{x} = 7.8, & m = 10.
\end{array}
$$

The first sample has mean 3 and median 3. Notice that the sample is symmetric about its mean. In the second sample, because of the outlying data point 20, the sample mean has increased to 6, while the median pays no attention to the outlier, retaining its value of 3. This phenomenon is the reason why some statisticians prefer the median to the mean: It is less affected by extreme observations, which may in fact have been obtained erroneously. The third sample contrasts with the first in that two rather large observations have replaced 4 and 5, destroying the symmetry of the original data about 3. Several observations are piled up on the left, and a couple of observations are far to the right of them. Again the effect is to pull up the mean above the median; $\bar{x} = 6.8$, while m is still 3. A random sample is said to be *skewed to the right* (or *positively skewed*) in the case that we have just described in which the mean exceeds the median. A histogram of such a data set will be shaped as in Fig. 7.1(a). In the fourth sample we have reversed the direction of the skewness; several large values are bunched and are accompanied in the sample by a couple of observations that are far to the left of them. This time, the few smaller values pull down the sample mean to 7.8, which is now less than the median of 10. When a random sample is such that the mean is less than the median, we say that it is *skewed to the left* (or *negatively skewed*). A left-skewed histogram will look like part (b) of Fig. 7.1.

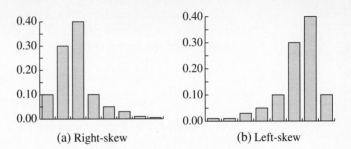

Figure 7.1   Skewness of a Sample

7.1.3   Multivariate Summary Statistics

Many data analysis problems involve taking several measurements on each of a group of subjects. The information pertaining to a single subject is therefore a vector, and the entire data set is a random sample of vectors. Statisticians usually refer to the ith component of the data vector as the ith measurement variable for a subject. We are then interested in summarizing information about the individual variables and about the dependence of each variable on the others.

For example, I taught a mathematical statistics class of 14 students a few years ago, in which I gave three midterm tests. The results are listed in Table 7.2. The subjects in the table are the students, for which there are three measurement variables, namely the scores on the three tests. A general mathematical model that applies to this data set uses the idea of a *data matrix*:

$$\mathbf{X} = \begin{bmatrix} X_{11} & X_{12} & \dots & X_{1n} \\ X_{21} & X_{22} & \dots & X_{2n} \\ \vdots & \vdots & \ddots & \ddots \\ X_{p1} & X_{p2} & \dots & X_{pn} \end{bmatrix} = \begin{bmatrix} \mathbf{X}_1 & \mathbf{X}_2 & \dots & \mathbf{X}_n \end{bmatrix}, \qquad (7.4)$$

where each column $\mathbf{X}_i$ is a multivariate observation with p component variables. For our test data, there are $p = 3$ variables and $n = 14$ observations, and the data matrix is of the same structure as Table 7.2, so that

$$\mathbf{X}_1 = \begin{bmatrix} 74 \\ 66 \\ 78 \end{bmatrix}, \qquad \mathbf{X}_2 = \begin{bmatrix} 85 \\ 61 \\ 90 \end{bmatrix}, \qquad \dots, \qquad \mathbf{X}_{14} = \begin{bmatrix} 51 \\ 68 \\ 82 \end{bmatrix}.$$

Table 7.2

Mathematical Statistics Test Scores

	1	2	3	4	5	6	7	8	9	10	11	12	13	14
Test 1	74	85	52	87	56	61	92	57	93	90	76	78	73	51
Test 2	66	61	76	86	72	76	70	52	85	91	90	92	88	68
Test 3	78	90	98	93	76	80	92	62	93	90	83	97	83	82

It is logical to define the *vector sample mean* $\overline{\mathbf{X}}$ as the column vector whose components are the row means of the data matrix; in other words, the ith component of the mean vector is

$$\overline{X}_i = \frac{1}{n} \sum_{j=1}^{n} X_{ij}. \tag{7.5}$$

Thus $\overline{X}_i$ is just the sample mean for the ith measurement variable. For the data in Table 7.2 we have

$$\overline{x}_1 = \frac{74 + 85 + \cdots + 51}{14} = 73.2,$$

$$\overline{x}_2 = \frac{66 + 61 + \cdots + 68}{14} = 76.6,$$

$$\overline{x}_3 = \frac{78 + 90 + \cdots + 82}{14} = 85.5.$$

Hence $\overline{\mathbf{x}} = [73.2 \ \ 76.6 \ \ 85.5]'$.

When we consider summary measures for the spread of multivariate data, there is more to think about than in the single-variable case. Each individual row in the data matrix $\mathbf{X}$ can be thought of as a sample in its own right, with corresponding sample variance

$$S_i^2 = \frac{1}{n-1} \sum_{j=1}^{n} (X_{ij} - \overline{X}_i)^2. \tag{7.6}$$

Another item of interest is how the row variables depend on one another. Consider the data in the ith and kth rows, $X_{i1}, X_{i2}, \ldots, X_{in}$ and $X_{k1}, X_{k2}, \ldots, X_{kn}$, which we view as measurements of two variables each on n subjects. We can ask for an estimate of the theoretical covariance and correlation of variables i and k. Recall that

$$\mathrm{Cov}(X_i, X_k) = E[(X_i - \mu_i)(X_k - \mu_k)], \qquad \mathrm{Corr}(X_i, X_k) = \frac{\mathrm{Cov}(X_i, X_k)}{\sigma_i \sigma_k}, \tag{7.7}$$

where μ_i and μ_k are the theoretical means and σ_i and σ_k are the theoretical standard deviations of the two random variables. The analogous *sample covariance* between the ith and kth variables is defined by

$$S_{ik} = \frac{1}{n-1} \sum_{j=1}^{n} (X_{ij} - \overline{X}_i)(X_{kj} - \overline{X}_k). \tag{7.8}$$

Also, just as the theoretical correlation is the theoretical covariance divided by the product of the standard deviations, the *sample correlation* between the ith and kth variables is

$$R_{ik} = \frac{S_{ik}}{S_i \cdot S_k} = \frac{\sum_{j=1}^{n}(X_{ij} - \overline{X}_i)(X_{kj} - \overline{X}_k)}{\sqrt{\sum_{j=1}^{n}(X_{ij} - \overline{X}_i)^2} \cdot \sqrt{\sum_{j=1}^{n}(X_{kj} - \overline{X}_k)^2}}. \tag{7.9}$$

Example 7.1.5

In the test data, for example, the sample variances for tests 1 and 3 are

$$s_1^2 = \frac{1}{13}((74 - 73.2)^2 + (85 - 73.2)^2 + \cdots + (51 - 73.2)^2) = 233.7,$$

$$s_3^2 = \frac{1}{13}((78 - 85.5)^2 + (90 - 85.5)^2 + \cdots + (82 - 85.5)^2) = 95.2.$$

The sample covariance between tests 1 and 3 is

$$s_{13} = \frac{1}{13}((74 - 73.2)(78 - 85.5) + (85 - 73.2)(90 - 85.5) + \cdots$$
$$+ (51 - 73.2)(82 - 85.5)) = 74.6.$$

Hence the sample correlation between tests 1 and 3 is

$$r_{13} = \frac{s_{13}}{s_1.s_3} = \frac{74.6}{\sqrt{233.7 \cdot 95.2}} = .500.$$

From the definition of the sample covariance as a sum of products, and the example calculation, it should be apparent that the sample covariance will be large and positive if the data values for variable i tend to be large (or small) precisely when the data values for variable k tend to be large (or small). The sample covariance will be large and negative if variable i tends to be large when variable k is small, and vice versa. The sample correlation has similar properties, but the sample correlation is also invariant under linear transformations of the data (see Exercise 16).

There is a matrix analogous to the theoretical covariance matrix that contains all of the sample covariance information. The *sample covariance matrix* is defined by

$$\mathbf{S} = \begin{bmatrix} S_1^2 & S_{12} & \cdots & S_{1p} \\ S_{21} & S_2^2 & \cdots & S_{2p} \\ \vdots & \vdots & \ddots & \vdots \\ S_{p1} & S_{p2} & \cdots & S_p^2 \end{bmatrix}. \tag{7.10}$$

Like the theoretical covariance matrix, it is a symmetric matrix with the variances on its diagonal, and the covariances in the off-diagonal positions. There is also a sample correlation matrix $\mathbb{R}$, whose $i - j$ entry is the sample correlation R_{ij}.

Example 7.1.6

We have already calculated some of the entries in the covariance matrix for the test data set. Similar computations can calculate the rest of the entries, which come out as follows:

$$\mathbf{s} = \begin{bmatrix} 233.7 & 78.4 & 74.6 \\ 78.4 & 154.9 & 71.3 \\ 74.6 & 71.3 & 95.2 \end{bmatrix}.$$

Notice that the sample covariance matrix is symmetric. (Why must this be so?)

Formula (7.9) and the foregoing covariance matrix yield the sample correlation matrix

$$r = \begin{bmatrix} 1 & .412 & .500 \\ .412 & 1 & .588 \\ .500 & .588 & 1 \end{bmatrix}.$$

?Question 7.1.6 Why should the diagonal entries of the sample correlation matrix be 1? (*Hint:* What would the sample covariance between a variable and itself be?)

The sample correlation matrix is an estimator of the theoretical correlation matrix; hence we would expect it to give information about the degree of linear dependence among the measurement variables. We do not yet know how to judge whether a sample correlation of .412 is significantly different from zero in a probabilistic sense, but we will consider that question later.

We close this section with an analogue to Proposition 4.3.2, which shows that the sample correlation between two variables is bounded, as is the theoretical correlation, and that it directly measures the degree of linear association in the sample between the two variables. The proof is almost identical to that of the earlier proposition, with expectation replaced by the sum over all data points, divided by $n - 1$.

PROPOSITION 7.1.1

Let $(X_{11}, X_{21}), (X_{12}, X_{22}), \ldots, (X_{1n}, X_{2n})$ be a random sample of pairs. Then,
  (a) $|R_{12}| \leq 1$;
  (b) the observed value of the sample correlation $|r_{12}| = 1$ if and only if the observed sample points $(x_{11}, x_{21}), (x_{12}, x_{22}), \ldots, (x_{1n}, x_{2n})$ all fall along a line.

EXERCISES 7.1

1. Comment on the possible differences between the responses that might be given by subjects to personal interview questions as opposed to anonymous mail surveys. In particular, is it possible that a sample obtained by interviewing subjects might have results very different from a mail survey sample of the very same subjects?

2. A survey of Protestant clergy in 1993–1994 by Hartford Seminary on divorce rates was performed by mailing 10,000 questionnaires to ministers of various denominations (Douthat, 1995). A total of 2458 women and 2086 men returned the forms. Some of the results are as follows: Among Unitarian-Universalists, 47% of the women and 44% of the men said that they had been divorced at some point; among Baptists, 19% of the women and 13% of the men reported a divorce; and among Presbyterians it was 25% of the women and 19% of the men. Nationwide, census data indicate that about 23% of women and 22% of men have been divorced.

Some of the more conservative denominations have explicit restrictions or implicit pressures against divorced ministers. Comment on what you think of the legitimacy of the survey results.

3. I have mentioned in earlier examples that my physician is conducting a survey to evaluate service and to target areas that might be improved. Respondents are to answer 24 questions on a scale of 1–4 with 1 being most satisfied and 4 being least satisfied. Examples of the wording of questions are "Appointment scheduled at a convenient time of day"; "The caring concern of our nurses"; "The professionalism of our technical staff"; "Effectiveness of our health information materials"; "The thoroughness of the examination"; and "Hours of operation convenient for you." Patients are given questionnaires as they check in for their appointments and are asked either to take a few minutes afterward to fill them out and leave them directly with the receptionist or to send them back by mail. Comment on potential problems with the survey.

4. As I was writing this, I had three students—Jun Zhu, Vivek Malhotra, and Ed Spencer—working on a project for the college librarian on space needs in the science-mathematics library. Books there are shelved according to the Library of Congress encoding system. The main subject classifications beginning with Q, for example, are QA: mathematics; QB: astronomy; QC: physics; QD: chemistry; QE: geology; QH: natural history; QK: botany; QL: zoology; QM: human anatomy; QP: physiology; QR: microbiology. The librarian has available yearly budget figures for the departments of mathematics, biology, chemistry, and physics and is interested in forecasting when shelf space will run out in each category. Discuss the kind of information that you would need to do such a forecast and the sampling procedures that you would design to obtain that information.

5. Suppose that you are surveying students about their overall satisfaction with their academic life. You would like information categorized by graduating class; the class sizes are 400 freshmen, 350 sophomores, 300 juniors, and 250 seniors. How might you go about obtaining separate random samples by class, and how would you use the data? How would you use only the information you have to create an estimate $\hat{p}$ of the overall proportion of students who are satisfied, which has the property that its expected value is the true proportion of satisfied students. Incidentally, a sample that consists of separate random samples from subgroups (called *strata*) of the whole population is called a *stratified sample*.

6. Suppose that a population of total size N is divided into a total of k strata (see Exercise 5) with sizes $N_1, N_2, \ldots, N_k$, respectively. A numerical characteristic denoted by X is of interest, and the ith stratum has mean μ_i and variance σ_i^2. Random samples of size $n_1, n_2, \ldots, n_k$ are drawn from the strata. The sample means $\overline{X}_1, \overline{X}_2, \ldots, \overline{X}_n$ for the samples are computed and combined into an estimate,

$$\overline{X} = \sum_{i=1}^{k} \frac{N_i}{N} \overline{X}_i,$$

of the overall mean μ of X. Show that the variance of the estimator $\overline{X}$ is

$$\sum_{i=1}^{k} \left(\frac{N_i}{N} \right)^2 \frac{\sigma_i^2}{n_i}$$

and then use Lagrange multipliers to find the sample sizes $n_1, n_2, \ldots, n_k$ that minimize this variance subject to the constraint that the total sample size is n.

7. In a uniform $[0, 1]$ pseudo–random number generator as discussed in the section, if *modulus* $= 65536$, $m = 25173$, and *inc* $= 13849$ (values suggested in Cooper and Clancy, 1985), and the initial value of *seed* $= 8$, produce the first ten simulated numbers and use them to simulate a random sample of ten integers from 0 to 9. (Include the number you get from the initial seed value.)

8. Find the expected value and variance of the number of primes in a random sample without replacement of size 5 taken from the set of integers $\{1, 2, \ldots, 20\}$. Repeat for the number of evens and the number of odds. (See Example 4.3.3.)

9. Show that the sample variance S^2 of a random sample $X_1, X_2, \ldots, X_n$ can also be written as

$$\frac{1}{n-1} \left[\sum_{i=1}^{n} X_i^2 - n\overline{X}^2 \right] = \frac{1}{n-1} \left[\sum_{i=1}^{n} X_i^2 - \frac{1}{n} \left(\sum_{i=1}^{n} X_i \right)^2 \right].$$

Discuss what this implies about statistical calculators with mean and variance keys. In particular, does the calculator have to store all of the individual data values entered by its user?

10. a. Let $X_1, X_2, \ldots, X_n$ be a random sample and consider the empirical probability mass function, which assigns probability $1/n$ to each of the observed sample values. Show that the mean of the empirical distribution is $\overline{x}$ and the variance of the empirical distribution is $(n-1)s^2/n$.

   b. Consider the following data set:

   0.1, 1.2, 0.5, 3.6, 2.1, 1.8, 0.5, 0.7, 2.4, 1.1, 0.6, 3.1, 2.8, 0.4, 1.4, 1.3, 2.2, 3.2

   Categorize the data into categories 0–1, 1–2, 2–3, and 3–4, and find the empirical distribution, that is, the proportion of observations in each category. Compute the empirical mean and variance by treating the categories as their midpoints .5, 1.5, 2.5, and 3.5. Do you get the same numbers as if you had computed the sample mean and variance directly from the uncategorized raw data? Would you expect the differences to be very large?

11. The following data, from the International Labour Organization, are national unemployment rates from several countries usually considered well developed. Find the 10th, 20th, 50th, 80th, and 90th percentiles and the range. Is the data set skewed right, left, or neither?

Canada	11.2	U.S.	6.9	Japan	2.5
Australia	10.9	New Zealand	9.8	Austria	4.8
Belgium	12.1	Denmark	12.1	Finland	18.2
France	11.7	Germany	8.9	Greece	10.0
Ireland	17.6	Italy	10.2	Luxembourg	2.2
Netherlands	8.3	Norway	6.0	Portugal	5.1
Spain	22.7	Sweden	8.2	Switzerland	4.5
U.K.	10.3	Turkey	8.6		

12. Suppose that all of the data $X_1, X_2, \ldots, X_n$ in a random sample are converted to a different system of units by a linear transformation $Y_i = aX_i + b$. What is the relationship between the summary statistics $\overline{X}$, S_X^2, M, and IQ_X (the interquartile range) and the corresponding summary statistics for the new Y sample?

13. Argue that by changing just one value in an observed random sample $x_1, x_2, \ldots, x_n$ the mean can be made to be at least C units larger than the median of the sample, where C is an arbitrary positive number.

14. Following are the 50 state populations plus the District of Columbia as of the 1990 U.S. Census, rounded to the nearest .1 million. Find the median, mean, variance, and interquartile range. Are the data roughly symmetric, or are they positively or negatively skewed?

    4.0, 0.6, 3.7, 2.4, 29.8, 3.3, 3.3, 0.7, 0.6, 12.9, 6.5, 1.1, 1.0,
    11.4, 5.5, 2.8, 2.5, 3.7, 4.2, 1.2, 4.8, 6.0, 9.3, 4.4, 2.6, 5.1,
    0.8, 1.6, 1.2, 1.1, 7.7, 1.5, 18.0, 6.6, 0.6, 10.8, 3.1, 2.8, 11.9,
    1.0, 3.5, 0.7, 4.9, 17.0, 1.7, 0.6, 6.2, 4.9, 1.8, 4.9, 0.5

15. The data in Table 7.3 (from *Illinois Academe*, 1991, the American Association of University Professors) gives average salaries of full-time faculty

Table 7.3

Average Faculty Salaries

University	Full Professors	Associate Professors	Assistant Professors
Chicago State	$41,200	$33,400	$29,500
Eastern Illinois	43,100	35,800	31,600
Northeastern Illinois	42,600	35,500	31,300
Western Illinois	44,200	38,500	32,800
Illinois State	50,700	39,400	33,200
Northern Illinois	53,000	40,800	33,600
Sangamon State	45,700	39,900	32,200
SIU–Carbondale	51,400	40,000	32,400
SIU–Edwardsville	48,600	42,100	34,900
U. of I.–Chicago	60,900	44,000	38,100
U. of I.–Urbana	63,700	45,600	40,300

by rank at 11 Illinois universities. Find the covariance and correlation matrix, if the three variables of interest are the salaries for the three ranks. What does the correlation matrix seem to say about salary decisions?

16. Show that the sample correlation between variables i and k does not change if the data for variable i are transformed by $Y_{ij} = aX_{ij} + b$, and the data for variable k are transformed by $Y_{kj} = cX_{kj} + d$, where $a, c > 0$.

17. Compute the sample covariance and correlation between U.S. and Mexican work times using the data in Example 3.4.1.

18. Table 7.4 contains 1993–94 total grade point averages among all students in each of Calculus I, Calculus II, and Calculus III at 20 universities, whose identities are suppressed for confidentiality. (The information was obtained by an e-mail survey of department chairs conducted by a faculty member at one of the responding universities.) Identify the data matrix, and compute the mean vector and covariance matrix of this vector sample.

19. What is the expected value of a covariance matrix?

Table 7.4

Calculus Grade Point Averages

Calculus I	Calculus II	Calculus III
1.93	2.45	2.78
2.25	2.19	2.31
1.98	2.26	2.23
2.53	1.95	1.89
1.8	1.5	1.7
2.28	2.82	2.9
2.46	2.39	2.41
1.87	2.07	2.14
2.29	2.66	2.81
2.23	1.96	2.26
2.67	2.59	2.71
1.82	1.8	1.62
2.07	2.5	2.5
2.0	2.5	2.3
2.52	2.52	2.82
2.368	2.47	2.51
2.237	2.58	2.133
2.69	2.84	3.47
2.05	2.32	2.34
2.73	2.81	3.01

20. Show that the sample covariance matrix can also be obtained as

$$\mathbf{S} = \frac{1}{n-1} \sum_{j=1}^{n} (\mathbf{X}_j - \overline{\mathbf{X}})(\mathbf{X}_j - \overline{\mathbf{X}})'.$$

Here $\mathbf{X}_j$ refers to the jth column of the data matrix and $\overline{\mathbf{X}}$ is the vector sample mean.

7.2 Graphical Data Analysis

Although the numerical data summaries that we discussed in the first section of this chapter are certainly useful, they do not always provide the best way of understanding the data. When presented properly in a picture, prominent features of a data set can leap out and tell a story with clarity and impact in a way that numbers alone cannot. In addition, a graph can preserve much of the detail of a data set that is lost when the data are summarized by some small collection of numbers. So, statistical graphics is a powerful technique for doing exploratory data analysis and can yield unexpected and interesting results.

In this section we will cover some of the standard graphical methods for displaying data: bar graphs, stem-and-leaf plots, box-and-whisker plots, dot plots, scattergrams, time series plots, and normal quantile plots. Each method has its own particular strength and domain of application. Sometimes, moreover, the use of more than one method gives information that one alone may not give.

Two wonderful books are available that reach a much greater depth than we will attempt here in the presentation of graphical data. One is a classic text by John Tukey (1977) that introduced many clever ways of tallying, displaying, and interpreting data. The other, by William Cleveland (1985), discusses many practical do's and don'ts of creating graphs and reports some interesting psychological results on human perception of pictures that lead to his recommendations.

7.2.1 Tallying Data: Stem-and-Leaf Diagrams

In many cases a problem yields a large set of data. To better understand the distribution of the data, we could tabulate the frequency of occurrence of values of the data and graph the frequencies in a histogram. We can use a device for the preliminary task of tallying data that produces a graph that helps us visualize the distribution of the data, similarly to a histogram. This graph type, called the *stem-and-leaf diagram*, is illustrated in the next two examples.

2	5 6 5 6 9 4 6 6 9 7 5
3	0 4 2 2 9 5 5 5 2 8 5 0 0 2 2 8 2 6 6 3 9 5 8 2 0 3
4	0 5 9 5 5 2 9 4 0 2
5	3
6	2 0

Figure 7.2 Days Inn Room Rates

Example 7.2.1

I chose a random sample of 50 Days Inn hotels from the 1990 directory and recorded the basic room rate for one adult. The data, in units of dollars, follow.

62, 40, 25, 26, 30, 34, 45, 32, 25, 32, 26, 39, 35, 49, 45, 60, 29, 35, 35, 24, 45, 26, 32, 42, 38, 35, 30, 30, 32, 32, 49, 44, 38, 32, 36, 26, 29, 36, 33, 39, 40, 27, 35, 38, 42, 32, 30, 33, 53, 25

To draw a stem-and-leaf diagram, begin by choosing main category headings, the *stems*. Each stem will have a line on the diagram, with a header indicating what the category is, and bits of more detailed information, the *leaves*, recorded on the line. In the case of numerical data of the sort that we have, we could choose our stems to be the tens digits of the room rates, which run from 2 through 6. Then, as we scan the data set we record in the appropriate stem lines the units digit of the data. In the stem 6 line we record 2, in the stem 4 line we record 0, in the stem 2 line we record 5, and so on. The complete stem-and-leaf diagram for the room rate data set is displayed in Fig. 7.2. It organizes the data in such a way that counting category frequencies is simplified, all of the information in the original data set is preserved, and the viewer can visualize the distribution of the data. Rates between $30 and $39 per night are clearly the most frequent. A few extremely high rates are present and lend a right-skewness to the distribution. (These outlying points correspond to Worcester, Massachusetts; Seminole, Florida; and Phoenix, Arizona.)

Example 7.2.2

The classified section of the February 26, 1993, issue of the *Peoria Journal-Star* newspaper yielded a data set pertaining to the age and make of used cars. I examined all of the used car ads and used a stem-and-leaf diagram to record, by year ranging from 1975 through 1993, the make of the car being advertised (see Fig. 7.3). As stems, I chose the years, and as symbols for the leaves, I used two-letter codes for the car manufacturers, which are translated at the bottom of the diagram. Thus the stem-and-leaf technique is easily adaptable to nonnumeric data. Two new features of the figure are helpful: The leaves are sorted to simplify spotting frequently advertised makes, and an additional column with the frequency counts is displayed next to the stems.

	freq	
'75	4	Ch Dt Fo Ol
'76	5	Bu Ch Ch Fo Po
'77	8	Cr Cr Fo Fo Me Ol Ol Po
'78	4	Ch Fo Po Po
'79	8	Ca Ch Do Fo Fo Fo Li Po
'80	3	Ca Ol Vw
'81	5	Bu Bu Bu Bu Bu
'82	4	Bu Bu Ca To
'83	4	Bu Li Ol Ol
'84	7	Cr Do Li Li Ol Pl Po
'85	16	Bu Ca Ch Ch Ch Ch Ch Do Do Fo Fo Fo Fo Fo Ol Ol
'86	10	Bu Ch Ch Do Fo Fo Li Ol Ol Po
'87	11	Ch Ch Ch Ch Fo Li Li Ni Ol Po Po
'88	11	Bu Bu Ca Ch Ch Fo Fo Je Li Ol Ol
'89	18	Bu Ca Ca Ch Ch Ch Ch Cr Do Fo Fo Fo Fo Ho Li Li Li Po
'90	8	Ca Ch Ch Ch Ch Fo Fo Fo
'91	14	Bu Bu Ca Ch Ch Do Fo Fo Fo Fo Ol Po Po Po
'92	18	Bu Bu Ch Ch Ch Ch Ch Ch Ch Cr Fo Fo Li Pl Po Po Po Vw
'93	1	Fo

Bu: Buick; Ca: Cadillac; Ch: Chevy; Cr: Chrysler; Do: Dodge; Dt: Datsun;
Fo: Ford; Ho: Honda; Je: Jeep; Li: Lincoln; Me: Mercedes; Ni: Nissan;
Ol: Olds; Pl: Plymouth; Po: Pontiac; To: Toyota; Vw: Volkswagen

Figure 7.3 Used Car Age and Make Distribution

7.2.2 Charting Data: Bar and Dotplots

A histogram of frequencies is a type of bar chart that is already familiar to us from Section 1.1, but there are some other kinds of charts to attend to, which are particularly helpful when multiple data sets are present.

Example 7.2.3

A recent study looked at the consumption of coffee at home by people who fall into various age groups (Dortch, 1995). The data that follow are from two years, 1985 and 1994, and the numbers are the percentages of people in the age group specified who drank any coffee at home in the six months prior to the survey.

	18–24	25–34	35–44	45–54	55–64	65+
1985	71%	75%	84%	90%	93%	90%
1994	64%	67%	74%	82%	85%	86%

Figure 7.4 gives a *grouped column graph* of these data, in which bars for the 1985 percentages are dark and bars for the 1994 percentages are light. Each age group is a *cluster*, and the progression of percentages with age group for a given year is a *series* in the language of graphics. The story that comes

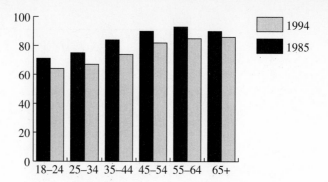

Figure 7.4 Percentage of Coffee Drinkers by Age

through from the graph is that in the ten years between the two surveys, coffee consumption is down by just under 10%, almost uniformly across age groups. The least decrease is in the 65 and older group, and the greatest decrease is in the 35–44 group.

?Question 7.2.1 What is the main story that would be told if clusters consisted of all percentage data for a given year and series were progressions from 1985 to 1994 for a given age group? Do you think that such a graph is more or less informative than the one in Fig. 7.4?

Example 7.2.4 The demographic distribution of the United States is changing. An article in the *U.S. News and World Report* (Friedman, 1995) cited figures on the 1995 racial makeup of the country and the projected 2050 distribution:

	1995	2050 (projected)
White	73.7%	52.5%
Black	12.6%	15.7%
Hispanic	10.2%	22.5%
Asian, Pacific Island	3.5%	10.3%
Amer. Indian, Eskimo	.9%	1.1%

A so-called *stacked bar chart* of the percentages for each year appears in Fig. 7.5. This graph uses subrectangle heights to represent the proportions of the total for a given year made up by each data point. With a little more effort, you can compare the two years; for instance, the main story that seems to be told here is that the share of the overall white population decreases rather drastically, with the largest increases going to the Hispanic and Asian groups.

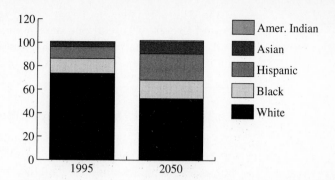

Figure 7.5 Projected Racial Makeup of the United States

—— **?Question 7.2.2** Exactly what is it that makes the main story of the last graph (Fig. 7.5) difficult to pick out very precisely? Can you design a different kind of graph that would tell the story better?

Let us now return to methods for displaying the distribution of a data set. A type of graph similar to a histogram is a *dotplot* (see Fig. 7.6). This graph displays the state space of the observations being plotted on a horizontal line, and it represents the data points themselves as dots located at the appropriate position on the line. When data points overlap, or there is not enough resolution to display them side by side, they are stacked on top of one another. Like the histogram, the dotplot is helpful for seeing how a set of data are dispersed; unlike the histogram, you nearly see every data point individually, because the extent to which data must be categorized by stacking dots is determined only by the width of the dots; the finer the dots, the less is the need to categorize by stacking. The next example illustrates the use of dotplots.

Example 7.2.5 In Example 5.5.2, we presented data on average salaries for assistant and full professors at 25 institutions. Figure 7.6 contains parallel dotplots of these data. The fact that the plots are done on the same horizontal scale helps the viewer to see the conclusions that we reached earlier numerically. The center of the assistant professor salary distribution seems to be at about $36,000, the data are rather symmetrically distributed around the center, and the spread

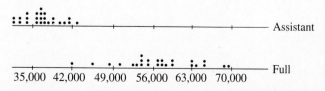

Figure 7.6 Salaries for Assistant and Full Professors

of the data is smaller than that of the full professor data, whose center is around $57,000. The extreme data points tend to stand out in a dotplot as well; here the data point at $42,000 and the two points near $70,000 in the full professor data are apart from the rest.

7.2.3 Graphical Data Summaries: Box-and-Whisker Plots

The preceding graphical methods attempted to preserve as much as possible of the original data and therefore produced very detailed pictures. We will now discuss a "quick-and-dirty" method of summarizing a data set graphically that plays on the numerical summary statistics discussed in the last section. Recall the median (the 50th percentile) and the quartiles (the 25th and 75th percentiles) of the sample. A *box-and-whisker plot* of a set of data (see, e.g., Fig. 7.7) consists of a box whose upper and lower boundaries are the quartiles, divided by a segment at the position of the sample median, together with a line segment protruding from both the top and bottom of the box. In our diagrams these line segments, called *whiskers*, extend to the smallest and largest order statistic. Other authors and computer programs may draw the whiskers differently and explicitly mark extreme outlying data. (For example, *Minitab*'s special slant on these diagrams is to use what are called *fences*. An *inner fence* length is computed as one and one-half times the width of the box, and an *outer fence* length is computed as three times the width of the box. Whiskers are drawn from the box edges to the most extreme observations that are still within the inner fences. Observations between the inner and outer fences are marked with an asterisk, and observations outside of the outer fences are marked with a circle.)

Example 7.2.6

Tukey (1977) reported on a very famous case study of graphical data analysis, Lord Rayleigh's experiments on the composition of air in the 1890s, which led him to discover the gaseous element argon. Rayleigh obtained what he thought was nitrogen from various sources, including air, by a purification process. He then measured the weights of standard volumes of the nitrogen. The data are given in Table 7.5.

Consider the box-and-whisker plot of Fig. 7.7(a). By itself it doesn't seem to indicate very much. It appears to possess a large spread, at least given the small axis scale, and the median is not very near to the midpoint of the box. This indicates that there are some large outlying values skewing the distribution. But now look at the dotplot of the same data in Fig. 7.7(b). Whereas boxplots excel at giving the viewer a quick and rough summary look at the distribution of data, dotplots preserve more of the detail and offer some interesting things to see in that detail. The data set divides into at least two clear groups, an upper and a lower group. This is the sort of picture that leads one to wonder whether something about the way the data were taken causes that grouping. A look at the numbers in the table shows

Table 7.5

Rayleigh's Nitrogen Weight Measurements

Date	Source	Weight
November 29, 1993	NO	2.30143
December 5, 1993	NO	2.29816
December 6, 1993	NO	2.30182
December 8, 1993	NO	2.29890
December 12, 1993	Air	2.31017
December 14, 1993	Air	2.30986
December 19, 1993	Air	2.31010
December 22, 1993	Air	2.31001
December 26, 1993	N_2O	2.29889
December 28, 1993	N_2O	2.29940
January 9, 1994	NH_4NO_2	2.29849
January 13, 1994	NH_4NO_2	2.29889
January 27, 1994	Air	2.31024
January 30, 1994	Air	2.31030
February 1, 1994	Air	2.31028

that the very cohesive upper group of observations were all derived from air.

Figure 7.8 contains parallel boxplots of the subgroups of data corresponding to air-based "nitrogen" and nitrogen obtained from the other sources. The split is very pronounced. The composition of the other sources would have been rather well known to Rayleigh, so that he could have been confident that what he had in the purified gas was nitrogen. The fact that an equivalent volume of air clearly weighed more suggested there was some unknown gas in the air that was not removed by purifying the air sample. This unknown gas turned out to be argon. The moral of the story is that it is often not enough to just look at one graph of a data set. Another kind of graph may

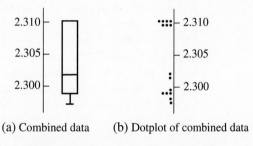

(a) Combined data   (b) Dotplot of combined data

Figure 7.7 Rayleigh's Nitrogen Data

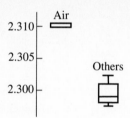

Figure 7.8 Rayleigh's Nitrogen Data, by Source

yield more helpful information than the first and may also suggest new ways to look at a data set and new graphs to draw.

?Question 7.2.3 If a data set was nearly symmetric, what would its boxplot look like? if it was extremely right-skewed? left-skewed?

7.2.4 Relationships Between Variables: Scattergrams

Many statistical problems involve taking two measurements on a subject. Usually one is interested in knowing whether the quantities being measured depend on one another, and the data available are a set of pairs (x_i, y_i), $i = 1, \ldots, n$ where the components refer to the two measurements on the particular subject numbered i. A simple plot of the pairs as points in the Cartesian plane can shed light on the dependence. A *scattergram* is such a plot. If more than two variables are being measured, then a collection of scattergrams two variables at a time can be drawn to illustrate pairwise dependence. [Computer tools are now becoming available to plot three measurement variables (x_i, y_i, z_i), over all subjects i, as points in three-dimensional space in order to illustrate multiple dependence.]

Example 7.2.7 Baseball has always been a statistics-heavy sport. Let us investigate the question of what factors contribute to a ball club's offensive production: power, hitting skill, or speed. The October 12, 1992, issue of *The Sporting News* contained the data in Table 7.6, which are the end-of-season figures on team batting average, home runs, stolen bases, and runs scored. We will use the first three variables as measures of hitting skill, power, and speed, respectively, and we would like to use scattergrams to study the dependence of runs scored, our measure of offensive production, on the three variables.

Scattergrams of runs against average, runs against home runs, and runs against stolen bases appear in Figs. 7.9(a) and (b) and 7.10. Runs scored do seem to increase with batting average, in a more or less linear way. In a later chapter on *regression*, we will see how to find a best fitting line through data points like those in the figure. Run production also seems to increase with

Table 7.6

Final 1992 Major League Statistics

Team	Average	Home Runs	Stolen Bases	Runs
Cubs	.254	104	77	593
Expos	.252	102	196	648
Mets	.235	93	129	599
Phils	.253	118	127	686
Pirates	.255	106	110	693
Cards	.262	94	208	631
Braves	.254	138	126	682
Reds	.260	99	125	660
Astros	.246	96	139	608
Dodgers	.248	72	142	548
Padres	.255	135	69	617
Giants	.244	105	112	574
Orioles	.259	148	89	705
Red Sox	.246	84	44	599
Indians	.266	127	144	674
Tigers	.256	182	66	791
Brewers	.268	82	256	740
Yankees	.261	163	78	733
Blue Jays	.263	163	129	780
Angels	.243	88	160	579
White Sox	.261	110	160	738
Royals	.256	75	131	610
Twins	.277	104	123	747
Athletics	.258	142	143	745
Mariners	.263	149	100	679
Rangers	.250	159	81	682

home runs in a linear way, and the slope seems steeper. (Be careful about how you interpret that result, however; see the next question.) You would intuitively expect such results. But what is perhaps surprising for the 1992 data is that runs do not seem to depend in a systematic way on stolen bases. There is nothing that these data can do to support the argument that speed helps a ball club score.

?Question 7.2.4 In Figs. 7.9(a) and (b), use a straightedge to try to estimate slope, and answer

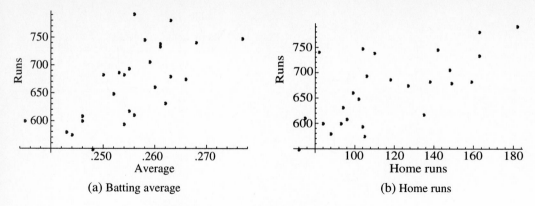

(a) Batting average (b) Home runs

Figure 7.9 Factors Influencing Runs Scored

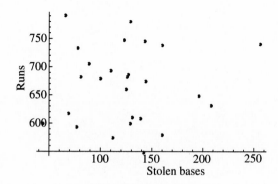

Figure 7.10 Lack of Dependence of Runs on Stolen Bases

the following question: If you were the manager of a team, which would you prefer: an extra 10 points in team batting average, or an extra 50 home runs?

7.2.5 Time Series Plots

In many data sets we are interested in changes of a variable with time. A graph in which the values of the variable are plotted as a function of time is called a *time series plot*.

Example 7.2.8

Terasaki and coworkers (1995) recently studied kidney grafts with the goal of comparing the survival rates of grafts from spousal and other living donors with grafts from deceased organ donors. The percentages of grafts that still survived at the end of the first, second, and third years were determined separately for each of four patient age groups and for each type of donor. The results are as follows (estimated from a graph in the paper):

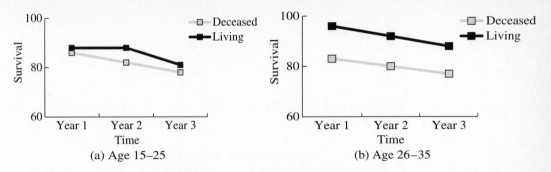

Figure 7.11 Kidney Graft Survival Rates

Living Donors

Age	Year 1	Year 2	Year 3
15–25	88%	88%	81%
26–35	96%	92%	88%
36–50	92%	87%	84%
>50	87%	80%	80%

Deceased Donors

Age	Year 1	Year 2	Year 3
15–25	86%	82%	78%
26–35	83%	80%	77%
36–50	80%	75%	70%
>50	75%	68%	62%

The series values are the yearly graft survival percentages, which appear on the *y*-axis of Fig. 7.11. Part (a) of the figure compares time series plots for living and deceased donors for recipients in the 15–25 age group, and part (b) does the same for the 26–35 age group. What story is told by these graphs? It appears as if patients who receive kidney grafts from living donors, especially those who are close to them, do better for the first few years than those who get their grafts via cadaveric transplant. Perhaps the perfection of the physical match of tissue takes second place to more psychological factors. Try producing similar plots for the other age groups. Is there an even more pronounced phenomenon for them? Do you see any evidence that as years pass, the survival rates for grafts from living and deceased donors are coming together?

7.2.6 Normal Quantile Plots

In many instances in the upcoming chapters, performing statistical analysis of a random sample will require that we assume the sample came from a normal distribution. An analytical procedure is available for checking this assumption that we will derive later: however, a graphical procedure also exists, which we will now study.

Let $X_1, X_2, \ldots, X_n$ be a random sample, with order statistics $X_{(1)}, X_{(2)}, \ldots, X_{(n)}$. Answer the following question before proceeding.

?Question 7.2.5   Show that the ith order statistic is the $i/(n+1) \cdot 100th$ percentile of the sample.

If the sample had come from the $N(\mu, \sigma^2)$ distribution, then it stands to reason that the order statistic $X_{(i)}$ should be a good estimate of the theoretical $i/(n+1) \cdot 100th$ percentile of the $N(\mu, \sigma^2)$ distribution, call it q_i. Therefore the pairs $(q_i, X_{(i)})$ should approximately fall on the line $y = x$ in the coordinate plane. The problem is that we don't know the normal parameters μ and σ^2 a priori, so that we can't compute q_i directly. But there is a very clever way to get around the problem using the technique of standardization.

Let z_i be the unique $i/(n+1) \cdot 100th$ percentile of the standard normal distribution, which satisfies the equation

$$\frac{i}{n+1} = P[Z \le z_i] = \int_{-\infty}^{z_i} \frac{1}{\sqrt{2\pi}} e^{-z^2/2} \, dz. \tag{7.11}$$

We can derive a relationship between q_i and z_i as follows:

$$\frac{i}{n+1} = P[X \le q_i] = P\left[\frac{X-\mu}{\sigma} \le \frac{q_i - \mu}{\sigma}\right] = P[Z \le \frac{q_i - \mu}{\sigma}]$$

$$\Rightarrow z_i = \frac{q_i - \mu}{\sigma}$$

$$\Rightarrow q_i = \sigma z_i + \mu. \tag{7.12}$$

Thus, if the sample is normal, since $q_i \approx X_{(i)}$ the points $(z_i, X_{(i)})$ should lie near the line $y = \sigma x + \mu$.

The discussion of the last paragraph motivates the following type of graph and its interpretation. A *normal quantile plot* (sometimes called a *normal scores plot*) of a data set $X_1, X_2, \ldots, X_n$ is a graph of the points $(z_i, X_{(i)})$, where $X_{(i)}$ is the ith order statistic of the data and z_i is the $i/(n+1) \cdot 100th$ percentile of the standard normal distribution. Significant departures from linearity indicate that the data are probably not a random sample from a normal distribution. If the normal quantile plot is nearly linear, then we conclude that the data could have come from a normal distribution $N(\mu, \sigma^2)$, and in addition the slope of the line is an estimate of the standard deviation and the y-intercept is an estimate of the mean.

Example 7.2.9   Consider the data set of nine sample values:

$$2.33, \ 5.69, \ 0.39, \ -2.28, \ -2.14, \ 3.31, \ -0.72, \ 1.63, \ -3.91.$$

Is it reasonable to suppose that this data set is a random sample from some normal distribution? Since $n = 9$ here, we need the normal quantiles

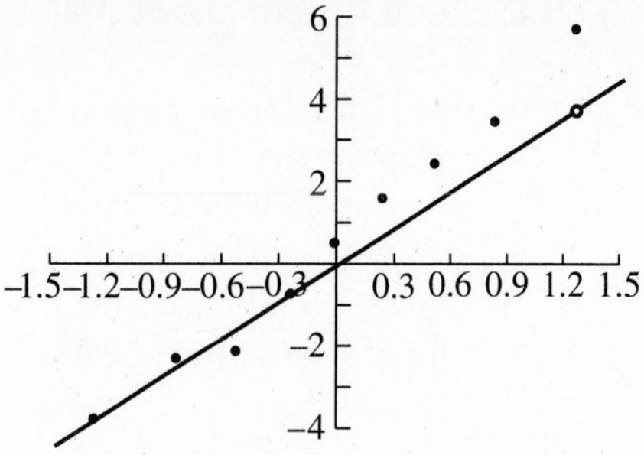

Figure 7.12 Normal Quantile Plot

$z_1, z_2, \ldots, z_9$ corresponding to the probabilities $i/(n+1) = i/10$, $i = 1, 2, \ldots,$ 9. For example, we can easily determine from Table 3 of Appendix A that the value of z_9, the point satisfying $P[Z \leq z_9] = 9/10$, is 1.28. Ordering the data and pairing them with the normal quantiles obtained in this way results in the following points:

$$(-1.28, -3.91), \quad (-.84, -2.28), \quad (-.52, -2.14), \quad (-.25, -.72), \quad (0, .39),$$
$$(.25, 1.63), \quad (.52, 2.33), \quad (.84, 3.31), \quad (1.28, 5.69).$$

The normal scores plot is contained in Fig. 7.12. It looks more or less linear, except that the rightmost observation is a bit higher than it should be. This tells us that it would be a good idea to check the source of that data point to see whether it was an error or just a valid but extreme outlying piece of data. The truth can now be told. I simulated the original data from the normal distribution with mean 0 and standard deviation 3, but I changed a simulated value of 3.69 to 5.69 to make the point about noticing outlying points in the normal scores plot. The real point is shown as an open circle on the plot, and the graph of the line $y = 3x + 0$ (recall that the ideal line is $y = \sigma x + \mu$) is superimposed to show the fit of the points to the theoretical line.

?Question 7.2.6 Verify the normal quantiles used in the last example: $-1.28, -.84, -.52, -.25,$ 0, .25, .52, .84, 1.28.

See Exercise 17 for a result relating skewness of the underlying data to the appearance of the normal quantile plot.

EXERCISES 7.2

1. From the classified ad section of your local paper, obtain a sample of listed rental prices for one-bedroom apartments and another sample of prices of two-bedroom apartments. Plot the two samples using parallel dotplots with the same scale.

2. Form a stem-and-leaf plot and a histogram of the state population data in Exercise 14 of Section 7.1. Note that the choices of stems and categories for the histogram are parts of the problem; they should be neither so broad that the detail of the distribution of the data is lost nor so narrow that patterns cannot be seen. Does the population distribution seem symmetric or skewed?

3. Exercise 11 of Section 7.1 contained a data set of unemployment rates. Draw a box-and-whisker plot and dotplot of the overall data set, then compare parallel box and dotplots of the non-European and European data. What story, if any, do these graphs tell?

4. Draw parallel dotplots and box-and-whisker diagrams of the test 1, test 2, and test 3 data in Table 7.2 of Section 7.1. Which test performance seems to have been best, and which was worst?

5. Following are ages at death for people in the obituary section of the *Peoria Journal-Star* newspaper on May 28, 1993:

$$100, \ 74, \ 83, \ 79, \ 71, \ 79, \ 78, \ 59, \ 59, \ 29, \ 54,$$
$$52, \ 79, \ 53, \ 74, \ 51, \ 87, \ 58, \ 56, \ 47$$

The following data are ages at death from the obituary section of the *Chicago Tribune* on the same day (this is actually a subset of the obituaries, since not all death ages were listed):

$$95, \ 93, \ 69, \ 77, \ 93, \ 40, \ 62, \ 96, \ 90, \ 74, \ 77, \ 47, \ 70, \ 90, \ 97, \ 71,$$
$$80, \ 94, \ 68, \ 19, \ 79, \ 69, \ 40, \ 91, \ 74, \ 2, \ 55$$

Pose your own questions about the data, and investigate them using graphical techniques. Do you think that the data are a random sample of all deaths? Why or why not? Do the pictures divulge information that you did not expect?

6. I monitored my home electric and gas consumption over a period of 28 months, together with the dollar costs of each energy form. The data appear in Table 7.7. What kind of questions can be asked about the data? Devise bar charts to answer those questions. What patterns do the graphs allow you to see in the data? What deviations from the patterns can you pick out? (*Hint:* The gas powers my heat and hot water, I have central air-conditioning, and in the fall of 1991 I was mostly out of town on a sabbatical.) Would time series graphs allow you to see certain aspects of the data better than bar charts?

7. The following data are the percentages of all stock shares sold on registered exchanges accounted for by the New York Stock Exchange (NYSE), the American Stock Exchange (ASE), and other exchanges between the years of 1971 and 1975. (Data obtained from Johnson, 1978.)

	1971	1972	1973	1974	1975
NYSE	72.1	71.4	75.7	79.0	81.1
ASE	17.7	17.5	12.9	9.8	8.7
Other	10.2	11.1	11.4	11.2	10.2

Table 7.7

Power Consumption

Month	Electricity (kwh)	Electricity Cost	Gas (therms)	Gas Cost
January '91	319	37.90	187	91.03
February	485	51.97	192	90.84
March	307	38.79	112	59.73
April	270	35.48	61	37.33
May	420	47.02	34	24.05
June	1049	108.88	20	17.74
July	631	70.09	18	16.79
August	650	71.79	20	17.50
September	174	28.44	14	14.70
October	125	23.42	24	19.17
November	164	26.33	65	37.51
December	308	38.35	194	85.56
January '92	398	44.70	194	92.11
February	358	44.64	155	77.96
March	387	49.52	147	66.92
April	323	43.84	91	47.16
May	480	55.53	29	21.33
June	570	71.29	19	16.79
July	598	73.90	18	16.42
August	658	76.18	22	17.78
September	352	47.62	31	21.75
October	299	40.21	68	37.31
November	421	49.39	170	76.13
December	344	43.56	181	79.40
January '93	398	46.82	192	85.41
February	443	50.85	206	92.72
March	406	48.68	144	71.39
April	483	55.22	97	52.62

Produce a grouped column graph that shows the progression of proportion of the market held by each exchange. Compare to a time series graph and to a stacked bar chart. Which do you find most informative, and why?

8. Use at least one appropriate graphical technique to compare full professors', associate professors', and assistant professors' salaries in Exercise 15 in Section 7.1.

9. A recent issue of *Advertising Age* magazine (Williamson, 1995) reported on 1994 and expected 1999 spending (in billions of dollars) by segments of the communications industry, as indicated in the following table. Produce a graph that tells a story about the data, then summarize that story in words.

	1994	1999
Newspapers	$48	$61.7
Film	29.5	39
Television	29	37.6
Book Publishers	23.8	33.2
Radio	10.3	14.5
Digital Media	5.8	14.2

10. A survey (Wilkie, 1995) of 20,000 people gave the following demographic information on people who have used perfume or cologne more than seven times in the week prior to the survey: 12.6% of women aged 18–24 had done so, as had 12.9% of women aged 25–34, 12.0% of women aged 35–44, 16.9% of women aged 45–54, 16.0% of women aged 55–64, and 8.0% of women aged 65 and older. Also, 13.1% of men aged 18–24 had used cologne more than seven times, as had 11.6% of men aged 25–34, 9.5% of men aged 35–44, 12.4% of men aged 45–54, 10.6% of men aged 55–64, and 9.1% of men aged 65 and older. Construct an appropriate graph to display this information, and discuss any conclusions you can draw with the help of the graph.

11. Plot scattergrams of the calculus grade point data in Table 7.4 of Section 7.1: Calculus I vs. Calculus II, Calculus I vs. Calculus III, and Calculus II vs. Calculus III. What conclusions can you draw?

12. Follow the common stock prices of Sears and Wal-Mart for at least two weeks, and create time series plots on the same graph. Do you see any interesting phenomena?

13. Topps is a company that produces baseball trading cards that are much sought after by hobbyists and speculators. Each year a set of cards is issued, and occasionally the number of cards in the set changes. Older sets tend to be more valuable than newer ones, and the condition of the cards also influences their market price. Table 7.8 has estimated prices in dollars by condition (taken from the *1983 Sport Americana Baseball Card Price Guide*) for the years 1960–1980, as well as the number of cards in the set for that year. Three conditions are possible: mint, very good to excellent, and fair to good. What kinds of questions can be asked about these data? Use appropriate graphical techniques to study them.

14. Do the runs scored that appear in Table 7.6 seem to be approximately normally distributed?

15. The following are 14 simulated uniform [0,1] observations. Produce a normal quantile plot, and comment on what you see. If you were to have simulated a much larger number of observations, what do you think you

Table 7.8

Baseball Card Prices

Year	Mint	Very Good to Excellent	Fair to Good	Number
1960	310	210	80	572
1961	650	450	200	589
1962	335	225	100	598
1963	500	325	140	576
1964	240	160	70	587
1965	260	180	80	598
1966	375	250	100	598
1967	500	350	140	609
1968	190	135	60	598
1969	210	140	60	664
1970	210	140	60	720
1971	210	130	50	752
1972	220	150	65	787
1973	120	80	35	660
1974	90	60	25	660
1975	100	65	30	660
1976	50	35	15	660
1977	40	26	11	660
1978	32	22	9	726
1979	28	20	8	726
1980	20	13	5	726

would have seen?

.46, .01, .67, .42, .78, .10, .41, .24, .53, .98, .99, .41, .80, .19

16. Economists estimate that tax-free municipal bonds need a rate of return of 7% in order to keep pace with a common stock whose return is 10% but which is taxed at a rate of 28% on its appreciation and 36% on its dividends (Forsyth, 1995). The following are data on 16 such municipal bonds. Does the distribution of the rates of return seem normal? Estimate the mean and variance, and decide what percentage of such bonds will exceed the cutoff value of 7%.

6.70, 6.86, 6.86, 6.93, 6.74, 7.09, 6.94, 6.83,
6.81, 7.05, 7.02, 6.88, 6.63, 6.74, 6.86, 6.97

17. Argue that when a sample of data is skewed to the right, the normal scores plot will be flat on its left side. What can you guess about the normal scores plot for a sample of data that is skewed to the left? What would the normal scores plot of a U-shaped collection of data look like?

7.3 | **Principles of Graph Construction**

In the last section we were mostly concerned with illustrating several popular kinds of statistical graphs. You saw the graphs as finished products, and perhaps you did not think about the underlying principles involved in designing them. Now we will concentrate on why one kind of graph might be preferable to another, drawing heavily from the ideas contained in Cleveland (1985). Three principles should be applied in graph construction:

1. A graph should successfully tell a story about the data.
2. It should be designed so as to make the viewer's task of understanding it as simple as possible.
3. It should be labeled helpfully and strategically so as to stand by itself.

The purpose of a graph is to convey information relatively quickly to the viewer. Typically a data set contains much information, so the grapher must decide what story to highlight. If the graph is well designed, then there may be secondary stories that come out, but the first responsibility of the grapher is to be sure that the main story is prominent.

This viewer-oriented philosophy forces us to pay attention to what is easy for a viewer to pick out of a diagram. Some recent research on human cognition bears on principles for drawing good graphs. To understand graphs, viewers frequently must make several kinds of judgments and estimates:

1. Relative positions
2. Lengths
3. Areas
4. Angles or slopes

We are beginning to gather experimental evidence (Cleveland, 1985, Chap. 4) that the foregoing tasks become more difficult as we move down the list; that is, it is easiest for the viewer to estimate or compare relative positions, and hardest for the viewer to estimate or compare angles. To make a graph more understandable, it is therefore beneficial to choose graph designs that require the easier tasks as opposed to the harder tasks.

Furthermore, a well-labeled graph facilitates the interpretation of the viewer by clarifying the details, such as a general description of what the graph is presenting, the meaning of the variables plotted on each axis, and the variable to which a symbol pertains.

We illustrate these principles using several examples.

Example 7.3.1

The Illinois State Lottery has a daily game in which a sequence of three digits (i.e., 0 through 9) is drawn at random. Table 7.9 contains data on the number of times during a year-long period ending June 12, 1993, that each digit was picked in each position.

Suppose that you are interested in the first digit only and will ignore the other data for the time being. The obvious question becomes: How do the

Table 7.9

Frequency of Digits in Illinois Pick 3 Lottery

Digit	1st	2nd	3rd
0	35	36	45
1	44	40	43
2	25	44	36
3	40	26	34
4	50	31	34
5	26	38	39
6	33	26	32
7	38	31	29
8	37	50	33
9	35	41	38

digit frequencies compare? Which digits are the most frequently drawn, and which are the least frequently drawn?

The main variable of interest is the digit frequency on the first draw, for digits from 0 to 9. The most obvious choice of graph type is a frequency histogram, in which the bar heights are proportional to the observed digit frequencies. The graph is shown in Fig. 7.13. The graph title indicates what the data set is, with very little background information demanded of the viewer. The two axes bear labels that tell what the x- and y-coordinates mean, and there are sufficiently many ticks for the viewer to estimate values visually, yet not so many as to produce a cluttered diagram.

What stories stand out in Fig. 7.13? Probably the most striking is that the maximum frequency of 50 (digit 4) is twice as great as the minimum frequency of 25 (digit 2). Other frequencies are scattered between the two extremes. Note that the viewer's visual-cognitive task in understanding the story is simply to compare the positions of the tops of the bars to the fixed

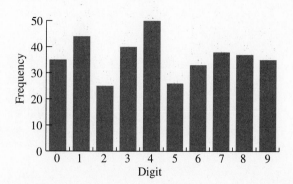

Figure 7.13   Frequency of Selection on First Draw of Illinois Pick 3 Lottery

vertical scale, or to one another, which is the easiest of our list of tasks. Despite the fact that frequencies are conceptualized as being proportional to lengths, it is not really necessary to make length judgments in this graph. The main story—the comparison of frequencies of digits—is therefore told in a way that is easy for the viewer to pick out.

The graph tells two substories that are not easy to appreciate. First, with the digits arranged in the usual ascending order, it is hard to read the complete progression of the most common to the least common digit (4, 1, 3, 7, etc.), because the eye must jump around. To make this substory stand out, it would be necessary to reorder the digits on the horizontal axis in decreasing order of frequency, which also would not interfere with the main story. The viewer could as a result more easily find the range of frequencies.

Second, we don't get a good idea from the graph of what the expected frequencies are, or how far these frequencies are from what is expected. Since there are a total of 363 plays of the Illinois Pick 3 Lottery in the data set (add down any column of Table 7.9), and for random drawing each digit has an equal likelihood of 1/10 of being selected, it follows that 36.3 is the expected frequency of each digit. If the graph is being presented to the casual viewer who does not have much mathematical background, however, it may be wiser to concentrate not on frequencies but on the percentage of the time that each digit was drawn. The viewer is more likely to reason that this percentage is expected to be 10% than to compute the expected frequency of 36.3 as we did. The graph is resketched (Fig. 7.14) with digit relative frequencies, from highest to lowest, on the vertical axis rather than absolute frequencies, and a line of reference is drawn at 10%. This graph allows the viewer to gauge the magnitude of the variation in digit frequencies more easily, by comparing lengths of bars (or relative positions of the tops of the bars).

?Question 7.3.1   Another possible design for the graph of digit frequencies would have dots at the appropriate vertical positions of frequencies in place of bars. Can you

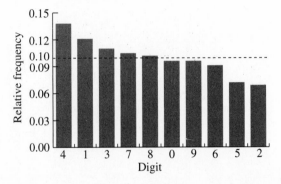

Figure 7.14 Relative Frequencies of Digits, Illinois Pick 3 Game

think of any advantages or disadvantages of this approach relative to the histogram approach?

Example 7.3.2

When the data that are to be graphed represent proportions of a total accounted for by each of several categories, *pie charts* are often used to depict the data. Each category gets a wedge of a circular pie whose angle θ is proportional to the category's fraction of the total. (To be specific, if the total is N and the category share is n, then θ satisfies $n/N = \theta/2\pi$.)

Figure 7.15 shows a pie chart for the following data (Fulkerson, 1995). A total of 1200 adults nationwide were asked whether the amount that they had given to charity in the last year was less than \$25, between \$25 and \$99, and so on, as indicated. The percentages who placed themselves in the six categories are given.

<\$25	\$25–\$99	\$100–\$499	\$500–\$999	≥\$1000	Not sure
7%	34%	39%	11%	6%	3%

In the figure, pie wedges are marked with the category to which they correspond. Can you tell from the pie chart that the \$25–\$99 category has a smaller share than the \$100–\$499 category? Can you tell that about a fifth as many people give under \$25 as \$25–\$99? These tasks require accurate estimation and comparison of angles, which, again, is difficult. It would have been helpful to annotate the categories with the percentages, but then the graph would not show very much that the raw data themselves did not show. Its chief value is in providing a visual substitute for a list of numbers, the rough pattern of which may be easier to see and remember than the raw numbers.

?Question 7.3.2

As we have seen, pie charts present the viewer with difficult cognitive tasks. Worse yet, some pie charts are published that use a pseudo-three-dimensional

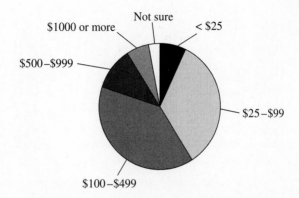

Figure 7.15 Distribution of Charity Donations

style and a sight perspective that can give misleading impressions about the data. This is a case where technology has made it possible to create fancy, aesthetically pleasing pictures that in the final analysis fail to do the job they were meant to do: to tell a story about the data. Find examples in the media of this sort of misguided use of graphics.

Example 7.3.3

The following is an example of a misguided graph the like of which I have seen in some very respectable places. (If you are ambitious, check out the financial page of the June 21, 1993, issue of the *Chicago Tribune* in which a similar graph related to the values of health-care bonds and total bond issues appears.) Exercise 6 of Section 7.2 contained data on electricity and gas costs for my home. For the months of January through June of 1991, bar charts of the total energy cost and the cost due to gas consumption are displayed side by side in Fig. 7.16. The legend indicates that the graph is meant to tell a story about gas cost as a fraction of the total.

The strategy of placing the two graphs side by side separates the bars whose lengths the viewer is to compare by a large distance, making that comparison more difficult than it needs to be. The larger problem is that in order to make the graphs equally tall, different scales were used on the vertical axes on the two graphs. The dual scales require us to imagine, for instance, the January gas bar shrunken by a factor of around 2/3 and then embedded inside the January total bar. Then we are to conclude that the shrunken January bar is 3/4 as large as the total bar. This is too difficult a mental task. In this case a stacked bar graph as in Fig. 7.17 would have been a much better choice. Since we have no direct interest in the electric cost we needn't do the more cognitively difficult task of comparing lengths of different electric bars, which do not emanate from the same baseline. The

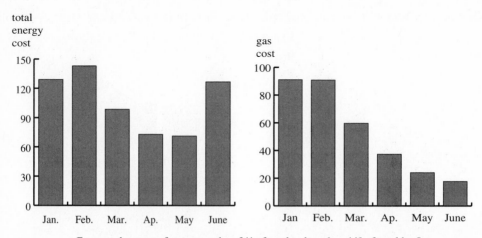

Gas cost decreases from more than 3/4 of total to less than 1/6 of total by June

Figure 7.16   A Confusing Juxtaposition

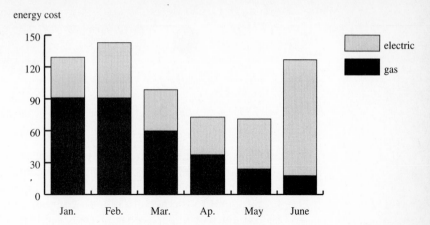

Figure 7.17   Gas Cost: Share of Total

primary story about the progression of gas percentages of the total stands out fairly well.

This example raises a general issue relative to comparing two data sets. We could produce two similar graphs, and we could then situate them side by side or one above the other. In either case we say that we are *juxtaposing* the graphs. Or, we could *superimpose* the graphs, that is, put them all in the same plot region. In that case though, we must take steps to ensure that the graph does not become too complicated for the viewer to figure out and that the data remain visible and distinguishable. It is not correct to say that one of these methods is always better than the other. The strategy of superimposition worked well here, but in Exercise 1 at the end of this section, which reprises the Illinois Lottery example, you may find otherwise. Superimposition brings the data to be compared closer together, which eases the cognitive task, but if the two data sets cross each other frequently, it can create a cluttered and confusing graph.

?Question 7.3.3   There is a very important tenet of graphing which states that if you want to illustrate the behavior of some quantity, then graph it directly, not indirectly. In view of this, do you think that the gas cost story of the last example is suitably told yet? What might be done better? (You could try Exercise 4 at the end of this section, which asks you to make a better graph.)

Example 7.3.4   The four inner planets of our solar system have equatorial diameters roughly as follows:

Mercury   3000 mi   Venus   7700 mi
Earth   7926 mi   Mars   4230 mi

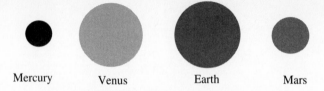

Mercury   Venus   Earth   Mars

Figure 7.18 Inner Planets of the Solar System

Suppose that we are principally interested in the comparison of planetary surface area. A diagram of the planets is contained in Fig. 7.18. It is drawn approximately to scale linearly, so that the equatorial radii in the picture are proportional to the actual radii. To compare surface areas, the viewer must be content to compare areas of cross sections in the plane of the page. (Is this a proper thing to do?). Can you do this very easily? For example, can you tell whether Earth has 1–1/2 times, 2 times, 3 times, or more than 3 times the surface area of Mars?

Though it must be admitted that the spheres in Fig. 7.18 have a certain eye-catching appeal over Fig. 7.19, the latter graph does convey the main story more easily and more directly to the viewer. In Fig. 7.19, the surface areas (computed by the formula $4\pi r^2 = \pi d^2$) are graphed directly using a dot graph variant of a bar chart. You probably will find that you underestimated the differences in surface area when you first looked at Fig. 7.18; Earth has more than three times the surface area of Mars, and around eight times the surface area of Mercury.

Note that to convey more information and to facilitate interpretation, there are two technical aspects in Fig. 7.19: The diameter is listed on a nonlinear right-hand scale, and reference lines are drawn across the graph to let the eye trace back from the data point to the vertical axes.

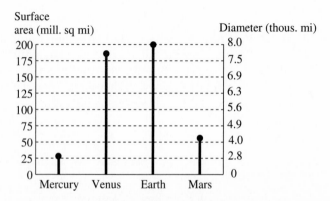

Figure 7.19 Planetary Surface Areas

Sometimes the creator of a graph is interested in telling a story about rates of change of variables. For a time series graph, the rate of change at a point (x_0, y_0) is the slope m of the segment connecting (x_0, y_0) with its neighboring point (x_1, y_1). Estimating a rate of change, or comparing two rates of change, is one of the most difficult visual-cognitive tasks that a viewer can be asked to do. If θ is the angle made by the segment with the horizontal, then the relationship

$$\tan \theta = \frac{y_1 - y_0}{x_1 - x_0} = m$$

between the angle and the slope is nonlinear, particularly for angles near $\pm \pi/2$. All angles are tricky to estimate, and for steep angles near these extremes, small changes in angles yield large changes in slope, which can easily lead a viewer astray. This notion is illustrated by the following example.

Example 7.3.5

Exercise 13 of Section 7.2 contained a data set on 1983 market prices of sets of baseball cards for the years 1960–1980. The sets could be in one of three conditions: mint, very good to excellent, or fair to good. Time series graphs for each of the three conditions are shown in Fig. 7.20. Several stories are nicely told by this graph, one that the price behaviors for the three conditions mirror one another as time goes by, with the mint cards being the most expensive and the fair cards the least expensive.

Suppose, however, that the main story we would like to tell involves the yearly *changes* in value. Since the horizontal variable changes in increments of exactly one year, we are interested in following the slopes of segments and comparing them for the three conditions. The visual impression created by Fig. 7.20 is that those slopes are nearly the same. But remember the main principle of graph design: choose a graph that tells the main story. Figure 7.20 does not directly highlight slopes, but rather prices themselves, and leads to the difficult perceptual task of slope comparison.

So, a better idea is to compute those slopes, which is easy to do from Table 7.8, and plot them against time. A graph appears in Fig. 7.21. This

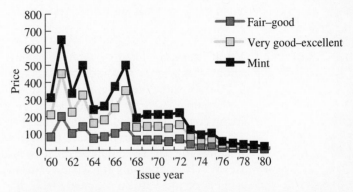

Figure 7.20   1983 Baseball Card Prices, Issue Years 1960–1980

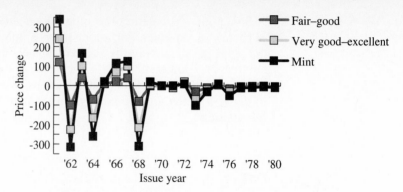

Figure 7.21   1983 Baseball Card Yearly Price Changes

graph tells a story rather different from what you may have expected when you looked at Fig. 7.20. The price changes for the three conditions follow a similar pattern, but the mint sets tend to change at a rate around 1 and 1/2 times as great as very good–excellent cards, and around 3 times the rate of fair–good cards. This large difference in the fluctuation is not apparent in the first graph. We observe that when the point of a graph is to show slopes or estimate slope differences, the best strategy is to graph those slopes directly in order to avoid angle estimation.

?Question 7.3.4   Another problem with graphs that are steeply sloped involves difference estimation. Focus your attention on the leftmost portion of Fig. 7.20, around the years 1960 and 1961, for mint and very good–excellent sets. Would you say that the two segments are close together? What is the visual impression left by the two steep, apparently parallel segments? To estimate differences in mint vs. very good–excellent price, what distance should you be looking at? Formulate in a sentence the perceptual problem of estimating differences when graphs are steep, and recommend a principle for graphing when the main story involves such differences.

Example 7.3.6   Finally we would like to look at a technical aspect of the graph design process about which there is some controversy. Consider parts (a) and (b) of Fig. 7.22. Both show the progression of kidney graft survival rates from the data of Example 7.2.8 for patients aged 36–50 who received their grafts from living donors. You might imagine the headline that would go along with part (a): "Kidney graft survival rate plunges with time!" On the other hand the headline for part (b) might be the less lurid: "Kidney graft survival rate stays steady." It is tempting for a viewer to evaluate the data by looking at the heights of the vertical segments from the time axis to the time series line. Then graph (a) suggests a drastic reduction in percentage of surviving graphs

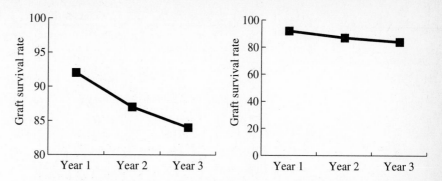

Figure 7.22   Kidney Graph Survival Rates, Ages 36–50, Living Donors

by the third year to about a third of the level of the first year. Of course, the vertical axis tick labels say otherwise, and the graph of part (b), which includes 0, gives the correct visual impression that the large majority of grafts that survived the first year lasted through the third year.

You may have noticed this phenomenon yourself in graphs of political support or trends of economic indicators. When 0 is included in the vertical scale, what may have otherwise appeared to be large changes look like small changes. Proponents of the strategy of setting the vertical axis range according to the maximum and minimum data values regardless of where zero is would argue that their method fills the plot region better and focuses the viewers' attention on where the data actually are. Those who favor the inclusion of zero say that it is in the interest of honesty to do so, because the viewer may not understand or look carefully at the axis labels, and may therefore be misled into exaggerating the magnitude of small changes. My own recommendation, viewer-oriented as always, is to think of who the viewer is and the story that you want the viewer to understand. If your viewer will not be confused by an axis whose origin is not 0, then exclude 0 and choose the scale so as to just enclose the data. Otherwise, if there is a real possibility that your viewer could be misled, include 0.

EXERCISES 7.3

1. Use juxtaposition to create a graph comparing the frequency distributions of the three draws in the Illinois Pick 3 Lottery game using the data in Table 7.9. Then try superimposing the graphs. Which do you prefer and why?

2. Table 7.10 gives U.S. Commerce Department data on 1992 average per capita income for the states. Design a graph to compare the income distributions for northern states east of the Mississippi and southern states also east of the Mississippi. (Essentially, the Ohio River is the border between north and south. Count Minnesota, West Virginia, and Delaware as north, and Maryland and Washington D.C. as south.)

3. Suppose that for the used car data in Example 7.2.2 (see the stem-and-leaf diagram in Fig. 7.3), we are interested in the distribution of car makes

Table 7.10

Average Per Capita Incomes

Alabama	16,220	Alaska	21,603	Arizona	17,119
Arkansas	15,439	California	21,278	Colorado	20,124
Connecticut	26,979	Delaware	21,451	Dist. Col.	26,360
Florida	19,397	Georgia	18,130	Hawaii	21,218
Idaho	16,067	Illinois	21,608	Indiana	18,043
Iowa	18,287	Kansas	19,376	Kentucky	16,534
Louisiana	15,712	Maine	18,226	Maryland	22,974
Massachusetts	24,059	Michigan	19,508	Minnesota	20,049
Mississippi	14,088	Missouri	18,835	Montana	16,062
Nebraska	19,084	Nevada	20,266	New Hampshire	22,934
New Jersey	26,457	New Mexico	15,353	New York	23,534
N. Carolina	17,667	N. Dakota	16,854	Ohio	18,624
Oklahoma	16,198	Oregon	18,202	Pennsylvania	20,253
Rhode Island	20,299	S. Carolina	15,989	S. Dakota	16,558
Tennessee	17,341	Texas	17,892	Utah	15,325
Vermont	18,834	Virginia	20,629	Washington	20,398
W. Virginia	15,065	Wisconsin	18,727	Wyoming	17,423

instead of ages. Design a graph with this focus. If possible, have your graph convey at least something about the age distribution.

4. Devise a sequence of juxtaposed pie charts to study the fraction of total utility expense due to the two sources electricity and gas from Table 7.7 for the first half of calendar year 1992. Then devise another graph to tell the same story that requires the viewer to do an easier perceptual task.

5. Use graphical techniques to investigate the *relative* price changes of sets of baseball cards of the three conditions: mint, very good–excellent, and fair–good using the data in Exercise 13 of Section 7.2. (The relative change for a time period is the ratio of the change in value over a period to the old value.) What main story does your graph tell?

6. I estimated the following data from a graph in a recent news article in *Chemical Week* ("Markets and Economics," 1995, p. 30). They involve the demand by three North American countries for polystyrene in three different years. (The units of the data are billions of pounds.) Design one graph whose main story is about the change in demand with time, and another whose main story is about the comparison between countries.

	1990	1994	1995
Mexico	.4	.4	.5
Canada	.4	.3	.3
U.S.	4.7	5.6	5.8

7. The following data are industrywide delivery volumes of several kinds of recreational vehicles for two successive Junes (Kisiel, 1995). Decide on

a story that you would like to tell about the data, and produce a graph and short discussion which tells that story.

	June 1995	**June 1994**
Conventional Trailers	6400	6800
5th Wheel Trailers	3400	4700
Folding Camping Trailers	4800	5500
Truck Campers	1300	1000

8. The next graph superimposes time series graphs for a company's revenue and cost in the same plot region. Design a graph that better tells a story about company profits than this one. Is it easy to see from the original graph when the largest profit occurred? Why or why not? When according to your graph did the largest profit occur?

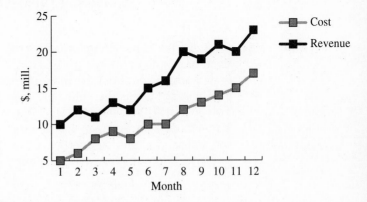

9. Stevens' Law from psychology states that a person's estimate of the magnitude of an attribute whose true magnitude is x is of the form $p(x) = cx^\beta$. Different kinds of attributes, such as length, area, and volume, have particular β values. If β happens to be about .7 for area estimation, argue that there is a conservatism in which small areas appear bigger than they are relative to large areas, and large areas appear smaller than they are relative to small areas.

10. In comparing average housing prices in an area over time, would you consider a graph in which the plot symbols are pentagonal houses whose heights are proportional to price? Why or why not?

11. If an angle θ that a line makes with the horizontal changes by an amount $d\theta$, by how much does the slope of the line change? What is the implication of this result to graphical perception?

12. Using the data from Example 7.2.9, which is graphed in Fig. 7.12, form a graph whose main story is about the *residuals*, the vertical deviations of the points from the line. Do the residuals follow any pattern? Where are the largest residuals?

13. Suppose that the graph in the accompanying figure represents the growth in child abductions in a certain area between the years 1990 and 1992. The actual numbers of abductions are noted in parentheses as annotation

for the graph. The heights of the child figures are drawn so as to be proportional to the number of cases reported. What impression is left on the eye by this diagram? Is this a good graph, or does it mislead, and why? Try to make a better graph.

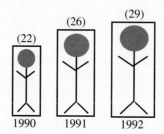

Numbers of Child Abductions

14. Consider the runs scored vs. home runs scattergram in Fig. 7.9(b). By sight, draw a line through the graph that seems to best fit the data. Now draw another graph whose main purpose is to show how well or how poorly your line fits the points of the scattergram.

15. Two political candidates are competing for a Senate seat. They both have access to the same independent poll showing their support in four successive weeks. The poll data follow. Draw a graph that candidate A would probably release to the press, and one that candidate B would prefer.

Candidate A:	30%, 31%, 29%, 27%
Candidate B:	25%, 26%, 28%, 29%
Undecided:	45%, 43%, 43%, 44%

CHAPTER 8

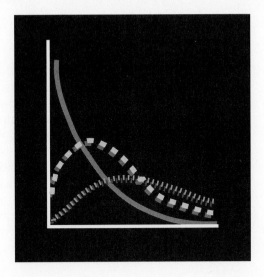

POINT AND INTERVAL ESTIMATION

8.1 | Unbiased Estimation

A significant part of the statistician's job is to pose a reasonable probabilistic model, then to use data to estimate unknown parameters that characterize that model. For example, a statistician dealing with a service system in which arrivals occur at random times may have reason to believe that the arrival stream forms a Poisson process, whose rate parameter λ must be estimated. Recall that the times $S_1, S_2, S_3, \ldots, S_n$ between the first n successive arrivals are independent exponential(λ) random variables. Thus these random variables form a random sample, and the problem would be to find a function of the sample to estimate λ.

One important question concerns how to create estimators, or functions of random samples. To address this question, later in this chapter we will study the method of *maximum likelihood*, which gives a systematic and reasonable way of deriving functional forms for estimators. Another question is, How do we evaluate the quality of our estimators? We would like to study the latter question in this section, focusing mostly on estimators that have no so-called *bias*, that is, no tendency to consistently overestimate or underestimate the desired parameter.

8.1.1 Bias and Standard Error

To set up the estimation problem, assume that $X_1, X_2, \ldots, X_n$ is a random sample of size n from a probability distribution with density (or mass function) $f(x; \theta)$. The X_i's and their possible values x may be vectors, as may the distributional parameter θ, although we will mostly deal with the one-dimensional case here. To understand what is to come, not only in this section but throughout the rest of the book, you must remember that a function $Z = g(X_1, X_2, \ldots, X_n)$ of random variables $X_1, X_2, \ldots, X_n$ is also a random variable. Therefore this function has a probability distribution, mean, and variance of its own. Typically, this distribution, mean, and variance will depend on the parameters of the distribution from which the X_i's were sampled; this allows us to gain information from the sample about the parameters. The following is the central definition of this section.

DEFINITION 8.1.1

(a) An estimator $Z = g(X_1, \ldots, X_n)$ of a parameter θ has *bias* equal to

$$\text{bias}(Z) = E[Z - \theta]. \qquad (8.1)$$

If Z has bias equal to zero, we call it an *unbiased estimator* of θ.

(b) The *standard error* of an estimator Z is its standard deviation

$$\sigma_z = \sqrt{\text{Var}(Z)}. \qquad (8.2)$$

?Question 8.1.1 Before you go on, try to determine why unbiasedness might be an important property for an estimator to have and why we would be concerned about an estimator's standard error.

The main qualities that we should expect of a good estimator are accuracy and precision. In other words, the random estimator Z should on the average be about θ, and it should not be highly variable. Otherwise, how could we have much confidence in an estimate obtained from a single random sample? To illustrate, consider Fig. 8.1(a), which depicts the probability distributions of two typical estimators Z_1 and Z_2 of a parameter θ. Now Z_1 appears to be symmetrically distributed about θ. Thus Z_1 would not systematically overestimate or underestimate θ, and its expected value is θ. On the other hand, Z_2 has a very high likelihood of overestimating θ, because its probability weight is mostly to the right of θ, and its mean is clearly more than θ. Of the two estimators, Z_1 is superior from the standpoint of accuracy; Z_1 has lower bias than Z_2. Part (b) of the figure shows a different contrast. There,

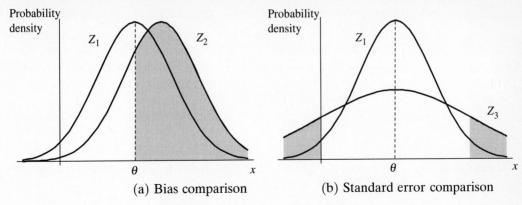

(a) Bias comparison (b) Standard error comparison

Figure 8.1 Comparison of Two Estimators

both Z_1 and another estimator, Z_3, seem to be about θ on average, meaning they have low bias. The trouble with Z_3 is that its probability distribution has such a large variance (and standard deviation) that it has a much greater likelihood of taking a value distant from θ than does Z_1. It is not very precise compared to Z_1, because of its larger standard error. So, we search for estimators that are unbiased and have small standard errors.

Example 8.1.1

The side of the box that contains the brand of garbage bags that I usually use claims that a product testing service established that their bags had a score of 250 on a tear resistance scale, as compared to 110 for other brands. How might such a claim be checked? Several bags may be selected at random from a particular manufacturing run and subjected to a standardized tearing test. Let $X_1, X_2, \ldots, X_n$ be the resulting tearing scores for n bags. Suppose that we adopt the model that $X_1, X_2, \ldots, X_n$ are a random sample from the $N(\mu, \sigma^2)$ distribution, where we know neither μ nor σ^2 a priori. We would like a good estimator of the population mean tearing score μ.

Common sense indicates that the sample average $\overline{X}_n = \frac{1}{n}(X_1 + X_2 + \cdots + X_n)$ should estimate μ well. We might also guess that if we did not use all of the data, we would not get as precise an estimate. Let us compare $\overline{X}_n$ to $\overline{X}_2$, the mean of only the first two observations. By linearity of expectation,

$$E[\overline{X}_n] = \frac{1}{n}(E[X_1] + E[X_2] + \cdots + E[X_n]) = \frac{1}{n} \cdot n\mu = \mu, \tag{8.3}$$

and similarly

$$E[\overline{X}_2] = \frac{1}{2} \cdot 2\mu = \mu. \tag{8.4}$$

Thus $\overline{X}_n$ and $\overline{X}_2$ are unbiased for μ. Also, since the X_i's are independent,

$$\text{Var}(\overline{X}_n) = \text{Var}\left[\frac{1}{n}(X_1 + X_2 + \cdots + X_n)\right] = \frac{1}{n^2} \cdot n \cdot \sigma^2 = \frac{\sigma^2}{n}, \tag{8.5}$$

and by a similar computation,

$$\text{Var}(\overline{X}_2) = \frac{\sigma^2}{2}. \tag{8.6}$$

Hence the standard error $\sigma_{\overline{X}_n} = \sigma/\sqrt{n}$ of $\overline{X}_n$ is less than the standard error $\sigma_{\overline{X}_2} = \sigma/\sqrt{2}$ of $\overline{X}_2$ as long as $n > 2$. The conclusion is that a sample average of a random sample of any size is an accurate estimator of the population mean, and the precision increases as the sample size increases. We will be able to see a more precise relationship between the sample size and the precision of the estimate later in the chapter.

?Question 8.1.2 In Example 8.1.1, was it really necessary to assume that the distribution was normal in order to derive the results?

In a comparative judgment of the precision of two estimators, it is the relative spreads of their distributions that matters. Hence we are led to make the following definition.

DEFINITION 8.1.2

The *relative efficiency* of an unbiased estimator Z_1 to an unbiased estimator Z_2 is

$$\frac{\text{Var}(Z_2)}{\text{Var}(Z_1)}. \tag{8.7}$$

Because a small standard error (hence a small variance) is a desirable quality in an estimator, the estimator Z_1 is better than Z_2 if the relative efficiency of Z_1 to Z_2 is more than 1. In Example 8.1.1, the relative efficiency of $\overline{X}_n$ to $\overline{X}_2$ is $n/2$, which exceeds 1 for $n > 2$.

Example 8.1.2 Again let $X_1, X_2, \ldots, X_n$ be a random sample from the $N(\mu, \sigma^2)$ distribution, and consider the problem of estimating the population variance σ^2. The sample variance

$$S^2 = \sum_{i=1}^{n} \frac{(X_i - \overline{X})^2}{n-1} \tag{8.8}$$

is one sensible estimator; another is the average square deviation

$$V = \sum_{i=1}^{n} \frac{(X_i - \overline{X})^2}{n}. \tag{8.9}$$

To check for bias, we can compute as follows:

$$E[S^2] = \frac{1}{n-1} E\left[\sum_{i=1}^{n} (X_i - \overline{X})^2\right]$$

$$= \frac{1}{n-1} E \left[\sum_{i=1}^{n} X_i^2 - n\overline{X}^2 \right]$$

$$= \frac{1}{n-1} \sum_{i=1}^{n} (\text{Var}(X_i) + E[X_i]^2) - n(\text{Var}(\overline{X}) + E[\overline{X}]^2)$$

$$= \frac{1}{n-1} \sum_{i=1}^{n} (\sigma^2 + \mu^2) - n(\sigma^2/n + \mu^2)$$

$$= \frac{1}{n-1} (n\sigma^2 + n\mu^2 - \sigma^2 - n\mu^2)$$

$$= \frac{1}{n-1} (n-1)\sigma^2 = \sigma^2. \tag{8.10}$$

So, S^2 is unbiased for σ^2. Notice that you can get V from S^2 just by multiplying S^2 by $(n-1)/n$. Therefore

$$E[V] = \frac{n-1}{n} E[S^2]$$

$$= \frac{n-1}{n} \sigma^2,$$

by (8.10). Thus V has bias

$$E[V - \sigma^2] = \frac{n-1}{n}\sigma^2 - \sigma^2 = -\frac{1}{n}\sigma^2.$$

This means that V on the average underestimates σ^2. However the bias of V does approach 0 as n goes to ∞.

To compare the standard errors, recall that $(n-1)S^2/\sigma^2$ has the $\chi^2(n-1)$ distribution. It follows that $(n-1)S^2/\sigma^2$ has variance $2(n-1)$. (Make sure that you know why.) Thus

$$\text{Var}(S^2) = \text{Var}\left(\frac{\sigma^2}{n-1} \cdot \frac{(n-1)S^2}{\sigma^2} \right)$$

$$= \frac{\sigma^4}{(n-1)^2} \text{Var}\left(\frac{(n-1)S^2}{\sigma^2} \right)$$

$$= \frac{\sigma^4}{(n-1)^2} \cdot 2(n-1) = \frac{2\sigma^4}{(n-1)}. \tag{8.11}$$

Therefore

$$\text{Var}(V) = \text{Var}\left(\frac{n-1}{n} S^2 \right)$$

$$= \left(\frac{n-1}{n} \right)^2 \text{Var}(S^2)$$

$$= \left(\frac{n-1}{n} \right)^2 \cdot \frac{2\sigma^4}{(n-1)} = \frac{2(n-1)\sigma^4}{n^2}.$$

The efficiency of V relative to S^2 is

$$\frac{\text{Var}(S^2)}{\text{Var}(V)} = \frac{2\sigma^4/(n-1)}{2(n-1)\sigma^4/n^2} = \frac{n^2}{(n-1)^2},$$

which is greater than 1 when n exceeds 1. So V, despite being biased, has a smaller standard error than S^2. Asymptotically, as $n \to \infty$ the relative efficiency of V to S^2 (and S^2 to V) approaches 1.

In summary, it is not clear which estimator of σ^2, S^2 or V, is better. The former is unbiased, hence it is more accurate, while the latter has smaller variance, hence it is more precise. For large sample size, it makes little difference. Most statisticians have accepted the unbiased estimator S^2.

Example 8.1.3

In the introduction to this section we referred to the estimation of the rate parameter λ in a Poisson process. Suppose that we are able to observe just two interarrival times S_1 and S_2 with values $s_1 = 4.0$ and $s_2 = 3.0$. Recall that λ is the parameter of the exponential distribution

$$f(s; \lambda) = \lambda e^{-\lambda s}, \qquad s > 0 \tag{8.12}$$

of S_1 and S_2, which is interpreted as the average number of arrivals per unit of time. Its reciprocal $\theta = 1/\lambda$ is therefore the average time interval between arrivals, that is, the expected value of both S_1 and S_2. To estimate λ, it suffices to estimate θ.

We can write the density function as

$$f(s; \theta) = \frac{1}{\theta} e^{-s/\theta}, \qquad s > 0. \tag{8.13}$$

The arithmetical average $\overline{S}$ of the two times is unbiased for θ (Why?). Its standard error is

$$\sigma_{\overline{S}} = \sqrt{\text{Var}(\overline{S})} = \sqrt{\text{Var}(S)/2} = \frac{\theta}{\sqrt{2}}. \tag{8.14}$$

(Check this.) Here $\overline{S}$ has observed value $(3.0 + 4.0)/2 = 3.5$, so we estimate the rate as $1/3.5 = .286$.

A competitive estimator of θ is the geometric mean $Z = \sqrt{S_1 S_2}$. Its expectation is

$$E[Z] = E[\sqrt{S_1}\sqrt{S_2}] = \int_0^\infty \int_0^\infty \sqrt{s_1}\sqrt{s_2}\frac{1}{\theta^2}e^{-(s_1+s_2)/\theta}\, ds_1 ds_2$$

$$= \left[\int_0^\infty \sqrt{s_1}\frac{1}{\theta}e^{-s_1/\theta}\, ds_1\right]^2$$

$$= \left[\sqrt{\theta}\int_0^\infty \sqrt{u}e^{-u}\, du\right]^2$$

$$= \left[\sqrt{\theta}\cdot\Gamma(3/2)\right]^2 = \theta \cdot \left(\frac{1}{2}\sqrt{\pi}\right)^2 = \frac{\pi\theta}{4}.$$

The third line arises from the substitution $u = s_1/\theta$. Then Z is biased, but the estimator defined by $Y = (4/\pi)Z$ is unbiased. To compare the standard errors of $\overline{S}$ and Y, we can compute:

$$\text{Var}(Y) = \frac{16}{\pi^2}\text{Var}(Z) = \frac{16}{\pi^2}\left[E[Z^2] - (E[Z])^2\right]$$

$$= \frac{16}{\pi^2}\left[E[S_1 S_2] - \left(\frac{\pi\theta}{4}\right)^2\right]$$

$$= \frac{16}{\pi^2}\left[(E[S_1])^2 - \left(\frac{\pi\theta}{4}\right)^2\right]$$

$$= \frac{16}{\pi^2}\left[\theta^2 - \left(\frac{\pi\theta}{4}\right)^2\right]$$

$$= \left(\frac{16}{\pi^2} - 1\right)\theta^2 \approx .62\theta^2.$$

(Why is the third line true?) The standard error of Y is about $\sqrt{.62\theta^2} \approx .79 \cdot \theta$ as compared to $\theta/\sqrt{2} \approx .71 \cdot \theta$ for $\overline{S}$. Therefore $\overline{S}$ is somewhat more efficient than Y.

Example 8.1.4

Since buses are scheduled forms of transportation, it is reasonable to assume that the time by which a bus will be late to a particular bus stop has both a lower and an upper limit, say 0 and θ for convenience, where θ is an unknown parameter. Suppose also that we believe that the distribution of $X = $ the number of minutes late is uniform. Let us investigate whether the largest order statistic Y_n of a random sample $X_1, X_2, \ldots, X_n$ from the uniform$(0, \theta)$ distribution is unbiased for θ, and if not, in what way it should be adjusted so as to be unbiased. You will study other estimators for θ in Exercise 10 at the end of this section.

The p.d.f. and c.d.f. of X are, respectively,

$$f(x; \theta) = \begin{cases} 1/\theta & \text{if } x \in (0, \theta), \\ 0 & \text{otherwise;} \end{cases} \qquad F(x; \theta) = \begin{cases} 0 & \text{if } x \leq 0, \\ x/\theta & \text{if } x \in (0, \theta), \\ 1 & \text{otherwise.} \end{cases}$$

Hence by Proposition 5.2.3, the distribution of Y_n is

$$f_{Y_n}(y) = n\left(\frac{y}{\theta}\right)^{n-1} \cdot 1/\theta = \frac{n}{\theta^n} \cdot y^{n-1}, \qquad y \in (0, \theta).$$

Therefore

$$E[Y_n] = \int_0^\theta y \cdot \frac{n}{\theta^n} \cdot y^{n-1}\, dy = \frac{n}{\theta^n} \cdot \frac{y^{n+1}}{n+1}\Big|_0^\theta = \frac{n}{n+1} \cdot \theta. \tag{8.15}$$

We can then compute the bias of Y_n as

$$E[Y_n - \theta] = \frac{n}{n+1} \cdot \theta - \theta = \frac{-1}{n+1} \cdot \theta.$$

Thus Y_n underestimates θ on the average, but the bias approaches 0 as the sample size n approaches ∞. (Statisticians sometimes use the phrase *asymptotically unbiased* for this behavior.) By linearity of expectation and (8.15), the estimator $Z = (n + 1)Y_n /n$ stretches out the largest observed sample value enough so as to be unbiased for θ.

—— **?Question 8.1.3**   Find the standard error of the estimator Z in the last example.

8.1.2   Absolute Efficiency: The Rao–Cramer Lower Bound

We have been comparing estimators to one another on the basis of precision. The next question to ask is whether there is a fixed standard to which to compare individual estimators, that is, a best possible precision. This question is answered in the affirmative in the next theorem, the *Rao–Cramer inequality*. We omit the proof on the grounds that it is based on a not particularly instructive trick. A proof can be found in (Larsen and Marx, 1986, p. 283) and other mathematical statistics texts.

PROPOSITION 8.1.1

Let $f(x; \theta)$ be a density (or probability mass function) with parameter θ, such that the set of points where $f(x; \theta) > 0$ does not depend on θ. Then if Z is any unbiased estimator of θ,

$$\text{Var}(Z) \geq \frac{1}{n \cdot E\left[\left(\dfrac{\partial \ln f(X; \theta)}{\partial \theta}\right)^2\right]}. \tag{8.16}$$

The quantity on the right side of (8.16) is called the *Rao–Cramer lower bound*. The formula for the bound is complicated enough to merit explanation. First, X is a random variable with density f. In the denominator, $f(X; \theta)$ is a function of X; hence it is a random variable as well. So is $\ln f(X; \theta)$, and in turn so is the partial derivative of this function with respect to θ. The square of this partial is yet another function of X, so that it makes sense to take its expectation, which is defined as the integral (or sum) of $[\partial \ln f(x; \theta)/\partial \theta]^2$ times the density (or p.m.f.) f over the state space of X. Normally, properties of expectation will shortcut the computation so that it is not necessary to evaluate an integral to find the Rao–Cramer (R–C) bound.

Before proceeding to the examples, let us derive an equivalent expression for the denominator of the R–C lower bound that is often easier to compute than the one in (8.16). Specifically, let us show that

$$E\left[\left(\frac{\partial \ln f(X; \theta)}{\partial \theta}\right)^2\right] = -E\left[\left(\frac{\partial^2 \ln f(X; \theta)}{\partial \theta^2}\right)\right]. \tag{8.17}$$

In full form, we have

$$E\left[\left(\frac{\partial \ln f(X; \theta)}{\partial \theta}\right)^2\right] = \int \left(\frac{\partial \ln f(x; \theta)}{\partial \theta}\right)^2 f(x; \theta)\, dx$$

$$= \int \left(\frac{\partial f(x; \theta)}{\partial \theta} \bigg/ f(x; \theta)\right)^2 \cdot f(x; \theta)\, dx. \tag{8.18}$$

Also, we can write

$$E\left[\left(\frac{\partial^2 \ln f(X;\theta)}{\partial\theta^2}\right)\right]$$

$$= \int \left(\frac{\partial^2 \ln f(x;\theta)}{\partial\theta^2}\right) f(x;\theta)\,dx$$

$$= \int \frac{\partial}{\partial\theta}\left(\frac{\partial f(x;\theta)}{\partial\theta}\Big/ f(x;\theta)\right) \cdot f(x;\theta)\,dx$$

$$= \int \left[\left(f(x;\theta)\frac{\partial^2 f(x;\theta)}{\partial\theta^2} - \left(\frac{\partial f(x;\theta)}{\partial\theta}\right)^2\right)\Big/ f(x;\theta)^2\right] \cdot f(x;\theta)\,dx$$

$$= \int \frac{\partial^2 f(x;\theta)}{\partial\theta^2}\,dx - \int \left(\frac{\partial f(x;\theta)}{\partial\theta}\Big/ f(x;\theta)\right)^2 \cdot f(x;\theta)\,dx$$

$$= 0 - \int \left(\frac{\partial f(x;\theta)}{\partial\theta}\Big/ f(x;\theta)\right)^2 \cdot f(x;\theta)\,dx. \tag{8.19}$$

The integral of $\partial^2 f(x;\theta)/\partial\theta^2$ in the next to last line vanishes because, interchanging derivative with integral, it is the second derivative of the integral of f, which is the constant 1. Combining this result with (8.18) yields the desired equality.

Example 8.1.5

Show that the sample mean $\overline{X}$ of a random sample of size n from the $N(\mu,\sigma^2)$ distribution has the smallest possible standard error among all unbiased estimators of μ.

Since by Proposition 8.1.1 no estimator can have a variance smaller than the R–C lower bound, if we show that the variance of $\overline{X}$ achieves the bound, then no other estimator can do strictly better. It will then follow that the standard error of $\overline{X}$ is the smallest possible.

From formula (8.5), the variance of $\overline{X}$ is σ^2/n. To compute the R–C bound for the normal distribution, note that

$$f(x;\mu) = \frac{1}{\sqrt{2\pi\sigma^2}}e^{-(x-\mu)^2/2\sigma^2}$$

$$\Rightarrow \ln f(x;\mu) = -\frac{1}{2}\ln(2\pi\sigma^2) - \frac{(x-\mu)^2}{2\sigma^2}$$

$$\Rightarrow \frac{\partial \ln f(x;\mu)}{\partial\mu} = \frac{-1}{2\sigma^2}\cdot 2(x-\mu)(-1) = \frac{x-\mu}{\sigma^2}.$$

At this point we could take one of two directions. Using the original formula (8.16) for the Rao–Cramer bound,

$$E\left[\left(\frac{\partial \ln f(X;\mu)}{\partial\mu}\right)^2\right] = E\left[\left(\frac{X-\mu}{\sigma^2}\right)^2\right] = \frac{1}{\sigma^4}E\left[(X-\mu)^2\right] = \frac{\sigma^2}{\sigma^4} = \frac{1}{\sigma^2}.$$

Alternatively, we could use expression (8.17). First, we compute:

$$\frac{\partial^2 \ln f(x;\mu)}{\partial\mu^2} = \frac{\partial}{\partial\mu}\left(\frac{x-\mu}{\sigma^2}\right) = \frac{-1}{\sigma^2};$$

hence

$$E\left[\left(\frac{\partial \ln f(X;\mu)}{\partial \mu}\right)^2\right] = -E\left[\left(\frac{\partial^2 \ln f(X;\mu)}{\partial \mu^2}\right)\right] = -E\left[\frac{-1}{\sigma^2}\right] = \frac{1}{\sigma^2}.$$

In either case, the R–C bound is σ^2/n. Since the variance of $\overline{X}$ equals the bound, we have accomplished our objective. Notice how the alternative form of the R–C bound simplified the expectation part of the computation.

DEFINITION 8.1.3

The *efficiency of an unbiased estimator* $Z = g(X_1, X_2, \ldots, X_n)$ of θ is the ratio

$$\frac{\text{R–C lower bound}}{\text{Var}(Z)},$$

where the Rao–Cramer lower bound is given by formula (8.16).

Therefore the efficiency of $\overline{X}$ in Example 8.1.5 is 1.

Example 8.1.6

Find, if possible, an unbiased estimate of the probability p of an event in some experiment that has the smallest possible variance among unbiased estimators.

Let n independent replications of the experiment be performed. Construct random variables $X_1, X_2, \ldots, X_n$ such that

$$X_i = \begin{cases} 1 & \text{if the event occurs on replication } i, \\ 0 & \text{otherwise.} \end{cases}$$

Then $X_1, X_2, \ldots, X_n$ is a random sample from the Bernoulli(p) distribution. The sample mean $\overline{X} = \sum_{i=1}^{n} X_i/n = \widehat{p}$, is the proportion of times that the event occurred in the sample, an intuitively appealing estimator of p. Since $E[X_i] = p$, $\widehat{p}$ is unbiased for p. Let us try to verify that $\widehat{p}$ is a best estimator by comparing its variance to the R–C lower bound.

?Question 8.1.4 Check that $\text{Var}(\widehat{p}) = p(1-p)/n$.

Example 8.1.6

Continued Now we can write the common mass function of the X_i's as

$$f(x;p) = p^x (1-p)^{1-x}, \qquad x = 0, 1.$$

Then

$$\frac{\partial^2 \ln f(X;p)}{\partial p^2} = \frac{\partial}{\partial p}\left(\frac{\partial}{\partial p}\left[x\ln(p) + (1-x)\ln(1-p)\right]\right)$$

$$= \frac{\partial}{\partial p}\left(\frac{x}{p} - \frac{1-x}{1-p}\right)$$

$$= -\left(\frac{x}{p^2} + \frac{1-x}{(1-p)^2}\right).$$

Hence,

$$-E\left[\left(\frac{\partial^2 \ln f(X;p)}{\partial p^2}\right)\right] = E\left[\frac{X}{p^2} + \frac{1-X}{(1-p)^2}\right] = \frac{p}{p^2} + \frac{1-p}{(1-p)^2} = \frac{1}{p(1-p)}.$$

The R–C lower bound, from (8.16) and (8.17), is $p(1-p)/n$. Therefore the sample proportion $\widehat{p}$ is a minimum variance unbiased estimator for p.

EXERCISES 8.1

1. Find two unbiased estimators of the parameter μ in the Poisson(μ) distribution based on a random sample $X_1, X_2, \ldots, X_n$. Show that they are indeed unbiased, and compute their standard errors.

2. Find an unbiased estimator for the parameter α in the $\Gamma(\alpha, 2)$ distribution, and an unbiased estimator for the parameter $\beta = 1/\lambda$ in the $\Gamma(2, 1/\beta)$ distribution. Show that they are indeed unbiased, and compute their standard errors.

3. Let $X_1, X_2, \ldots, X_n$ be a random sample from some distribution whose mean is μ. Fill in the blank (with justification): A linear combination $\sum_{i=1}^n a_i X_i$ is unbiased for μ if and only if _____.

4. If an estimator Z is unbiased for a population parameter θ, does it necessarily follow that a function $h(Z)$ is unbiased for $h(\theta)$? If not, give a counterexample, and find an extra condition on the function h such that the statement is true.

5. Show that the sample median of a random sample of size 5 from the exponential distribution with parameter λ is not an unbiased estimator of the population median of that distribution.

6. Devise an unbiased estimator of the 25th percentile of the uniform$(0, \theta)$ distribution based on the order statistics of a random sample of size 7. Calculate the value of the estimator if the observed sample values are .50, .13, 1.25, .68, 1.53, .31, and 1.10.

7. Suppose that copies of a limited edition collector's plate are issued with serial numbers $1, 2, \ldots, \theta$. A random sample of plates whose numbers are $X_1, X_2, \ldots, X_n$ is available, and you would like to estimate how many plates were actually made. Show that an approximately unbiased estimator of θ is $Y = (^{n+1}/_n)(X_{(n)} + 1)$, where $X_{(n)}$ is the largest order statistic of the sample. (*Hint:* To approximate the expected value, view the sum as a Riemann sum of a continuous function, and approximate it by an integral.)

8. Let X_1, X_2, X_3 be a random sample of size 3 from some distribution whose mean μ we want to estimate. Show that among all unbiased linear combinations $\sum_{i=1}^{3} a_i X_i$ of the sample values, the linear combination with $a_1 = a_2 = a_3 = 1/3$ has the smallest standard error.

9. Find the value of c such that $E[(cS^2 - \sigma^2)^2]$ is minimized, where S^2 is the sample variance of a random sample of size n from the $N(\mu, \sigma^2)$ distribution. Interpret the result in terms of the estimation of σ^2.

10. Let $X_{(1)}$ be the first order statistic of a sample of size n from the uniform$(0, \theta)$ distribution. Find a function of $X_{(1)}$ that is unbiased for θ. Show that $2\overline{X}$ is also an unbiased estimator of θ. Which of these two estimators is more efficient?

11. Find a constant c such that cS is unbiased for σ, where S^2 is the sample variance of a random sample of size n from the $N(\mu, \sigma^2)$ distribution. (*Hint:* Do a preliminary computation of $E[\sqrt{Y}]$, where Y has the $\chi^2(r)$ distirbution.)

12. Find the Rao–Cramer lower bound for estimation of the parameter μ of the Poisson distribution, and find a minimum variance unbiased estimator for μ.

13. Let S^2 be the sample variance of a random sample of size n from the $N(\mu, \sigma^2)$ distribution. Then S^2 is an unbiased estimator of σ^2. Does S^2 achieve the Rao–Cramer lower bound?

14. Show that the estimator $(n+1)/n \cdot Y_n$ of the parameter θ in the uniform$(0, \theta)$ distribution discussed in Example 8.1.4 has variance *less* than the R–C lower bound. Explain how this is possible.

15. Show that the estimator $\overline{X}$ achieves the R–C lower bound for estimation of the parameter $\mu = 1/\lambda$ in the exponential(λ) distribution.

16. Show that the estimator $\overline{X}$ achieves the R–C lower bound for estimation of the parameter $q = 1/p$ in the geometric(p) distribution:

$$f(k; p) = (1 - p)^{k-1} p, \qquad k = 1, 2, \ldots .$$

8.2 | Maximum Likelihood Estimation

We have studied criteria for judging the quality of statistical estimators of population parameters, but we postponed the question of where estimators come from. Is there some routine from which we can take the functional form $f(x; \theta)$ of the distribution being sampled from and obtain an estimator $Z = g(X_1, X_2, \ldots, X_n)$ of θ? We will examine one such method in this section, called the *maximum likelihood method*. It will be designed to be intuitively appealing, and it can also be proven to yield estimators with good properties. At the end of this section we will also remark briefly on another procedure, called the *method of moments*, that frequently involves less computation but still produces reasonable estimates.

Let us first develop some intuition. Answer the following question before going on.

—— **?Question 8.2.1**   Suppose that you had just two observations $x_1 = 6.5$, $x_2 = 7.5$ from a normal population whose mean and variance were unknown to you. Does it seem likelier that the true mean is $\mu = 0$, $\mu = 7$, or $\mu = 13$?

In the absence of any other information you should have answered $\mu = 7$, because you know that the normal density has most of its probability weight near its mean. Hence x_1 and x_2 are likelier to have been observations from $N(7, \sigma^2)$ than from $N(0, \sigma^2)$, which has most of its weight near 0, or from $N(13, \sigma^2)$, which has most of its weight near 13. To be more precise, among the three listed choices of μ, and perhaps among all real μ, the joint distribution of X_1 and X_2, which is bivariate normal $(7, 7, \sigma^2, \sigma^2, 0)$ should have highest probability density at the point where $x_1 = 6.5$, $x_2 = 7.5$. We will now quantitatively define and justify this approach to the search for good estimators.

DEFINITION 8.2.1

Let $X_1, X_2, \ldots, X_n$ be a random sample from a distribution characterized by $f(x; \theta)$, where θ is an unknown parameter. The *likelihood function* of the sample is the joint density (or p.m.f. in the discrete case) of the random variables $X_1, X_2, \ldots, X_n$ viewed as a function of θ, that is,

$$L(\theta) = L(\theta; x_1, x_2, \ldots, x_n) = f(x_1, x_2, \ldots, x_n; \theta) = \prod_{i=1}^{n} f(x_i; \theta). \quad (8.20)$$

A *maximum likelihood estimator (MLE)* $\widehat{\theta}$ of θ achieves the maximum of $L(\theta)$ subject to any constraints that may be present on permissible values of θ.

To find an MLE, you must write the function $L(\theta)$ and find its maximum with respect to θ, which usually entails a calculus optimization. Since L also depends on the observed sample values $x_1, x_2, \ldots, x_n$, the maximizing argument $\widehat{\theta}$ will be some function g of the x_i's. When that function is applied to the random sample variables, you obtain an estimator $Z = \widehat{\theta} = g(X_1, X_2, \ldots, X_n)$.

Why does the maximum likelihood method make sense? Consider the case where $f(x; \theta)$ is a discrete p.m.f. Then the likelihood function is the joint p.m.f. of the sample random variables:

$$L(\theta) = f(x_1, x_2, \ldots, x_n; \theta) = P[X_1 = x_1, X_2 = x_2, \ldots, X_n = x_n; \theta],$$

where the probability expression on the right assumes that θ is the true parameter. Therefore the maximization finds the θ value $\widehat{\theta}$ that maximizes the joint probability that the observed data $x_1, x_2, \ldots, x_n$ occurred. Loosely speaking, such a θ value best explains the observed sample. The motivation in the continuous case is similar, except that we maximize the joint probability density associated with the observed sample instead of the actual joint probability.

The following fact is helpful. As you are asked to show rigorously in Exercise 11, the MLE $\hat{\theta}$ also is the value of θ that maximizes $\ln(L(\theta))$, which is often simpler to maximize, as we will see in the following examples.

Example 8.2.1

Find the maximum likelihood estimator of the parameter μ in the Poisson(μ) distribution.

The Poisson p.m.f. has the form

$$f(x; \mu) = \frac{\mu^x e^{-\mu}}{x!}, \qquad x = 0, 1, 2, \ldots,$$

and the possible values of μ are $\mu \in (0, \infty)$. The likelihood function is

$$L(\mu; x_1, x_2, \ldots, x_n) = f(x_1, x_2, \ldots, x_n; \mu) = \prod_{i=1}^{n} f(x_i; \theta)$$

$$= \prod_{i=1}^{n} \frac{\mu^{x_i} e^{-\mu}}{x_i!}$$

$$= \frac{\mu^{\sum_{i=1}^{n} x_i} e^{-n\mu}}{\prod_{i=1}^{n} x_i!}. \qquad (8.21)$$

By itself the function in (8.21) would be relatively difficult to maximize. By the comment above, however, it suffices to maximize

$$\ln(L(\mu)) = \ln\left(\frac{\mu^{\sum_{i=1}^{n} x_i} e^{-n\mu}}{\prod_{i=1}^{n} x_i!}\right) = \left(\sum_{i=1}^{n} x_i\right) \ln(\mu) - n\mu - \ln\left(\prod_{i=1}^{n} x_i!\right). \qquad (8.22)$$

Differentiating,

$$\frac{\partial \ln(L(\mu))}{\partial \mu} = \frac{\sum_{i=1}^{n} x_i}{\mu} - n. \qquad (8.23)$$

Since the second derivative is $-\sum_{i=1}^{n} x_i/\mu^2 < 0$, and there is only one critical point, that critical point is a local and global maximum. Setting the expression in (8.23) equal to 0 yields

$$\hat{\mu} = \frac{\sum_{i=1}^{n} x_i}{n} = \bar{x}. \qquad (8.24)$$

Thus the estimator $\hat{\mu} = \sum_{i=1}^{n} X_i/n = \bar{X}$ is the MLE for μ.

Example 8.2.2

Suppose that a device consists of an original part and a backup that is automatically put into use when the original fails. Both parts fail at exponentially distributed times with a common parameter λ, and the failure times are independent of one another. Five copies of the device are observed, which fail at times 2.1, 3.2, 2.5, 4.6, and 3.8. Find the value of a maximum likelihood estimator for λ.

Because the backup part begins to operate when the original part breaks, the time T to failure of the device is the sum of the failure times of the parts. Recall from Proposition 5.3.2 that the sum of two i.i.d. exponential random variables with parameter λ has the $\Gamma(2, \lambda)$ density:

$$f(t; \lambda) = \frac{\lambda^2}{\Gamma(2)} t^{2-1} e^{-\lambda t} = \lambda^2 t e^{-\lambda t}, \qquad t > 0. \tag{8.25}$$

We assume that the five copies operate independently of one another, and so the problem supplies us with observed values t_1, t_2, t_3, t_4, and t_5 of a random sample from the distribution specified by $f(t; \lambda)$ in (8.25). We now must find the form of the MLE and compute its value for the given sample information.

The likelihood function is

$$L(\lambda) = \prod_{i=1}^{5} f(t_i; \lambda) = \prod_{i=1}^{5} \lambda^2 t_i e^{-\lambda t_i} = \lambda^{10} \left(\prod_{i=1}^{5} t_i \right) e^{-\lambda \sum_{i=1}^{5} t_i}.$$

Its graph for the given data values appears in Fig. 8.2(a), and the graph of its natural logarithm is in part (b) of the figure. Note that, as expected, the point where the maximum value is taken on is the same for both functions. We can find the MLE of λ by computing as follows:

$$\frac{\partial \ln(L(\lambda))}{\partial \lambda} = \frac{\partial}{\partial \lambda} \left(10 \ln(\lambda) + \sum_{i=1}^{5} \ln(t_i) - \lambda \sum_{i=1}^{5} t_i \right)$$

$$= \frac{10}{\lambda} - \sum_{i=1}^{5} t_i.$$

Setting the last expression to zero, the MLE comes out to be $\widehat{\lambda} = 10 / \sum_{i=1}^{5} T_i$, which has the value $10/(2.1 + 3.2 + 2.5 + 4.6 + 3.8) = .617$ for the given sample.

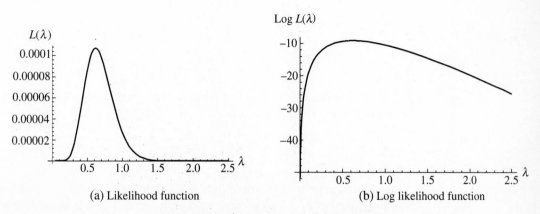

(a) Likelihood function   (b) Log likelihood function

Figure 8.2 Maximum Likelihood Estimator

—— **?Question 8.2.2**   Check that $\hat{\lambda}$ is indeed a global maximizer over the set of possible values for λ.

Example 8.2.3

The problem of finding the MLE of θ in the uniform$[0, \theta]$ distribution requires a slightly different analysis. Let $X_1, X_2, \ldots, X_n$ be a random sample from this population, whose p.d.f. is

$$f(x; \theta) = \begin{cases} 1/\theta & \text{if } x \in (0, \theta), \\ 0 & \text{otherwise.} \end{cases} \tag{8.26}$$

Notice that since each X_i must take a value less than or equal to θ, θ is constrained to be at least as large as all of the X_i's; that is, $\theta \geq \max X_i = X_{(n)}$. The likelihood is

$$L(\theta) = \begin{cases} 1/\theta^n & \text{if } \theta \geq x_i \quad \text{for all } i, \\ 0 & \text{otherwise,} \end{cases} \tag{8.27}$$

which clearly decreases as a function of θ. The maximum value of L is then to be found at the leftmost point of the set of possible values $[x_{(n)}, \infty)$; hence the MLE $\hat{\theta}$ of θ is the largest order statistic $X_{(n)}$.

Example 8.2.4

The maximum likelihood method applies as well to the estimation of a vector parameter. To illustrate, let us find the joint MLEs of μ and σ^2 in the normal distribution.

The density is

$$f(x; \mu, \sigma^2) = \frac{1}{\sqrt{2\pi\sigma^2}} e^{-(x-\mu)^2/2\sigma^2}, \tag{8.28}$$

and we wish to maximize the likelihood as a function of the vector parameter $\boldsymbol{\theta} = (\mu, \sigma^2)$. The likelihood function is

$$L(\boldsymbol{\theta}; x_1, x_2, \ldots, x_n) = \prod_{i=1}^{n} f(x_i; \boldsymbol{\theta})$$

$$= \prod_{i=1}^{n} \frac{1}{\sqrt{2\pi\sigma^2}} e^{-(x_i-\mu)^2/2\sigma^2}$$

$$= (2\pi\sigma^2)^{-n/2} e^{-\sum_{i=1}^{n}(x_i-\mu)^2/2\sigma^2}. \tag{8.29}$$

To better conceptualize what we are doing, consider the Calculus I grade point data from Table 7.4, reproduced here for your convenience:

1.93, 2.25, 1.98, 2.53, 1.8, 2.28, 2.46, 1.87, 2.29, 2.23, 2.67,
1.82, 2.07, 2.0, 2.52, 2.368, 2.237, 2.69, 2.05, 2.73

There are $n = 20$ observed sample values $x_1 = 1.93, x_2 = 2.25, \ldots, x_{20} = 2.73$. These values may be substituted into the formula for L in (8.29), and a two-variable function of μ and σ^2 results. It is shown in Fig. 8.3. We look for the point $(\hat{\mu}, \hat{\sigma}^2)$ at which that function hits its maximum.

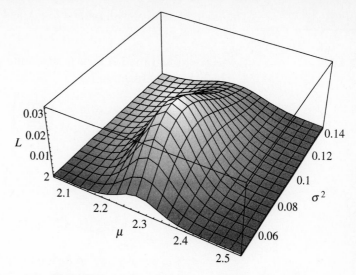

Figure 8.3 A Two-Variable Likelihood Function

As usual, it helps to consider instead the log of the likelihood function:

$$\ln(L(\mu, \sigma^2)) = -\frac{n}{2}(\ln(2\pi) + \ln(\sigma^2)) - \frac{1}{2\sigma^2}\sum_{i=1}^{n}(x_i - \mu)^2. \qquad (8.30)$$

To locate the exact optimum values, set the partials of $\ln(L)$ with respect to μ and σ^2 equal to zero. Doing this, we obtain

$$\frac{\partial \ln(L(\mu, \sigma^2))}{\partial \mu} = -\frac{1}{2\sigma^2}\sum_{i=1}^{n}(x_i - \mu) \cdot (-2) = 0$$

$$\implies \hat{\mu} = \overline{X}, \qquad (8.31)$$

$$\frac{\partial \ln(L(\mu, \sigma^2))}{\partial \sigma^2} = -\frac{n}{2\sigma^2} + \frac{1}{2} \cdot \frac{1}{(\sigma^2)^2}\sum_{i=1}^{n}(x_i - \mu)^2 = 0$$

$$\implies \hat{\sigma}^2 = \frac{1}{n}\sum_{i=1}^{n}(X_i - \hat{\mu})^2 = \frac{1}{n}\sum_{i=1}^{n}(X_i - \overline{X})^2. \qquad (8.32)$$

[You should fill in the computational details of (8.31) and (8.32) yourself.]

The MLE for μ is $\overline{X}$ as you probably would have expected, but the MLE for σ^2 is not the unbiased estimator S^2 but rather $V = \frac{n-1}{n}S^2$. For the foregoing calculus data, the sample mean turns out to be 2.239 and the value of V is .0826, which is consistent with Fig. 8.3.

?Question 8.2.3 Review the second partials test for categorization of critical points in multivariate optimization. Apply it to show that $(\hat{\mu}, \hat{\sigma}^2)$ in (8.31) and (8.32) is indeed a local maximum.

Example 8.2.5

The Weibull density with parameters λ and β is

$$f(t; \lambda, \beta) = \beta \lambda^\beta t^{\beta-1} e^{-(\lambda t)^\beta}, \qquad t > 0. \tag{8.33}$$

Let us try to find the MLE of β, assuming that it is known that $\lambda = 2$.

The likelihood function is

$$L(\beta) = \beta^n 2^{n\beta} \left(\prod_{i=1}^{n} t_i \right)^{\beta-1} e^{-\sum_{i=1}^{n} (2t_i)^\beta}.$$

Hence,

$$\frac{\partial \ln(L(\beta))}{\partial \beta} = \frac{\partial}{\partial \beta} \left(n \ln(\beta) - n\beta \ln(2) - (\beta - 1) \ln \left(\prod_{i=1}^{n} t_i \right) - \sum_{i=1}^{n} (2t_i)^\beta \right)$$

$$= \frac{n}{\beta} - n \ln(2) - \ln \left(\prod_{i=1}^{n} t_i \right) - \sum_{i=1}^{n} (2t_i)^\beta \ln(2t_i). \tag{8.34}$$

In this case we are not so fortunate as to have an analytically solvable form for the MLE. If we had particular observed values $t_1, t_2, \ldots, t_n$ sampled from the Weibull$(2, \beta)$ distribution, then we would have to substitute them into (8.34), set the expression equal to 0, and solve numerically for β. You should be aware that maximum likelihood estimation does sometimes result in equations that force the use of numerical techniques, but we will pursue these cases no further here.

The estimators that we have produced in these examples, together with other examples that you will study in the exercise set, are statistics that we have previously seen to be unbiased estimators of minimum variance. This should give you even more confidence in the use of the maximum likelihood technique to construct estimators. But a sophisticated theory also exists to give support to the use of this method. We will not go into detail here, but you can find elaboration in Arnold (1990, pp. 273 ff). One very important result is that if an estimator Z of a parameter θ exists that is unbiased and also achieves the Rao–Cramer lower bound, then that estimator must be a function of the MLE. Thus, except in cases where no best estimator exists at all, there is no need to look for any estimator other than the MLE.

We close this section with a brief nod to another procedure for obtaining estimators, the *method of moments*. Consider a case where there are two parameters to be estimated. The sample moments $M_1 = \overline{X}$ and $M_2 = \sum_{i=1}^{n} X_i^2 / n$ should be close to the population moments $m_1 = \mu$ and $m_2 = E[X^2]$. The first moment m_1 is some function $m_1 = m_1(\theta)$ of the unknown parameters, as is the second moment $m_2 = m_2(\theta)$. Setting

$$m_1(\theta) = M_1 = \overline{X}, \qquad m_2(\theta) = M_2 = \sum_{i=1}^{n} \frac{X_i^2}{n}, \tag{8.35}$$

and solving for θ gives the method of moments estimate $\widetilde{\theta}$. If θ is only a one-dimensional parameter, then only the first of the two equations in (8.35) is necessary. In the case that θ has three or more components, the third-order and higher population moments $m_k = E[X^k]$ would be set equal to the third-order and higher sample moments $M_k = \sum_{i=1}^{n} X_i^k/n$ as necessary in order to produce enough equations to solve uniquely for all components of θ.

For example, for the method of moments estimator of the parameter λ in the exponential (λ) distribution, recall that the population mean is $m_1 = 1/\lambda$, hence we set $1/\lambda = \overline{X}$, which implies $\tilde{\lambda} = 1/\overline{X}$. Back in Example 3.3.1 on the distribution of the delay until a new word is reached in "The Road Not Taken," I used this technique to estimate λ.

As another example, in the $\Gamma(\alpha, \lambda)$ distribution, the mean is α/λ and the variance is α/λ^2, so that the second moment is $m_2 = \alpha/\lambda^2 + (\alpha/\lambda)^2$. The method of moments equations for the two components α and λ are thus

$$\frac{\alpha}{\lambda} = \overline{X}, \qquad \frac{\alpha + \alpha^2}{\lambda^2} = \frac{1}{n} \sum_{i=1}^{n} X_i^2 = M_2.$$

Solving for α in the first equation and substituting into the second gives

$$\alpha = \lambda \overline{X} \Rightarrow \frac{\lambda \overline{X} + \lambda^2 \overline{X}^2}{\lambda^2} = M_2 \Rightarrow \tfrac{1}{\lambda} \overline{X} = M_2 - \overline{X}^2$$

$$\Rightarrow \tilde{\lambda} = \frac{\overline{X}}{M_2 - \overline{X}^2}, \qquad \tilde{\alpha} = \frac{\overline{X}^2}{M_2 - \overline{X}^2}.$$

EXERCISES 8.2

1. Suppose that you are watching a Poisson process whose rate you do not know. In one time interval of length 1 you count 2 arrivals. In another, disjoint from the first, you count 4 arrivals. Which of the following possible values of the rate parameter seems likeliest: $\lambda = 1, 2, 3, 4,$ or 5? Interpret in terms of the likelihood function.

2. Produce a graph of the likelihood function for the normal example with which we began the section, for each of the σ^2 values 1, 4, and 9. In each case, where is the maximum of the likelihood obtained?

3. On eight successive days the price of a stock is observed and whether the stock gained or lost is recorded. The results are gain, gain, loss, gain, loss, gain, loss, gain. Pose a probabilistic model for the experiment, and find the numerical value of a maximum likelihood estimator for the unknown parameter.

4. Compute the maximum likelihood estimator for the parameter λ in the exponential (λ) distribution. Is it the same as the method of moments estimator?

5. a. Cars arrive to a car wash, some of which get a hot wax. Suppose that we repeat n independent times the experiment of counting the number of cars that pass through the car wash until the first car gets a hot wax. Formulate a probabilistic model that lets you find a maximum likelihood estimator of the probability that a single car requests a hot wax.

   b. Repeat part (a), this time using as the observed data the number of cars until the second hot wax customer.

6. Find the maximum likelihood estimator of the mean μ of a normal distribution, assuming that the variance σ^2 is known.

7. Find the maximum likelihood estimator of the variance σ^2 of a normal distribution, assuming that the mean μ is known.

8. Suppose that the fraction of an hour during which a phone booth is unoccupied has the density function $f(x; \theta) = \theta x^{\theta-1}$, $x \in (0, 1)$. Find a maximum likelihood estimator of θ.

9. A random variable X has the uniform distribution on the interval $(\theta, 1)$. Find the maximum likelihood estimator of θ based on a random sample of size n. Is the MLE unbiased?

10. The little country of Lower Statistovia knows that its enemy Upper Statistovia numbers its tanks serially starting with 1 and ending with the number θ of tanks that it has produced. Spies from Lower Statistovia have observed tanks numbered 4, 16, 21, 43, and 58. Find the maximum likelihood estimator of the number of Upper Statistovian tanks. Be clear about the assumptions you are making in your probabilistic model.

11. Show that if $\widehat{\theta}$ is the point of maximization for a real-valued function of a real variable $h(\theta)$, then it is also the point of maximization for $\ln(h(\theta))$.

12. Let $X_1, X_2, \ldots, X_n$ be a random sample of truck tire wearout mileages from the Weibull$(\lambda, 1.5)$ distribution. Find the form of the MLE of λ.

13. Set up, but do not try to solve, the equation for the maximum likelihood estimator of the parameter α in the $\Gamma(\alpha, 2)$ distribution.

14. In an opinion poll, a sequence of n independently selected respondents answer yes, no, or don't know to the question of interest. We can model the state space for a random variable associated with one respondent as the set of three points $(1, 0)$ for yes, $(0, 1)$ for no, and $(0, 0)$ for don't know. Find the maximum likelihood estimator of the vector parameter $\boldsymbol{\theta} = (p_1, p_2)$, where $p_1 = P[\text{yes}]$ and $p_2 = P[\text{no}]$. (*Hint:* Write the p.m.f. of the vector random variable (X_1, X_2) whose state space is $E = \{(1, 0), (0, 1), (0, 0)\}$.)

15. Find the method of moments estimator of the parameter in Exercise 5(a).

16. Find the method of moments estimators of the parameters θ_1 and θ_2 in the uniform (θ_1, θ_2) distribution.

17. Find the method of moments estimator (a) for μ in the Poisson (μ) distribution; (b) for μ and σ^2 (simultaneously) in the $N(\mu, \sigma^2)$ distribution.

8.3 | Sufficient Statistics

In our work on parameter estimation, you have undoubtedly noticed the same estimators coming up again and again in different contexts. In particular, statistics such as the sum of sample values, the sum of squares of sample values, and functions of them such as the sample mean and sample variance appear as maximum likelihood estimators, and also as unbiased, minimum variance estimators. You may have been wondering whether there is some underlying cause for this, and whether in general maximum likelihood statistics capture all that is needed in the raw data for estimation purposes.

This section uses the concept of *sufficiency* to tie together the ideas of parameter estimation. The material is theoretical in nature and so is deep in the background of the practice of statistics. Hence it may be omitted at first reading by those more interested in applications or in the other topics to come. But if you are intrigued by the underlying theory of the subject and if you appreciate the aesthetic beauty of mathematics, this section is for you. We will not attempt complete coverage here, some of which requires the study of measure theory to understand. If you wish to study sufficiency further, two places to begin are Hogg and Craig (1978) and Arnold (1990), upon which much of this section is based.

Here is a brief roadmap through the important results of this section. A *sufficient statistic* is an estimator such that if its value is known, then the details of the sample can shed no more light on the true value of the parameter. By factoring the likelihood function strategically, all at once the form of the sufficient statistic can be discovered, and its sufficiency can be proved. For distributions belonging to a certain class called the *exponential class* (to which many of the distributions we have seen do belong), the sufficient statistic is easy to read off of the density itself, and it takes the form of a sum of elementary functions of the sample values. Probably most important for our purposes is that if minimum variance unbiased estimators exist, then they must be functions of sufficient statistics. Furthermore, under mild regularity conditions, maximum likelihood estimators are functions of sufficient statistics, and hence if an unbiased transformation of an MLE can be found, the resulting estimator will be a best estimator.

DEFINITION 8.3.1

Let $X_1, X_2, \ldots, X_n$ be a random sample from a distribution with density (or p.m.f.) $f(x; \theta)$. A statistic $Z = g(X_1, X_2, \ldots, X_n)$ is called *sufficient for* θ if the conditional joint distribution of $X_1, X_2, \ldots, X_n$ given Z,

$$f(x_1, x_2, \ldots, x_n \,|\, z) = \frac{f(x_1, x_2, \ldots, x_n, z)}{f_z(z)}, \qquad (8.36)$$

does not depend on θ.

The fact that for a sufficient statistic $f(x_1, x_2, \ldots, x_n \,|\, z)$ does not depend on θ says that the conditional distribution of any estimator $Y = h(X_1, X_2, \ldots, X_n)$ given $Z = z$ has nothing to do with θ; hence Y can provide no more statistical information about θ. Roughly speaking, a sufficient statistic carries in it all of the useful information for estimating a parameter.

Example 8.3.1

Let us illustrate the idea of sufficiency by trying to show that the sum $Z = \sum_{i=1}^{n} X_i$ of random sample values from the $N(\mu, \sigma^2)$ distribution is sufficient for μ, where for simplicity we assume that σ^2 is known.

The joint density of $X_1, X_2, \ldots, X_n$ and $Z = \sum_{i=1}^{n} X_i$ vanishes unless $z = \sum_{i=1}^{n} x_i$, and in the latter case it just agrees with the joint density of the random variables $X_1, X_2, \ldots, X_n$. Recall also Proposition 5.3.4, which implies that Z has the $N(n\mu, n\sigma^2)$ distribution. Thus formula (8.36) becomes

$$
\begin{aligned}
f(x_1, x_2, \ldots, x_n \mid z) &= \frac{f(x_1, x_2, \ldots, x_n)}{f_z(z)} \\
&= \frac{\prod_{i=1}^{n} f(x_i; \mu)}{f_z(z)} \\
&= \frac{e^{-\sum_{i=1}^{n}(x_i - \mu)^2/2\sigma^2} / \left(\sqrt{2\pi\sigma^2}\right)^n}{e^{-(z-n\mu)^2/2n\sigma^2}/\sqrt{2\pi n\sigma^2}}, \qquad z = \sum_{i=1}^{n} x_i \\
&= \left(2\pi\sigma^2\right)^{(-n+1)/2} \sqrt{n} \, e^{-\frac{\sum(x_i - \mu)^2}{2\sigma^2} + \frac{\left(\sum x_i - n\mu\right)^2}{2n\sigma^2}} \\
&= \left(2\pi\sigma^2\right)^{(-n+1)/2} \sqrt{n} \, e^{\frac{1}{2n\sigma^2}\left(\left(\sum x_i\right)^2 - n\sum x_i^2\right)}, \qquad (8.37)
\end{aligned}
$$

after expanding and simplifying the exponent. The resulting conditional density does not depend on μ. Since σ^2 is known, $Z = \sum_{i=1}^{n} X_i$ is a sufficient statistic for μ. You would probably guess that if σ^2 had not been known, so that $\boldsymbol{\theta} = (\mu, \sigma^2)$ is a vector parameter, Z would not have been sufficient for $\boldsymbol{\theta}$. More than just the information of the sum of the sample values is necessary to estimate $\boldsymbol{\theta}$ in this case. (What do you think is also needed?)

?Question 8.3.1 We just found that when σ^2 is known, $Z = \sum_{i=1}^{n} X_i$ is sufficient for μ in $N(\mu, \sigma^2)$. Do you think that $\overline{X}$ would also be sufficient? Why?

Fortunately it is not necessary to use the definition of sufficiency to check whether a given statistic is sufficient, nor do we have to make burnt offerings to the statistical gods to find candidates for sufficient statistics. The next theorem gives two alternative characterizations of sufficiency that are very useful. Condition (b) is called the *Fisher–Neyman criterion*, which is really just a stepping stone to condition (c), the *factorization criterion*. Some authors adopt the factorization criterion as the original definition of sufficiency.

PROPOSITION 8.3.1

Let $X_1, X_2, \ldots, X_n$ be a random sample from a distribution characterized by density (or p.m.f.) $f(x; \theta)$, whose support set (i.e., set of points x of nonzero density) does not depend on θ. Let $Z = g(X_1, X_2, \ldots, X_n)$ be an estimator of θ. The following are equivalent:

(a) Z is sufficient for θ.

(b) The likelihood function L of the sample factors into the product

$$L(\theta; x_1, x_2, \ldots, x_n) = b(x_1, x_2, \ldots, x_n) \cdot f_z(z; \theta) \qquad (8.38)$$

of some function b that does not depend on θ and the marginal density (or p.m.f.) of Z.

(c) The likelihood function L of the sample factors into the product

$$L(\theta; x_1, x_2, \ldots, x_n) = b(x_1, x_2, \ldots, x_n) \cdot k(g(x_1, x_2, \ldots, x_n); \theta) \qquad (8.39)$$

of some function b that does not depend on θ and some function k of $z = g(x_1, x_2, \ldots, x_n)$ and θ.

■ **Proof** We will prove the chain of implications (a)$\Rightarrow$(b)$\Rightarrow$(c)$\Rightarrow$(a).

(a)$\Rightarrow$(b). Suppose that Z is sufficient for θ. The likelihood is simply the joint density $f(x_1, x_2, \ldots, x_n; \theta)$. Also, as noted already, the joint density of $X_1, X_2, \ldots, X_n$ and Z is equal to $f(x_1, x_2, \ldots, x_n; \theta)$ when $z = g(x_1, x_2, \ldots, x_n)$, and it equals zero otherwise. Then

$$L(\theta) = f(x_1, x_2, \ldots, x_n; \theta) = f(x_1, x_2, \ldots, x_n \mid z) \cdot f_z(z; \theta),$$

and by (8.36), $f(x_1, x_2, \ldots, x_n \mid z) = b(x_1, x_2, \ldots, x_n)$ is a function of the observed sample values which does not depend on θ.

(b)$\Rightarrow$(c). This implication follows trivially by taking

$$k(g(x_1, x_2, \ldots, x_n), \theta) = f_z(g(x_1, x_2, \ldots, x_n); \theta).$$

(c)$\Rightarrow$(a). Suppose the factorization criterion (8.39) holds. We do the proof (following Hogg and Craig, 1978, p. 344) in the continuous case only; the discrete case is left as Exercise 11.

Apply the transformation

$$u_1 = z = g(x_1, \ldots, x_n), \quad u_2 = g_2(x_1, \ldots, x_n), \quad \ldots, \quad u_n = g_n(x_1, \ldots, x_n),$$

where the g_i's are any convenient functions that, together with g, form a 1–1 transformation. If the inverse transformation is

$$x_1 = h_1(z, u_2, \ldots, u_n), \quad x_2 = h_2(z, u_2, \ldots, u_n), \ldots, \quad x_n = h_n(z, u_2, \ldots, u_n),$$

then we obtain that the joint p.d.f. of $Z, U_2, \ldots, U_n$, is

$$\widetilde{f}(z, u_2, \ldots, u_n) = f(h_1, h_2, \ldots, h_n) \cdot |J|,$$
$$= b(h_1, h_2, \ldots, h_n) \cdot k(z; \theta) \cdot |J|, \qquad (8.40)$$

where J is the determinant of the matrix of partials $(\partial h_i / \partial u_j)_{i,j=1,\ldots,n}$. The marginal density of Z can be found by integrating the right side of (8.40) with respect to all of $u_2, u_3, \ldots, u_n$:

$$f_z(z) = \int \int \cdots \int \widetilde{f}(z, u_2, \ldots, u_n) \, du_2 du_3 \cdots du_n$$
$$= \int \int \cdots \int b(h_1(z, u_2, \ldots, u_n), \ldots, h_n(z, u_2, \ldots, u_n)) \cdot k(z; \theta)$$
$$\cdot |J| \, du_2 du_3 \cdots du_n$$
$$= k(z; \theta) \cdot c(z). \qquad (8.41)$$

In the third line of (8.41) we have recognized that $k(z; \theta)$ factors out of the integral, all variables $u_2, \ldots, u_n$ are integrated out of the remaining integral, and by assumption neither b nor J, nor the limits of integration, depend on θ; hence the integral, which we call c, does not depend on θ.

Therefore, the conditional density of $X_1, X_2, \ldots, X_n$ given $Z = z$ is

$$f(x_1, x_2, \ldots, x_n \mid z) = \frac{f(x_1, x_2, \ldots, x_n, z)}{f_z(z)}$$

$$= \frac{b(x_1, x_2, \ldots, x_n) \cdot k(z; \theta)}{k(z; \theta) \cdot c(z)}$$

$$= \frac{b(x_1, x_2, \ldots, x_n)}{c(z)},$$

which does not depend on θ. Thus Z is sufficient and the proof is complete.

Since the likelihood function $L(\theta; x_1, x_2, \ldots, x_n)$ is easy to find from the population distribution, the factorization criterion (8.39) allows us to discover sufficient statistics. Simply factor out of L any factors that depend only on $x_1, x_2, \ldots, x_n$ (i.e. b), and look at the residual factor (k). In it will be a function of the sample variables (g) and terms involving the parameter θ. Then the estimator $Z = g(X_1, X_2, \ldots, X_n)$ is sufficient for θ.

Example 8.3.2

Find a sufficient statistic for the parameter p in the geometric distribution

$$f(x; p) = p(1 - p)^{x-1}; \qquad x = 1, 2, 3, \ldots.$$

The likelihood function is

$$L(p) = \prod_{i=1}^{n} f(x_i; p) = p^n (1 - p)^{\sum_{i=1}^{n} x_i - n}. \qquad (8.42)$$

Setting $b = 1$, $k = L$, and $g(x_1, x_2, \ldots, x_n) = \sum_{i=1}^{n} x_i$ in (8.39) shows us that $Z = \sum_{i=1}^{n} X_i$ is sufficient for p.

You may remember Exercise 16 in Section 8.1, which proved that $\overline{X}$ was a minimum variance unbiased estimator of the parameter $q = 1/p$ in this context. Setting $p = 1/q$ in (8.42) and reorganizing the exponent gives

$$L(q) = \left(\frac{1}{q}\right)^n \left(1 - \frac{1}{q}\right)^{n \cdot \sum_{i=1}^{n} x_i / n - n},$$

which implies by the factorization criterion that $\overline{X} = \sum_{i=1}^{n} X_i / n$ is sufficient for q. We will be saying more about the connection between sufficient statistics and minimum variance unbiased estimators later in the section.

?Question 8.3.2 In Example 8.3.2, is $\overline{X}$ sufficient for p? Is $Z = \sum X_i$ sufficient for q? Do you think anything can be said about the uniqueness of sufficient statistics, or of the parameters they estimate?

Example 8.3.3

Show that $Z = \sum_{i=1}^{n} T_i$ is sufficient for λ, where $T_1, T_2, \ldots, T_n$ is a random sample from the $\Gamma(\alpha, \lambda)$ distribution and we assume that α is known.

The density is

$$f(t, \lambda) = \frac{\lambda^\alpha}{\Gamma(\alpha)} t^{\alpha-1} e^{-\lambda t}, \qquad t > 0,$$

which implies that the likelihood function is

$$L(\lambda; t_1, t_2, \ldots, t_n) = \prod_{i=1}^{n} f(t_i, \lambda)$$

$$= \frac{\lambda^{n\alpha}}{(\Gamma(\alpha))^n} \left(\prod_{i=1}^{n} t_i \right)^{\alpha-1} e^{-\lambda \sum_{i=1}^{n} t_i}$$

$$= \left(\frac{\left(\prod_{i=1}^{n} t_i \right)^{\alpha-1}}{(\Gamma(\alpha))^n} \right) \cdot \left(\lambda^{n\alpha} e^{-\lambda \sum_{i=1}^{n} t_i} \right). \qquad (8.43)$$

Let the left factor in (8.43) be the function b in (8.39) (which is valid because α is known), and let the right factor be the function k. Then $g(t_1, t_2, \ldots, t_n) = \sum_{i=1}^{n} t_i$ and the estimator $Z = g(T_1, T_2, \ldots, T_n) = \sum_{i=1}^{n} T_i$ is sufficient for λ by the factorization criterion.

?Question 8.3.3

In Example 8.3.3, is the MLE of λ a function of the sufficient statistic? Is $\sum_{i=1}^{n} T_i$ also sufficient for the parameter $\beta = 1/\lambda$? If so, can you find a function of the sufficient statistic that is unbiased for β and also has minimum variance among unbiased estimators?

The following is a simple result that you may have guessed as you have been reading. Intuitively it must be true, because knowledge of the value $z = g(x_1, x_2, \ldots, x_n)$ of a sufficient statistic is equivalent to knowledge of a 1–1 function of z.

PROPOSITION 8.3.2

If $Z = g(X_1, X_2, \ldots, X_n)$ is a sufficient statistic for θ, and h is an invertible function of z, then the estimator $h(Z)$ is also sufficient for θ.

■ **Proof**   The proof is left for Exercise 4.

We frequently see $\sum_{i=1}^{n} X_i$ as a sufficient statistic for a parameter; Proposition 8.3.2 tells us, for example, that $\overline{X} = \sum_{i=1}^{n} X_i/n$ or, in fact, any 1–1 function of $\sum_{i=1}^{n} X_i$ is also sufficient.

Since the likelihood function is obtained by multiplying copies of the density (or p.m.f.) $f(x; \theta)$, it stands to reason that the factorization criterion may imply that conditions on the density itself will allow us to recognize sufficient statistics. To develop this idea, we introduce the next definition and prove a result.

DEFINITION 8.3.2

A density (or p.m.f.) is said to be of the *exponential class* if its support is independent of θ, and if it can be written in the form

$$f(x;\theta) = e^{K(x)p(\theta)+S(x)+q(\theta)} \tag{8.44}$$

for some functions K, p, S, and q.

PROPOSITION 8.3.3

If the distribution $f(x;\theta)$ is of the exponential class, then the statistic $Z = \sum_{i=1}^{n} K(X_i)$ is sufficient for θ.

■ **Proof** The appropriate likelihood is

$$
\begin{aligned}
L(\theta;x_1,x_2,\ldots,x_n) &= \prod_{i=1}^{n} f(x_i;\theta) \\
&= \prod_{i=1}^{n} e^{K(x_i)p(\theta)+S(x_i)+q(\theta)} \\
&= e^{\sum_{i=1}^{n} K(x_i)p(\theta)+\sum_{i=1}^{n} S(x_i)+nq(\theta)} \\
&= e^{\sum_{i=1}^{n} S(x_i)} \cdot e^{p(\theta)\sum_{i=1}^{n} K(x_i)+nq(\theta)}. \tag{8.45}
\end{aligned}
$$

Let the left factor of the last expression be b, let the right factor be k, and let the sum in the exponent of the right factor be the function g. Then by (8.39), $Z = \sum_{i=1}^{n} K(X_i)$ is sufficient.

Since so many distributions can be put into the form (8.44), Proposition 8.3.3 explains why the sum of simple functions of the sample values X_i arises so frequently as a sufficient statistic. Since, as we establish later, unbiased functions of sufficient statistics have optimal properties, such sums as $\sum K(X_i)$ often come up as minimum variance unbiased estimators as well.

—— **?Question 8.3.4** The gamma density can be written

$$f(t;\lambda) = \frac{\lambda^{\alpha}}{\Gamma(\alpha)} t^{\alpha-1} e^{-\lambda t}, \qquad t > 0$$

$$= \exp[\alpha \ln \lambda - \ln \Gamma(\alpha) + (\alpha - 1)\ln t - \lambda t]. \tag{8.46}$$

Use this formula to show in another way that $\sum_{i=1}^{n} T_i$ is sufficient for λ if α is known.

Example 8.3.4

Consider the estimation of the normal variance σ^2 assuming a known mean μ. The normal density can be written as

$$f(x;\sigma^2) = \frac{1}{\sqrt{2\pi\sigma^2}}e^{-(x-\mu)^2/2\sigma^2}$$

$$= \exp\left[-\frac{1}{2}\ln(2\pi\sigma^2) - \frac{(x-\mu)^2}{2\sigma^2}\right]. \tag{8.47}$$

Let $K(x) = (x-\mu)^2$, $p(\sigma^2) = -1/2\sigma^2$, $S(x) = 0$, and $q(\sigma^2) = -\frac{1}{2}\ln(2\pi\sigma^2)$. Then $\sum_{i=1}^{n} K(X_i) = \sum_{i=1}^{n}(X_i - \mu)^2$ is sufficient for σ^2, by Proposition 8.3.3.

In the exercises you will see more examples of members of the exponential class. The examples of the normal distribution and gamma distribution may have made you wonder about sufficient statistics in the case where the parameter has multiple components. Exercise 12 investigates this issue.

We close this section with one of the triumphant achievements of mathematical statistics, the *Rao–Blackwell theorem*, and a corollary that has to do with maximum likelihood estimators.

PROPOSITION 8.3.4

Rao–Blackwell   Let $Z = g(X_1, X_2, \ldots, X_n)$ be a sufficient statistic for θ based on a random sample of size n from a distribution $f(x;\theta)$. If Y is any unbiased estimator of θ, then there is a function $\phi(Z)$ of the sufficient statistic that is also unbiased and has variance smaller than or equal to the variance of Y. Hence if a minimum variance unbiased estimator of θ exists, then there must also be a function of the sufficient statistic Z that is a minimum variance unbiased estimator.

■ **Proof**   We claim that the function $\phi(Z)$ in the statement of the theorem is

$$\phi(Z) = E[Y \mid Z]. \tag{8.48}$$

Since Z is sufficient for θ, by Definition 8.3.1 the distribution of Y given Z does not depend on θ. Also, since Y is unbiased for θ, by Proposition 4.2.1 we have

$$E[\phi(Z)] = E[E[Y \mid Z]] = E[Y] = \theta; \tag{8.49}$$

therefore $\phi(Z)$ is also unbiased for θ.

To show that $\mathrm{Var}(Y) \geq \mathrm{Var}(\phi(Z))$, we can compute as follows:

$$\begin{aligned}
\mathrm{Var}(Y) &= E[(Y-\theta)^2]\\
&= E[((Y-\phi(Z))+(\phi(Z)-\theta))^2]\\
&= E[(Y-\phi(Z))^2]+2E[(Y-\phi(Z))(\phi(Z)-\theta)]+E[(\phi(Z)-\theta)^2]\\
&= E[(Y-\phi(Z))^2]+2E[E[(Y-\phi(Z))(\phi(Z)-\theta)\mid Z]]+\mathrm{Var}(\phi(Z))\\
&= E[(Y-\phi(Z))^2]+2E[(\phi(Z)-\theta)(E[Y\mid Z]-\phi(Z))]+\mathrm{Var}(\phi(Z))\\
&= E[(Y-\phi(Z))^2]+\mathrm{Var}(\phi(Z)). \tag{8.50}
\end{aligned}$$

In the second line we simply add and subtract $\phi(Z)$; then we expand the square in the third line. Line four follows by conditioning and unconditioning on Z (Proposition 4.2.1 again), and by applying the fact that θ is the mean of $\phi(Z)$. In the crucial fifth line we remove from the conditional expectation the factor $(\phi(Z) - \theta)$ involving Z alone and use linearity to split the remaining expectation. By our choice of $\phi(Z)$, $E[Y \mid Z] - \phi(Z) = 0$, which produces the final line. But since the leading term $E[(Y - \phi(Z))^2]$ on the right side of (8.50) is nonnegative, $\text{Var}(Y) \geq \text{Var}(\phi(Z))$ as desired.

Finally, if Y is a minimum variance unbiased estimator, then by what we have shown, $\phi(Z)$ is unbiased and $\text{Var}(Y) \geq \text{Var}(\phi(Z))$. Consequently $\phi(Z)$ is also a minimum variance unbiased estimator, which proves the last statement.

The Rao–Blackwell theorem implies that in searching for optimal estimators, it is only necessary to look at unbiased functions of sufficient statistics. With some extra regularity conditions on the distribution, it turns out that for a given distribution there is only one unbiased function of a sufficient statistic for the parameter, which must therefore be a best estimator.

We conclude this section by relating the concept of maximum likelihood to sufficiency. This result shows why the maximum likelihood method so often produces estimators with optimal properties. Its proof, which you should fill in yourself, follows directly from the factorization criterion.

PROPOSITION 8.3.5

If $Z = g(X_1, X_2, \ldots, X_n)$ is a sufficient statistic for θ, and $\widehat{\theta}$ is an MLE for θ, then $\widehat{\theta}$ is a function of Z.

Consequently, an unbiased function of a maximum likelihood estimator is an unbiased function of a sufficient statistic, which therefore has minimum variance (under the regularity conditions just mentioned; see the sources mentioned at the start of the section).

EXERCISES 8.3

1. Use the definition of sufficiency to show that $\sum_{i=1}^{n} X_i$ is a sufficient statistic for the parameter μ in the Poisson (μ) distribution.
2. Use the definition of sufficiency to show that $\sum_{i=1}^{n} X_i$ is a sufficient statistic for the parameter p in the Bernoulli (p) distribution [i.e., the $b(1, p)$ distribution].
3. Show using the definition of sufficiency rather than the factorization criterion that $Z = \sum_{i=1}^{n} T_i$ is sufficient for λ, where $T_1, T_2, \ldots, T_n$ is a random sample from the $\Gamma(\alpha, \lambda)$ distribution and α is known.
4. Prove Proposition 8.3.2.
5. Find a sufficient statistic for the parameter λ in the exponential(λ) distribution.
6. Use the Factorization Theorem to show that $\overline{X} = \sum_{i=1}^{n} X_i / n$ is a sufficient statistic for the parameter μ in the $N(\mu, \sigma^2)$ distribution, where σ^2 is known.

7. Find a sufficient statistic for the parameter θ in the distribution with density

$$f(x; \theta) = \theta x^{\theta-1}; \qquad 0 < x < 1.$$

Show that the MLE of θ is a function of the sufficient statistic and is itself sufficient.

8. Write the Pareto density

$$f(x; \theta) = \frac{\theta}{(1+x)^{\theta+1}}, \qquad 0 < x < \infty$$

as a member of the exponential class, and find a sufficient statistic for θ.

9. Use the definition of sufficiency to show that the largest order statistic Y_n of a random sample of size n from the uniform $(0, \theta)$ distribution is sufficient for θ. (Note that the support of the density is *not* free of θ.)

10. Is the Weibull(λ, β) density a member of the exponential class? If so, use Proposition 8.3.3 to find a sufficient statistic for λ, assuming a known value of β.

11. Prove the implication (c)$\Rightarrow$(a) of Proposition 8.3.1 in the case of discrete distributions. (*Hint:* Make use of the sets $A_z = \{(x_1, x_2, \dots, x_n) : Z = g(x_1, x_2, \dots, x_n) = z\}$.)

12. Nothing in Definition 8.3.1 forbids θ from being a vector parameter $\boldsymbol{\theta} = (\theta_1, \theta_2, \dots, \theta_l)$. The factorization theorem Proposition 8.3.1 follows with only notational changes. Use it to show that $\mathbf{Z} = (\sum_{i=1}^{n} X_i, \sum_{i=1}^{n} X_i^2)$ is jointly sufficient for $\boldsymbol{\theta} = (\mu, \sigma^2)$ in the normal distribution. Also, prove that $\mathbf{Z} = (\overline{X}, S^2)$ is sufficient for $\boldsymbol{\theta}$.

13. Argue, using the fourth line of (8.45), that for exponential class densities the MLE will be a function of the sufficient statistic.

8.4 | Confidence Intervals for One-Sample Problems

Thus far in our study of estimation we have spent all of our time on estimation of parameters by single numerical values. We have also been developing good mathematical rationale for using particular point estimates, yet the full power of probability has not appeared. It begins to emerge when we consider measures of uncertainty of the point estimates.

There are two main facets to uncertainty that result from the fact that an estimator Z of a parameter θ is a random variable: (1) Z will not in general equal θ, but rather will fall within some distance ϵ of θ, called the *margin of error*; (2) Z is not even absolutely certain to be in the interval $[\theta - \epsilon, \theta + \epsilon]$; rather, for a well chosen ϵ it has some high probability that we will call the *confidence level* of falling into that interval. (See Fig. 8.4.)

To reflect this dual nature of uncertainty in estimation, we pose the following definition.

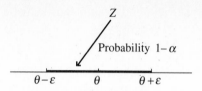

Figure 8.4 A Confidence Interval

DEFINITION 8.4.1

Let A and B be random variables such that $A \leq B$ with probability one. Then $[A, B]$ is a $(1 - \alpha) \times 100\%$ *confidence interval* for a parameter θ if

$$P[\theta \in [A, B]] = 1 - \alpha. \tag{8.51}$$

In many cases the confidence interval is centered on a point estimate Z, so that it has the form $[Z - \epsilon, Z + \epsilon]$, where the interval radius (or margin of error) ϵ may also be a random variable. To tie this back to the introductory discussion, note that $\theta \in [Z - \epsilon, Z + \epsilon]$ if and only if $Z \in [\theta - \epsilon, \theta + \epsilon]$, so that our estimation of θ is bound not only by the constraint that we can only produce an interval estimate but also by the fact that an interval estimate only has some high probability of containing the parameter.

To find a confidence interval for a parameter θ based on a random sample $X_1, X_2, \ldots, X_n$, the approach is to find a function $h(X_1, X_2, \ldots, X_n; \theta)$ of the sample that depends functionally on the parameter but whose probability distribution does not depend on the parameter. To see that this approach is not as contradictory as it seems, consider the standardized mean of a normal random sample:

$$W = \frac{\overline{X} - \mu}{\sigma / \sqrt{n}}. \tag{8.52}$$

The random variable $W = h(X_1, X_2, \ldots, X_n; \mu)$ has the $N(0, 1)$ distribution, independent of $\theta = \mu$, yet W contains μ in its defining formula.

Continuing with the normal example, let $z_{\alpha/2}$ and its opposite $-z_{\alpha/2}$ be standard normal critical values obtained from Table 3 of Appendix A as shown in Fig. 8.5, between which lies $(1 - \alpha) \times 100\%$ of the total normal curve area of 1. For instance, for $(1 - \alpha) = .90$, the corresponding $z_{\alpha/2} = z_{.05} = 1.645$, and for $(1 - \alpha) = .95$, the corresponding $z_{\alpha/2} = z_{.025} = 1.96$. Then we can write the probabilistic inequalities

$$1 - \alpha = P\left[-z_{\alpha/2} \leq W \leq z_{\alpha/2}\right]$$

$$= P\left[-z_{\alpha/2} \leq \frac{\overline{X} - \mu}{\sigma / \sqrt{n}} \leq z_{\alpha/2}\right]$$

$$= P\left[-z_{\alpha/2} \cdot \frac{\sigma}{\sqrt{n}} \leq \overline{X} - \mu \leq z_{\alpha/2} \cdot \frac{\sigma}{\sqrt{n}}\right]$$

$$= P\left[\overline{X} - z_{\alpha/2} \cdot \frac{\sigma}{\sqrt{n}} \leq \mu \leq \overline{X} + z_{\alpha/2} \cdot \frac{\sigma}{\sqrt{n}}\right]. \tag{8.53}$$

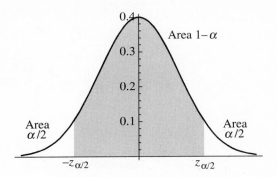

Figure 8.5   Standard Normal Density

Thus the random interval

$$\left[\overline{X} - z_{\alpha/2} \cdot \frac{\sigma}{\sqrt{n}}, \ \overline{X} + z_{\alpha/2} \cdot \frac{\sigma}{\sqrt{n}}\right] \tag{8.54}$$

is a $(1 - \alpha) \times 100\%$ confidence interval for μ. For this interval to be useful, however, the variance σ^2 must be known, which it rarely is.

Although it may not have wide applicability, the example of the normal mean with known variance is useful to crystalize the ideas. Consider a sample of size $n = 25$, with $\sigma^2 = 9$ and confidence level 95%, that is, $1 - \alpha = .95$. Thus $\alpha = .05$ and $z_{\alpha/2} = 1.96$. Then by (8.54) the confidence interval has the form

$$\left[\overline{X} - 1.96 \cdot \frac{3}{5}, \ \overline{X} + 1.96 \cdot \frac{3}{5}\right] = [\overline{X} - 1.176, \overline{X} + 1.176].$$

This means that $\overline{X}$ will fall within a margin of error of 1.176 of μ with a level of confidence of 95%.

Let us consider the meaning of the last sentence more carefully. Each time a sample of 25 is observed, $\overline{X}$ has a different value; hence the confidence interval has different endpoints. Most of the observed confidence intervals will contain the true parameter, and a few will not. In fact, if samples of size 25 were taken indefinitely, the proportion of them that contain the true μ would converge to 95%, by the Law of Large Numbers. To display this phenomenon, I simulated 20 samples of size 25 from the $N(0,9)$ distribution, computed $\overline{X}$ and $\overline{X} \pm 1.176$ for each of the 20 samples, and plotted the confidence intervals in Fig. 8.6. As the figure shows, 17 of the intervals included the point $y = 0$—and hence contained the true $\mu = 0$—and 3 intervals—namely numbers 1, 6, and 20—did not. Thus in these few replications we observed only $17/20 = 85\%$ rather than 95% trapping of the true parameter by the confidence intervals.

—— **?Question 8.4.1**   For fixed confidence level, what is the effect of increasing the sample size on the margin of error in the normal mean problem just discussed? For

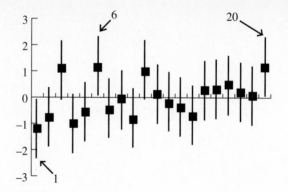

Figure 8.6 Twenty Simulated $N(0, 9)$ Confidence Intervals

fixed margin of error, what is the effect of increasing the sample size on the confidence level? Do you think that these principles would generalize to problems other than the estimation of the normal mean?

Example 8.4.1

As part of a class project to forecast library space needs, my students Jun Zhu, Vivek Malhotra, and Ed Spencer became interested in the variability of departmental book purchases. Following are expenditures (in dollars) by the Department of Mathematics and Computer Science for new library book purchases from the academic years of 1982–83 to 1991–92:

$$4710, \quad 2451, \quad 4413, \quad 3947, \quad 4856, \quad 5128, \quad 6904, \quad 3565, \quad 6181, \quad 2521$$

As the normal scores plot of Fig. 8.7 confirms, there is no strong evidence in the data against assuming that book expenditures are normally distributed. Let us derive an 80% confidence interval for the normal variance σ^2.

We think immediately of the point estimate S^2 and recall that the random variable $Y = (n-1)S^2/\sigma^2$ has the $\chi^2(n-1)$ distribution, from Proposition 5.4.5. Notice that Y depends functionally on the parameter σ^2, but its probability distribution does not. As in Fig. 8.8, select $a = \chi^2_{1-\alpha/2}$ and $b = \chi^2_{\alpha/2}$

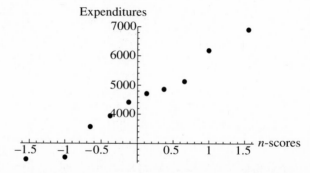

Figure 8.7 Normal Scores Plot of Expenditures

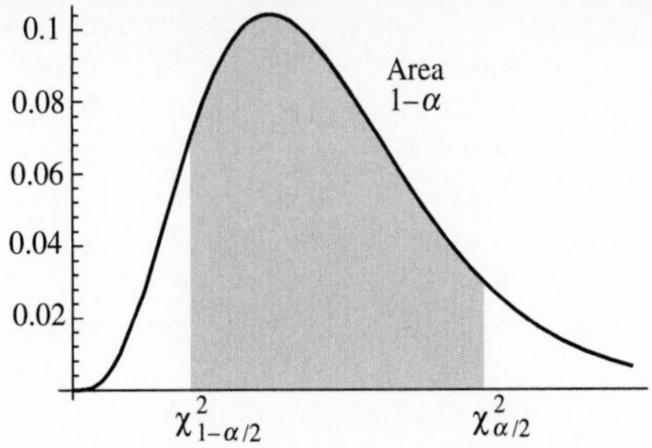

Figure 8.8   Critical Values for χ^2 Distribution

from the $\chi^2(n-1)$ table (Table 4 in Appendix A) to satisfy

$$\frac{\alpha}{2} = P[Y \leq a]; \qquad \frac{\alpha}{2} = P[Y \geq b]. \tag{8.55}$$

Then

$$1 - \alpha = P[\chi^2_{1-\alpha/2} \leq Y \leq \chi^2_{\alpha/2}]$$

$$= P\left[\chi^2_{1-\alpha/2} \leq \frac{(n-1)S^2}{\sigma^2} \leq \chi^2_{\alpha/2}\right]$$

$$= P\left[\frac{(n-1)S^2}{\chi^2_{\alpha/2}} \leq \sigma^2 \leq \frac{(n-1)S^2}{\chi^2_{1-\alpha/2}}\right]. \tag{8.56}$$

So the last set of inequalities on the right describes a $(1-\alpha) \times 100\%$ confidence interval for the normal variance. In our book expenditure problem, $n = 10$, $s^2 = 2{,}050{,}624$, $\alpha = .2$, $\alpha/2 = .1$, $\chi^2_{1-\alpha/2} = 4.168$, and $\chi^2_{\alpha/2} = 14.68$, so the observed confidence interval for σ^2 is

$$\left[\frac{9 \cdot 2{,}050{,}624}{14.68}, \; \frac{9 \cdot 2{,}050{,}624}{4.168}\right] = [1{,}257{,}195, \; 4{,}427{,}930].$$

?Question 8.4.2   In the last example, what do you conclude about the variability of the data and about the reliability of predictions of yearly expenditures? How would you find an interval estimate of the standard deviation?

Example 8.4.2   In Example 7.3.1 we looked at data on frequency of digits drawn in the Illinois Pick 3 Lottery. For example (see Table 7.9), on the first of the three draws, over a period of 363 days, the digit 4 was drawn 50 times. Since there are 10 possible digits, the expected proportion of 4s drawn during this period is $10\% = .10$ (or 36.3 total 4s). Is the observed proportion of $50/363 \approx .138$ larger than would reasonably be accounted for by random variation?

We can formulate the question in another way: Is the probability of 4, .1 under randomness, within a suitable margin of error of the sample proportion of 4s? To answer the question, we can try to find a confidence interval for the parameter p =probability that 4 is drawn, based on the estimator $\widehat{p}$ = sample proportion of 4s. Then we will consider the theoretical probability p =.1 to be incompatible with the data if it falls outside of the computed confidence interval.

To form a probabilistic model for the experiment, assume that the 363 draws were conducted independently and under identical conditions. Define random variables $X_1, X_2, \ldots, X_{363}$ to have value 1 or 0, respectively, according to whether the draw corresponding to the subscript is or isn't a 4. Then $X_1, X_2, \ldots, X_{363}$ is a random sample from the Bernoulli (p) distribution $[b(1, p)]$:

$$f(x; p) = p^x (1 - p)^{1-x}, \qquad x = 0, 1. \tag{8.57}$$

The random variable $\widehat{p}$ can be standardized by subtracting its mean p and dividing by its standard deviation $\sqrt{p(1 - p)/n}$. (Recall why these are the correct moments.) Then the Central Limit Theorem implies that the resulting random variable Z, whose functional form contains p, has an approximately $N(0, 1)$ distribution, independently of p. By imitating the normal mean computation (8.53), we can write

$$1 - \alpha \approx P[-z_{\alpha/2} \leq Z \leq z_{\alpha/2}]$$

$$= P\left[-z_{\alpha/2} \leq \frac{\widehat{p} - p}{\sqrt{p(1 - p)/n}} \leq z_{\alpha/2}\right]$$

$$= P[\widehat{p} - z_{\alpha/2} \cdot \sqrt{p(1 - p)/n} \leq p \leq \widehat{p} + z_{\alpha/2} \cdot \sqrt{p(1 - p)/n}]. \tag{8.58}$$

The only remaining difficulty is that the radius of the confidence interval $\epsilon = z_{\alpha/2} \cdot \sqrt{p(1 - p)/n}$ depends on the unknown parameter p.

One way of proceeding is to suppose that because the sample size is large, $\widehat{p}$ will approximate p closely and we lose little in replacing p by $\widehat{p}$ in the margin of error ϵ, since the confidence level $1 - \alpha$ is only approximate. Then the approximate $(1 - \alpha) \times 100\%$ confidence interval for p is

$$[\widehat{p} - z_{\alpha/2} \cdot \sqrt{\widehat{p}(1 - \widehat{p})/n}, \ \widehat{p} + z_{\alpha/2} \cdot \sqrt{\widehat{p}(1 - \widehat{p})/n}]. \tag{8.59}$$

For the lottery data, if we use a confidence level of $(1 - \alpha) = .90$, then we find from the standard normal table that $z_{\alpha/2} = z_{.05} = 1.645$. Therefore by (8.59) the computed confidence interval is

$$\left[.138 \pm 1.645 \cdot \sqrt{\frac{.138(1 - .138)}{363}}\right] \approx [.108, .168].$$

Since the theoretical probability of .1 lies outside of this interval, we conclude with 90% confidence that .1 is not the true probability of drawing a 4.

An alternative to estimating p in (8.58) by $\widehat{p}$ is explored in Exercise 13.

?Question 8.4.3 Think of the confidence interval concept relative to political polling. Do news media truly report confidence intervals? If not, what is missing? Do you think that this omission is a serious problem?

Example 8.4.3 The following data (USDA, 1991, 1976) represent the percentage of fat in milk produced by all cows in the United States between the years 1963–1990. Let us derive a 95% confidence interval estimate of the mean percentage of milk fat.

3.71, 3.70, 3.70, 3.69, 3.69, 3.67, 3.67, 3.66, 3.66, 3.68,
3.66, 3.67, 3.68, 3.66, 3.65, 3.67, 3.66, 3.65, 3.64, 3.65,
3.66, 3.66, 3.67, 3.67, 3.65, 3.67, 3.68, 3.65

If Y is the random variable representing the fat content in a given year, then we have the problem of finding a confidence interval for μ_y. To have a probabilistic model from which to proceed, suppose that the data form a random sample from the $N(\mu_y, \sigma_y^2)$ distribution. A frequency histogram of the data, given in Fig. 8.9, does not have drastic departures from the normal bell shape, although there is a suspicious indication of some right skewness. (Look hard at the sequence of data. Do you have any other reservations about the assumptions we have made?) Even if the population distribution is not normal, we have seen in the Central Limit Theorem that the distribution of the sample mean $\overline{Y}$ is approximately normal.

So we can find a 95% confidence interval for μ_y analogously to the derivation in (8.53), but we must compensate for the fact that the variance σ_y^2 is not known. To do this, recall from Proposition 5.5.1 that the random variable

$$T = \frac{\overline{Y} - \mu_y}{S_y/\sqrt{n}} \qquad (8.60)$$

has the $t(n-1)$ distribution, independent of μ_y, where S_y^2 is the sample variance. Let $t_{\alpha/2} = t_{\alpha/2}(n-1)$ be the critical value from the table of the

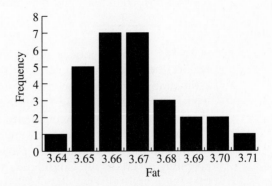

Figure 8.9   Histogram of Milk Fat Percentages

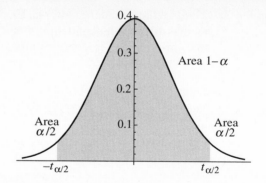

Figure 8.10 Critical Values for the t-Distribution

t-distributions (Table 5 of Appendix A) satisfying $P[T \geq t_{\alpha/2}] = \alpha/2$, as displayed in Fig. 8.10. Then by the symmetry of the t-density,

$$1 - \alpha = P[-t_{\alpha/2} \leq T \leq t_{\alpha/2}]$$

$$= P\left[-t_{\alpha/2} \leq \frac{\overline{Y} - \mu_y}{S_y/\sqrt{n}} \leq t_{\alpha/2}\right]$$

$$= P\left[\overline{Y} - t_{\alpha/2} \cdot \frac{S_y}{\sqrt{n}} \leq \mu_y \leq \overline{Y} + t_{\alpha/2} \cdot \frac{S_y}{\sqrt{n}}\right]. \tag{8.61}$$

Hence the random interval

$$\left[\overline{Y} - t_{\alpha/2} \cdot \frac{S_y}{\sqrt{n}}, \ \overline{Y} + t_{\alpha/2} \cdot \frac{S_y}{\sqrt{n}}\right] \tag{8.62}$$

is a $(1 - \alpha) \times 100\%$ confidence interval for the normal mean μ_y of Y in the case that the variance σ_y^2 is unknown.

For the milk fat problem, the observed value of the sample mean is $\overline{y} = 3.67$, and the standard deviation s_y is .01728. The sample size n is 28, and the $t(27)$ critical value for level 95% is $t_{.025}(27) = 2.05$. Then the confidence interval is $[3.67 \pm 2.05 \cdot (.01728)/\sqrt{28}] = [3.663, 3.677]$. Notice that since the data were very concentrated and the sample size was fairly large, we found a very sharp confidence interval estimate.

Earlier, we asked a question about the relation between sample size, confidence level, and margin of error, that is, confidence interval radius. Since estimation in general becomes more precise as sample size increases, we would expect that, all else being fixed, the effect of increasing sample size should be to (a) increase confidence level and (b) decrease margin of error. Our four main examples were (1) normal μ, assuming σ^2 known; (2) normal σ^2; (3) Bernoulli (p), assuming large sample; and (4) normal μ, assuming σ^2 unknown. Considering the first case first, formula (8.54) shows that the margin of error for a $(1 - \alpha) \times 100\%$ confidence interval for μ is $z_{\alpha/2} \cdot \sigma/\sqrt{n}$. Thus the margin of error decreases with n at a rate of $1/\sqrt{n}$.

If, on the other hand, we view the margin of error as fixed, then since σ is constant, increasing n requires an increase in the critical value $z_{\alpha/2}$ to keep a constant margin of error, which results in a smaller α value (refer to Fig. 8.5) and a larger confidence level $1 - \alpha$.

—— **?Question 8.4.4** Argue that similar effects should be expected to take place in case (4) where the variance is unknown, but they are not certain. What role does the critical value $t_{\alpha/2}(n - 1)$ play in your argument?

For the Bernoulli (p) interval, we can make an interesting observation about the maximum possible margin of error. By (8.58), the margin of error is

$$z_{\alpha/2} \cdot \sqrt{p(1 - p)/n}.$$

But the quantity $p(1 - p)$ can never be greater than $1/4$. (Justify this property.) Hence the maximum margin of error is

$$z_{\alpha/2} \cdot \sqrt{1/4n} = \frac{1}{2} \cdot \frac{z_{\alpha/2}}{\sqrt{n}}.$$

The same sample size effects on the margin of error and the confidence level noted earlier for the normal mean occur here for the Bernoulli success parameter. If the level is fixed, the margin of error decreases; if the margin is fixed, then the confidence level increases as sample size increases.

You are asked to discuss the effect of increasing sample size on the estimation of the normal variance σ^2 in Exercise 11.

EXERCISES 8.4

1. (Data from Milton and Arnold, 1986, p. 210) In a study of the greenhouse effect, the following 30 atmospheric CO_2 measurements were made (the units are parts per million). Check to see whether it is reasonable to suppose that the data come from a normal distribution. Produce a 95% confidence interval estimate of the average CO_2 concentration.

    319, 325, 330, 320, 326, 339, 338, 340, 330, 343, 349, 335, 337, 331, 321, 350, 341, 338, 339, 341, 327, 322, 338, 333, 328, 336, 337, 334, 332, 334

2. In a political poll of 500 randomly selected respondents, 212 people expressed support for candidate A. Give an approximate margin of error for the support for A at a confidence level of 90%.

3. The numbers that follow are average wage rates for farm field workers in 18 regions of the United States in July 1990 (USDA, 1991). Suppose that this is a random sample from a normal distribution with variance .7225. Find a 95% confidence interval for the mean wage rate.

    5.38, 5.20, 4.40, 4.75, 4.10, 5.32, 5.06, 5.18, 4.49, 4.41, 5.05, 4.32, 4.58, 4.71, 4.51, 5.25, 5.68, 7.97

4. Find 95% confidence intervals for each of the variances of the statistics test data in Table 7.2. Do the tests seem to differ significantly in variability?

5. A study in Georgia of teenage driving habits (Centers for Disease Control and Prevention, 1995a) found that among 64 young people involved in vehicle crashes during 1993 in a certain Georgia county, 28 had driven at least 20 miles per hour over the speed limit six or more times in a three-month period. Find an approximate 90% confidence interval for the proportion of drivers of this kind who exceed the speed limit. Do the same for a group of teenage drivers who were not involved in crashes, 65 of 227 of whom sped six or more times. Does there seem to be a real difference between the two groups?

6. A medical study (Franklin et al., 1995, p. 881) looked at several measures of exertion during snow removal. One measure was heart rate (beats per minute). Ten men were assigned to use manual snow shoveling and ten others to use a snowblower to clear a specific area. The subjects were monitored by a mobile ECG machine. The study reported a sample mean of 175 for the manual group and 124 for the snowblower group, with sample standard deviations of 15 and 18, respectively. Devise an approach that would let you quantify the evidence for a significant difference in average heart rate for the two groups. (By the way, the study lists its data for manual shoveling in the form 175 ± 15 and for snowblowing as 124 ± 18. Is this misleading in any way?)

7. A certain home pregnancy test is imperfect: It shows positive on women who are not pregnant 5% of the time, and it shows negative on women who are pregnant 1% of the time. In a group of 100 women who take this test, 12 show positive. Find an approximate 95% confidence interval for the number of the 100 women who are pregnant. (*Hint:* Think of the Law of Total Probability.)

8. Following are 1990 data (USDA, 1991) from the 50 U.S. states on average number of eggs laid in the state per laying hen (i.e., the per hen laying rate per year).

   a. Produce a graph to check whether an assumption of normality is reasonable.

   b. Find a 90% confidence interval for the mean laying rate.

   c. Find a 90% confidence interval for the variance of laying rate.

$$232, \quad 187, \quad 241, \quad 232, \quad 250, \quad 250, \quad 271, \quad 228, \quad 245, \quad 245,$$
$$233, \quad 255, \quad 255, \quad 263, \quad 260, \quad 240, \quad 250, \quad 219, \quad 272, \quad 259,$$
$$269, \quad 264, \quad 259, \quad 240, \quad 247, \quad 261, \quad 250, \quad 172, \quad 270, \quad 272,$$
$$241, \quad 267, \quad 232, \quad 236, \quad 264, \quad 231, \quad 258, \quad 268, \quad 256, \quad 261,$$
$$260, \quad 227, \quad 240, \quad 250, \quad 301, \quad 250, \quad 265, \quad 238, \quad 264, \quad 194$$

9. A *two-dimensional confidence region* of level $1 - \alpha$ for a vector parameter $\boldsymbol{\theta} = (\theta_1, \theta_2)$ is a random set $A \subseteq \mathbb{R}^2$ such that

$$P[\boldsymbol{\theta} \in A] = 1 - \alpha.$$

Recall the data in Table 7.6 on runs scored and bases stolen. Find a confidence region of level .9025 for the parameter (μ_r, μ_s), where μ_r is the mean runs scored and μ_s is the mean stolen bases. What assumptions are you making? Does the data set support those assumptions?

10. The number of workers absent in a day at a particular large retail establishment is believed to have a Poisson distribution. The actual numbers of absentees were recorded on 23 days, and it was found that there were 1 day with 0 absences, 2 days with 1 absence, 3 days with 2 absences, 5 days with 3 absences, 4 days with 4 absences, 3 days with 5 absences, 2 days with 6 absences, 1 day with 7 absences, 1 day with 8 absences, and 1 day with 9 absences. Devise an approximate 95% confidence interval for the parameter μ of the Poisson distribution.

11. Discuss the effects of increasing sample size on the problem of confidence interval estimation of the normal variance σ^2. (To do this thoroughly, you may want to use a computer to carefully study the behavior of chi-square critical values.)

12. Construct an approximate large sample $(1 - \alpha) \times 100\%$ confidence interval for the parameter p of the geometric (p) distribution.

13. Using formula (8.58), algebraically rearrange the inequalities to characterize a confidence interval for p that does not require the estimation of p by $\hat{p}$ in the expression $\sqrt{p(1 - p)/n}$.

14. Derive a $(1 - \alpha)$ level confidence interval for the parameter θ in the uniform $[0, \theta]$ distribution based on the largest order statistic Y_n of a random sample of size n. (*Hint:* Start by writing expressions for $P[Y_n < c]$ and $P[Y_n > d]$ in terms of θ.)

15. Derive a general two-dimensional confidence region (see Exercise 9) for the vector parameter $\boldsymbol{\theta} = (\mu, \sigma^2)$ in the $N(\mu, \sigma^2)$ distribution.

16. a. Use moment-generating functions to show that if $X_1, X_2, \ldots, X_n$ is a random sample from the exponential distribution with parameter λ, then the random variable $2 \cdot \lambda \sum_{i=1}^{n} X_i$ has the $\chi^2(2n)$ distribution.

   b. Use the result in part (a) to find a 90% confidence interval for the probability that a Poisson process has no arrivals in a time interval of unit length.

17. Suppose that you want to estimate a normal mean to within .1, and the variance of the distribution is known to be 2.1. At least how many samples are necessary, if 95% confidence is required?

18. Give a conservative estimate of the number of samples necessary to estimate the Bernoulli success probability p to within a tolerance of .05 at a confidence level of 90%.

8.5 | Confidence Intervals for Two-Sample Problems

Many interesting data analysis problems involve comparisons between two populations or distributions. Here are just a few examples:

■ Do athletes trained by one regimen have faster times in the 200-meter dash than athletes trained by another regimen?

■ Do more people like the new formulation of a soft-drink than the old one?

■ Does a new manufacturing process reduce variability of the output product?

(Try to think of a few other examples of problems involving comparisons.)

To answer questions like these, we can obtain a random sample from each of the two distributions involved and use the samples to estimate some suitable function of the two population parameters that would allow us to make a comparison. Stop for a moment to consider the following question.

—— **?Question 8.5.1** What probabilistic model pertains to the 200-meter dash question; in other words, what distribution might be assumed for the two samples of runners, and what comparison of parameters would we need to make? Try to answer analogous questions for the soft-drink preference problem and the manufacturing variability problem.

One of the most frequent comparisons that experimenters and researchers need to make is between population averages. For example, in the problem of the 200-meter runners, we might have some hypothesized distribution f_x of times for one training regimen, with mean μ_x, and another hypothesized distribution f_y of times for the other regimen, with its own mean μ_y. To determine whether the first method produces faster average times than the second, we would check whether $\mu_x < \mu_y$, in other words, $\mu_x - \mu_y < 0$. The function of the parameters of interest here is therefore $\mu_x - \mu_y$. In the soft-drink problem, we might be interested in the difference of two Bernoulli success parameters $p_x - p_y$, where p_x is the proportion of people who like the new version of the drink and p_y is the proportion who like the old one. In the manufacturing problem it could be the ratio σ_y^2/σ_x^2 between the variances of the new and the old processes that is suitable for comparing the variabilities; if this ratio is less than 1, then the new process Y is less variable. Other problems could require other functions of the parameters, but we will be concentrating on these three cases in this section, which illustrate the thought process well.

It is easy to concoct point estimates of such functions of parameters. Simply replace each distributional parameter in the function by its individual point estimate, so that $\overline{X} - \overline{Y}$, $\hat{p}_x - \hat{p}_y$, and S_y^2/S_x^2 are natural estimates of the functions of parameters discussed in the last paragraph. We will devote this section to the construction of confidence intervals, which makes use of the theory of transformations of random variables that we developed earlier.

An issue surfaces involving the randomness of the sampling process that did not arise when we studied one-sample problems. A brief introduction will suffice for now; we will study it more thoroughly when we reach *experimental design*, in Chapter 11. There will be two samples of data $X_1, X_2, \ldots, X_n$ and $Y_1, Y_2, \ldots, Y_m$, one from each population, on which our interval estimates will be based. The issue is, Are these samples independent of one another?

Sometimes the data gathering process determines the answer, and other times the experimenter can design the experiment so as to produce independence or dependence. To illustrate, consider an after-school study program to prepare high school juniors to take the ACT exam. The question is, Does this program increase the average score? The universe of discourse is all high school juniors, or perhaps the more restricted universe of all juniors at a particular high school. Two scenarios for the gathering of data arise:

1. A single group of n students is selected for the program. They are given a practice version of the ACT before they begin to study, and a second version at the end of the program. The data are X_i = score of ith student before study, Y_i = score of ith student after study, $i = 1, 2, \ldots, n$.
2. Two groups of students are randomly selected from the junior class. The first group of, say, n students does not participate in the study program, and the second group of, say, m students does participate. The exam is given to all of the $n + m$ students at the end of the program. The data are X_i = score of ith student in first group, $i = 1, 2, \ldots, n$, and Y_j = score of jth student in second group, $j = 1, 2, \ldots, m$.

In the first design, the data structure is a sample of pairs (X_1, Y_1), (X_2, Y_2), $\ldots$, (X_n, Y_n). Although it may be reasonable to suppose that the performance of one student (X_i, Y_i) is independent of the performance of another (X_j, Y_j), it is certainly not reasonable to suppose that for a single student X_i is independent of Y_i. In the second design, however, the two groups were independently selected and have no students in common, so we can assume not only that the X_i's are independent of each other and the Y_j's are independent of each other, but also that all of the X_i's are independent of all of the Y_j's. As we will see, because of the different independence assumptions, each design requires a different analysis.

A statistician may well be called on to analyze the data after the experiment has already been set up and performed, in which case there may be no choice as to which method is used. Still, both designs appear to be viable ways of comparing mean performance with and without the study program, and it is quite an interesting statistical problem to decide beforehand which approach is better. It is not even clear which criteria we should use to define "better." These are the kinds of ideas that we will study in detail in Chapter 11. For now, just be aware that confidence interval estimates will depend on whether the data are structured as pairs, as in design (1), or whether the data are completely random as in design (2).

?Question 8.5.2   If we are trying to estimate $\mu_x - \mu_y$, which desirable criteria might there be to help us choose between the two experimental designs?

A sequence of exercises (10–12) leads you through the derivation and applications of confidence intervals for the difference $\mu_x - \mu_y$ when the paired data $(X_1, Y_1), (X_2, Y_2), \ldots, (X_n, Y_n)$ are a random sample from the bivariate normal distribution. Be sure to do these problems. The main result is that if S_d

is the sample standard deviation of the sample of differences $D_1 = X_1 - Y_1$, $D_2 = X_2 - Y_2, \ldots, D_n = X_n - Y_n$, then the random interval

$$(\overline{X} - \overline{Y}) \pm t_{\alpha/2}(n - 1) \cdot \frac{S_d}{\sqrt{n}} \tag{8.63}$$

is a $(1 - \alpha) \times 100\%$ confidence interval for $\mu_x - \mu_y$, where $t_{\alpha/2}(n - 1)$ is the critical value from the t-table in Appendix A for the $t(n - 1)$ distribution such that $P[T > t_{\alpha/2}] = \alpha/2$. We treat the independent samples design in the next example.

Example 8.5.1

In an effort to understand the life cycle of sea bass off the mid-Atlantic coast, some researchers counted the frequency of lengths (mm) of larvae caught and measured during each of the months of August and October for the years 1977–1987 (Able et al., 1995). The frequency table follows. Estimate the difference in average bass larvae length by a 90% confidence interval.

Length	2	3	4	5	6	7	8	**Total**
August	15	66	36	18	13	3	1	152
October	3	23	12	11	10	2	0	61

We have two samples: $X_1, X_2, \ldots, X_{152}$ for August and $Y_1, Y_2, \ldots, Y_{61}$ for October. We do not know the distributions from which the samples came; however, by the Central Limit Theorem the sample means $\overline{X}$ and $\overline{Y}$ are approximately normally distributed with means μ_x and μ_y, respectively, and variances $\sigma_x^2/152$ and $\sigma_y^2/61$, respectively. Also since we assume that the X and Y samples are independent of one another, so are $\overline{X}$ and $\overline{Y}$. By Proposition 5.3.4,

$$\overline{X} - \overline{Y} \approx N\left(\mu_x - \mu_y, \frac{\sigma_x^2}{152} + \frac{\sigma_y^2}{61}\right).$$

Therefore

$$\frac{\left(\overline{X} - \overline{Y}\right) - (\mu_x - \mu_y)}{\sqrt{\sigma_x^2/152 + \sigma_y^2/61}} \approx N(0, 1). \tag{8.64}$$

It is straightforward to show that if $z_{\alpha/2}$ is the usual standard normal critical value at level $\alpha/2$, then

$$1 - \alpha \approx P\left[\mu_x - \mu_y \in \left[\overline{X} - \overline{Y} \pm z_{\alpha/2} \cdot \sqrt{\frac{\sigma_x^2}{152} + \frac{\sigma_y^2}{61}}\right]\right]. \tag{8.65}$$

Since we don't know the variances σ_x^2 and σ_y^2, the best that we can do is to replace them by their unbiased estimators S_x^2 and S_y^2. Thus an approximate $(1 - \alpha)$ level confidence interval for $\mu_x - \mu_y$ is

$$\left[\overline{X} - \overline{Y} - z_{\alpha/2} \cdot \sqrt{\frac{S_x^2}{152} + \frac{S_y^2}{61}}, \ \overline{X} - \overline{Y} + z_{\alpha/2} \cdot \sqrt{\frac{S_x^2}{152} + \frac{S_y^2}{61}}\right]. \tag{8.66}$$

For this data set, $\bar{x} = 3.74$, $\bar{y} = 4.13$, $s_x^2 = 1.54$, and $s_y^2 = 2.01$, and the appropriate normal critical value for level 90% is $z_{\alpha/2} = 1.645$. Substituting these data into (8.66), we find that the interval has numerical endpoints $[-.731, -.048]$. Notice that the point 0 is not within the 90% confidence margin of error of the estimate $\bar{x} - \bar{y} = -.39$; hence we have evidence that $\mu_x - \mu_y$ is less than 0, as it should be if the age distribution (and hence the larvae lengths) is moving toward higher values in the two months between August and October.

$\rule{3cm}{0.4pt}$

—— **?Question 8.5.3**   Verify the confidence interval formula (8.65).

In the last example we found a large sample, approximate $(1 - \alpha)$ level confidence interval (8.66) for the difference $\mu_x - \mu_y$ when the X and Y variances are unknown. When the variances are unknown *but equal* and the samples come from normal populations, an exact confidence interval of an interesting form can be derived. Let $X_1, X_2, \ldots, X_n$ and $Y_1, Y_2, \ldots, Y_m$ be independent random samples drawn from the $N(\mu_x, \sigma_x^2)$ and $N(\mu_y, \sigma_y^2)$ distributions, where $\sigma_x^2 = \sigma_y^2 = \sigma^2$. Then by Proposition 5.3.4,

$$\bar{X} - \bar{Y} \approx N\left(\mu_x - \mu_y, \frac{\sigma^2}{n} + \frac{\sigma^2}{m}\right).$$

Also, the sample variances are independent and satisfy

$$\frac{(n-1)S_x^2}{\sigma^2} \sim \chi^2(n-1), \qquad \frac{(m-1)S_y^2}{\sigma^2} \sim \chi^2(m-1).$$

Hence by Proposition 5.3.3,

$$\frac{(n-1)S_x^2}{\sigma^2} + \frac{(m-1)S_y^2}{\sigma^2} \sim \chi^2(n-1+m-1) = \chi^2(n+m-2).$$

This last random variable is independent of $\bar{X} - \bar{Y}$. Therefore, by the definition of the *t*-distribution, the random variable

$$T = \frac{\dfrac{(\bar{X} - \bar{Y}) - (\mu_x - \mu_y)}{\sqrt{\sigma^2/n + \sigma^2/m}}}{\sqrt{\left(\dfrac{(n-1)S_x^2}{\sigma^2} + \dfrac{(m-1)S_y^2}{\sigma^2}\right)\Big/(n+m-2)}}$$

$$= \frac{(\bar{X} - \bar{Y}) - (\mu_x - \mu_y)}{\sqrt{1/n + 1/m} \cdot \sqrt{\dfrac{(n-1)S_x^2 + (m-1)S_y^2}{n+m-2}}} \tag{8.67}$$

has the $t(n+m-2)$ distribution.

The quantity $[(n-1)S_x^2 + (m-1)S_y^2]/(n+m-2)$ in the denominator of T is a weighted average of the two sample variances, and therefore it is sometimes called the *pooled variance* and is denoted by S_p^2. Then S_p is the *pooled standard deviation*. By simple algebra, it follows from (8.67) that

$$\left[(\bar{X} - \bar{Y}) \pm t_{\alpha/2}(n + m - 2) \cdot \sqrt{\frac{1}{n} + \frac{1}{m}} \cdot S_p \right] \tag{8.68}$$

is a $(1 - \alpha) \times 100\%$ confidence interval for $\mu_x - \mu_y$ in the case of equal unknown variances. Here, as usual, the quantity $t_{\alpha/2}(n + m - 2)$ is the critical value from the t-table with $(n + m - 2)$ degrees of freedom, such that $P[T > t_{\alpha/2}] = \alpha/2$.

Example 8.5.2

To illustrate another kind of two-sample confidence interval, let us return to the setting of Exercise 6 in Section 8.4, where two groups of ten men each were checked for heart rate during snow removal activity. The first group shoveled an area manually, and the second group used snowblowers. The exercise stated that the sample mean heart rates were $\bar{x} = 175$ for the manual group and $\bar{y} = 124$ for the snowblower group, and the sample standard deviations were $s_x = 15$ and $s_y = 18$, respectively. In the interest of checking whether it is reasonable to use confidence interval (8.68) to estimate the difference of mean heart rates, let us devise a confidence interval for the ratio of variances. If that interval contains 1, then we can be comfortable saying that we don't have significant evidence against the assumption that $\sigma_x^2 = \sigma_y^2$.

We will proceed under the assumption that both samples $X_1, X_2, \ldots, X_m$ for manual shoveling and $Y_1, Y_2, \ldots, Y_n$ for snowblowing came from normal populations and that all random variables in these lists are independent of all others. The sample sizes here are $m = n = 10$, but we will derive our results generally.

To construct a confidence interval for the ratio of variances σ_y^2/σ_x^2, note that

$$\frac{(m - 1)S_x^2}{\sigma_x^2} \sim \chi^2(m - 1) \quad \text{and} \quad \frac{(n - 1)S_y^2}{\sigma_y^2} \sim \chi^2(n - 1).$$

By the definition of the F-distribution,

$$F = \frac{(m - 1)S_x^2/[\sigma_x^2 \cdot (m - 1)]}{(n - 1)S_y^2/[\sigma_y^2 \cdot (n - 1)]} = \frac{S_x^2/\sigma_x^2}{S_y^2/\sigma_y^2} \sim f(m - 1, n - 1). \tag{8.69}$$

Let $f_{\alpha/2}(m - 1, n - 1)$ and $f_{1-\alpha/2}(m - 1, n - 1)$ be critical values from the F-table with $(m - 1)$ and $(n - 1)$ degrees of freedom in Table 6 of Appendix A such that

$$\frac{\alpha}{2} = P[F > f_{\alpha/2}], \qquad \frac{\alpha}{2} = P[F < f_{1-\alpha/2}].$$

(See Fig. 8.11.) Then

$$1 - \alpha = P[f_{1-\alpha/2} \leq F \leq f_{\alpha/2}]$$

$$= P[f_{1-\alpha/2} \leq \frac{S_x^2/\sigma_x^2}{S_y^2/\sigma_y^2} \leq f_{\alpha/2}]$$

$$= P\left[\frac{S_y^2}{S_x^2} \cdot f_{1-\alpha/2} \leq \frac{\sigma_y^2}{\sigma_x^2} \leq \frac{S_y^2}{S_x^2} \cdot f_{\alpha/2} \right]. \tag{8.70}$$

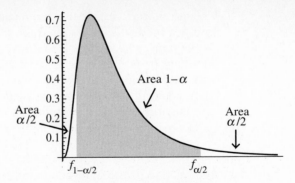

Figure 8.11 F-Critical Values

The last inequality on the right side of (8.70) describes a $(1 - \alpha) \times 100\%$ confidence interval for the ratio of variances σ_y^2 / σ_x^2.

For a confidence level of .90, and $m - 1 = n - 1 = 9$ for example, $f_{\alpha/2}(9, 9) = f_{.05}(9, 9) = 3.18$ and $f_{1-\alpha/2}(9, 9) = f_{.95}(9, 9) = 1/f_{.05} = 1/3.18 = .314$. (Verify this.) Therefore the observed confidence interval for σ_y^2 / σ_x^2 is

$$\left[\frac{18^2}{15^2}(.314), \; \frac{18^2}{15^2}(3.18) \right] = [.452, 4.58].$$

Notice that this interval does contain 1, meaning that $\sigma_y^2 / \sigma_x^2 = 1$ is within a margin of error of the observed ratio of sample variances that corresponds to 90% confidence. Thus we can feel pretty well assured that we are not doing an outrageous thing by supposing that $\sigma_x^2 = \sigma_y^2$, or that the heart rate variances of the two groups are the same.

?Question 8.5.4   Finish the last example by computing a 95% confidence interval for the difference in means using the equal unknown variance form (8.68). Do you feel confident in saying that the use of snowblowers reduces heart rate?

Example 8.5.3   As a final example of the development of confidence intervals in the two-sample case, let us derive a confidence interval for the difference $p_x - p_y$ of Bernoulli success parameters of two independent experiments in the following context. The Centers for Disease Control and Prevention (1995b), besides monitoring infectious diseases, also amass statistics on injuries in accidents. In 1991, among 2015 crashes in Wisconsin known to have involved unhelmeted motorcycle riders, 55 of those riders died. In 994 motorcycle crashes in which the rider wore a helmet, 19 of the riders died. Does there seem to be a significant lessening in mortality when riders wear helmets?

In the one-sample proportion problem of the last section, we used the Central Limit Theorem to obtain an approximate large sample confidence interval for the Bernoulli success parameter based on the sample proportion

$\hat{p}$ of successes. Recall that $\hat{p}$ has mean p and variance $p(1-p)/n$. We now have two random samples $X_1, X_2, \ldots, X_n$ and $Y_1, Y_2, \ldots, Y_m$ from Bernoulli distributions, where $X_i = 1$ or 0 according to whether the ith unhelmeted rider died or not, and $Y_j = 1$ or 0 according to whether the jth helmeted rider died or not. Then for large n and m, assuming that the crashes are independent of one another, the sample proportions of successes $\hat{p}_x$ and $\hat{p}_y$ have the approximate distributions:

$$\hat{p}_x \approx N\left(p_x, \frac{p_x(1-p_x)}{n}\right), \qquad \hat{p}_y \approx N\left(p_y, \frac{p_y(1-p_y)}{m}\right).$$

Also, $\hat{p}_x$ and $\hat{p}_y$ are independent of each other. In this case we can say that approximately:

$$\hat{p}_x - \hat{p}_y \approx N\left(p_x - p_y, \frac{p_x(1-p_x)}{n} + \frac{p_y(1-p_y)}{m}\right), \tag{8.71}$$

which implies that

$$Z = \frac{(\hat{p}_x - \hat{p}_y) - (p_x - p_y)}{\sqrt{p_x(1-p_x)/n + p_y(1-p_y)/m}} \approx N(0, 1). \tag{8.72}$$

Deriving in the usual way, and replacing p_x and p_y in the denominator by the estimates $\hat{p}_x$ and $\hat{p}_y$, gives the large sample approximate confidence interval for $p_x - p_y$, the difference between mortality probabilities for unhelmeted and helmeted riders:

$$1 - \alpha \approx P[-z_{\alpha/2} \leq Z \leq z_{\alpha/2}]$$

$$= P\left[-z_{\alpha/2} \leq \frac{(\hat{p}_x - \hat{p}_y) - (p_x - p_y)}{\sqrt{p_x(1-p_x)/n + p_y(1-p_y)/m}} \leq z_{\alpha/2}\right]$$

$$\approx P\left[p_x - p_y \in \left[\hat{p}_x - \hat{p}_y \pm z_{\alpha/2} \cdot \sqrt{\frac{\hat{p}_x(1-\hat{p}_x)}{n} + \frac{\hat{p}_y(1-\hat{p}_y)}{m}}\right]\right].$$

$$\tag{8.73}$$

The critical value for an interval with 95% confidence is $z_{\alpha/2} = 1.96$, and for these data $\hat{p}_x = 55/2015 = .027$, $\hat{p}_y = 19/994 = .019$, $n = 2015$, $m = 994$. Substituting these values into (8.73) gives an interval of $[-.003, .019]$. Since 0 is in this interval—that is, it is within the 95% margin of error .011 of the observed difference $\hat{p}_x - \hat{p}_y = .008$—we see that statistical variability could account for this observed difference even if $p_x = p_y$. However, we are demanding a high confidence level here, which has the effect of widening the interval. Try recomputing the interval for lower levels to find the level of confidence such that the left endpoint exceeds 0. (You shouldn't have to go far below 90%.) The observed proportions are indeed relatively unlikely if unhelmeted motorcyclists had the same mortality rate as helmeted ones.

EXERCISES 8.5

1. An agricultural researcher is interested in comparing the effect of two herbicides on yield of corn. A test field is available that can be divided into subfields called plots. Conditions are rather homogeneous within a

plot but nonhomogeneous between different plots. Describe two experimental designs for comparing average corn yield, one that takes into account the possible differences between plots and the other that does not, analogous to the ACT experiment described in the section.

2. Referring again to the Illinois State Lottery data in Table 7.9, does it seem as if the digit 8 is significantly more likely to occur on the 2nd than the 3rd draw? Use an approximate 95% level of confidence.

3. Grenade Cola Company is interested in whether its customers can tolerate an even higher concentration of caffeine in its product. In a blind taste test between the original formulation of Grenade and a competitive brand, Nervous Tension Cola, 54% of the 500 people tested preferred Grenade. In a subsequent test using a new formula, 60% of 200 people preferred Grenade. Find an approximate 90% confidence interval for the difference between the proportion of people who prefer the original formula to Nervous Tension and the proportion who prefer the new formula to Nervous Tension. Can you reach a conclusion?

4. Exercise 5 in Section 7.2 contained information about death ages of 20 people who died on a particular day in Peoria and 27 people who died on the same day in Chicago. Estimate the difference in average death age by a 90% confidence interval.

5. Suppose that in April a political candidate draws 25% support in a poll of 500 randomly selected voters, and in May the candidate has 30% support of 500 other randomly selected voters. Use confidence intervals to decide whether the candidate can say with high probability that his true support has increased.

6. In an experiment to assess the possibility that paranormal abilities might help students improve academically, ESP researchers Cashen and Ramseyer (1970) gave an ESP diagnostic test to 32 students at the beginning of a first-year psychology class. The results separated the group into 15 high and 17 low ESP students. Before each test, the students were asked to predict the questions that would be on it. The experimenters then counted the number of correct predictions by each person. The data follow. Ask and answer an interesting question about the data.

High ESP: 79, 62, 95, 83, 59, 51, 78, 78, 63, 55, 24, 69, 64, 55, 52

Low ESP: 82, 64, 44, 29, 50, 43, 41, 30, 38, 46, 36, 42, 23, 49, 43, 31, 24

7. An advertisement in a recent issue of the *Journal of the American Medical Association* ("Cardizem," 1995) contained data on possible side effects of a drug called Cardizem, used to treat hypertension or angina. It reported on 607 subjects who had been tested with Cardizem, and 301 control subjects who had taken a harmless placebo. Of these, 5.4% of the Cardizem subjects suffered headaches, compared to 5.0% of the controls, and 2.6% of the Cardizem subjects suffered edema (fluid buildup), compared to 1.3% of the controls. Do you think that the pharmaceutical company advertising this drug can claim legitimately that there is no strong evidence of an appreciable side effect of either kind?

8. In Example 5.5.1 and Exercise 6 of Section 5.5 there are data from the months of January and February on number of users of a computer lab. Find a 90% confidence interval for the ratio of variances. Would you recommend using formula (8.68) to derive a confidence interval for the difference of means?

9. While studying statistics at the University of Iowa, I came upon a group of researchers in the Department of Psychology who were studying environmental versus genetic influences in the characteristics of twins. Among many other things, they were attempting to detect the strength of genetic influences by separately treating so-called *monozygotic* twin pairs, who share more genes, and *dizygotic* twin pairs, who share fewer genes. The problem with the data was that the twin sets could not be genetically tested to determine for certain which category they belonged to. Instead, the experimenters needed to classify the twins as monozygotic or as dizygotic based on the mother's responses to a questionnaire. This questionnaire was scored on the basis of 40 points and was designed so that those twin pairs who were monozygotic would generally receive higher scores than those who were not. Fortunately, there was a pilot sample of twin pairs who had been genetically tested and whose mothers had responded to the questionnaire. The order statistics of the scores of 30 known dizygotic twin pairs are as follows:

0, 0, 2, 4, 4, 4, 4, 6, 6, 6, 6, 6, 8, 8, 9, 9, 10, 10, 10, 10,
12, 12, 12, 14, 14, 18, 22, 28, 29, 30

Order statistics of the scores of 21 known monozygotic twin pairs were

23, 24, 25, 26, 26, 26, 26, 26, 26, 26, 28, 28, 28,
28, 28, 30, 30, 30, 30, 36, 38

a. Construct a dotplot of the two samples of zygosity scores. Does there appear to be a difference in the distributions of scores? Find a 95% confidence interval for the difference in mean score in order to give a reasoned answer.

b. There is some overlap in possible scores, specifically, some dizygotic twins can receive higher scores than some monozygotic twins. Still, it is necessary to use a boundary score x^* to classify new twin pairs as dizygotic if their score is x^* or less, and monozygotic if their score is greater than x^*. Find an expression for the *Total Probability of Misclassification* (TPM) of a twin pair, under the assumption that 2/3 of all twin pairs are dizygotic and scores of each twin type are normally distributed. Use this expression and the preceding data to find the value of x^* that minimizes the TPM. (You will likely need a symbolic algebra-graphical package such as *Mathematica* to complete the calculation.)

10. Derive the confidence interval (8.63) for the difference in normal means for paired data.

11. In Section 7.1 we looked at a data set of student scores on three statistics tests (see Table 7.2).

Table 8.1

Cotton Yield, 1988–1989

Nation	1988	1989	Nation	1988	1989
Angola	202	202	Mozambique	315	300
Cameroon	612	477	Nigeria	114	108
Cent. Af. Rep	167	173	South Africa	374	363
Chad	276	313	Sudan	443	456
Egypt	718	683	Tanzania	147	131
Kenya	204	171	Uganda	100	98
Malawi	179	131	Zaire	117	119
Morocco	508	638	Zimbabwe	371	294

    a. Using a 95% confidence interval, does there appear to be a significant difference in average performance on tests 1 and 2?

    b. Does there appear to be a significant difference in average performance on tests 2 and 3? Again use a 95% confidence interval.

12. Table 8.1 gives 1988 and 1989 results on the cotton crop in several African nations (USDA, 1991). The units of measurement are kilograms/hectare. Use the paired data form (8.63) of the confidence interval for difference in means to make a judgment as to whether the yield changed significantly from one year to the next.

13. The number of cars arriving to an intersection from the south was counted for 16 independent 1-minute periods:

$$1, \ 2, \ 5, \ 3, \ 3, \ 2, \ 4, \ 1, \ 4, \ 3, \ 5, \ 2, \ 4, \ 3, \ 3, \ 6$$

The number arriving to the same intersection from the west was counted for 16 additional, independent 1-minute periods:

$$2, \ 1, \ 2, \ 3, \ 2, \ 2, \ 1, \ 3, \ 4, \ 1, \ 2, \ 1, \ 0, \ 0, \ 1, \ 2$$

Find an approximate 95% confidence interval for the difference in average arrival rates. What probabilistic model underlies your analysis?

14. The number of employee absences at two branches of a retail store was tracked over a period of 100 days. The frequencies for the numbers of observed absences are recorded as follows. Does there seem to be a difference between stores in their average absentee rates?

Store 1

# Absentees	0	1	2	3	4	5	6	7	8
# Days	11	29	20	21	10	6	2	1	0

Store 2

# Absentees	0	1	2	3	4	5	6	7	8
# Days	1	8	15	19	20	18	9	6	4

15. Many data sets follow the *1-factor linear model with 2 levels*. In this, there are data points X_{ij}, $i = 1, 2$; $j = 1, 2, \ldots, n$, such that

$$X_{ij} = \mu + \alpha_i + \epsilon_{ij},$$

where μ is constant, α_1 and α_2 are constants satisfying $\alpha_1 + \alpha_2 = 0$, and ϵ_{ij} are i.i.d. normal random variables with mean 0 and known variance.
a. Derive maximum likelihood estimators for μ, α_1, and α_2.
b. Derive the form of a confidence interval for the difference $\alpha_1 - \alpha_2$.

CHAPTER 9

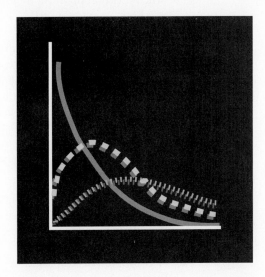

HYPOTHESIS TESTING

9.1 | Basic Ideas of Hypothesis Testing

When we studied confidence interval problems in the last chapter, we often adopted a decision-making point of view. That is, we decided that an hypothesized parameter value was reasonable based on whether a confidence interval contained the parameter value. In this chapter we will be taking a direct approach to the decision-making problem by developing statistical procedures to choose between competing hypotheses about parameters. More than just decision-making is involved, however; we also want to estimate the probability of making an erroneous decision. The smaller is this probability, the more effective is our decision rule. Just as in Chapter 8 when we introduced confidence intervals, we will use tests for means, variances, and proportions as examples, and we will be interested in the beneficial effect of increasing sample size.

9.1.1 Definitions

We begin the chapter by describing the basic concepts and language of hypothesis testing. A *statistical hypothesis* is a statement of the form "The unknown distribution has property *P*." Most often, an hypothesis has the special form "The unknown distributional parameter θ belongs to a set Ω." There will be two competing hypotheses of this form:

$$H_0 : \theta \in \Omega_0 \quad \text{vs.} \quad H_1 : \theta \in \Omega_1,$$

where Ω_0 and Ω_1 are disjoint sets of possible values of the parameter θ whose union makes up the entire set of possible values for θ. Hypothesis H_0 is called the *null hypothesis*, and H_1 is called the *alternative hypothesis*. (You will sometimes see the notation H_a used for the alternative hypothesis instead). The reason is that H_0 is usually chosen to be a kind of default or no change hypothesis, which the statistician tends to believe unless given strong evidence otherwise.

Here are a few examples of hypothesis testing problems that fit the foregoing description:

- A new cancer therapy designed to halt the growth of tumors has an effect beyond that of a placebo.
- Average global temperature has increased as compared to ten years ago.
- A computer disk manufacturing process produces disks with no more than one bad spot in every 240 sectors, on the average.
- A herbicide reduces the variation in soybean yield in a test field.

—— **?Question 9.1.1** Think of some other hypothesis testing situations.

In the first example we can use a Bernoulli model. There is a placebo effect: Some baseline proportion p_0 of cancerous tumors will stop growing without treatment. Let X be 1 or 0, respectively, according to whether a randomly selected tumor treated by the new therapy does or does not halt its growth. Then the success parameter p of the distribution of X is to be compared with the baseline p_0. Either $p \leq p_0$ (or perhaps it is more sensible to say $p = p_0$) if the therapy is ineffective, or $p > p_0$ if the therapy is effective. Symbolically, we have the two hypotheses

$$H_0 : p \in \Omega_0 = [0, p_0] \quad \text{vs.} \quad H_1 : p \in \Omega_1 = (p_0, 1],$$

or more concisely,

$$H_0 : p \leq p_0 \quad \text{vs.} \quad H_1 : p > p_0.$$

—— **?Question 9.1.2** Set up a probabilistic model appropriate for the hypothesis testing scenario in the second example on global warming.

We will sometimes use the following terminology. When the set of possible values of the parameter specified by an hypothesis has one element, such as $H_0 : p = p_0$, we say that the hypothesis is *simple*; otherwise, as in $H_1 : p > p_0$, the hypothesis is called *composite*. When an hypothesis consists of two pieces, such as, $H_1 : p \neq p_0$—that is, $H_1 : p > p_0$ or $p < p_0$—we call it a *two-sided hypothesis*; otherwise it is *one-sided*.

Like confidence interval decisions, the decision that we make when testing competing hypotheses is based on a random sample of data from the population of interest. In a comparative test between two parameters θ_1 and θ_2 of two distributions, we would need two random samples, one from each distribution. For the time being, however, let us consider only single-parameter, single-sample hypothesis testing in order to introduce the ideas.

So, let there be a random sample $X_1, X_2, \ldots, X_n$ from a distribution $f(x; \theta)$, and consider the hypotheses $H_0 : \theta \in \Omega_0$ vs. $H_1 : \theta \in \Omega_1$. A *decision rule* is any unambiguous criterion based on the sample that leads to a selection of either H_0 or H_1 for all possible outcomes of the random sample. A decision rule is characterized by a set called a *critical region* C; H_0 is rejected in favor of H_1 in the event that the data vector $[X_1, X_2, \ldots, X_n]'$ falls into the set C. For the cancer therapy problem for instance, we would have success-failure data $X_1, X_2, \ldots, X_n$ where $X_i = 1$ or 0, respectively, according to whether or not patient i was successfully treated. An intuitively reasonable decision rule is to reject $H_0 : p \leq p_0$ if the sample proportion $\hat{p}$ of successes exceeds some constant c, that is, if the data vector $[X_1, X_2, \ldots, X_n]'$ falls into the region $C = \{(x_1, x_2, \ldots, x_n) : \hat{p} = \sum x_i / n > c\}$.

Because the data are random, for some random samples $[X_1, X_2, \ldots, X_n]'$ will fall into C and the null hypothesis will be rejected, and for others $[X_1, X_2, \ldots, X_n]'$ will not fall into C and H_0 will be accepted. Only one of H_0 or H_1 is actually true, so for some samples the decision will be correct, and for others it will be in error. A good decision rule will be one that is not very likely to be in error. The diagram in Fig. 9.1 summarizes the possible outcomes of an hypothesis test.

The decision procedure is subject to two possible errors: incorrect rejection of H_0 (i.e., acceptance of H_1), which is called *type I error*, and incorrect acceptance of H_0 (i.e., rejection of H_1), which is *type II error*. Since we commit a type I error when the data fall into the critical region but H_0 is true, type I error is an event whose probability we can determine. Denote

$$\alpha = P[\text{type I error}] = P[H_0 \text{ rejected}; H_0 \text{ true}]. \qquad (9.1)$$

Similarly, denote

$$\beta = P[\text{type II error}] = P[H_0 \text{ accepted}; H_1 \text{ true}]. \qquad (9.2)$$

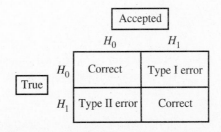

Figure 9.1   Outcomes of an Hypothesis Test

(I will presently try to convince you that α and β need to be defined more carefully, but for now just go on.) A good test will have small error probabilities α and β.

In the cancer problem again, with

$$H_0 : p = p_0 \quad \text{vs.} \quad H_1 : p > p_0,$$

the type I error probability would be

$$\alpha = P[[X_1, X_2, \ldots, X_n]' \in C; \ H_0 \text{ true}]$$
$$= P\left[\sum \frac{X_i}{n} > c; \ p = p_0\right]. \tag{9.3}$$

Since the distribution of $\sum X_i$ is binomial with parameters n and p_0 if $p = p_0$, α can be computed using a table of the binomial distribution. Note, however, that had we not specified a particular single parameter value p_0, using instead the original $H_0 : p \leq p_0$, we could not have completed the computation of α, because we would not know the value of the Bernoulli success parameter for $\sum X_i$. Thus $\alpha = \alpha(p_0)$ should really be viewed as a function of the particular member of the H_0 set of parameter values. Similarly, the type II error probability β can be determined only for particular single values $p = p_1$ in the set $\Omega_1 = \{p : p > p_0\}$. For such a simple alternative,

$$\beta = P[[X_1, X_2, \ldots, X_n]' \notin C; \ H_1 \text{ true}]$$
$$\beta(p_1) = P\left[\sum \frac{X_i}{n} \leq c; \ p = p_1\right]. \tag{9.4}$$

The complementary probability to the type II error probability—that is, the probability of correctly rejecting the null hypothesis—is called the *power* of the test. A good decision rule should have high power when H_1 is true. Because β is a function of the parameter, so is the power. Denote this function by

$$K = 1 - \beta = P[[X_1, X_2, \ldots, X_n]' \in C; \ H_1 \text{ true}]$$
$$K(p_1) = P\left[\sum \frac{X_i}{n} > c; \ p = p_1\right]. \tag{9.5}$$

Both functions β and K can be extended to operate on the parameter set Ω_0 as well. The power $K(p)$ for p belonging to $\Omega_0 = \{p : p \leq p_0\}$ would then equal the probability of rejecting H_0 given that p is the true parameter value; that is, it would be the type I error probability for the particular member p of Ω_0.

───── **?Question 9.1.3**   What do you think the relationship is between α and power? In other words, as the critical region changes so as to increase α, does the power increase or decrease?

The *significance level* of a test of hypotheses $H_0 : \theta \in \Omega_0$ vs. $H_1 : \theta \in \Omega_1$ is the maximum probability of type I error over all parameter values θ in

Ω_0. Often, H_0 is a simple hypothesis $\theta = \theta_0$, in which case the significance level and type I error probability are identical, and both equal the following probability:

$$\alpha = P[[X_1, X_2, \ldots, X_n]' \in C; \ \theta = \theta_0].$$

In some problems, Ω_0 has many elements. If in the cancer example the null hypothesis H_0 is of the form $p \leq p_0$, then the significance level is

$$\max_{p \leq p_0} P\left[\sum \frac{X_i}{n} > c; \ p \text{ is true}\right]. \tag{9.6}$$

It is fairly obvious that $\sum X_i$ is likeliest to exceed the constant nc when p is as large as it can be (see Exercise 8 at the end of this section), and therefore the maximum in (9.6) is taken on at p_0. Then the significance level is just the type I error probability α in (9.3).

Let us review these concepts and terms in the context of another example.

Example 9.1.1

A reporter from a local newspaper once called me to ask how to interpret the results of a small poll conducted prior to the 1992 presidential election. Of those responding to the poll, 46 favored George Bush, 50 favored Bill Clinton, 22 favored Ross Perot, and 29 were undecided. Does it seem that Perot had significantly more than 10% popularity?

We can formulate a test of hypotheses as

$$H_0 : p \leq .1 \quad \text{vs.} \quad H_1 : p > .1,$$

where p is the proportion of the population being sampled who favor Perot, which is the same as the probability that a randomly selected individual supports Perot. To continue, we must assume that the polling was done randomly, producing a random sample $X_1, X_2, \ldots, X_n$, $n = 46 + 50 + 22 + 29 = 147$, from the Bernoulli (p) distribution. Here, $X_i = 1$ or 0, respectively, according to whether the ith person polled favored or did not favor Perot. Since each of these X_i's has mean p and variance $p(1 - p)$, the sample proportion $\hat{p} = \overline{X} = \sum X_i/n$ has mean p and variance $p(1 - p)/n$. Therefore by the Central Limit Theorem,

$$Z = \frac{\hat{p} - p}{\sqrt{p(1 - p)/n}} \approx N(0, 1). \tag{9.7}$$

An intuitively reasonable critical region for the preceding hypotheses is

$$C = \left\{(x_1, x_2, \ldots, x_n) : \hat{p} = \sum x_i/n > k\right\}$$

because large values of $\hat{p}$ indicate that H_1 is true, not H_0. What value of k will create a critical region of approximate significance level .05? It is clear that the maximum probability that $\hat{p}$ falls into C among values of $p \leq .1$ occurs for $p = .1$. Then we have

$$.05 = \alpha = P[\text{type I error}] = P[\hat{p} > k; \ p = .1]$$

$$= P\left[\frac{\hat{p} - .1}{\sqrt{.1(.9)/n}} > \frac{k - .1}{\sqrt{.1(.9)/n}}\right]$$

$$\approx P\left[Z > \frac{k - .1}{\sqrt{.1(.9)/n}}\right]. \tag{9.8}$$

Hence from the $N(0, 1)$ table,

$$1.645 = \frac{k - .1}{\sqrt{.1(.9)/147}} \Rightarrow k = .1 + 1.645\sqrt{.1(.9)/147} = .14.$$

The critical region appropriate to level .05 is therefore $\hat{p} > .14$. The observed value of $\hat{p}$ from the data set is $22/147 = .149$. Therefore H_0 is rejected at level .05; that is, we have made the decision that Perot does have more than 10% support, and the probability of type 1 error is just .05.

In the computation (9.8) you may have noticed an alternative, simple way of performing the test. Instead of solving for k, express the critical region equivalently as $Z > 1.645$, where Z is the standardized version of $\hat{p}$ in (9.7) (using the hypothesized $p_0 = .1$ for p). Compute the observed value of Z from the sample:

$$z = \frac{\hat{p} - .1}{\sqrt{.1(.9)/147}} = \frac{.149 - .1}{\sqrt{.1(.9)/147}} = 1.98.$$

Because 1.98 exceeds the critical value 1.645, H_0 is rejected. You may prefer this method if type I error is all that is important in a problem, but as you will see in the next paragraphs, knowledge of the critical value k assists in computing power and type II error probability.

We should further investigate the effectiveness of our test. What is the approximate type II error probability when the true value of p is .15? In other words, what proportion of the time will the data lead us to conclude that Perot has no more than 10% support when in fact he has 15% support? Recall that a type II error is committed when H_0 is incorrectly accepted. Because of the decision rule that we constructed,

$$\beta(.15) = P[\text{type II error}; \ p = .15] = P[\hat{p} \leq .14; \ p = .15]$$

$$= P\left[\frac{\hat{p} - .15}{\sqrt{.15(.85)/147}} \leq \frac{.14 - .15}{\sqrt{.15(.85)/147}}\right]$$

$$\approx P[Z \leq -.34]$$

$$= .37. \tag{9.9}$$

This is a fairly large probability of error of the second type. What has happened is that in choosing to design the critical region so as to reduce type I error to a small probability of .05, we have made it difficult in general to reject H_0. Thus the test is relatively unlikely to reject H_0 when indeed it should be rejected, when the true p is .15. Do you think that the β value would be lower if we had taken a more extreme value for p, such as .20? Yes, as p grows, it becomes less and less likely that the data will fall into the acceptance region for H_0. You might want to rework the computation for $p = .20$ and some other values of p in the H_1 region to convince yourself of this.

A computation like the preceding one shows that the approximate power function for this test is

$$K(p) = P[\text{reject } H_0; \ p \text{ true}]$$

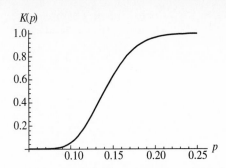

Figure 9.2   Power Function of the Large Sample Test for Bernoulli p

$$= P[\hat{p} > .14;\ p \text{ true}]$$

$$= 1 - \Phi\left(\frac{.14 - p}{\sqrt{p(1-p)/147}}\right), \tag{9.10}$$

where Φ is the standard normal c.d.f. A graph of $K(p)$ appears in Fig. 9.2. Notice that $K(.1) = .05$ (why?) and $K(.15) = 1 - \beta(.15) = .63$. Figure 9.2 shows that the power is very high for values of p of at least .2, which is far from the boundary .10 between the two regions $\Omega_0 = [0, .1]$, $\Omega_1 = (.1, 1]$, but the power is not particularly high near .10. When the true alternative is close to the null hypothesized value, however, we shouldn't expect the test to be too successful in making sharp distinctions.

After your experience with confidence intervals, you may well be led to ask: What is the effect of increasing sample size on the power of a test? The next example, which continues the political poll example, sheds some light on the question.

Example 9.1.2

Return to the beginning of the 1992 presidential campaign poll of Example 9.1.1, where a pollster wishes to test $H_0 : p = .1$ vs. $H_1 : p > .1$ using a random sample $X_1, X_2, \ldots, X_n$ of Bernoulli (p) random variables, but suppose that n is not known. Working as in (9.8) to obtain a level .05 test, k and n together would satisfy

$$.05 \approx P\left[Z > \frac{k - .1}{\sqrt{(.1)(.9)/n}}\right],$$

so that

$$1.645 = \frac{k - .1}{\sqrt{(.1)(.9)/n}} \Rightarrow k = .1 + 1.645\sqrt{(.09)/n}. \tag{9.11}$$

Notice that as n increases, k decreases. Thus, for fixed α, growth in sample size makes the critical region $\hat{p} > k$ larger, in turn making it more likely to reject H_0 when H_0 is false. Thus the power increases as n increases.

How large a sample is required to reject H_0 with high probability, say .9, when the true p is in the H_1 region, say $p = .15$? This requires the power $K(.15)$ to be at least .9, which yields

$$.9 \le K(.15) = P[\hat{p} > k; \; p = .15]$$

$$= P[\hat{p} > .1 + 1.645\sqrt{(.09)/n}; \; p = .15]$$

$$\approx P\left[Z > \frac{(.1 + 1.645\sqrt{(.09)/n}) - .15}{\sqrt{(.15)(.85)/n}}\right],$$

where Z is $N(0,1)$. (Why is the last line true?) From the standard normal table,

$$-1.28 \ge \frac{(.1 + 1.645\sqrt{(.09)/n}) - .15}{\sqrt{(.15)(.85)/n}} = \frac{-.05}{\sqrt{(.15)(.85)/n}} + 1.645\frac{\sqrt{.09}}{\sqrt{(.15)(.85)}}$$

$$\Rightarrow -1.28 - 1.645\frac{\sqrt{.09}}{\sqrt{(.15)(.85)}} \ge \sqrt{n}\frac{-.05}{\sqrt{(.15)(.85)}}$$

$$\Rightarrow n \ge 361.4.$$

Therefore, $n = 362$ people would have to be polled in order to design a test of level .05 whose power is at least .9 at $p = .15$. From (9.11), we would then reject H_0 that Perot's support is no more than 10% if

$$\hat{p} > .1 + 1.645\sqrt{\frac{.09}{362}} = .126.$$

Practitioners frequently use an approach to hypothesis testing that differs a little from what we have been doing. Our procedure has been to choose a nominal significance level α (.01, .05, and .1 are popular levels depending on one's aversion to type I error) and to use it to find an appropriate critical value specifying the values of a test statistic for which H_0 is rejected. Then the data are taken, the test statistic is computed and compared with the critical region, and a decision between H_0 and H_1 is made. The other way of proceeding is to observe data first, then compute the value of the test statistic. Let this be the critical value of the test, thereby building in rejection of H_0. The only error possible is type I error. Compute the probability of this type I error, usually called the *p-value* of the test. If this probability is very small, then it is likely that the decision to reject H_0 was the correct one.

In the political poll, this approach would dictate rejection of H_0 if the proportion $\hat{p}$ of people in the sample favoring Perot is greater than $22/147 = .149$. Then the *p*-value would be

$$P[\hat{p} > .149; \; p = .1] \approx P\left[Z > \frac{.149 - .1}{\sqrt{(.1)(.9)/147}}\right]$$

$$= P[Z > 1.98] \approx .034. \tag{9.12}$$

Since this probability is small, we can confidently reject H_0. Note that the p-value is in fact the smallest significance level at which H_0 can be rejected, so that reporting a p-value actually gives more information than reporting that H_0 is rejected at some single, nominal level.

—— **?Question 9.1.4** Without reference to the text, try to recite the definitions of the key concepts in hypothesis testing: decision rule, critical region, types I and II error, significance level, power, and p-value.

9.1.2 Sign Test

This section will close by giving you an early taste of an inference method that belongs to a different class of methods than those we have seen so far. They are the so-called *nonparametric* (or *distribution-free*) procedures, so named because they either are not linked to numerical parameters or make minimal assumptions about the underlying distribution from which the sample comes. They are intuitive and simple to carry out, and in most cases the test statistics have small sample distributions that are easy to derive. They are also competitive in power to the classical parametric techniques in cases where both are applicable. We will introduce such methods from time to time when they fit the development.

Here let us consider a test for population percentiles of continuous distributions. By definition, the $p \times 100$th percentile κ_p satisfies

$$P[X \leq \kappa_p] = p. \tag{9.13}$$

(You should now draw a picture to illustrate this equation.)

Suppose we have a random sample $X_1, X_2, \ldots, X_n$ from a continuous distribution with density f, and we are interested in testing the two-sided alternative

$$H_0 : \kappa_p = \kappa_{p0} \quad \text{vs.} \quad H_a : \kappa_p \neq \kappa_{p0}, \tag{9.14}$$

where κ_{p0} is some proposed value for the pth percentile (p is given). The most natural thing to do is to observe that a proportion p of the sample values is expected to be less than or equal to κ_{p0} if the latter is the correct value of the percentile. Neither too many nor too few of the sample values should be less than or equal to κ_{p0}; otherwise we should disbelieve H_0. If too many sample values are less than or equal to κ_{p0}, then the distribution is more heavily weighted to the left than κ_{p0} indicates, so that we would tend to believe that the true κ_p is less than κ_{p0}. On the other hand, if too few sample values are less than or equal to κ_{p0}, then we would have evidence that the true κ_p is greater than κ_{p0}.

The sign test of hypotheses (9.14) works as follows. Form the differences $X_i - \kappa_{p0}$, $i = 1, \ldots, n$. Count the number of negative signs Y among these differences, that is, the number of X_i's less than κ_{p0}. Reject H_0 in favor of H_a if $Y \leq c_0$ or $Y \geq c_1$ where we will choose c_0 and c_1 such that

$$\frac{\alpha}{2} \geq P[Y \leq c_0] \quad \text{and} \quad \frac{\alpha}{2} \geq P[Y \geq c_1] \tag{9.15}$$

in order to make a test of level no more than α. (What do you think is the procedure for one-sided alternatives?)

The distribution of the test statistic Y under H_0 follows from the fact that the X_i's are i.i.d. and all have a likelihood of p of being $\leq \kappa_{p0}$ if $\kappa_p = \kappa_{p0}$. The X_i's therefore form a sequence of n Bernoulli trials of success probability p. Since Y is the number of successes, Y has the $b(n, p)$ distribution under H_0, independently of the form of the underlying distribution of the X_i's. This is what makes the sign test a distribution-free procedure. Since p is known (but not κ_p), and the sample size n is known, the binomial table or a statistical program lets us find c_0 and c_1 satisfying (9.15).

—— **?Question 9.1.5** What difficulties are there in this analysis if the underlying distribution of the X_i's is discrete rather than continuous? What would you do in this case?

Example 9.1.3 Exercise 5 in Section 7.2 contained the data set, displayed as follows, of ages at death of people on a certain day in Chicago. We can test the hypothesis that the median death age is 65 using the sign test.

$$95, \ 93, \ 69, \ 77, \ 93, \ 40, \ 62, \ 96, \ 90, \ 74, \ 77, \ 47, \ 70, \ 90,$$
$$97, \ 71, \ 80, \ 94, \ 68, \ 19, \ 79, \ 69, \ 40, \ 91, \ 74, \ 2, \ 55$$

The median is of course the 50th percentile of the death age distribution. Since we know nothing in advance about whether the likely alternative is an older or a younger death age, the competing hypotheses can be formulated in two-sided fashion:

$$H_0 : \kappa_{.5} = 65 \quad \text{vs.} \quad H_a : \kappa_{.5} \neq 65.$$

The parameters are $n = 27$ and $p = 1/2$. To construct an $\alpha = .05$ level critical region, we would like to choose c_0 and c_1 such that

$$.025 \geq P[Y \leq c_0] = \sum_{k=0}^{c_0} \binom{27}{k} \left(\frac{1}{2}\right)^k \left(\frac{1}{2}\right)^{n-k} \tag{9.16}$$

and

$$.025 \geq P[Y \geq c_1] = \sum_{k=c_1}^{n} \binom{27}{k} \left(\frac{1}{2}\right)^k \left(\frac{1}{2}\right)^{n-k}. \tag{9.17}$$

Exact equality of these two probabilities to .025 may not be achievable because of the discrete nature of the binomial distribution. The best we can do is find values of the critical numbers c_0 and c_1 such that the probabilities $P[Y \leq c_0]$ and $P[Y \geq c_1]$ are close to, but do not exceed, .025. From *Minitab*'s cdf command, we find that

$$P[Y \leq 7] \approx .01, \qquad P[Y \geq 20] \approx .01,$$

so the critical region can be designed to reject H_0 if the number Y of negative differences between death ages X_i and 65 is ≤ 7 or ≥ 20. From the given data,

$y = 7$, so the conclusion is that H_0 is rejected at level .02. For a study of the power of the sign test, do Exercise 17.

?Question 9.1.6   How might you use a normal approximation to find c_0 and c_1 in the last example?

EXERCISES 9.1

1. For both the computer disk example and the herbicide example that are listed at the start of the section, propose a probabilistic model and appropriate hypotheses that should be tested in order to test the claim.

2. For the Illinois Pick 3 Lottery data in Table 7.9, the digit 5 was observed 26 times on the first selection of the 363 plays tallied. Test at the .05 significance level whether 5 is less likely than it would be under randomness assumptions. Approximate the type II error probability of the test if the true probability of selecting a 5 is .06.

3. A civic researcher is investigating the distribution of household sizes in her community. She suspects that there are no more than 45% of all households that have two or fewer members. Twenty-five households are selected at random, among which are fourteen that have two or fewer members.
   a. Is the researcher's null hypothesis rejected at level .05? What is the approximate p-value of this test? (Use a normal approximation in this part.)
   b. Use the binomial distribution to devise an exact critical region of level close to, but no more than, .05. What is the power of this test if the true proportion of households with two or fewer members is 55%?

4. A 1994 study (Douthat, 1995) mentioned previously in Exercise 2 of Section 7.1 reported that among 2086 male clergy responding to a questionnaire, 20% had been divorced at least once. Previous 1985 census data indicated that nationwide about 22% of male clergy had been divorced. Test at the .01 level the hypothesis that the divorce rate has decreased in the nine years between the two studies. What is the approximate p-value of this test?

5. A recent article (Fulkerson, 1995) gave statistics on the value of charitable donations. Among 1200 adults surveyed, 80% had contributed less than $500 over the course of a year. Test at level .05 the null hypothesis that the proportion who give under $500 is 75% against the alternative that this proportion exceeds 75% using this data. Find the approximate type II error probability if the proportion is 78%.

6. Argue that for a test of the Bernoulli parameter $H_0 : p = p_0$ vs. $H_1 : p > p_0$ using a fixed random sample $X_1, X_2, \ldots, X_n$, as α decreases β must increase for any alternative value of p.

7. Residents of an unincorporated area are voting on whether they wish to be annexed by a neighboring village. A simple majority will pass the annexation. In a poll of 50 of these residents, 28 favor annexation. Test

the hypothesis that the area will be annexed at significance level .05. Write an expression for the approximate power function of this test, and use a computer graphing program to graph it.

8. Use a normal approximation to verify that the maximum in formula (9.6) occurs at $p = p_0$, for fixed values of p_0, n, and $c \in (p_0, 1]$.

9. In this exercise we relate a large sample confidence interval for the Bernoulli parameter p to the hypothesis test: $H_0 : p = p_0$ vs. $H_a : p \neq p_0$. Show that a $(1 - \alpha) \times 100\%$ confidence interval for p is precisely the set of all hypothesized p_0 values for which we would accept H_0 at level α. [Use the form (8.58) of the confidence interval, in which the p in the variance term is not replaced by its estimate $\hat{p}$.]

10. This exercise previews the topic of the next section: one sample hypothesis tests for the mean. Let $X_1, X_2, \ldots, X_{25}$ be a random sample from the $N(\mu, \sigma^2)$ distribution, where we assume that σ^2 is known to be 4. Devise a critical region of level .01 for the test $H_0 : \mu = 0$ vs. $H_1 : \mu > 0$. Find the power function of the test, and use a computer graphing program to sketch it.

11. Suppose that X_1, X_2, X_3 is a random sample of size 3 from the uniform $(0, \theta)$ distribution, and we are interested in the hypotheses: $H_0 : \theta = 1$ vs. $H_a : \theta < 1$. A reasonable critical region is $Y_3 < .6$, where Y_3 is the largest order statistic of the sample. Find the significance level of this test and the probability of type II error when $\theta = .8$.

12. Consider the computer disk example at the start of the section, and recall your response to Exercise 1. If $n = 5$ disks are inspected, devise an exact critical region of level no more than .10 for the test. What is the power of the test at 1 bad spot per 240 sectors? at 1.5? at 2?

13. Discuss how you would perform a large sample approximate test of $\lambda = \lambda_0$ vs. $\lambda \neq \lambda_0$ based on a random sample $X_1, X_2, \ldots, X_n$ from the Poisson(λ) distribution.

14. Recall the data on ages at death for 20 people in Peoria (Exercise 5 in Section 7.2). Test at a level not exceeding .10 the hypotheses $H_0 : \kappa_{25} = 35$ vs. $H_1 : \kappa_{25} > 35$, where κ_{25} is the 25th percentile of the death age distribution. What is the p-value of this test?

15. Following is part of the data set on ozone concentrations mentioned in Exercise 14 of Section 3.3, for a time period beginning in early August 1990. If m is the median of the concentration distribution, use the sign test to test $H_0 : m = 50$ vs. $H_1 : m > 50$. Give the p-value of the test.

$$27, \ 46, \ 47, \ 88, \ 44, \ 52, \ 46, \ 70, \ 69, \ 77,$$
$$41, \ 43, \ 75, \ 70, \ 37, \ 55, \ 56, \ 62, \ 28, \ 58$$

16. Use a normal approximation to the exact distribution of the sign test statistic to perform the following test at approximate level .05. The following data are 1989 average market prices (dollars per bushel) for rye in several states (USDA, 1991). Does the 75th percentile of the price distribution appear to be $3 per bushel, or less than that?

$$1.65, \ 2.85, \ 2.26, \ 2.10, \ 2.10, \ 1.80, \ 1.70, \ 1.71, \ 2.85, \ 1.70, \ 1.66,$$
$$2.80, \ 1.90, \ 3.35, \ 2.55, \ 2.90, \ 1.67, \ 3.50, \ 2.00, \ 1.90, \ 3.25, \ 2.05,$$
$$1.70, \ 2.05, \ 2.40, \ 2.15$$

17. Consider a sign test of the hypotheses $H_0 : \kappa_p = \kappa_{p0}$ vs. $H_1 : \kappa_p < \kappa_{p0}$. Devise the general form of the rejection region and the distribution of the test statistic. Write an expression for the power function of the test. If $p = 1/2$, $n = 8$, $\kappa_{p0} = 3$, the test is to have level no more than .05, and the underlying distribution being sampled from is exponential(λ), then simplify the expression for the power function as much as possible. Compute its value when the true $\kappa_{.5} = 1$.

9.2 | One-Sample Location Tests

In the last section you learned about the fundamentals of hypothesis testing by way of tests for proportions. Now we will concentrate on tests for the horizontal location of a distribution, which could be measured by its mean or its median. You had a chance to see two examples of location tests in Section 9.1. If you did Exercise 10 there, you constructed a test of $H_0 : \mu = 0$ vs. $H_1 : \mu > 0$, where μ was the mean of the $N(\mu, \sigma^2)$ distribution and σ^2 was known. Since the value of μ determines the center of symmetry of the normal distribution, this test is a test for the horizontal location of that distribution. You also saw the sign test, which permits you to test hypotheses about the median or 50th percentile of a continuous distribution. The purpose of this section is to solidify and expand upon the idea of hypothesis testing for the normal mean and to refine the sign test into a new nonparametric procedure called the Wilcoxon test for testing hypotheses about the median of a continuous, symmetric distribution.

—— **?Question 9.2.1**   When are the mean and median of a distribution the same?

9.2.1   Tests for the Mean

Suppose that we would like to perform one of three hypothesis tests,

$$(1) \ H_0 : \mu = \mu_0 \text{ vs. } H_1 : \mu \neq \mu_0,$$
$$(2) \ H_0 : \mu = \mu_0 \text{ vs. } H_2 : \mu > \mu_0,$$
$$(3) \ H_0 : \mu = \mu_0 \text{ vs. } H_3 : \mu < \mu_0,$$

using a random sample $X_1, X_2, \ldots, X_n$ from $N(\mu, \sigma^2)$ with known σ^2. The sample mean $\overline{X}$ has the $N(\mu_0, \sigma^2/n)$ distribution if H_0 is true, and so we can construct critical regions based on $\overline{X}$ that are intuitively reasonable for the two-sided alternative H_1, or either of the one-sided alternatives H_2 or H_3.

To illustrate, consider H_1. We should reject H_0 if $\overline{X}$ is too far from μ_0 in either direction; in other words the critical region should have two pieces of the form

$$\overline{X} > c_1 \quad \text{or} \quad \overline{X} < c_2. \tag{9.18}$$

For a level α test we have

$$\alpha = P[\overline{X} > c_1 \text{ or } \overline{X} < c_2; \ \mu = \mu_0]$$

$$= P\left[Z = \frac{\overline{X} - \mu_0}{\sigma/\sqrt{n}} > \frac{c_1 - \mu_0}{\sigma/\sqrt{n}}\right] + P\left[Z < \frac{c_2 - \mu_0}{\sigma/\sqrt{n}}\right], \qquad (9.19)$$

hence

$$\frac{c_1 - \mu_0}{\sigma/\sqrt{n}} = z_{\alpha/2} \quad \text{and} \quad \frac{c_2 - \mu_0}{\sigma/\sqrt{n}} = -z_{\alpha/2}$$

$$\Rightarrow c_1 = \mu_0 + z_{\alpha/2} \cdot \frac{\sigma}{\sqrt{n}} \quad \text{and} \quad c_2 = \mu_0 - z_{\alpha/2} \cdot \frac{\sigma}{\sqrt{n}}. \qquad (9.20)$$

Notice that the critical values c_1 and c_2 are symmetric about μ_0. The critical distance that $\overline{X}$ must be from μ_0 in order to reject H_0 is $z_{\alpha/2} \cdot \sigma/\sqrt{n}$ for the two-sided alternative H_1.

Look at the last statement carefully. Because $H_0 : \mu = \mu_0$ is accepted if and only if μ_0 and $\overline{X}$ are within the critical distance of each other, the null hypothesis is accepted at level α for precisely those μ_0 that lie in a $(1 - \alpha) \times 100\%$ confidence interval for μ. As we continue our study of hypothesis testing, you should continue to note this connection between two-sided tests and confidence intervals. See Exercise 8 for a note on one-sided tests and confidence intervals.

The test of H_0 vs. H_1 can be performed by carrying out the computation of c_1 and c_2 in (9.20) and comparing them to the observed sample mean $\overline{x}$. If either inequality in (9.18) is true, then H_0 is rejected. Equivalently, you can compute the observed value of Z:

$$z = \frac{\overline{x} - \mu_0}{\sigma/\sqrt{n}}. \qquad (9.21)$$

If $z > z_{\alpha/2}$ or $z < -z_{\alpha/2}$, then H_0 is rejected. You can see easily that if n is large, then the Central Limit Theorem produces an approximate level α test identical to the preceding one even if the underlying distribution is not normal.

?Question 9.2.2 What form do the critical regions have for alternatives H_2 and H_3? Write them in terms of both $\overline{x}$ and z.

Example 9.2.1 Exercise 15 in Section 7.1 contained a data set on faculty salaries (see Table 7.3). Full professors at several universities in Illinois had the following average salaries:

$$41{,}200, \ 43{,}100, \ 42{,}600, \ 44{,}200, \ 50{,}700, \ 53{,}000,$$
$$45{,}700, \ 51{,}400, \ 48{,}600, \ 60{,}900, \ 63{,}700$$

Suppose that an advocacy group knows that the national average salary for full professors is 55,200, and the standard deviation is 7000. Does this group have a case on statistical grounds that full professors are underpaid in Illinois?

You can compute that the sample average for the $n = 11$ data points is 49,554, a somewhat lower figure than the national average. To see whether it is significantly so, we test

$$H_0 : \mu = 55,200 \quad \text{vs.} \quad H_3 : \mu < 55,200,$$

where μ is the average Illinois salary. Assuming that the salary distribution is normal, we reject H_0 at level .05 if

$$Z = \frac{\overline{X} - \mu_0}{\sigma/\sqrt{n}} < -z_{.05} = -1.645.$$

The observed z-value is

$$z = \frac{49,554 - 55,200}{7000/\sqrt{11}} = -2.67.$$

Since the z-value is less than the critical value -1.645, we reject H_0. In terms of the original problem context, the advocacy group can claim that the sample average of 49,554 is unlikely to be so far below the national mean if Illinois' average is the same as the national average. They can try to use this as evidence that Illinois professors are being shortchanged.

The p-value of the test is useful here. Recall that the p-value is the probability that a type I error is made by barely rejecting H_0 based on the observed data. Therefore,

$$p\text{-value} = P[\overline{X} < 49,554; \ \mu = 55,200] = P[Z < -2.67] = .0038.$$

Thus we can reject H_0 at a level as small as .0038, which is impressive evidence against the null hypothesis. Notice how the p-value expresses well the gist of the advocacy group's argument. It is only about .4% likely for the Illinois sample mean to be 49,554 or less if Illinois is truly comparable to the national norms.

Equations (9.18) and (9.20) allow us to study the power of the normal mean test. For the two-sided test H_0 vs. H_1, Exercise 3(a) asks you to show that

$$K(\mu) = \Phi\left(\frac{c_2 - \mu}{\sigma/\sqrt{n}}\right) - \Phi\left(\frac{c_1 - \mu}{\sigma/\sqrt{n}}\right) + 1, \qquad (9.22)$$

where Φ is the standard normal c.d.f. For example, if $n = 25$, $\mu_0 = 0$, and $\sigma = 3$, the power function of a level .05 test is sketched in Fig. 9.3. It achieves its minimum value of .05 at $\mu = 0$ (why?) and takes on values near 1 at a distance of about 2 from 0. It is also symmetric about 0 (see Exercise 3b).

Before we leave the subject of the normal mean test with known variance, a graphical interpretation may be useful to strengthen your understanding of the concepts. Consider $H_0 : \mu = \mu_0$ vs. $H_2 : \mu > \mu_0$. Now $\overline{X}$ has the $N(\mu, \sigma^2/n)$ distribution. The question is, Is μ equal to μ_0 or to something else larger than μ_0? The choice that we are making is between a particular normal density centered about μ_0, as on the left in Fig. 9.4, and some

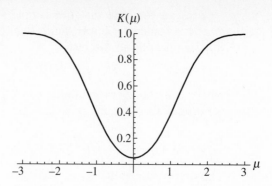

Figure 9.3 Power Function of the Normal Mean Test

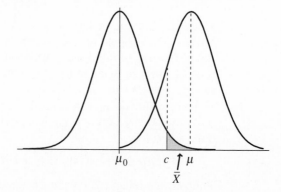

Figure 9.4 Geometric Meaning of Normal Mean Test

other normal density whose center is at $\mu > \mu_0$. A value of $\overline{X}$ such as the one shown, which is a good deal greater than μ_0, is not likely to happen if $\mu = \mu_0$. It seems likelier that $\overline{X}$ is an observation from the distribution on the right. Our decision rule picks a critical value c such that the area to its right under the μ_0 density is a small value α, the type I error probability. If $\overline{X}$ is greater than this c, it is "too far" from μ_0 for us to reasonably believe H_0; hence we reject H_0. By setting α small and hence c large, we are taking the point of view that H_0 is "innocent until proven guilty" by a very extreme value of the test statistic $\overline{X}$. To check your understanding, answer the next question.

—— **?Question 9.2.3** In Fig. 9.4, what area would represent the type II error probability for the alternative μ shown? What happens to this area as the density on the right moves closer to the density on the left?

Thus far we have made an assumption that is probably not realistic, that the variance σ^2 of the underlying normal distribution is known. When the variance is unknown, the most intuitively obvious thing to do is to replace

σ^2 in the test statistic Z by its unbiased estimate S^2. But then the probability law of the test statistic changes, and we produce a new test called the *t-test* for the mean.

Recall that, since $V = (n-1)S^2/\sigma^2$ has the $\chi^2(n-1)$ distribution and is independent of $U = (\overline{X} - \mu)/(\sigma/\sqrt{n})$,

$$T = \frac{U}{\sqrt{V/(n-1)}} = \frac{(\overline{X} - \mu)/(\sigma/\sqrt{n})}{\sqrt{S^2/\sigma^2}} = \frac{\overline{X} - \mu}{S/\sqrt{n}} \sim t(n-1). \tag{9.23}$$

Then, for example, a level α *t*-test of $H_0 : \mu = \mu_0$ vs. $H_3 : \mu < \mu_0$ could be constructed to reject H_0 if and only if

$$T = \frac{\overline{X} - \mu_0}{S/\sqrt{n}} < -t_\alpha(n-1), \tag{9.24}$$

where $t_\alpha(n-1)$ is the critical point for the $t(n-1)$ distribution such that α probability lies to its right. By symmetry, α probability lies to the left of $-t_\alpha(n-1)$. The critical region for $H_2 : \mu > \mu_0$ is analogous: $T > t_\alpha$. For the two-sided alternative $H_1 : \mu \neq \mu_0$, the critical region is two-sided: $T < -t_{\alpha/2}$ or $T > t_{\alpha/2}$.

Example 9.2.2

Exercise 16 in Section 7.2 estimated the investment performance of some municipal bonds. To be competitive with common stocks with 10% rates of return, these bonds need to give at least a 7% rate of return, in light of the extra tax savings. It would be important for an investor who is leaning toward purchasing these bonds to know whether, among the population of all such bonds that could be selected, the average is truly more than 7% or in fact is less. The data given in the exercise (from Forsyth, 1995) are reproduced here for convenience.

6.70, 6.86, 6.86, 6.93, 6.74, 7.09, 6.94, 6.83,
6.81, 7.05, 7.02, 6.88, 6.63, 6.74, 6.86, 6.97

If the investor's point of view is to give the bonds the benefit of the doubt unless strong evidence indicates otherwise, then the appropriate hypotheses to use are

$$H_0 : \mu \geq 7 \quad \text{vs.} \quad H_a : \mu < 7.$$

Since $n = 16$, the degrees of freedom parameter for the *t* critical value is 15. Using significance level .1 for example, $t_{.1}(15) = 1.34$. The rejection region for these hypotheses would then be

$$T = \frac{\overline{X} - 7}{S/\sqrt{n}} < -t_{.1}(15) = -1.34.$$

You can compute that $\overline{x} = 6.87$ and $s = .13$. Thus the observed *t*-value is

$$t = \frac{6.87 - 7}{.13/\sqrt{16}} = -4.0 < -1.34.$$

The conclusion of the test is to reject H_0 at level .1 in favor of H_a, which implies that the bonds are on the average not a competitive investment with

Table 9.1

Comparison of Normal and t-Critical Values

	Level		
	.1	*.05*	*.025*
Normal	1.282	1.645	1.960
$t(29)$	1.311	1.699	2.045
$t(49)$	1.299	1.677	2.010
$t(99)$	1.290	1.660	1.984

common stocks. Actually, since the computed t-value was so extreme, we have very strong evidence of a difference from 7. The p-value of the test is around .0006, which I obtained using *Mathematica's CDF* command to evaluate $P[T < -4.0]$.

One last comment on the t-test for the mean: When the sample size n is large, t critical values are very close to standard normal critical values. Table 9.1 shows normal critical values z_α at levels $\alpha = .1, .05,$ and $.025,$ and t-critical values at the same levels for 29 degrees of freedom. Even for samples as small as 30 t and normal critical values are similar, and for sample sizes of at least 100 they are nearly equal. These results are not surprising, since the t-distribution converges to the standard normal distribution as the degrees of freedom approaches ∞. Even better, an extended Central Limit Theorem shows that the distribution of T converges to $N(0, 1)$ even when the underlying distribution from which the sample is taken is nonnormal (see Hogg and Craig, 1978, p. 198). So for large samples we can confidently use T as a test statistic and obtain critical values from the $N(0, 1)$ table.

9.2.2 Wilcoxon Test

When we studied the sign test in the last section, it may have struck you that the test did not make full use of the data. In a sign test for a median $H_0 : \kappa_{.5} \equiv m = m_0$ vs. $H_1 : m \neq m_0$, only the signs of the differences $X_i - m_0$ are taken into account, not their magnitudes. For example, in a test of $H_0 : m = 3$ vs. $H_1 : m \neq 3$, using the data set

$$1, \quad 2, \quad 10, \quad 11, \quad 15$$

the sign test would not notice that several values in the sample are very much larger than 3. The fact that two sample values among five are less than 3 would lead to acceptance of H_0 at any reasonable level. This example makes us suspect that the power of the sign test—its ability to reject H_0 when necessary—might not be the best possible.

The problem before us is to test the null hypothesis $H_0 : m = m_0$, where m is the median of a distribution that we assume is continuous and symmetric about m. This includes such cases as the uniform, normal, t-, and Cauchy distributions, as well as others. In the nonparametric test known as the *Wilcoxon test*, we sort the absolute differences

$$|X_1 - m_0|, |X_2 - m_0|, \ldots, |X_n - m_0|$$

into increasing order, keeping track of which differences are negative and which are positive. The smallest $|X_j - m_0|$ is given rank 1, the next smallest rank 2, and so on, and the largest absolute difference is given rank n. Let R_j be the rank of $|X_j - m_0|$ in this list, and let $I_j = +1$ or -1, respectively, according to whether the difference $X_j - m_0$ is positive or negative. The *Wilcoxon test statistic* is the sum of the signed ranks:

$$W = \sum_{j=1}^{n} I_j \cdot R_j. \tag{9.25}$$

Large negative values of W indicate that many of the data values were much less than the hypothesized m_0, because the large ranks contribute to W with negative sign. Similarly large positive values of W suggest that many of the data values were much larger than m_0. In the former case, we have evidence that the true median is less than m_0, and in the latter, the true median is larger than m_0. The forms of critical regions for one and two-sided alternatives follow naturally from this observation:

Alternative	Critical Region
$m > m_0$	$W > c$
$m < m_0$	$W < -c$
$m \neq m_0$	$W < -d$ or $W > d$

In this little example, the ranked differences would be as in Table 9.2. The observed value of W is then

$$w = (-1)(2) + (-1)(1) + (1)(3) + (1)(4) + (1)(5) = 9,$$

which is the dot product of the sign indicator vector in the last row of the table and the rank vector in the third row. The idea is to produce a sum of indicators weighted by their ranks, in order to take more advantage of the magnitude of the data values than the sign test does. In essence, the sign test does not weight the sign indicators at all. The question remains, Is this value of W particularly unlikely if the null hypothesis is true?

Table 9.2

Computation of W

$X_j - m_0$	-2	-1	7	8	12
$|X_j - m_0|$	2	1	7	8	12
R_j	2	1	3	4	5
I_j	-1	-1	1	1	1

—— **?Question 9.2.4**   Could two of the absolute differences $|X_i - m_0|$ and $|X_j - m_0|$ be tied, theoretically? In practice, if data are recorded only to fixed accuracy, could they be tied? (The usual convention in the case of ties is to give each tied absolute difference the average of the two ranks for which they are competing.)

To find an upper critical value w_α such that $P[W > w_\alpha; H_0] = \alpha$, or a lower critical value u_α such that $P[W < u_\alpha; H_0] = \alpha$, we must be able to compute the distribution of W under H_0. Unfortunately there is no closed form for it, although it is tabulated in various texts and is available on statistical computer packages. This distribution can be enumerated by brute force, as we will see in the next example, by using the fact that if H_0 is true, then the jth ranked absolute difference $|X_j - m_0|$ is as likely to be a positive as a negative signed difference, independently of whether any of the other differences was positive or negative. This follows from the symmetry of the distribution about its median and from the independence of the sample values. Then W has the same distribution under H_0 as another random variable defined by

$$W' = \sum_{j=1}^{n} S_j \cdot j, \tag{9.26}$$

where the random variables S_j are independent and take on the two possible values ± 1 each with probability $1/2$. Since this is true regardless of the underlying distribution being sampled from, the Wilcoxon test falls into the category of nonparametric techniques.

Example 9.2.3

Four of the biggest home purchasing months tend to be June, July, August, and September. Data taken in 1992 by the Peoria Area Association of Realtors showed that the median among these four months of the average dollar value per transaction for the month was \$72,283. In 1993 the four monthly average values per transaction were \$77,372, \$80,104, \$78,490, and \$77,783. To shed light on the question of whether housing prices are going up, let us ask, What is the p-value of the Wilcoxon test of $H_0 : m = \$72,283$ vs. $H_2 : m > \$72,283$, where m is the theoretical median value per transaction in 1993?

Remember that the distribution of W under H_0 is the same as that of W' in (9.26). We can just enumerate all 16 possible combinations of signs S_j for $j = 1$ to 4, and the w' values they generate, in order to list out this distribution.

This listing is shown in Table 9.3. For instance, when each of the S_j's is positive, $w' = (1)(1) + (1)(2) + (1)(3) + (1)(4) = 10$, and this is the only combination among the 16 that yields a value of $w' = 10$; hence $P[W' = 10] = P[W = 10] = 1/16$. Note that ± 4, ± 2, and 0 each occur as w' values in two ways. Therefore the probability mass function of W is

$$g(w) = P[W = w] = \begin{cases} 1/16 & \text{if } w = 10, 8, 6, -6, -8, -10, \\ 2/16 & \text{if } w = 4, 2, 0, -2, -4, \\ 0 & \text{otherwise.} \end{cases} \tag{9.27}$$

Table 9.3

Distribution of W'

S_1	S_2	S_3	S_4	w'
1	1	1	1	10
1	1	1	−1	2
1	1	−1	1	4
1	1	−1	−1	−4
1	−1	1	1	6
1	−1	1	−1	−2
1	−1	−1	1	0
1	−1	−1	−1	−8
−1	1	1	1	8
−1	1	1	−1	0
−1	1	−1	1	2
−1	1	−1	−1	−6
−1	−1	1	1	4
−1	−1	1	−1	−4
−1	−1	−1	1	−2
−1	−1	−1	−1	−10

Notice that this distribution is symmetric about 0. (In Exercise 18 you are asked to show that this is true in general.) For the given data, all of the four observations are greater than the hypothesized median \$72,283. Hence the sample value of the test statistic is $w = 10$, just as above. For the one-sided alternative specified, the appropriate critical region would be of the form $W > c$. Therefore, the p-value of the test is

$$p\text{-value} = P[W > 9;\ H_0\ \text{true}] = P[W = 10;\ H_0\ \text{true}] = 1/16 = .0625.$$

In other words, such data are little more than 6% likely if the median value for 1993 home purchases is \$72,283, which suggests that there has indeed been an increase in value over the year.

?Question 9.2.5 In the little data set 1,2,10,11,15 at the start of the section, the value of the Wilcoxon statistic was 9. Using the technique of the last example, but enumerating only the largest possible values of the statistic, find the probability under H_0 that $W \geq 9$.

Although the random variables $j \cdot S_j$ being added in (9.26) are independent under H_0, they are not identically distributed. (Why not?) However, a stronger version of the Central Limit Theorem (see Parzen, 1960, p. 430) yields that for large sample size n,

$$\frac{W' - E[W']}{\sigma(W')} \approx N(0,1). \tag{9.28}$$

Since W and W' have the same distribution under H_0, the approximation in (9.28) holds for W as well.

Putting large sample approximate Wilcoxon procedures into practice requires that we find $E[W']$ and $\sigma(W')$, which are the same as the mean and standard deviation of W. The mean is simple:

$$E[W'] = E\left[\sum_{j=1}^{n} S_j \cdot j\right] = \sum_{j=1}^{n} E[S_j] \cdot j = \sum_{j=1}^{n} \left(\tfrac{1}{2} \cdot 1 + \tfrac{1}{2} \cdot (-1)\right) \cdot j = 0. \tag{9.29}$$

As a result, the variance is the expected square of W', which is computed as follows:

$$
\begin{aligned}
\text{Var}(W') = E[(W')^2] &= E\left[\left(\sum_{j=1}^{n} S_j \cdot j\right)\left(\sum_{k=1}^{n} S_k \cdot k\right)\right] \\
&= E\left[\sum_{j=1}^{n} S_j^2 \cdot j^2\right] + E\left[\sum\sum_{j \neq k} S_j \cdot S_k \cdot j \cdot k\right] \\
&= \sum_{j=1}^{n} 1 \cdot j^2 + 0 = \frac{n(n+1)(2n+1)}{6}.
\end{aligned}
\tag{9.30}
$$

[Be sure that you understand why the second and third lines of (9.30) are true.] Therefore the random variable

$$Z = \frac{W}{\sqrt{n(n+1)(2n+1)/6}} \approx N(0,1) \tag{9.31}$$

under $H_0 : m = m_0$ for large n.

Example 9.2.4

My student Chris Najim, who was studying portfolio theory, tracked 30 weekly rates of return (change in weekly value/value at start of week) for several stocks, including Sears. The Sears data (in %) is presented, sorted in order of increasing magnitude for our convenience.

0.00, −0.28, −0.35, 0.63, 0.68, −0.70, 0.92, −1.14, −1.88, −1.91, 2.03, −2.23, −2.36, 2.38, 2.60, 2.85, −3.31, −3.53, −3.58, −3.77, 4.03, −4.22, −4.23, 4.50, 4.78, −4.98, 5.00, 5.18, 5.76, 8.03

Does the stock seem to be growing in value?

The question can be translated into an hypothesis test about the median of the distribution of the rate of return random variable. Either the median rate of return is positive if Sears stock is growing in value, or it is 0 if the stock is not growing. Thus we will test $H_0 : m = 0$ vs. $H_a : m > 0$ using a large sample Wilcoxon test of level, say, .1.

The null hypothesis would be rejected if Z in (9.31) exceeds $z_{.1} = 1.28$. The observed value of W can be found by adding the ranks of the data in the list times the sign of the elements:

$$w = (1)(1) + (-1)(2) + (-1)(3) + \cdots + (1)(29) + (1)(30) = 49.$$

Since $n = 30$, the observed value of Z is $z = 49/\sqrt{30(31)(61)/6} = .50$. Thus H_0 is easily accepted; there is no statistical evidence that Sears has a positive median rate of return.

EXERCISES 9.2

1. A report by the Pakistani government (Government of Pakistan, 1982) gave the following figures on average yearly earnings (rupees) in 1977 by factory workers in various industries. In 1976, the average for all industries was 4660.71. Is there strong evidence that there has been an increase in average earnings by factory workers? Use (a) a normal test that assumes that the standard deviation is the same as the sample standard deviation for the same industries in previous year, which was 1202.72; (b) a t-test. Do you reach different conclusions? Is there a circumstance in which you might reach different conclusions?

Textile	5283.96	*Paper and Printing*	5837.13
Cotton	7028.48	*Wood, Stone, and Glass*	5913.30
Engineering	6807.07	*Skin and Hides*	6275.54
Mineral Metals	4842.37	*Mints*	6870.40
Chemical and Dyes	7216.71		

2. Find the p-value of the test $H_0 : \mu = 0$ vs. $H_3 : \mu < 0$ based on the data $-1, 2, -1.5, -2.5, -3, -2.75, -2, -1.25$, assuming that the data are a random sample from the normal distribution with standard deviation 1.2.

3. a. Verify formula (9.22) for the power function of the two-sided test of $H_0 : \mu = \mu_0$.
   b. Show that the power function in formula (9.22) is symmetric about μ_0.

4. Derive the power function $K(\mu)$ for a normal mean test $H_0 : \mu = \mu_0$ vs. $H_2 : \mu > \mu_0$ of level α. Assume that the variance σ^2 is known.

5. A consumer group is attempting to gather evidence for a legal case against a cigarette manufacturer who claims in an advertisement that a brand that it markets contains no more than 14 mg of tar per cigarette. The group pretests a small sample of cigarettes and finds a standard deviation of about .8 mg. How many cigarettes should the consumer group test in order to have no more than a .01 probability of falsely accusing the company, but have at least a 90% probability of making its case if it believes that the true content is 14.5 mg?

6. Show that for any alternative μ in the two-sided test of $H_0 : \mu = \mu_0$ in the case of known variance, the power at μ approaches 1 as the sample size n approaches ∞.

7. It has been said that the probability that a newborn baby is a boy is actually not exactly .5. Conservatively, this probability is closer to .512, and it could be more. The following data (U.S. Dept. of Health and Human Services, 1991) are 1970–1991 estimates of the numbers of male births per thousand female births in the United States. Formulate appropriate hypotheses and use a t-test to test the hypotheses at level .025.

1055, 1052, 1051, 1052, 1055, 1054, 1053, 1053, 1053, 1052, 1053, 1052, 1051, 1052, 1050, 1052, 1051, 1050, 1050, 1050, 1050, 1046

8. A *confidence ray* of type $[A, \infty)$ and level $1 - \alpha$ for a parameter θ is a random interval such that

$$1 - \alpha = P[\theta \in [A, \infty)] = P[\theta \geq A].$$

A similar definition can be given for a confidence ray of type $(-\infty, A]$.
a. Develop a confidence ray of the form $[A, \infty)$ for the normal mean based on a random sample of size n, assuming known variance.
b. Show that a number μ_0 belongs to this confidence ray if and only if a one-sided test of $H_0 : \mu = \mu_0$ is accepted.

9. (From Milton and Arnold, 1986, p. 233) Automotive engineers are considering a design for a certain model of car that utilizes more aluminum than the present design does, in order to reduce the weight of the car and consequently to improve the gas mileage. The current design averages 26 mpg in highway driving tests. Thirty-six cars with the extra aluminum design are driven, and the following are the observed mpg figures:

33.8, 24.3, 18.8, 23.7, 25.3, 29.6, 24.9, 31.5, 34.4, 28.0, 20.5, 36.7, 30.3, 33.5, 27.4, 27.6, 22.5, 30.7, 28.6, 27.1, 28.8, 16.5, 32.7, 25.2, 33.1, 37.5, 25.1, 34.5, 29.5, 26.8, 30.0, 28.4, 25.6, 19.8, 28.9, 27.7

Bearing in mind that, because full implementation of a new design is costly, the engineers should be conservative about claims of improved mileage, set up and execute an appropriate hypothesis test. Use graphics to decide whether a normality assumption about gas mileage is reasonable.

10. Following are measurements of concentrations of pollutants (mg/l) in certain rivers in India, taken during the years 1984–1988 (United Nations, 1994). Test at level .05 the hypotheses $H_0 : \mu \leq 2.6$ vs. $H_a : \mu > 2.6$. Use the t-table to estimate the p-value of the test. Produce a normal quantile plot to check the fit to normality of the data.

6.52, 2.21, 1.22, 3.89, 1.08, 1.32, 5.57, 1.14, 3.50, 2.40, 2.95, 3.24, 3.67, 2.28, 6.55, 1.89, 3.11, 1.90, 2.01, 1.33, 2.06

11. Using the Sears rate of return data in Example 9.2.4 and a t-test, check whether Sears rate of return is significantly different from 0. Does the normal theory test reach the same conclusion as the Wilcoxon test?

12. What is the critical distance that $\overline{X}$ must be from an hypothesized μ_0 in order to reject $H_0 : \mu = \mu_0$ at level α in favor of $H_3 : \mu < \mu_0$? Assume that the distribution being sampled from is $N(\mu, \sigma^2)$ with unknown σ^2. What factors does this distance depend on? Discuss how changes to these factors affect the critical distance.

13. State and prove a theorem about the relation of confidence intervals to hypothesis tests for the case of the normal mean with unknown variance.

14. As in Exercise 10, pollution measurements were taken during 1988–1992 in some Japanese rivers (United Nations, 1994). The concentrations follow. By enumerating only the most extreme few outcomes, devise a small sample Wilcoxon test of level no more than .01 for the hypotheses $H_0 : m = 2$ vs. $H_a : m < 2$. Does the test accept or reject H_0 using these data?

$$.73, \ 1.40, \ 2.56, \ 0.83, \ 1.66, \ 1.71, \ 1.39, \ 0.52, \ 0.65$$

15. Find the complete distribution under H_0 of the Wilcoxon statistic W for sample size $n = 5$. Use this to execute a two-sided test of level no more than .2 of $H_0 : m = 6$ vs. $H_1 : m \neq 6$ for the data set $-1, 3, 10, 5, 12$.

16. Following are percentages of the voting population who did vote in the 1984 presidential elections in several small cities in the states of Georgia, Alabama, and Tennessee (Thomas, 1992). Use a large sample Wilcoxon test to test at level .05 the hypotheses $H_0 : m = 33.3$ vs. $H_a : m < 33.3$. (Average the ranks of tied observations to compute the value of W.)

$$30.6, \ 26.2, \ 24.9, \ 25.8, \ 14.3, \ 27.0, \ 30.9, \ 22.5, \ 27.0, \ 32.4,$$
$$36.0, \ 37.6, \ 39.3, \ 30.1, \ 31.1, \ 29.7, \ 32.4, \ 30.5, \ 32.6$$

17. Use the data on Calculus I grade point averages in Exercise 18 of Section 7.1 and the large sample Wilcoxon procedure to test at level .05 whether the median GPA is 2.0 or something larger than 2.0. (Average the ranks of tied observations and delete the data value identically equal to 2.0 in computing the observed value of W.) Does a t-test give the same conclusion?

18. Show that the distribution of the Wilcoxon test statistic W must be symmetric about 0 under the null hypothesis.

19. A moment-generating function approach can be used to find the small sample distribution of the Wilcoxon test statistic W under H_0. First recall that for a discrete random variable X taking on values $a_1, a_2, \ldots, a_n$ with probabilities $p_1, p_2, \ldots, p_n$, respectively, the m.g.f. of X would be

$$M_x(t) = E[e^{tX}] = \sum_{i=1}^{n} e^{ta_i} \cdot p_i.$$

   a. Find the m.g.f. of the random variable $j \cdot S_j$ in (9.26).
   b. Write a general expression for the m.g.f. of W' in (9.26).
   c. Apply your expression to the case where $n = 3$ to find the m.g.f. of W' and, from it, the probability mass function of W.

9.3 | Tests for Dispersion

Another distributional parameter of concern in testing hypotheses is the variance, a measure of the spread or dispersion of the distribution. For example, in quality control problems a process may be operating correctly if the variation in its output does not exceed some tolerable level. Or, there may be competitive versions of a process, the less variable of which may be the more effective. We will also see later that there are comparative statistical procedures whose applicability depends on equality of variance of two or more populations. For all of these reasons, the study of tests for dispersion is worthwhile. Here we will cover the classical normal theory tests for a single variance and for a ratio of variances, and we also introduce a nonparametric test for comparing variability of two populations.

9.3.1 Normal Variance Tests

To begin, recall from Section 5.4 that the random variable $(n-1)S^2/\sigma^2$ has the $\chi^2(n-1)$ distribution, where S^2 is the sample variance of a random sample $X_1, X_2, \ldots, X_n$ from the $N(\mu, \sigma^2)$ distribution. For this reason, for normal populations it is easy to develop tests for the null hypothesis $H_0 : \sigma^2 = \sigma_0^2$ against one- and two-sided alternatives. For instance, if S^2 is significantly larger than σ_0^2, then one tends to favor the alternative $H_2 : \sigma^2 > \sigma_0^2$ over the null hypothesis. To judge what constitutes "significantly larger," note that if H_0 is true, then

$$P\left[U = \frac{(n-1)S^2}{\sigma_0^2} > \chi_\alpha^2(n-1)\right] = \alpha. \tag{9.32}$$

A rejection region of the form inside brackets in (9.32) therefore produces a test of level α for this one-sided alternative.

?Question 9.3.1 Write out in full the rejection regions suitable for level α tests of $H_0 : \sigma^2 = \sigma_0^2$ vs. $H_1 : \sigma^2 \neq \sigma_0^2$, and $H_3 : \sigma^2 < \sigma_0^2$.

Example 9.3.1 Silver Medal Company is interested in controlling the quality of its 5-lb sacks of flour. Not only must the fill weight of the sacks be 5 lb on the average, but also sacks must be of very uniform weight. Suppose that the uniformity requirement is that the standard deviation of weight be less than or equal to $1/20 = .05$ lb. A sample of 20 sacks is weighed, and the following weights are observed. Is there any evidence that the filling process is out of control?

$$5.008, \ 4.969, \ 5.037, \ 4.949, \ 5.036, \ 4.985, \ 4.992,$$
$$4.981, \ 4.986, \ 4.970, \ 5.064, \ 5.030, \ 4.998, \ 5.076,$$
$$5.067, \ 4.966, \ 4.841, \ 5.006, \ 5.049, \ 4.978$$

Since there are two aspects to controlled behavior here, we will test two hypotheses. Our probabilistic model is that $X_1, X_2, \ldots, X_{20}$ is a random sample from $N(\mu, \sigma^2)$. A histogram of these data, shown in Fig. 9.5, supports an assumption of normality, except for one unusually low outlying observation. We need to test hypotheses about the mean fill weight:

$$H_0 : \mu = 5 \quad \text{vs.} \quad H_1 : \mu \neq 5,$$

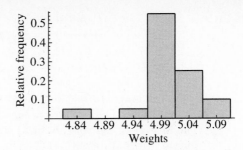

Figure 9.5   Histogram of Fill Weights

and the standard deviation of fill weight:

$$H_0 : \sigma \le .05 \text{ (i.e., } \sigma^2 \le .0025) \quad \text{vs.} \quad H_a : \sigma > .05 \text{ (i.e., } \sigma^2 > .0025).$$

If either hypothesis is rejected, then there is cause to believe that the process is not under control.

Treating the mean hypotheses first, the data yields $n = 20$, $\overline{X} = 4.9995$, and $S = .0525$. Hence the observed value of the T-statistic of the last section is

$$t = \frac{4.9995 - 5}{.0525/\sqrt{20}} = -.04.$$

This result is safely between $+t_{\alpha/2}$ and $-t_{\alpha/2}$ for any reasonably small level α, so we accept the null hypothesis that the filling machine is producing the proper average weight.

As for the variance hypotheses, the observed value of S is .0525, which is slightly larger than the desired .05. Is this slight excess statistically significant? The appropriate critical region for a level α test would be

$$\frac{(n-1)S^2}{\sigma_0^2} > \chi_\alpha^2(n-1). \tag{9.33}$$

For a level $\alpha = .05$ test, $\chi_{.05}^2(19) = 30.143$. We can compute the value of the test statistic as $19(.0525)^2/.05^2 = 20.9$, which does not exceed the critical value. Thus we again accept the null hypothesis that the filling process is in control.

?Question 9.3.2   Use statistical software to find the p-value of the test in Example 9.3.1. What intuitive evidence does this p-value give?

The type II error probability is not hard to find for the normal variance test. To see why, return to the setting of Example 9.3.1, and consider the quantity $\beta(.06) = P[\text{accept } H_0; \sigma = .06]$. By (9.33),

$$\beta(.06) = P\left[\frac{(n-1)S^2}{\sigma_0^2} \le \chi_\alpha^2(n-1); \sigma = .06\right]$$

$$= P[(n-1)S^2 \le (.05)^2 \cdot (30.143); \sigma = .06]$$

$$= P\left[\frac{(n-1)S^2}{(.06)^2} \le \frac{(.05)^2(30.143)}{(.06)^2}; \sigma = .06\right]$$

$$= P[V \le 20.93] = .66. \tag{9.34}$$

In the last line of (9.34) the transformed random variable $V = (n-1)S^2/(.06)^2$ has the $\chi^2(n-1)$ distribution, because its denominator is the correct variance. The answer of .66 was obtained from *Minitab*'s CDF command for the chi-square distribution with 19 degrees of freedom. The choice of a small value of α has made the critical region small enough that its complement, the acceptance region, has large probability under the alternative $\sigma = .06$. This has led to a very large type II error probability.

It is probably more common in applications of statistics to compare variances of two populations rather than to test hypotheses about a single population variance. There is sometimes a control group and a treated group, or two competing treated groups, whose dispersions may be different. Such differences can be detected by testing the null hypothesis that the ratio of variances σ_x^2/σ_y^2 of the two groups X and Y equals 1 against the usual three alternatives.

So we continue our study by developing a test of $H_0 : \sigma_x^2/\sigma_y^2 = 1$ vs. one- and two-sided alternatives, based on two independent random samples $X_1, X_2, \ldots, X_m$ and $Y_1, Y_2, \ldots, Y_n$ from $N(\mu_x, \sigma_x^2)$ and $N(\mu_y, \sigma_y^2)$ respectively. The approach is similar to the confidence interval problem for two variances. The following random variables have known distributions:

$$U = \frac{(m-1)S_x^2}{\sigma_x^2} \sim \chi^2(m-1), \qquad V = \frac{(n-1)S_y^2}{\sigma_y^2} \sim \chi^2(n-1). \tag{9.35}$$

Furthermore, since the samples are independent, U and V are independent. It is logical to base a test of σ_x^2/σ_y^2 on the ratio of unbiased estimators S_x^2/S_y^2; if the latter ratio is significantly greater than or significantly less than 1, there is evidence that σ_x^2/σ_y^2 is greater than or less than 1.

Formulas (9.35) allow us to create a test statistic that has a known distribution under the null hypothesis. If $H_0 : \sigma_x^2/\sigma_y^2 = 1$ (i.e., $\sigma_x^2 = \sigma_y^2$) is true, then

$$F = \frac{\dfrac{(m-1)S_x^2}{\sigma_x^2(m-1)}}{\dfrac{(n-1)S_y^2}{\sigma_y^2(n-1)}} = \frac{S_x^2}{S_y^2} \sim f(m-1, n-1). \tag{9.36}$$

A level α critical region for the two-sided alternative $H_1 : \sigma_x^2/\sigma_y^2 \ne 1$ (i.e., $\sigma_x^2 \ne \sigma_y^2$) is

$$F < f_{1-\alpha/2}(m-1, n-1) \quad \text{or} \quad F > f_{\alpha/2}(m-1, n-1), \tag{9.37}$$

where the critical points are points for the $F(m-1, n-1)$ distribution to the left and right of which lies probability $\alpha/2$, as shown in Fig. 9.6. One-sided alternatives have corresponding critical regions.

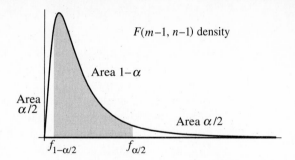

Figure 9.6   Two-Sided F Critical Values

Example 9.3.2

Average life spans of women in several South American and North American countries (including some island nations) are listed (United Nations, 1994).

South America

73, 55, 68, 75, 69, 68, 66, 69, 67, 71, 75, 73

North America

75, 72, 72, 76, 80, 77, 76, 68, 64, 66, 72, 59, 56, 66,
75, 74, 66, 76, 68, 74, 79, 71, 75, 72, 79

We will perform the F-test discussed earlier on the hypotheses $H_0 : \sigma_1^2 = \sigma_2^2$ vs. $H_a : \sigma_1^2 \neq \sigma_2^2$, where σ_1^2 and σ_2^2 are, respectively, the variances of the South and North American lifespans.

We must assume that the distributions of such life spans are normal. The sample standard deviations are $s_1 = 5.38$ and $s_2 = 6.10$. Therefore the test statistic has the value

$$f = \frac{s_1^2}{s_2^2} = \frac{5.38^2}{6.10^2} = .778.$$

Since $m = 12$ and $n = 25$, the appropriate critical values are taken from the $F(11, 24)$ distribution. For a test of level .1, for example,

$$f_{.95}(11, 24) = 1/f_{.05}(24, 11) = 1/2.61 = .38 \quad \text{and} \quad f_{.05}(11, 24) = 2.22.$$

Because the sample f-value is strictly between the two critical points, H_0 is accepted at level .1. Any preconceived notions you may have had about the effects of uniform economic conditions and availability of medical treatment in this setting are not supported by the data.

?Question 9.3.3

How would you go about computing type II error probabilities for the test of equality of two variances? Is it necessary to specify both alternative value σ_1^2 and σ_2^2?

9.3.2   Mood's Nonparametric Dispersion Test

As we have observed, nonparametric statistical methods make minimal assumptions about the distribution from which a sample is taken. When you are reasonably confident about the normality of the underlying distribution, you can employ the classical normal theory test procedures that we have studied. But it is also important to have nonparametric methods for dispersion testing to use when normality is questionable and/or sample sizes are not large. There are a number of similar techniques (see Gibbons, 1971, Ch. 10) based on ranks. One of these, called the *Mood test*, is described in the next paragraphs.

Suppose that there are samples $X_1, X_2, \ldots, X_m$ and $Y_1, Y_2, \ldots, Y_n$ from two populations whose dispersions are to be compared. Throughout the following, we will denote $N = m + n$. If we combine the samples and sort them, we obtain a sequence of N observations such as

$$\ldots Y X Y X Y X \ldots \quad \text{or} \quad \ldots Y Y X X X Y \ldots,$$

in which we keep track only of relative order and of whether the observations came from the X or the Y sample. We would like to create a statistic that would conclude from the first arrangement that the X and Y populations have the same dispersion and would conclude from the second arrangement that the Y's are more variable than the X's.

—— **?Question 9.3.4**   Suppose that the six observations just shown constitute the complete random sample of size $N = m + n = 6$. In each case find the sum of the differences between the X ranks and the average rank in the combined sample. What does your result say about the usefulness of this sum as a statistic for a dispersion test?

To answer this question, you must compute the average rank in the combined sample,

$$\frac{1 + 2 + \cdots + N}{N} = \frac{N(N+1)/2}{N} = \frac{N+1}{2},$$

which is $7/2$ when $N = 6$. The sum of deviations of X ranks from this average is, for the first arrangement, $(2 - 7/2) + (4 - 7/2) + (6 - 7/2) = 3/2$; and for the second arrangement, $(3 - 7/2) + (4 - 7/2) + (5 - 7/2) = 3/2$. Since this statistic is unable to distinguish between arrangements that show differences in variability, it is of no use.

However, our detour should motivate a better approach, especially if you remember the idea behind the sample variance. For a random sample $X_1, X_2, \ldots, X_n$, the sum of deviations from the sample mean $\sum_{i=1}^{n} (X_i - \overline{X})$ is always 0. (Be sure that you can show this.) This causes us to think of squaring those deviations before adding them. Analogously, define the *Mood statistic*,

$$M = \sum_{i=1}^{N} \left(i - \frac{N+1}{2} \right)^2 Z_i, \tag{9.38}$$

where $Z_i = 1$ or 0, respectively, according to whether the ith ranked member of the combined sample is an X or a Y. Thus M is the sum of the squared deviations of the X ranks from the average rank in the combined sample.

To see how M gives information about the comparative spread of the X and Y populations, we must first assume that the medians of these populations are the same. If, say, X has the smaller variation, then X's will tend to be ranked near the average rank in the combined sample. If Y has the smaller variation, then X's will have both very low and very high ranks, which are distant from the average rank. Therefore a small value of M indicates that X has smaller dispersion than Y, and a large value of M indicates that X has larger dispersion than Y.

To go further, we must now find the distribution of M under the null hypothesis that the X and Y samples actually came from the same distribution (as opposed to two distributions with common median and different spread). Under this H_0, any arrangement of m X symbols and n Y symbols is as likely as any other. The problem of computing the probability of the event that $M > m_1$ or $M < m_2$, where m_1 and m_2 are to be critical values for the test, reduces to the combinatorial problem of counting the number of arrangements of X's and Y's that yield such extreme M values, and dividing by the total number of possible arrangements $\binom{N}{m}$. (This reduction to a combinatorial problem is a typical situation in nonparametric testing.)

Unfortunately, this is difficult to do in general. But when n and m are small enough, we can enumerate the possibilities for M, as in the next example.

Example 9.3.3

Find the distribution of M under the null hypothesis that the X and Y samples come from the same distribution for the case $m = n = 3$. Use this result to construct a critical region of level no more than .1 for the alternative that the X population is more disperse than the Y population.

There are $\binom{3+3}{3} = \binom{6}{3} = 20$ possible arrangements of 3 X's and 3 Y's, as displayed in Table 9.4 with their associated M values. A couple of example calculations are:

$$XXXYYY : (1 - 7/2)^2 + (2 - 7/2)^2 + (3 - 7/2)^2 = 35/4,$$
$$XXYXYY : (1 - 7/2)^2 + (2 - 7/2)^2 + (4 - 7/2)^2 = 35/4.$$

You should have no difficulty in verifying the rest.

Combining probabilities of like values of M from the table, the probability mass function of M under the null hypothesis is

$$g(x) = P[M = x] = \begin{cases} 2/20 & \text{if } x = 11/4, 19/4, 27/4, 43/4, 51/4, 59/4, \\ 8/20 & \text{if } x = 35/4, \\ 0 & \text{otherwise.} \end{cases}$$

The alternative that the X's are more disperse would be favored by large values of M. Therefore a critical region of exact level $2/20 = .1$ is $M = 59/4$, which consists of the two arrangements $XYYYXX$ and $XXYYYX$.

Table 9.4

Possible Sequences and Values of M

Arrangement	M-Value	Arrangement	M-Value
$XXXYYY$	35/4	$YXYYXX$	43/4
$XXYXYY$	35/4	$YYXYXX$	35/4
$XXYYXY$	43/4	$YYYXXX$	35/4
$XXYYYX$	59/4	$YYXXYX$	27/4
$XYXYYX$	51/4	$YYXXXY$	11/4
$XYYXYX$	51/4	$YXYXYX$	35/4
$XYYYXX$	59/4	$YXYXXY$	19/4
$XYXXYY$	27/4	$YXXYYX$	35/4
$XYXYXY$	35/4	$YXXYXY$	19/4
$XYYXXY$	35/4	$YXXXYY$	11/4

—— **?Question 9.3.5** How many arrangements of X and Y symbols are possible when $m = n = 2$? Enumerate them, calculate the associated M-values, and write out the null probability distribution of M.

A large sample approximately standard normal version of the test statistic exists:

$$Z = \frac{M - E[M]}{\sqrt{\text{Var}(M)}} \approx N(0, 1) \tag{9.39}$$

(see Gibbons, 1971, p. 179), where

$$E[M] = \frac{m(N^2 - 1)}{12} \tag{9.40}$$

and

$$\text{Var}(M) = \frac{mn(N + 1)(N^2 - 4)}{180}. \tag{9.41}$$

The test can be performed by comparing Z to standard normal critical values ($\pm z_\alpha$ for the two possible one-sided tests, $\pm z_{\alpha/2}$ for the two-sided test).

Example 9.3.4 In a rating guide to small U.S. cities (Thomas, 1992), it is reported that several cities in Ohio and North Carolina had the following per capita spending (in dollars) on police-related activities:

Ohio: 36, 35, 22, 26, 29, 42, 39, 38, 32, 22, 49, 38, 37, 37, 29

North Carolina: 44, 28, 35, 31, 39, 27, 36, 38, 46, 29, 32, 39

Test at level .1 the equality of the dispersions of the two populations using the large sample normal approximation to the Mood statistic.

The combined ordered sample follows, with parenthetical notation of the state of origin (*o* for Ohio, *n* for North Carolina).

22(*o*), 22(*o*), 26(*o*), 27(*n*), 28(*n*), 29(*n*), 29(*o*), 29(*o*), 31(*n*), 32(*n*), 32(*o*), 35(*o*), 35(*n*), 36(*o*), 36(*n*), 37(*o*), 37(*o*), 38(*o*), 38(*o*), 38(*n*), 39(*o*), 39(*n*), 39(*n*), 42(*o*), 44(*n*), 46(*n*), 49(*o*)

For these data $m = 15, n = 12$, and so $N = 27$, and the average rank in the combined sample is $(27 + 1)/2 = 14$. We will replace observations by ranks, and tied observations by the average of the ranks that the observations are competing for. In view of the ties, the list of ranks is therefore

1.5, 1.5, 3, 4, 5, 7, 7, 7, 9, 10.5, 10.5, 12.5, 12.5, 14.5, 14.5, 16.5, 16.5, 19, 19, 19, 22, 22, 22, 24, 25, 26, 27

For example the three 29s in ranks 6, 7, and 8 are each given rank 7. Then the ranks of the Ohio observations are 1.5, 1.5, 3, 7, 7, 10.5, 12.5, 14.5, 16.5, 16.5, 19, 19, 22, 24, 27. Subtracting 14 from each member of this list and squaring yields the value $m = 941.75$ for the Mood statistic. From (9.40) and (9.41), we obtain

$$E[M] = \frac{15 \cdot (27^2 - 1)}{12} = 910; \qquad \text{Var}(M) = \frac{15(12)(28)(27^2 - 4)}{180} = 20300.$$

Hence

$$z = \frac{941.75 - 910}{\sqrt{20,300}} = .22.$$

A two-sided test would reject H_0 at level .1 if this standardized value of the Mood statistic was less than -1.645 or greater than 1.645; hence we very safely accept the null hypothesis of no difference between dispersions at this level.

We close this section by deriving formula (9.40) for the mean of M. You are guided through a derivation of the variance (9.41) in Exercise 16.

By the construction of the Mood statistic (9.38) and linearity of expectation, the expected value of M is

$$E[M] = E\left[\sum_{i=1}^{N} \left(i - \frac{N+1}{2}\right)^2 \cdot Z_i\right] = \sum_{i=1}^{N} \left(i - \frac{N+1}{2}\right)^2 \cdot E[Z_i]. \qquad (9.42)$$

Since Z_i only takes on the values 0 and 1,

$$E[Z_i] = 1 \cdot P[Z_i = 1] + 0 = P[i\text{th ranked value is an } X]. \qquad (9.43)$$

Under the null hypothesis, the likelihood that the ith ranked value is an X value is simply the proportion m/N of X values in the combined sample. Therefore we can finish the calculation (9.42) as follows:

$$E[M] = \sum_{i=1}^{N} \left(i - \frac{N+1}{2}\right)^2 \cdot \frac{m}{N}$$

$$= \frac{m}{N} \cdot \sum_{i=1}^{N} (i^2 - (N+1)i + (N+1)^2/4)$$

$$= \frac{m}{N} \cdot \left(\frac{N(N+1)(2N+1)}{6} - \frac{N(N+1)^2}{2} + \frac{N(N+1)^2}{4} \right)$$

$$= m(N+1) \left(\frac{2N+1}{6} - \frac{N+1}{4} \right)$$

$$= m(N+1) \frac{N-1}{12}$$

$$= \frac{m(N^2 - 1)}{12}.$$

EXERCISES 9.3

1. Our old friend Fast Eddie is now in the hamburger business. Being quality conscious, he is concerned that the service times of his customers are reasonable on the average and do not vary greatly so that the large majority of his customers go away satisfied. The goal is to serve the customer in an average of no more than 4 minutes (including waiting time in line) with a standard deviation of no more than 1 minute. A sample of 14 times follows:

   5.3, 3.7, 3.8, 4.0, 2.3, 4.2, 6.3, 1.2, 4.3, 2.7,
   3.9, 7.2, 5.4, 4.6

   Test at the .1 level whether his quality goals are being achieved. Use the χ^2-table (Table 4 in Appendix A) to estimate the power of the test if the true variance is 4.

2. Following are data (USDA, 1976, 1991) on average number of pounds of cotton harvested per acre devoted to cotton in the United States. The data are broken into two groups, 1959–1975 and 1976–1990, and these yields are in time order. You can compute that the mean and standard deviation for the first group are about 473 and 36, respectively. Does the second group appear to have a significantly higher standard deviation than 36? Look at the data, and give a commonsense explanation of what seems to be happening.

   1959–1975: 461, 446, 438, 457, 517, 517, 527, 480, 447, 516,
   434, 438, 438, 507, 520, 441, 453

   1976–1990: 465, 520, 420, 547, 404, 542, 590, 508, 600, 630,
   552, 706, 619, 614, 640

3. Explore the connection between confidence intervals and two-sided hypothesis tests for the normal variance.

4. Derive the general form of the power function of a level α test of $H_0: \sigma^2 = \sigma_0^2$ vs. $H_a: \sigma^2 > \sigma_0^2$, where σ^2 is the variance of an $N(\mu, \sigma^2)$ distribution. Use a symbolic algebra-graphics program to plot it for $\sigma_0^2 = 1$, $\alpha = .05$, and values of $n = 20, 50$, and 100.

5. A group of researchers studied the effect of a new kind of individualized care for very low birth weight premature babies (Als et al., 1995).

A control group of 18 infants received the usual treatment in neonatal intensive care including constant noise, bright lights, and medical procedures, and an experimental group of 20 infants received a different treatment meant to be stress-reducing. The researchers listed means and standard deviations on several outcome variables, including number of days of mechanical ventilation, number of days before bottle feeding, and hospital charges (1000s of dollars). These data are given next. For which of these variables can you detect a difference in population variances at significance level .05?

	Control		Experimental	
	Mean	*St. D.*	*Mean*	*St. D.*
Ventilation	63.8	72.9	28.3	23.3
Feeding	104.1	85.8	59.2	25.8
Charges	189	174	98	37

6. Below are two data sets from the February 26, 1993, issue of the *Peoria Journal-Star*. The first is a sample of monthly rental prices for one-bedroom apartments and the second, rental prices for two-bedroom apartments.

   - 350, 400, 375, 300, 450, 375, 355, 185, 410, 350, 310, 225, 335, 195, 210,
   - 350, 510, 700, 680, 365, 620, 390, 420, 425, 450, 425, 475, 365, 375, 275, 325, 250, 285, 525

   Test at level .05 the hypotheses $H_0 : \sigma_1^2 = \sigma_2^2$ vs. $H_a : \sigma_1^2 \neq \sigma_2^2$, where σ_1^2 and σ_2^2 are, respectively, the variances of the populations of one- and 2-bedroom apartment rents in Peoria.

7. Perform a two-sample test of equality of variance on the data on cotton yields in Exercise 2. Use level .05 and a one-sided alternative.

8. Reread Exercise 6 in Section 8.4 on heart rates of men doing snow removal. Devise an appropriate hypothesis test about variances. Give a careful write-up of your assumptions, procedures, results, and conclusions.

9. Devise a general expression for the type II error probability of a level α test of $H_0 : \sigma_x^2 / \sigma_y^2 = 1$ vs. $H_2 : \sigma_x^2 / \sigma_y^2 > 1$, as a function of the true ratio $r = \sigma_x^2 / \sigma_y^2$.

10. A *test for scale parameter* is one in which the null hypothesis is that one population c.d.f. F_X equals another F_Y, with the alternative that F_Y is just a rescaled version of F_X. If m denotes the common median of the X and Y distributions, then the random variables $X - m$ and $Y - m$ both have median 0, and the competing hypotheses can be expressed as

$$H_0 : F_{Y-m}(y) = F_{X-m}(y) \quad \text{vs.} \quad H_a : F_{Y-m}(y) = F_{X-m}(\theta y),$$

where θ, the *scale parameter*, is a positive number. Show that the two-sample normal variances test $H_0 : \sigma_x^2 = \sigma_y^2$ vs. $H_a : \sigma_x^2 \neq \sigma_y^2$ is a special case of a scale parameter test, when it is assumed that the two means are equal. What is the scale parameter for this test in terms of the two variances?

11. The Mood test is an example of a *two-sample linear rank procedure*—that is, it is based on a statistic of the form

$$\sum_{i=1}^{N} a_i Z_i,$$

where Z_i is 1 or 0 respectively according to whether the ith ranked member of the combined X and Y samples is an X or a Y. The coefficients a_i are called *weights* and are chosen by the statistician in an intuitively reasonable way for the purpose of a particular test. Show that the Mood weights are symmetric about the average rank in the combined sample (you should treat the even and odd cases separately). Can you devise another system of weights symmetric about the average rank that would be useful for the dispersion problem?

12. Enumerate the possible sequences of X's and Y's and the corresponding values of the Mood statistic M when the sample sizes are $m = 3$ and $n = 2$. Use this to find the p.m.f. of M under the null hypothesis.

13. Suppose that one sample consists of the observations 5.2, 4.9, 6.0 and another consists of the observations 4.7, 5.0, 8.0, 9.0. What would be the *p*-value of a Mood test with the one-sided alternative that the first population is less disperse than the second?

14. Two catalysts A and B are being compared relative to the yield that they produce in a chemical reaction. Twenty observed yields for catalyst A are

9.20, 9.86, 11.83, 12.78, 11.61, 7.25, 11.80, 9.28, 5.58, 11.79,
7.99, 8.95, 11.00, 6.72, 10.19, 9.62, 11.91, 11.89, 11.81, 9.29

and seventeen observed yields for catalyst B are

10.46, 14.66, 17.10, 14.34, 14.06, 18.77, 12.07, 6.51, 11.33,
9.21, 10.29, 10.56, 14.62, 6.07, 13.78, 4.71, 12.02

Perform a large sample approximate level .05 Mood test of the null hypothesis of equality of dispersion vs. the alternative that the catalyst A yield is less disperse than the yield for catalyst B. (*Note:* The first data set was simulated from the $N(10, 4)$ distribution, and the second from the $N(10, 16)$ distribution. This problem lets you check whether the Mood test is powerful enough to pick out this difference in variances.)

15. Use the data in Table 7.6 and a large sample Mood test to test for a difference in dispersion of National League home runs and American League home runs. (The American League teams begin in the table with the Orioles.) Perform the test at level .05.

16. Derive formula (9.41) for the variance of the Mood statistic by doing the following:
    a. Show that under H_0, $\text{Var}(Z_i) = mn/N^2$, where Z_i is as in (9.38).
    b. Show that $\text{Cov}(Z_i, Z_j) = -mn/N^2(N - 1)$.
    c. Show by induction that

$$\sum_{i=1}^{N} i^3 = \left(\frac{N(N+1)}{2}\right)^2; \quad \sum_{i=1}^{N} i^4 = \frac{N(N+1)(2N+1)(3N^2+3N-1)}{30}.$$

    d. Complete the calculation of $\text{Var}(M)$.

9.4 | Two-Sample Location Tests

You will recall that we dealt with comparisons of normal means via confidence intervals in Section 8.5. After what you have seen in this chapter, it should not surprise you that there are corresponding normal theory hypothesis tests for differences in means that allow you to compare the locations of two distributions. In this section we will quickly cover the normal theory for independent samples and for paired samples, which you have mostly seen before, and then go on to describe a nonparametric test for location comparisons that is in the same spirit as earlier nonparametric methods. This is known as the *Mann–Whitney test*, or the *Wilcoxon rank-sum test*.

9.4.1 Normal Means Tests

Let $X_1, X_2, \ldots, X_m$ and $Y_1, Y_2, \ldots, Y_n$ be random samples from $N(\mu_x, \sigma_x^2)$ and $N(\mu_y, \sigma_y^2)$ distributions, respectively. We will develop an hypothesis test for

$$H_0 : \mu_x - \mu_y = 0 \quad \text{vs.} \quad \begin{cases} H_1 : \mu_x - \mu_y \neq 0 \text{ or} \\ H_2 : \mu_x - \mu_y > 0 \text{ or} \\ H_3 : \mu_x - \mu_y < 0. \end{cases} \tag{9.44}$$

There are two main cases to consider depending on the experimental design used to gather the data. Recall the example of the students training for the ACT exam which we considered early in Section 8.5. In one design, a single test group was used, which gave rise to paired data $(X_i, Y_i)_{i=1,2,\ldots,n}$ of scores before and after the training program. Here the two samples were of the same size, and one student's scores were assumed independent of another's, but the two scores X_i and Y_i for a particular student could be correlated. In the other design, the two samples contained entirely different students, the sample sizes m and n could be different, and every X_i was independent of every Y_j. We will refer to the first case as the *paired samples design*, and the second as the *independent samples design*.

Let us look first at the independent samples design. The most intuitively obvious way to construct a critical region for a test of $\mu_x - \mu_y$ is to base it on the difference of sample means $\overline{X} - \overline{Y}$, which is an unbiased estimate of $\mu_x - \mu_y$. Rejection regions for alternatives H_1, H_2, and H_3, respectively, would take the form

$$\overline{X} - \overline{Y} > k_1 \quad \text{or} \quad \overline{X} - \overline{Y} < k_2$$
$$\overline{X} - \overline{Y} > c_1$$
$$\overline{X} - \overline{Y} < c_2.$$

It remains to see how to choose the critical values in order to achieve a test of the desired level α.

As always, the sample means have normal distributions, and since the samples, hence the sample means, are independent we have

$$\overline{X} - \overline{Y} \sim N\left(\mu_x - \mu_y, \frac{\sigma_x^2}{m} + \frac{\sigma_y^2}{n}\right). \tag{9.45}$$

In the unusual case that the two variances σ_x^2 and σ_y^2 are known we are already done, because under the null hypothesis of equal population means we have

$$Z = \frac{\overline{X} - \overline{Y}}{\sqrt{\sigma_x^2/m + \sigma_y^2/n}} \sim N(0,1), \tag{9.46}$$

and so critical values for Z ($\pm z_{\alpha/2}$ in the two-sided case H_1, z_α and $-z_\alpha$ in the two one-sided cases H_2 and H_3) can be found from the standard normal table. It is then a simple matter to express the rejection region in terms of $\overline{X} - \overline{Y}$ if necessary. If the variances are unknown and if the sample sizes m and n are both large, one can replace σ_x^2 and σ_y^2 in (9.46) by their unbiased estimates S_x^2 and S_y^2 to produce an approximate level α test. For the three alternatives H_1, H_2, and H_3, respectively, the critical regions would be

$$Z = \frac{\overline{X} - \overline{Y}}{\sqrt{S_x^2/m + S_y^2/n}} < -z_{\alpha/2} \text{ or } Z > z_{\alpha/2}$$

$$Z > z_\alpha$$

$$Z < -z_\alpha. \tag{9.47}$$

Example 9.4.1

This example introduces a problem from a study by a group of developmental psychologists (Kasser et al., 1995) that we will return to at several points later. It is a very rich illustration of the application of a number of statistical techniques, including how the practitioner must think about how to organize a large amount of data so as to draw conclusions.

The researchers were interested in discovering connections between the methods by which mothers raised their children and the relative valuation of financial success with three other kinds of aspirations by the children: self-acceptance, affiliation with friends and family, and community feeling. The theory to be tested was that children tend to take on the values of those who raise them, and that a warm family environment tends to free the child from the kind of insecurity that leads to heavy emphasis on making money.

Surveys were administered to more than 100 18-year-olds and to their mothers. The psychologists used the survey data to create an index of maternal nurturance for each mother. The higher that index, the warmer and more supportive the mother was judged to be. The teenage children were then classified into two groups according to whether their preferred value was financial success or self-acceptance. Among 72 of the teens for whom financial success was preferred, the mean and standard deviation of their mother's nurturance were -0.88 and 3.15, respectively. For the other 57 teens who preferred self-acceptance, the mean and standard deviation of the mother's nurturance were $.97$ and 3.60. Are there significant differences in the characteristics of the mothers between the two groups?

We test whether the mean mother's nurturance μ_x for the first group is equal to, or rather is less than, the mean mother's nurturance μ_y for the second group. Using the large sample normal approximation in (9.47),

$$z = \frac{-0.88 - .97}{\sqrt{3.15^2/72 + 3.60^2/57}} = -2.93.$$

The approximate *p*-value of the test is $P[Z < -2.93] = .0017$. This gives strong evidence that the mothers of the teens who valued financial success more had lower nurturance indices than the mothers of the teens who preferred self-acceptance.

Interestingly, for small sample sizes and unknown, unequal variances a suitable test statistic with a known distribution is not available in general. An attempt to construct a *t*-statistic can be done (see Exercise 3), but an algebraic roadblock arises. But we can continue in the special case of unknown but equal variances.

Suppose once again that $X_1, X_2, \ldots, X_m$ and $Y_1, Y_2, \ldots, Y_n$ are independent random samples from $N(\mu_x, \sigma_x^2)$ and $N(\mu_y, \sigma_y^2)$ distributions, respectively, where $\sigma_x^2 = \sigma_y^2 = \sigma^2$. Then as above, under the null hypothesis

$$Z = \frac{\overline{X} - \overline{Y}}{\sqrt{\sigma^2/m + \sigma^2/n}} \sim N(0, 1). \tag{9.48}$$

Also, we know that

$$U = \frac{(m-1)S_x^2}{\sigma^2} \sim \chi^2(m-1) \quad \text{and} \quad V = \frac{(n-1)S_y^2}{\sigma^2} \sim \chi^2(n-1). \tag{9.49}$$

Hence

$$U + V \sim \chi^2(m + n - 2).$$

Since both of the sample variances are independent of both of the sample means, $U + V$ is independent of Z. By the definition of the *t*-distribution,

$$T = \frac{Z}{\sqrt{(U + V)/(m + n - 2)}} = \frac{\overline{X} - \overline{Y}}{S_p\sqrt{1/m + 1/n}} \sim t(m + n - 2), \tag{9.50}$$

where

$$S_p^2 = \frac{(m-1)S_x^2 + (n-1)S_y^2}{m + n - 2} \tag{9.51}$$

is the pooled estimate of the common variance σ^2. (Make sure you can simplify the first expression for T in (9.50) into the form of the second.) Suitable rejection regions of level α are then based on critical values obtained from the $t(m + n - 2)$ table:

$$T < -t_{\alpha/2}(m + n - 2) \quad \text{or} \quad T > t_{\alpha/2}(m + n - 2) \text{ for } H_1,$$
$$T > t_{\alpha}(m + n - 2) \quad \text{for } H_2,$$
$$T < -t_{\alpha}(m + n - 2) \quad \text{for } H_3. \tag{9.52}$$

—— **?Question 9.4.1** We derived the hypotheses tests for $H_0 : \mu_x - \mu_y = 0$. What alterations must be made in order to test $H_0 : \mu_x - \mu_y = c$ for some $c \neq 0$?

Example 9.4.2

A student named Brian Kendall once sought my help while researching ethanol metabolization by Drosophila melanogaster (fruit flies). One of the specific things that Brian was interested in was the effect of a hormone called JH (juvenile hormone) on this process. A control group of 36 batches of flies was given a sucrose-ethanol diet, and a treatment group of 24 batches was given JH in addition to the sucrose-ethanol diet. He measured the activity level of a key enzyme called ADH, which is known to be associated with the metabolization process. For the flies in the plain ethanol group, the mean ADH activity level was 170.313 and the standard deviation was 34.206. The juvenile hormone group produced a mean of 260.389 and a standard deviation of 38.084. Is there a significant difference in ADH activity due to the presence of JH?

The closeness of the standard deviations gives us some confidence in using the equal variance version of the test statistic. You can compute that

$$s_p^2 = \frac{(m-1)s_x^2 + (n-1)s_y^2}{m+n-2} = \frac{35(34.206)^2 + 23(38.084)^2}{36+24-2} = 1281.2,$$

which results in the observed $s_p = 35.79$. The t-statistic therefore has the value

$$t = \frac{\bar{x} - \bar{y}}{s_p\sqrt{1/m + 1/n}} = \frac{170.313 - 260.389}{35.79\sqrt{1/36 + 1/24}} = -9.5.$$

The rejection region for a level α test of $H_0: \mu_x - \mu_y = 0$ vs. $H_3: \mu_x - \mu_y < 0$ is $T < -t_\alpha(m+n-2)$, and as we have remarked before, for a large value of the degree of freedom parameter such as 58 here, t critical values are close to standard normal critical values. The observed test statistic value of -9.5 is then outrageously small under the null hypothesis; its p-value is essentially 0. Therefore we have very strong evidence in favor of a JH effect on alcohol metabolization.

?Question 9.4.2

Perform the test for equality of normal variances on the fruit fly groups in Example 9.4.2. Does it support our assumption of equal variances?

Next we consider the paired samples design. In it, we have n independent random vectors $(X_1, Y_1), (X_2, Y_2), \ldots, (X_n, Y_n)$, which we assume constitute a random sample from the bivariate normal distribution. If (X, Y) is a generic bivariate normal vector with mean vector (μ_x, μ_y), variances σ_x^2 and σ_y^2, and correlation coefficient ρ, then by Proposition 5.4.6, with $\mathbf{X} = [X\ Y]'$ and $A = [1\ -1]$,

$$X - Y \sim N(\mu_x - \mu_y, \sigma_x^2 + \sigma_y^2 - 2\rho\sigma_x\sigma_y). \tag{9.53}$$

(Verify this.) If we form the sample of differences $D_i = X_i - Y_i$, $i = 1, \ldots, n$, we then have a single random sample $D_1, D_2, \ldots, D_n$ from a normal distribution with mean $\mu_x - \mu_y$ and variance $\sigma_d^2 = \sigma_x^2 + \sigma_y^2 - 2\rho\sigma_x\sigma_y$. It does not matter that we do not know what ρ, σ_x or σ_y are individually; we can still do a one-sample t-test on the sample of differences, estimating the standard deviation σ_d of the population of differences by the sample standard deviation S_d of the D_i's. To test $H_0 : \mu_x - \mu_y = \mu_d = 0$ in this case, you can compare the test statistic

$$T = \frac{\overline{D}}{S_d/\sqrt{n}} = \frac{\overline{X} - \overline{Y}}{S_d/\sqrt{n}} \tag{9.54}$$

to t-critical values of the desired level with $n - 1$ degrees of freedom. This is illustrated in the next example.

Example 9.4.3

Exercise 8 in Section 8.4 listed a data set of yearly egg laying rates per hen for the 50 states in 1990. The same government report (USDA, 1991) gave laying rates for the year 1989. The state-by-state differences of 1990 minus 1989 are given here. Does there seem to be a significant difference between the two years?

1, 0, −17, 0, 7, −4, 19, 1, 1, 5, −16, 5, 6, 2, 8, −9, 9, 1, 8, 3, 3, 7, 3, −1, 7, 5, −3, −4, −4, 14, −15, −1, −7, 9, 5, −6, −9, 0, 16, −1, 5, −14, −2, 2, 37, 0, −5, 5, 8, 21

To check for change in the laying rates, we test $H_0 : \mu_x - \mu_y = 0$ vs. $H_a : \mu_x - \mu_y \neq 0$, where the X population is for 1990 and the Y is for 1989. The numbers are the observed values of the difference variables $D_i = X_i - Y_i$, $i = 1, \ldots, 50$, and our test is equivalent to the test $H_0 : \mu_d = 0$ vs. $H_a : \mu_d \neq 0$. It can be computed that $\overline{D} = 2.1$, and $S_d = 9.5$. The value of the t-statistic is therefore

$$t = \frac{\overline{d}}{s_d/\sqrt{n}} = \frac{2.1}{9.5/\sqrt{50}} = 1.56.$$

For 49 degrees of freedom, the t critical value for a two-sided test at level .1 is 1.677 (see Table 9.1), so we cannot reject the null hypothesis at this level. The data give no firm proof that the egg laying rate has changed from 1989 to 1990.

Now that you have studied two designs that lead to tests for $\mu_x - \mu_y$, it is natural to wonder which one to use. For tests of equal significance level α, one test is better than another if it has more power. It is only fair to compare the two tests for difference of normal means if their total sample sizes $m + n$ and $2n$ are the same, so let the sample sizes for the independent samples design both equal n, which is the number of pairs in the paired design. Also, for simplicity, suppose that a friendly genie (or a pilot sample) has told us that $\sigma_x^2 = \sigma_y^2 = \sigma^2$ is known, and the population correlation coefficient ρ is also known. Then the two test statistics are

$$Z = \frac{\overline{X} - \overline{Y}}{\sqrt{\sigma^2/n + \sigma^2/n}}, \qquad Z_d = \frac{\overline{X} - \overline{Y}}{\sigma_d/\sqrt{n}} \tag{9.55}$$

for the independent and paired sample cases, respectively, where

$$\sigma_d^2 = \sigma^2 + \sigma^2 - 2\rho\sigma\sigma = 2\sigma^2(1-\rho). \tag{9.56}$$

The test statistics Z and Z_d have the $N(0,1)$ distribution under the null hypothesis, and H_0 would be rejected for, say, $H_2 : \mu_x - \mu_y > 0$ iff the observed value of the test statistic exceeds z_α.

If the true difference in means $\mu_x - \mu_y$ is some nonzero d, then we can express the critical regions in terms of $\overline{X} - \overline{Y}$ and restandardize as follows:

$$(\text{independent samples}): \quad \overline{X} - \overline{Y} > z_\alpha \cdot \sigma \cdot \sqrt{2/n}$$

$$\Rightarrow \frac{\overline{X} - \overline{Y} - d}{\sigma \cdot \sqrt{2/n}} > z_\alpha - \frac{d}{\sigma \cdot \sqrt{2/n}}, \tag{9.57}$$

$$(\text{paired samples}): \quad \overline{X} - \overline{Y} > z_\alpha \cdot \sigma_d \cdot \sqrt{1/n}$$

$$\Rightarrow \frac{\overline{X} - \overline{Y} - d}{\sigma_d \cdot \sqrt{1/n}} > z_\alpha - \frac{d}{\sigma \cdot \sqrt{2/n} \cdot \sqrt{1-\rho}}. \tag{9.58}$$

The last part of (9.58) follows from the expression for σ_d in (9.56). The test with the higher power has the larger critical region, hence the smaller right-hand side. As long as $\rho > 0$, the right-hand side will be smaller for the paired samples test. In other words, for equal level and total sample size, if the variances of the two populations are known and equal, the paired sample design has higher power if the two variables X and Y are positively correlated and lower power if they are negatively correlated. This occurs because of the smaller variance of the estimator $\overline{X} - \overline{Y}$ in the paired case. (Make sure that you see why this is so.)

Of course, the assumption of known variances is unrealistic, and it is more difficult to draw definitive conclusions when they are not known. Then we must use the t-statistics (9.50) for the independent case and (9.54) for the paired case. The critical values would then be $t_\alpha(2n-2)$ in the independent case and $t_\alpha(n-1)$ in the paired case. The lower degrees of freedom in the paired case increases the critical value, which partially offsets the benefit derived from reducing the variance estimate. See Exercise 14.

9.4.2 Mann–Whitney–Wilcoxon Test

Several nonparametric methods are available for comparing locations. Two of the oldest of them are the *Mann–Whitney test* and the *Wilcoxon rank-sum test*. It turns out that the test statistics differ by a constant and hence are equivalent. So, after a short description of the Mann–Whitney procedure, we focus our attention on the Wilcoxon test.

Suppose that two continuous distributions F_X and F_Y are identical with the possible exception that the F_Y distribution is shifted to the right by an amount θ from the F_X distribution. Then $F_Y(x) = F_X(x - \theta)$. The null hypothesis that the two distributions are actually the same therefore translates to $H_0 : \theta = 0$, which can be tested against both one- and two-sided alternatives. The alternative that $\theta < 0$ means that F_Y is shifted to the left from F_X.

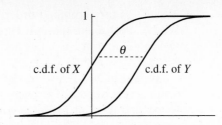

Figure 9.7   c.d.f.'s in Two-Sample Location Test

—— **?Question 9.4.3**   If θ as described in the preceding paragraph is greater than zero, then how do the medians of the two populations X and Y compare? What if $\theta < 0$?

An example of the situation that we are testing is displayed in Fig. 9.7. If the Y population distribution is shifted to the right of that of X, as shown in the figure, then for each fixed x, it is likelier that $X \leq x$ than $Y \leq x$. In other words, X values are more likely to be smaller than Y values. Thus, if random samples $X_1, X_2, \ldots, X_m$ and $Y_1, Y_2, \ldots, Y_n$ are obtained from the two distributions, in the combined ordered sample, the X's will tend to be smaller than the Y's. Similarly, if θ is negative—the Y population is shifted to the left—then the Y's will tend to be smaller than the X's. Both the Mann–Whitney and the Wilcoxon rank-sum tests are founded on this observation.

We can begin by combining the X and Y samples and sorting the combined sample into increasing order. To form the Mann–Whitney test statistic, consider the mn random variables

$$D_{ij} = \begin{cases} 1 & \text{if } Y_j < X_i, \\ 0 & \text{otherwise,} \end{cases}$$

$i = 1, \ldots, m; j = 1, \ldots, n$. Define

$$U = \sum_{i=1}^{m} \sum_{j=1}^{n} D_{ij}, \tag{9.59}$$

that is, define the Mann–Whitney statistic U to be the number of times in the combined sample that Y's precede X's. Then H_0 would be rejected in favor of $H_2 : \theta > 0$ if $U < c_1$, and H_0 would be rejected in favor of $H_3 : \theta < 0$ if $U > c_2$ for suitable critical values.

The Wilcoxon statistic is similar in form to the Mood statistic for dispersion. If the combined ordered sample is formed and the X ranks are low, then there is an indication that the Y distribution is shifted to the right of the X distribution. If the X ranks are added, then the sum will be small under the alternative $\theta > 0$, and large under the alternative that the Y distribution is shifted to the left, that is, $\theta < 0$. So, define

$$W = \sum_{i=1}^{N} i \cdot Z_i, \tag{9.60}$$

where $N = m + n$, and $Z_i = 1$ if the ith ranked observation in the combined sample is an X, and $Z_i = 0$ otherwise. In other words, W is the sum of the

X ranks in the combined ordered sample. Reject H_0 in favor of $H_2 : \theta > 0$ if $W < c_1$, and reject the null hypothesis in favor of $H_3 : \theta < 0$ if $W > c_2$. The two-sided alternative is favored if either extremely low or high values of W occur.

At first glance it may seem that the Mann–Whitney test and Wilcoxon test are very similar. This remains true at second and third glance too (even fourth and fifth), because as you are asked to show in Exercise 18,

$$U = W - \frac{m}{2}(m + 1). \qquad (9.61)$$

The two tests therefore give equivalent conclusions, and we can focus on just one of them, the Wilcoxon test.

As with the Mood statistic, the small sample distribution of W can be enumerated by listing the arrangements of X and Y symbols that are possible when m and n are given. If the two underlying distributions are the same, then each such sequence of X's and Y's is equally likely. The probability that a particular possible value of W is assumed is therefore the number of arrangements that generate that value divided by the total number of possible arrangements $\binom{N}{m}$.

Example 9.4.4

Find the distribution of W under the null hypothesis in the case $m = n = 3$. Use it to construct a critical region of level no more than .1 for the alternative that the Y population is shifted to the right of the X population.

To find the null distribution of the test statistic, it suffices to list the possible arrangements as already described, and calculate the values of W that are associated with them. I have done this in Table 9.5. A couple of sample calculations are shown, which should be self-explanatory.

Table 9.5

Possible Sequences and Values of W

Arrangement	W-Value	Arrangement	W-Value
XXXYYY	$1 + 2 + 3 = 6$	YXYYXX	13
XXYXYY	$1 + 2 + 4 = 7$	YYXYXX	14
XXYYXY	8	YYYXXX	15
XXYYYX	9	YYXXYX	13
XYXYYX	10	YYXXXY	12
XYYXYX	11	YXYXYX	12
XYYYXX	12	YYXYXX	11
XYXXYY	8	YXXYYX	11
XYXYXY	9	YXXYXY	10
XYYXXY	10	YXXXYY	9

Thus the distribution of W is

$$f_W(w) = \begin{cases} 3/20 & \text{if } w = 9, 10, 11, 12, \\ 2/20 & \text{if } w = 8, 13, \\ 1/20 & \text{if } w = 6, 7, 14, 15, \\ 0 & \text{otherwise.} \end{cases}$$

The alternative that Y is shifted to the right is favored if the sum of the X ranks is small. Therefore if we reject H_0 when $W < 8$ (when $W = 6$ or 7), we have a rejection region of exact level $1/20 + 1/20 = 1/10 = .1$. Reading from Table 9.5, this critical region rejects the null hypothesis if either of the arrangements $XXXYYY$ or $XXYXYY$ appear.

A large sample normal approximation holds (Gibbons, 1971, p. 165). To use it, we must compute the mean and variance of W, then base the test on the standardized W:

$$Z = \frac{W - E[W]}{\sqrt{\text{Var}(W)}}. \tag{9.62}$$

Critical values are obtained in the usual way from the standard normal table, with the direction of the inequality in the rejection region determined by the direction that would be appropriate for W itself, depending on the alternative hypothesis. The computations of the expectation and variance are similar to the computations for the Mood statistic (see in Section 9.3 the end discussion and Exercise 16). It is rather easy (see Exercise 19) to check that

$$E[W] = \frac{m(N + 1)}{2}. \tag{9.63}$$

We will show next that

$$\text{Var}(W) = \frac{mn(N + 1)}{12}. \tag{9.64}$$

Since W is a sum, but the components of the sum are correlated, the variance of W is the sum of the variances of all terms plus the covariances of all pairs of terms. (See Proposition 4.3.3.) To prepare for the computation, we note that under H_0 the likelihood that the ith ranked value is an X is m/N, and the likelihood that both the jth and kth ranked values are X's is $m(m - 1)/(N(N - 1))$. Hence

$$\text{Var}(Z_i) = E[Z_i^2] - E[Z_i]^2 = E[Z_i^2] - E[Z_i]^2$$
$$= \frac{m}{N} - \frac{m^2}{N^2}$$
$$= \frac{mn}{N^2}, \tag{9.65}$$

$$\text{Cov}(Z_j, Z_k) = E[Z_j Z_k] - E[Z_j]E[Z_k]$$
$$= \frac{m(m - 1)}{N(N - 1)} - \frac{m^2}{N^2}$$
$$= -\frac{mn}{N^2(N - 1)}. \tag{9.66}$$

[Verify the details of (9.65) and (9.66).] Therefore the variance of W is

$$\text{Var}(W)$$

$$= \text{Var}\left(\sum_{i=1}^{N} i \cdot Z_i\right)$$

$$= \sum_{i=1}^{N} i^2 \cdot \text{Var}(Z_i) + \sum\sum_{j \neq k} j \cdot k \cdot \text{Cov}(Z_j, Z_k)$$

$$= \frac{mn}{N^2} \sum_{i=1}^{N} i^2 - \frac{mn}{N^2(N-1)} \sum_{j=1}^{N} j \cdot \left(\sum_{k=1, k \neq j}^{N} k\right)$$

$$= \frac{mn}{N^2} \cdot \frac{N(N+1)(2N+1)}{6} - \frac{mn}{N^2(N-1)} \sum_{j=1}^{N} j \cdot \left(\frac{N(N+1)}{2} - j\right)$$

$$= \frac{mn}{N^2}\left[\frac{N(N+1)(2N+1)}{6} - \frac{1}{N-1} \cdot \frac{N(N+1)}{2} \sum_{j=1}^{N} j + \frac{1}{N-1} \sum_{j=1}^{N} j^2\right]$$

$$= \frac{mn}{N^2}\left[\frac{N(N+1)(2N+1)}{6} - \frac{1}{N-1} \cdot \frac{N^2(N+1)^2}{4} + \frac{1}{N-1} \cdot \frac{N(N+1)(2N+1)}{6}\right]$$

$$= \frac{mn(N+1)}{12}, \tag{9.67}$$

after algebraic simplification of the sixth line.

The next example illustrates the application of the large sample Wilcoxon rank-sum test.

Example 9.4.5

Spring in the Midwest seems to take interminably long to arrive. But I comfort myself by saying, "By the time March comes it is becoming more springlike, so winter is essentially over." Is it really? Let us use the Wilcoxon test to check for an upward shift in the temperature distribution from February to March. Following are National Weather Service data on average daily temperatures in the city of Peoria for 1993 during the month of February:

32, 30, 38, 38, 37, 33, 37, 34, 36, 42, 34, 31, 28, 21,
21, 21, 5, 5, 16, 27, 32, 18, 6, 8, 16, 22, 21, 21

March average temperatures were

34, 36, 36, 34, 36, 36, 36, 37, 36, 34, 27, 27, 18, 16, 35, 35,
22, 26, 30, 36, 36, 41, 41, 40, 46, 50, 49, 48, 51, 59, 49

Figure 9.8 gives parallel dotplots of the temperatures for these two months. March does seem to be a little warmer.

If you combine the $m = 28$ February temperatures with the $n = 31$ March temperatures and order the resulting sample of $N = 59$ observations, you

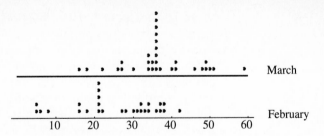

Figure 9.8 February and March Temperatures

obtain the following arrangement:

5F, 5F, 6F, 8F, 16F, 16F, 16M, 18F, 18M, 21F, 21F, 21F, 21F,
21F, 22F, 22M, 26M, 27F, 27M, 27M, 28F, 30F, 30M, 31F, 32F,
32F, 33F, 34F, 34F, 34M, 34M, 34M, 35M, 35M, 36F, 36M, 36M,
36M, 36M, 36M, 36M, 36M, 36M, 37F, 37F, 37M, 38F, 38F,
40M, 41M, 41M, 42F, 46M, 48M, 49M, 49M, 50M, 51M, 59M

There are a lot of ties, which I resolved by the usual device of averaging the ranks that the tied observations occupy. The sum of the February ranks is $(1.5 + 1.5 + 3 + \cdots + 47.5 + 47.5 + 52) = 606$. From formulas (9.63) and (9.64) the mean and variance of W are $E[W] = m(N + 1)/2 = 28(60)/2 = 840$, and $\text{Var}(W) = mn(N + 1)/12 = 28(31)(60)/12 = 4340$. Thus the value of the standardized test statistic is

$$z = \frac{606 - 840}{\sqrt{4340}} = -3.55$$

We would reject the null hypothesis of equal temperature distribution for the alternative that March temperatures are higher if the sum of the February ranks is too small, equivalently, if Z is too small. This observed value z is less than the critical value -3.08 for a .001 level test; hence these data give very strong evidence that March temperatures are higher than February temperatures.

You might be wondering whether there is a nonparametric correspondent to the paired t-test. In fact there is, and it is just an adaptation of a procedure that we have discussed previously. If the data are pairs (X_i, Y_i), then if the distributions are the same, the difference $D_i = X_i - Y_i$ should have a distribution whose median is 0. The one-sample Wilcoxon signed rank test of Section 9.2 can be applied to the sample of differences to check this null hypothesis.

EXERCISES 9.4

1. Another thing that the psychologists of Example 9.4.1 did was to compute an index of socioeconomic advantage for each of the 18-year-olds. The higher that index, the more well-to-do is the person. The teenagers were classified into two groups according to whether their preferred value was

financial success or family affiliation. Among 61 of the teens for whom financial success was preferred, the mean and standard deviation of their socioeconomic index were −1.08 and 3.55. The other 68 teens who preferred self-acceptance had mean and standard deviation of socioeconomic status of 1.28 and 3.74, respectively. Use both a large sample normal test and a *t*-test with the equal unknown variance assumption to test for differences. Interpret what the results could mean.

2. From (Casella and Berger, 1990, p. 396) A Byzantine church was constructed in stages. A group interested in comparing the ages of the core and the periphery of the church used dating techniques on samples of wood from each area, which gave ages as follows:

Core: 1294, 1279, 1274, 1264, 1263, 1254, 1251, 1251, 1248, 1240, 1232, 1220, 1218, 1210

Periphery: 1284, 1274, 1272, 1264, 1256, 1256, 1254, 1250, 1242

Use the large sample form of the independent samples *t*-test to test the null hypothesis at level .05 that the core is the same age as the periphery against the two-sided alternative.

3. Attempt to derive a test statistic for the independent samples normal means problem in the case that the samples are small, and the two variances are unequal and unknown. Where does the difficulty lie?

4. Using the data in Table 7.6, test at level .025 whether, as has been claimed, the National League is a faster league (i.e., steals more bases) than the American League. Recall that in the table, the National League teams begin with the Cubs and end with the Giants, and the Orioles through the Rangers are American League teams. Check to see whether the assumption of equal variances is reasonable.

5. Consider a two-sample means test based on random samples $X_1, X_2, \ldots, X_n$ and $Y_1, Y_2, \ldots, Y_n$. Find a specific expression for the error that you make in computing the value of the test statistic if you assume that the samples are independent when in fact they are paired. (You may suppose that the population variances are equal.) Discuss the effect of the correlation between X and Y on this error.

6. Exercise 6 in Section 8.5 reported a study of paranormal phenomena (Cashen and Ramseyer, 1970), that pretested 32 students in order to classify them into a group that had high latent ESP abilities and a group that did not. Throughout a semester the researchers asked the students to predict the nature of exam questions, and the researchers counted how many questions each student was able to predict. The data follows. Perform an appropriate hypothesis test and interpret the results.

High E.S.P.: 79, 62, 95, 83, 59, 51, 78, 78, 63, 55, 24, 69, 64, 55, 52

Low E.S.P.: 82, 64, 44, 29, 50, 43, 41, 30, 38, 46, 36, 42, 23, 49, 43, 31, 24

7. Derive an expression for the power function of the two-sided level α independent samples normal means test, assuming that the two popula-

tion variances are equal and known. Write it as a function of the true difference $d = \mu_x - \mu_y$.

8. Example 8.5.1 cited a data set of sea bass larvae length for specimens caught in the months of August and October. In that example we used a confidence interval to get information about the difference $\mu_a - \mu_o$ between August and October mean lengths. Devise an hypothesis test of $H_0 : \mu_a - \mu_o = -.3$ vs. $H_3 : \mu_a - \mu_o < -.3$, and carry it out at level .05.

9. Exercise 1 in Section 9.2 gave data on 1977 average earnings by Pakistani workers in several industries. Following are similar data for 1976 (Government of Pakistan, 1982). Test at level .05 $H_0 : \mu_{1977} - \mu_{1976} = 1000$ vs. $H_2 : \mu_{1977} - \mu_{1976} > 1000$.

Textile	4325.05	*Paper and Printing*	6121.58
Cotton	2580.05	*Wood, Stone, and Glass*	3095.35
Engineering	4730.86	*Skin and Hides*	5894.89
Mineral Metals	4619.38	*Mints*	4539.83
Chemical and Dyes	5669.11		

10. Sometimes literary investigations are conducted on works of unknown authorship. They employ statistical methods to determine whether an hypothesized author could have written the work under consideration. Design and carry out an experiment to see whether average sentence length can be used to distinguish a work of Hemingway from a work of Dostoyevsky.

11. Suppose that you are designing an independent samples test of $H_0 : \mu_x - \mu_y = 0$ vs. $H_2 : \mu_x - \mu_y > 0$. A pilot sample gave evidence that the two population variances were not significantly different from 1. For simplicity, you will make the same number of observations from each population. At least how many observations should you make so that a .05 level test has a probability of at least .90 of detecting a true difference in means of .2?

12. In the past, I have noticed my bowling scores going up remarkably between the first and the second game that I bowl in a given evening. The question is not so much whether they rise, but by how much. Test the null hypothesis that the theoretical average score on the second game is no more than 15 points better than the first against the alternative that the second game average is more than 15 points better than the first. Use level .01 and the following data:

Date	1	2	3	4	5	6	7	8	9	10
Game 1	102	95	110	105	115	90	98	111	120	92
Game 2	110	123	118	120	135	124	140	146	133	116

13. Develop a large sample approximate test for the difference in Bernoulli success parameters p_x and p_y based on independent random samples of sizes m and n, respectively, from the two populations.

14. For the problem of testing for $d = \mu_x - \mu_y = 0$ against $d > 0$ in the independent and paired samples cases with equal numbers of observations and equal, unknown variance, assume that the true difference $d = 2$, the

common standard deviation is $\sigma = 3$, the significance level is $\alpha = .05$, and $n = 21$. Taking into account the degrees of freedom of the critical values, is it still true that the power of the paired t-test exceeds that of the independent samples t-test as long as $\rho > 0$? (*Hint:* Write the power functions and suppose that S_p will be nearly equal to σ and S_d will be nearly equal to σ_d.)

15. We commented at the end of the section on a nonparametric analogue of the paired samples t-test, which transforms the data and applies the one-sample Wilcoxon signed rank test. Devise and carry out a nonparametric test of level no more than .1 for location difference on the bowling scores in Exercise 12. You may use a normal approximation.

16. a. Find the distribution of the Wilcoxon test statistic W under the null hypothesis when $m = 4$, and $n = 3$.

    b. Use it to construct a critical region of level no more than .1 for the alternative that the Y population is shifted to the left of the X population.

    c. A sample of 3 batteries of brand Y was tested, and their lifetimes were 100, 115, and 135 hours of use. Another independent sample of 4 brand X batteries gave lifetimes of 89, 95, 102, and 109 hours. Perform a one-sided Wilcoxon rank-sum test of level no more than .1 to decide whether there is a difference in location of the lifetime distributions.

17. Following are percentages of voters who turned out for the 1984 presidential elections from several small cities in New England, and several others from Western states (Thomas, 1992). Use a two-sided Wilcoxon test of level no more than .05 to check for a difference in location of the voter turnout distributions. (You may simply itemize the extreme cases instead of finding the whole probability distribution.)

    New England: 47.5, 45.4, 37.3, 41.7, 40.3
    West: 38.3, 29.2, 39.5, 36.1, 39.0

18. Prove formula (9.61) relating the Mann–Whitney statistic to the Wilcoxon statistic. You may assume for simplicity that there are no ties in the data.

19. Derive formula (9.63) for the mean of the Wilcoxon statistic.

20. The following data are yearly divorce rates for a number of countries in Asia, and in Europe (United Nations, 1994). Use a large sample normal approximation to the Wilcoxon test statistic to check whether the distribution of European divorce rates is shifted to the right of the distribution of Asian divorce rates. Use approximate level .01. Resolve ties by rank averaging.

    Asia: 1.3, 1.6, 1.2, 0.8, 0.4, 1.4, 1.0, 0.8, 0.6, 0.1, 1.3,
             1.3, 0.9, 2.8, 0.8, 1.5, 1.9, 0.2, 0.5, 0.7, 1.6, 0.2
    Europe: 0.8, 2.1, 3.4, 1.9, 1.3, 2.7, 2.9, 2.5, 2.4, 3.0, 3.7,
             1.1, 2.9, 1.9, 1.9, 3.0, 3.1, 0.9, 2.4, 2.2, 2.8, 0.5

21. The probability mass function of the Wilcoxon rank-sum statistic can also be generated recursively in the way we describe in this problem. Denote

$$f_{m,n}(w) = P[W = w; \ X \text{ sample has size } m, \ Y \text{ sample has size } n].$$

a. Find the functions $f_{0,0}(w)$, $f_{0,1}(w)$, $f_{1,0}(w)$, and $f_{1,1}(w)$.

b. Show that

$$f_{m,n}(w) = \frac{m}{N} \cdot f_{m-1,n}(w - N) + \frac{n}{N} \cdot f_{m,n-1}(w).$$

c. Use the results of parts (a) and (b) to find the function $f_{2,2}(w)$.

22. Perform an approximate level .05 large sample Wilcoxon test for a difference in location of the death age distributions for Peoria and Chicago using the data in Exercise 5 in Section 7.2. Give tied data the average of the ranks they occupy.

23. This problem describes another way of constructing a nonparametric test for comparison of locations of distributions called the *median test*. Let $X_1, X_2, \ldots, X_m$ and $Y_1, Y_2, \ldots, Y_n$ be random samples from distributions F_X and F_Y, and suppose that we are interested in testing the null hypothesis that $F_X = F_Y$ against the alternative that the Y distribution is shifted to the left of the X distribution. Let M be the number of X's in the combined sample that are less than the median of the combined sample (For simplicity, assume that $N = m + n$ is even.) Argue that the statistic M has the following probability mass function under the null hypothesis:

$$P[M = k] = \frac{\binom{N/2}{k} \cdot \binom{N/2}{m-k}}{\binom{N}{m}}, \quad k = 0, \ldots, \min\{m, N/2\}.$$

Explain how to use this result to execute the test we have described at a level of no more than some desired α.

9.5 | Likelihood Ratio Tests

When we studied parameter estimation, we found that intuitively reasonable statistics like $\overline{X}$, $\hat{p}$, and $V = (n-1)S^2/n$ came out as those functions of the random sample that maximized the *likelihood function*, that is, the joint p.m.f. or p.d.f. of the sample values. This satisfying theoretical development has a counterpart here in the domain of hypothesis testing. Tests for means, variances, and proportions that we have constructed, supported only by common sense, also have a desirable property with regard to the likelihood function. This short section shows you that many of our procedures belong to the family called *likelihood ratio tests*.

Recall that if $X_1, X_2, \ldots, X_n$ is a random sample from a population whose p.m.f. or p.d.f. is $f(x; \theta)$, then the likelihood function is

$$L(\theta) = L(\theta; x_1, x_2, \ldots, x_n) = \prod_{i=1}^{n} f(x_i; \theta). \tag{9.68}$$

Consider a test of $H_0 : \theta \in \Omega_0$ vs. $H_a : \theta \in \Omega \setminus \Omega_0$, where Ω in this context denotes the set of all possible values for the parameter θ. To illustrate the ideas, think of the very special case of $\Omega_0 = \{\theta_0\}$ and $\Omega = \{\theta_0, \theta_1\}$, where the competing hypotheses are $H_0 : \theta = \theta_0$ vs. $H_a : \theta = \theta_1$. If the likelihood value at θ_0 is small in comparison to the likelihood value at θ_1, then we would tend to believe H_a rather than H_0, because θ_1 makes the observed data more likely than θ_0. The same can be said if the likelihood value at θ_0 is small in comparison to the maximum of the likelihood over all members of Ω—in our example the maximum over both θ_0 and θ_1. This motivates the following definition.

DEFINITION 9.5.1

The *likelihood ratio test* of $H_0 : \theta \in \Omega_0$ vs. $H_a : \theta \in \Omega \setminus \Omega_0$ based on a random sample $X_1, X_2, \ldots, X_n$ rejects H_0 if and only if

$$Y \equiv \frac{\max_{\theta \in \Omega_0} L(\theta)}{\max_{\theta \in \Omega} L(\theta)} < k. \tag{9.69}$$

The denominator in (9.69) is often just the likelihood function L evaluated at the maximum likelihood estimator $\widehat{\theta}$, unless there are other restrictions such as one-sidedness that Ω imposes. If H_0 is a simple hypothesis $\theta = \theta_0$, then there is nothing more to do to calculate the numerator of the likelihood ratio test statistic Y, because it is just $L(\theta_0)$. If either of these pleasant situations fail, then you have to call on your calculus to find the maximum values.

?Question 9.5.1 Argue that the critical region in (9.69) makes sense only if $0 < k < 1$.

Also, the inequality $Y < k$ frequently can be rewritten in an equivalent form in order to match the form of a critical region we derived earlier in the chapter, as you will see in the next examples.

Example 9.5.1 Let us derive the likelihood ratio test for $H_0 : \mu = \mu_0$ vs. $H_a : \mu > \mu_0$, where μ is the mean of a normal population with known variance σ^2. The numerator of Y in (9.69) is

$$\max_{\mu \in \{\mu_0\}} L(\mu) = L(\mu_0) = \prod_{i=1}^{n} \frac{1}{\sqrt{2\pi\sigma^2}} e^{-\frac{(x_i - \mu_0)^2}{2\sigma^2}}.$$

We know from Chapter 8 that the maximum of $L(\mu)$ over all $\mu \in \mathbb{R}$ is taken on at $\overline{X}$, which is the lone critical point of L. But in the denominator of (9.69) the maximum is to be taken over $\Omega = \{\mu : \mu \geq \mu_0\}$, because H_a is one-sided here. Then if $\overline{X} \geq \mu_0$, the maximum of L over Ω is at $\overline{X}$, otherwise the maximum is at μ_0 itself, as shown in Fig. 9.9.

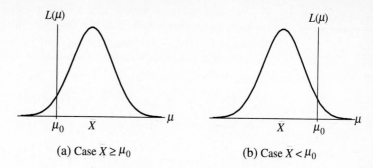

$L(\mu)$ $L(\mu)$

     μ_0 X μ X μ_0 μ

    (a) Case $X \geq \mu_0$ (b) Case $\bar{X} < \mu_0$

Figure 9.9   Normal Likelihood Function

Thus

$$Y = \begin{cases} \dfrac{L(\mu_0)}{L(\mu_0)} = 1 & \text{if } \bar{x} < \mu_0, \\[2ex] \dfrac{L(\mu_0)}{L(\bar{x})} & \text{if } \bar{x} \geq \mu_0. \end{cases} \tag{9.70}$$

In Question 9.5.1 you showed that in the critical region $Y < k$, k must be less than 1. Hence $Y < k$ cannot happen when $\bar{x} < \mu_0$. The likelihood ratio test criterion therefore rejects H_0 if and only if both $\bar{x} \geq \mu_0$ and

$$Y < k \Leftrightarrow \frac{L(\mu_0)}{L(\bar{x})} < k$$

$$\Leftrightarrow \prod_{i=1}^{n} \frac{1}{\sqrt{2\pi\sigma^2}} e^{-\frac{(x_i - \mu_0)^2}{2\sigma^2}} \bigg/ \prod_{i=1}^{n} \frac{1}{\sqrt{2\pi\sigma^2}} e^{-\frac{(x_i - \bar{x})^2}{2\sigma^2}} < k$$

$$\Leftrightarrow \exp\left[-\frac{1}{2\sigma^2} \sum_{i=1}^{n} (x_i - \mu_0)^2 - (x_i - \bar{x})^2\right] < k$$

$$\Leftrightarrow \sum_{i=1}^{n} (x_i - \mu_0)^2 - (x_i - \bar{x})^2 > c. \tag{9.71}$$

In the last line of (9.71) we have taken the log of both sides of the inequality, multiplied by $-2\sigma^2$, and just called the resulting right side by a new name c. The difference of squares in the last line can be expanded out and simplified to the form $n(\bar{x} - \mu_0)^2$, and because $\bar{x} \geq \mu_0$ the critical region has the form

$$n(\bar{x} - \mu_0)^2 > c \Leftrightarrow \bar{x} - \mu_0 > d.$$

This is exactly the decision rule we used earlier for the one-sided normal mean test.

Example 9.5.2

Next consider the two-sided test for the normal variance $H_0: \sigma^2 = \sigma_0^2$ vs. $H_a: \sigma^2 \neq \sigma_0^2$. We will derive the likelihood ratio test under the assumption that the mean μ is unknown.

What is different here is that the sets of parameter values Ω_0 and Ω must acknowledge the presence of the auxiliary parameter μ, even though it is not being tested for. Otherwise the test statistic Y would have the unknown μ in it, and hence Y would not be computable from the observed sample.

So, we set

$$\Omega_0 = \{(\mu, \sigma^2): \mu \in \mathbb{R}, \sigma^2 = \sigma_0^2\} \text{ and } \Omega = \{(\mu, \sigma^2): \mu \in \mathbb{R}, \sigma^2 > 0\}.$$

The likelihood function is

$$L(\mu, \sigma^2) = \prod_{i=1}^{n} \frac{1}{\sqrt{2\pi\sigma^2}} e^{-\frac{(x_i-\mu)^2}{2\sigma^2}}.$$

From Exercise 6 in Section 8.2, when $\sigma^2 = \sigma_0^2$ the maximizing value for μ is $\bar{x}$, and from Example 8.2.4, the joint maximizers over Ω are $\bar{x}$ and $v = \sum_{i=1}^{n}(x_i - \bar{x})^2/n$. Therefore the likelihood ratio statistic is

$$Y = \frac{\max_{\theta \in \Omega_0} L(\theta)}{\max_{\theta \in \Omega} L(\theta)} = \frac{L(\bar{x}, \sigma_0^2)}{L(\bar{x}, v)}$$

$$= \prod_{i=1}^{n} \frac{1}{\sqrt{2\pi\sigma_0^2}} e^{-\frac{(x_i-\bar{x})^2}{2\sigma_0^2}} \Big/ \prod_{i=1}^{n} \frac{1}{\sqrt{2\pi v}} e^{-\frac{(x_i-\bar{x})^2}{2v}}$$

$$= \left(\frac{v}{\sigma_0^2}\right)^{n/2} \exp\left(-\sum_{i=1}^{n} \frac{(x_i-\bar{x})^2}{2\sigma_0^2} + \sum_{i=1}^{n} \frac{(x_i-\bar{x})^2}{2v}\right)$$

$$= \left(\frac{v}{\sigma_0^2}\right)^{n/2} \exp\left(-\sum_{i=1}^{n} \frac{(x_i-\bar{x})^2}{2\sigma_0^2} + \frac{n}{2}\right)$$

$$= Cv^{n/2} \exp(-Dv). \tag{9.72}$$

In the last line we have grouped constants together into two positive constants C and D. Thus the likelihood ratio statistic is expressible as a certain function of the MLE V of σ^2. The null hypothesis is rejected when Y is small.

Figure 9.10 displays typical behavior of Y as a function of v for specific constants. Notice that the functional value is small when v is either too small or too large. Thus the rejection region for H_0 has the form $V > c_1$ or $V < c_2$, which is equivalent to the two-sided region using $S^2 = nV/(n-1)$ that we saw earlier.

Several other likelihood ratio tests are examined in the exercises.

In a more advanced course (see for example Bickel and Doksum, 1977), you will see proofs of some interesting properties of likelihood ratio tests. Under a few conditions, the random variable $W = -2\log Y$ is approximately chi-squared for large samples. The degrees of freedom parameter is the difference between the dimension of the Ω space and the dimension of the Ω_0 space. Also, it is a theorem called the *Neyman–Pearson lemma* that likelihood ratio tests have higher power than any other tests of the same significance

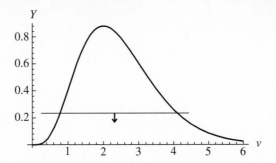

Figure 9.10 Behavior of Y as a Function of v

level when both hypotheses are simple, and in some more general cases as well.

EXERCISES 9.5

1. Derive the likelihood ratio test of $H_0 : \mu \le \mu_0$ vs. $H_a : \mu > \mu_0$ in the case of the normal mean with known variance. Simplify the test, and show that it is equivalent to the earlier one-sided test in Section 9.2.

2. Derive the likelihood ratio test of $H_0 : \mu = \mu_0$ vs. $H_a : \mu \ne \mu_0$ in the case of the normal mean with unknown variance. Simplify the test, and show that it is equivalent to the earlier one-sided test in Section 9.2.

3. Derive the likelihood ratio test of $H_0 : \sigma^2 = \sigma_0^2$ vs. $H_a : \sigma^2 < \sigma_0^2$ in the case of the normal variance with known mean. Simplify the test.

4. Derive the likelihood ratio test of $H_0 : p = p_0$ vs. $H_a : p \ne p_0$ in the case of the Bernoulli success parameter. Simplify the test and show that it is equivalent to the earlier two-sided test in Section 9.1.

5. Derive the likelihood ratio test of $H_0 : \lambda = \lambda_0$ vs. $H_a : \lambda > \lambda_0$ in the case of the parameter λ of the exponential distribution. Express the test in an equivalent form involving $\overline{X}$.

6. Derive the likelihood ratio test of $H_0 : \mu = \mu_0$ vs. $H_a : \mu \ne \mu_0$ in the case of the parameter μ of the Poisson distribution. Express the test in an equivalent form involving $\overline{X}$.

CHAPTER 10

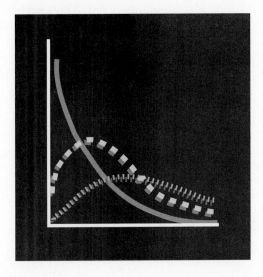

REGRESSION AND CORRELATION

10.1 | Least Squares Estimation

One of the most common statistical problems is to model the functional relationship between one variable X and another Y using a sample of pairs $(X_1, Y_1), (X_2, Y_2), \ldots, (X_n, Y_n)$. For instance, we might be interested in yield of corn in a field as a function of rainfall, or gas consumption of a car as a function of speed at which the car is driven, or height of a tree as a function of its girth, or chemical reaction rate as a function of reaction temperature, or myriad other possible functional relationships. In all of these examples, the presence of randomness leads the data points (X_i, Y_i) to scatter imperfectly about the curve relating the two variables. What we must try to do is use the scattered data to make inferences about the functional form of the relationship of Y to X. (The point of view usually taken is that Y is a function of X, although it might as well be the other way around, and we make no special claims about causal dependence of Y on X.)

So in this chapter we study the area of statistics known as *regression*. Section 10.1 concentrates on the estimation aspects of the regression problem in the special case where Y is linearly related to X and both X and Y are one-dimensional. The problem in which Y is a linear, single-variable function of a vector variable $\mathbf{X}$ is considered in Section 10.2. In later sections we look at hypothesis tests and confidence intervals for regression model parameters (Section 10.3), graphical tools for validating model assumptions (Section 10.4), and the related problem of correlation testing (Section 10.5).

Example 10.1.1

Consider the following data set of expected life spans for men (X) and for women (Y) for $n = 12$ nations in North America (United Nations, 1994). What is the relationship between the two variables? The scattergram in Fig. 10.1 suggests that the life spans for women depend roughly linearly on the male life spans, although the points do not fall perfectly on a line. How can we estimate the slope and intercept of a "best-fitting" line? What would you predict for the average life span of a woman in a country in which the male life span is 70?

$$(68, 75), \quad (67, 72), \quad (70, 72), \quad (69, 76), \quad (73, 80), \quad (72, 77),$$
$$(73, 76), \quad (64, 68), \quad (51, 64), \quad (60, 66), \quad (66, 72), \quad (55, 59)$$

After discussing some of the theory, we will return to this example.

Before we can progress, we must propose a statistical model that is amenable to analysis. The data are a random sample of pairs of random variables $(X_1, Y_1), (X_2, Y_2), \ldots, (X_n, Y_n)$. Assume that this sample comes from a population represented by a generic pair (X, Y) such that, conditioned on the observed value x of X,

$$Y = b_0 + b_1 x + \epsilon, \tag{10.1}$$

where b_0 and b_1 are the intercept and slope parameters to be estimated, and ϵ is a random variable with the $N(0, \sigma^2)$ distribution for some σ^2. For the

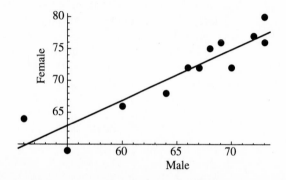

Figure 10.1 Scattergram of Female vs. Male Life Spans

entire sample of pairs then, we are assuming that

$$Y_1 = b_0 + b_1 x_1 + \epsilon_1,$$
$$Y_2 = b_0 + b_1 x_2 + \epsilon_2,$$
$$\vdots$$
$$Y_n = b_0 + b_1 x_n + \epsilon_n, \tag{10.2}$$

where $x_1, x_2, \ldots, x_n$ are the observed values of the X_i's, and the ϵ_i's are i.i.d. $N(0, \sigma^2)$ random variables.

It follows that for each $i = 1, 2, \ldots, n$, conditioned on $X_i = x_i$, $Y_i \sim N(b_0 + b_1 x_i, \sigma^2)$. Remember that one case in which the conditional mean of Y is linear in x is when the pair (X, Y) has the bivariate normal distribution discussed in Section 4.4. The exact form of the conditional mean is

$$\mu_{Y|x} = \mu_y + \frac{\rho \sigma_y}{\sigma_x}(x - \mu_x). \tag{10.3}$$

Hence a random sample from the bivariate normal distribution gives rise to experimental data that fits model (10.2).

For the next few sections we will adopt the convenient assumption that the observed values $x_1, x_2, \ldots, x_n$ are actually known. Any distributional results about estimators should then be interpreted as conditional upon $X_1 = x_1, X_2 = x_2, \ldots, X_n = x_n$. When we talk about correlation later, we will see that this simplification is not an injurious one. One of the most important tasks in regression is to test for significance of the slope coefficient b_1, and this test turns out to be the same regardless of whether we assume that the X_i's are random or known constants.

Look again at the model (10.1). The line $y = b_0 + b_1 x$, expressing the relationship between x and the mean of Y, is called the *theoretical regression line*. We will now find the maximum likelihood estimators $\widehat{b}_0$ and $\widehat{b}_1$ of the intercept and slope parameters, so as to produce an *estimated regression line* $y = \widehat{b}_0 + \widehat{b}_1 x$. Henceforth we will use the notation

$$\widehat{Y} = \widehat{b}_0 + \widehat{b}_1 x \tag{10.4}$$

for the predicted y value using this estimated regression equation for a given x. Some authors refer to this y value (especially when it is the predicted $\widehat{Y}_i = \widehat{b}_0 + \widehat{b}_1 x_i$ for one of the data points x_i) as the *fitted value* of y for the given x.

PROPOSITION 10.1.1

The maximum likelihood estimators of the coefficients b_0 and b_1 in the single variable linear regression model (10.2) are

$$\widehat{b}_1 = \frac{n \sum_{i=1}^{n} x_i y_i - \left(\sum_{i=1}^{n} x_i\right) \cdot \left(\sum_{i=1}^{n} y_i\right)}{n \sum_{i=1}^{n} x_i^2 - \left(\sum_{i=1}^{n} x_i\right)^2}$$
$$\widehat{b}_0 = \overline{y} - \widehat{b}_1 \overline{x}. \tag{10.5}$$

Moreover, the MLEs are the values that achieve the minimum of the *least squares function*:

$$\min_{b_0,b_1} l(b_0, b_1) = \min_{b_0,b_1} \sum_{i=1}^{n} (y_i - [b_0 + b_1 x_i])^2. \tag{10.6}$$

■ **Proof** Since the ϵ_i's are independent, so are the Y_i's. By our assumption $Y_i \sim N(b_0 + b_1 x_i, \sigma^2)$. Therefore the joint density of the Y_i's is

$$f(y_1, y_2, \ldots, y_n) = \prod_{i=1}^{n} \frac{1}{\sqrt{2\pi\sigma^2}} e^{-(y_i - [b_0 + b_1 x_i])^2 / 2\sigma^2}$$

$$= (2\pi\sigma^2)^{-n/2} e^{-\frac{1}{2\sigma^2} \sum_{i=1}^{n} (y_i - [b_0 + b_1 x_i])^2}. \tag{10.7}$$

Viewed as a function of b_0 and b_1 (and σ^2), this joint density expression is just the likelihood function $L(b_0, b_1, \sigma^2)$. Since the function $e^{-u/2\sigma^2}$ is a decreasing function of u for $u > 0$, to maximize the likelihood, it suffices to minimize the sum in the exponent, that is, $l(b_0, b_1) = \sum_{i=1}^{n} (y_i - [b_0 + b_1 x_i])^2$. This proves the second claim of the proposition.

A little calculus completes the process of estimating b_0 and b_1. The partial derivatives of the least squares function l in (10.6) with respect to b_0 and b_1 are

$$\frac{\partial l}{\partial b_0} = \sum_{i=1}^{n} 2(y_i - [b_0 + b_1 x_i])(-1)$$

$$\frac{\partial l}{\partial b_1} = \sum_{i=1}^{n} 2(y_i - [b_0 + b_1 x_i])(-x_i). \tag{10.8}$$

Equating both partials to 0 produces the so-called *normal equations* for $\widehat{b}_0$ and $\widehat{b}_1$:

$$\begin{cases} \sum_{i=1}^{n} y_i = \widehat{b}_0 \cdot n + \widehat{b}_1 \cdot \sum_{i=1}^{n} x_i \\[2mm] \sum_{i=1}^{n} x_i y_i = \widehat{b}_0 \cdot \sum_{i=1}^{n} x_i + \widehat{b}_1 \cdot \sum_{i=1}^{n} x_i^2. \end{cases} \tag{10.9}$$

The solution of this system of linear equations is the pair of expressions in (10.5). This finishes the proof.

Look carefully at the least squares criterion (10.6) to which the maximum likelihood method led, and refer to Fig. 10.2. The MLEs $\widehat{b}_0$ and $\widehat{b}_1$ are values of b_0 and b_1 such that the sum of the square differences between the data y_i values and their predicted values $\widehat{y}_i = b_0 + b_1 x_i$ is minimized. In other words, the maximum likelihood method says that we should choose $\widehat{b}_0$ and $\widehat{b}_1$ to minimize the total squared vertical distance of the data points to the line that $\widehat{b}_0$ and $\widehat{b}_1$ determine.

Because of our observation following (10.7), we were able to sidetrack the problem that in maximizing the likelihood function we need also to maximize

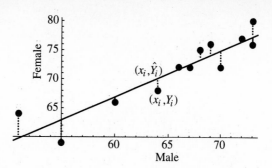

Figure 10.2   Graphical Description of the Least Squares Method

over the third unknown parameter σ^2. However the maximum likelihood estimator of σ^2 will have both practical and theoretical importance later. You are to derive it in Exercise 7 at the end of the section.

—?Question 10.1.1   Fill in the details of the computations in (10.8) and (10.9). Verify that the stationary point determined by equating the partials of l to 0 is indeed a minimum point.

Example 10.1.1

Continued   For the life span data, the regression summary statistics that make up formulas (10.5) are computed as

$$n = 12, \quad \sum x_i = 788, \quad \sum y_i = 857, \quad \sum x_i^2 = 52294, \quad \sum x_i y_i = 56713.$$

Substitution of these values into (10.5) gives

$$\widehat{b}_1 = \frac{12 \cdot 56713 - 788 \cdot 857}{12 \cdot 52294 - 788^2} \approx .796, \qquad \widehat{b}_0 = \frac{857}{12} - .796 \cdot \frac{788}{12} \approx 19.15.$$

The line in Fig. 10.1 is the estimated regression line $y = \widehat{b}_0 + \widehat{b}_1 x$. With the equation of the estimated line in hand, we can predict values of the life span of a woman given the male life span (or vice versa). To answer the question posed in the original example, the predicted average life span of a woman in a country in which the male life span is 70 would be $19.15 + .796(70) \approx 74.87$. We will investigate the precision of estimates like this one later in this chapter.

   Figure 10.3 contains the so-called *residuals* $e_i = y_i - \widehat{y}_i$ plotted in a time series. The residuals are the differences between the observed y-values and the fitted y-values. You are asked something about them in the next question.

—?Question 10.1.2   The residuals are plotted in order of their subscript in Fig. 10.3. Bearing in mind the original model, what do the e_i's estimate? What features would you

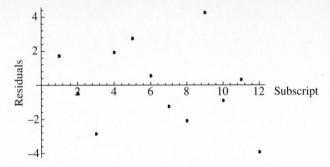

Figure 10.3 Residual Plot for Life Span Data

expect to see in the plot if the model assumptions are correct? We will return to this subject in Section 10.4.

Computing regression estimates by hand is unnecessary when it is so simple to let a machine do the dirty work. A screen dump of part of a *Minitab* session for our life span data is shown in Fig. 10.4. This output is typical of the way in which statistics programs handle regression analysis. The top line shows how to request regression estimates (see Appendix B on *Minitab* for more information). We are regressing the dependent variable *female*, whose values are in column 2 of the *Minitab* data worksheet, on one predic-

```
MTB>regress 'female' on 1 predictor 'male';
SUBC>residuals in c20;
SUBC>fits in C21.

The regression equation is
female = 19.2 + 0.796 male

Predictor    Coef     Stdev    t-ratio      p
Constant    19.155    7.095      2.70     0.022
male         0.7959   0.1075     7.40     0.000

s = 2.518    R-sq = 84.6%    R-sq(adj) = 83.0%

Analysis of Variance

SOURCE        DF     SS       MS       F       p
Regression     1   347.53   347.53   54.83   0.000
Error         10    63.39     6.34
Total         11   410.92

Unusual Observations
Obs.   male   female    Fit    Stdev.Fit   Residual   St.Resid
  9    51.0   64.000   59.744     1.736      4.256      2.33R

R denotes an obs. with a large st. resid.
```

Figure 10.4 *Minitab* Regression Output

tor variable named *male*, located in column 1. (The *residuals* subcommand is optional. It saves the values $e_i = y_i - \hat{y}_i$ into the designated column. The optional "fits" subcommand saves the fitted values $\hat{y}_i$ into the column requested.) The estimators can be read off the displayed linear equation, or from the coefficient column in the table below it. We will discuss what some of the other *Minitab* output means later.

The slope estimator $\hat{b}_1$ can be written in several equivalent ways, all of which convey different but useful information. The form (10.5) has the value of being computationally efficient. You can also check that

$$\hat{b}_1 = \frac{\sum_{i=1}^n (x_i - \bar{x})(Y_i - \bar{Y})}{\sum_{i=1}^n (x_i - \bar{x})^2} = \frac{\sum_{i=1}^n (x_i - \bar{x})Y_i}{\sum_{i=1}^n (x_i - \bar{x})^2}. \tag{10.10}$$

The middle form of $\hat{b}_1$ suggests an interesting parallel to the bivariate normal conditional mean formula, as I would like you to explore in the next question. The last form in (10.10) indicates that $\hat{b}_1$ is a linear function of the Y_i's; hence it is normally distributed under our model assumptions.

—— **?Question 10.1.3**   Verify (10.10). Show that the middle form is equal to RS_y/s_x, where R is the sample correlation between x and Y. Does this remind you of anything?

By the result in Question 10.1.3, the estimated regression line has the form

$$y = \hat{b}_0 + \hat{b}_1 x = \bar{y} - \hat{b}_1 \bar{x} + \hat{b}_1 x$$
$$= \bar{y} + \hat{b}_1 (x - \bar{x})$$
$$= \bar{y} + R\frac{S_y}{s_x}(x - \bar{x}). \tag{10.11}$$

Thus the estimated line has a form satisfyingly similar to the formula for $E[Y|X=x]$, with theoretical parameters replaced by their natural estimates (see formula (10.3)).

The *Minitab* output of Fig. 10.4 references the R^2 value 84.6% = .846, which is a measure of how well the regression model predicts the observed data, or more specifically, how much of the total variability of the data is accounted for by the model. This R^2 is calculated as follows, using sums of squares. Define the *total sum of squares* of the Y_i data by

$$SST = \sum_{i=1}^n (Y_i - \bar{Y})^2, \tag{10.12}$$

and the *sum of squares due to error* by

$$SSE = \sum_{i=1}^n (Y_i - \hat{Y}_i)^2, \tag{10.13}$$

where, as before, $\hat{Y}_i = \hat{b}_0 + \hat{b}_1 x_i$. Then SSE measures the deviation of the actual Y's from the values predicted for them by the model, and SST measures the total deviation of the Y's from their sample mean. Finally define

$$SSR = \sum_{i=1}^n (\hat{Y}_i - \bar{Y})^2, \tag{10.14}$$

called the *regression sum of squares*. The regression and error sums of squares are components of the total sum of squares, as expressed in the following decomposition formula, which you are asked to prove in Exercise 12 at the end of this section:

$$SST = SSE + SSR. \tag{10.15}$$

The intuition is that the Y_i's deviate from their mean $\overline{Y}$, partly because they have different expected values. After accounting for the predictable deviation SSR of Y values from $\overline{Y}$, what is left from the total sum of squares is unexplained deviation between the predicted and actual values of the Y_i's. The larger is this unexplained deviation, the poorer is the fit of the model to the data. It then makes sense to define the regression statistic R^2 by

$$R^2 = \frac{SSR}{SST} = 1 - \frac{SSE}{SST}, \tag{10.16}$$

the proportion of total variability explained by the model. The number R^2 will be close to 1 when the fit is good, and close to 0 otherwise.

You can acquire an intuitive sense through the experience of working with scattergrams of what a numerical value of R^2 between 0 and 1 actually means; for instance, for the life span data set, $R^2 = .846$, which seems to indicate a good fit, with little deviation of points from the line. You would see data points that are more widely scattered about the estimated regression line if $R^2 = .5$, and you would see almost no linear dependence between the two variables if $R^2 = .1$. For another important characterization of R^2 as a squared correlation (thus the notation), see Exercise 14 at the end of this section.

Minitab calculates the sums of squares and displays them in the analysis of variance table (see Fig. 10.4) under the *SS* column. For the life span example, $SSR = 347.53$, $SSE = 63.39$, and $SST = 410.92$, which yields the reported value $R^2 = SSR/SST = 347.53/410.92 = .846$.

The error sum of squares has another important role. We now show that a function of SSE is unbiased for the error variance σ^2, and we also show the unbiasedness of our estimators of the slope and intercept parameters.

PROPOSITION 10.1.2

The statistics $\widehat{b}_0$ and $\widehat{b}_1$, are unbiased estimators of b_0 and b_1, respectively. Also,

$$E[SSE] = (n - 2)\sigma^2, \tag{10.17}$$

and hence the statistic

$$MSE = \frac{SSE}{n - 2} \tag{10.18}$$

is unbiased for σ^2. (We refer to MSE as the *mean square for error*.)

■ **Proof** The computation of $E[\widehat{b}_1]$ is long but straightforward, using the rightmost form of $\widehat{b}_1$ in (10.10):

$$E[\widehat{b}_1] = E\left[\frac{\sum_{i=1}^{n}(x_i - \overline{x})Y_i}{\sum_{i=1}^{n}(x_i - \overline{x})^2}\right]$$

$$= \frac{1}{\sum_{i=1}^{n}(x_i - \overline{x})^2} \cdot \sum_{i=1}^{n}(x_i - \overline{x})E[Y_i]$$

$$= \frac{1}{\sum_{i=1}^{n}(x_i - \overline{x})^2} \cdot \sum_{i=1}^{n}(x_i - \overline{x})(b_0 + b_1 x_i)$$

$$= \frac{1}{\sum_{i=1}^{n}(x_i - \overline{x})^2} \cdot \left[b_0 \sum_{i=1}^{n}(x_i - \overline{x}) + b_1 \sum_{i=1}^{n}(x_i - \overline{x})x_i \right]$$

$$= \frac{1}{\sum_{i=1}^{n}(x_i - \overline{x})^2} \cdot \left[0 + b_1 \sum_{i=1}^{n}(x_i - \overline{x})(x_i - \overline{x} + \overline{x}) \right]$$

$$= \frac{1}{\sum_{i=1}^{n}(x_i - \overline{x})^2} \cdot \left[b_1 \sum_{i=1}^{n}(x_i - \overline{x})^2 + b_1\overline{x} \sum_{i=1}^{n}(x_i - \overline{x}) \right]$$

$$= \frac{1}{\sum_{i=1}^{n}(x_i - \overline{x})^2} \cdot \left[b_1 \sum_{i=1}^{n}(x_i - \overline{x})^2 + b_1\overline{x} \cdot 0 \right] = b_1.$$

Hence $\widehat{b}_1$ is an unbiased estimator of b_1.

You are asked to check in Exercise 8 that $\widehat{b}_0$ is unbiased for b_0.

To show (10.17), we will do the lengthy (but not difficult) computation in small chunks. First, by the definition of SSE, we can express the desired expectation in terms of the expectations of the squares of the Y_i's, the square of $\overline{Y}$, and the square of $\widehat{b}_1$, by expanding out the square in SSE:

$$E[SSE] = E\left[\sum_{i=1}^{n}(Y_i - \widehat{Y}_i)^2 \right]$$

$$= E\left[\sum_{i=1}^{n}(Y_i - [\overline{Y} + \widehat{b}_1(x_i - \overline{x})])^2 \right]$$

$$= E\left[\sum_{i=1}^{n}(Y_i - \overline{Y})^2 - 2\sum_{i=1}^{n}(Y_i - \overline{Y})\widehat{b}_1(x_i - \overline{x}) + \sum_{i=1}^{n}\widehat{b}_1^2(x_i - \overline{x})^2 \right]$$

$$= E\left[\sum_{i=1}^{n}(Y_i - \overline{Y})^2 - \widehat{b}_1^2 \sum_{i=1}^{n}(x_i - \overline{x})^2 \right]$$

$$= E\left[\sum_{i=1}^{n}Y_i^2 - n\overline{Y}^2 - \widehat{b}_1^2 \sum_{i=1}^{n}(x_i - \overline{x})^2 \right]. \tag{10.19}$$

The expression is expanded out in the third line by grouping the Y_i and $\overline{Y}$ terms together. The second and third terms on the right in line 3 are brought together in the fourth line using formula (10.10) for $\widehat{b}_1$. Expansion and simplification of $(Y_i - \overline{Y})^2$ produces the expression in line 5.

Now the expected square of any random variable W is the sum of the variance of W and the squared expectation of W. We just have to apply this to each of Y_i^2, $\overline{Y}^2$, and $\widehat{b}_1^2$ in order to complete the computation. We have

$$\text{Var}(Y_i) = \sigma^2, \qquad E[Y_i] = b_0 + b_1 x_i;$$

$$\text{Var}(\overline{Y}) = \frac{1}{n^2} \sum_{i=1}^{n} \text{Var}(Y_i) = \frac{\sigma^2}{n},$$

$$E[\overline{Y}] = \frac{1}{n} \sum_{i=1}^{n} E[Y_i] = \frac{1}{n} \sum_{i=1}^{n} (b_0 + b_1 x_i) = b_0 + b_1 \overline{x};$$

$$\text{Var}(\widehat{b}_1) = \text{Var}\left(\frac{\sum_{i=1}^{n}(x_i - \overline{x})Y_i}{\sum_{i=1}^{n}(x_i - \overline{x})^2} \right)$$

$$= \left[\frac{1}{\sum_{i=1}^{n}(x_i - \overline{x})^2} \right]^2 \text{Var}\left[\sum_{i=1}^{n}(x_i - \overline{x})Y_i \right]$$

$$= \left[\frac{1}{\sum_{i=1}^{n}(x_i - \overline{x})^2} \right]^2 \sum_{i=1}^{n}(x_i - \overline{x})^2 \sigma^2 = \frac{\sigma^2}{\sum_{i=1}^{n}(x_i - \overline{x})^2}$$

$$E[\widehat{b}_1] = b_1.$$

Substituting all these into (10.19) and simplifying gives

$$E[SSE] = \sum_{i=1}^{n} [\text{Var}(Y_i) + (E[Y_i])^2] - n(\text{Var}(\overline{Y}) + (E[\overline{Y}])^2)$$

$$- \sum_{i=1}^{n}(x_i - \overline{x})^2 (\text{Var}(\widehat{b}_1) + (E[\widehat{b}_1])^2)$$

$$= \sum_{i=1}^{n} [\sigma^2 + (b_0 + b_1 x_i)^2] - n[\sigma^2/n + (b_0 + b_1 \overline{x})^2]$$

$$- \sum_{i=1}^{n}(x_i - \overline{x})^2 \left[\frac{\sigma^2}{\sum_{i=1}^{n}(x_i - \overline{x})^2} + b_1^2 \right]$$

$$= n\sigma^2 - \sigma^2 - \sigma^2 + \sum_{i=1}^{n}(b_0 + b_1 x_i)^2 - n(b_0 + b_1 \overline{x})^2$$

$$- b_1^2 \sum_{i=1}^{n}(x_i - \overline{x})^2$$

$$= (n - 2)\sigma^2.$$

(Check the cancellation in the last line.) This completes the proof.

The mean square for error $MSE = SSE/(n - 2)$ will play a key role in the hypothesis testing and confidence interval procedures that we will consider later in the chapter.

Example 10.1.2

I have coordinated our college mathematics placement system for several years. We administer a 30-question multiple-choice test to entering freshmen, and we use that information, together with college board mathematics test scores, to advise students on the most appropriate mathematics course. I became interested in the ability of test scores to predict each other and to predict students' performance in later courses. Table 10.1 contains data for 25 students, with variables as follows: placement test score (0–30), SAT math score, ACT math score, and Calculus I grade (encoded as 4.0 = A, 3.7 = A−, 3.3 = B+, 3.0 = B, etc.).

We will use the complete data set in the next section when we discuss the problem of regressing a dependent variable on several predictor variables. For now, let us see how to predict the calculus grade by the placement exam score.

Figure 10.5 displays the scattergram of grade vs. placement, which shows a great deal of variability. From *Minitab*, the unbiased estimates of the re-

Table 10.1

Student Scores and Grades

Placement	SAT	ACT	Calculus
15	530	26	B+ (3.3)
19	530	21	F (0)
22	650	26	B+ (3.3)
23	570	30	A (4)
22	520	29	B (3)
22	510	24	F (0)
24	540	25	B (3)
21	640	23	C (2)
21	710	29	B (3)
21	580	26	C (2)
19	590	21	C (2)
21	550	26	C (2)
17	490	28	B (3)
9	310	17	C− (1.7)
23	690	32	B− (2.7)
24	690	30	A− (3.7)
23	560	21	C (2)
17	580	26	A (4)
22	690	29	A− (3.7)
20	610	22	B (3)
22	690	25	B (3)
22	610	27	B− (2.7)
21	540	30	C (2)
19	570	25	B− (2.7)
22	700	30	A− (3.7)

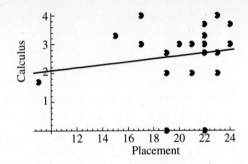

Figure 10.5   Scattergram of Calculus Grades vs. Placement Scores

gression coefficients are $\hat{b}_0 = 1.58$ and $\hat{b}_1 = .0511$, which gives an estimated regression line of

$$y = 1.58 + .0511x,$$

shown on the scattergram. For example, a student who scores 20 on the placement exam would be predicted to get a grade of $1.58 + .0511(20) = 2.60$ in Calculus I. Also, *Minitab* tells us that $SSE = 25.337$ and $SSR = .663$, which implies that the observed R^2-value is $.663/(25.337 + .663) = .025$. Thus only about 2.5% of the variability in the data is explained by the regression model; the remaining 97.5% is error variance. The unbiased estimate of the error variance σ^2 is $MSE = SSE/23 = 1.102$, which is large in comparison with the kind of grade point codes (0–4) that are taken on in this problem. All of this corresponds with the fact that the data are so widely scattered about the regression line and confirms that we should be cautious about drawing definitive conclusions on the basis of the placement test alone. Perhaps better predictions are possible if the other test variables are used as predictors. We will find out in the next section.

— **?Question 10.1.4**   How would you answer the question, Is the ACT a better predictor of calculus grades than our homegrown placement exam? Go to a statistics program and find the answer.

EXERCISES 10.1   1. The first coordinates in the following pairs are labels for years (beginning in 1985), and the second coordinates are the numbers of tourists (millions) who visited Australia in those years (United Nations, 1994). Use a regression model to answer these questions: What is the tourism in 1993 predicted to be? When will the tourism exceed three million?

$$(1, 1.138),\ (2, 1.425),\ (3, 1.779),\ (4, 2.249),$$
$$(5, 2.080),\ (6, 2.215),\ (7, 2.370),\ (8, 2.603)$$

2. Some authors present the regression model and estimators in a different, but equivalent form. Let $\bar{x}$ be the mean of the x_i's. The linear model
$$Y_i = a + b(x_i - \bar{x}) + \epsilon_i, \qquad i = 1, 2, \ldots, n$$
is an alternative model to (10.2), related to it by the change of parameters $b_0 = a - b\bar{x}$ and $b_1 = b$. Derive the maximum likelihood estimators of the parameters a and b in this revised regression model. Show that the equation of the estimated regression line is the same as it is using the maximum likelihood estimators $\widehat{b}_0$ and $\widehat{b}_1$ in (10.5).

3. The following pairs give first the x-coordinate, the approximate number of milk cows (in thousands) in the United States during each of the years 1976–1990, and then the y-coordinate, the average number of pounds of milk given during the year per cow (USDA, 1991). Find the equation of the estimated regression line relating the two variables, and use it to estimate the number of cows that would maximize the total yield of milk. Comment on the reasonableness of your mathematical analysis in the real situation.

(11,032, 10,894),   (10,945, 11,206),   (10,803, 11,243),   (10,734, 11,492),
(10,799, 11,891),   (10,898, 12,183),   (11,011, 12,306),   (11,059, 12,622),
(10,793, 12,541),   (10,981, 13,024),   (10,773, 13,285),   (10,327, 13,819),
(10,262, 14,145),   (10,126, 14,244),   (10,127, 14,642)

4. Find the equations of the regression lines relating runs scored to (a) home runs and (b) team batting average for the baseball data set in Table 7.6. What do you think of these two variables as predictors of runs scored? If you were a manager, would you rather have an extra 10 points in team batting average or an extra 25 home runs?

5. Recall the data set on bowling scores in Exercise 12 in Section 9.4. Produce a regression model relating the score on the second game to the score on the first game. Plot the estimated regression line together with the data. Comment on how well your model fits the data. Estimate the variance σ^2 of the error terms ϵ_i, and predict the score on the second game if the score on the first is 100.

6. The following data pairs are percentages of the gross national product spent by several Asian nations and percentages spent for defense-related purposes by the same nations in 1990 (United Nations, 1994). Produce a scatterplot with the estimated regression line superimposed. Do you see a strange data point? Redo the problem without this point, and comment on how the results change, and why.

(13.5, .9),   (11.6, 3.3),   (11.1, 3.1),   (9.2, .9),   (10.8, 3.7),
(14.2, 3.4),   (35.4, 28.3),   (13.7, 3.2),   (9.4, 3.4)

7. Derive the maximum likelihood estimator for the error variance in the linear regression model. Did you see it in this section in another context?

8. Show that the maximum likelihood estimator $\widehat{b}_0$ is unbiased for b_0.

9. Recall the statistics test data in Table 7.2. Do the data show a strong linear relationship between the test 1 and test 3 scores? Find the estimated regression equation. What would a student who received a 70 on test 1 be expected to score on test 3?

10. Suppose we impose the restriction on the linear model that the intercept parameter must be 0. Find the new maximum likelihood estimator of the slope parameter.

11. The model $Y = ae^{bx}$, where ϵ is a random variable with some mean and variance, is not linear. Could a change of variables produce a linear model? What conditions would ϵ have to satisfy in order to apply the analysis of this section? Describe carefully how estimates of the parameters a and b could be obtained.

12. Verify the sum of squares decomposition (10.15).

13. Argue that in order to pick an even number of values $x_1, x_2, \ldots, x_n$ on a closed interval $[p, q]$ so as to minimize the variance of $\widehat{b}_1$, you should set half of the x_i's equal to p and the other half equal to q.

14. Prove that $R^2 = 1 - SSE/SST$ is also the square of the sample correlation between the observed Y_i's and their predicted values $\widehat{Y}_i$.

15. For the bowling score data in Exercise 5, suppose that the true value of the slope coefficient β_1 is .5, and the true error variance $\sigma^2 = 121$. Find the probability that the estimated slope $\widehat{\beta}_1$ comes within .1 of the true β_1. (*Hint:* Use what was shown in Proposition 10.1.2.)

16. There is a nonparametric estimation procedure for the slope coefficient in a linear model. Suppose that $Y_i = b_0 + b_1 x_i + \epsilon_i$, $i = 1, 2, \ldots, n$ where the errors ϵ_i are not necessarily normal but are uncorrelated with mean 0 and common variance. The *Thiel estimator* of b_1 is the median of the set of all possible slope values $m_{ik} = (Y_i - Y_k)/(x_i - x_k)$ of line segments connecting two of the data points at a time. Note that we must assume that no two points share a common x value. Discuss the possible advantages of this estimator over the least squares estimator. (*Hint:* What if there were a strange outlying data point in the sample?) Compute the Thiel estimator of b_1 for the data (1.2,6.3), (1.3,7.0), (1.4,6.8), (1.5,7.3), (1.6,7.5).

10.2 | Multiple Regression

We will go on in this section to study least squares estimation in multiple regression models, in which Y has a mean that is linear in many variables $x_1, x_2, \ldots, x_k$. The best way to approach this problem is to use a matrix formulation of the model, which specializes to the one-variable case.

Looking, for instance, at the test score and calculus grade data in Table 10.1, you can see that the calculus grades constitute a column vector $\mathbf{Y}$, whose ith component Y_i is the grade of the ith student. Similarly, the placement exam scores form a column vector $\mathbf{x}_p = [x_{1,p}\, x_{2,p} \ldots x_{25,p}]'$. Let $\boldsymbol{\beta} = [b_0\, b_p]'$, $\boldsymbol{\epsilon} = [\epsilon_1\, \epsilon_2 \ldots \epsilon_{25}]'$, and let $\mathbf{1}$ stand for a column vector consisting of 25 1s. Then the single variable regression model can be written

$$\mathbf{Y} = \begin{bmatrix} \mathbf{1} & \mathbf{x}_p \end{bmatrix} \cdot \boldsymbol{\beta} + \boldsymbol{\epsilon}, \tag{10.20}$$

or in full form

$$\begin{bmatrix} Y_1 \\ Y_2 \\ \vdots \\ Y_{25} \end{bmatrix} = \begin{bmatrix} 1 & x_{1,p} \\ 1 & x_{2,p} \\ \vdots & \vdots \\ 1 & x_{25,p} \end{bmatrix} \cdot \begin{bmatrix} b_0 \\ b_p \end{bmatrix} + \begin{bmatrix} \epsilon_1 \\ \epsilon_2 \\ \vdots \\ \epsilon_{25} \end{bmatrix}.$$

The matrix $X = [\mathbf{1} \, \mathbf{x}_p]$ on the right side of (10.20) is called the *design matrix*. To extend the regression model to include the SAT and ACT variables as well, it is necessary to put a column into the design matrix for each of these, that is, $X = [\mathbf{1} \, \mathbf{x}_p \, \mathbf{x}_s \, \mathbf{x}_a]$ and to enlarge the vector of parameters so that each variable has its own coefficient, say, $\boldsymbol{\beta} = [b_0 \, b_p \, b_s \, b_a]'$. The model $\mathbf{Y} = X\boldsymbol{\beta} + \boldsymbol{\epsilon}$ in full form is

$$\begin{bmatrix} Y_1 \\ Y_2 \\ \vdots \\ Y_{25} \end{bmatrix} = \begin{bmatrix} 1 & x_{1,p} & x_{1,s} & x_{1,a} \\ 1 & x_{2,p} & x_{2,s} & x_{2,a} \\ \vdots & \vdots & \vdots & \vdots \\ 1 & x_{25,p} & x_{25,s} & x_{25,a} \end{bmatrix} \cdot \begin{bmatrix} b_0 \\ b_p \\ b_s \\ b_a \end{bmatrix} + \begin{bmatrix} \epsilon_1 \\ \epsilon_2 \\ \vdots \\ \epsilon_{25} \end{bmatrix}. \qquad (10.21)$$

—— **?Question 10.2.1** For the single-variable model in which calculus grades are related to placement scores only, show that the least squares function $l(b_0, b_p)$ is the same as $(\mathbf{Y} - X\boldsymbol{\beta})' \cdot (\mathbf{Y} - X\boldsymbol{\beta}) = \|\mathbf{Y} - X\boldsymbol{\beta}\|^2$.

As the last paragraph suggested, a *multiple regression model* with k predictor variables $x_1, x_2, \ldots, x_k$ and n data points can be written in matrix form as

$$\mathbf{Y} = X\boldsymbol{\beta} + \boldsymbol{\epsilon}, \qquad (10.22)$$

or

$$\mathbf{Y} = \begin{bmatrix} Y_1 \\ Y_2 \\ \vdots \\ Y_n \end{bmatrix}, \quad X = \begin{bmatrix} 1 & x_{11} & x_{12} & \cdots & x_{1k} \\ 1 & x_{21} & x_{22} & \cdots & x_{2k} \\ \vdots & \vdots & \vdots & \ddots & \vdots \\ 1 & x_{n1} & x_{n2} & \cdots & x_{nk} \end{bmatrix}, \quad \boldsymbol{\beta} = \begin{bmatrix} \beta_0 \\ \beta_1 \\ \beta_2 \\ \vdots \\ \beta_k \end{bmatrix}, \quad \boldsymbol{\epsilon} = \begin{bmatrix} \epsilon_1 \\ \epsilon_2 \\ \vdots \\ \epsilon_n \end{bmatrix},$$

$$(10.23)$$

where $\epsilon_1, \epsilon_2, \ldots, \epsilon_n$ are independent and identically distributed $N(0, \sigma^2)$ random variables. By multiplying out the matrices, the model in full form is

$$Y_1 = \beta_0 + \beta_1 x_{11} + \cdots + \beta_k x_{1k} + \epsilon_1,$$
$$Y_2 = \beta_0 + \beta_1 x_{21} + \cdots + \beta_k x_{2k} + \epsilon_2,$$
$$\vdots$$
$$Y_n = \beta_0 + \beta_1 x_{n1} + \cdots + \beta_k x_{nk} + \epsilon_n. \qquad (10.24)$$

In the matrix X each column 1 through k represents a variable; the column of 1s can be thought of as an additional constant variable. A row corresponds to a particular multivariate observation $(1, x_{i1}, x_{i2}, \ldots, x_{ik}, Y_i)$ among the n sample observations. For the placement test example we have $k = 3$ predictor variables (plus one for the constant), and $n = 25$ observations. The estimation problem of multiple regression is to find estimators for the unknown parameters $\beta_0, \beta_1, \ldots, \beta_k$.

—— **?Question 10.2.2** What distribution does the vector $\boldsymbol{\epsilon}$ have? In light of this, what distribution does $\mathbf{Y}$ have?

The least squares criterion again gives the maximum likelihood estimates (see Exercise 5), which means that we could simply do the appropriate calculus minimization of the sum of squares

$$l(\beta_0, \beta_1, \ldots, \beta_k) = \sum_{i=1}^{n} (Y_i - [\beta_0 + \beta_1 x_{i1} + \cdots + \beta_k x_{ik}])^2 \qquad (10.25)$$

in order to derive the optimal values of $\beta_0, \beta_1, \ldots, \beta_k$ (see Exercise 2). Instead we will take a matrix approach, which sets up the distributional results that we will prove later for the least squares estimators $\widehat{\beta}_0, \widehat{\beta}_1, \ldots, \widehat{\beta}_k$.

We need the following results. If $\mathbf{a}$ is a constant vector and $f(\mathbf{x}) = \mathbf{a}' \cdot \mathbf{x} = \mathbf{x}' \cdot \mathbf{a}$ is a linear function of a vector $\mathbf{x} = [x_1 \ \cdots \ x_n]'$, then the gradient of f is

$$\nabla f(\mathbf{x}) = \begin{bmatrix} \partial f / \partial x_1 \\ \vdots \\ \partial f / \partial x_n \end{bmatrix} = \mathbf{a}. \qquad (10.26)$$

(Verify this.) Also, if $\mathbf{A} = (a_{ij})$ is a constant symmetric matrix and the function $g(\mathbf{x})$ is the quadratic form $g(\mathbf{x}) = \mathbf{x}'\mathbf{A}\mathbf{x}$, then

$$\nabla g(\mathbf{x}) = 2\mathbf{A}\mathbf{x}. \qquad (10.27)$$

—— **?Question 10.2.3** Show (10.27) by expressing the quadratic form $\mathbf{x}'\mathbf{A}\mathbf{x}$ in expanded form as $\sum_j \sum_k a_{jk} x_j x_k$.

As Question 10.2.1 indicated, the expression in (10.25) is really a dot product of the row vector $(\mathbf{Y} - X\boldsymbol{\beta})'$ with the column vector $(\mathbf{Y} - X\boldsymbol{\beta})$, which is the same as the norm squared $\|\mathbf{Y} - X\boldsymbol{\beta}\|^2$. The least squares function can therefore be written

$$
\begin{aligned}
l(\boldsymbol{\beta}) &= (\mathbf{Y} - X\boldsymbol{\beta})'(\mathbf{Y} - X\boldsymbol{\beta}) \\
&= (\mathbf{Y}' - \boldsymbol{\beta}'X')(\mathbf{Y} - X\boldsymbol{\beta}) \\
&= \mathbf{Y}'\mathbf{Y} - \boldsymbol{\beta}'X'\mathbf{Y} - \mathbf{Y}'X\boldsymbol{\beta} + \boldsymbol{\beta}'X'X\boldsymbol{\beta}.
\end{aligned}
$$

The gradient of l, by (10.26) and (10.27), is

$$\nabla l(\boldsymbol{\beta}) = -2X'\mathbf{Y} + 2X'X\boldsymbol{\beta}.$$

(Check that $X'X$ is indeed symmetric, so that we can use (10.27).) Setting this gradient equal to zero gives

$$X'X\boldsymbol{\beta} = X'\mathbf{Y}. \qquad (10.28)$$

Formula (10.28) is the matrix version of the *normal equations*. If $X'X$ is invertible, then the solution to the normal equations—the least squares estimate of the vector parameter $\boldsymbol{\beta}$—is

$$\hat{\boldsymbol{\beta}} = (X'X)^{-1}X'\mathbf{Y}. \tag{10.29}$$

?Question 10.2.4 Suppose that the multiple regression model is $Y_i = \beta_0 + \beta_1 x_{i1} + \beta_2 x_{i2} + \epsilon_i$, $i = 1, 2, \ldots, n$. Find $X'X$. Write out the normal equations (10.28).

In full form, the normal equations become

$$n\beta_0 + \left(\sum_i x_{i1}\right)\beta_1 + \left(\sum_i x_{i2}\right)\beta_2 + \cdots + \left(\sum_i x_{ik}\right)\beta_k = \sum_i Y_i$$

$$\left(\sum_i x_{i1}\right)\beta_0 + \left(\sum_i x_{i1}^2\right)\beta_1 + \left(\sum_i x_{i1}x_{i2}\right)\beta_2 + \cdots + \left(\sum_i x_{i1}x_{ik}\right)\beta_k = \sum_i x_{i1}Y_i$$

$$\left(\sum_i x_{i2}\right)\beta_0 + \left(\sum_i x_{i1}x_{i2}\right)\beta_1 + \left(\sum_i x_{i2}^2\right)\beta_2 + \cdots + \left(\sum_i x_{i2}x_{ik}\right)\beta_k = \sum_i x_{i2}Y_i$$

$$\vdots$$

$$\left(\sum_i x_{ik}\right)\beta_0 + \left(\sum_i x_{i1}x_{ik}\right)\beta_1 + \left(\sum_i x_{i2}x_{ik}\right)\beta_2 + \cdots + \left(\sum_i x_{ik}^2\right)\beta_k = \sum_i x_{ik}Y_i. \tag{10.30}$$

(Verify this.) The solutions $\hat{\beta}_0, \hat{\beta}_1, \hat{\beta}_2, \ldots, \hat{\beta}_k$, if they exist, are the least squares estimators of the regression parameters $\beta_0, \beta_1, \beta_2, \ldots, \beta_k$.

Example 10.2.1

Returning to the placement exam data, with variables x_p, x_s, and x_a standing for the placement, SAT, and ACT scores respectively, and Y defined as the calculus grade, the relevant data are

$$\sum_{i=1}^{n} x_{ip} = 511, \qquad \sum_{i=1}^{n} x_{ip}^2 = 10699, \qquad \sum_{i=1}^{n} x_{ip}x_{is} = 304180,$$

$$\sum_{i=1}^{n} x_{ip}x_{ia} = 13397, \qquad \sum_{i=1}^{n} x_{is} = 14650, \qquad \sum_{i=1}^{n} x_{is}^2 = 8773300,$$

$$\sum_{i=1}^{n} x_{is}x_{ia} = 384020, \qquad \sum_{i=1}^{n} x_{ia} = 648, \qquad \sum_{i=1}^{n} x_{ia}^2 = 17116,$$

$$\sum_{i=1}^{n} y_i = 65.5, \qquad \sum_{i=1}^{n} x_{ip}y_i = 1351.8, \qquad \sum_{i=1}^{n} x_{is}y_i = 39366,$$

$$\sum_{i=1}^{n} x_{ia}y_i = 1748.6.$$

The matrix normal equations are

$$\begin{bmatrix} 25 & 511 & 14650 & 648 \\ 511 & 10699 & 304180 & 13397 \\ 14650 & 304180 & 8773300 & 384020 \\ 648 & 13397 & 384020 & 17116 \end{bmatrix} \begin{bmatrix} \beta_0 \\ \beta_p \\ \beta_s \\ \beta_a \end{bmatrix} = \begin{bmatrix} 65.5 \\ 1351.8 \\ 39366 \\ 1748.6 \end{bmatrix}.$$

I have computed the values of the coefficients of the unknown β's in (10.30) from the data only to emphasize the fact that the problem of finding the least squares estimators is one of solving a matrix linear equation. Statistics packages are well able to do all of the computational work for you, and you should rely on them.

Figure 10.6 gives an example of *Minitab* output for this problem. The estimated coefficients given by the REGRESS command are the solutions of the preceding matrix equation, which turn out to be

$$\widehat{\beta}_0 = -1.61, \qquad \widehat{\beta}_p = -0.137, \qquad \widehat{\beta}_s = 0.00513, \qquad \widehat{\beta}_a = 0.155.$$

Note that to format the data correctly for the REGRESS command to work, each variable $Y, x_1, x_2, x_3, \ldots$ must be in a column, with observations in consistent order, so that a row of the worksheet corresponds to a multivariate observation $(x_{i1}, x_{i2}, x_{i3}, \ldots, Y_i)$. As an example of the use of the regression equation, a student with placement score 15, math SAT score 500, and math

```
MTB>Regress 'calculus' 3 'place' 'sat' 'act'.

The regression equation is
calculus = -1.61 - 0.137 place + 0.00513 sat + 0.155 act

Predictor      Coef     Stdev     t-ratio     p
Constant     -1.605     1.370      -1.17     0.254
place       -0.13727    0.07476    -1.84     0.081
sat          0.005128   0.002789    1.84     0.080
act          0.15532    0.05833     2.66     0.015

s = 0.8408     R-sq = 42.9%     R-sq(adj) = 34.8%

Analysis of Variance

SOURCE        DF      SS        MS       F       p
Regression     3    11.1559   3.7186   5.26    0.007
Error         21    14.8441   0.7069
Total         24    26.0000

SOURCE        DF    SEQ SS
place          1    0.6629
sat            1    5.4820
act            1    5.0110

Unusual Observations
Obs.   place   calculus    Fit     Stdev.Fit   Residual   St.Resid
  2    19.0     0.000     1.766     0.292       -1.766      -2.24R
  6    22.0     0.000     1.718     0.345       -1.718      -2.24R
 14     9.0     1.700     1.389     0.652        0.311       0.58X

R denotes an obs. with a large st. resid.
X denotes an obs. whose X value gives it large influence.
```

Figure 10.6 *Minitab* Output for Multiple Regression

ACT score 20 would be predicted to have a calculus grade point average of

$$\hat{y} = \hat{\beta}_0 + \hat{\beta}_p x_p + \hat{\beta}_s x_s + \hat{\beta}_a x_a$$
$$= -1.61 + (-0.137) \cdot 15 + (0.00513) \cdot 500 + (0.155) \cdot 20 = 2.0.$$

Such a student would probably not be a good risk to take calculus. A strange phenomenon has occurred in this regression analysis: The placement exam score actually has a negative coefficient. For the equation to best fit the data, the part of the predicted GPA contributed by the SAT and the ACT overshoots the expected grade point, demanding a compensatory negative term to suppress the grade point. But this produces the counterintuitive result that, SAT and ACT scores being equal, the student with the lower placement score is predicted to have the greater success. This is not exactly what I was shooting for when I made up this exam. Fortunately, a more extensive analysis with many more data points does not show this phenomenon; students with higher placement scores are predicted to have higher calculus GPAs.

⎿

As you can see from Fig. 10.6, *Minitab*'s multiple regression output again gives R^2 values and sums of squares. They are computed in the same way as before. The total sum of squares

$$SST = \sum_{i=1}^{n}(Y_i - \overline{Y})^2$$

is partitioned into the sum of the regression sum of squares

$$SSR = \sum_{i=1}^{n}(\hat{Y}_i - \overline{Y})^2$$

plus the error sum of squares

$$SSE = \sum_{i=1}^{n}(Y_i - \hat{Y}_i)^2,$$

so that $SST = SSR + SSE$. The fitted values $\hat{Y}_i$ are computed as

$$\hat{Y}_i = \hat{\beta}_0 + \hat{\beta}_1 x_{i1} + \cdots + \hat{\beta}_k x_{ik}.$$

The R^2 value for a multiple regression problem is again the ratio SSR/SST, which is a measure of the goodness of the fit of the estimated regression equation to the data. In the placement exam example, the inclusion of the SAT and ACT has lifted the R^2 value from 2.5% to 42.9%, which makes it more reasonable to try to predict student performance by test scores. It is still a risky business, however.

A special case of multiple regression is *polynomial regression*, in which the model is of the form

$$Y = \beta_0 + \beta_1 x + \beta_2 x^2 + \cdots + \beta_k x^k + \epsilon. \tag{10.31}$$

All that is necessary to treat such problems is to regress the dependent variable Y on the k variables $x_1 = x, x_2 = x^2, x_3 = x^3, \ldots, x_k = x^k$. The normal equations are the same, with x_{ij} replaced by the jth power x_i^j. To set up the data set for *Minitab* computations, it is sufficient to place the observed Y values in a column, to place the x values in another column, and then to use the LET command to install the powers $x^2, x^3, \ldots$ into their own columns. For example, if x is in column 2, then LET C3 = C2**2 would put the values of the variable x^2 into column 3.

Example 10.2.2

I once assigned a project in which the students were to find the relationship between the tape counter of a VCR or audio cassette machine and actual time. You may find it interesting to use geometry to derive a quadratic formula relating the two, which depends on some constants (like tape thickness) that may be inconvenient to measure. One project group came up with the data in Table 10.2 for a machine that they used. For counter readings in increments of 50 from 0 to 1000, they recorded the clock time that had passed in seconds. A plot of clock time versus counter reading submitted to me by students Chris Katholi and Andy King is given in Fig. 10.7. There is a slight curvature in the data, suggesting that a quadratic function may be a good model.

Thus, we are proposing that if $Y =$ time and $x =$ counter reading, then $Y = \beta_0 + \beta_1 x + \beta_2 x^2 + \epsilon$, where $\epsilon \sim N(0, \sigma^2)$. I used *Minitab* to regress time on counter and the square of counter, as described above. The resulting polynomial regression equation was

$$y = -0.861 + 1.85x + .000323x^2,$$

and the R^2 value, found by dividing $SSR = 9,086,431$ by $SST = 9,086,433$ was very nearly 100%. Therefore our fitted model is extremely good, which perhaps is to be expected in a mechanical system subject to definite physical laws and only small random disturbances and errors in measurement. Our earlier example on test scores of human students shows the different nature of predictions of human behavior.

Table 10.2

VCR Counter Readings and Real Time

Count	Time	Count	Time	Count	Time
0	0	50	92.1	100	186.9
150	283	200	381.2	250	481
300	582.6	350	685.6	400	790
450	895.9	500	1004	550	1113.3
600	1224.1	650	1336.9	700	1450.7
750	1566.4	800	1684	850	1803.4
900	1923.6	950	2046.3	1000	2169.8

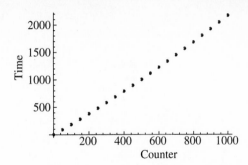

Figure 10.7 Scatterplot of Time vs. Tape Counter Reading

EXERCISES 10.2

1. Consider the baseball data set in Table 7.6. Regress the runs scored on all of the three predictor variables home runs, average, and stolen bases. Also regress the runs on each pair of other variables. Are there any variables that do not seem to contribute significantly to runs? Predict the number of runs scored by a team that hits 100 home runs, steals 150 bases, and has a team batting average of .258.

2. Suppose that the regression model is

$$Y_i = \beta_0 + \beta_1 x_{i1} + \beta_2 x_{i2} + \beta_3 x_{i3} + \epsilon_i, \qquad i = 1, 2, \ldots, n.$$

Using simple calculus on the least squares function, derive the form of the normal equations.

3. Like many women, my wife Lynn experienced gestational diabetes during her pregnancy with our daughter Emily. As part of her treatment, she had to monitor her blood glucose level before and after eating and at bedtime. Table 10.3 gives the data for breakfast, lunch, and bedtime for ten days.
   a. Does there seem to be a significant linear and/or quadratic relationship between the before- and after-breakfast readings?
   b. Answer the same question for the before- and after-lunch readings.
   c. Try to regress the glucose level at bedtime on all of the other four variables, then only on the after-meal variables. What conclusions can you draw?

4. In previous exercises (Exercise 10 in Section 4.4, Exercise 3 in Section 5.5) we have referred to a data set of performance by Peoria students. Table 10.4 gives values of several variables: years of experience of teaching staff, teacher salary, expense per pupil, reading score, and math score (all averages for the individual school) for a number of elementary schools.
   a. Which, if any, of the first three variables seem to be influential predictors of the reading score? What score would be predicted for a school whose teaching staff averaged 15 years' experience and a $30,000 salary?
   b. Which, if any, of the first three variables seem to be influential predictors of the math score? What score would be predicted for a school whose teaching staff averaged 12 years' experience and that spent $3200 per pupil?

Table 10.3

Blood Glucose Levels

Before Breakfast	After Breakfast	Before Lunch	After Lunch	Bedtime
90	124	74	112	90
81	142	86	103	110
82	108	78	99	105
90	153	85	125	88
84	120	74	92	104
94	134	80	120	90
89	142	67	119	102
92	132	96	105	100
91	134	96	103	132
102	128	76	120	85

Table 10.4

Peoria Elementary School Data

Years Experience	Teacher Salary	Expense per Pupil	Reading Score	Math Score
16.6	31463	3177	276	272
15.2	30009	3174	296	328
11.7	25877	3148	247	200
11.5	28639	3390	286	307
11.2	25804	2902	239	258
17.7	29404	3047	297	281
19.2	31416	3813	346	387
17.5	31553	7174	292	293
18.9	31223	3927	289	300
14.1	29542	3049	348	320
17.1	31273	3276	307	344
13.6	26512	3618	244	264
13.3	29810	3792	285	298
16.5	31781	4120	286	295
15.2	24772	3304	257	320
11.1	26179	3649	288	266
17.9	31787	3751	285	308
13.2	28277	3814	268	297
17.3	25552	6590	348	326
17.4	31036	3357	291	284
10.6	25900	3554	285	310
13.9	28918	2952	278	272

5. Show that the least squares estimators for multiple regression are also the maximum likelihood estimators of the parameters.

6. In Exercise 9 in Section 8.5 I mentioned a group of psychologists with whom I had worked for a short while. They were studying twin children in an effort to understand genetic and environmental causes of behavior. Following is a small part of their data set, in which 25 pairs of twin girls were given individual scores on a scale for their degree of activeness. The triples are the age of the twins, the activity level of twin 1 (arbitrarily chosen between the two twins in the pair), and the activity level of twin 2.

$$(11, 81, 41), \ (12, 46, 41), \ (9, 46, 46), \ (6, 56, 51), \ (6, 56, 61),$$
$$(8, 46, 46), \ (5, 61, 56), \ (10, 41, 41), \ (8, 51, 56), \ (12, 41, 41),$$
$$(6, 51, 66), \ (8, 46, 46), \ (5, 66, 41), \ (6, 41, 41), \ (7, 56, 56),$$
$$(8, 46, 46), \ (7, 41, 41), \ (6, 51, 46), \ (6, 41, 41), \ (6, 41, 41),$$
$$(6, 51, 46), \ (5, 41, 41), \ (5, 46, 51), \ (6, 41, 41), \ (7, 66, 41)$$

a. Find the individual regression equations expressing the activity level of twin 2 as a function of each of the other variables alone. Note the R^2 values. Produce scatter plots of twin 2 versus age and twin 2 versus twin 1.

b. Find the multiple regression equation expressing the activity level of twin 2 as a function of both of the two other variables together. What conclusions can you draw?

c. Observations 1, 7, 14, 18, 19, 20, 21, and 25 scored low on a scale of genetic similarity. Delete these, and investigate how well the remaining data fits a linear model.

7. Verify that the matrix normal equations $X'X\beta = X'Y$ yield the same estimates $\widehat{\beta}_0$ and $\widehat{\beta}_1$ as in Section 10.1 for the special case of regression on a single predictor variable.

8. Place yourself in the position of a physicist of many years ago studying a great new breakthrough in mechanics, the inverse square law of gravitation, which is thought to be

$$F = \frac{GmM}{r^2},$$

where m and M are the masses of the two objects, F is the force of gravitation between them, r is the distance between them, and G is a constant to be estimated. Suppose that you could measure the force (with some measurement error) for a given fixed distance r, and an assortment of pairs of masses. Explain how multiple regression could be used to estimate the constant G.

9. The following list of numbers is the work force of the former Soviet Union devoted to agriculture (in millions of people) for the years 1970–1987 (Pockney, 1991). Use a quadratic regession model to predict the work force in 1989. Does a quadratic model give a better fit to the data than a linear model?

$$23.8, \ 23.3, \ 23.5, \ 23.6, \ 23.6, \ 23.5, \ 23.3, \ 23.3, \ 23.1, \ 22.9,$$
$$22.7, \ 22.9, \ 23.0, \ 22.8, \ 22.4, \ 22.0, \ 21.3, \ 20.4$$

10. Because of economies of scale in costs, a company theorizes that the relationship between the total cost of manufacture and the number of items manufactured may not be linear, but instead quadratic. Investigate the company's theory using the following data, in which the first component is the number of items and the second is the cost. Does a quadratic model seem to give a better fit than a linear model?

$(78, 70)$, $(97, 74)$, $(123, 129)$, $(132, 81)$, $(165, 149)$, $(201, 133)$, $(268, 218)$, $(280, 221)$, $(294, 182)$, $(317, 182)$, $(345, 202)$, $(365, 219)$, $(367, 228)$, $(386, 219)$, $(399, 197)$

11. A graph of data issued by the New Jersey Casino Control Commission (*Chicago Tribune*, July 4, 1993) yielded the following approximate gross revenues for Atlantic City casinos for the years 1978–1992 (in billions of dollars):

.15, .3, .6, 1.1, 1.5, 1.8, 1.9, 2.1, 2.3, 2.5, 2.7, 2.8, 2.9, 2.95, 3.2

a. Fit a parabolic relationship between the variables. Use it to predict the revenue in 1993.

b. Try fitting a cubic relationship. Does the presence of the cubic term explain much more of the model variability?

12. Derive the estimators of the coefficients in a quadratic regression model directly from the least squares criterion, if there is no constant parameter in the model, that is, if

$$Y_i = \beta_1 x_i + \beta_2 x_i^2 + \epsilon_i, \qquad i = 1, 2, \ldots, n.$$

13. Derive the computational formula

$$SSE = \sum_{i=1}^{n} Y_i^2 - \widehat{\beta}_0 \sum_{i=1}^{n} Y_i - \widehat{\beta}_1 \sum_{i=1}^{n} x_{i1} Y_i - \cdots - \widehat{\beta}_k \sum_{i=1}^{n} x_{ik} Y_i$$

for the error sum of squares in the multiple regression model.

10.3 | Statistical Inference for Regression

So far, we have merely derived point estimators of the regression parameters in the linear model $Y = \beta_0 + \beta_1 x_1 + \beta_2 x_2 + \cdots + \beta_k x_k + \epsilon$. Although point estimation is certainly useful, it doesn't give any indication of how precise the estimates are. For that we need confidence interval estimates. Also, we have no way of testing hypotheses about the parameters such as $H_0 : \beta_j = 0$, which allows us to check whether the variable Y depends on x_j at all. (If we conclude that the coefficient β_j is not zero, then we say that the variable x_j is a *significant predictor* of Y.)

The purpose of this section is to derive inference procedures for regression parameters based on the maximum likelihood estimators $\widehat{\beta}_j$. As preparation, we must first study the probability distributions of the $\widehat{\beta}_j$'s. The following theorem sets the stage. It is stated for the multiple regression model, but it specializes to the one-dimensional setting as well.

PROPOSITION 10.3.1

Let $\mathbf{Y} = X\boldsymbol{\beta} + \boldsymbol{\epsilon}$ be a multiple regression model, with parameter vector $\boldsymbol{\beta} = [\beta_0 \, \beta_1 \ldots \beta_k]'$ and error vector $\boldsymbol{\epsilon} = [\epsilon_1 \, \epsilon_2 \, \ldots \, \epsilon_n]'$ which is multivariate normal with mean vector $\mathbf{0}$ and covariance matrix $\sigma^2 \cdot I$. Assume that $X'X$ is invertible. Then the least squares estimators $\widehat{\beta}_0, \widehat{\beta}_1, \ldots, \widehat{\beta}_k$ have the multivariate normal distribution with mean $\boldsymbol{\beta}$ and covariance matrix

$$\sum = \sigma^2 \, (X'X)^{-1}. \tag{10.32}$$

■ **Proof** Since $X'X$ is invertible, we have from the normal equations $\widehat{\boldsymbol{\beta}} = (X'X)^{-1} X'\mathbf{Y}$. But by assumption, $\mathbf{Y}$ is multivariate normal with mean vector $E[\mathbf{Y}] = E[X\boldsymbol{\beta} + \boldsymbol{\epsilon}] = X\boldsymbol{\beta}$ and covariance matrix $\sum_Y = \mathrm{Cov}(X\boldsymbol{\beta} + \boldsymbol{\epsilon}) = \mathrm{Cov}(\boldsymbol{\epsilon}) = \sigma^2 I$. Hence by Proposition 5.4.6, $\widehat{\boldsymbol{\beta}}$ is multivariate normal, with mean vector

$$
\begin{aligned}
E\left[\widehat{\boldsymbol{\beta}}\right] &= E\left[(X'X)^{-1} X'\mathbf{Y}\right] \\
&= (X'X)^{-1} X' E[\mathbf{Y}] \\
&= (X'X)^{-1} X'X\boldsymbol{\beta} \\
&= \boldsymbol{\beta}
\end{aligned}
\tag{10.33}
$$

and covariance matrix

$$
\begin{aligned}
\sum = \mathrm{Cov}\left(\widehat{\boldsymbol{\beta}}\right) &= \mathrm{Cov}\left[(X'X)^{-1} X'\mathbf{Y}\right] \\
&= (X'X)^{-1} X' \cdot \sigma^2 I \left((X'X)^{-1} X'\right)' \\
&= \sigma^2 \cdot (X'X)^{-1} X'X (X'X)^{-1} \\
&= \sigma^2 \cdot (X'X)^{-1}.
\end{aligned}
\tag{10.34}
$$

Example 10.3.1

In the simple linear regression case we have $\boldsymbol{\beta} = [b_0 \, b_1]'$, $\widehat{\boldsymbol{\beta}} = \left[\widehat{b}_0 \, \widehat{b}_1\right]'$ and

$$
X = \begin{bmatrix} 1 & x_1 \\ 1 & x_2 \\ \vdots & \vdots \\ 1 & x_n \end{bmatrix}, \qquad X'X = \begin{bmatrix} n & \sum_{i=1}^{n} x_i \\ \sum_{i=1}^{n} x_i & \sum_{i=1}^{n} x_i^2 \end{bmatrix}.
$$

Then the covariance matrix of $\widehat{\boldsymbol{\beta}}$ is

$$
\mathrm{Cov}(\widehat{\boldsymbol{\beta}}) = \sigma^2 \, (X'X)^{-1} = \frac{\sigma^2}{n \sum_{i=1}^{n} x_i^2 - (\sum_{i=1}^{n} x_i)^2} \begin{bmatrix} \sum_{i=1}^{n} x_i^2 & -\sum_{i=1}^{n} x_i \\ -\sum_{i=1}^{n} x_i & n \end{bmatrix}.
\tag{10.35}
$$

Since $\widehat{\boldsymbol{\beta}}$ is multivariate normal, the marginal distributions of $\widehat{b}_0$ and $\widehat{b}_1$ are normal. As we already know, the means are b_0 and b_1, respectively. The variance of $\widehat{b}_0$ is the entry in row 1 and column 1 of $\sigma^2 \cdot (X'X)^{-1}$. After some algebraic rearrangement, this is

$$\mathrm{Var}(\widehat{b}_0) = \frac{\sigma^2}{n}\left[1 + \frac{n\bar{x}^2}{\sum_{i=1}^{n}(x_i - \bar{x})^2}\right]. \tag{10.36}$$

Also, the variance of $\widehat{b}_1$ is the entry in row 2 and column 2 of $\sigma^2 \cdot (X'X)^{-1}$, which is

$$\mathrm{Var}(\widehat{b}_1) = \frac{\sigma^2 \cdot n}{n\sum_{i=1}^{n}x_i^2 - (\sum_{i=1}^{n}x_i)^2} = \frac{\sigma^2}{\sum_{i=1}^{n}(x_i - \bar{x})^2}. \tag{10.37}$$

(We actually derived this formula earlier in the process of proving Proposition 10.1.2.) Note that $\widehat{b}_0$ and $\widehat{b}_1$ cannot be independent, since their covariance, the off-diagonal entry of $\sigma^2 \cdot (X'X)^{-1}$, is nonzero.

Now consider the predicted value of y for a given new x value, $\widehat{Y} = \widehat{b}_0 + \widehat{b}_1 x$. Since $\widehat{Y}$ is a linear combination of normal random variables, it is also normally distributed. Its mean is $E[\widehat{Y}] = E[\widehat{b}_0 + \widehat{b}_1 x] = b_0 + b_1 x$; that is, the mean of the predicted $\widehat{Y}$ is the (conditional) mean of Y given x. The variance of $\widehat{Y}$ is

$$\mathrm{Var}(\widehat{Y}) = \frac{\sigma^2}{n} + \sigma^2 \cdot \frac{(x - \bar{x})^2}{\sum_{i=1}^{n}(x_i - \bar{x})^2}. \tag{10.38}$$

You are asked to prove formula (10.38) in Exercise 14.

?Question 10.3.1   Describe how the variability of $\widehat{Y}$ depends on the error variance, sample size, and the spread of the x-values.

When we were discussing confidence intervals and hypothesis tests for the normal mean in the case of unknown variance, we were able to replace the unknown σ^2 in the standardized sample mean by its unbiased estimator S^2. The distribution of the resulting test statistic was $t(n-1)$. We would like to be able to do the same thing here with the normally distributed estimators $\widehat{b}_0$ and $\widehat{b}_1$; in other words, we would like to begin by forming random variables

$$\frac{\widehat{b}_0 - b_0}{\sqrt{\mathrm{Var}(\widehat{b}_0)}}, \qquad \frac{\widehat{b}_1 - b_1}{\sqrt{\mathrm{Var}(\widehat{b}_1)}}. \tag{10.39}$$

The expressions for the variances of $\widehat{b}_0$ and $\widehat{b}_1$ in (10.36) and (10.37) depend on the unknown σ^2. But we would like to replace σ^2 in these expressions by its unbiased estimate, the mean square error $MSE = SSE/(n-2)$. We will need to know that a normalized form of SSE has the χ^2 distribution and is independent of the $\widehat{b}$'s.

This plan demands that we have knowledge about the statistical properties of the sums of squares. We now state a result applicable to multiple regression as well as the one-variable case.

PROPOSITION 10.3.2

Let $\mathbf{Y} = \mathbf{X}\boldsymbol{\beta} + \boldsymbol{\epsilon}$ be a multiple regression model with $k+1$ parameters $\beta_0, \beta_1, \ldots, \beta_k$ satisfying the assumptions of Proposition 10.3.1. Then

a. $SSE/\sigma^2 \sim \chi^2(n - k - 1)$.
b. If all parameters except possibly β_0 are zero, then $SSR/\sigma^2 \sim \chi^2(k)$.
c. SSE and SSR are independent; also $\widehat{\boldsymbol{\beta}}$ is independent of SSE.

—— **?Question 10.3.2**   Before reading on, try to construct a t-statistic for b_0 and b_1 in one-variable linear regression. Compare your answer with what appears shortly.

Unfortunately, the amount of detail in the proof of Proposition 10.3.2 makes it prohibitive for us to include here. Exercises 16–18 lead you through enough of it to be convincing. The proof depends on results on quadratic forms of normal random vectors (see Propositions 5.4.7 and 5.4.8). The main thrust of the proof is to show that the error sum of squares $\|\mathbf{Y} - X\widehat{\boldsymbol{\beta}}\|^2$ is the quadratic form

$$SSE = \mathbf{Y}' \left(I - X \left(X'X \right)^{-1} X' \right) \mathbf{Y}, \tag{10.40}$$

and the matrix $Q_e = I - X \left(X'X \right)^{-1} X'$ of this quadratic form is idempotent with rank $n - k - 1$. The regression sum of squares can be written as another quadratic form, and under the hypothesis of the theorem the matrix $Q_r = \left(X \left(X'X \right)^{-1} X' - Q_0 \right)$ of this quadratic form is idempotent with rank k. Here, Q_0 is an $n \times n$ matrix whose every entry is $1/n$. In addition, $Q_e \cdot Q_r = 0$. Hence, by Proposition 5.4.8, SSE and SSR are independent. The quadratic form $(\widehat{\boldsymbol{\beta}} - \boldsymbol{\beta})'X'X(\widehat{\boldsymbol{\beta}} - \boldsymbol{\beta})$ is also expressible as a form in $\mathbf{Y} - X\boldsymbol{\beta}$ with a matrix which products to zero with Q_e; hence SSE and $\widehat{\boldsymbol{\beta}}$ are also independent.

What this all means in the case of simple one-variable linear regression is that one can form t-statistics for the parameters as follows:

$$T_{b_0} = \frac{(\widehat{b}_0 - b_0)/\sqrt{\text{Var}(\widehat{b}_0)}}{\sqrt{\dfrac{SSE/\sigma^2}{n-2}}}$$

$$= \frac{(\widehat{b}_0 - b_0) \Big/ \sqrt{\dfrac{\sigma^2}{n} \left[1 + \dfrac{n\overline{x}^2}{\sum_{i=1}^{n}(x_i - \overline{x})^2} \right]}}{\sqrt{MSE/\sigma^2}}$$

$$= \frac{(\widehat{b_0} - b_0)}{\sqrt{MSE}\sqrt{\dfrac{1}{n} + \dfrac{\overline{x}^2}{\sum_{i=1}^{n}(x_i - \overline{x})^2}}} \sim t(n-2), \qquad (10.41)$$

and

$$T_{b_1} = \frac{(\widehat{b_1} - b_1)/\sqrt{\text{Var}(\widehat{b_1})}}{\sqrt{\dfrac{SSE/\sigma^2}{n-2}}}$$

$$= \frac{(\widehat{b_1} - b_1)\bigg/\sqrt{\dfrac{\sigma^2}{\sum_{i=1}^{n}(x_i - \overline{x})^2}}}{\sqrt{MSE/\sigma^2}}$$

$$= \frac{(\widehat{b_1} - b_1)}{\sqrt{MSE}\bigg/\sqrt{\sum_{i=1}^{n}(x_i - \overline{x})^2}} \sim t(n-2). \qquad (10.42)$$

Confidence intervals and hypothesis tests for the parameters b_0 and b_1 can then be built analogously to those for the normal mean in the unknown variance case.

Also, in both the single and multiple regression problems under the null hypothesis that the entire regression model is insignificant—that is, all coefficients $\beta_1, \beta_2, \ldots, \beta_k$, are zero—the ratio of mean squares

$$F = \frac{SSR/k}{SSE/(n-k-1)} = \frac{MSR}{MSE} \qquad (10.43)$$

has the $F(k, n-k-1)$ distribution. The sums of squares, their degrees of freedom, the mean squares, and this F statistic are all given in the *Minitab* analysis of variance table. If the F-statistic is unreasonably large, then the regression model explains more of the total variation than it is expected to if the regression coefficients were all zero; hence we would conclude that at least one of the coefficients must be significant. *Minitab* even reports a p-value for this test of model significance in the analysis of variance table.

We illustrate in the next example.

Example 10.3.2

Let us look again at the life span data of Example 10.1.1. Figure 10.4 contains the *Minitab* output necessary for us to (a) perform a test of the significance of male life spans as a predictor of female life spans; (b) test the significance of the intercept term b_0 in the model; and (c) devise a 90% confidence interval for b_1, the rate of change of female life spans with respect to male life spans.

A linear model $Y = b_0 + b_1 x + \epsilon$ is assumed, where Y is the female life span, x is the male life span, and ϵ is a normally distributed error. The sample size is $n = 12$, so the degrees of freedom for error is $n - 2 = 10$. The hypotheses for the test referred to in (a) are

$$H_0 : b_1 = 0 \quad \text{vs.} \quad H_a : b_1 \neq 0.$$

We know that under H_0, by (10.42), the random variable

$$T_{b_1} = \frac{\hat{b}_1}{\sqrt{MSE}/\sqrt{\sum_{i=1}^{n}(x_i - \bar{x})^2}} \qquad (10.44)$$

has the $t(n-2)$ distribution. The estimated b_1 is $\hat{b}_1 = .7959$, and from the analysis of variance table, $MSE = 6.34$. The standard deviation s_x of the male data can be calculated directly to be 7.062. Thus

$$7.062 = \sqrt{\sum_{i=1}^{n}(x_i - \bar{x})^2/(n-1)} \Rightarrow \sqrt{\sum_{i=1}^{n}(x_i - \bar{x})^2} = 7.062 \cdot \sqrt{11} = 23.42.$$

Thus the sample value of the test statistic is

$$t_{b_1} = \frac{.7959}{\sqrt{6.34}/23.42} = \frac{.7959}{.1075} = 7.40.$$

Look closely at the regression parameter table in Fig. 10.4. Under the "Stdev" column in the "male" row you will observe the number 0.1075, the denominator of the t-statistic for b_1. In other words, *Minitab* knows how to calculate the estimated standard deviation of $\hat{b}_1$ (its *standard error*). But even better, the t-ratio column contains the value of t_{b_1}, so that it is not necessary to do any calculating. Since the observed t is more than the critical value $t_{.01}(10) = 2.76$ for a two-sided .02 level test, we reject the null hypothesis at level .02. In fact, *Minitab* also gives us the p-value of the test in the column titled "p." Apparently the p-value is less than .001, which is very strong evidence that the male life span is a significant predictor of the female life span. Also, the p-value for the regression model in the analysis of variance table is less than .001 due to the high magnitude of the observed F-statistic.

Now that we know what *Minitab* is telling us, it is easy to do part (b). The constant term b_0 has a t-value of 2.70 and a p-value of .022, which is also strong evidence that the parameter b_0 belongs in the model.

The confidence interval for b_1 is easy to derive. At a level of 90% and 10 degrees of freedom, the t-critical values are ± 1.81. Denote the standard error of $\hat{b}_1$, the denominator in (10.42), by $SE(\hat{b}_1)$. Its numerical value is .1075. Then

$$.90 = P[-t_{.05}(10) \le T_{b_0} \le t_{.05}(10)]$$
$$= P[\hat{b}_1 - t_{.05}(10) \cdot SE(\hat{b}_1) \le b_1 \le \hat{b}_1 + t_{.05}(10) \cdot SE(\hat{b}_1)]. \quad (10.45)$$

The numerical values of the confidence interval endpoints in (10.45) are therefore

$$[.7959 - 1.81\,(.1075),\, .7959 + 1.81\,(.1075)] = [.6013, .9905].$$

?Question 10.3.3   How would you obtain a confidence interval for b_0 in the last example?

Example 10.3.3

Remember that in formula (10.38) we derived the variance of a predicted value $\widehat{Y} = \widehat{b}_0 + \widehat{b}_1 x$ of a new observation for a given new x-value. Let us construct a random interval called a *prediction interval* for the new y-value $Y = b_0 + b_1 x + \epsilon$. Then, let us apply it in the context of the calculus grade data set from Example 10.1.2 to give a 90% interval estimate of the calculus grade of a new student who receives a 28 on the ACT.

Now Y is normal with mean $b_0 + b_1 x$ and variance σ^2, and $\widehat{Y} = \widehat{b}_0 + \widehat{b}_1 x$ is also normal with mean $b_0 + b_1 x$ and variance as in (10.38). Since $\widehat{Y}$ is based on the data Y_i's that are independent of the new observation Y, the variance of $Y - \widehat{Y}$ is

$$\text{Var}\left(Y - \widehat{Y}\right) = \text{Var}(Y) + \text{Var}(\widehat{Y})$$

$$= \sigma^2 + \frac{\sigma^2}{n} + \sigma^2 \cdot \frac{(x - \bar{x})^2}{\sum_{i=1}^{n}(x_i - \bar{x})^2}$$

$$= \sigma^2 \left(1 + \frac{1}{n} + \frac{(x - \bar{x})^2}{\sum_{i=1}^{n}(x_i - \bar{x})^2}\right).$$

Also, $Y - \widehat{Y}$ has mean 0. (Why?) Because of the independence of the estimated coefficients $\widehat{b}_0$ and $\widehat{b}_1$ (and Y) from SSE, we can conclude that the following has the $t(n-2)$ distribution:

$$T_Y = \frac{(Y - \widehat{Y})/\sqrt{\text{Var}(Y - \widehat{Y})}}{\sqrt{MSE/\sigma^2}} = \frac{Y - \widehat{Y}}{\sqrt{MSE}\sqrt{1 + \frac{1}{n} + \frac{(x - \bar{x})^2}{\sum_{i=1}^{n}(x_i - \bar{x})^2}}}. \quad (10.46)$$

Thus,

$$1 - \alpha = P[-t_{\alpha/2} \le T_Y \le t_{\alpha/2}]$$

$$= P\left[Y \in \left[\widehat{Y} \pm t_{\alpha/2} \cdot \sqrt{MSE}\sqrt{1 + \frac{1}{n} + \frac{(x - \bar{x})^2}{\sum_{i=1}^{n}(x_i - \bar{x})^2}}\right]\right]. \quad (10.47)$$

Formula (10.47) gives the general result for a $(1 - \alpha) \times 100\%$ prediction interval for a new Y value corresponding to x.

In the calculus grade data set from Table 10.1, we can regress the dependent variable $Y = $ calculus grade on one independent variable $x = $ ACT score. Figure 10.8 contains *Minitab* output.

We have $n = 25, MSE = .7791$ (note that $\sqrt{MSE} = .8827$ appears as s in the table), $\widehat{b}_0 = -1.5$, and $\widehat{b}_1 = 0.15895$. For a prediction interval of level 90%, the critical values are $\pm t_{.05}(23) = \pm 1.71$. It can be computed from the ACT data in the table that the mean and variance of the x_i's are 25.92 and 3.65^2, respectively. Then for $x = 28$,

$$(x - \bar{x})^2 = (28 - 25.92)^2 = 4.3264 \quad \text{and} \quad \sum_{i=1}^{n}(x_i - \bar{x})^2 = 24 \cdot 3.65^2 = 319.74.$$

```
MTB>Regress 'calculus' 1 'act'.

The regression equation is
calculus = -1.50 + 0.159 act

Predictor     Coef      Stdev     t-ratio      p
Constant    -1.500      1.291     -1.16      0.257
act          0.15895    0.04935    3.22      0.004

s = 0.8827      R-sq = 31.1%      R-sq(adj) = 28.1%

Analysis of Variance

SOURCE       DF      SS        MS         F         p
Regression    1    8.0812    8.0812    10.37     0.004
Error        23   17.9188    0.7791
Total        24   26.0000
```

Figure 10.8 *Minitab* Output for Regression of Grade on ACT

The predicted y-value is $y = -1.5 + 0.15895 \cdot 28 = 2.95$. Substituting into (10.47), the 90% prediction interval becomes

$$2.95 \pm 1.71 \cdot .8827 \sqrt{1 + \frac{1}{25} + \frac{4.3264}{319.74}} = [1.4, 4.5].$$

— **?Question 10.3.4** Look closely at the preceding computation. The interval radius is so large that the interval has limited practical value in predicting the student's calculus performance. What factors made the radius so large?

Statistical inference for multiple regression follows along the same lines. Proposition 10.3.1 contained the general expression for $\sum$, the covariance matrix of the vector $\widehat{\boldsymbol{\beta}}$ of regression estimators. Because this formula is available, one can find the variance of any of the estimated parameters $\widehat{\beta}_j$ by inspecting the jth diagonal element of the covariance matrix. This variance will have σ^2 as a factor by (10.32). Calculations just like that of (10.41) and (10.42) can be done, leading to t-statistics

$$T_{\beta_j} = \frac{\widehat{\beta}_j - \beta_j}{SE(\widehat{\beta}_j)}. \tag{10.48}$$

The degrees of freedom of these t-random variables are those of the sum of squares due to error, namely $n - k - 1$.

Minitab knows how to find the standard error in the denominator. It is reported in the "Stdev" column of the regression table for the estimator of interest; furthermore, the value of the t-statistic $\widehat{\beta}_j/SE(\widehat{\beta}_j)$ for the test of $H_0 : \beta_j = 0$ is reported in the t-column. Thus drawing conclusions about the significance of variables in the multiple regression model is easy, as our next example illustrates.

Example 10.3.4

Consider again the *Minitab* output of Fig. 10.4 for the calculus grade prediction problem. The value of the F-statistic is 5.26, with a p-value of .007, which gives strong evidence that the model has predictive value. The constant term β_0 has a t-value of -1.17, a value that has probability .254 of being exceeded even if in reality $\beta_0 = 0$. (Why?) The placement, SAT, and ACT parameters have somewhat more extreme t-values, -1.84, 1.84, and 2.66, respectively. In the case of the ACT, the fact that the p-value is 0.015 implies that the data could only have produced such a high t-value 1-1/2% of the time if the ACT coefficient truly was zero, which is strong evidence that the ACT exam is a significant predictor of calculus performance for this data set. The home-grown placement exam and the SAT are somewhat less significant, but the p-values of about .08 are small enough to persuade us to include them in the linear model.

Example 10.3.5

Recall Example 9.4.1 in which the team of psychologists gathered data on maternal behavior and relative valuation of several kinds of life goals by the children. That example analyzed the data by dividing the teenage children into disjoint groups according to whether they valued financial success or a more benevolent quality more highly, and a t-test was applied to contrast the maternal nurturance of the two groups.

The researchers came up with another way to look at their data. For each of 95 children they prepared three contrast variables Y^1, Y^2, and Y^3 that measured the differences between the child's valuation of self-acceptance, affiliation, and community feeling and the child's valuation of financial success (a positive observation of a Y^j would indicate that financial success was less highly valued). Then each of these dependent variables was regressed on three independent variables: mother's social values, maternal nurturance, and socioeconomic advantage of the family, to see which if any of these three factors might contribute to the child's attitudes.

For example, for the self-acceptance versus financial success contrast, the reported regression coefficients for maternal values, nurturance, and socioeconomics were .26, .04, and .23, respectively. Nurturance in this case was not a significant predictor of the child's values. Mother's values was significant at a level less than .05, and socioeconomic advantage was significant at a level less than .1. The regression model F-value was 6.89, which was significant at a level less than .01. Still, R^2 only came out to .19, so that although the model had nonnegligible predictive power, error variability still dominated. The conclusion was that maternal values were most influential in predicting the child's preference for self-acceptance over financial success. For family affiliation, socioeconomic advantage was the only barely significant predictor, and for community feeling, maternal nurturance was the most significant predictor, and maternal values were also barely significant. It is interesting that the three different kinds of social values in the children seem to be most influenced by three different environmental variables.

EXERCISES 10.3

1. The following list of numbers indicates the volume of domestic air passenger traffic (millions of passengers) in Pakistan from the years 1970–1981 (Government of Pakistan, 1982). Would you prefer a linear, quadratic, or cubic model of passenger volume as a function of time? Explain carefully. Find a 90% confidence interval for the yearly growth rate for a linear model.

   1.231, 1.075, 1.139, 1.498, 1.904, 2.364, 2.760, 3.278, 3.358, 3.534, 3.574, 3.534

2. Table 10.5 contains 1990 data from the Illinois Department of Health on 14 cities in Illinois. The table gives the number of babies born to married teenaged mothers and for the same town the number of births to unwed mothers. Plot a scattergram in which the number of unwed births is on the vertical axis and the number of married births is on the horizontal axis. Does a linear model seem to fit? Test the hypothesis that the number of births to married teenage mothers is a significant predictor of the number of births to unwed mothers. Find an 80% prediction interval for the number of unwed births in a city with 30 births to married teenage mothers. What does your interval say about this data set?

Table 10.5

Births to Teenagers in Illinois Cities, 1990

City	Births, Wed	Births, Unwed
Bloomington	12	69
Canton	11	19
Decatur	54	222
East Peoria	13	36
East St. Louis	15	411
Galesburg	18	50
Kewanee	6	21
LaSalle	6	10
Macomb	7	5
Morton	5	5
Pekin	26	55
Peoria	40	394
Springfield	45	262
Washington	4	15

Table 10.6

Car Thefts, 1992

City	Population	No. Car Thefts
Houston	16	30.9
Dallas	10	20.5
New York	73	126.9
Philadelphia	16	22.1
San Diego	11	20.2
Chicago	28	44.9
Detroit	10	27.3
Los Angeles	36	7.9
Beaumont, TX	1.19	1.7
Tallahassee, FL	1.30	2.0
Jacksonville, FL	6.64	8.8
Baton Rouge, LA	2.37	4.1
Ft. Lauderdale, FL	1.56	2.8
Buffalo, NY	3.30	5.5
Tampa, FL	2.91	7.3
Jackson, MS	2.00	4.1

3. Data were gathered from law enforcement agencies in 16 cities on auto thefts and population for the year 1992. The data are displayed in Table 10.6, with the population in units of 100,000 people and the theft incidents in units of thousands. Test at an appropriate level whether population is a significant predictor of the number of auto thefts. Is the constant term in the linear model significantly different from 0? Find a prediction interval of level 90% for the number of auto thefts in a city of 500,000 people.

4. Look back at Exercise 3 in Section 10.2 on the blood glucose readings. Write a paragraph that essentially answers the same questions but uses the results of this section.

5. Consider a two-variable model $Y_i = \beta_0 + \beta_i x_{i1} + \beta_2 x_{i2} + \epsilon_i$, $i = 1, \ldots, 10$ with 10 pairs of values of (x_1, x_2) observations as follows:

$$(11.3, 6.5), \quad (9.5, 3.2), \quad (15.7, 8.4), \quad (9.3, 2.1), \quad (12.1, 6.9),$$
$$(8.4, 3.0), \quad (13.7, 6.8), \quad (12.2, 5.1), \quad (6.0, 1.2), \quad (10.6, 4.2).$$

Compute the covariance matrix of the estimate $\hat{\boldsymbol{\beta}} = (\hat{\beta}_0, \hat{\beta}_1, \hat{\beta}_2)$ in terms of σ^2.

6. Suppose that the error variance σ^2 in a linear regresssion model $Y = b_0 + b_1 x + \epsilon$ is equal to 1. Find the variance of the predicted Y-value $\hat{Y}$ for an x-value of 4 if the data on which the regression estimates are based are the following:

$$(1, 3.2), \quad (3, 5.0), \quad (5, 7.6), \quad (7, 9.1), \quad (9, 10.8).$$

7. Using the data in Exercise 1 in Section 10.1 on Australian tourism, find a prediction interval of level 90% for the 1993 tourism, and a 90% confidence interval for the yearly rate of growth.

8. In the baseball data (see Table 7.6):
   a. Test for significance of the three possible predictors of runs scored: home runs, batting average, and stolen bases.
   b. Find an 80% prediction interval for runs scored by a team hitting 100 home runs.
   c. Find an 80% prediction interval for runs scored by a team batting .275.
   d. For a regression of runs scored on home runs, find a 90% confidence interval for b_0, which we can interpret as the expected number of runs that could be scored by a completely punchless team who hits no home runs.

9. In Exercise 3 in Section 10.1 on milk cows, test to see whether the model is significantly improved by including a quadratic factor. Find an 80% confidence interval for the parameter b_0 in the linear model. (By the way, if you enjoy absurdity, give an intuitive interpretation of your interval estimate in this problem context.)

10. In the setting of the statistics test score data of Table 7.2 answer the following questions.
    a. Is test 1 a significant predictor of test 3 at level .05? What is an 80% prediction interval for the test 3 score of a student who scored 70 on test 1?
    b. Is test 2 a significant predictor of test 3 at level .05? What is an 80% prediction interval for the test 3 score of a student who scored 75 on test 2?

11. Recall the Peoria school data in Exercise 4 in Section 10.2 (see Table 10.4).
    a. Test for significance of the variables years of experience, teacher salary, and expenditure per pupil on both the reading and the math score.
    b. Find an 80% prediction interval for the reading score in a school whose teachers have 15 years of experience.

12. Show that

$$\widehat{b}_0 + \widehat{b}_1 x \pm t_{a/2}(n-2) \cdot \sqrt{MSE} \cdot \sqrt{\frac{1}{n} + \frac{(x-\overline{x})^2}{\sum_{i=1}^{n}(x_i - \overline{x})^2}}$$

is a $(1-\alpha) \times 100\%$ confidence interval for the mean of Y for a given x, that is, $b_0 + b_1 x$.

13. Show that a two-sided test of $H_0 : b_1 = 0$ in one-variable regression is equivalent to a test that rejects H_0 if and only if

$$F = \frac{SSR}{SSE/(n-2)} > f_\alpha(1, n-2),$$

where $f_\alpha(1, n-2)$ is the upper tail level α critical point for the f-distribution with degrees of freedom 1 and $n-2$.

14. Verify that

$$\mathrm{Var}(\widehat{Y}) = \frac{\sigma^2}{n} + \sigma^2 \cdot \frac{(x-x)^2}{\sum_{i=1}^{n}(x_i - \overline{x})^2},$$

where $\widehat{Y} = \widehat{b}_0 + \widehat{b}_1 x$ is the fitted y-value for a given x-value in one-variable regression.

15. Discuss in general terms how one might obtain a prediction interval for a new Y-value for given new x_1- and x_2-values in a two-variable linear model $Y = \beta_0 + \beta_1 x_1 + \beta_2 x_2 + \epsilon$ that is analogous to the one-variable case.

16. a. Show formula (10.40) for SSE.

   b. Prove that the matrix Q_e of the quadratic form in (10.40) is idempotent, that is, $Q_e^2 = Q_e$.

   c. Show that SSE also equals

$$(\mathbf{Y} - X\boldsymbol{\beta})' \left(I - X\,(X'X)^{-1} X'\right) (\mathbf{Y} - X\boldsymbol{\beta}).$$

   Why can you now conclude that SSE/σ^2 has a χ^2-distribution?

17. Show that the rank of the matrix $Q_e = I - X(X'X)^{-1}X'$ in (10.40) is $n - k - 1$.

   [*Hint:* See Appendix E, on linear algebra. It is true that if a symmetric matrix Q is idempotent and if λ is an eigenvalue for Q, then λ must be either 0 or 1. Use the fact that the trace $\mathrm{tr}(A)$ of a matrix A is not only the sum of its diagonal entries but also the sum of its eigenvalues. Recall also that $\mathrm{tr}(AB) = \mathrm{tr}(BA)$ and $\mathrm{tr}(A - B) = \mathrm{tr}(A) - \mathrm{tr}(B)$. Using these facts, show that the trace of Q_e is $n - k - 1$. Finally, diagonalize Q_e.]

18. a. Show that $SSR = \mathbf{Y}'Q_r\mathbf{Y}$, where

$$Q_r = \left(X\,(X'X)^{-1} X' - Q_0\right).$$

   Recall that we denoted by Q_0 the $n \times n$ matrix whose every entry is $1/n$. [*Hint:* As a lemma for this part and the others, show that the row sums of the matrix $X\,(X'X)^{-1} X'$ are one by multiplying it by $X \cdot \mathbf{c}$, where $\mathbf{c}$ is a vector with 1 in its first component and zeros elsewhere. Now what is the product $Q_0 \cdot X\,(X'X)^{-1} X'$?]

   b. Show that Q_r is idempotent.

   c. Show that $Q_e \cdot Q_r = 0$.

   d. Finally, show that if all of the $\beta_j = 0$ except possibly β_0, then SSR is also equal to

$$(\mathbf{Y} - X\boldsymbol{\beta})'Q_r(\mathbf{Y} - X\boldsymbol{\beta}).$$

   Why can you now conclude that under the null hypothesis of no significant variables, SSR/σ^2 has a χ^2-distribution and is independent of SSE?

10.4 | Diagnostic Checking

For a regression model of the form $Y = f(\mathbf{x}) + \epsilon$ with observed data $(\mathbf{x}_1, Y_1)$, $(\mathbf{x}_2, Y_2), \ldots, (\mathbf{x}_n, Y_n)$, several assumptions deserve checking. In studying the

R^2-value and the significance of coefficients we have already shown some ways of checking the model's basic structure. If R^2 is near zero, then we are led to think that the variables we have may not be good predictors; perhaps we should look for others. A special case occurs when the model is underfit; for example, a linear model was chosen when in fact the combination of the R^2-value with curvature in the scatterplot of the data indicates that a quadratic model would be better.

Thus, we have already looked at diagnostic checking for the proper $f(\mathbf{x})$. The other main issues involve the error terms ϵ. Recall the model assumptions:

1. The errors $\epsilon_1, \epsilon_2, \ldots, \epsilon_n$ in the data sample are normally distributed with mean 0.
2. The errors are mutually independent.
3. The variances $\mathrm{Var}(\epsilon_i)$ are all equal to some constant σ^2.

This section will begin by talking briefly and in generality about methods for checking these assumptions. We will then look at some examples to illustrate the specifics.

Our diagnostic methods will make heavy use of the residuals, which were defined earlier to be

$$e_i = Y_i - \widehat{Y}_i, \tag{10.49}$$

where $\widehat{Y}_i$ is the ith fitted value. Geometrically, the residuals are the lengths of the vertical segments shown in Fig. 10.2, connecting the data points to the estimated regression curve. Since the random errors satisfy $\epsilon_i = Y_i - f(\mathbf{x}_i)$ and the fitted values $\widehat{Y}_i$ estimate $f(\mathbf{x}_i)$, formula (10.49) implies that we can use the residuals as estimates of the unobservable errors ϵ_i. Their empirical properties should mirror the theoretical properties of the ϵ_i's listed above.

— **?Question 10.4.1**   Devise a geometric interpretation of the errors ϵ_i.

The normality assumption is easily checked graphically by plotting a histogram of the residuals, which should have the characteristic bell shape. An alternative is to produce a normal quantile plot of the residuals as described in Chapter 7. In the normal quantile plot, departures from linearity indicate possible nonnormality of the error terms. Transformations of variables can sometimes correct this (see later).

Independence of the errors is a bit harder to check and also harder to correct when it is in doubt. Some kinds of data sets, especially when the data are gathered in time order, show a tendency for response variables to be correlated with preceding response variables in the data. This indicates correlation of the error terms ϵ_i. When such correlation is present, it should be reflected in some pattern in the residuals. A scatterplot of residuals e_{i+1} against their predecessors e_i, which should be without pattern, may in fact show a trend if independence is violated. You may also be able to pick up such a pattern in a time series plot of the residuals, in which you might see a lot of positive residuals successively or high residuals bunched. A time series plot of residuals should show no pattern if the error terms are indeed independent. (However, see Exercise 17, which shows that residuals are not independent of one another.) When lack of independence is suggested, the whole regression approach is in jeopardy, and the next recourse might be to turn to a statistical model called a *time series*. We will not study time series in this book, but see Appendix D, on Research Projects, for a quick introduction.

The equal variance assumption is called *homoscedasticity*. *Heteroscedasticity* is the term used for unequal variance. Again, the residuals provide helpful information. For instance, a common case of heteroscedasticity occurs when larger Y_i's are more variable than smaller ones. Then the residuals for those large values of the response will be more scattered than the residuals for smaller values. This produces a "fanning out" effect in a scatterplot of residuals against Y_i values. As you look from left to right on such a plot, you would see the points becoming more widely spread. Since Y_i is estimated by $\widehat{Y_i}$, the same sort of information is provided by a plot of residuals against corresponding fitted values, which has become the traditional approach in regression model checking.

— **?Question 10.4.2**   Would you also be able to check for heteroscedasticity by a plot of residuals versus x-values for a linear model? Why?

Frequently a simple transformation of the y-variable, the x-variable, or both can eliminate heteroscedasticity, and surprisingly often it can also correct nonnormality of the error terms. When large y-values need to be damped down, resulting in a decrease in residuals for those large y-values, a power function y^r for a constant exponent $r \in (0, 1)$ can be applied to the y's. In more extreme cases the natural log of the y's can be taken.

Let us apply the ideas to several sample data sets.

Example 10.4.1   To see what should appear under ideal conditions, I simulated a data set that satisfies the assumptions. First, I generated 33 x-values between 0 and 1, using a uniform random number generator. Then to produce a known theoretical regression line, I computed $y_i = 2x_i + 1$ for each of the 33 x_i's. To generate observed random Y_i's whose deviations ϵ_i from the theoretical average are i.i.d. normal random variables, I simulated 33 such ϵ_i's from the $N(0, .2^2)$ distribution, and computed $Y_i = 2x_i + 1 + \epsilon_i$. So, the resulting (x_i, Y_i) pairs, $i = 1, 2, \ldots, 33$ should satisfy the linear regression hypotheses. Therefore the data set should pass our diagnostic tests with flying colors.

The scattergram of the data appears in Fig. 10.9(a), and the fit to the theoretical regression line is visually good. *Minitab* reports an R^2-value of 93.1%, which indicates that the model structure is good, and the single variable x provides good predictive power. Both coefficients are highly significant at levels less than .001.

A plot of residuals versus fits is shown in Fig. 10.9(b). It is rather featureless, as it should be, although there are a couple of large residuals for fits near

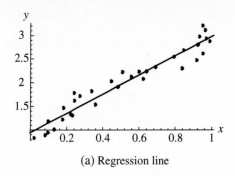

(a) Regression line

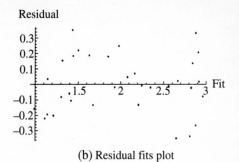

(b) Residual fits plot

Figure 10.9   Diagnostics for Simulated Data

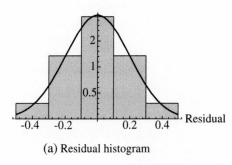

(a) Residual histogram

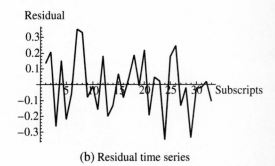

(b) Residual time series

Figure 10.10   Diagnostics for Simulated Data, Continued

3. A relative frequency histogram of residuals shows a near perfect normal bell shape (see Fig. 10.10a). And finally, a time series plot of residuals in Fig. 10.10(b) shows no eccentricities.

?Question 10.4.3   Given the way I produced the x- and Y-values in the last example, what should a plot of the actual errors versus residuals look like?

Example 10.4.2   The following data give average yearly prices per bushel of rye in 26 states, for the years 1988 and 1989 (USDA, 1991). (The states are in alphabetical order from one year to the next, so that, for instance, the first members of the two lists (2.15, 1.65) form a price pair for the state of Colorado, (2.05, 2.85) is the price pair for Delaware, etc.)

1988: 2.15, 2.05, 2.61, 2.90, 2.65, 2.36, 2.15, 2.35, 2.01,
2.70, 1.97, 3.00, 2.40, 3.25, 2.35, 2.95, 1.77, 4.00,
3.35, 2.34, 3.55, 2.85, 2.12, 2.70, 2.70, 2.60
1989: 1.65, 2.85, 2.26, 2.10, 2.10, 1.80, 1.70, 1.71, 2.85,
1.70, 1.66, 2.80, 1.90, 3.35, 2.55, 2.90, 1.67, 3.50,
2.00, 1.90, 3.25, 2.05, 1.70, 2.05, 2.40, 2.15

Let us perform a regression analysis of 1989 price on 1988 price and do model checking.

Using a statistical program, you should be able to obtain the following equation:

$$price89 = .460 + .687 \cdot price88.$$

The p-values for the constant and linear term are .32 and .0005, respectively, and R^2 comes out to be about .40. So we have a highly significant slope coefficient, but not a great deal of the variability is explained by the model. A scatterplot of the data with the estimated regression line superimposed appears in Fig. 10.11(a).

I have produced a residual versus fit graph in part (b) of Fig. 10.11. Since it shows no striking pattern, we have no cause for alarm about the homoscedasticity hypothesis. But the normal quantile plot and histogram in Figs. 10.12(a) and (b) should catch our eye. The flattened shape of the quantile plot in the middle and its curvature at the tails arise from the overabundance of data in the middle and the slight right-skew seen better in the histogram. (Why?) Although the lack of fit to normality by the residuals is not too severe, it is worth investigating transformations of the data.

After trying a few alternatives, including transforming both the 1988 and the 1989 data, I decided to leave the 1988 data alone and to regress the natural logarithm of the 1989 data. The resulting regression equation was

$$\log(price89) = .0419 + .284 \cdot price88.$$

The p-values were about .83 and .0007 for the constant and linear term, and R^2 was about .39, nearly the same as before. The scatterplot and the residuals versus fits graph looked sufficiently like the original ones in Fig. 10.11 that it is not worth displaying them again. Figure 10.13(a) shows the new residual histogram, which looks somewhat better, so this particular transformation has made marginal improvement with regard to the model assumptions. The data were not collected in time order, so that we would not expect a time series graph of the residuals to show any correlation between residuals, but I have included this graph for completeness sake in Fig. 10.13(b). Any correlations between residuals might take place for states that are close to one another geographically, which could be checked.

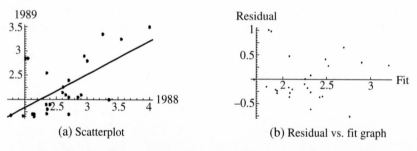

(a) Scatterplot   (b) Residual vs. fit graph

Figure 10.11 Regression and Residual Analysis of Rye Price Data

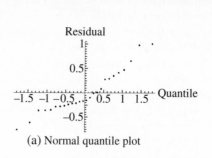

Residual

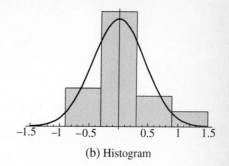

(a) Normal quantile plot

(b) Histogram

Figure 10.12   Distribution of Residuals for Rye Price Data

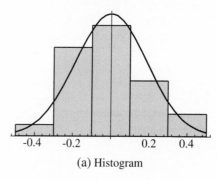

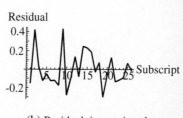

(a) Histogram

(b) Residual time series plot

Figure 10.13   Model-Checking of Log Transformed Rye Price Data

——— **?Question 10.4.4**   How would you go about predicting values of the 1989 rye prices given 1988 prices using the modified model in Example 10.4.2?

Example 10.4.3   Let us analyze the data from Exercise 2 of Section 10.3 on births to teenage mothers. We had a list, by town, of numbers of births to married and to unmarried young women (see Table 10.5). The question is, Can the number of births to married teenagers effectively predict the number of births to unmarried teenagers?

If you did that problem, you know that there are some strange things about this data set. A scattergram shows a great deal of variability, which is borne out by the R^2-value in the following results produced by *Minitab*:

$$\text{unwed} = 2.1 + 5.89 \cdot \text{wed}$$

	t	*p*
Constant	.04	.965, $R^2 = 43.2\%$
Wed	3.02	.011

Also, *Minitab* reports that the observation (54,222) for Decatur has a large influence on the regression equation and that the observation (15,411) for East St. Louis is an outlier. Being a rather large city, Decatur may have characteristics different from smaller towns like Kewanee and Canton. It would be interesting to know how the results of the regression would change if we excluded this data point. Also, economic conditions in East St. Louis have been very bad, which makes it a more exceptional case than the other towns in the list. While acknowledging that cases like East St. Louis exist, we can obtain information about a more restricted universe if we perform the analysis again without this point. (When outliers occur, deleting them automatically is not the thing to do, because they could be legitimately obtained observations that give information about the variables under study. Sometimes, however, such outliers are in fact errors in data gathering and reporting or are by-products of some extremely rare, even unrepeatable phenomenon. In that case it is reasonable to do a second analysis without the outlier, to try to get a sense of a more accurate, or more typical model. You should always report the deletion and your reasons for it, in the interest of honesty in research.)

The residuals versus fits graph for the reduced data set is shown in Fig. 10.14(a), and the histogram of residuals is in Fig. 10.14(b). The first graph indicates a fanning out phenomenon, with larger residuals corresponding to larger fitted values; thus we may have heteroscedasticity. The second graph shows a possible asymmetry of the distribution of residuals. Therefore a transformation of the data may be helpful. Let us try transforming the unwed values by natural log. (You might experiment with a less severe power transformation (unwed)^r for $r < 1$.) The regression results follow:

$$\log(\text{unwed}) = 2.02 + .0892 \cdot \text{wed}$$

	t	p	
Constant	6.85	.000,	$R^2 = 79.8\%$
Wed	6.29	.000	

We see a much better R^2-value than before, and we see that both coefficients are very significant. The residuals versus fits graph and residuals histogram for the transformed data are shown in Figs. 10.15(a) and (b). Notice how the fanning out phenomenon has been corrected (perhaps even overcorrected) although the histogram is not particularly better. So we have succeeded in some ways, but bear in mind that we reduced the data set so

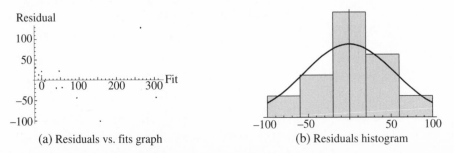

(a) Residuals vs. fits graph (b) Residuals histogram

Figure 10.14   Diagnostic Graphs for Teenage Mother Data

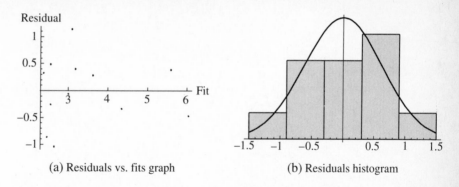

(a) Residuals vs. fits graph   (b) Residuals histogram

Figure 10.15 Diagnostic Graphs for Transformed, Reduced Teenage Mother Data

as to get a better result, and whatever conclusions we draw apply only to a narrower universe.

Example 10.4.4

Exercise 9 in Section 10.2 dealt with the number of people in the work force of the former Soviet Union who were involved in agriculture between 1970 and 1987. If you did that problem, you found that a quadratic model of work force as a function of time provided a good fit. Using 1970 as year 1, the estimated regression equation is

$$work\ force = 23.30 + .126 \cdot year - 0.14 \cdot year^2,$$

where the units of measurement of work force are in millions of people. The R^2-value is over .88, which seems to indicate that we have a very good model.

But consider the scattergram of the data with the regression curve in Fig. 10.16(a) and the residuals versus fits plot in part (b) of the figure. There are some strange, regular bounces in the data that are enough to convince us to look immediately at possible time correlation of the residuals.

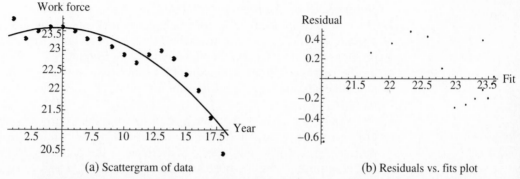

(a) Scattergram of data   (b) Residuals vs. fits plot

Figure 10.16 Regression Curve and Residual Analysis for Russian Work Force Data

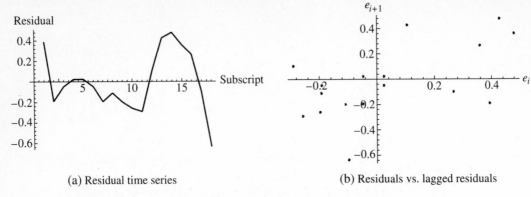

(a) Residual time series   (b) Residuals vs. lagged residuals

Figure 10.17   Residual Correlation Analysis for Russian Work Force Data

A time series of residuals from this regression is graphed in Fig. 10.17(a). It is smoother than one would expect, showing a tendency for a residual to be positive (resp. negative) when its preceding residual was positive (resp. negative). It is easier to see this phenomenon in Fig. 10.17(b), which displays the pairs (e_i, e_{i+1}) for $i = 1, \ldots, 17$ that is, the residuals against the same residuals lagged by one time unit. It appears that residuals are rather substantially correlated with their predecessors, which is contrary to the assumptions of the regression model. I even did a linear regression of the y-coordinates e_{i+1} against the x-coordinates e_i, and came out with a slope coefficient significant at level .02.

We can proceed no further, since the apparent lack of independence invalidates the simple regression model. This is a situation where the special methods of time series forecasting would come into play, which I encourage you to pursue in later study.

EXERCISES 10.4

1. Recall the blood glucose data in Exercise 3 of Section 10.2 (see Table 10.3). Do a set of model diagnostic checks for the regression of bedtime glucose level on the after-lunch level. Can you spot an outlier?

2. Do a set of diagnostic checks for the regression of runs scored on batting average for the baseball data set in Table 7.6. Are there any apparent problems? What effect do the two points (.235, 599) and (.277, 747) have on the regression line? What variables other than average, home runs, and stolen bases might an experimenter look at as predictors of runs scored?

3. In Exercise 4 in Section 10.2 we looked again at the Peoria school data. Do a set of diagnostic checks for the regression problem in which reading score is the response and years of experience is the predictor. Report on anything unusual, and if necessary transform the data and recheck the model.

4. Do a set of diagnostic checks on the car theft data of Exercise 3 of Section 10.3. Report any problems, and fix them if possible.

5. The data in Table 10.7 (from McClave and Dietrich, 1994, p. 791) deal with selling prices of antiques at an auction as a function of the age of the item and the number of bids offered on the item.

   a. Do a regression analysis using a two-variable linear model

$$price = \beta_0 + \beta_1 \cdot age + \beta_2 \cdot numbid + \epsilon,$$

   including appropriate checks on the model assumptions. [Delete one outlier, the point $(170, 14, 1131)$.]

   b. Plot residuals against number of bids. Do you find anything unusual?

   c. Use a new regression model that is quadratic instead of linear in number of bids. Pay particular attention to the distribution of the residuals.

   d. Try regressing several powers of the price on age, number of bids, and the square of number of bids to see the effect on the regression analysis, particularly on the normal scores plot of residuals. Confine your experimentation to powers $(price)^r$ for r between .5 and 1.5. What power seems to work best?

6. The following is a set of data points (x, y) for which a linear model has been proposed. Do the model assumptions check? If not, does a power or log transformation on the y-variable help? If the data were created by $y = 2x_i + 1 + \epsilon_i$, where ϵ_i were simulated $N(0, (1 + 0.3(i - 1))^2)$ observations, would your answers to the preceding questions be reasonable?

Table 10.7

Prices of Auctioned Items

Age	Numbids	Price	Age	Numbids	Price
127	13	1235	170	14	1131
115	12	1080	182	8	1550
127	7	845	162	11	1884
150	9	1522	184	10	2041
156	6	1047	143	6	854
182	11	1979	159	9	1483
156	12	1822	108	14	1055
132	10	1253	175	8	1545
137	9	1297	108	6	729
113	9	946	179	9	1792
137	15	1713	111	15	1175
117	11	1024	187	8	1593
137	8	1147	111	7	785
153	6	1092	115	7	744
117	13	1152	194	5	1356
126	10	1336	168	7	1262

$(2.0, 4.61)$, $(2.5, 6.25)$, $(3.0, 7.78)$, $(3.5, 8.76)$, $(4.0, 11.95)$, $(4.5, 11.81)$, $(5.0, 3.15)$, $(5.5, 13.65)$, $(6.0, 13.05)$, $(6.5, 17.83)$, $(7.0, 16.74)$, $(7.5, 19.94)$, $(8.0, 9.28)$, $(8.5, 18.14)$, $(9.0, 19.96)$, $(9.5, 24.00)$, $(10.0, 17.39)$, $(10.5, 20.31)$, $(11.0, 12.28)$, $(11.5, 18.82)$

7. Do the following data show signs of heteroscedasticity and/or nonlinearity? Can you find a power transformation y^r to apply to the y-values that would correct the problem?

$(6.0, 276.6)$, $(6.2, 348.8)$, $(6.4, 403.9)$, $(6.6, 387.3)$, $(6.8, 428.3)$, $(7.0, 441.6)$, $(7.2, 413.8)$, $(7.4, 604.4)$, $(7.6, 613.3)$, $(7.8, 541.2)$, $(8.0, 578.8)$, $(8.2, 518.2)$, $(8.4, 757.1)$, $(8.6, 747.1)$, $(8.8, 745.8)$

8. The following data were reported in a National Institutes of Health study (Johnston, et al., 1993). The first component of each pair is the midpoint of an age group classification, and the second is the percentage of people in that age group who have ever used marijuana. First use a linear function to model the dependence of lifetime prevalence of marijuana use on age group. Check appropriate residuals graphs, then try another function. Are there still potential problems with your revised model?

$(18, 33)$, $(19.5, 42)$, $(21.5, 53)$, $(23.5, 60)$, $(25.5, 64)$, $(27.5, 68)$, $(29.5, 70)$, $(31.5, 71)$

9. The data set on Atlantic City casino revenues was introduced in Exercise 11 of Section 10.2. If you did that problem, you found a quadratic relationship between revenues and time. Do a set of diagnostic checks, paying particular attention to the time series plot and residuals versus lagged residuals plot. Can you find any evidence of a correlation of residuals?

10. In Exercise 1 in Section 10.1 we saw data on Australian tourism. Do diagnostic checks on a quadratic model of tourism as a function of time. Does the distribution of the residuals seem good? Try transforming both variables by logarithms and repeating the checks. Report on the results.

11. Discuss the following: Could an incorrect linear model applied to a relationship that is actually quadratic make itself known by skewing the distribution of the residuals?

12. The radioactive decay model $A(t) = A_0 e^{-kt}$ applies to substances that decay at a rate proportional to the amount present. Here $A(t)$ refers to the amount remaining at time t, A_0 is the initial amount, and k is the proportionality constant in the aforementioned decay condition. If you attempted a linear regression of A on t using experimental data, what would you expect to see in the residual graphs? Are there transformed variables that should bear a linear relationship to one another?

13. In economics a classic model called the *Cobb–Douglas production model* proposes that the productivity P of an organization is related to its capital investment I and its labor force L by

$$P = kI^\alpha L^{1-\alpha}, \qquad \alpha \in (0, 1).$$

In a multiple linear regression of P on I and L, what would plots of residuals against each predictor variable look like? Are there transformed variables that should bear a linear relationship?

14. The filling process of five-pound bags of sugar is subject to random errors in fill weight. What sort of regression model should relate the fill weight Y to its order X of production in the line (1st, 2nd, 3rd, ...)? Using the following data, is there any evidence that the filling process may be going astray? Be sure to check a time series graph of residuals.

> 5.10, 5.31, 4.66, 4.19, 3.91, 4.22, 4.46, 4.40, 3.95, 4.24,
> 4.74, 4.18, 4.70, 4.58, 4.20, 3.93, 3.98, 4.53, 4.52, 4.50

15. Recall the data in Example 9.2.4 in which Chris Najim followed the weekly rate of return (in %) for Sears stock for 30 consecutive weeks. The data, in time order, follows. Is there any evidence of lack of independence of error terms in a linear regression model of rate of return on week?

> −3.77, −4.22, 0.63, −1.88, −2.23, −3.58, −2.36, 4.50,
> −3.31, 0.68, 2.38, −4.98, −0.35, −0.70, −3.53, 4.03,
> −4.23, 4.78, 8.07, 2.60, 2.85, 0.00, 0.92, 5.18,
> 2.03, −0.28, −1.14, 5.76, −1.91, 5.00

16. Show that the variance of the ith residual in a single-variable linear regression model is

$$\text{Var}(e_i) = \sigma^2 \left[1 - \frac{1}{n} - \frac{(x_i - \bar{x})^2}{\sum_{j=1}^{n}(x_j - \bar{x})^2} \right].$$

Therefore the residuals do not have the same variance as the errors ϵ_i.

17. Show that the correlation between two residuals e_i and e_j in a single-variable regression model is nonzero; hence the residuals are not independent random variables.

18. Suppose that in the single variable linear regression model $Y_i = b_0 + b_1 x_i + \epsilon_i$, $i = 1, 2, \ldots, n$, the variance of ϵ_i is some constant c_i^2 times σ^2. Show that the transformed model

$$\frac{1}{c_i} Y_i = \frac{1}{c_i} b_0 + \frac{1}{c_i} b_1 x_i + \frac{1}{c_i} \epsilon_i$$

satisfies a linear multiple regression model with no constant with the usual assumptions on error terms. Show that the least squares method for the new problem is equivalent to the *weighted least squares method* in which the estimates are obtained by minimizing the quantity

$$\sum_{i=1}^{n} (Y_i - [b_0 + b_1 x_i])^2 \cdot \frac{1}{c_i^2}.$$

10.5 | Correlation

Several times in this text we have encountered the concept of correlation, which measures the strength of the linear relationship between two variables.

We have seen correlation in the context of a set of bivariate data and in the context of probability distributions. This section helps tie together the two ideas and explores the intimate connection of correlation testing with the linear regression problem.

Let us trace back through the development of the idea of correlation. Section 4.3 introduced correlation for joint probability distributions. The correlation $\rho = \mathrm{Cov}(X, Y)/\sigma_x\sigma_y$ between two random variables X and Y is near 1 in magnitude when the joint density concentrates most of its mass near a line. One of the interesting things we found about the bivariate normal density is that, given the value x taken on by X, the conditional mean of the random variable Y is

$$\mu_{y|x} = \mu_y + \rho\frac{\sigma_y}{\sigma_x}\left(x - \mu_x\right). \tag{10.50}$$

It is of interest to construct a test statistic for the bivariate normal correlation coefficient, and that is where the sample correlation comes in.

For random samples, the magnitude of the sample correlation $R = S_{XY}/S_X S_Y$ is near 1 when the data points lie near to a line, as in Fig. 10.1 on the male and female life spans, and it is near 0 when the x and y coordinates of the data points do not bear a very strong linear relationship to one another, as in Fig. 10.5 on the calculus grade-placement exam data.

In this chapter we discovered that the estimated regression line relating the mean of the response variable Y to the value of the predictor variable x is

$$Y = \widehat{b}_0 + \widehat{b}_1 x = \overline{Y} + \widehat{b}_1\left(x - \overline{x}\right). \tag{10.51}$$

Furthermore we can write

$$\widehat{b}_1 = R_x \cdot \frac{S_y}{S_x}, \tag{10.52}$$

where

$$R_x = \frac{\sum_{i=1}^{n}(x_i - \overline{x})(Y_i - \overline{Y})}{\sqrt{\sum_{i=1}^{n}(x_i - \overline{x})^2} \cdot \sqrt{\sum_{i=1}^{n}(Y_i - \overline{Y})^2}} \tag{10.53}$$

is the sample correlation between x- and Y-values. Thus the estimated regression line is

$$Y = \overline{Y} + R_x \cdot \frac{S_y}{S_x}\left(x - \overline{x}\right), \tag{10.54}$$

which parallels (10.50). The correlation R_x in (10.53) is really a conditioned version of the actual sample correlation written in full as follows, given observed values of the predictor variable X:

$$R = \frac{\sum_{i=1}^{n}(X_i - \overline{X})(Y_i - \overline{Y})}{\sqrt{\sum_{i=1}^{n}(X_i - \overline{X})^2} \cdot \sqrt{\sum_{i=1}^{n}(Y_i - \overline{Y})^2}}. \tag{10.55}$$

Despite this difference, the ideas of correlation and regression seem to be connected in a deep way.

We would like to construct an hypothesis test of $H_0 : \rho = 0$ for a bivariate normal correlation coefficient ρ based on the sample correlation R by exploiting this connection to regression and the known distributional results for regression estimators. The next theorem shows how.

PROPOSITION 10.5.1

Let $(X_1, Y_1), (X_2, Y_2), \ldots, (X_n, Y_n)$ be a random sample from a bivariate normal distribution with parameters $\mu_x, \mu_y, \sigma_x^2, \sigma_y^2$, and ρ. Let R be the sample correlation between X and Y. Then the random variable

$$T = \frac{\sqrt{n-2} \cdot R}{\sqrt{1-R^2}} \qquad (10.56)$$

has the $t(n-2)$ distribution under the hypothesis that $\rho = 0$.

■ **Proof** Write S_{XY} for the numerator of (10.55) divided by $n-1$, or the sample covariance. Write S_X^2 for the sample variance of X, and S_Y^2 for the sample variance of Y. We will use similar notations with lowercase x's: S_{xY}, s_x^2 for the sample covariance and variance, respectively, conditioned on the observed x_i's. Then from (10.53) and (10.55),

$$R_x = \frac{S_{xY}}{s_x S_Y} \quad \text{and} \quad R = \frac{S_{XY}}{S_X S_Y}. \qquad (10.57)$$

From Section 10.3 we know that conditioned on the observed X_i's,

$$T = \frac{\hat{b}_1 - b_1}{\sqrt{SSE/(n-2)}/\sqrt{s_x^2 \cdot (n-1)}} \sim t(n-2),$$

and since $\rho = 0$ by hypothesis, formula (10.50) implies that also $b_1 = 0$. By (10.52) we can write

$$T = \frac{\sqrt{n-2} \cdot R_x \cdot S_Y/s_x}{\sqrt{SSE}/\sqrt{(n-1)s_x^2}} \sim t(n-2). \qquad (10.58)$$

In Exercise 4 you are asked to verify algebraically the identity

$$\frac{R_x S_Y/s_x}{\sqrt{SSE}/\sqrt{(n-1)s_x^2}} = \frac{R_x}{\sqrt{1-R_x^2}}. \qquad (10.59)$$

This result implies that, conditioned on the observed x_i's, the random variable T of (10.56) has the $t(n-2)$ distribution under $H_0 : \rho = 0$. But the form of the distribution has no dependence on the x_i's, so the unconditional distribution of T must be the same as its conditional distribution given the X_i's (see Exercise 5). Thus, $T \sim t(n-2)$, which completes the proof.

To test $H_0 : \rho = 0$, against one- or two-sided alternatives, it suffices to compare the computed $t = \sqrt{n-2} \cdot r/\sqrt{1-r^2}$ to critical values of the desired

484 *Chapter 10: Regression and Correlation*

level obtained from the $t(n-2)$ table (Table 5 in Appendix A), and to reject H_0 if the magnitude of the computed t is too large. Notice that since the slope coefficient b_1 is the same as $\rho\sigma_y/\sigma_x$ in the bivariate normal setting, and the variances are positive, testing for $\rho = 0$ is the same as testing for $b_1 = 0$, that is, for the significance of X as a linear predictor of Y. This particular regression inference problem has therefore been extended to the case where X as well as Y is random.

?Question 10.5.1 Verify that r is large and positive (respectively negative) if and only if t is large and positive (respectively negative).

Example 10.5.1 Exercise 15 in Section 7.1 listed data on average faculty salary by rank for 11 universities in Illinois. Scattergrams of full professor salaries against associate professor salaries and against assistant professor salaries are shown in Figs. 10.18(a) and (b). A fairly high degree of linear dependence is suggested. Since it is reasonable to suppose that large salaries in one rank correspond to large salaries in another, we will test the null hypothesis $H_0 : \rho = 0$ against the one-sided alternative $H_a : \rho > 0$ for full and associate professors.

The sample correlation between full and associate professors comes out as .909. Hence the value of the t-statistic is

$$t = \frac{\sqrt{n-2} \cdot r}{\sqrt{1-r^2}} = \frac{\sqrt{9} \cdot (.909)}{\sqrt{1-(.909)^2}} = 6.54.$$

Since the level .001 critical value for the $t(9)$ distribution is 4.297, we easily reject the null hypothesis of no correlation at level .001.

?Question 10.5.2 The correlation observed between full and assistant professors is .929. Compute the value of the t-statistic. Use a statistics package to find the p-value of the test of $H_0 : \rho = 0$ based on these data.

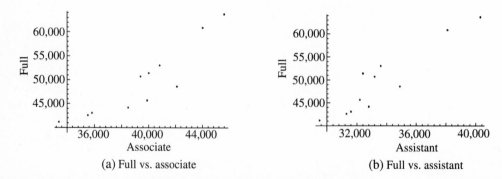

(a) Full vs. associate   (b) Full vs. assistant

Figure 10.18 Salaries of College Faculty

Example 10.5.2

Exercise 3 in Section 10.3 contained data on number of car thefts and population sizes in a sample of 16 cities. Is there a significant correlation between these two variables?

Figure 10.19 shows the scattergram for these data. Notice the outlier at $(36, 7.9)$, which corresponds to Los Angeles. The sample correlation between thefts and city size for the full data set is .901, and without the outlier it is a huge .994. The calculated t-values are

$$t = \frac{\sqrt{14} \cdot .901}{\sqrt{1 - (.901)^2}} = 7.77$$

for the complete data and

$$t = \frac{\sqrt{13} \cdot .994}{\sqrt{1 - (.994)^2}} = 32.76$$

for the data with Los Angeles deleted. Both values are highly significant, and it is apparent that Los Angeles is an exceptional case to a human phenomenon that is surprisingly predictable.

We conclude this brief section by paying homage to another test statistic, one used occasionally for correlation testing. It can be shown (see Bickel and Doksum, 1977, p. 222) that if R is the sample correlation of a sample of pairs (X_i, Y_i) of size n with theoretical correlation ρ, then the random variable

$$Z = \frac{(1/2)\ln\big((1+R)/(1-R)\big) - (1/2)\ln\big((1+\rho)/(1-\rho)\big)}{1/\sqrt{n-3}} \qquad (10.60)$$

has an approximate normal distribution with mean 0 and variance 1 for large sample size. The Z-statistic is explored in several of the exercises. Since Z increases in magnitude as R does (see Exercise 13), it is intuitively reasonable to reject the null hypothesis $H_0 : \rho = 0$ for Z-values of large magnitude. In the computation of Z we would use $\rho = 0$, the null hypothesized value, which

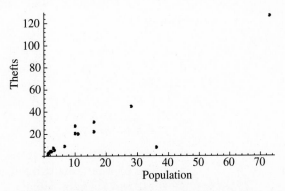

Figure 10.19   Scatterplot of Car Theft Data

simplifies the expression in (10.60). One advantage that this method has over the t-test is that it applies to null hypotheses of the form $H_0 : \rho = \rho_0$ as well. The appropriate test statistic Z would be as in (10.60) with ρ replaced by ρ_0. Since the exact meaning of an hypothesis like $H_0 : \rho = .5$ isn't too clear, it is probably more important that Z can be used to construct an approximate level $1 - \alpha$ confidence interval for ρ. We leave this as Exercise 12.

Example 10.5.3

The team of psychologist-researchers (Kasser et al. 1995), whose work on attitudes of mothers and their teenage children we have now looked at several times, also had occasion to do some correlation analysis. Among several of their computations was the sample correlation of the value placed by the mother and the child on each of the financial success, self-acceptance, and familial affiliation scales. They found these correlations to be .21, .22, and .37, respectively, for $n = 129$ sets of mothers and children. The observed value of the z-statistic for financial success correlation is then

$$z = \frac{(1/2)\ln\big((1+r)/(1-r)\big)}{1/\sqrt{n-3}} = \sqrt{126} \cdot \ln\left(\frac{1+.21}{1-.21}\right) \Big/ 2 = 2.39.$$

From the normal table, the approximate p-value for a one-sided test would be $P[Z > 2.39] = .008$. Similar calculations for the self-acceptance variable give $z = 2.51$ and approximate p-value .006. For the family affiliation variable, which one might expect to be the most highly correlated, the observed z is 4.36, which gives a p-value of nearly 0.

EXERCISES 10.5

1. Exercise 18 in Section 7.1 contained data on grade point averages in Calculus I, II, and III at 20 universities. Can you find significant correlations between any of the variables? Use level .05 and a one-sided test.

2. My student Chris Najim obtained rate of return data on several stocks, including Sears and Wal-Mart, over the same period of time. In the following list, the first component is the Sears rate of return for a particular day, and the second is the Wal-Mart rate of return for the same day. Are the two variables significantly correlated at level .05? Would you expect them to be?

   $(-3.77, 2.79)$, $(-4.22, -0.74)$, $(0.63, -1.99)$, $(-1.88, 0.76)$, $(-2.23, -3.78)$, $(-3.58, -2.88)$, $(-2.36, 1.35)$, $(4.50, 4.52)$, $(-3.31, -5.85)$, $(0.68, -0.54)$, $(2.38, 6.52)$, $(-4.98, -3.32)$, $(-0.35, -0.53)$, $(-0.70, 3.71)$, $(-3.53, 6.14)$, $(4.03, 1.45)$, $(-4.23, 1.43)$, $(4.78, 5.85)$, $(8.07, 2.21)$, $(2.60, -1.95)$, $(2.85, 0.88)$, $(0.00, -0.88)$, $(0.92, -4.86)$, $(5.18, 1.16)$, $(2.03, -2.06)$, $(-0.28, -2.34)$, $(-1.14, 3.60)$, $(5.76, -2.08)$, $(-1.91, -0.95)$, $(5.00, 1.67)$

3. Consider the statistics test scores from Table 7.2 of Section 7.1. Do there appear to be significant correlations between any pair of tests? Use both the t-test and the approximate Z-test in (10.60).

4. Verify formula (10.59).

5. To complete the proof of Proposition 10.5.1, show the following. Suppose that X and Y are jointly distributed continuous random variables with the property that the conditional density $f_{y|x}$ of Y given $X = x$ has no dependence on x. Then the marginal density of Y is the same as the conditional density.

6. Obtain the residuals from the quadratic regression of casino revenues on year using the data in Exercise 11 in Section 10.2 (see also Exercise 9 in Section 10.4). Test at level .1 for correlation between residuals and their predecessors.

7. Verify the computational formula for the sample correlation:

$$r = \frac{\sum_{i=1}^{n} x_i y_i - n\overline{x}\overline{y}}{\sqrt{\sum_{i=1}^{n} x_i^2 - n\overline{x}^2} \cdot \sqrt{\sum_{i=1}^{n} y_i^2 - n\overline{y}^2}}.$$

8. In Exercise 1 in Section 3.4 we presented a data set of 28 pairs in which one member of the pair was the birth rate of a country in which there had recently been mass forced migration, and the other was an index of human suffering for that country. Use the large sample approximate normal test to check for significant correlation. Report the p-value, and give a conclusion.

9. The following list of data are average yields per acre (in pounds) of cotton (x) and tobacco (y) for the years 1976–1989 (USDA, 1991). Produce a scattergram. Test for a significant correlation between the variables and comment on the results.

   (465, 2041), (520, 1982), (420, 2101), (547, 1844), (404, 1939),
   (542, 2113), (590, 2185), (508, 1811), (600, 2183), (630, 2197),
   (552, 2001), (706, 2028), (619, 2160), (614, 2016)

10. I have mentioned before (see Exercise 6 in Section 10.2, for example) that I worked with a group of psychologists studying the behavior of twins. Another little part of the data set follows. In the triples, the first score is the aggression level, the second is the activity level, and the third measures the frequency of sleep problems for some of the children in the study. Can you find any significant correlations among the variables? Comment on your results.

    (43, 41, 65), (46, 66, 43), (46, 46, 50), (39, 41, 50), (50, 41, 50),
    (39, 56, 50), (78, 56, 57), (53, 41, 50), (64, 46, 57), (50, 41, 43),
    (39, 61, 50), (53, 61, 72), (61, 41, 50), (53, 51, 43), (39, 41, 43),
    (39, 41, 57), (50, 46, 43), (61, 51, 50), (43, 41, 43), (39, 41, 43)

11. How large would the correlation coefficient for a sample of 17 pairs have to be for the theoretical correlation to be judged significantly different from zero at level .05? (Assume a one-sided test.)

12. Use the transformation (10.60) to construct an approximate level $1 - \alpha$ confidence interval for ρ. Apply this to find an approximate 90% confidence interval for the correlation between car thefts and population in Example 10.5.2. Delete Los Angeles from the data set to do this.

13. Show that the random variable Z in (10.60) increases in magnitude as R becomes further from ρ. Thus, under the null hypothesis $H_0 : \rho = 0$, Z increases in magnitude as R does.

14. A nonparametric version of the correlation coefficient is as follows. Let $(X_1, Y_1), (X_2, Y_2), \ldots, (X_n, Y_n)$ be a random sample of pairs, let Q_i^x be the rank of X_i in the X sample, and let Q_i^y be the rank of Y_i in the Y sample. Then the *rank correlation coefficient* of the sample is

$$\tilde{R} = 1 - \frac{6 \cdot \sum_{i=1}^{n} (Q_i^x - Q_i^y)^2}{n(n^2 - 1)}.$$

a. Explain why this statistic measures the association between X and Y.

b. Compute $\tilde{R}$ for the data:

$$(3.65, 8.02), \quad (2.16, 7.50), \quad (2.50, 7.30), \quad (4.04, 9.50),$$
$$(1.23, 4.86), \quad (3.41, 7.94), \quad (4.25, 7.81), \quad (5.16, 9.08).$$

c. Find by listing all possibilities the probability distribution of $\tilde{R}$ in the case $n = 4$ under the null hypothesis that X and Y are independent.

CHAPTER 11

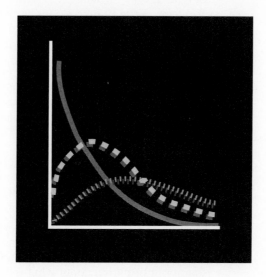

EXPERIMENTAL DESIGN AND ANALYSIS OF VARIANCE

11.1 | Experimental Design and the Linear Model

This chapter discusses the design and analysis of a certain well-defined class of experiments. In these experiments, as in regression models, a response variable that may be influenced by other variables or conditions is observed. This time, however, the predictor variables, called *factors*, can only take on finitely many values called *levels*. We are interested in estimating and testing the effect of the factor levels on the mean of the response and in finding possible interactions between different factors. By the latter, we mean that the effect of level 1 of factor *A* and level 2 of factor *B* on the mean response may be more than just the total of separate effects due to combining the two factors.

To illustrate the ideas involved in these experiments, consider a problem in which we are trying to compare the gas mileages that cars achieve when driven on full tanks of regular (87 octane), special blend (89 octane), and premium gasoline (92 octane). Then gas mileage, Y, is the response variable, and the single predicting factor is the grade of gasoline, which has the three possible levels: regular (1), special (2), and premium (3). Imitating regression problems, we may take as our starting point a model in which Y is the sum of a nonrandom term dependent on the gasoline grade, which determines Y's mean, and a random error term that is normally distributed with mean 0. In symbols,

$$Y = \begin{cases} \mu_1 + \epsilon & \text{if } \text{level} = 1, \\ \mu_2 + \epsilon & \text{if } \text{level} = 2, \\ \mu_3 + \epsilon & \text{if } \text{level} = 3, \end{cases} \tag{11.1}$$

where $\epsilon \sim N(0, \sigma^2)$.

The usual questions of interest in such a model are comparisons between means $\mu_1, \mu_2, \mu_3, \ldots$. In the gasoline example, if the grade of gasoline made no difference in the mileage achieved, then it would follow that $\mu_1 = \mu_2 = \mu_3$. So, if we are interested in testing whether a factor does indeed have a significant influence on the response variable, then we would want to test the null hypothesis $H_0 : \mu_1 = \mu_2 = \mu_3 = \cdots$ against the complementary hypothesis that two or more of the means are unequal. To obtain sharper information about what the inequalities are, if any, we might further be interested in pairwise mean comparisons if this H_0 is rejected.

This leads to the question of how to collect data that can be used in tests like these. The present section concentrates solely on the design of experiments to gather the data and on how the design relates to the proposed linear model. Subsequent sections discuss the analysis of the data for a few of the simplest models. Appendix D, Research Projects, points you in the direction of more complicated models.

—— **?Question 11.1.1** Try to think of other examples of the kind of experiment that we have described. What data would you propose to gather?

Consider the process of collecting data for the gas mileage problem. It doesn't require a lot of imagination to conclude that each grade of gasoline should be run on several cars and we should record the gas mileage obtained, thus producing a sample of regular mileages, say $Y_{11}, Y_{12}, \ldots, Y_{1n_1}$, a sample of special blend mileages, say $Y_{21}, Y_{22}, \ldots, Y_{2n_2}$, and a sample of premium mileages, say $Y_{31}, Y_{32}, \ldots, Y_{3n_3}$, not necessarily of the same sample sizes n_1, n_2, and n_3. There are some subtleties, however. Do we use the same test vehicles for one grade as for another, or do we use different ones? Do we run each car for the same number of miles? Do we record just one gas mileage, or do we run the car for several trips, record each trip mileage, and average the trip mileages?

—— **?Question 11.1.2** What do you think is the most important effect of running each car for a very large number of miles as opposed to a few miles? What possible problem could arise if some cars were run for a greater distance than others?

As you thought about Question 11.1.2, you probably decided that consistency is desirable, otherwise the variability of one piece of random data might be inherently greater than the variability of another, which could complicate the analysis. Also, the larger the number of miles the better, for greater precision of the estimate of mean miles per gallon. Practical considerations of time and expense enter this decision. Exercise 8 of this section sheds light on the question of averaging trip mileages, which we will say no more about. This leaves the main question of whether to use the same cars or different cars for the gasoline grades. The two possibilities illustrate the main types of single-factor design, the *completely randomized design* and the *randomized block design*.

The completely randomized design randomly assigns a given number of *experimental units*, here the cars, to the levels of the factor. The design is called *balanced* if each level receives the same number of units, and *unbalanced* otherwise. For simplicity in notation we will only consider balanced designs for the time being. Some number $3n$ of cars are chosen, then divided randomly into three subsets of n cars each. The first subset uses regular gasoline for a fixed number of miles, the second subset uses special blend for the same number of miles, and the third uses premium. Independent random samples of gas mileages are obtained:

$$\begin{aligned}
&Y_{11}, Y_{12}, Y_{13} \ldots, Y_{1n} && \text{(regular)}, \\
&Y_{21}, Y_{22}, Y_{23}, \ldots, Y_{2n} && \text{(special)}, \\
&Y_{31}, Y_{32}, Y_{33}, \ldots, Y_{3n} && \text{(premium)}.
\end{aligned} \tag{11.2}$$

The probabilistic model that we use to describe the data is

$$\begin{aligned}
Y_{1j} &= \mu_1 + \epsilon_{1j}; && j = 1, 2, \ldots, n, \\
Y_{2j} &= \mu_2 + \epsilon_{2j}; && j = 1, 2, \ldots, n, \\
Y_{3j} &= \mu_3 + \epsilon_{3j}; && j = 1, 2, \ldots, n,
\end{aligned} \tag{11.3}$$

or more concisely

$$Y_{ij} = \mu_i + \epsilon_{ij}; \qquad i = 1, 2, \ldots, a, \quad j = 1, 2, \ldots, n, \tag{11.4}$$

where ϵ_{ij} are i.i.d. $N(0, \sigma^2)$ random variables and $a = 3$ for the gas mileage example. An equivalent model has the mean μ_i broken down into an overall mean μ plus a residual effect α_i for the ith level of the factor:

$$Y_{ij} = \mu + \alpha_i + \epsilon_{ij}; \qquad i = 1, 2, \ldots, a, \quad j = 1, 2, \ldots, n. \tag{11.5}$$

It is usually assumed in this case that the residual effects sum to zero, that is, $\sum_{i=1}^{a} \alpha_i = 0$. (See Exercise 6.)

The randomized block design uses a total of b experimental units, and tests each unit with each level of the factor. For the gas mileage example, each of b cars would be driven with each grade of gasoline. It is probably best to also randomize the order in which the grades are used within each test car; that is, car 6 may use regular, then premium, then special blend, while car 9 may use premium, then special, then regular. The idea is not to confuse an actual effect of the grade of gasoline with some artifact dependent on when the cars were driven. (A warm-up factor could indeed confound the results in this problem.) The data can still be listed as in (11.2), but whereas columns had no meaning in the completely randomized case, now columns are *blocks* of data with some common characteristic, as illustrated in Fig. 11.1. In our example a block consists of data obtained from the same test car. The reason to choose a block design is to try to eliminate sources of variation in gas mileage due to the extraneous factor of the test car. If the analysis can be performed somehow car by car, perhaps it will be easier for us to find real differences between grades.

Clearly we will have a different probabilistic model and different statistical analysis in the blocked case than in the randomized case. Let b denote the number of blocks and again let a be the number of levels of the factor. Since the blocks are meaningful and may have an effect on the mean response, we add a term β_j for the jth block to (11.5) to obtain the model for the randomized block design.

$$Y_{ij} = \mu + \alpha_i + \beta_j + \epsilon_{ij}; \qquad i = 1, \ldots, a, \quad j = 1, 2, \ldots, b. \qquad (11.6)$$

Again it is usually assumed that the residual effects due to blocks total to zero, that is, $\sum_{j=1}^{b} \beta_j = 0$. In both the completely randomized and randomized block experiments modeled by (11.5) and (11.6), respectively, we are primarily concerned with testing whether $\alpha_1 = \alpha_2 = \cdots = \alpha_a = 0$. In the randomized block experiment we are secondarily interested in whether there is an effect due to blocks, that is, whether $\beta_1 = \beta_2 = \cdots = \beta_b = 0$.

The problem can be extended to permit the study of two predicting factors at once. For a balanced two-factor experiment, which we will study in Section 11.3, the layout of the data is as in the table in Fig. 11.2. Each factor has several levels, say a levels for factor A and b levels for factor B. In each combination (i, j) of levels there is a sample of responses:

$$\text{sample } ij = Y_{ij1}, Y_{ij2}, \ldots, Y_{ijn};$$

hence there are $n \cdot a \cdot b$ observations in total.

For a completely randomized two-factor model there would be $n \cdot a \cdot b$ experimental units, assigned randomly to treatment combinations (i, j) with n

Blocks

		1	2		b
Levels	1	Y_{11}	Y_{12}	$\cdots$	Y_{1b}
of	2	Y_{21}	Y_{22}	$\cdots$	Y_{2b}
factor	$\vdots$				
A	a	Y_{a1}	Y_{a2}	$\cdots$	Y_{ab}

Figure 11.1 Data Layout for Randomized Block Design

Levels of factor B

Levels of factor A	1	2	$\cdots$	b
1	$Y_{111}\ Y_{112}\cdots Y_{11n}$	$Y_{121}\ Y_{122}\cdots Y_{12n}$		$Y_{1b1}\ Y_{1b2}\cdots Y_{1bn}$
2	$Y_{211}\ Y_{212}\cdots Y_{21n}$	$Y_{221}\ Y_{222}\cdots Y_{22n}$		$Y_{2b1}\ Y_{2b2}\cdots Y_{2bn}$
$\vdots$				
a	$Y_{a11}\ Y_{a12}\cdots Y_{a1n}$	$Y_{a21}\ Y_{a22}\cdots Y_{a2n}$		$Y_{ab1}\ Y_{ab2}\cdots Y_{abn}$

Figure 11.2 Data Layout for Balanced, Randomized Two-Factor Design

units to each combination. An extension of our gas mileage example would be to consider not only the gasoline grade but the speed at which the vehicle in the test is driven. There might be speed categories, say 0–40 mph, 41–55 mph, and over 55 mph. Thus there are nine grade–speed combinations to each of which n cars are assigned at random. The random variable Y_{ijk} is the gas mileage observed for the kth car in the ith gas grade and jth speed category. Both factors may influence the mean of the mileage and therefore should contribute to the probabilistic model.

It is also conceivable that the contribution of a treatment combination (i, j) to the mean mileage may be more than just the total of separate contributions from level i of the gas grade factor and level j of the speed factor. Something special about the combination of i and j could introduce a new *interaction term* $\alpha\beta$ into the model as follows:

$$y_{ijk} = \mu + \alpha_i + \beta_j + (\alpha\beta)_{ij} + \epsilon_{ijk}; \qquad i = 1, 2, \ldots, a, \\ j = 1, 2, \ldots, b, \\ k = 1, 2, \ldots, n. \tag{11.7}$$

Again ϵ_{ijk} are i.i.d. $N(0, \sigma^2)$ random variables. The mean of the response Y_{ijk} is an overall mean term μ, plus terms α_i and β_j for the levels of factor A and B, respectively, used on the experimental unit, plus an interaction term $(\alpha\beta)_{ij}$ for the (i, j) treatment combination. We could be interested not only in the existence of significant factor effects as characterized by nonzero α's or β's, but in *interaction* between factors characterized by nonzero $(\alpha\beta)$'s. We usually assume that $\sum_i (\alpha\beta)_{ij} = 0\ \forall\, j, \sum_j (\alpha\beta)_{ij} = 0\ \forall\, i$, together with the zero sum conditions from before on the α's and β's.

?Question 11.1.3 Express in your own words the meaning of an interaction term in the gas mileage example.

To this point, we have supposed that the (finitely many) levels of a factor are the only ones possible, or of interest, and are determined in advance by the experimenter. Another situation has levels essentially taken at random from some universe of possible levels. Consider, for example, a manufacturing study in which equipment can be run at two speeds to produce an item with some measurable response characteristic such as breaking strength. The experimenter may be interested in not only the effect of speed, but the effect of the particular machine used for production. Some number of machines might be selected, say machines $1, 2, \ldots, 5$, and each might be run at each of the two speeds (in random order) to produce n items. The data layout is as in Fig. 11.2, with two levels of factor A, speed, and 5 levels of factor B, machine. So what is different?

The machines used are only five selected randomly from many possible machines. They make *random* contributions B_1, B_2, B_3, B_4, and B_5, respectively, to the observed breaking strength. The mean contribution μ_b by all machines may be assumed to be absorbed into μ, the overall mean, so that we may suppose the random variables B_i have mean 0. What is left as a measure of machine effect is the variance of the B's. The appropriate probabilistic model (without interaction) is

$$Y_{ijk} = \mu + \alpha_i + B_j + \epsilon_{ijk}; \qquad i = 1, 2; \quad j = 1, 2, 3, 4, 5; \quad k = 1, 2, \ldots, n,$$

$$(11.8)$$

where the random variables $B_j \sim N(0, \sigma_b^2)$ are independent of the i.i.d. normal errors ϵ_{ijk}. The null hypothesis regarding the effect of B would be $H_0 : \sigma_b^2 = 0$; in other words, the machines contribute no significant variability to the strength. A similar model with random interaction terms $(\alpha B)_{ij} \sim N(0, \sigma_{\alpha B}^2)$ is possible. Factor B here is called a *random* factor, and factor A is a *fixed* factor. A model such as (11.8) that includes random factors is called a *random effects model*. We will look at the analysis of single factor random effects models in Section 11.2, and two-factor random effects models in Section 11.3.

— **?Question 11.1.4**   Try to think of another experimental situation in which a random factor appears.

You can imagine a variety of models like those mentioned here, for example, experiments with more than two-factors, unbalanced designs, or combinations of fixed and random factors.

For models with fixed effects, there is a common structure that is most easily seen by referring to matrices. All of these models are additive in terms of the constant parameters making up the mean response. This makes it possible to write the observed values of the response variable in a vector, which is then set equal to a matrix of 0s and 1s, times a vector of parameters, plus a vector of errors corresponding to the response values. For the gas mileage model (11.5), for example, we could define

$$
\mathbf{Y} = \begin{bmatrix} Y_{11} \\ Y_{12} \\ \vdots \\ Y_{1n} \\ \hline Y_{21} \\ Y_{22} \\ \vdots \\ Y_{2n} \\ \hline Y_{31} \\ Y_{32} \\ \vdots \\ Y_{3n} \end{bmatrix}, \quad
X = \begin{bmatrix} \overset{\mu}{1} & \overset{\alpha_1}{1} & \overset{\alpha_2}{0} & \overset{\alpha_3}{0} \\ 1 & 1 & 0 & 0 \\ \vdots & \vdots & \vdots & \vdots \\ 1 & 1 & 0 & 0 \\ \hline 1 & 0 & 1 & 0 \\ 1 & 0 & 1 & 0 \\ \vdots & \vdots & \vdots & \vdots \\ 1 & 0 & 1 & 0 \\ \hline 1 & 0 & 0 & 1 \\ 1 & 0 & 0 & 1 \\ \vdots & \vdots & \vdots & \vdots \\ 1 & 0 & 0 & 1 \end{bmatrix}, \quad
\boldsymbol{\theta} = \begin{bmatrix} \mu \\ \alpha_1 \\ \alpha_2 \\ \alpha_3 \end{bmatrix}, \quad
\boldsymbol{\epsilon} = \begin{bmatrix} \epsilon_{11} \\ \epsilon_{12} \\ \vdots \\ \epsilon_{1n} \\ \hline \epsilon_{21} \\ \epsilon_{22} \\ \vdots \\ \epsilon_{2n} \\ \hline \epsilon_{31} \\ \epsilon_{32} \\ \vdots \\ \epsilon_{3n} \end{bmatrix}.
$$

$$(11.9)$$

Then,

$$\mathbf{Y} = X\boldsymbol{\theta} + \boldsymbol{\epsilon}. \tag{11.10}$$

The vector $\mathbf{Y}$ is the *data vector* for this model. The matrix X is called the *design matrix*, which captures more than anything else the structure of the experiment. Since the overall mean μ is present in every formula for Y_{ij}, X has 1s in every entry in its first column. The first group of n Y's are treated at factor level 1, the second at level 2, and the third at level 3, which accounts for the blocks of 1s and 0s in the $\alpha_1, \alpha_2,$ and α_3 columns of X. The vector $\boldsymbol{\theta}$ contains the unknown parameters, and the error vector $\boldsymbol{\epsilon}$ has an entry for each Y_{ij}. The linear form of the relationship between $\mathbf{Y}$ and $\boldsymbol{\theta}$ in (11.10) motivates the term *linear model* for the problems we are studying in this chapter. The matrix representation of the linear model will facilitate the derivations of distributional results necessary for statistical inference. It also is a unifying idea that glues together the seemingly disparate methods and models of experimental design.

Many of those methods are very intuitive, even if the justification of them is sometimes difficult. As we will see in the next sections, the analysis is based on comparisons of sample means. To get a headstart now on the notation so that you may concentrate more on the ideas later, we introduce the conventional dot and bar notation. In problems of experimental design a response Y is subscripted by one, two, three, or even more indicators that identify the factor–level combination the response pertains to. For instance, in the two-factor randomized design we have seen Y_{ijk}, which is member k of a random sample of observations at level i of factor A and level j of factor B. When a dot is used in place of a subscript, it means that the observations in that group of data are being summed; for example

$$Y_{ij\cdot} = \sum_{k=1}^{n} Y_{ijk} \quad \text{and} \quad Y_{i\cdot\cdot} = \sum_{j=1}^{b}\sum_{k=1}^{n} Y_{ijk}. \tag{11.11}$$

Furthermore, when the Y has a bar over it, it means that an average is taken; that is, the sum indicated by the dots is divided by the number of elements being summed. For the foregoing examples, the averages would be written:

$$\overline{Y}_{ij\cdot} = \frac{1}{n}\sum_{k=1}^{n} Y_{ijk} \quad \text{and} \quad \overline{Y}_{i\cdot\cdot} = \frac{1}{nb}\sum_{j=1}^{b}\sum_{k=1}^{n} Y_{ijk}. \tag{11.12}$$

For the overall average of all Y's, we will simplify the notation and omit the dot subscripts, preferring instead just $\overline{Y}$. We will use the dot and bar notation freely in what is to come.

?Question 11.1.5 What would be the meaning of $Y_{i\cdot}$ in the one-factor completely randomized experiment? What is $\overline{Y}_{\cdot j}$ in the two-factor randomized experiment?

EXERCISES 11.1

1. Suppose you were interested in comparing four reputed weight loss plans. Describe in detail such an experiment, specifying the data you would gather and the probabilistic model you would propose. Does a completely randomized or a randomized block experiment seem more appropriate here?

2. Recall the statistics test data from Table 7.2. These data were not the result of a designed experiment. Nevertheless, discuss whether it is possible to apply a probabilistic model of the kind detailed in this section to test whether the expected performance on the three tests is the same.

3. A cereal manufacturer is concerned that there may be factors out of its control that influence sales of its products. Particularly, it believes that the shelf height at which a box of cereal is displayed at the grocery store may affect sales volume. Design an experiment to test this factor. If the manufacturer is also simultaneously interested in whether the predominant color of its boxes has a significant effect on sales, design an experiment that tests both the shelf height and color factors. Carefully write the probabilistic model for this experiment.

4. Ten fields at different locations are available for testing how the total yield in bushels of a corn crop depends upon the brand of seed used. Four brands are under consideration. Design a randomized block experiment to test this situation.

5. This question was inspired by a consultation with a student, Brian Kendall, doing research on alcohol metabolism by fruit flies. Suppose each of two varieties of flies are given each of three diets: pure sucrose, sucrose with ethanol, and sucrose with ethanol and a substance called juvenile hormone. The activity level of an enzyme (called ADH) suspected to be associated with alcohol metabolization was measured for aggregates of 100 flies, with six replications for each diet–variety pair. Write a probabilistic model for the experiment, and discuss possible hypothesis tests.

6. Show that the form (11.5) with the side condition $\sum_{i=1}^{a} \alpha_i = 0$ follows by setting μ equal to the average of the μ_i's in the original linear model for the completely randomized one-factor experiment, and $\alpha_i = \mu_i - \mu$.

7. The design of a one-factor completely randomized experiment entails the random assignment of n experimental units into k test groups of sizes $n_1, n_2, \ldots, n_k$. In how many ways can this be done?

8. In the gas mileage example of this section, show that if a car is driven for a number N of miles, the overall average gas mileage for the N travel miles is *not* the same as the average of the trip gas mileages computed as if it had been driven for k trips of N/k miles each. For which method is it more reasonable to assume normality of the gas mileage data?

9. What is the probability distribution of the ith sample mean $\overline{Y}_{i\cdot} = \frac{1}{n} \sum_{j=1}^{n} Y_{ij}$ in the balanced completely randomized one-factor experiment?

10. What is the expectation and variance of the difference between the ith and kth sample means (see Exercise 9) in the balanced completely randomized one-factor experiment?

11. Find the expected value and variance of the average

$$\overline{Y}_{\cdot j \cdot} = \frac{1}{an} \sum_{i=1}^{a} \sum_{k=1}^{n} Y_{ijk}$$

of observations at the jth level of factor B in a two-factor randomized experiment without interaction term.

12. Display the general randomized block model in matrix form.

13. In a single-factor balanced completely randomized model with a random factor, what is the probability distribution of $\overline{Y}_{i\cdot} = \frac{1}{n} \sum_{j=1}^{n} Y_{ij}$?

14. Write the design matrix for a two-factor experiment with three replications for each of two levels of factor A and four levels of factor B.

11.2 | One-Factor Analysis of Variance

In this section we will derive the inferential procedures necessary to test for significance of a factor in the mean of a response variable in experiments of the kind we described in Section 11.1. The heart of the analysis is the decomposition of the total variation in the data into contributions from various sources, much as we did in regression. Thus the term *analysis of variance* has been attached to these and other related procedures. For now we will confine our attention to the case of one factor. Two or more factors are considered in Section 11.3.

11.2.1 Completely Randomized Design

Recall that in the completely randomized experiment, the model for the observed values of the response variable is

$$Y_{ij} = \mu + \alpha_i + \epsilon_{ij}; \qquad i = 1, 2, \ldots, a, \quad j = 1, 2, \ldots, n, \tag{11.13}$$

where we are assuming that there are a levels of the factor and n observations are taken at each factor level. The random variables ϵ_{ij} are i.i.d. $N(0, \sigma^2)$ random variables, which implies that the random variables Y_{ij} are independent of one another, and $Y_{ij} \sim N(\mu + \alpha_i, \sigma^2)$. Also, we will suppose that $\sum_{i=1}^{a} \alpha_i = 0$. For simplicity we are considering the balanced design, but the case where there are n_i observations taken at factor level i requires only minor modifications. We will point out some of the changes as the development proceeds.

Formulas (11.9) and (11.10) show how to express the model in matrix form. For notational ease let us denote:

$$\mathbf{1}_n = \text{column vector of 1s with } n \text{ components,}$$
$$\mathbf{0}_n = \text{column vector of 0's with } n \text{ components,}$$
$$\mathbf{Y}_i = [Y_{i1}, Y_{i2}, \ldots, Y_{in}]',$$
$$\boldsymbol{\epsilon}_i = [\epsilon_{i1}, \epsilon_{i2}, \ldots, \epsilon_{in}]'. \tag{11.14}$$

Then (11.13) becomes

$$
\begin{bmatrix} \mathbf{Y}_1 \\ \mathbf{Y}_2 \\ \vdots \\ \mathbf{Y}_a \end{bmatrix}
=
\begin{bmatrix}
\mathbf{1}_n & \mathbf{1}_n & \mathbf{0}_n & \cdots & \mathbf{0}_n \\
\mathbf{1}_n & \mathbf{0}_n & \mathbf{1}_n & \cdots & \mathbf{0}_n \\
\vdots & \vdots & \vdots & \ddots & \vdots \\
\mathbf{1}_n & \mathbf{0}_n & \mathbf{0}_n & \cdots & \mathbf{1}_n
\end{bmatrix}
\begin{bmatrix} \mu \\ \alpha_1 \\ \alpha_2 \\ \vdots \\ \alpha_a \end{bmatrix}
+
\begin{bmatrix} \boldsymbol{\epsilon}_1 \\ \boldsymbol{\epsilon}_2 \\ \vdots \\ \boldsymbol{\epsilon}_a \end{bmatrix}, \tag{11.15}
$$

or

$$\mathbf{Y} = X\boldsymbol{\theta} + \boldsymbol{\epsilon}. \tag{11.16}$$

As we mentioned in Section 10.1, the matrix X of 1s and 0s that multiplies the parameter vector $\boldsymbol{\theta} = [\mu \; \alpha_1 \; \alpha_2 \; \ldots \; \alpha_a]'$ is called the *design matrix*.

— **?Question 11.2.1**   Verify by multiplication that (11.15) is equivalent to (11.13).

Example 11.2.1

As a native midwesterner I am frequently confronted by incredulous people from other areas of the country wondering how I can stand the temperature extremes in this area. Recently I came across data (Thomas, 1992) from several small cities in various states on the total number of days in a year in which the temperature either rose to 90° or more, or fell to 32° or less. The data follow. Is really a difference in temperature extremes by region?

Illinois, Indiana (Midwest):   152, 154, 164, 157, 156, 149, 155, 149, 168, 153, 160, 156

Texas, Arkansas (South):   144, 117, 143, 148, 133, 143, 149, 123, 144, 140, 140, 134

Oregon, Washington (Northwest):   70, 209, 70, 121, 182, 72, 48, 62, 41, 146, 100, 151

Under model (11.13), the Midwest data form the random vector $\mathbf{Y}_1$, the components of which are normally distributed with mean $\mu + \alpha_1$. Similarly

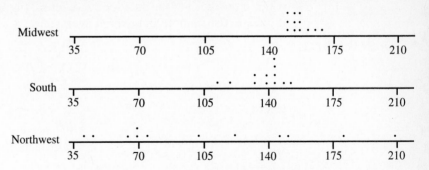

Figure 11.3   Dotplots of Temperature Extreme Data

the South data form the random vector $\mathbf{Y}_2$, whose components are normal with mean $\mu + \alpha_2$, and the Northwest data form the random vector $\mathbf{Y}_3$, whose components are normal with mean $\mu + \alpha_3$. The data vector $\mathbf{Y}$ would consist of the 36 points above stacked in a column with the Midwest data first, then the South data, and the Northwest data last. The design matrix is

$$
X = \begin{array}{c} \\ M \\ S \\ N \end{array}
\begin{array}{cccc} \mu & \alpha_1 & \alpha_2 & \alpha_3 \end{array}
\left[\begin{array}{cccc}
\mathbf{1}_{12} & \mathbf{1}_{12} & \mathbf{0}_{12} & \mathbf{0}_{12} \\
\mathbf{1}_{12} & \mathbf{0}_{12} & \mathbf{1}_{12} & \mathbf{0}_{12} \\
\mathbf{1}_{12} & \mathbf{0}_{12} & \mathbf{0}_{12} & \mathbf{1}_{12}
\end{array} \right].
$$

We have labeled the columns with the parameters to which they correspond and the row blocks with the regions to which they correspond.

The test of whether the regions differ in average number of extreme temperature days per year becomes the test of the null hypothesis

$$ H_0 : \alpha_1 = \alpha_2 = \alpha_3 = 0 $$

against the complementary alternative. Parallel dot plots of the data are shown in Fig. 11.3.

?**Question 11.2.2**   Do the data in Example 11.2.1 indicate differences among regions? Intuitively, what test statistics would you look at to make such a determination?

Your answer to Question 11.2.2 points out the direction we will be taking; the test statistic will be based on differences between sample means corresponding to the factor levels (sample means for regions in the last example). Denote as in Section 11.1

$$ \overline{Y}_{i\cdot} = \sum_{j=1}^{n} \frac{Y_{ij}}{n}, \qquad (11.17) $$

which is the mean of the data for factor level i (in the unbalanced case replace n by n_i). Also, denote the overall average by

$$\overline{Y} = \sum_{i=1}^{a} \sum_{j=1}^{n} \frac{Y_{ij}}{N}, \tag{11.18}$$

where

$$N = \begin{cases} n \cdot a & \text{balanced case,} \\ \sum_{i=1}^{a} n_i & \text{unbalanced case.} \end{cases} \tag{11.19}$$

Intuitively, if the factor levels differ in their mean parameters, then the $\overline{Y}_{i\cdot}$'s should differ significantly from one another. Equivalently, the following sum of squares should be large if the means $\mu + \alpha_i$ are not all equal:

$$\sum_{i=1}^{a} (\overline{Y}_{i\cdot} - \overline{Y})^2. \tag{11.20}$$

So, H_0 will be rejected when this quantity is large.

It is necessary to modify the intuitively appealing statistic in (11.20) in order to produce a random variable with a known probability distribution. To prepare the way, recall the idea of the projection of a vector $\mathbf{y}$ on the linear space spanned by a collection of vectors $\mathbf{x}_1, \mathbf{x}_2, \ldots, \mathbf{x}_k$:

$$P_x \mathbf{y} = \sum_{i=1}^{k} \frac{\mathbf{x}_i' \mathbf{y}}{\|\mathbf{x}_i\|^2} \cdot \mathbf{x}_i = \left(\sum_{i=1}^{k} \frac{\mathbf{x}_i \mathbf{x}_i'}{\|\mathbf{x}_i\|^2} \right) \mathbf{y}. \tag{11.21}$$

The matrix in parentheses on the right of (11.21), which accomplishes the projection of $\mathbf{y}$ by matrix multiplication, is called the *projection matrix*. A depiction of the geometric meaning of projection is given in Fig. 11.4.

What do projections have to do with the analysis of variance? It turns out that the vector of responses $\mathbf{Y}$ can be projected onto subspaces spanned by groups of columns of the design matrix. Moreover the squared lengths of the projected components of $\mathbf{Y}$ give a sum of squares decomposition via the Pythagorean theorem, which we can use to show that the sums of squares have chi-square distributions.

The previous paragraph is a kind of roadmap of the proof of the next theorem. Some of the straightforward computational details will be left to you as exercises.

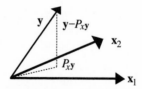

Figure 11.4 Projection of a Vector onto a Space Spanned by Vectors

PROPOSITION 11.2.1

(a) In the completely randomized one-factor model (11.13) we have the following sum of squares decomposition:

$$\sum_{i=1}^{a}\sum_{j=1}^{n} Y_{ij}^2 = an\overline{Y}^2 + n\cdot\sum_{i=1}^{a}(\overline{Y}_{i\cdot} - \overline{Y})^2 + \sum_{i=1}^{a}\sum_{j=1}^{n}(Y_{ij} - \overline{Y}_{i\cdot})^2. \tag{11.22}$$

Equivalently

$$SST = SSF + SSE,$$

where

$$SST = \sum_{i=1}^{a}\sum_{j=1}^{n} Y_{ij}^2 - an\overline{Y}^2 = \sum_{i=1}^{a}\sum_{j=1}^{n}(Y_{ij} - \overline{Y})^2, \tag{11.23}$$

$$SSF = n\cdot\sum_{i=1}^{a}(\overline{Y}_{i\cdot} - \overline{Y})^2, \tag{11.24}$$

$$SSE = \sum_{i=1}^{a}\sum_{j=1}^{n}(Y_{ij} - \overline{Y}_{i\cdot})^2. \tag{11.25}$$

The sums of squares SST, SSF, and SSE are called the *total, factor,* and *error sums of squares*, respectively.

(b) Furthermore, under H_0

$$\frac{SSF}{\sigma^2} \sim \chi^2(a-1) \quad \text{and} \quad \frac{SSE}{\sigma^2} \sim \chi^2(N-a), \tag{11.26}$$

and these random variables are independent of one another.

■ **Proof**   We will project the data vector **Y** onto (1) the space spanned by the first column in the design matrix X (corresponding to μ); and (2) the space spanned by the last a columns of X (corresponding to $\alpha_1, \alpha_2, \ldots, \alpha_a$). Then **Y** will be the sum of the two projected vectors plus a residual.

The projection matrix for the μ column from (11.21) is

$$P_\mu = \frac{1}{na}\cdot\begin{bmatrix}\mathbf{1}_n\\\mathbf{1}_n\\\vdots\\\mathbf{1}_n\end{bmatrix}[\mathbf{1}'_n\ \mathbf{1}'_n\ \cdots\ \mathbf{1}'_n] = \frac{1}{na}\cdot[\mathbf{1}_{na\times na}], \tag{11.27}$$

where $\mathbf{1}_{na\times na}$ is an $na \times na$ matrix consisting entirely of 1s. The jth column in the α-group in the design matrix has blocks of 0s everywhere but in the jth row block, in which there are n 1s. Then the contribution that this column makes to the total projection matrix in (11.21) is

$$P_{\alpha_j} = \frac{1}{n}\begin{bmatrix}\mathbf{0}_n\\\vdots\\\mathbf{1}_n\\\vdots\\\mathbf{0}_n\end{bmatrix}[\mathbf{0}'_n\ldots\mathbf{1}'_n\ldots\mathbf{0}'_n] = \frac{1}{n}\begin{bmatrix}\mathbf{0}_{n\times n} & \cdots & \mathbf{0}_{n\times n} & \cdots & \mathbf{0}_{n\times n}\\\vdots & \ddots & \vdots & & \vdots\\\mathbf{0}_{n\times n} & \cdots & \mathbf{1}_{n\times n} & \cdots & \mathbf{0}_{n\times n}\\\vdots & & \vdots & \ddots & \vdots\\\mathbf{0}_{n\times n} & \cdots & \mathbf{0}_{n\times n} & \cdots & \mathbf{0}_{n\times n}\end{bmatrix},$$

where $\mathbf{0}_{n\times n}$ and $\mathbf{1}_{n\times n}$ are matrices of size $n \times n$ of all 0s and all 1s, respectively. The projection matrix for the α-columns is therefore

$$P_\alpha = \sum_{i=1}^{a} P_{\alpha_i} = \frac{1}{n} \begin{bmatrix} \mathbf{1}_{n\times n} & \cdots & \mathbf{0}_{n\times n} & \cdots & \mathbf{0}_{n\times n} \\ \vdots & \ddots & \vdots & & \vdots \\ \mathbf{0}_{n\times n} & \cdots & \mathbf{1}_{n\times n} & \cdots & \mathbf{0}_{n\times n} \\ \vdots & & \vdots & \ddots & \vdots \\ \mathbf{0}_{n\times n} & \cdots & \mathbf{0}_{n\times n} & \cdots & \mathbf{1}_{n\times n} \end{bmatrix}. \tag{11.28}$$

Notice that P_μ and P_α are symmetric matrices.

Now by adding and subtracting $P_\mu\mathbf{Y}$ and $P_\alpha\mathbf{Y}$ to and from $\mathbf{Y}$, we can write

$$\mathbf{Y} = P_\mu\mathbf{Y} + (P_\alpha - P_\mu)\mathbf{Y} + (I - P_\alpha)\mathbf{Y}, \tag{11.29}$$

where I is an $na \times na$ identity matrix. The formula in full form is

$$\begin{bmatrix} Y_{11} \\ \vdots \\ Y_{1n} \\ Y_{21} \\ \vdots \\ Y_{2n} \\ \vdots \\ Y_{a1} \\ \vdots \\ Y_{an} \end{bmatrix} = \begin{bmatrix} \overline{Y} \\ \vdots \\ \overline{Y} \\ \overline{Y} \\ \vdots \\ \overline{Y} \\ \vdots \\ \overline{Y} \\ \vdots \\ \overline{Y} \end{bmatrix} + \begin{bmatrix} \overline{Y}_{1\cdot} - \overline{Y} \\ \vdots \\ \overline{Y}_{1\cdot} - \overline{Y} \\ \overline{Y}_{2\cdot} - \overline{Y} \\ \vdots \\ \overline{Y}_{2\cdot} - \overline{Y} \\ \vdots \\ \overline{Y}_{a\cdot} - \overline{Y} \\ \vdots \\ \overline{Y}_{a\cdot} - \overline{Y} \end{bmatrix} + \begin{bmatrix} Y_{11} - \overline{Y}_1 \\ \vdots \\ Y_{1n} - \overline{Y}_1 \\ Y_{21} - \overline{Y}_2 \\ \vdots \\ Y_{2n} - \overline{Y}_2 \\ \vdots \\ Y_{a1} - \overline{Y}_a \\ \vdots \\ Y_{an} - \overline{Y}_a \end{bmatrix}. \tag{11.30}$$

The rest of the proof will follow from some tedious but easy to verify facts about the matrices P_μ and P_α, most of which we leave to the exercises. First,

$$P_\mu \cdot P_\alpha = \frac{1}{na} \cdot [\mathbf{1}_{na\times na}] \cdot \frac{1}{n} \begin{bmatrix} \mathbf{1}_{n\times n} & \cdots & \mathbf{0}_{n\times n} & \cdots & \mathbf{0}_{n\times n} \\ \vdots & \ddots & \vdots & & \vdots \\ \mathbf{0}_{n\times n} & \cdots & \mathbf{1}_{n\times n} & \cdots & \mathbf{0}_{n\times n} \\ \vdots & & \vdots & \ddots & \vdots \\ \mathbf{0}_{n\times n} & \cdots & \mathbf{0}_{n\times n} & \cdots & \mathbf{1}_{n\times n} \end{bmatrix}$$

$$= \frac{1}{n^2 a} \cdot \begin{bmatrix} n\mathbf{1}_{n\times n} & \cdots & n\mathbf{1}_{n\times n} \\ \vdots & \ddots & \vdots \\ n\mathbf{1}_{n\times n} & \cdots & n\mathbf{1}_{n\times n} \end{bmatrix} = P_\mu. \tag{11.31}$$

You can also check that $P_\alpha P_\mu = P_\mu$, $P_\alpha^2 = P_\alpha$, and $P_\mu^2 = P_\mu$ (Exercise 13) But then

$$(P_\mu \mathbf{Y})'(P_\alpha - P_\mu)\mathbf{Y} = \mathbf{Y}'P_\mu'(P_\alpha \mathbf{Y} - P_\mu \mathbf{Y})$$
$$= \mathbf{Y}'P_\mu P_\alpha \mathbf{Y} - \mathbf{Y}'P_\mu P_\mu \mathbf{Y}$$
$$= \mathbf{Y}'P_\mu \mathbf{Y} - \mathbf{Y}'P_\mu \mathbf{Y}$$
$$= 0.$$

This means that the projected vector $P_\mu \mathbf{Y}$ and the projected vector $(P_\alpha - P_\mu)\mathbf{Y}$ are orthogonal. In a similar way you can check that $P_\mu \mathbf{Y}$ and $(I - P_\alpha)\mathbf{Y}$ are orthogonal, and that $(P_\alpha - P_\mu)\mathbf{Y}$ and $(I - P_\alpha)\mathbf{Y}$ are orthogonal. Thus equation (11.29) expresses $\mathbf{Y}$ as the sum of orthogonal vectors. It follows from the Pythagorean theorem that

$$\|\mathbf{Y}\|^2 = \|P_\mu \mathbf{Y}\|^2 + \|(P_\alpha - P_\mu)\mathbf{Y}\|^2 + \|(I - P_\alpha)\mathbf{Y}\|^2. \tag{11.32}$$

The sum of squares decomposition asserted in part (a) is established by applying (11.32) to the full forms of the vectors in (11.30).

To prove part (b), first recall that we said above that P_μ was a symmetric, idempotent matrix (i.e., $P_\mu^2 = P_\mu$). Also $P_\alpha - P_\mu$ is symmetric, and

$$(P_\alpha - P_\mu)(P_\alpha - P_\mu) = P_\alpha^2 - P_\mu P_\alpha - P_\alpha P_\mu + P_\mu^2$$
$$= P_\alpha - P_\mu - P_\mu + P_\mu$$
$$= P_\alpha - P_\mu.$$

Hence $P_\alpha - P_\mu$ is idempotent. You can show (Exercise 14) similarly that $(I - P_\alpha)$ is idempotent.

Since the norm squared of a vector is the dot product of the transpose of the vector with itself, the sum of squares decomposition (11.32) can be written:

$$\mathbf{Y}' \cdot I \cdot \mathbf{Y} = \mathbf{Y}'P_\mu'P_\mu \mathbf{Y} + \mathbf{Y}'(P_\alpha - P_\mu)'(P_\alpha - P_\mu)\mathbf{Y} + \mathbf{Y}'(I - P_\alpha)'(I - P_\alpha)\mathbf{Y}.$$

But since the matrices P_μ, $P_\alpha - P_\mu$, and $I - P_\alpha$ are each symmetric and idempotent, this identity can be rewritten:

$$\mathbf{Y}'(I - P_\mu)\mathbf{Y} = \mathbf{Y}'(P_\alpha - P_\mu)\mathbf{Y} + \mathbf{Y}'(I - P_\alpha)\mathbf{Y}. \tag{11.33}$$

Under the null hypothesis $H_0 : \alpha_1 = \alpha_2 = \cdots = \alpha_a = 0$, $\mathbf{Y}$ is a vector of independent normal random variables each of which has mean μ and variance σ^2. Now it is easy to check that the row sums of each of the matrices I, P_μ, and P_α are 1. Therefore the row sums of $(I - P_\mu)$, $P_\alpha - P_\mu$, and $I - P_\alpha$ are all zero. By Proposition 5.4.7 all of the sums of squares in (11.33) have χ^2 distributions when divided by σ^2. You will show in Exercise 15 that $I - P_\mu$ has rank $N - 1$, $P_\alpha - P_\mu$ has rank $a - 1$, and $I - P_\alpha$ has rank $N - a$. These observations prove the claim that SSF/σ^2 and SSE/σ^2 have $\chi^2(a - 1)$ and $\chi^2(N - a)$ distributions, and furthermore that the total sum of squares $SST/\sigma^2 \sim \chi^2(N - 1)$.

Lastly, since

$$(P_\alpha - P_\mu) \cdot (I - P_\alpha) = P_\alpha - P_\alpha^2 - P_\mu + P_\mu P_\alpha = P_\alpha - P_\alpha - P_\mu + P_\mu = 0,$$

Proposition 5.4.8 proves the independence of SSF and SSE. This completes the proof.

—— **?Question 11.2.3** Check by multiplying out the matrices that (11.33) is just a matrix version of the sum of squares decomposition $SST = SSF + SSE$.

The last proposition is true in the unbalanced case as well, with suitable alterations in the sums of squares

$$SST = \sum_{i=1}^{a} \sum_{j=1}^{n_i} (Y_{ij} - \overline{Y})^2,$$

$$SSF = \sum_{i=1}^{a} n_i (\overline{Y}_{i\cdot} - \overline{Y})^2,$$

$$SSE = \sum_{i=1}^{a} \sum_{j=1}^{n_i} (Y_{ij} - \overline{Y}_{i\cdot})^2. \tag{11.34}$$

The only change in the proof is the loss of perfect balance in the P_α matrix.

We can now describe the F-test for the null hypothesis of no factor effect in the completely randomized one-factor experiment. As in regression, let the sums of squares divided by their degrees of freedom be called *mean squares*. To test

$$H_0 : \alpha_1 = \alpha_2 = \cdots = \alpha_a = 0,$$

compute the statistic

$$F = \frac{SSF/(a-1)}{SSE/(N-a)} = \frac{MSF}{MSE}, \tag{11.35}$$

which has the $F(a-1, N-a)$ distribution under H_0, by Proposition 11.2.1. Reject H_0 at level α if $F \geq f_\alpha(a-1, N-a)$ where $f_\alpha(a-1, N-a)$ is the point to the right of which lies probability α for the $f(a-1, N-a)$ distribution.

Statistical computer software does the tedious work of calculating the sums of squares, mean squares, and F-statistic. A table of information called an ANOVA table (an acronym for *An*alysis *o*f *Va*riance) is output, which usually looks something like this:

Source	Sum Sq	df	Mean Sq	F
Factor	#	$a-1$	#	#
Error	#	$N-a$	#	
Total	#	$N-1$		

The factor, error, and total sums of squares for the data set being analyzed are listed in a column, with their appropriate degrees of freedom in an adjacent column. The mean square column takes the quotient of the sums of squares with their degrees of freedom. The F-statistic is the quotient of MSF and MSE from the mean square column. A p-value may also be reported by the software.

Example 11.2.1

Continued For the regional temperature extreme data, *Minitab* reports the ANOVA table as follows:

Source	Sum Sq	df	Mean Sq	F
Factor	15456	2	7728	7.30
Error	34955	33	1059	
Total	50411	35		

Since there are three levels of the region factor and 36 total data points, the factor sum of squares has $3 - 1 = 2$ degrees of freedom, error has $36 - 3 = 33$ degrees of freedom, and total has $36 - 1 = 35$ degrees of freedom. The F-statistic takes on the very large value of 7.30, with a corresponding p-value of around .002. Thus the regions do appear to differ significantly, confirming the visual impression given by the dotplots in Fig. 11.3. Actually, we may be in trouble here because our model requires errors to have equal variance, and the Pacific Northwest data seem to be suspiciously spread out. Transformation methods similar to those for regression can correct heteroscedasticity, but we will not discuss them here.

We can gain more insight into what the F-test does by looking at the expected values of the mean squares.

PROPOSITION 11.2.2

In the completely randomized one-factor experiment, MSE is always unbiased for σ^2. Furthermore, in general

$$E[MSF] = \sigma^2 + \frac{n}{a-1} \sum_{i=1}^{a} \alpha_i{}^2. \tag{11.36}$$

When H_0 is true, MSF is unbiased for σ^2, otherwise $E[MSF]$ is strictly greater than σ^2.

■ **Proof**   Consider first the error mean square

$$E[MSE] = E[SSE/(N-a)]$$

$$= \frac{1}{N-a} \cdot E\left[\sum_{i=1}^{a}\sum_{j=1}^{n}(Y_{ij} - \overline{Y}_{i\cdot})^2\right]$$

$$= \frac{1}{N-a} \cdot \sum_{i=1}^{a} E\left[\sum_{j=1}^{n}(Y_{ij} - \overline{Y}_{i\cdot})^2\right].$$

The sum over j in brackets is just $n-1$ times the sample variance S_i^2 for the ith factor level. Regardless of whether H_0 is true, this sample variance is unbiased for σ^2. So,

$$E[MSE] = \frac{1}{N-a} \cdot \sum_{i=1}^{a} E[(n-1)S_i^2]$$

$$= \frac{1}{N-a} \cdot \sum_{i=1}^{a} (n-1) \cdot \sigma^2$$

$$= \frac{\sigma^2}{N-a}(n-1)a = \frac{\sigma^2}{N-a} \cdot (na-a) = \sigma^2.$$

Thus MSE is an unbiased estimate of the error variance σ^2. (See Exercise 5 for the extension of the result to the unbalanced case.)

As for MSF, we have

$$E[MSF]$$

$$= \frac{1}{a-1}E\left[n\sum_{i=1}^{a}(\overline{Y}_{i\cdot} - \overline{Y})^2\right]$$

$$= \frac{n}{a-1}E\left[\sum_{i=1}^{a}\overline{Y}_{i\cdot}^2 - a \cdot \overline{Y}^2\right]$$

$$= \frac{n}{a-1}\left[\sum_{i=1}^{a}(\text{Var}(\overline{Y}_{i\cdot}) + (E[\overline{Y}_{i\cdot}])^2) - a \cdot (\text{Var}(\overline{Y}) + (E[\overline{Y}])^2)\right]$$

$$= \frac{n}{a-1}\left[\sum_{i=1}^{a}\frac{\sigma^2}{n} + ((\mu + \alpha_i)^2) - a \cdot \left(\frac{\sigma^2}{na} + \left(\frac{1}{na}\cdot\sum_{i=1}^{a}\sum_{j=1}^{n}(\mu + \alpha_i)\right)^2\right)\right]$$

$$= \frac{n}{a-1}\left[\frac{a}{n}\sigma^2 + \sum_{i=1}^{a}(\mu^2 + 2\mu\alpha_i + \alpha_i^2) - \frac{\sigma^2}{n} - \frac{1}{a}\cdot\left(\sum_{i=1}^{a}(\mu + \alpha_i)\right)^2\right]$$

$$= \frac{n}{a-1}\left[\frac{a-1}{n}\sigma^2 + a\mu^2 + \sum_{i=1}^{a}\alpha_i^2 - \frac{1}{a}\cdot a^2\mu^2\right]$$

$$= \sigma^2 + \frac{n}{a-1}\sum_{i=1}^{a}\alpha_i^2. \tag{11.37}$$

If H_0 is true, then the sum on the right of the final expression in (11.37) vanishes and MSF is also unbiased for σ^2. If H_0 is not true, then the sum is positive, causing MSF to have a mean greater than σ^2.

By this proposition, it is intuitively reasonable to reject H_0 if $F = MSF/MSE$ is large because if H_0 is true then the numerator and the denominator of the F-statistic are expected to be about the same thing. We will see again in our study of random effects models and the two-factor model the importance of looking at expected mean squares in order to determine the appropriate F-ratio.

Here is one last comment on the one-factor completely randomized model. The need for the extra constraint $\sum_{i=1}^{a}\alpha_i = 0$ can be seen as a need for an extra condition to uniquely estimate the model parameters $\mu, \alpha_1, \alpha_2, \ldots, \alpha_a$ by the least squares criterion. As in regression, the parameter estimates for the model $\mathbf{Y} = X\boldsymbol{\theta} + \boldsymbol{\epsilon}$ that minimize $\|\mathbf{Y} - X\boldsymbol{\theta}\|^2$ satisfy

$$X'X\boldsymbol{\theta} = X'\mathbf{Y}. \tag{11.38}$$

You can verify that

$$X'X = \begin{bmatrix} N & n & n & \cdots & n \\ n & n & 0 & \cdots & 0 \\ n & 0 & n & \cdots & 0 \\ \vdots & \vdots & & \ddots & \\ n & 0 & 0 & \cdots & n \end{bmatrix}, \qquad X'Y = \begin{bmatrix} an\overline{Y} \\ n\overline{Y}_{1.} \\ \vdots \\ n\overline{Y}_{a.} \end{bmatrix}. \qquad (11.39)$$

(See Exercise 18.) But the first row of $X'X$ is the sum of the other a rows, so $X'X$ is not invertible, and the least squares equations do not have a unique solution. One way of adjoining a condition to provide a unique solution is to drop the first row of $X'X$ and add a row $[0\,1\,1\,\ldots 1]$, and correspondingly add an entry of 0 to the vector $X'Y$, in effect adjoining the condition $\sum \alpha_i = 0$ to the system. Why do we use this particular side condition? You saw in Exercise 6 in Section 11.1 that changing the original parameterization $Y_{ij} = \mu_i + \epsilon_{ij}$ to $Y_{ij} = \mu + \alpha_i + \epsilon_{ij}$ was merely a matter of defining μ to be the average of the μ_i's and setting $\alpha_i = \mu_i - \mu$. The condition $\sum_{i=1}^{a} \alpha_i = 0$ follows easily. Note that in reparameterizing, we have added an extra, nonestimable parameter (where there were a parameters μ_i, there are now $a+1$ parameters μ and $\alpha_1, \ldots \alpha_a$), which accounts for the lack of invertibility of $X'X$.

11.2.2   Randomized Block Design

Recall that in the randomized block model the observations $Y_{1j}, Y_{2j}, \ldots, Y_{aj}$ form a block of data having some common characteristic, such as having come from the same machine operator, or batch of product, or plot of ground. The model to describe the data is

$$Y_{ij} = \mu + \alpha_i + \beta_j + \epsilon_{ij}, \qquad i = 1, 2, \ldots, a, \quad j = 1, 2, \ldots, b. \qquad (11.40)$$

In matrix form this can be expressed as

$$\mathbf{Y} = X\boldsymbol{\theta} + \boldsymbol{\epsilon}, \qquad (11.41)$$

where

$$\mathbf{Y} = \begin{bmatrix} Y_{11} \\ \vdots \\ Y_{1b} \\ Y_{21} \\ \vdots \\ Y_{2b} \\ \vdots \\ Y_{a1} \\ \vdots \\ Y_{ab} \end{bmatrix}, \quad X = \begin{bmatrix} \overset{\mu}{1} & \overset{\alpha_1}{1} & \overset{\alpha_2}{0} & \overset{\cdots}{\cdots} & \overset{\alpha_a}{0} & \overset{\beta_1}{1} & \overset{\cdots}{\cdots} & \overset{\beta_b}{0} \\ \vdots & \vdots & \vdots & & \vdots & \vdots & \ddots & \vdots \\ 1 & 1 & 0 & \cdots & 0 & 0 & \cdots & 1 \\ 1 & 0 & 1 & \cdots & 0 & 1 & \cdots & 0 \\ \vdots & \vdots & \vdots & & \vdots & \vdots & \ddots & \vdots \\ 1 & 0 & 1 & \cdots & 0 & 0 & \cdots & 1 \\ \vdots & \vdots & \vdots & & \vdots & \vdots & & \vdots \\ 1 & 0 & 0 & \cdots & 1 & 1 & \cdots & 0 \\ \vdots & \vdots & \vdots & & \vdots & \vdots & \ddots & \vdots \\ 1 & 0 & 0 & \cdots & 1 & 0 & \cdots & 1 \end{bmatrix}, \quad \boldsymbol{\theta} = \begin{bmatrix} \mu \\ \alpha_1 \\ \alpha_2 \\ \vdots \\ \alpha_a \\ \beta_1 \\ \vdots \\ \beta_b \end{bmatrix},$$

$$(11.42)$$

and $\boldsymbol{\epsilon}$ is a $N(\mathbf{0}, \sigma^2 I)$ distributed random vector of ab components. The block structure of X is appealing: The μ column is a vector of ab 1s, each α_i column has a block $\mathbf{1}_b$ of ones in the ith block position and blocks $\mathbf{0}_b$ of zeros in the other positions, and the β columns consist of a identical blocks of $b \times b$ identity matrices $I_{b \times b}$.

The analysis for this design parallels what we did for the completely randomized experiment, but there is an extra vector in the decomposition of $\mathbf{Y}$ resulting from the extra group of parameters $\beta_1, \beta_2, \ldots, \beta_b$.

PROPOSITION 11.2.3

(a) In the randomized block model (11.40) we have the following sum of squares decomposition:

$$\sum_{i=1}^{a} \sum_{j=1}^{b} Y_{ij}^2 = a \cdot b \cdot \overline{Y}^2 + b \sum_{i=1}^{a} (\overline{Y}_{i \cdot} - \overline{Y})^2 + a \sum_{j=1}^{b} (\overline{Y}_{\cdot j} - \overline{Y})^2$$

$$+ \sum_{i=1}^{a} \sum_{j=1}^{b} (Y_{ij} - \overline{Y}_{i \cdot} - \overline{Y}_{\cdot j} + \overline{Y})^2. \tag{11.43}$$

Equivalently

$$SST = SSF + SSB + SSE,$$

where

$$SST = \sum_{i=1}^{a} \sum_{j=1}^{b} Y_{ij}^2 - ab\overline{Y}^2 = \sum_{i=1}^{a} \sum_{j=1}^{b} (Y_{ij} - \overline{Y})^2, \tag{11.44}$$

$$SSF = b \cdot \sum_{i=1}^{a} (\overline{Y}_{i \cdot} - \overline{Y})^2, \tag{11.45}$$

$$SSB = a \sum_{j=1}^{b} (\overline{Y}_{\cdot j} - \overline{Y})^2, \tag{11.46}$$

$$SSE = \sum_{i=1}^{a} \sum_{j=1}^{b} (Y_{ij} - \overline{Y}_{i \cdot} - \overline{Y}_{\cdot j} + \overline{Y})^2. \tag{11.47}$$

The sums of squares SST, SSF, SSB, and SSE are called the *total, factor, block,* and *error sums of squares,* respectively.

(b) Under the null hypothesis that all α_i and β_j equal zero,

$$\frac{SST}{\sigma^2} \sim \chi^2(ab-1), \qquad \frac{SSF}{\sigma^2} \sim \chi^2(a-1),$$

$$\frac{SSB}{\sigma^2} \sim \chi^2(b-1), \qquad \frac{SSE}{\sigma^2} \sim \chi^2((a-1)(b-1)) \tag{11.48}$$

and $SSF, SSB,$ and SSE are mutually independent.

■ **Proof**   We will skim the details of the matrix approach to the sum of squares decomposition and the associated distributional results, leaving you to fill in the gaps in the exercises.

As in the completely randomized experiment, let P_μ and P_α be the projection matrices for the spaces spanned by the μ column and the group of α columns in the design matrix X in (11.42). Their explicit forms are in (11.27) and (11.28). Furthermore, let P_β be the projection matrix for the group of β column vectors in X. Exercise 23 asks you to use (11.21) to write out P_β explicitly. Now decompose the data vector $\mathbf{Y}$ by adding and subtracting $P_\alpha\mathbf{Y}$, $P_\beta\mathbf{Y}$, and two copies of $P_\mu\mathbf{Y}$. Then,

$$\mathbf{Y} = P_\mu\mathbf{Y} + (P_\alpha - P_\mu)\mathbf{Y} + (P_\beta - P_\mu)\mathbf{Y} + (I - P_\alpha - P_\beta + P_\mu)\mathbf{Y}. \quad (11.49)$$

To get a feeling for what this decomposition does, recall that the vector $\mathbf{Y}$ consists of a blocks each of length b. Using (11.49), the jth data value in the ith block turns out to be

$$Y_{ij} = \overline{Y} + (\overline{Y}_{i\cdot} - \overline{Y}) + (\overline{Y}_{\cdot j} - \overline{Y}) + (Y_{ij} - \overline{Y}_{i\cdot} - \overline{Y}_{\cdot j} + \overline{Y}). \quad (11.50)$$

In addition to the properties $P_\mu^2 = P_\mu$, $P_\alpha^2 = P_\alpha$, and $P_\alpha P_\mu = P_\mu P_\alpha = P_\mu$ that we already had for P_μ and P_α, it can be shown that $P_\beta^2 = P_\beta$, $P_\beta P_\mu = P_\mu P_\beta = P_\mu$, and $P_\beta P_\alpha = P_\alpha P_\beta = P_\mu$. (See Exercise 23.) With these computational formulas in hand, it is easy to show that the vectors on the right side of (11.49) are orthogonal, which therefore implies that their norms squared add to the square of the norm of $\mathbf{Y}$. In matrix form this gives

$$\mathbf{Y}'\mathbf{Y} = \mathbf{Y}'P_\mu\mathbf{Y} + \mathbf{Y}'(P_\alpha - P_\mu)\mathbf{Y} + \mathbf{Y}'(P_\beta - P_\mu)\mathbf{Y}$$
$$+ \mathbf{Y}'(I - P_\alpha - P_\beta + P_\mu)\mathbf{Y}$$

$$\Leftrightarrow$$

$$\mathbf{Y}'(I - P_\mu)\mathbf{Y} = \mathbf{Y}'(P_\alpha - P_\mu)\mathbf{Y} + \mathbf{Y}'(P_\beta - P_\mu)\mathbf{Y}$$
$$+ \mathbf{Y}'(I - P_\alpha - P_\beta + P_\mu)\mathbf{Y}. \quad (11.51)$$

The sum of squares decomposition (11.43) follows by expanding out the products.

Also, the matrices $P_\alpha - P_\mu$, $P_\beta - P_\mu$, and $I - P_\alpha - P_\beta + P_\mu$ are idempotent, and their row sums are equal to 0. Therefore under the null hypothesis, by Proposition 5.4.7, the sums of squares divided by σ^2 have χ^2 distributions. An example of the computation of one of the ranks—the degrees of freedom—is in Exercise 24. It is also straightforward to verify that these idempotent matrices have zero products with one another; hence by Proposition 5.4.8 the sums of squares are independent. This finishes the sketch of the proof.

The sums of squares can be used to form F-statistics

$$\begin{aligned} F_a &= \frac{MSF}{MSE} = \frac{SSF/(a-1)}{SSE/((a-1)(b-1))}, \\ F_b &= \frac{MSB}{MSE} = \frac{SSB/(b-1)}{SSE/((a-1)(b-1))}. \end{aligned} \quad (11.52)$$

The null hypothesis of no factor effect is rejected at level α if $F_a > f_\alpha(a-1, (a-1)(b-1))$, and the null hypothesis of no block effect is rejected at level α if $F_b > f_\alpha(b-1, (a-1)(b-1))$.

Example 11.2.2

The following data are adapted from an experiment run by a student, Brian Kendall, who studied the dependence of the action of an enzyme called ADH in fruit flies on diet. For three of the diets—plain sucrose, sucrose with ethanol, and sucrose with ethanol and a substance called juvenile hormone (JH)—measurements of ADH activity for aggregated specimens were made for four experimental runs.

	Run 1	Run 2	Run 3	Run 4
Sucrose	155.939	98.843	44.386	115.462
Ethanol	232.314	161.198	134.645	162.327
Juvenile Hormone	294.294	262.609	213.436	271.216

Note that in each run the activity levels are lowest for sucrose and highest for JH, but there is considerable run-to-run variability. Run 3 has the smallest observations, and run 1 the largest. So a randomized block model $Y_{ij} = \mu + \alpha_i + \beta_j + \epsilon_{ij}$ seems reasonable, with the diet as the factor of principal interest and runs as blocks. *Minitab* reports sums of squares, mean squares, and F-values as follows:

Source	DF	SS	MS	F
Diet	2	49365	24683	150.5
Run	3	14141	4714	28.7
Error	6	986	164	
Total	11	64492		

Because $f_{.01}(2, 6) = 10.9$ and $f_{.01}(3, 6) = 9.78$, both the blocking factor of experimental run and the treatment factor of diet are very significant.

?Question 11.2.4

Why are the degrees of freedom as they are in the ANOVA table in the last example? Granting the correctness of the sums of squares, check the mean squares and the computation of the F-values.

As in the completely randomized experiment, under the extra conditions that $\sum_i \alpha_i = 0$ and $\sum_j \beta_j = 0$, the least squares estimates of the model parameters exist uniquely and take the form

$$\hat{\mu} = \overline{Y}, \qquad \hat{\alpha}_i = \overline{Y}_{i.} - \overline{Y}, \qquad \hat{\beta}_j = \overline{Y}_{.j} - \overline{Y}. \qquad (11.53)$$

Many experiments can be designed as either completely randomized experiments or as randomized block experiments. To block or not to block, that is the question (to paraphrase another great author). In both cases you can test $H_0 : \alpha_1 = \alpha_2 = \cdots = \alpha_a = 0$, the hypothesis of no effect of factor A. Let us look carefully at the criteria for rejection of H_0:

Completely
randomized: $\quad F = \dfrac{SSF/(a-1)}{SSE/(ab-a)} > f(a-1, ab-a)$

Randomized block: $\quad F' = \dfrac{SSF/(a-1)}{SSE'/(a-1)(b-1)} > f(a-1, (a-1)(b-1)).$

Now $SSF = b \sum_{i=1}^{a} (\overline{Y}_i. - \overline{Y})^2$ is the same for both experiments, so the differences are in the error sums of squares and in the degrees of freedom $ab - a = a(b - 1)$ and $(a - 1)(b - 1)$. Since the error sum of squares for the randomized block experiment is $SST - SSF - SSB$ and the error sum of squares for the completely randomized experiment is $SST - SSF$, blocking makes SSE smaller (intuitively, removing some of the variability of the data). This tends to make the test statistic F' for the randomized block larger, leading to more power in this experimental design. However the degrees of freedom $(a - 1)(b - 1)$ for the randomized block experiment are also smaller (by exactly $b - 1$), which damps down F', and which also means that the critical value on the right side is somewhat larger for the randomized block, which partially counteracts the SSE effect. So the choice between the two designs is not clear-cut. But our analysis has shown that blocking is most likely to increase the power of the test when SSB is large; that is, there is a high degree of variability between blocks. Exercise 25 asks you to analyze the data of Example 11.2.2 without the blocking. The question is, Would the test still have picked up the significance of the diet effect had we not accounted for blocks?

11.2.3 Random Effects Model

In the single-factor random effects model, the factor levels are randomly chosen from a universe of possible levels, and they contribute a normally distributed term with mean 0 to the response. The model is

$$Y_{ij} = \mu + A_i + \epsilon_{ij}, \qquad i = 1, 2, \ldots, a, \quad j = 1, 2, \ldots, n, \qquad (11.54)$$

where $A_i \sim N(0, \sigma_a^2)$ and $\epsilon_{ij} \sim N(0, \sigma^2)$ are independent. The null hypothesis of interest is $H_0 : \sigma_a^2 = 0$.

When we began this chapter, we motivated the one-factor problem with a gas mileage example, which is easy to convert to a random effects model to illustrate the ideas. Suppose we are interested in whether the model of car driven in the mileage test has an effect on the variance of gas mileage. We might select three car models at random and run eight different test cars each from these models. The response value Y_{ij} $i = 1, 2, 3$, and $j = 1, 2, \ldots, 8$, is the miles per gallon for the jth car of the ith model. The fact that the car is from the ith model contributes a random term A_i to the mileages of each of the cars of that model type, in addition to which is the error term ϵ_{ij} specific to car j.

The method of performing the hypothesis test of $H_0 : \sigma_a^2 = 0$ is exactly the same as in the fixed effects case. This is because when the null hypothesis is true, the model is $Y_{ij} = \mu + \epsilon_{ij}$, just as it is in the fixed effects problem. The sums of squares have the same distribution under H_0 as before, and the statistic

$$F = \frac{MSF}{MSE}$$

again has the $F(a - 1, N - 1)$ distribution.

The expected mean squares come out differently, however, and they shed some light on why the rejection region $F > f_\alpha(a - 1, N - 1)$ still is intuitively sensible. The next proposition computes them.

PROPOSITION 11.2.4

For the single-factor random effects model (11.54),

$$E[SSF] = E\left[n \sum_{i=1}^{a} (\overline{Y}_{i\cdot} - \overline{Y})^2 \right] = (a - 1)(\sigma^2 + n\sigma_a^2), \tag{11.55}$$

$$E[SSE] = E\left[\sum_{i=1}^{a} \sum_{j=1}^{n} (Y_{ij} - \overline{Y}_{i\cdot})^2 \right] = (N - a)\sigma^2. \tag{11.56}$$

Hence, $E[MSF] = \sigma^2 + n\sigma_a^2$ and $E[MSE] = \sigma^2$.

■ **Proof**   It is easy to verify that

$$SSF = \frac{1}{n} \sum_{i=1}^{a} Y_{i\cdot}^2 - \frac{1}{N} Y_{\cdot\cdot}^2, \tag{11.57}$$

where $Y_{\cdot\cdot} = \sum_{i=1}^{a} \sum_{j=1}^{n} Y_{ij}$. Since the expected square of a random variable is its variance plus its mean squared, we have

$$E[SSF] = \frac{1}{n} \sum_{i=1}^{a} (\text{Var}(Y_{i\cdot}) + (E[Y_{i\cdot}])^2) - \frac{1}{N}(\text{Var} Y_{\cdot\cdot} + (E[Y_{\cdot\cdot}])^2)$$

$$= \frac{1}{n} \sum_{i=1}^{a} \left(\text{Var}\left(n\mu + nA_i + \sum_{j=1}^{n} \epsilon_{ij} \right) + (n\mu)^2 \right)$$

$$\qquad - \frac{1}{N} \left(\text{Var}\left(na\mu + \sum_{i=1}^{a} nA_i + \sum_{i=1}^{a} \sum_{j=1}^{n} \epsilon_{ij} \right) + (na\mu)^2 \right)$$

$$= \frac{1}{n} \sum_{i=1}^{a} (n^2 \sigma_a^2 + n\sigma^2 + n^2 \mu^2) - \frac{1}{N}(n^2 a \sigma_a^2 + na\sigma^2 + n^2 a^2 \mu^2)$$

$$= na\sigma_a^2 + a\sigma^2 + na\mu^2 - \frac{1}{na}(n^2 a \sigma_a^2 + na\sigma^2 + n^2 a^2 \mu^2)$$

$$= na\sigma_a^2 - n\sigma_a^2 + a\sigma^2 - \sigma^2$$

$$= n(a - 1)\sigma_a^2 + (a - 1)\sigma^2 = (a - 1)(\sigma^2 + n\sigma_a^2).$$

The third line takes advantage of the independence of the As and ϵs and the usual elementary properties of variance. The rest of the computation is straightforward.

We leave the computation of $E[SSE]$ as Exercise 26.

So, the numerator of the F-statistic has mean $\sigma^2 + n\sigma_a^2$ (or σ^2 when H_0 is true) and the denominator has mean σ^2. If the factor contributes significant variability to the response, then the F-statistic is expected to be large, otherwise it should be near 1. This justifies the form of the critical region.

Example 11.2.3

Suppose the gas mileage data are as follows:

Model 1: 27.2 25.1 25.4 28.6 30.1 27.2 25.6 26.9
Model 2: 24.2 25.3 27.1 24.8 22.2 28.7 25.4 26.1
Model 3: 22.0 24.2 23.6 25.4 24.5 24.0 26.8 25.1

Minitab reports the following analysis of variance information.

Source	Sum Sq	df	Mean Sq	F
Factor	26.62	2	13.31	4.62
Error	60.50	21	2.88	
Total	87.12	23		

The value of F for the car model factor is 4.62, with an associated p-value of .022, which gives strong evidence that there is excess variance due to the car model.

?Question 11.2.5 Treat the data in the last example as if the three car models were levels of a fixed factor. How would you go about deciding exactly where the differences between models were?

EXERCISES 11.2

1. The delay (in months) between submission and publication of articles in four journals is listed for 14 articles sampled randomly from each journal. Test for a difference in average delay time at significance level .1.

   American Statistician: 14, 19, 31, 24, 17, 14, 14, 18, 34, 25, 19, 25, 32, 30

   Operations Research: 12, 37, 38, 20, 33, 40, 50, 29, 27, 28, 34, 43, 26, 36

   SIAM Journal on Computing: 22, 24, 33, 34, 17, 21, 20, 19, 41, 21, 32, 18, 25, 24

   Bulletin of AMS: 16, 11, 15, 16, 8, 8, 9, 18, 11, 7, 9, 34, 5, 14

2. Given are divorce rates in several countries, grouped by continent (United Nations, 1994). Suppose that the completely randomized model applies. Test at level .05 for a difference in divorce rate by continent.

   Oceania: 1.1, 2.5, 1.0, 2.7, 0.9, 2.9, 0.3, 0.6
   Africa: 7.0, 1.4, 0.6, 0.7, 1.3, 0.9, 1.3, 1.6
   Asia: 1.3, 1.6, 1.2, 0.8, 0.4, 1.4, 1.0, 0.8
   Europe: 0.8, 2.1, 3.4, 1.9, 1.3, 2.7, 2.9, 2.5

3. The percentages of voters in several small cities who turned out for the 1984 presidential elections is given below (Thomas, 1992). Cities from the states of Georgia, Alabama, and Tennessee are grouped into a category labeled South; those from Maine, New Hampshire, Vermont, and Rhode Island are grouped and labeled New England; and those from

Colorado, Wyoming, Idaho, and Utah are labeled West. Check for regional differences in voter turnout. Where do the differences seem to be, if any?

South:	30.6, 26.2, 24.9, 25.8, 14.3, 27.0, 30.9, 22.5, 27.0, 32.4, 36.0, 37.6, 39.3, 30.1
New England:	47.5, 45.4, 37.9, 37.3, 40.5, 41.7, 40.3
West:	38.3, 29.2, 39.5, 41.8, 36.1, 41.8, 39.0, 40.4

4. Verify the following computational formulas for the total and factor sums of squares:

$$SST = \sum_{i=1}^{a} \sum_{j=1}^{n} Y_{ij}^2 - \frac{(Y_{..})^2}{na};$$

$$SSF = \frac{1}{n} \sum_{i=1}^{a} Y_{i\cdot}^2 - \frac{(Y_{..})^2}{na}.$$

5. Show that for a general completely randomized experiment $E[MSE] = \sigma^2$.

6. In Exercise 4 in Section 4.1 we listed data (Reisinger et al., 1994) on a survey of people from Russia, Ukraine, and Lithuania about their response to the statement "Party competition will make the political system stronger." Using those data, devise a statistical approach to answering the question of whether significant differences exist between the countries in their attitudes toward this issue. Document any assumptions you are making. (*Hint:* You can compute the sums of squares if you know the group means and variances.)

7. A study on infant Hib immunization looked at potential differences in scheduling and in variants of the vaccine (Anderson et al., 1995). The 242 subjects in the study were divided into five groups as in the following table. The notation M-N-R was used by the investigators to indicate what vaccines the subjects received at age 2 months, 4 months, and 6 months; P is an abbreviation for one variety of the vaccine, H is another, and S stands for saline, that is no active agents were present (in order to study schedules involving only two vaccinations). The group average antibody concentrations (micrograms/milliliter) at age 7 months were reported and are listed in the Mean column. Since the exact standard deviations of the groups were not reported, I have made educated guesses based on other information. Do an appropriate analysis of variance to check for possible group differences at level .01. (*Hint:* You can compute the sums of squares from the group means and standard deviations.)

Group	Sample Size	Mean	Std. Dev.
P–P–S	39	0.85	.20
H–H–H	96	2.97	.55
P–S–P	36	1.19	.25
H–P–P	35	2.43	.50
P–H–H	36	4.28	1.10

8. Show that in the balanced completely randomized experiment, the value of the F-ratio does not change if the data are measured in different units, that is, if each observed Y response is rescaled and translated by $Y' = cY + d$.

9. The Federal Reserve Bank in Chicago maintains data on the average return on assets of banks throughout Illinois. Returns on assets for banks in three central Illinois counties for 1993 are shown. If you treat the problem as a random effects model, what hypothesis is tested? What conclusion do the data lead you to?

   Peoria County: 1.47, 1.27, 1.25, 1.25, 1.25, 1.24, 1.15, 1.03, .93, .91, .74, .72, .69, .46, .37

   Tazewell County: 1.63, 1.42, 1.41, 1.29, 1.16, .99, .96, .88, .87, .65, .55, .49

   Knox County: 1.52, 1.18, .97, .94, .91, .22

10. Exercise 3 in Section 10.2 contained some of my wife's blood glucose readings for several days while she was experiencing gestational diabetes. In this problem we will be interested in comparing the glucose levels after meals. Table 11.1 repeats the after-breakfast and the after-lunch readings and includes the after-dinner readings for the same period of days. (One data point from earlier is omitted because she didn't do an after-dinner check that day.) First do the analysis of variance on meal differences using a completely randomized model. Explain why it could be advantageous to block by days. Then use a randomized block model in which you block by days. Do you get different conclusions?

11. The Illinois Department of Employment Security gathered unemployment percentages for December 1993, November 1993, and December 1992 by county for 16 counties, which appear below. What statistical model can be applied to check for changes in unemployment over time? Test the null hypothesis of no change in unemployment at level .05.

Table 11.1

Blood Glucose Levels

After Breakfast	After Lunch	After Dinner
124	112	89
142	103	116
108	99	120
153	125	124
120	92	102
134	120	114
132	105	122
134	103	111
128	120	133

	Bureau	**Cass**	**Fulton**	**Henry**	**Knox**	**McDonough**
Dec. 1993	6.7	6.4	7.7	5.3	6.8	4.1
Nov. 1993	5.3	4.8	7.0	4.4	6.2	4.0
Dec. 1992	7.3	8.8	10.5	7.6	7.8	4.8

	McLean	**Marshall**	**Mason**	**Peoria**	**Putnam**	**Schuyler**
Dec. 1993	4.7	6.1	7.9	5.6	8.3	7.0
Nov. 1993	3.5	4.8	8.0	5.5	5.1	6.2
Dec. 1992	4.7	7.6	11.0	7.1	10.6	7.8

	Stark	**Tazewell**	**Warren**	**Woodford**
Dec. 1993	5.2	6.1	6.2	3.3
Nov. 1993	5.4	5.4	5.9	2.7
Dec. 1992	9.2	7.3	8.3	4.3

12. In a research consultation with some members of the Department of Psychology at the University of Iowa, I encountered a project involving twins. The level of aggression for each twin in a pair was estimated in a survey of the twins' mother. A subset of the data follows. Let Y_{ij} be the aggression level of the jth twin ($j = 1, 2$) in the ith twin pair, $i = 1, 2, \ldots, 20$. Explain how the random effects model $Y_{ij} = \mu + A_i + \epsilon_{ij}$ can be used to help make a judgment about whether aggression has a genetic component. Perform the test at level .05.

    Twin 1: 50, 50, 53, 39, 57, 46, 50, 57, 57, 50, 46, 61, 61, 57, 43, 39, 50, 57, 53, 68
    Twin 2: 43, 46, 46, 39, 50, 39, 78, 53, 64, 50, 39, 53, 61, 53, 39, 39, 50, 61, 43, 39

13. In the notation of the proof of Proposition 11.2.1 show that
    a. $P_\alpha P_\mu = P_\mu$.
    b. $P_\alpha^2 = P_\alpha$.
    c. $P_\mu^2 = P_\mu$.

14. In the notation of the proof of Proposition 11.2.1, show that the matrix $I - P_\alpha$ is symmetric and idempotent.

15. Show that the ranks of the matrices $I - P_\mu, P_\alpha - P_\mu$, and $I - P_\alpha$ in Proposition 11.2.1 are $N - 1$, $a - 1$, and $N - a$, respectively.

16. Table 11.2 has data on wool production (average pounds per fleece) in the United States in each of three years, broken down by state for 16 states (USDA, 1991). Do an appropriate analysis of variance to check for a difference in production by years.

17. Table 11.3 lists the yield per acre (pounds) of tobacco in several producing states for each of the years 1988–1990 (USDA, 1991). Do an analysis of variance using a randomized block model to check for a difference in yield by years. Then try a completely randomized model. Comment on your findings.

18. Verify formulas (11.39) for $X'X$ and $X'Y$, and use the side condition $\sum \alpha_i = 0$ to derive the formula $\overline{Y}_{i.} - \overline{Y}$ for the least squares estimator of α_i in the one-factor completely randomized experiment.

Table 11.2

Wool Production

State	1988	1989	1990
AK	9.0	8.4	8.8
AZ	6.6	6.8	6.8
CA	7.4	7.7	7.5
CO	6.6	7.7	7.4
CT	7.5	7.5	7.3
ID	9.9	10.0	10.2
IL	6.8	6.9	6.9
IN	6.8	6.8	6.8
IA	7.0	6.9	6.6
KS	6.9	6.7	7.1
KY	6.7	6.6	6.5
LA	6.7	6.7	6.3
ME	8.0	7.2	7.6
MD	7.0	7.6	6.3
MA	7.2	7.2	7.3
MI	7.6	7.4	7.2

Table 11.3

Tobacco Yields

State	1988	1989	1990
CT	1641	1641	1648
FL	2680	2650	2720
GA	2260	2180	2400
IN	1990	2170	2100
KY	2247	2059	2248
MD	1330	1110	1330
MA	1475	1554	1651
MO	2010	2180	2250
NC	2211	2029	2240
OH	1854	1750	2050
PA	1913	1887	1978
SC	2225	2160	2155
TN	1920	1754	2031
VA	1973	1892	2055
WV	1600	1300	1600
WI	2002	2045	1967

19. A *contrast* in the parameters $\mu_i = \mu + \alpha_i$ is a linear combination $\sum_{i=1}^{a} c_i \mu_i$ such that $\sum_{i=1}^{a} c_i = 0$. Notice that because of the latter condition, $\sum_{i=1}^{a} c_i \mu_i = \sum_{i=1}^{a} c_i \alpha_i$. The corresponding contrast in factor level sums is $C = \sum_{i=1}^{a} c_i Y_i$ (the contrast in means is similar). It can be shown that if the null hypothesis that the contrast $\sum c_i \mu_i = 0$ is true, then the *contrast sum of squares* $SSC = C^2/(n \sum_{i=1}^{a} c_i^2)$ divided by σ^2 has the $\chi^2(1)$ distribution and is independent of SSE. With this information, special comparisons of means in a completely randomized one-factor experiment can be done. Explain how to test the hypothesis that a contrast in the underlying means is 0, and use your reasoning to test the hypothesis that in Exercise 3 the mean turnout for the third group is the average of the mean turnouts for the first two groups.

20. Show that the variance of $\overline{Y}_{i.} - \overline{Y}$ in the randomized block experiment is $((a-1)/N)\sigma^2$.

21. Show that the least squares estimate of β_j in the randomized block experiment is $\overline{Y}_{.j} - \overline{Y}$, if the side conditions $\sum_{i=1}^{a} \alpha_i = 0$ and $\sum_{j=1}^{b} \beta_j = 0$ are true.

22. Verify by brute force calculation the randomized block sum of squares decomposition (11.43).

23. Write out the matrix P_β in the randomized block experiment and show that
    a. $P_\beta^2 = P_\beta$.
    b. $P_\mu P_\beta = P_\beta P_\mu = P_\mu$.
    c. $P_\beta P_\alpha = P_\alpha P_\beta = P_\mu$.

24. In the randomized block experiment, verify that the rank of the matrix $P_\beta - P_\mu$ is $b - 1$.

25. In Example 11.2.2, suppose that you ignore the presence of the experimental runs as a blocking factor. Do you still reach the same conclusion as in the example about the significance of the factor levels? Explain what happened.

26. Verify formula (11.56) for $E[SSE]$ in the random effects model.

27. Find an unbiased estimator for the factor variance σ_a^2 in the random effects model.

28. Show that the independent samples t-test for the difference between two normal means assuming equal but unknown population variances is equivalent to the ANOVA F-test for the balanced completely randomized experiment.

11.3 | Two-Factor Analysis of Variance

The ideas of one-factor analysis of variance carry over directly to experiments in which the response variable depends on two factors. In addition we now encounter the possibility that the two factors do not make independent contributions to the mean response; there may also be an interaction term that is dependent on the combination of factor levels used to generate a particular value of the response.

11.3.1   Fixed Effects Model

To restate the probabilistic model for a balanced experiment, let there be a levels of factor A and b levels of factor B. At each combination (i, j) of levels of factor A and B there are n replications. The defining equation for the fixed effects completely randomized model is

$$Y_{ijk} = \mu + \alpha_i + \beta_j + (\alpha\beta)_{ij} + \epsilon_{ijk};$$
$$i = 1, 2, \ldots, a; \qquad j = 1, 2, \ldots, b; \qquad k = 1, 2, \ldots, n. \qquad (11.58)$$

Here μ is the overall mean, α_i is the effect due to the ith level of factor A, β_j is the effect due to the jth level of factor B, $(\alpha\beta)_{ij}$ is the interaction effect for the pair (i, j) of levels, and ϵ_{ijk} is the $N(0, \sigma^2)$ distributed error term.

Example 11.3.1

Craik and coworkers (1994), interested in memory phenomena, designed an experiment in which subjects were presented either visually or auditorially with a long list of words, and for each word either a shallow or a deep mnemonic cue was supplied. An example of a shallow cue might be the number of syllables a word has. A deep cue would be more conceptual and hit closer to the word's meaning, for instance "found in China" might be a deep cue for *pagoda*. The researchers then gave the subjects a new long list of words, in which only half had been present in the study list and the other half were new. They gathered data on the proportion of words that the subjects were able to recognize correctly as being on the study list. The format of the data would be as follows:

		Cue	
	$i \setminus j$	*Shallow*	*Deep*
Presentation	*Visual*	$Y_{111}, \ldots, Y_{11n}(.56)$	$Y_{121}, \ldots, Y_{12n}(.84)$
	Auditory	$Y_{211}, \ldots, Y_{21n}(.57)$	$Y_{221}, \ldots, Y_{22n}(.83)$

In our model, factor A would be the presentation mode, which has two levels, and factor B would be the cue style, which also has two levels. In the display, Y_{ijk} is the proportion of words recognized by subject k using presentation mode i and cue style j. We would be interested in testing for differences in factor A levels; in other words, are subjects who are given their words visually better or worse recognizers than those who hear their words? Also we would be interested in differences in the levels of factor B; that is, do more conceptual cues assist the subjects' memories better than less conceptual ones? Interactions between how the information is presented and the semantic content of the information would also be important to know.

For your interest I have included the average success rates reported by the investigators among subjects in the four categories in parentheses. They suggest significant differences in cue style but not in presentation mode, which is what was found when the analysis was performed using the methods that we are about to learn. In addition, no significant interaction between the two factors was found.

Let us make progress toward the analysis, which makes use of matrix results again. As in the single-factor case, certain averages are important. Dot and bar notation is used just as before:

$$\overline{Y}_{ij\cdot} = \frac{1}{n}\sum_{k=1}^{n} Y_{ijk}, \qquad \overline{Y}_{i\cdot\cdot} = \frac{1}{bn}\sum_{j=1}^{b}\sum_{k=1}^{n} Y_{ijk},$$

$$\overline{Y}_{\cdot j\cdot} = \frac{1}{an}\sum_{i=1}^{a}\sum_{k=1}^{n} Y_{ijk}, \qquad \overline{Y} = \frac{1}{abn}\sum_{i=1}^{a}\sum_{j=1}^{b}\sum_{k=1}^{n} Y_{ijk}.$$

(11.59)

Note that $\overline{Y}_{ij\cdot}$ is the mean in cell $i-j$, $\overline{Y}_{i\cdot\cdot}$ is the mean in row i, $\overline{Y}_{\cdot j\cdot}$ is the mean in column j, and $\overline{Y}$ is the overall mean of all data. You should expect that the test for factor A effect would be based on sums of squares of $\overline{Y}_{i\cdot\cdot} - \overline{Y}$, and the test for factor B effect would involve sums of squares of $\overline{Y}_{\cdot j\cdot} - \overline{Y}$.

We can stack the response observations on one another to form a matrix version of (11.58) as follows.

$$\mathbf{Y} = X\boldsymbol{\theta} + \boldsymbol{\epsilon}.$$

(11.60)

To fix the ideas, consider a small model in which factor A has just two levels, factor B has two levels, and there are $n = 2$ replications for each level combination. Then equation (11.60) would look like

$$
\begin{bmatrix}
Y_{111} \\
Y_{112} \\
\hline
Y_{121} \\
Y_{122} \\
\hline
Y_{211} \\
Y_{212} \\
\hline
Y_{221} \\
Y_{222}
\end{bmatrix}
=
\begin{bmatrix}
1 & 1 & 0 & 1 & 0 & 1 & 0 & 0 & 0 \\
1 & 1 & 0 & 1 & 0 & 1 & 0 & 0 & 0 \\
\hline
1 & 1 & 0 & 0 & 1 & 0 & 1 & 0 & 0 \\
1 & 1 & 0 & 0 & 1 & 0 & 1 & 0 & 0 \\
\hline
1 & 0 & 1 & 1 & 0 & 0 & 0 & 1 & 0 \\
1 & 0 & 1 & 1 & 0 & 0 & 0 & 1 & 0 \\
\hline
1 & 0 & 1 & 0 & 1 & 0 & 0 & 0 & 1 \\
1 & 0 & 1 & 0 & 1 & 0 & 0 & 0 & 1
\end{bmatrix}
\begin{bmatrix}
\mu \\
\alpha_1 \\
\alpha_2 \\
\beta_1 \\
\beta_2 \\
\alpha\beta_{11} \\
\alpha\beta_{12} \\
\alpha\beta_{21} \\
\alpha\beta_{22}
\end{bmatrix}
+
\begin{bmatrix}
\epsilon_{111} \\
\epsilon_{112} \\
\hline
\epsilon_{121} \\
\epsilon_{122} \\
\hline
\epsilon_{211} \\
\epsilon_{212} \\
\hline
\epsilon_{221} \\
\epsilon_{222}
\end{bmatrix}.
$$

(11.61)

In the design matrix, the first column corresponds to μ, the next two columns to α_1, α_2, the next two columns to β_1, β_2, and the last four columns to $\alpha\beta_{11}, \alpha\beta_{12}, \alpha\beta_{21}, \alpha\beta_{22}$. In general in the design matrix for the two-factor model, the μ column consists of $a \cdot b$ copies of $\mathbf{1}_n$, a column vector of 1s with n components. The α_1 column has b copies of $\mathbf{1}_n$ followed by zeros, and similarly other α columns will have b $\mathbf{1}_n$'s consecutively in row blocks corresponding to the Y's in which their levels are used. The β columns will have a cycle of $\mathbf{1}_n$'s and $\mathbf{0}_n$'s, with the $\mathbf{1}_n$'s appearing for column β_j in row blocks in which j is the level of factor B used for the response data.

— **?Question 11.3.1**   Write out the design matrix for a model with three levels of factor A, two of factor B, and three replications each.

The approach to analyzing two-factor experiments is very similar to the one-factor case, but the algebra is more extensive and tedious. We will provide an outline of the proof of the following theorem, leaving a few of the details to you in the exercises.

PROPOSITION 11.3.1

(a) In the balanced two-factor fixed effects model (11.58) we have the following sum of squares decomposition:

$$SST = SSA + SSB + SSAB + SSE, \tag{11.62}$$

where

$$SST = \sum_{i=1}^{a}\sum_{j=1}^{b}\sum_{k=1}^{n} Y_{ijk}^2 - abn\overline{Y}^2 = \sum_{i=1}^{a}\sum_{j=1}^{b}\sum_{k=1}^{n}(Y_{ijk} - \overline{Y})^2,$$

$$SSA = bn\sum_{i=1}^{a}(\overline{Y}_{i..} - \overline{Y})^2, \qquad SSB = an\sum_{j=1}^{b}(\overline{Y}_{.j.} - \overline{Y})^2,$$

$$SSAB = n \cdot \sum_{i=1}^{a}\sum_{j=1}^{b}(\overline{Y}_{ij.} - \overline{Y}_{i..} - \overline{Y}_{.j.} + \overline{Y})^2,$$

$$SSE = \sum_{i=1}^{a}\sum_{j=1}^{b}\sum_{k=1}^{n}(Y_{ijk} - \overline{Y}_{ij.})^2. \tag{11.63}$$

(b) Furthermore, under $H_{0A}: \alpha_1 = \alpha_2 = \cdots = \alpha_a = 0$, $H_{0B}: \beta_1 = \beta_2 = \cdots = \beta_b = 0$, and $H_{0AB}: \alpha\beta_{ij} = 0 \ \forall i, j$, the following random variables are independent and have chi-square distributions: $SSA/\sigma^2 \sim \chi^2(a-1)$, $SSB/\sigma^2 \sim \chi^2(b-1)$, $SSAB/\sigma^2 \sim \chi^2((a-1)(b-1))$, $SSE/\sigma^2 \sim \chi^2(ab(n-1))$.

■ **Proof**   Let P_μ be the projection matrix for the μ column in the design matrix, P_α for the α columns, P_β for the β columns, and $P_{\alpha\beta}$ for the $\alpha\beta$ columns. By adding and subtracting like terms we obtain the vector decomposition

$$\mathbf{Y} = P_\mu\mathbf{Y} + (P_\alpha - P_\mu)\mathbf{Y} + (P_\beta - P_\mu)\mathbf{Y} + (P_{\alpha\beta} - P_\alpha - P_\beta + P_\mu)\mathbf{Y}$$
$$+ (I - P_{\alpha\beta})\mathbf{Y}. \tag{11.64}$$

Using the identities

$$
\begin{aligned}
P_\mu P_\alpha &= P_\alpha P_\mu = P_\mu, \\
P_\mu P_\beta &= P_\beta P_\mu = P_\mu, \\
P_\mu P_{\alpha\beta} &= P_{\alpha\beta} P_\mu = P_\mu, \\
P_\alpha P_\beta &= P_\beta P_\alpha = P_\mu, \\
P_\alpha P_{\alpha\beta} &= P_{\alpha\beta} P_\alpha = P_\alpha, \\
P_\beta P_{\alpha\beta} &= P_{\alpha\beta} P_\beta = P_\beta,
\end{aligned}
\tag{11.65}
$$

(see Exercise 12) and the idempotency of the projection matrices (see Exercise 13), you can show that the vectors on the right of (11.64) are orthogonal. The sum of squares decomposition (11.62) is just the Pythagorean theorem applied to (11.64):

$$
\begin{aligned}
\mathbf{Y}'\mathbf{Y} = \mathbf{Y}'P_\mu \mathbf{Y} + \mathbf{Y}'(P_\alpha - P_\mu)\mathbf{Y} + \mathbf{Y}'(P_\beta - P_\mu)\mathbf{Y} \\
+ \mathbf{Y}'(P_{\alpha\beta} - P_\alpha - P_\beta + P_\mu)\mathbf{Y} + \mathbf{Y}'(I - P_{\alpha\beta})\mathbf{Y}.
\end{aligned}
\tag{11.66}
$$

The matrices in the quadratic forms in (11.66) are idempotent (see Exercise 13) and their row sums are zero (see Exercise 14). Furthermore, their pairwise products are zero (see Exercise 15). Therefore under the null hypotheses, the quadratic forms divided by σ^2 are independent χ^2 random variables. It can be checked that the ranks of $P_\alpha - P_\mu$, $P_\beta - P_\mu$, $(P_{\alpha\beta} - P_\alpha - P_\beta + P_\mu)$, and $(I - P_{\alpha\beta})$, respectively, are $a - 1, b - 1, (a - 1)(b - 1)$, and $ab(n - 1)$. Since these ranks are the corresponding degrees of freedom, the proof is complete.

It should not come as a great surprise that the quantities $\overline{Y}_{i\cdot\cdot} - \overline{Y}, \overline{Y}_{\cdot j\cdot} - \overline{Y}$, and $\overline{Y}_{ij\cdot} - \overline{Y}_{i\cdot\cdot} - \overline{Y}_{\cdot j\cdot} + \overline{Y}$ in the sums of squares are the least squares estimates of the parameters α_i, β_j, and $\alpha\beta_{ij}$ respectively, under the side conditions $\sum_i \alpha_i = 0, \sum_j \beta_j = 0, \sum_i (\alpha\beta)_{ij} = 0, \sum_j (\alpha\beta)_{ij} = 0$. (See Exercise 7.)

Statistics packages compute sums of squares, mean squares (sums of squares divided by degrees of freedom) and F-statistics (quotients of mean squares with error mean square) for both factors and the interaction, and display them in ANOVA tables like those we have seen before. (See Appendix B, on *Minitab*, for information about how to proceed.) Then the appropriate level α rejection regions for the null hypotheses of no interaction $H_{0AB} : (\alpha\beta)_{ij} = 0, \forall i, j$, no factor B effect $H_{0B} : \beta_j = 0, \forall j$, and no factor A effect $H_{0A} : \alpha_i = 0, \forall i$ are respectively

$$
F_{AB} = \frac{MSAB}{MSE} = \frac{\dfrac{SSAB}{(a - 1)(b - 1)}}{\dfrac{SSE}{ab(n - 1)}} > f_\alpha((a - 1)(b - 1), ab(n - 1)),
\tag{11.67}
$$

$$
F_B = \frac{MSB}{MSE} = \frac{\dfrac{SSB}{(b - 1)}}{\dfrac{SSE}{ab(n - 1)}} > f_\alpha(b - 1, ab(n - 1)),
\tag{11.68}
$$

$$F_A = \frac{MSA}{MSE} = \frac{\frac{SSA}{(a-1)}}{\frac{SSE}{ab(n-1)}} > f_\alpha(a-1, ab(n-1)). \qquad (11.69)$$

Example 11.3.2

The following data on barley yields (metric tons per hectare) in certain countries are classified according to two factors: A, continent, and B, year (USDA, 1991). Test at level .05 the hypotheses of no interaction effect, no factor A effect, and no factor B effect.

		Year		
		1988	*1989*	*1990*
	Europe	2.69, 4.15, 3.04	2.69, 4.72, 3.33,	2.69, 5.45, 3.59,
		4.00, 2.77	4.43, 2.90	3.60, 2.82
		(3.33)	(3.61)	(3.63)
Continent	*Africa*	0.58, 2.95, 1.47,	0.70, 3.09, 1.06,	0.66, 3.09, 1.30,
		1.38, 1.71	1.25, 3.00	0.89, 2.62
		(1.62)	(1.82)	(1.71)
	Asia	1.67, 2.31, 1.32,	1.74, 2.36, 0.90,	1.73, 1.62, 1.04,
		3.50, 2.50	3.28, 4.62	3.26, 5.33
		(2.26)	(2.58)	(2.60)

The cell means are displayed in parentheses to give you a feel for the data. The row means for Europe, Africa, and Asia, respectively, are 3.52, 1.72, and 2.48, and the column means for 1988, 1989, and 1990 are, respectively, 2.40, 2.67, and 2.65. There is a mild indication of an increase in productivity between 1988 and 1989, but it is hard to tell whether it is significant. It appears that there is a substantial effect of continent on yield. Since each continent experiences roughly the same change in barley yield as years go by, we would not expect significant interaction.

The ANOVA table obtained from *Minitab* follows.

Source	*df*	*SS*	*MS*	*F*
Cont	2	24.718	12.359	9.67
Year	2	.660	.330	.26
Interaction	4	.086	.022	.02
Error	36	45.990	1.277	
Total	44	71.454		

(Make sure you know how the df, MS, and F columns were obtained.) We see from this table that because of the large number of degrees of freedom for error, the mean square error is small. The $f(2, 36)$ critical value for level .05 is 3.26, and the $f(4, 36)$ critical value is 2.63. Since the observed F-value for the continent factor of 9.67 exceeds the critical value, continent is significant at this level. Neither year nor the continent–year interaction is significant at any reasonable level.

—— **?Question 11.3.2** Use the data given in the last example to check by hand the computation of the sum of squares for continent.

—— **?Question 11.3.3** In the last example, compute the least squares estimators of the parameters α_i and β_j.

Models with more than two factors can be built and analyzed similarly (see Montgomery, 1991 for details). For instance, for a three-factor model, the possible interactions are between all pairs of factors and also among all three factors simultaneously. Some or all factors and interactions may not be significant and so can be eliminated. The full three-factor model would look like

$$Y_{ijkl} = \mu + \alpha_i + \beta_j + \gamma_k + \alpha\beta_{ij} + \alpha\gamma_{ik} + \beta\gamma_{jk} + \alpha\beta\gamma_{ijk} + \epsilon_{ijkl}. \qquad (11.70)$$

Factor means of all data for fixed i $(\overline{Y}_{i\ldots})$, fixed j $(\overline{Y}_{\cdot j\cdot})$, and fixed k $(\overline{Y}_{\cdot\cdot k\cdot})$ can be computed. Subtracting the overall mean $\overline{Y}$, squaring, and summing over the subscript again gives the sums of squares for the factors. Interaction sums of squares and the error sum of squares are formed analogously to the two-factor case. The degrees of freedom for all terms in the model are similar to the two-factor case. Once again, statistical programs compute all of this information, together with the mean squares and the F statistics, and one rejects the null hypotheses that the model terms are zero if the observed f values exceed the critical values of the desired level.

11.3.2 Random Effects Model

When one or both factors are random—that is, when the levels of the factor are not fixed but rather are randomly selected from a universe of possible levels—the model, null hypotheses, and analysis change somewhat. In random effects models, if a factor is random, then its contribution to the model equation is a random variable and we assume as well that any interaction term that includes the factor is random. For concreteness let us consider only the case where factor A is fixed and factor B (therefore the interaction factor) is random. The model equation is

$$Y_{ijk} = \mu + \alpha_i + B_j + (\alpha B)_{ij} + \epsilon_{ijk},$$
$$i = 1, \ldots, a, \quad j = 1, \ldots, b, \quad k = 1, \ldots, n, \qquad (11.71)$$

where ϵ_{ijk} are i.i.d. $N(0, \sigma^2)$ random variables, $B_1, B_2, \ldots, B_b$ are i.i.d. $N(0, \sigma_b^2)$ random variables independent of the ϵ's, and for fixed i the $(\alpha B)_{ij}$'s are i.i.d. $N(0, \sigma_{ab}^2)$ random variables independent of both the B's and the ϵ's. For the sum of the excess effects due to factor A to be zero, we require that $\sum_i \alpha_i = 0$ and also $\sum_i (\alpha B)_{ij} = 0$ for each $j = 1, 2, \ldots, b$. Notice that this implies a correlation between $(\alpha B)_{ij}$ and $(\alpha B)_{kj}$ for $i \neq k$. The null hypotheses of no effect become

$$H_{0AB} : \sigma_{ab}^2 = 0, \qquad H_{0B} : \sigma_b^2 = 0, \qquad H_{0A} : \alpha_i = 0 \, \forall i. \qquad (11.72)$$

Example 11.3.3

In analyzing the performance of a complicated computer algorithm, a computer scientist may resort to simulation testing. Suppose that four particular sorting algorithms are being compared and that the computer randomly generates three data sets on which to run each of them. Due to changing conditions in the computer's environment, non-determinism enters into the running times in a small way. Two replications for the experiment of running an algorithm on a data set are done for each algorithm–data set pair, yielding the running time results shown in the table. We are interested in testing for differences in the running times produced by the four algorithms, for effect of the data set, and for interactions between the algorithms and the data sets.

		Data Set		
		1	*2*	*3*
	1	2.52004, 1.92923	1.53198, .89860	2.40773, 1.73537
	2	2.62897, 2.70122	1.80498, 1.68569	2.30763, 2.14115
Algorithm	*3*	.82177, .44678	2.03421, 2.30390	1.10039, .86721
	4	2.88669, 2.99239	.21713, .45465	2.17922, 1.73152

The algorithms are predetermined and appear to be the only ones of interest in the investigation. It makes sense to treat the algorithm factor, factor A say, as a fixed factor with four levels. The data sets, however, are only three among many possible inputs to the algorithms; therefore the data set factor B and the interaction term should both be random. Model (11.71) with $a = 4$, $b = 3$, and $n = 2$ describes the situation. We will analyze the data shortly.

Under the null hypotheses the sums of squares again have χ^2-distributions with the same parameters, but we must be careful to correctly set up the F-statistics to be sensitive to the alternative hypotheses for the new model. The next result shows how to do this. Its proof is a lengthy but routine process of computing means and variances of sums.

PROPOSITION 11.3.2

For the random effects model (11.72) with A fixed and B random, we have

$$E[MSA] = \sigma^2 + n \frac{a}{a-1} \sigma_{ab}^2 + \frac{nb}{a-1} \sum_{i=1}^{a} \alpha_i^2, \qquad (11.73)$$

$$E[MSB] = \sigma^2 + na\sigma_b^2, \qquad (11.74)$$

$$E[MSAB] = \sigma^2 + n\frac{a}{a-1}\sigma_{ab}^2, \tag{11.75}$$

$$E[MSE] = \sigma^2. \tag{11.76}$$

■ **Proof** It is easy to verify the following equivalent versions of the sums of squares:

$$SSA = \frac{1}{bn}\sum_{i=1}^{a} Y_{i..}^2 - \frac{1}{abn}Y_{...}^2, \tag{11.77}$$

$$SSB = \frac{1}{an}\sum_{j=1}^{b} Y_{.j.}^2 - \frac{1}{abn}Y_{...}^2, \tag{11.78}$$

$$SSAB = \frac{1}{n}\sum_{i=1}^{a}\sum_{j=1}^{b} Y_{ij.}^2 - \frac{1}{bn}\sum_{i=1}^{a} Y_{i..}^2 - \frac{1}{an}\sum_{j=1}^{b} Y_{.j.}^2 + \frac{1}{abn}Y_{...}^2, \tag{11.79}$$

$$SSE = \sum_{i=1}^{a}\sum_{j=1}^{b}\sum_{k=1}^{n} Y_{ijk}^2 - \frac{1}{n}\sum_{i=1}^{a}\sum_{j=1}^{b} Y_{ij.}^2. \tag{11.80}$$

To compute the expected values of these random variables, recall that $E[W^2] = \text{Var}(W) + (E[W])^2$ for any random variable W. By linearity, it suffices to calculate the mean and variance of each of: $Y_{ijk}, Y_{ij.}, Y_{i..}, Y_{.j.},$ and $Y_{...}$. First,

$$E[Y_{ijk}] = E[\mu + \alpha_i + B_j + (\alpha B)_{ij} + \epsilon_{ijk}] = \mu + \alpha_i + 0 + 0 + 0 = \mu + \alpha_i, \tag{11.81}$$

and

$$\begin{aligned}
\text{Var}(Y_{ijk}) &= \text{Var}(\mu + \alpha_i + B_j + (\alpha B)_{ij} + \epsilon_{ijk}) \\
&= \text{Var}(B_j + (\alpha B)_{ij} + \epsilon_{ijk}) \\
&= \sigma_b^2 + \sigma_{ab}^2 + \sigma^2.
\end{aligned} \tag{11.82}$$

Next, by (11.81)

$$E[Y_{ij.}] = E\left[\sum_{k=1}^{n} Y_{ijk}\right] = \sum_{k=1}^{n} E[Y_{ijk}] = n(\mu + \alpha_i). \tag{11.83}$$

Also,

$$\begin{aligned}
\text{Var}(Y_{ij.}) &= \text{Var}\left(\sum_{k=1}^{n} Y_{ijk}\right) \\
&= \text{Var}\left(n(\mu + \alpha_i + B_j + (\alpha B)_{ij}) + \sum_{k=1}^{n} \epsilon_{ijk}\right) \\
&= n^2\text{Var}(B_j) + n^2\text{Var}((\alpha B)_{ij}) + \sum_{k=1}^{n}\text{Var}(\epsilon_{ijk}) \\
&= n^2\sigma_b^2 + n^2\sigma_{ab}^2 + n\sigma^2.
\end{aligned} \tag{11.84}$$

Now by (11.83),

$$E[Y_{i..}] = E\left[\sum_{j=1}^{b} Y_{ij.}\right] = \sum_{j=1}^{b} n(\mu + \alpha_i) = nb(\mu + \alpha_i), \tag{11.85}$$

and

$$\text{Var}(Y_{i..}) = \text{Var}\left(\sum_{j=1}^{b} Y_{ij.}\right)$$

$$= \text{Var}\left[\sum_{j=1}^{b}\left(n(\mu + \alpha_i + B_j + (\alpha B)_{ij}) + \sum_{k=1}^{n} \epsilon_{ijk}\right)\right]$$

$$= \text{Var}\left[nb(\mu + \alpha_i) + n\sum_{j=1}^{b} B_j + n\sum_{j=1}^{b}(\alpha B)_{ij} + \sum_{j=1}^{b}\sum_{k=1}^{n} \epsilon_{ijk}\right]$$

$$= n^2 b \cdot \sigma_b^2 + n^2 b \cdot \sigma_{ab}^2 + nb \cdot \sigma^2. \tag{11.86}$$

Continuing to $Y_{.j.}$ we have

$$E[Y_{.j.}] = E\left[\sum_{i=1}^{a} Y_{ij.}\right] = E\left[\sum_{i=1}^{a}\left(n(\mu + \alpha_i + B_j + (\alpha B)_{ij}) + \sum_{k=1}^{n} \epsilon_{ijk}\right)\right]$$

$$= E\left[na\mu + 0 + na \cdot B_j + 0 + \sum_{i=1}^{a}\sum_{k=1}^{n} \epsilon_{ijk}\right]$$

$$= na\mu + 0 + 0 = na\mu, \tag{11.87}$$

by the conditions $\sum \alpha_i = 0$ and $\sum_i (\alpha B)_{ij} = 0$. The same conditions let us compute

$$\text{Var}(Y_{.j.}) = \text{Var}\left(na\mu + na \cdot B_j + \sum_{i=1}^{a}\sum_{k=1}^{n} \epsilon_{ijk}\right)$$

$$= n^2 a^2 \sigma_b^2 + na\sigma^2. \tag{11.88}$$

Finally for $Y\dots$ we have

$$E[Y\dots] = E\left[\sum_{j=1}^{b} Y_{.j.}\right] = E\left[\sum_{j=1}^{b}\left(na\mu + naB_j + \sum_{i=1}^{a}\sum_{k=1}^{n} \epsilon_{ijk}\right)\right]$$

$$= E\left[nab\mu + na\sum_{j=1}^{b} B_j + \sum_{i=1}^{a}\sum_{j=1}^{b}\sum_{k=1}^{n} \epsilon_{ijk}\right]$$

$$= nab\mu + na \cdot 0 + 0 = nab\mu \tag{11.89}$$

and

$$\text{Var}(Y\dots) = \text{Var}\left(nab\mu + na\sum_{j=1}^{b} B_j + \sum_{i=1}^{a}\sum_{j=1}^{b}\sum_{k=1}^{n} \epsilon_{ijk}\right)$$

$$= (na)^2 b\sigma_b^2 + nab\sigma^2. \tag{11.90}$$

Inserting these results into (11.77) gives

$$E[SSA] = \frac{1}{bn} \sum_{i=1}^{a} [(n^2 b \sigma_b^2 + n^2 b \sigma_{ab}^2 + nb\sigma^2) + (nb)^2(\mu + \alpha_i)^2]$$

$$- \frac{1}{nab}[(na)^2 b \sigma_b^2 + nab\sigma^2 + (nab\mu)^2]$$

$$= na\sigma_b^2 + na\sigma_{ab}^2 + a\sigma^2 + nb \sum_{i=1}^{a} (\mu + \alpha_i)^2 - na\sigma_b^2 - \sigma^2 - nab\mu^2$$

$$= na\sigma_{ab}^2 + (a-1)\sigma^2 + nb \sum_{i=1}^{a} (\mu^2 + 2\mu\alpha_i + \alpha_i^2) - nab\mu^2$$

$$= (a-1)\sigma^2 + na\sigma_{ab}^2 + nb \sum_{i=1}^{a} \alpha_i^2, \tag{11.91}$$

from which the formula for $E[MSA]$ follows. Turning to SSB, from (11.87) and (11.88),

$$E[SSB] = \frac{1}{an} \sum_{j=1}^{b} [n^2 a^2 \sigma_b^2 + na\sigma^2 + (na\mu)^2]$$

$$- \frac{1}{nab}[(na)^2 b \sigma_b^2 + nab\sigma^2 + (nab\mu)^2]$$

$$= nab\sigma_b^2 + b\sigma^2 + nab\mu^2 - na\sigma_b^2 - \sigma^2 - nab\mu^2$$

$$= na(b-1)\sigma_b^2 + (b-1)\sigma^2. \tag{11.92}$$

The formula for $E[MSB]$ follows by dividing by the degrees of freedom $b - 1$. You are asked to finish the computations of the expected interaction mean square and error mean square in Exercise 22.

The expected mean squares (11.73)–(11.76) indicate which mean square ratios to form for which tests. To test $H_{0AB}: \sigma_{ab}^2 = 0$, note that when the alternative holds, $MSAB$ is expected to be significantly greater than MSE due to the presence of the $n \cdot \frac{a}{a-1} \sigma_{ab}^2$ term. The appropriate rejection region for H_{0ab} is

$$F_{AB} = \frac{MSAB}{MSE} > f_\alpha[(a-1)(b-1), ab(n-1)]. \tag{11.93}$$

For similar reasons, the rejection region for $H_{0B}: \sigma_b^2 = 0$ is

$$F_B = \frac{MSB}{MSE} > f_\alpha(b-1, ab(n-1)). \tag{11.94}$$

So far there are no surprises. But look at the form of $E[MSA]$ in (11.73). If MSA is larger than MSE there is no way of telling whether the reason is that some α_i's are nonzero or that σ_{ab}^2 is nonzero. All is not lost, however, because if you compare $E[MSA]$ to $E[MSAB]$ in (11.75), you see that if

MSA is significantly greater than $MSAB$, there is evidence that it is the α_i's that are nonzero, since both expected mean squares have the same σ_{ab}^2 term. Therefore the rejection region for $H_{0A} : \alpha_i = 0$ for all i is

$$F_A = \frac{MSA}{MSAB} > f_\alpha[a - 1, (a - 1)(b - 1)].\qquad (11.95)$$

In general, a rule of thumb for determining the mean square to use in the denominator of an F-ratio for a test of a factor is to use the nearest random interaction factor containing the factor being tested, or MSE in the absence of any such factor.

—— **?Question 11.3.4**   Suppose factor B is fixed and A is random. What do you think the critical regions would be? What if both factors were random?

Example 11.3.3

Continued   For the computer algorithm data, *Minitab* supplies us with the information we need to draw conclusions.

Source	df	SS	MS	F
Algorithm	3	2.7465	.9155	
Data Set	2	2.2714	1.1357	15.27
Interaction	6	9.2606	1.5434	20.75
Error	12	.8926	.0744	
Total	23	15.1711		

Because their observed f-values are so large, both the data set and interaction terms have nonzero variance at reasonable levels of significance. I purposely left the F column for the algorithm factor blank, because *Minitab* reports a value of 12.31, computed by taking the ratio of the algorithm mean square to the error mean square. This is not the correct thing to do, as we saw earlier. The proper ratio to form is $MSA/MSAB = .9155/1.5434$, which comes out to .593. Thus the algorithms here are not significantly different at any reasonable level. *Minitab*'s F-value would have misled us had we not acknowledged the fact that the data set factor is random.

■ **EXERCISES 11.3**

1. In Section 11.1 we introduced the design of factorial experiments with an example on gas mileage of autos. A two-factor model was proposed in which a group of cars was divided randomly into $a \cdot b = 3 \cdot 3 = 9$ subgroups according to the combination of levels of two factors: gasoline grade (regular, special, premium), and driving speed (0–40, 41–55, over 55). Suppose that we observe the following data. Test at level .05 the hypotheses of no interaction effect, no factor A effect, and no factor B effect.

		Speed		
		0–40	*41–55*	*Over 55*
	Regular	17.6, 19.6, 22.2	23.3, 22.6, 21.7	24.7, 27.4, 27.3
Grade	Special	19.2, 21.9, 22.6	22.3, 25.0, 20.6	27.1, 27.3, 29.4
	Premium	24.0, 20.6, 24.1	23.6, 22.7, 25.1	27.8, 29.0, 28.1

2. A test field is broken up into 18 plots in order to test the effect on soybean yield of the variety of seed bean planted, of which two are of interest, and the herbicide applied to the soil, of which there are three types. The plots are randomly assigned to the six treatment combinations, three to one combination. Bean yields in bushels are given in the table. Test at level .05 for interaction between seed variety and herbicide and for effects of seed variety and herbicide on average yield.

		Herbicide		
		1	*2*	*3*
	1	45, 50, 48	50, 52, 55	38, 42, 26
Variety	2	42, 47, 46	48, 49, 54	39, 40, 41

3. Write the design matrix for a general two-factor completely randomized model. For efficiency you might want to adopt the notation $\mathbf{1}_n$ for a vector of 1s of length n.

4. Kemph and Kasser (forthcoming), studying gender and attitudes toward male homosexuality performed an experiment in which an interviewer asked subjects to rate on a five-point scale their attitudes toward several scenarios relating to male homosexuality. High ratings correspond to more negative attitudes. The subjects were then given a composite test score. The data were classified according to the the gender of the subjects and also according to how the interviewer (who was male) presented himself. The interviewer was dressed either in cowboy boots and a cap with a feed store logo and a rodeo poster in the background (to suggest heterosexuality of the interviewer) or in a sweatshirt with "Gay Pride" written on it and a poster about homosexuality in the background (to suggest homosexuality of the interviewer). The questions of interest are: Do the subjects respond differently according to their perception of what the interviewer wants to hear? Do the responses depend significantly on the gender of the subject? and Do male responses change according to the interviewer condition in a way different from the change of female responses?

   Formulate a statistical model carefully, and use it and the following data to answer the questions. The category means reported by the experimenters were male-cowboy mean, 3.22; male-gay mean, 1.83; female-cowboy mean, 2.07; female-gay mean, 1.75. The raw data were unavailable, and it appears that the design was unbalanced, so I have created the following information that will lead to nearly the same conclusions that the researchers had: The common sample size for the four categories was 4, and the sum of squares due to error was 1.86.

5. In the two-factor experiment as we have described it, each level of factor *A* is combined with each level of factor *B*. Discuss the following issue (for concreteness suppose that each factor has two levels): What is gained or lost by combining level 1 of factor *A* with levels 1 and 2 of factor *B*, and level 2 of factor *A* with only level 1 of factor *B*?

6. A research group in the psychology department at Knox College headed by Professor Heather Hoffman and including students Kemeko Miller and Pamela Skoubis investigated a conditioning problem. The following data are extracted from their larger data set. A group of rats was divided into subgroups according to two treatments. The first treatment involved brain depletion; some rats had a toxic drug injected into their brains, others were given a harmless drug, and a control group was left intact. The second treatment was a stimulus presentation. Some rats (the *paired* group) were presented with a liquid food (Grape Kool Aid), paired immediately with the drug lithium chloride, which made them sick, while others were given the sickening drug three hours later (the *unpaired* group). The response measured was the rats' percent body weight gain after being presented with the food for a second time (an indirect measurement of how much they consumed and therefore their ability to learn about poisonous foods). Test at level .05 for a significant effect of the treatments and for interaction between the treatments.

	Paired	Unpaired
Toxic	.22, .62, 1.32, 1.08, 1.07, .68, .16, 1.04	.48, 4.46, 3.45, 1.50, 3.14, 2.55, .76, 3.33
Nontoxic	2.13, 1.27, .17, .21, 1.43, .75, 1.55, 1.26	3.35, 3.58, 4.66, 1.79, 2.64, 1.28, 3.96, 2.50
Control	1.21, .48, .89, 1.71, 3.04, 1.13, .15, .72	4.66, 3.51, 1.47, 3.43, 3.55, 1.60, 1.49, 2.24

7. In the model expressed by (11.61) with $a = 2$, $b = 2$, and $n = 2$, write out the least squares equation $X'X\theta = X'Y$. Argue that the least squares estimators of the βs and αs are $\hat{\beta}_j = \overline{Y}_{.j.} - \hat{\mu}$ and $\hat{\alpha}_i = \overline{Y}_{i..} - \hat{\mu}$ under the side conditions $\sum_i \alpha_i = 0$, $\sum_j \beta_j = 0$, and $\sum_i(\alpha\beta)_{ij} = \sum_j(\alpha\beta)_{ij} = 0$. (As in the one-factor case $\hat{\mu} = \overline{Y}$, which you may use without proof.)

8. A study (Johnston et al., 1993) of drug use among young people arrived at the following table of numbers, which are the estimated percentages of people at four different grade levels who used three types of drugs during each of the years 1991 and 1992. Formulate appropriate hypotheses, test them, and discuss your conclusions. (Note that actually the year is a possible blocking factor, but carry out the analysis as if the data in each cell is a random sample of size 2 anyway.)

	Cocaine	Stimulants	Tranquilizers
8th Grade	1.1, 1.5	6.2, 6.5	1.8, 2.0
10th Grade	2.2, 1.9	8.2, 8.2	3.2, 3.5
12th Grade	3.5, 3.1	8.2, 7.1	3.6, 2.8
College	3.6, 3.0	3.9, 3.6	2.4, 2.9

9. Compare a single-factor randomized block model with a two-factor completely randomized model with just one replication per treatment combination. How are the experiments different? Despite the differences, show the equivalence of the sum of squares decompositions for the two experiments.

10. As another variation on the Peoria school data analysis (see Exercises 4, 11, and 3 in Sections 10.2, 10.3, and 10.4, respectively), consider the following. Samples of three school average reading scores from each of two counties and each of two economic conditions reported in the district served by the school (under 15% poor and over 15% poor) have been tabulated. Formulate a model, and ask and answer interesting questions about it.

	Under 15% Poor	**Over 15% Poor**
Peoria County	297, 346, 292	276, 247, 286
Tazewell County	348, 285, 257	244, 288, 268

11. In Proposition 11.3.1, write out the matrices P_μ, P_α, P_β, and $P_{\alpha\beta}$.

12. In Proposition 11.3.1, show that $P_\alpha P_{\alpha\beta} = P_\alpha$, and $P_\beta P_{\alpha\beta} = P_\beta$ (see Exercise 11).

13. In Proposition 11.3.1, show that
    a. $P_{\alpha\beta}$ is idempotent.
    b. $P_{\alpha\beta} - P_\alpha - P_\beta + P_\mu$ is idempotent. [Use the identities (11.65).]

14. Show that the rows of the matrix $I - P_{\alpha\beta}$ in Proposition 11.3.1 sum to zero.

15. Show that all pairwise products of the matrices in the quadratic forms in (11.66) are zero and hence the forms are independent. [You may use the identities (11.65).]

16. In the university calculus GPA data mentioned several times earlier (see, e.g., Exercise 18 in Section 7.1), several institutions provided information on two offerings of each calculus course. Treating the institution as a random factor in the following data, test for differences in grade performance by course, for variability due to institution, and for interaction variability. Use level .05.

	Calc I	**Calc II**	**Calc III**
School 1	2.25, 1.98	2.19, 2.26	2.31, 2.23
School 2	2.28, 2.22	2.82, 2.74	2.90, 2.90
School 3	2.00, 1.82	2.03, 1.80	2.04, 1.62

17. Five randomly chosen machines are each run at two speeds to make a product. Suppose that the breaking strengths of 3 samples are measured for each machine–speed combination. For the data in the following table, test at level .05 for significant machine, speed, and interaction effects. Carefully state the probabilistic model, and the hypotheses you are testing.

	Machine				
	1	*2*	*3*	*4*	*5*
High	8.55, 9.17, 8.38	5.56, 7.06, 6.82	8.31, 7.70, 6.86	6.40, 6.77, 8.56	7.34, 7.12, 6.18
Speed					
Low	12.22, 11.35, 9.67	11.68, 10.01, 9.45	14.24, 12.08, 11.41	12.15, 10.31, 15.52	12.21, 10.86, 13.37

18. In each of four different time periods during 1990, the average wage rates of farm workers were estimated and broken down by their type (field versus livestock) and by region as indicated in the following table (USDA, 1991). The regions are a randomly chosen few among many. Formulate a probabilistic model, and carry out appropriate hypothesis tests. (As in Exercise 8, the time period is actually a potential blocking factor, but ignore that.)

	Field	**Livestock**
Northeast II	4.66, 5.45, 5.20, 5.45	3.93, 3.85, 4.65, 4.54
Southeast	4.10, 4.31, 4.10, 4.97	5.02, 5.02, 5.05, 4.63
S. Plains	4.56, 4.74, 4.32, 4.74	4.69, 4.87, 4.26, 4.78
Mountain III	4.96, 4.74, 4.51, 5.04	5.20, 4.88, 5.16, 4.90

19. This problem was inspired by the work of Jennifer Robinson, a student in the biology department of Knox College, in collaboration with Dr. Billy Geer. The data were not yet compiled at the time of this writing, so the following numbers are merely simulated. The experimenters were interested in the effects of heat shock and dietary ethanol on the fatty acid profiles of the Drosophila melanogaster fly. Diets of three different contents were applied to flies raised at three different temperatures: 18°C, 23°C, and 28°C. Measurements taken of one particular fatty acid (as a percent of the total of all fatty acids) appear in the table. Treating temperature as a random factor, test for main factor effects and interaction effect at level .05. Is the conclusion of the test for a temperature effect different from what it would have been had you considered temperature to be a fixed effect? How about the diet effect?

		Diet		
		1	*2*	*3*
	18°C	16.4, 16.5, 17.2, 16.9	14.6, 14.5, 15.4, 15.5	14.6, 14.5, 16.5, 15.7
Temperature	**23°C**	15.0, 17.1, 17.3, 17.3	19.1, 16.4, 16.8, 17.3	15.1, 16.7, 14.9, 18.1
	28°C	15.5, 17.0, 15.2, 14.8	16.6, 18.7, 15.2, 14.4	13.8, 15.1, 17.4, 16.4

20. Find $\text{Cov}(\overline{Y}_{ij\cdot}, \overline{Y}_{ik\cdot})$ in the two-factor model with factor A fixed and factor B random.

21. In a two-factor model with both factors random, find the expected mean square error.

22. a. Show that for the interaction sum of squares in the random effects model (11.71),

$$E[SSAB] = (a-1)(b-1)\sigma^2 + na(b-1)\sigma_{ab}^2,$$

from which formula (11.75) follows.

   b. Show that for the error sum of squares in the random effects model (11.71),

$$E[SSE] = ab(n-1)\sigma^2,$$

from which formula (11.76) follows.

23. Describe the conduct of, and write a linear model for, a two-factor fixed effects model with a separate blocking factor, that is, a two-factor randomized block experiment.

24. An *interaction diagram* for a two-factor experiment is a plot of sample means for treatment combinations, against, say, the levels of factor A. The points with coordinates (factor A level, sample mean) would be plotted for each factor A and B level combination, and points with the same level of factor B would be connected. So if there were levels 1, 2, 3, and 4 for factor A, and levels 1, 2, and 3 for factor B, one such diagram might look like the one on the left, another as in the figure on the right.

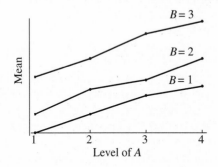 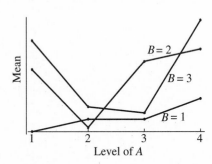

   a. What do you think that it says about the factors, if the polygonal graphs are nearly parallel to one another, as on the left? Similarly, what would the crossing pattern displayed on the right suggest about the factors?

   b. Do an interaction diagram for the data set in Exercise 1, and interpret it in light of your response to part (a).

CHAPTER 12

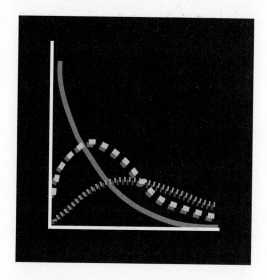

GOODNESS-OF-FIT

12.1 | Chi-Square Tests

Many times in this text we have had models or data sets in which we supposed that certain probability distributions applied. Frequently, for example, we have assumed that a set of data $X_1, X_2, \ldots, X_n$ was a random sample from the normal distribution with some mean and variance. You should have been wondering, How can this assumption be checked? This chapter shows methods for assessing how well data fit an hypothesized distribution and whether two or more distributions are significantly different. It turns out that similar methods can be applied to the apparently different question of whether events are independent of one another.

Consequently we are hitting at the heart of probabilistic modeling in this last chapter, which emerges as a very important part of statistical analysis. Yet, you will probably not find the ideas and techniques to be very difficult. They are based on the intuitive idea that if an hypothesized distribution leads to a theoretical set of probabilities for categories, then empirical probabilities based on a random sample should be nearly equal to the theoretical probabilities. The larger the overall discrepancy between hypothesized category probabilities and empirical probabilities, the more significant is the evidence against the hypothesis.

The analysis, as in other contexts we have seen, depends on sums of square deviations. In this case the square differences $(p_{i0} - \widehat{p}_i)^2$ seem to be most relevant, where $\widehat{p}_i$ is the proportion of sample values in category i and p_{i0} is the hypothesized probability for category i. The larger is the total sum of these square differences, the stronger should be your disbelief of the p_{i0}'s. An example of the kind of problem we are concerned with follows.

Example 12.1.1

Recall the data on the Illinois State Lottery presented in Chapter 7. In 363 games the frequencies of occurrence of the digits 0–9 on the first draw of three are given in Table 12.1.

Are all digits equally likely to be chosen? If so, then $p_{i0} = P[\text{draw } 1 = i] = .1$ for all $i = 0, \ldots, 9$. The test we will construct will compare these hypothesized probabilities to the empirical proportions obtained by dividing the frequencies in Table 12.1 by the total number of trials, 363. These empirical probabilities are shown on the third line of Table 12.2.

To know whether the discrepancies between the hypothesized and empirical probabilities are significantly large, we must look for a way of standardizing them to produce a measure that has some absolute meaning. We will continue this example when we have developed the approximate chi-square statistic for goodness-of-fit testing.

To set up the general multinomial goodness-of-fit situation, suppose that each member of a random sample $X_1, X_2, \ldots, X_n$ can fall into one of r categories, labeled $c_1, c_2, \ldots, c_r$. Denote

$$p_i = P[X_j = c_i], \tag{12.1}$$

Table 12.1

Frequencies of Digits, Illinois State Lottery

0	1	2	3	4	5	6	7	8	9
35	44	25	40	50	26	33	38	37	35

Table 12.2

Comparison of probabilities

0	1	2	3	4	5	6	7	8	9
.1	.1	.1	.1	.1	.1	.1	.1	.1	.1
.096	.121	.069	.110	.138	.072	.091	.105	.102	.096

and

$$Y_i = \text{number of } X_j\text{'s equal to } c_i. \tag{12.2}$$

Then $(p_1 + p_2 + \cdots + p_r) = 1$ and $(Y_1 + Y_2 + \cdots + Y_r) = n$. We will be testing the null hypothesis

$$H_0 : p_i = p_{i0} \ \forall i = 1, 2, \ldots, r \tag{12.3}$$

against its complementary alternative. Notice that the random variables Y_i follow the multinomial distribution with parameters n and $p_1, p_2, \ldots, p_r$. The following theorem gives birth to the *chi-square goodness-of-fit test*.

PROPOSITION 12.1.1

Under the assumptions of the last paragraph, the random variable

$$Q = \sum_{i=1}^{r} \frac{(Y_i - np_i)^2}{np_i} \tag{12.4}$$

has a limiting $\chi^2(r-1)$ distribution as $n \to \infty$.

The proof that Q has an approximate $\chi^2(r-1)$ distribution for large n is a little technical, so we will omit it in the interest of space. (Kendall 1946, p. 290) gives a rough argument. We can at least give a good motivation without excessive effort in the case of two categories. Since the random variables Y_i have the multinomial distribution, the marginal distribution of each Y_i is binomial, $Y_i \sim b(n, p_i)$. Thus Y_i has mean np_i and standard deviation $\sqrt{np_i(1 - p_i)}$. Therefore for large n,

$$\frac{Y_i - np_i}{\sqrt{np_i(1 - p_i)}.} \approx N(0, 1). \tag{12.5}$$

In the simple case $r = 2$, we can rewrite the random variable Q as follows.

$$\begin{aligned}
Q &= \frac{(Y_1 - np_1)^2}{np_1} + \frac{(Y_2 - np_2)^2}{np_2} \\
&= \frac{(Y_1 - np_1)^2}{np_1} + \frac{(n - Y_1 - n(1 - p_1))^2}{n(1 - p_1)} \\
&= \frac{(1 - p_1)(Y_1 - np_1)^2}{np_1(1 - p_1)} + \frac{p_1(Y_1 - np_1)^2}{np_1(1 - p_1)} \\
&= \frac{(Y_1 - np_1)^2}{np_1(1 - p_1)}. \tag{12.6}
\end{aligned}$$

The last expression is the square of an approximate $N(0, 1)$ random variable; therefore it has an approximate $\chi^2(1)$ distribution. In the general case of r categories, the basic idea is to use the fact that the r random variables Y_i are each approximately normal, but there is one constraint $\sum Y_i = n$ inhibiting their independence, so that Q, a quadratic form of appropriately standardized Y_i's, has an approximate χ^2 distribution, with one degree of freedom less than the full r.

Here is how the chi-square goodness-of-fit test proceeds. If H_0 is true, then p_i in (12.4) can be replaced by p_{i0}. The null hypothesis would be rejected at level α if

$$Q = \sum_{i=1}^{r} \frac{(Y_i - np_{i0})^2}{np_{i0}} > \chi_\alpha^2(r-1), \tag{12.7}$$

where $\chi_\alpha^2(r-1)$ is the percentage point of the $\chi^2(r-1)$ distribution to the right of which lies probability α. This test criterion is very intuitive. Since np_{i0} is the expected value of Y_i under H_0, we may describe Q as

$$\sum_{\text{categories}} \frac{(\text{observed-expected})^2}{\text{expected}}. \tag{12.8}$$

The test statistic totals square differences between observed category frequencies and hypothesized expected category frequencies, divided by an expression meant to measure those square differences according to a standard yardstick. If the squared differences total to a large positive number, we tend to disbelieve the hypothesized category probabilities.

—— **?Question 12.1.1**   Factor n out of the square in the numerator of the general term in (12.7) to see that the test criterion captures the earlier intuition that the test should be based on standardized comparisons between $\hat{p}_i$ and p_{i0}.

Example 12.1.1   **Continued**   For the lottery example, we can now test $H_0 : p_{00} = 1/10, p_{10} = 1/10, \ldots, p_{90} = 1/10$. The expected frequencies are all $n \cdot p_{i0} = 363 \cdot 1/10 = 36.3$. A display of the relevant calculations follows.

Category	0	1	2	3	4
Observed	35	44	25	40	50
Expected	36.3	36.3	36.3	36.3	36.3
$(obs.-exp.)^2$	1.69	59.29	127.69	13.69	187.69
$(obs.-exp.)^2/exp.$	.047	1.63	3.52	.38	5.17

Category	5	6	7	8	9
Observed	26	33	38	37	35
Expected	36.3	36.3	36.3	36.3	36.3
$(obs.-exp.)^2$	106.09	10.89	2.89	0.49	1.69
$(obs.-exp.)^2/exp.$	2.92	.3	.080	.013	.047

The sum of the numbers in the last row is Q, which comes out to 14.1. Notice that the heaviest contributors to Q are categories 4, 2, and 5. Since the observed Q is less than the critical value $\chi_{.05}^2(9) = 16.92$, there is insufficient evidence to reject H_0. We must conclude that the digits are equally likely to be drawn, but the fit to uniform probabilities is not particularly good.

Example 12.1.2

In Example 9.1.1 when we first studied hypothesis testing we devised a way to test an hypothesis about Ross Perot's support during the 1992 presidential campaign. It is interesting that we now have a way to simultaneously test hypotheses about all of the candidates.

Recall that in a poll, 46 respondents favored Bush, 50 favored Clinton, 22 favored Perot, and 29 were undecided. Four possible categories present themselves: Bush, Clinton, Perot, and undecided. We can test an hypothesis such as

$$H_0 : p_b = .30, \qquad p_c = .25, \qquad p_p = .25, \qquad p_u = .2,$$

where the p's are probabilities that a randomly selected individual prefers the candidate indicated by the subscript.

A total of 147 people were sampled, and we can easily compute the expected frequencies under H_0. For Bush for example it would be $n p_{b0} = 147(.30) = 44.1$. The complete list is in the following table.

Candidate	Bush	Clinton	Perot	Undecided
Observed	46	50	22	29
Expected	44.1	36.75	36.75	29.4

The approximate χ^2-statistic is

$$Q = \frac{(46 - 44.1)^2}{44.1} + \frac{(50 - 36.75)^2}{36.75} + \frac{(22 - 36.75)^2}{36.75} + \frac{(29 - 29.4)^2}{29.4} = 10.78.$$

Since the critical value at level .05 is $\chi^2_{.05}(3) = 7.815$, we have sufficient evidence to reject the null hypothesis. The main reason that this happened is that the null hypothesis so badly underestimated the Clinton support and overestimated the Perot support.

We have not mentioned the conditions under which the χ^2-approximation fits the distribution of Q well. The general consensus among statisticians is that the approximation is good when the expected category frequencies $n \cdot p_{i0}$ are all at least 5, as they have been in our examples. The approximation has been known to be good for even smaller expected frequencies, however, especially if the frequencies are well balanced among categories.

The analysis of the problem of testing equality of a single set of multinomial probabilities with an hypothesized set extends to the case where there are two multinomial populations of interest whose category probabilities are to be compared. To set the stage, consider the following example.

Example 12.1.3

Parallel surveys of 506 women in 1993 and 295 women in 1994 (IEEE, *Peoria Journal Star*, Aug. 1, 1995), who were in technical and scientific professions asked whether various family-oriented benefits existed in their workplaces. One of the benefits of interest was family leave. In 1993, 63% of the women said yes, they did have this benefit, 19% said no, and 18% said that they were not sure. In 1994 the distribution of response was 83% yes, 8% no, and 9% not sure. Is there a significant difference between years in the response offered by the professional women? We will answer the question below.

We would like to construct a test of the null hypothesis

$$H_0 : p_{11} = p_{21}, \ p_{12} = p_{22}, \ p_{13} = p_{23}, \ldots,$$

where p_{1j} is the probability that a member of population 1 falls into category j and p_{2j} is the corresponding probability for population 2. Let $X_{11}, X_{12}, \ldots,$ X_{1n_1} and $X_{21}, X_{22}, \ldots, X_{2n_2}$ be independent random samples from the two populations being compared, and let Y_{1j} and Y_{2j} be the observed numbers of observations in the two samples falling into category j. In comparing two multinomial populations we hit a snag that did not occur earlier when we were testing for equality to specific hypothesized category probabilities. That is, under the null hypothesis we do not know the probability that an observation falls into category j; we only know that this probability is common to both populations. If there are r categories, then $r - 1$ such category probabilities must be estimated, and the last is complementary to their sum.

If the true category probabilities were known, then each random variable

$$Q_1 = \sum_{j=1}^{r} \frac{(Y_{1j} - n_1 p_{1j})^2}{n_1 p_{1j}} \quad \text{and} \quad Q_2 = \sum_{j=1}^{r} \frac{(Y_{2j} - n_2 p_{2j})^2}{n_2 p_{2j}}$$

would have an approximate $\chi^2(r - 1)$ distribution. They are also independent; hence their sum

$$Q_1 + Q_2 = \sum_{i=1}^{2} \sum_{j=1}^{r} \frac{(Y_{ij} - n_i p_{ij})^2}{n_i p_{ij}} \tag{12.9}$$

would have an approximate $\chi^2(2(r - 1))$ distribution. But under H_0, the $r - 1$ parameters $p_{11}(= p_{21}), p_{12}(= p_{22}), \ldots, p_{1,r-1}(= p_{2,r-1})$ must be estimated, let's say by the pooled estimates of the category probabilities:

$$\widehat{p}_j = \frac{Y_{1j} + Y_{2j}}{n_1 + n_2}. \tag{12.10}$$

Note that this statistic is just the total number of observations in category j over the total sample size, that is, the marginal proportion in category j. It can be shown that this estimation causes a loss in rank of the matrix in the quadratic form in (12.9) of $r - 1$, therefore the test statistic

$$Q = \sum_{i=1}^{2} \sum_{j=1}^{r} \frac{(Y_{ij} - n_i \widehat{p}_j)^2}{n_i \widehat{p}_j} \tag{12.11}$$

has an approximate $\chi^2(2(r - 1) - (r - 1)) = \chi^2(r - 1)$ distribution. We reject the null hypothesis of equality of category probabilities for the populations if

$$Q \geq \chi^2_\alpha(r - 1). \tag{12.12}$$

—— **?Question 12.1.2**   How does the test statistic (12.11) relate to the intuitive test criterion of comparing squared differences of category frequencies Y_{1j} and Y_{2j}?

Example 12.1.3

Continued   For the family leave survey problem, the given information is $n_1 = 506$ for 1993, $n_2 = 295$ for 1994, and

$$Y_{1,yes} = 506(.63) = 319, \quad Y_{1,no} = 506(.19) = 96, \quad Y_{1,notsure} = 506(.18) = 91,$$
$$Y_{2,yes} = 295(.83) = 245, \quad Y_{2,no} = 295(.08) = 24, \quad Y_{2,notsure} = 295(.09) = 26.$$

Thus

$$\widehat{p}_{yes} = \frac{319 + 245}{506 + 295} = .70, \quad \widehat{p}_{no} = \frac{96 + 24}{506 + 295} = .15, \quad \widehat{p}_{notsure} = \frac{91 + 26}{506 + 295} = .15$$

are the pooled estimates of the category probabilities. The observed frequencies and expected frequencies follow. For example, the expected frequency of the yes response for 1993 is $n_1 \widehat{p}_{yes} = 506(.70) = 354$.

		Response		
		Yes	*No*	*Not Sure*
1993	*obs.*	319	96	91
	exp.	354	76	76
1994	*obs.*	245	24	26
	exp.	207	44	44

The statistic Q has value

$$q = \frac{(319 - 354)^2}{354} + \frac{(96 - 76)^2}{76} + \frac{(91 - 76)^2}{76}$$
$$+ \frac{(245 - 207)^2}{207} + \frac{(24 - 44)^2}{44} + \frac{(26 - 44)^2}{44}$$
$$= 3.46 + 5.26 + 2.96 + 6.98 + 9.09 + 7.36 = 35.1.$$

This result substantially exceeds the level .01 $\chi^2(2)$ critical value of 9.21; hence we can reject the null hypothesis that the categorical distribution of responses is the same for 1993 as for 1994. Notice that the biggest contributors to the high q-value were the categories no and not sure in 1994, which had lower frequencies than they should have if the distribution had not changed. It appears that at least the perception is that family leave is becoming a more common opportunity with time.

EXERCISES 12.1

1. Companies A, B, C, and D compete for shares of the market in the cocktail weenie business. Traditionally, A has held 10%, B has held 25%, C has held 50%, and D has held the remaining 15%. Recently company A has launched an agressive ad campaign (slogan: Buy our weenies: we know where you live) in an attempt to increase its share. To test the effectiveness of the campaign, company A polls 100 customers and finds that 15 prefer their brand, 20 prefer brand B, 40 prefer brand C, and 25 prefer brand D. Draw conclusions about whether the ad campaign has succeeded.

2. The researchers mentioned in Example 7.1.3 (Romana and Jose, 1991) who were studying Filipino attitudes toward Japan also gathered data on the fields of specialization of Filipino recipients of a Japanese study scholarship (the Monbusho scholarship). The following table gives the numbers of scholars falling into each of six categories and historical data on percentages who go into these fields in U.S. scholarship programs. Are there significant differences in the field chosen by the scholars between the Monbusho and U.S. programs?

Field	Monbusho Program # Recipients	U.S. Programs Proportion
Social Sciences	89	.279
Education/Counseling	89	.168
Engineering	70	.030
Medical	64	.056
Applied Sciences	61	.031
Agriculture	28	.052

(Be careful: there are other U.S. scholarship fields that are not listed.)

3. Example 7.3.2 listed the results of a survey of 1200 adults that distributed the respondents into categories by the amount they had given to charity in the last year. For your convenience the numbers are repeated below. Test the goodness-of-fit of the data to the set of hypothesized category probabilities: $< \$25, 8\%$; $\$25-\$99, 32\%$; $\$100-\$499, 40\%$; $\$500-\$999, 10\%$; $\geq 1000, 5\%$; not sure, 5%.

< \$25	\$25–\$99	\$100–\$499	\$500–\$999	≥ \$1000	Not Sure
7%	34%	39%	11%	6%	3%

4. An important application of χ^2 testing is in assessing the fit of genetic data to Mendel's laws. In situations involving two characteristics, both of which are determined by a single gene with two alleles that can be either recessive or dominant, there can be four physical types of offspring. They occur with theoretical proportions $9:3:3:1$ under assumptions of random genetic contributions by the parents. If observed offspring proportions in a particular experiment with 500 individuals are $11:4:3:2$, test at level .01 the goodness-of-fit of the empirical to the theoretical proportions.

5. The U.S. government monitors births nationwide (U.S. Dept. of Health and Human Services, 1995). Among 4,110,907 births in 1991, the numbers that occurred on each day of the week were as follows. Ask and answer an interesting goodness-of-fit question about this information.

Sunday	Monday	Tuesday	Wednesday
466,706	601,244	651,952	626,733

Thursday	Friday	Saturday
628,656	635,814	499,802

6. Simplify the formula for the test statistic Q in the case of r categories with equal hypothesized category probabilities. Find a simplified form for the joint probability mass function of the number of observations in each category among the total of n observations.

7. In a first-year level course that I taught some time ago I administered standardized class evaluations. The forms were then tabulated, and a report was sent back to me in which the responses of my class were compared to long-term past faculty averages for all courses. The questions in the survey have to do with several kinds of topics, including the methodology of the instructor. The aggregate of six questions and 138 total responses to those questions produced the following distribution of responses (NA = not applicable, SA = strongly agree, A = agree, D = disagree, SD = strongly disagree). Since all questions are phrased so that agreement means good things, I was somewhat concerned about the matchup between my methodology ratings and the college average. Test at level .05 for significant differences between my distribution and the college average.

	NA	SA	A	D	SD
My Class	2.9%	14.5%	61.6%	17.4%	3.6%
Faculty Average	2.9%	44.1%	42.9%	8.5%	1.7%

8. In Example 7.1.2, we looked at a small study in which subjects were asked for random numbers in order to check whether they had a good idea of what randomness meant. Carefully state an appropriate goodness-of-fit hypothesis using the information in Table 7.1. Then perform the test at level .05. (To improve the approximation, you may have to combine categories strategically.)

9. When epidemiologists (Redfield and Burke, 1988) at Walter Reed Hospital developed a six-stage classification procedure for patients infected with the HIV virus, they monitored a group of patients to estimate the proportions who moved from each stage to each more severe stage. Among 96 patients who began in stage 3 of the disease (a sharp drop in T4 cell count) 67% stayed in that stage at a follow-up 14 months later, 11% went to stage 4 (impairment of immune system operation), 15% to stage 5 (threat by viral infections), and 7% to stage 6 (systemwide immune deficiency). Are these observed proportions consistent with hypothesized proportions $p_{30} = 75\%$, $p_{40} = 15\%$, $p_{50} = 8\%$, $p_{60} = 2\%$?

10. What, if any, is the effect on the χ^2-test of increasing the sample size? To be more specific, suppose that the number of observations doubles, triples, and so on, in general is multiplied by a factor of c, and the category frequencies are controlled to bear the same proportion to one another as in the smaller sample size data sets. What change occurs to the χ^2-statistic? Say something about the power of the test using larger as opposed to smaller sample size.

11. Information about a drug for hypertension called Cardizem was given in Exercise 7 of Section 8.5. Among 607 subjects who had taken it, 5.4% had headaches, 2.6% had edema (assume that these are exclusive conditions),

and the rest had neither of the two. In a control group of 301 subjects who took a placebo instead, 5.0% had headaches and 1.3% had edema. Does it appear as if the distribution of side effects is the same for Cardizem as it is for the placebo?

12. Using the data in Exercise 7 of Section 7.3, test for a difference between years of the distributions of types of recreational vehicles sold.

13. In the following table, the data are numbers of thousands of people in the United States in 1980 who lived in three kinds of environments: metropolitan, over 1 million population; metropolitan, under 1 million population, and non-metropolitan (Schick, 1986). The data are broken into two populations, those under 65 years of age and those over 65 years of age. Test for a difference between these age groups in terms of their preferred living environment.

	Metro>1 mill	Metro<1 mill	Nonmetro
Under 65	75402	67375	61581
Over 65	8440	6644	8658

14. A collegewide grade survey for the academic year 1990–91 from a number of liberal arts colleges produced the following percentages of grades for two institutions:

	A	B	C	D	F	Other
Inst. 1	29.0	51.0	16.0	3.0	.5	.5
Inst. 2	27.7	50.2	17.0	3.1	1.1	0.9

Assume that there were 4000 grades given at each institution from which these percentages came. Test at level .05 for a difference in grade distribution between the two colleges.

15. In the section we developed a test for comparing hypothesized sets of category probabilities for two populations. Develop an analogous test for comparing some number p of populations. What do you think are the degrees of freedom of the test statistic?

16. In the case of $r = 3$ categories, show that the test statistic Q of formula (12.4) is equal to the quadratic form

$$[Y_1 - np_1 \ Y_2 - np_2]R \cdot \begin{bmatrix} Y_1 - np_1, \\ Y_2 - np_2 \end{bmatrix},$$

where

$$R = \frac{1}{n(1 - p_1 - p_2)} \begin{bmatrix} \frac{1-p_2}{p_1} & 1 \\ 1 & \frac{1-p_1}{p_2} \end{bmatrix}.$$

Explain why this is convincing evidence that Proposition 12.1.1 is true in the case $r = 3$.

12.2 | Goodness-of-Fit to Distributions

12.2.1 Chi-Square Tests

The techniques that we have developed for testing the fit of observed proportions to hypothesized multinomial probabilities can be adapted to the more general problem of testing the goodness-of-fit of a random sample to an hypothesized distribution, that is, testing $H_0 : F = F_0$. If an hypothesized distribution F_0 for a random variable X is true, then the probabilities

$$p_{i0} = P[X \in C_i] \tag{12.13}$$

can be computed, where C_i is a category among several mutually exclusive and exhaustive catgories within the state space of X. One can construct the categories, classify a random sample $X_1, X_2, \ldots, X_n$ into them, and then use a chi-square test to decide whether the observed frequencies are consistent with the category probabilities. If the category probabilities are rejected, then the hypothesized distribution must be false (subject to type I statistical error, of course). However, even if the category probabilities are accepted as correct, these probabilities do not determine the distribution completely, because other distributions could have the same category probabilities. So we can only say that if we accept the category probabilities, we do not have evidence against the null hypothesis.

Consider the following question before reading on.

—— **?Question 12.2.1**   The strategy of dividing the state space into categories is somewhat arbitrary. Can you think of a potential disadvantage in choosing a large number of categories?

In response to Question 12.2.1, if there are too many categories, the theoretical probability that an observation falls into one may become so small that the expected frequency np_{i0} may be smaller than it should be in order to have confidence in the χ^2 approximation. On the other hand, too few categories may well prevent you from detecting departures in the data from the hypothesized distribution. Think, for example, of a normal distribution and an exponential distribution with equal medians m and the extreme case of just two categories $(-\infty, m)$ and $[m, \infty)$. If the hypothesis is that the sample comes from the normal distribution but in actuality it comes from the exponential distribution, it still will be true that about half of the data will fall into each category. So, H_0 will probably be accepted erroneously. It is certainly reasonable to suppose based on this argument that the power of the test increases as the number of categories increases.

The overall guidelines are to select as many categories as possible given the size n of the sample so that the expected category frequencies are relatively large (at least 5 is best) and well balanced. This improves the accuracy of the χ^2-approximation.

One last comment before we proceed to examples. Frequently it is not a particular distribution to which you try to fit the data, but a family of distributions like gamma, Poisson, or normal, that are characterized by unknown parameters. Although it is beyond the scope of this text to prove, for reasons similar to the ones we saw in the last section, *for each parameter that you must estimate (by maximum likelihood) using the sample, a degree of freedom is lost from the χ^2-statistic.* If there are p parameters ard r categories, then the rejection region for the null hypothesis $H_0 : F = F_0$ is

$$Q = \sum_{i=1}^{r} \frac{(Y_i - n\widehat{p}_{i0})^2}{n\widehat{p}_{i0}} > \chi_{\alpha}^2(r - 1 - p), \qquad (12.14)$$

where Y_i as usual is the observed frequency in category i and $\widehat{p}_{i0}$ is the estimated value of the category i probability p_{i0} in (12.13).

?Question 12.2.2   What would be the degrees of freedom for a goodness-of-fit test to the gamma distribution if there are 5 categories? to the exponential distribution and 6 categories?

Example 12.2.1   Sometime ago my wife, Lynn, and I paneled our family room with car siding, which is tongue-in-groove pine. I noticed at the time a number of knotholes, distributed randomly on the boards. A common model for the number of occurrences of a phenomenon in a planar or spatial region is the Poisson distribution. Sadly, I painted over the data set. But a representative, fictitious sample follows. Each datum is the number of knotholes on a particular board. Does the distribution of the number of knotholes seem to be Poisson?

$$0,\ 2,\ 3,\ 1,\ 3,\ 4,\ 2,\ 1,\ 2,\ 2,\ 1,\ 0,\ 3,\ 3,\ 3,\ 2,$$
$$4,\ 1,\ 2,\ 4,\ 2,\ 2,\ 1,\ 5,\ 1,\ 3,\ 2,\ 0,\ 3,\ 2,\ 2,\ 1$$

There are $n = 32$ items in this sample. Recall that the maximum likelihood estimator of the parameter μ of the Poisson distribution is $\widehat{\mu} = \overline{X}$. For this sample, its value comes out to be $\widehat{\mu} = 2.1$. To decide on categories, note that we would like to have category probabilities p_j such that the expected frequencies $np_j = 32 \cdot p_j$ are at least 5. Therefore our category probabilities should be at least $5/32$, which is around .15. Using the estimate of μ, we can compute the Poisson probabilities for $k = 0, 1, 2, 3, 4$, as

$$\widehat{p}_{00} = P[X = 0; H_0 \text{ true}] = \frac{e^{-2.1}(2.1)^0}{0!} = .122$$

and similarly

$$\widehat{p}_{10} = .257,\ \widehat{p}_{20} = .270,\ \widehat{p}_{30} = .189,\ \widehat{p}_{40} = .099.$$

Also

$$P[X \geq 5] = 1 - P[X \leq 4] = 1 - (.122 + .257 + .270 + .189 + .099) = .063.$$

Category 0 does not quite have our desired category probability, but it is close enough that we can tolerate that. Categories 1, 2, and 3 satisfy the requirement. If we set our last category as 4 or more, the category probability of the new combined category is .099 + .063 = .162. So, the category frequencies of each of the values 0–3 are tallied next, together with the last category defined as all boards with four or more knotholes. Multiply the preceding estimated category probabilities by $n = 32$, to calculate the estimated expected frequencies in the third line of the table.

	0	**1**	**2**	**3**	**4 or more**
Observed	3	7	11	7	4
Expected	3.9	8.2	8.6	6.0	5.2

Since one parameter, μ, was estimated and we are using five categories, the approximate χ^2-statistic Q has $5 - 1 - 1 = 3$ degrees of freedom. The critical value for a level .05 test is therefore $\chi^2_{.05}(3) = 7.81$. You can calculate Q in the usual way from the frequencies in the table. Its value turns out to be about 1.5, and so we accept H_0 based on this data. We have little reason not to believe that the Poisson (2.1) distribution is a good model for the number of knots on a board. The probability histogram for Poisson (2.1) and the observed relative frequencies from the data set are displayed together in Fig. 12.1.

The chi-square statistic is a good analytical tool for checking normality assumptions, as the next example shows.

Example 12.2.2

In Sections 10.1 and 10.2 I mentioned that I am responsible for administering and analyzing the results of the placement exam in mathematics that we give to entering freshmen. Following are 75 scores selected from the class entering in the fall of 1992. Is it reasonable to suppose that placement scores are

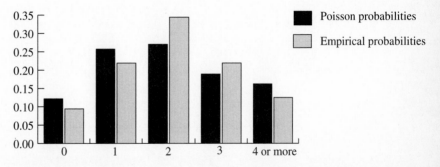

Figure 12.1 Histograms of Hypothesized Poisson and Empirical Probabilities

approximately normally distributed?

$$16, \ 13, \ 9, \ 13, \ 19, \ 15, \ 22, \ 22, \ 14, \ 10, \ 18, \ 13, \ 6, \ 14, \ 28,$$
$$15, \ 16, \ 16, \ 19, \ 15, \ 13, \ 11, \ 11, \ 22, \ 29, \ 15, \ 22, \ 19, \ 22, \ 17,$$
$$15, \ 9, \ 24, \ 15, \ 20, \ 20, \ 18, \ 23, \ 13, \ 26, \ 16, \ 26, \ 15, \ 10, \ 29,$$
$$7, \ 18, \ 12, \ 12, \ 22, \ 14, \ 10, \ 5, \ 12, \ 17, \ 16, \ 23, \ 9, \ 14, \ 20,$$
$$16, \ 18, \ 22, \ 22, \ 19, \ 11, \ 20, \ 20, \ 19, \ 15, \ 17, \ 7, \ 19, \ 13, \ 15$$

A normal scores plot of the sample is in Fig. 12.2. It appears reasonably straight, which is a good indication of normality. The sample mean and standard deviation are $\bar{x} = 16.5$, and $s = 5.3$. So, we will try a goodness-of-fit test to the $N(16.5, (5.3)^2)$ distribution, and we will make note when we select the critical value that two parameters, μ and σ^2 have been estimated.

Since the sample size is rather large $(n = 75)$ we can use a large number of categories, such as ten. In that case, if the category probabilities are chosen to be perfectly balanced at .1 each, the expected frequency in each category is exactly $(.1)(75) = 7.5$, which is safely more than 5.

How can we find ten equally likely categories? Using the standard normal table, we find that the 10th, 20th, 30th and so on, percentiles of the $N(0, 1)$ distribution are roughly

$$- 1.28, \ -.84, \ -.52, \ -.25, \ 0, \ .25, \ .52, \ .84, \ 1.28. \tag{12.15}$$

Since the percentiles for $N(\mu, \sigma^2)$ are σ times the $N(0, 1)$ percentiles plus μ, we obtain the following for the 10th, 20th, 30th and so on, percentiles of the $N(16.5, (5.3)^2)$ distribution:

$$9.7, \ 12.0, \ 13.7, \ 15.2, \ 16.5, \ 17.8, \ 19.3, \ 20.9, \ 23.3. \tag{12.16}$$

(See Question 12.2.3, below.)

We can count the frequencies in each category using the preceding data set.

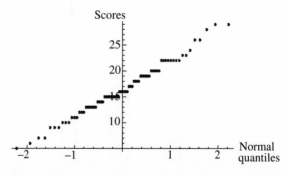

Figure 12.2 Normal Scores Plot of Placement Data

Category	$(-\infty, 9.7]$	$(9.7, 12.0]$	$(12.0, 13.7]$	$(13.7, 15.2]$	$(15.2, 16.5]$
Observed	7	9	6	13	6
Expected	7.5	7.5	7.5	7.5	7.5

Category	$(16.5, 17.8]$	$(17.8, 19.3]$	$(19.3, 20.9]$	$(20.9, 23.3]$	$(23.3, \infty)$
Observed	3	10	5	10	6
Expected	7.5	7.5	7.5	7.5	7.5

The value of the χ^2–statistic is $(7 - 7.5)^2/7.5 + \cdots + (6 - 7.5)^2/7.5 = 10.5$. There are $10 - 1 - 2 = 7$ degrees of freedom, and the χ^2 critical value for a .1 level test is 12.02; therefore we accept the null hypothesis of normality at that level.

?Question 12.2.3   Why is it true that $\pi_x = \sigma\pi_z + \mu$, where π_x and π_z are percentiles for $N(\mu, \sigma^2)$ and $N(0, 1)$, respectively?

It is also possible to use the techniques developed in Section 12.1 to test for equality of two distributions by categorizing the data. We illustrate in the next example.

Example 12.2.3   The following lists of numbers represent the production of wool (number of pounds per fleece) in 24 states for the years 1989 and 1990. (*Agricultural Statistics*, 1991)

> 1989: 8.4, 6.8, 7.7, 7.7, 7.5, 10.0, 6.9, 6.8, 6.9,
> 6.7, 6.6, 6.7, 7.2, 7.6, 7.2, 7.4, 7.0, 7.1, 10.0,
> 7.4, 9.5, 7.4, 6.6, 9.8

> 1990: 8.8, 6.8, 7.5, 7.4, 7.3, 10.2, 6.9, 6.8, 6.6,
> 7.1, 6.5, 6.3, 7.6, 6.3, 7.3, 7.2, 7.1, 7.3, 9.9,
> 7.2, 9.7, 7.3, 6.4, 9.8

Suppose that we are interested in testing whether the two data sets come from the same continuous distribution, that is, $H_0 : F_{1989}(x) = F_{1990}(x)$ for all x, where the F's are the two population c.d.f.'s. We can divide the real line into categories, say, $[0, 7)$, $[7, 7.5)$, $[7.5, 8)$, and $[8, \infty)$ and perform a multinomial test of equal category probabilities as described in Section 12.1. If the multinomial test leads to rejection, then we must also reject the equality of the distributions, although as before it is not so clear what acceptance of the revised null hypothesis of equal category probabilities really means.

The next table displays the observed frequencies for both samples in all categories together with estimated expected frequencies, which are the same for both samples because of the equal sample sizes of 24.

		[0, 7)	**[7, 7.5)**	**[7.5, 8)**	**[8, ∞)**
1989	*Observed*	8	7	4	5
	Expected	8	8	3	5

		[0, 7)	**[7, 7.5)**	**[7.5, 8)**	**[8, ∞)**
1990	*Observed*	8	9	2	5
	Expected	8	8	3	5

The chi-square statistic has the value $q = 2(0^2/8 + 1^2/8 + 1^2/3 + 0^2/5) = .92$, which is safely less than the critical value $\chi_\alpha^2(3)$ for any reasonable level α. We have no evidence here against the equality of the two wool production distributions.

12.2.2   Kolmogorov–Smirnov Test

A nonparametric test for goodness-of-fit to an hypothesized distribution is available and widely used. The beautiful thing about this test is that it makes use of the full data set without losing information in the categorization process.

Recall that the empirical distribution function $\widehat{F}_n$ of a sample $X_1, X_2, \ldots, X_n$ is a step function that begins at 0 and jumps by exactly $\frac{1}{n}$ at each of the order statistics of the sample. The value of $\widehat{F}_n(x)$ is $\frac{1}{n}(\#X_i\text{'s} \le x)$. The e.d.f. serves as an estimator of the cumulative distribution function F of the distribution of X, and it can also be used to judge how well an hypothesized c.d.f. F_0 fits the data. The farther is the e.d.f. from F_0, the more doubt is cast on the hypothesized distribution. This is the underlying motivation for the *Kolmogorov–Smirnov test*, which we now describe.

— **?Question 12.2.4**   Use what you know about binomial experiments to show that $E[\widehat{F}_n(x)] = F(x)$ and $\text{Var}(\widehat{F}_n(x)) = F(x)(1 - F(x))/n$. Comment on what these results say about $\widehat{F}_n(x)$ as an estimator of $F(x)$.

Consider a test of the null hypothesis,

$$H_0 : F = F_0 \text{ vs. } H_a : F \ne F_0, \tag{12.17}$$

where F is the c.d.f. of a continuous random variable. The *Kolmogorov–Smirnov statistic* is defined by

$$\tilde{D} = \sup_x \left| \widehat{F}_n(x) - F_0(x) \right|. \tag{12.18}$$

The associated level α test rejects H_0 if $\tilde{D} \ge d_\alpha$, where d_α is chosen so that $P[\tilde{D} \ge d_\alpha; H_0] = \alpha$.

To actually find the value of $\tilde{D}$ for a given sample, one must find where the vertical distance between the graph of F_0 and the graph of $\widehat{F}_n$ is greatest.

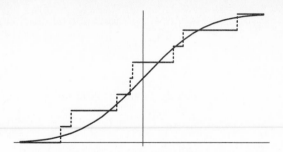

Figure 12.3   Empirical and Hypothesized c.d.f.'s

(See Fig. 12.3.) You are asked to argue in Exercise 17 that the maximum must occur at one of the observed data points $X_1, X_2, \ldots, X_n$ at which the e.d.f. jumps. Consequently, it suffices to find the largest value of all of the differences $|\widehat{F}_n(X_i) - F_0(X_i)|$ and $|\widehat{F}_n(X_i^-) - F_0(X_i)|$ by direct computation. [Here the notation $f(x^-)$ stands for the left-hand limit of f at the point x.]

It remains to show how to find critical values d_α. The next theorem does this and establishes moreover that the distribution of $\widetilde{D}$ under H_0 does not depend on the form of $F_0(x)$. This latter fact is what puts the Kolmogorov–Smirnov test in the class of distribution-free procedures.

PROPOSITION 12.2.1

The distribution of $\widetilde{D}$ in (12.18) under H_0 is the same as the distribution of $D = \sup_{u \in (0,1)} |\widehat{G}_n(u) - u|$, where $\widehat{G}_n$ is the empirical distribution function of a random sample $U_1, U_2, \ldots, U_n$ from the uniform $(0, 1)$ distribution.

■ **Proof**   Define the *indicator random variable* of an event A by $1_A(\omega) = 1$ if $\omega \in A$ and 0 otherwise. Since the empirical distribution function evaluated at a point x is the proportion of sample values X_i that are less than or equal to x, $\widehat{F}_n(x)$ can be written as

$$\widehat{F}_n(x) = \frac{1}{n} \sum_{i=1}^{n} 1_{\{X_i \le x\}}. \tag{12.19}$$

Denote $u = F_0(x)$ and introduce random variables $U_i = F_0(X_i)$, $i = 1, 2, \ldots, n$. Since F_0 is continuous, the U_i's form a random sample from the uniform $(0, 1)$ distribution under H_0 by Proposition 5.1.1. Since F_0 is nondecreasing,

$$\{X_i \le x\} \Leftrightarrow \{F_0(X_i) \le F_0(x)\} \Leftrightarrow \{U_i \le u\}.$$

Then $\widehat{F}_u(x)$ can be reexpressed as

$$\widehat{F}_n(x) = \frac{1}{n} \sum_{i=1}^{n} 1_{\{U_i \le u\}} = \widehat{G}_n(u),$$

where $\widehat{G}_n$ is the empirical distribution function of a uniform random sample of size n. Thus, since $u = F_0(x)$ achieves all values in $(0,1)$,

$$\tilde{D} = \sup_x \left| \widetilde{F}_n(x) - F_0(x) \right|$$

has the same distribution as

$$D = \sup_{u \epsilon (0,1)} \left| \widehat{G}_n(u) - u \right|,$$

which completes the proof.

Example 12.2.4

The distribution of D can be derived, although it is a messy job in all but small cases. Let us look at how it might be done for $n = 2$.

Let U_1, U_2 be a random sample from the uniform $(0,1)$ distribution. If $Y_1 \leq Y_2$ are the corresponding order statistics, then by the comment prior to the proposition, we would need to find the distribution of

$$D = \max \left\{ \left| \widehat{G}_n(Y_1) - Y_1 \right|, \left| \widehat{G}_n(Y_1^-) - Y_1 \right|, \left| \widehat{G}_n(Y_2) - Y_2 \right|, \left| \widehat{G}_n(Y_2^-) - Y_2 \right| \right\}.$$

But the function $\widehat{G}_n$ is very simple. It begins at 0, jumps to $1/2$ at Y_1, jumps to 1 at Y_2, and stays at 1 thereafter. (See Fig. 12.4.)

Therefore the value of D will be the largest of the four dashed vertical segments shown in the figure, that is

$$D = \max\{|1/2 - Y_1|, |Y_1|, |1 - Y_2|, |Y_2 - 1/2|\}. \tag{12.20}$$

We can now apply the c.d.f. technique to find the distribution of D:

$$\begin{aligned} G_D(y) &= P[D \leq y] \\ &= P[|1/2 - Y_1| \leq y, |Y_1| \leq y, |1 - Y_2| \leq y, |Y_2 - 1/2| \leq y]. \end{aligned} \tag{12.21}$$

The joint density of Y_1 and Y_2, by the results in Section 5.2, is

$$f(y_1, y_2) = \begin{cases} 2 & \text{if } 0 \leq y_1 \leq y_2 \leq 1, \\ 0 & \text{otherwise.} \end{cases} \tag{12.22}$$

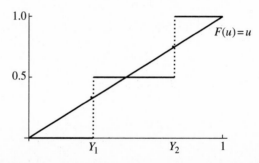

Figure 12.4   Kolmogorov–Smirnov Statistic When $n = 2$

You are asked in Exercise 18 to finish the computation of the probability in (12.21) by integrating this density over the region of (y_1, y_2) values specified by the event.

Critical values d_α such that $\alpha = P[D > d_\alpha]$ are widely available. An abridged list (obtained from Miller, Freund, and Johnson, 1990) is shown in Table 12.3. Notice that as the sample size grows, the critical value goes down, so that smaller distances between observed and theoretical distributions are required in order to reject the null hypothesis. This coincides with intuition.

Example 12.2.5

The following data are times in minutes between arrivals to a pretzel vendor's cart on Broad Street in Philadelphia during a randomly selected lunch hour. Check to see whether it is credible that customer arrivals form a Poisson process with rate $\lambda = 1$ by using the Kolmogorov–Smirnov goodness-of-fit test. Note that this means that the hypothesized interarrival distribution is $\exp(1)$.

2.82, 0.03, 1.52, 2.06, 0.15, 0.74, 0.90, 0.16, 2.05, 1.39
0.33, 0.44, 1.09, 3.72, 0.41, 0.65, 0.21, 0.58, 0.13, 0.71

?Question 12.2.5   Plot a histogram of the interarrival data. Does it make you believe the null hypothesis that the $\exp(1)$ distribution is correct?

Example 12.2.5

Continued   The c.d.f. of the $\exp(1)$ distribution is $F_0(x) = 1 - e^{-x}$, $x > 0$. Using Table 12.3, we will reject $H_0 : F = F_0$ at level .05 if the value of the test statistic $\widetilde{D}$ exceeds .294, since the sample size is $n = 20$.

Table 12.3

Kolmogorov–Smirnov Critical Values

Sample Size	$d_{.10}$	$d_{.05}$	Sample Size	$d_{.10}$	$d_{.05}$
5	.510	.565	13	.325	.361
6	.470	.521	14	.314	.349
7	.438	.486	15	.304	.338
8	.411	.457	16	.295	.328
9	.388	.432	17	.286	.318
10	.368	.410	18	.278	.309
11	.352	.391	19	.272	.301
12	.338	.375	20	.264	.294

To compute $\tilde{D}$, we must find the largest absolute difference between the e.d.f. and $F_0(x)$. To do this, arrange the data in order. The e.d.f. $\hat{F}_n(x)$ changes value at each order statistic; at the kth order statistic $X_{(k)}$ it jumps to the value k/n. It suffices to find the largest among the differences $|\hat{F}_n(X_{(k)}^-) - F_0(X_{(k)})|$ and $|\hat{F}_n(X_{(k)}) - F_0(X_{(k)})|$, $k = 1, 2, \ldots, 20$. Note that $\hat{F}_n(X_{(k)}^-) = \hat{F}_n(X_{(k-1)})$. Table 12.4 can assist us. The ordered data values are in the second column (the parentheses in the subscripts were dropped for legibility), followed by the hypothesized c.d.f. evaluated at those points. The e.d.f. is in the fourth column; note that it jumps by $1/20 = .05$ at each data value.

Using a spreadsheet it was easy to fill in the table by taking absolute differences between the third and fourth columns and between the third column and the fourth column lagged by one cell to treat the left-hand limits. The largest absolute difference occurs for $k = 12$ in the sixth column; hence the value of $\tilde{D}$ is .077 and we accept the null hypothesis that the distribution of interarrival times is $\exp(1)$.

Table 12.4

Computation of Kolmogorov–Smirnov Statistic

k	X_k	$F_0(X_k)$	$\hat{F}_n(X_k)$	$|\hat{F}_n(X_k^-) - F_0(X_k)|$	$|\hat{F}_n(X_k) - F_0(X_k)|$
1	0.03	.0296	.05	.0296	.0204
2	0.13	.1219	.10	.0719	.0219
3	0.15	.1393	.15	.0393	.0107
4	0.16	.1479	.20	.0021	.0521
5	0.21	.1894	.25	.0106	.0606
6	0.33	.2811	.30	.0311	.0189
7	0.41	.3364	.35	.0364	.0136
8	0.44	.3600	.40	.0060	.0440
9	0.58	.4401	.45	.0401	.0099
10	0.65	.4780	.50	.0280	.0220
11	0.71	.5084	.55	.0084	.0416
12	0.74	.5229	.60	.0271	.0771
13	0.90	.5934	.65	.0066	.0566
14	1.09	.6638	.70	.0138	.0362
15	1.39	.7509	.75	.0509	.0009
16	1.52	.7813	.80	.0313	.0187
17	2.05	.8713	.85	.0713	.0213
18	2.06	.8725	.90	.0225	.0275
19	2.82	.9404	.95	.0404	.0096
20	3.72	.9758	1.0	.0258	.0242

EXERCISES 12.2

1. Suppose that in a survey of 200 households taken in a town, 57 of them contained just one person, 69 contained two people, 31 contained three people, 27 contained four people, and 16 contained five or more people. Is it feasible that the distribution of the random variable X = household size -1 could be Poisson with parameter 1.25? (These data are consistent with information that was recently published in my county about the distribution of household size. I have just reduced the sample size.)

2. In Section 3.3 we saw a graphical comparison between an exponential density and a histogram of the number of words required to reach the next new word in Robert Frost's "The Road Not Taken." The authors of the study that generated this example (Badalamente et al., 1994) did a word count on Byron's "She Walks in Beauty." This poem has 120 words, each of which was given a unique number. The sequence of words that appear are listed below; for example the first repeated word is after word number 13, and that word was the same word as in position number 11. Test whether the distribution of delays until new words fits an exponential distribution using a χ^2-test. Use three categories: 0, 1, and 2 or more. Estimate the parameter of the distribution and adjust the degrees of freedom of the test accordingly.

1	2	3	4	5	6	7	8	9	10	11	12	13	11	14
15	16	8	17	11	18	19	3	20	21	11	20	22	23	24
25	26	27	28	29	30	25	31	32	33	34	35	6	36	34
37	6	38	39	40	41	6	42	43	29	44	3	45	46	47
48	49	50	51	20	52	53	54	55	56	57	58	59	58	60
61	62	63	11	64	26	65	11	51	26	66	67	68	67	69
70	71	6	72	26	73	6	74	26	75	76	77	8	32	3
78	79	80	81	82	83	84	14	85	80	86	87	88	89	90

3. A much cited data set (from Rinaman, 1993, p. 534) concerns the number of fatalities per year due to horse kicks in the Prussian Army during a 200 year time span. Are the following data consistent with a Poisson distribution with parameter .6? Plot an empirical histogram of relative frequencies with a histogram of the proposed Poisson mass function.

No. Fatalities per Year	0	1	2	3	4	5
Frequency	109	65	22	3	1	0

4. The following data are yearly divorce rates from countries in Europe (United Nations, 1994). Test whether the data could have come from a normal distribution using a χ^2-test with four categories.

0.8, 2.1, 3.4, 1.9, 1.3, 2.7, 2.9, 2.5, 2.4, 3.0, 3.7,
1.1, 2.9, 1.9, 1.9, 3.0, 3.1, 0.9, 2.4, 2.2, 2.8, 0.5,
4.2, 1.0, 4.1, 1.1, 2.2, 1.9, 2.4, 0.9, 1.0, 2.9, 1.4,
3.9, 1.0, 0.6, 2.2, 2.0, 3.7, 2.9, 0.8

5. Combine the data on ozone levels in Pennsylvania forests from Exercises 4 and 14 of Section 5.5 into a single data set of 42 observations. Test the

fit of the data to a gamma distribution with α parameter equal to 20 at level .05 using four categories: $[0, 40)$, $[40, 50)$, $[50, 60)$, $[60, \infty)$.

6. Suppose that a law of genetics predicts that genotypes AA, Aa, and aa should occur in a population in proportions $p_1 = r^2$, $p_2 = 2r(1 - r)$ and $p_3 = (1 - r)^2$, respectively, where r is an unknown parameter. If a sample of individuals is taken, 50 of whom are type AA, 40 are type Aa, and 30 are type aa, what is the value of a maximum likelihood estimator for r? Does a χ^2-test support the fit of the data to the prediciton?

7. A simple assumption about a baseball batter's performance is that his successive at bats are independent, and he has a constant probability of getting a hit on each at bat. Suppose that the following data are the numbers of base hits by a particular batter in 25 randomly selected games in which the batter had 4 official at-bats. Do the data support the assumption? What distribution should you hypothesize?

> 0, 1, 2, 2, 1, 0, 0, 0, 1, 1, 1, 0, 2, 1, 2, 2, 3, 1, 0, 1,
> 0, 4, 1, 1, 1

8. In the experiment mentioned several times before in the Department of Psychology at the University of Iowa in which I had a small part, data were gathered on the behavior of twins. Given is a subsample of the data, which are twenty survey scores from mothers of twin boys regarding the level of aggressiveness of one of the twins. Is an assumption of normality of aggression scores reasonable? Use a χ^2-test with four categories.

> 61, 57, 46, 39, 46, 50, 64, 46, 61, 43, 57, 46, 46, 50, 39, 64,
> 43, 64, 46, 71

9. To check the performance of the χ^2-test in a case where the true hypothesis is known, simulate 50 observations from the geometric distribution $f(k) = (1 - p)p^k, k = 0, 1, 2, \ldots$, with parameter $p = 1/2$. Divide them into three categories $k = 0$, $k = 1$, and $k = 2$ or more, and then apply the goodness-of-fit test to the geometric distribution with hypothesized parameters $\frac{1}{4}$, $\frac{1}{2}$, and $\frac{3}{4}$. Does your test result in the correct decision in all three cases?

10. Simulate 40 data points from the continuous distribution whose density is $f(x) = 2x$ if $x \in (0, 1)$, and $f(x) = 0$ otherwise. Then use a χ^2-test to check the goodness-of-fit of your data to the true distribution.

11. Test for significant differences in the egg-laying distributions for the years 1989–1990 given in Exercise 8 of Section 8.4 and Example 9.4.3. (Remember in the latter example, the information given was the list of differences between the 1990 egg laying rates and the 1989 rates, from which you can get the raw data on 1989 rates.)

12. Example 11.2.1 introduced data on temperature extremes in various regions of the country. Following is a larger data set that includes the previous one; recall that the numbers recorded here are the total numbers of days of temperatures either above 90° or below 32° in several small cities. Test for a difference by region of the distribution of temperature extremes.

Illinois, Indiana, Iowa: 152, 154, 164, 157, 156, 149, 155, 149, 168, 153, 160, 156, 160, 147, 157, 181, 162, 160, 176, 171, 177, 155

Texas, Arkansas, Mississippi: 144, 117, 143, 148, 133, 143, 149, 123, 144, 140, 140, 134, 135, 149, 149, 150, 133, 128, 139, 134, 132, 146, 106

13. The following data are response times by an emergency medical team to 911 calls. Use the Kolmogorov–Smirnov test to check whether the exponential distribution with mean 5 minutes would provide a reasonable model.

$$11.0, \ 2.4, \ 3.7, \ 3.6, \ 0.5, \ 0.3, \ 2.9, \ 10.4, \ 0.6, \ 0.6,$$
$$2.1, \ 2.8, \ 10.0, \ 1.3, \ 3.6, \ 0.3, \ 0.1, \ 7.9$$

14. There is a theorem which says that conditioned on the number of Poissonian arrivals in a given time interval, the time instants of arrivals have the distribution of the order statistics from a uniform random sample. If the following 15 arrival instants were observed during a time interval [0.10], does the Kolmogorov–Smirnov test support the hypothesis that arrivals form a Poisson process?

$$0.2, \ 0.5, \ 1.4, \ 1.5, \ 2.6, \ 3.1, \ 4.9, \ 5.2, \ 5.4, \ 7.0, \ 7.3, \ 8.0, \ 8.5, \ 9.1, \ 9.8$$

15. Use the Kolmogorov–Smirnov test to check for normality of the following data set. Use hypothesized mean 0 and variance 16.

$$-3.0, \ -2.8, \ -1.4, \ 0, \ .8, \ 2.2, \ 4.6, \ 5.1, \ 8.7, \ 9.2$$

16. Following is a list of 20 weekly rates of return for Wal-Mart stock obtained by my student Chris Najim. Use a Kolmogorov–Smirnov test to check whether the data fit a normal distribution with mean 1 and standard deviation 3.

$$2.79, \ -0.74, \ -1.99, \ 0.76, \ -3.78, \ -2.88, \ 1.35, \ 4.52, \ -5.85, \ -0.54,$$
$$6.52, \ -3.32, \ -0.53, \ 3.71, \ 6.14, \ 1.45, \ 1.43, \ 5.85, \ 2.21, \ -1.95$$

17. Argue that the supremum in $\tilde{D}$ in (12.18) must occur among the set of values $|\widehat{F}_n(X_i^-) - F_0(X_i)|, \quad i = 1, 2, \ldots, n$ or $|\widehat{F}_n(X_i) - F_0(X_i)|, \quad i = 1, 2, \ldots, n.$

18. Find the c.d.f. and density of D in the case $n = 2$ begun in Example 12.2.4. (*Hint:* In the computation of $P[D \le y]$ consider the cases $y \in (\frac{i-1}{4}, \frac{i}{4})$ separately for $i = 1, 2, 3, 4$.)

19. Use the Kolmogorov–Smirnov test statistic to construct a level $1 - \alpha$ confidence band for the true c.d.f. $F(x)$, by which we mean a lower function $L(x)$ and an upper function $U(x)$ such that

$$P[L(x) \le F(x) \le U(x), \forall x] = 1 - \alpha.$$

12.3 | Tests for Independence and Randomness

We have seen tests that use a sample of data to check one kind of model assumption, that is, the hypothesized distribution. Other kinds of assumptions are also made in statistical modeling, for example, that one random phenomenon is independent of another or that a sampling process or error distribution is random. This section concentrates on procedures for testing such conditions.

12.3.1 Chi-Square Tests

Consider a population, each of whose members can be classified according to the particular level of each of two factors that applies to them, among finitely many possible levels. Examples are many, including:

1. Social values of parent and child
2. Opinion about a political issue and political party affiliation
3. Job type and gender

The question of interest is the independence of the two factors, by which we mean the independence of each level of one factor of each level of the other. Before examining the precise probabilistic model, let us look at an example that we will analyze in detail later.

Example 12.3.1

The professional societies of mathematics gather employment information on new doctorates in the mathematical sciences. A recent report (A.M.S., 1994) contained the following information on the type of hiring institution and gender of individuals hired.

	I Math	II Math	III Math	IV Statistics	V Applied Math
Doctoral Hires, Male	140	43	119	43	23
Doctoral Hires, Female	23	22	33	19	7

Types I, II, and III are groups of mathematics departments ranked according to criteria based on research, with I the highest ranked group and III the lowest. Types IV and V are departments of statistics and applied mathematics, respectively. The counts are numbers of new doctoral hires in each category; for example, there were 19 new Ph.D.s who were female and were hired by statistics departments. You might ask, Is the type of employer independent of the gender of the person hired? We will try to answer shortly.

Here is a way of posing the independence question formally. Let there be two factors A and B, and suppose that factor A has h mutually exclusive and exhaustive levels or *categories* and factor B has k such categories. A random sample $X_1, X_2, \ldots, X_n$ is available such that each X_i is in exactly one of the categories of A and exactly one of the categories of B. The null hypothesis of independence of the factors is

$$H_0: \{X \in A_i\} \quad \text{and} \quad \{X \in B_j\} \quad \text{are independent events for all} \quad i = 1, 2, \ldots, h; \quad j = 1, 2, \ldots, k.$$

Denote

$$p_{ij} = P[X \in A_i \cap X \in B_j], \tag{12.23}$$
$$p_{i\cdot} = P[X \in A_i], \tag{12.24}$$
$$p_{\cdot j} = P[X \in B_j]. \tag{12.25}$$

Then the null hypothesis becomes

$$H_0: p_{ij} = p_{i\cdot} \cdot p_{\cdot j} \quad \text{for all } i = 1, 2, \ldots, h; \qquad j = 1, 2, \ldots, k. \tag{12.26}$$

?Question 12.3.1 Can you devise a way in which the null hypothesis (12.26) can be applied to the special case of the independence of two discrete random variables? We will show the method shortly.

In words, we would like to know whether the event that an observation falls into the ith category of one factor is independent of the event that it falls into the jth category of the other factor for all pairs of categories i and j. The data available to make this evaluation are the collection of frequencies Y_{ij}, where Y_{ij} is the number of observations among $X_1, X_2, \ldots, X_n$ that fall into A_i and B_j simultaneously. For convenience, denote

$$Y_{i\cdot} = \sum_{j=1}^{k} Y_{ij}; \qquad Y_{\cdot j} = \sum_{i=1}^{h} Y_{ij}, \tag{12.27}$$

which are respectively the ith row and jth column frequencies (see Fig. 12.5).

You may have realized that we already have the tools necessary to test this hypothesis. View the level combination (A_i, B_j) as one of $h \cdot k$ possible exclusive and exhaustive categories in which an observation can fall. Then

Factor A \ Factor B	1	2	. . .	k	
1	Y_{11}	Y_{12}		Y_{1k}	$Y_{1\cdot}$
2	Y_{21}	Y_{22}		Y_{2k}	$Y_{2\cdot}$
. . .					
h	Y_{h1}	Y_{h2}		Y_{hk}	$Y_{h\cdot}$
	$Y_{\cdot 1}$	$Y_{\cdot 2}$		$Y_{\cdot k}$	

Figure 12.5 Data Layout for Test of Independence of Two Factors

the frequencies Y_{ij} together form a random vector with the multinomial distribution with sample size n and category probabilities $p_{ij}, i = 1, 2, \ldots, h; \; j = 1, 2, \ldots, k$. We can simply use a χ^2-test statistic to check the goodness-of-fit of the observed frequencies to the expected frequencies under H_0.

If H_0 is true, then the only free parameters that must be estimated are the $h - 1$ probabilities $p_{1\cdot}, p_{2\cdot}, \ldots, p_{h-1\cdot}$ and the $k - 1$ probabilities $p_{\cdot 1}, p_{\cdot 2}, \ldots, p_{\cdot k-1}$. The degrees of freedom $hk - 1$ (the number of categories minus one) that would apply in the case of known parameters must be reduced by $(h - 1) + (k - 1)$. The resulting number of degrees of freedom is

$$hk - 1 - [(h - 1) + (k - 1)] = hk - h - k + 1 = (h - 1)(k - 1).$$

To compute expected frequencies, you can use the maximum likelihood estimators (see Exercise 5) of $p_{i\cdot}$ and $p_{\cdot j}$;

$$\widehat{p}_{i\cdot} = \frac{1}{n} Y_{i\cdot} ; \qquad \widehat{p}_{\cdot j} = \frac{1}{n} Y_{\cdot j} . \qquad (12.28)$$

Respectively, these statistics represent the proportion of all observations in level i of factor A and the proportion in level j of factor B. Therefore under H_0 the estimate for the probability that an observation falls into category (i, j) is

$$\widehat{p}_{ij} = \widehat{p}_{i\cdot} \widehat{p}_{\cdot j},$$

and the expected frequency for category (i, j) is

$$n\widehat{p}_{ij} = n\widehat{p}_{i\cdot}\widehat{p}_{\cdot j} = n \cdot \left(\frac{1}{n} \sum_j Y_{ij} \right) \left(\frac{1}{n} \sum_i Y_{ij} \right) = \frac{Y_{i\cdot} Y_{\cdot j}}{n}. \qquad (12.29)$$

If the frequency data are displayed in a table as in Fig. 12.5, then (12.29) says that the expected frequency for category (i, j) is the row i marginal frequency times the column j marginal frequency, divided by the sample size. Such a table is often called a *contingency table*.

The critical region for the independence test at level α is therefore

$$Q = \sum_{i=1}^{h} \sum_{j=1}^{k} \frac{(Y_{ij} - n\widehat{p}_{i\cdot}\widehat{p}_{\cdot j})^2}{n\widehat{p}_{i\cdot}\widehat{p}_{\cdot j}} > \chi_\alpha^2[(h - 1)(k - 1)], \qquad (12.30)$$

where as usual χ_α^2 is the critical value for the $\chi^2[(h - 1)(k - 1)]$ distribution to the right of which is probability α. Note that Q has the usual form of the sum over all categories of observed minus estimated expected frequencies squared divided by the expected frequencies.

Example 12.3.1

Continued   The employment data together with row and column totals are given here. The estimated expected frequencies under H_0 are shown in parentheses.

	I	II	III	IV	V	Row Totals
Male	140(127)	43(51)	119(118)	43(48)	23(23)	368
Female	23(36)	22(14)	33(33)	19(14)	7(7)	104
Column Totals	163	65	152	62	30	472

For example, by (12.29) the expected frequency of females in type III institutions would be computed as $104 \cdot (152/472) \approx 33$. The value of the χ^2-statistic is

$$q = \frac{(140 - 127)^2}{127} + \cdots + \frac{(7 - 7)^2}{7} = 14.2.$$

Since the degrees of freedom are $(h - 1)(k - 1) = (2 - 1)(5 - 1) = 4$, the q-value is compared to $\chi^2_{.05}(4) = 9.5$. Since $q > \chi^2_{.05}(4)$, the null hypothesis of independence of gender and department type is rejected at level .05. You can check that it would also be rejected at level .01, hence we have strong evidence against H_0. The most significant contributors to Q here are the female I and II categories. A disproportionately low number of women were hired by type I institutions, and a higher than expected number was hired by type II institutions. While this is not by itself direct evidence of an undervaluing of female Ph.D.s at the highest caliber universities, the analysis has raised red flags.

Example 12.3.2

Remember our psychologist team who studied value transmission from mothers to their teenage children (Kasser et al., 1995). In yet another way of looking at their data, they classified the mothers according to their preferred value, financial success or one of the others, and they classified the children of the same mothers by their own preferred value. The most significant results they obtained are in the following tables:

		Mother Financial Success	Self-Acceptance
Teen	*Financial Success*	50	22
	Self-Acceptance	20	37

		Mother Financial Success	Community Feeling
Teen	*Financial Success*	41	31
	Community Feeling	22	35

They computed values of Q as in (12.30) for both comparisons; for self-acceptance it was 15.13 and for community feeling it was 4.29. Since the degrees of freedom parameter is 1, the .05 level critical value is 3.84 and the .01 critical value is 6.63, the independence of mother and teen relative to community feeling can be rejected at level .05, and the independence of mother and teen relative to self-acceptance can be rejected at a level much smaller than .01. In general, mothers for whom financial success was the priority tended to have children for whom the same was true.

Earlier we asked about the problem of testing the independence of random variables. The χ^2-test can be applied here if the factors and levels are defined properly. Consider two random variables X_1 and X_2 that form a random vector $\mathbf{X} = (X_1\ X_2)'$. For the time being suppose that X_1 and X_2 are discrete, with possible values $z_1, z_2, \ldots, z_m$ and $w_1, w_2, \ldots, w_k$, respectively. Each observation $\mathbf{X}$ can be categorized according to the value z_i of its first component and the value w_j of its second component. Then,

$$p_{i\cdot} = P[X_1 = z_i], \qquad p_{\cdot j} = P[X_2 = w_j] \qquad \text{and} \qquad p_{ij} = P[X_1 = z_i,\ X_2 = w_j].$$
$$(12.31)$$

The χ^2-test of $H_0 : p_{ij} = p_{i\cdot} p_{\cdot j} \forall i, j$ becomes a test of the independence of X_1 and X_2.

?Question 12.3.2   What strategy would you use to test independence of two continuous random variables?

As you thought about Question 12.3.2, you probably arrived at the idea of breaking the state spaces of X_1 and X_2 into mutually exclusive and exhaustive subsets $A_1, \ldots, A_h$ for X_1 and $B_1, \ldots, B_k$ for X_2. Count the number of observations $\mathbf{X} = (X_1\ X_2)$ such that $X_1 \in A_i$ and $X_2 \in B_j$ for each i and j, which is the frequency Y_{ij} for the two-way classification. The null hypothesis is

$$H_0 : P[X_1 \in A_i, X_2 \in B_j] = P[X_1 \in A_i] \cdot P[X_2 \in B_j],\ \forall i, j. \qquad (12.32)$$

The A_i's and B_j's are only a few of the infinitely many sets A and B for which a factorization criterion similar to (12.32) must hold in order for X_1 and X_2 to be independent. If H_0 is rejected, however, then we are rejecting independence because factorization (12.32) must in particular hold for these A_i's and B_j's under independence. If H_0 is accepted, we can say only that we do not have statistically significant evidence against independence.

Example 12.3.3   A device consists of two components connected in parallel such that the components supposedly fail independently of each other. Twenty-eight such devices are tested and the failure times of each component are recorded. Do these data suggest a lack of independence of components?

$(20.1, 24.2)$ $(18.7, 23.1)$ $(15.6, 18.9)$ $(22.0, 16.1)$ $(25.2, 25.7)$
$(18.4, 20.2)$ $(17.5, 23.0)$ $(24.4, 21.6)$ $(23.2, 24.4)$ $(18.5, 21.5)$
$(20.3, 18.8)$ $(20.5, 24.1)$ $(16.2, 19.0)$ $(22.4, 20.1)$ $(18.7, 19.4)$
$(18.0, 21.4)$ $(22.3, 21.7)$ $(20.1, 22.6)$ $(19.5, 16.4)$ $(20.8, 23.2)$
$(18.8, 21.5)$ $(20.3, 17.9)$ $(25.6, 19.9)$ $(18.5, 19.7)$ $(21.4, 21.9)$
$(17.8, 19.2)$ $(21.4, 23.2)$ $(25.1, 22.0)$

Let us use two categories for each random failure time, as displayed in the following table. I chose rather arbitrarily a convenient cutoff value of 20. The frequencies are counted, and the expected frequencies are again computed as (row sum) · (column sum)/n.

		X_2	
		<20	*≥20*
X_1	*<20*	5 (3.9)	6 (7.1)
	≥20	5 (6.1)	12 (10.9)

You can compute that Q takes the value .79, which is clearly less than the $\chi^2((2-1)\cdot(2-1)) = \chi^2(1)$ critical value for any reasonable level; thus we do not have significant evidence to reject the null hypothesis of independence of components.

12.3.2   Runs Tests

Many sampling situations can be made to produce sequences of symbols of two kinds, say X's and Y's. For example, a list of data purported to be a random sample $X_1, \ldots, X_n$ from some distribution might give rise to a sequence of X's and Y's by noting

a. whether each X_i was above the sample median (if so write X in the ith position), or below the median (if so, write Y), or
b. whether each X_i was greater than its predecessor X_{i-1} for $i \geq 2$ (if so, write an X) or less than X_{i-1} (if so, write Y).

In both of these situations, if the sample is random, there is no preference for any particular arrangement of the X's and Y's over any other arrangement. (See, for example, Exercise 19.) Yet, most arrangements will not "look unusual" as the following do.

$$XXXXYYY, \quad YYYYXXXX, \quad YYXXXXYY$$

The first list suggests that something about how the X symbols were generated makes them tend to be earlier in the sequence than the Y's. This might arise in case (a) if there is a downward drift as the data sample is being gathered, or in case (b) if the sample values rise for a while and then fall. In the second sequence above the trends reverse. Notice that what links these unusual sequences together is the relatively small number of changes in symbol as one reads left to right.

— ?**Question 12.3.3**   What pattern is suggested in cases (a) and (b) for the third preceding sequence?

Therefore hypotheses about randomness can be tested in many experiments by forming sequences of two kinds of symbols from the data list involved in the problem and by counting the number of *runs* in the sequence.

By a *run*, we mean a sequence of one or more like symbols. The sequence $X\,Y\,XX\,Y\,X\,YY$ has 6 runs. The three preceding sequences have 2, 2, and 3 runs, respectively. A lack of randomness is indicated by a small number of runs, and can also be indicated by a large number of runs if a pairing phenomenon is possible. For instance, if X represents an increase in an animal population from the previous year and Y represents a decrease, extreme grouping of X's and Y's as in $XXXYY$ could result from a positive correlation between one year's growth and the growth in the next year, but the sequence $XYXYXY$ also indicates a cyclical behavior of population growth that is a departure from randomness. In the former case we have a small number of runs and in the latter case a large number. Hence a *runs test* consists of forming a sequence of X's and Y's from the data that is hypothesized to be random, and rejecting the randomness hypothesis if the total number of runs is too large or too small.

To use the idea of runs in hypothesis testing, we need the distribution of the number of runs under the null hypothesis of randomness of ordering of X's and Y's in the sequence. This is provided by the next theorem.

PROPOSITION 12.3.1

Suppose an experiment consists of selecting a random arrangement of m indistinguishable X symbols and n indistinguishable Y symbols. Then the random variable $R =$ total number of runs in the arrangement has state space $\{2, 3, \ldots, m + n\}$ and probability mass function given for the even and odd numbered states by

$$P[R = 2k] = \frac{2\binom{m-1}{k-1}\binom{n-1}{k-1}}{\binom{m+n}{m}}, \tag{12.33}$$

$$P[R = 2k + 1] = \frac{\binom{m-1}{k}\binom{n-1}{k-1} + \binom{m-1}{k-1}\binom{n-1}{k}}{\binom{m+n}{m}}. \tag{12.34}$$

■ **Proof** There are $\binom{m+n}{m}$ total possible arrangements of the m X's and n Y's. (You should make sure that you remember why.) Under the randomness assumption, all are equally likely. It therefore suffices to check that the numerators of (12.33) and (12.34) are the proper counts for the numbers of arrangements leading to an even number $2k$ of runs and an odd number $2k + 1$ of runs.

In the even case, since runs of X's and Y's alternate, the event of having $2k$ runs consists of the union of two disjoint subevents: (1) the sequence begins with an X run, ends with a Y run, and has k runs of X's and k runs of Y's; (2) the sequence begins with a Y run, ends with an X run, and has k runs of X's and k runs of Y's. It is easy to see that the number of ways of doing the second subevent is the same as the number of ways of doing the first; hence it suffices to show that there are $\binom{m-1}{k-1}\binom{n-1}{k-1}$ ways of doing the first subevent.

$$XX \;\square\; X \;\square\; XXX \;\square\; \cdots \;\square\; X.$$

To determine a complete sequence of the first type, it is necessary and sufficient to do two things: (1) determine the lengths of all X runs and (2) determine the lengths of all Y runs. The k X runs may be determined by inserting $k - 1$ markers indicated by the boxes in the diagram into the $m - 1$ possible slots between X symbols, after the first and before the last, with no two markers in the same slot. Since this is equivalent to sampling $k - 1$ marker positions without order or replacement from $m - 1$ slots, there are $\binom{m-1}{k-1}$ ways of subdividing the X symbols into runs. A similar argument yields that there are $\binom{n-1}{k-1}$ ways of subdividing the Y symbols into runs. Having done so, it remains only to insert the Y runs into the marker positions, with the final Y run on the right of the rightmost X run, and the sequence of X's and Y's is fully determined. Hence our assertion follows from the fundamental counting principle.

We leave the odd case as Exercise 18.

Example 12.3.4

Recall that in linear regression, the error terms ϵ_i are supposed to be independent and identically distributed. The residuals $e_i = Y_i - \widehat{Y}_i$ that estimate them should show no large dependence on one another if the model assumptions are true. As a check on model assumptions, suppose we record an X in the ith position of a list if residual e_i is greater than or equal to zero, and a Y otherwise. If we are worried that successive errors are positively correlated, for example, we would be disturbed by a sequence in which many X's occurred in succession and many Y's in succession, producing very few runs. If ten residuals are examined and the sequence

$$XXX\ YYYYY\ XX$$

is observed, can the null hypothesis that the regression model is correct be rejected at level .10?

To recast the question, we would like to know whether the likelihood is less than .10 that a random sequence of five X's and five Y's has three or fewer runs. By formulas (12.33) and (12.34), we have

$$P[R = 2] = \frac{2 \cdot \binom{4}{0}\binom{4}{0}}{\binom{10}{5}} = \frac{2}{\binom{10}{5}} = .008,$$

$$P[R = 3] = \frac{\binom{4}{1}\binom{4}{0} + \binom{4}{1}\binom{4}{0}}{\binom{10}{5}} = \frac{8}{\binom{10}{5}} = .032.$$

Since $P[R \le 3] = .040$ under randomness, we conclude that H_0 should be rejected at level .10. The event that the sequence contains as few as three runs is unlikely enough to make us disbelieve the i.i.d. assumption about the errors.

The existence of a test for randomness of the regression residuals is very attractive, but usually the data sets are larger than the one in the last example, and the p.m.f. of R is cumbersome to work with for large m and n. So, we now describe a large sample approximate normal procedure.

By clever use of indicator variables (see Exercise 20) it can be shown that under the null hypothesis of randomness,

$$E[R] = 1 + \frac{2mn}{m+n}, \tag{12.35}$$

$$\text{Var}(R) = \frac{2mn(2mn - m - n)}{(m+n)^2(m+n-1)}. \tag{12.36}$$

A brutal calculation using the mass function of R can be done, yielding that the limiting distribution of R as m and n approach infinity is normal (see Gibbons, 1971, p. 57 for an outline). Therefore the standardized variable

$$Z = \frac{R - E[R]}{\sqrt{\text{Var} R}} \tag{12.37}$$

has an approximate $N(0,1)$ distribution for large m and n. One can then reject H_0 if the value of Z is extremely low or high.

Example 12.3.5

Exercise 9 in Section 10.2 contained data on the number of people in the former Soviet Union between the years 1970 and 1987 who were employed in agriculture. The problem was followed up in Example 10.4.4, where we observed some suspicious behavior in the residuals from the estimated regression model $workforce = 23.30 + .126 \cdot year - .14 \cdot year^2$ (see Fig. 10.17). The list of residuals follows. Let us see whether the large sample runs test can pick up a departure from randomness.

0.392139, −0.191968, −0.048121, 0.02368, 0.023435, −0.048856, −0.193193, −0.109576, −0.198005, −0.25848, −0.291001, 0.104432, 0.427819, 0.47916, 0.358455, 0.265704, −0.099093, −0.635936

As in case (b) in the opening discussion, form a sequence of X's and Y's by observing whether each residual is greater than its predecessor (if so write an X) or less than its predecessor (if so write Y). The resulting sequence is

$$Y X X Y Y Y X Y Y Y X X X Y Y Y.$$

We are testing the null hypothesis that the combined list is a random arrangement of the two symbols, with $m = 6, n = 11$, against the general alternative. If this null hypothesis is rejected, then so must be the hypothesis that the residuals are i.i.d. If the hypothesis is accepted, we can only say that this way of testing gave no evidence against the model properties of the residuals.

There are seven runs in the observed list. The mean and variance of R from (12.35) and (12.36) are

$$E[R] = 1 + \frac{2(6)(11)}{6+11} = 8.76,$$

$$\text{Var}(R) = \frac{2 \cdot 6 \cdot 11(2 \cdot 6 \cdot 11 - 6 - 11)}{(6+11)^2(6+11-1)} = 3.28.$$

Thus the standardized test statistic has the value

$$z = \frac{7 - 8.76}{\sqrt{3.28}} = -.97.$$

For a two-sided .1 level test, the critical values $\pm z_{.05}$ are ± 1.645. Since our observed z falls between them, we are led to accept H_0; that is, we have insufficient evidence to conclude that there is a violation of the randomness assumptions.

EXERCISES 12.3

1. We can now be more precise about the issue of judging independence from data that we encountered early in the text (Sections 1.4, 1.5, 4.1, and 4.2). For instance, in Fig. 4.2 we classified 100 trapping days according to the numbers of mice and voles trapped on those days. The data are reproduced below. Use a χ^2-test of level .05 to test for the independence of the two random variables $M =$ number of mice and $V =$ number of voles.

		Voles			
		0	*1*	*2*	*3*
	0	16	8	7	6
	1	20	10	4	2
Mice	*2*	5	5	2	1
	3	9	2	2	1

2. Using the data given in Exercise 1(b) of Section 4.1, test for independence of the gender and the preference regarding mixed vs. single sex gym classes (Nelligan, 1994). For convenience, the data are repeated here.

	Mixed	**Single**	**Don't Care**
Male	33	6	6
Female	17	21	11

3. It is sometimes said that the ACT exam measures the level of mastery of high school material, while the SAT exam measures potential for advanced learning. Using the data in Table 10.1, test at level .05 for independence of the SAT and ACT scores of an individual. Carefully formulate the hypothesis being tested and the strategy for computing the test statistic.

4. Exercise 1 in Section 4.2 gave survey data on earthquake insurance status and county for four counties in California (Palm and Hodgson, 1992). Test at level .05 for independence of the two variables. The data are repeated here.

	Contra Costa	Santa Clara	Los Angeles	San Bernardino
Have Insurance	117	222	133	109
Previously Insured	28	26	10	14
Never Insured	376	307	193	249

5. Show that $\hat{p}_{i\cdot}$ and $\hat{p}_{\cdot j}$ of formula (12.28) are the maximum likelihood estimators of $p_{i\cdot}$ and $p_{\cdot j}$ based on the observed frequencies Y_{ij}.

6. The following data are consistent with percentages published in Dortch (1995) on coffee consumption in 1994. Suppose that 100 people in each of the following age categories were asked whether they had consumed any coffee at home in the last 6 months. Test at level .05 for independence of age group and response to this question.

		18–24	25–34	35–44	45–54	55–64	≥65
				Age Group			
Response	Yes	64	67	74	82	85	86
	No	36	33	26	18	15	14

7. A census of military service data (Schick, 1986) broke down the number of thousands of servicemen and women who served during recent wars by racial category. In the Vietnam era there were about 7826 thousand military personnel, and in the era of the Korean War there were about 4216 thousand. The percentages in each category are given here. Test for independence of the two factors of war and race of those serving at level .10.

	Korea	Vietnam
Caucasian	90.9	90.0
African-American	8.0	8.5
Other	1.1	1.5

8. Is there a difference between the χ^2-test of equality of two distributions that we studied in Section 12.1 and a χ^2-test of independence constructed so that factor A is the population to which an observation belongs and factor B is the category of states used in the test for distributional equality?

9. The American Mathematical Society survey on employment of new doctorates that was referred to in Example 12.3.1 also contained data on the type of employer hiring the doctorates, broken down by the five types of degree-granting institutions. The data are given in the following table. Does the employer type seem to be independent of the degree-granting institution? If not, where do the dependencies appear to be?

Institution

		I	II	III	IV	V
	Academic	235	101	102	91	45
Employer	*Gov't*	6	3	6	14	5
	Industry	26	16	20	49	36

10. Propose a chi-square test for the independence of three factors. How many degrees of freedom does your test statistic have?

11. Example 8.5.3 contained information on deaths of motorcycle riders in crashes and whether they were wearing a helmet at the time. Look at the data in another way using the ideas of this section, formulate and test an appropriate hypothesis, and give conclusions.

12. A chi-square test can be used to check for trends in data, especially time series data, by using the technique of grouping. For example, suppose that the following list of 24 data points was collected in time order but is supposed to be a random sample. Group, say, the first eight points, the middle eight, and the last eight. Then each point can be classified according to both its group and whether it is at least as big as the sample median or is less than the median. Finish the description of this procedure, discuss clearly how it enables you to test the hypothesis of randomness of the sample, and carry out the test on the following data at level .05.

$$-2.1 \ 1.1 \ -1.6 \quad 0.4 \ 2.1 \ 0.1 \ -1.8 \ -0.5$$
$$0.0 \ 1.2 \quad 3.4 \ -1.3 \ 1.5 \ 2.0 \quad 2.1 \quad 0.8$$
$$0.7 \ 3.1 \quad 2.3 \quad 1.8 \ 1.1 \ 2.2 \quad 1.6 \quad 1.5$$

13. Do a runs test of level no more than .05 for independence of the errors in the Pakistani airline passenger data set given in Exercise 1 of Section 10.3. Form the sequence of symbols by noting the signs of the residuals, and use a cubic regression model.

14. Do a runs test of level no more than .05 for independence of the errors in the casino revenue data set given in Exercise 11 of Section 10.2. Form the sequence of symbols by noting when data elements are greater or less than their predecessors, and use a quadratic regression model.

15. A romantic owner of a local entertainment establishment hopes that his customers enter randomly by gender, then exit after having found a person of the opposite gender. Use an exact small sample runs test to check his belief, based on the sequence of entries,

$$XXYXYYYXY,$$

and departures,

$$XYXYXYYXY,$$

where X indicates a female and Y a male.

16. Consider again the Peoria school data (see Table 10.4, Exercise 2, in Section 10.2). Perform a linear regression of reading score on years of teacher experience, and use an approximate large sample runs test to check the model assumptions on the errors. Form the sequence of symbols by noting the signs of the residuals.

17. Bags of flour are to be filled by a machine to a weight of five pounds. Use a runs test to check whether the filling process may be going out of control, if the following twelve successive observations of fill weight are made.

    4.98, 4.96, 5.02, 4.98, 5.04, 5.13, 5.14, 4.96, 5.20, 5.17, 5.05, 5.10

18. Complete the proof of Proposition 12.3.1 by proving formula (12.34).
19. If three independent observations X_1, X_2, and X_3 are taken from a continuous population, show that $P[X_i < X_j < X_k] = 1/6$ for each permutation of subscripts (i, j, k) from the set $\{1, 2, 3\}$.
20. Let $X_1, \ldots, X_n$ be a list of X's and Y's as in the runs test. Define

$$I_k = \begin{cases} 1 & \text{if } X_k \neq X_{k-1}, \\ 0 & \text{otherwise,} \end{cases}$$

    for $k = 2, 3, \ldots, n$. Use these indicator random variables to derive formula (12.35) for the expected number of runs.
21. Discuss how runs tests might be applied to test for a difference in location of two continuous distributions.

APPENDIX A

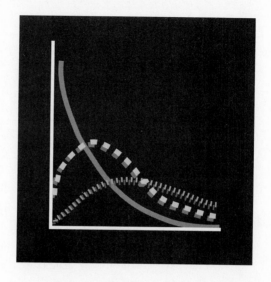

STATISTICAL TABLES

Table A.1

Binomial Mass Functions

$$q(k) = P[X = k] = \binom{n}{k} p^k (1-p)^{n-k}$$

n	k	p = 1/10	p = 1/5	p = 1/4	p = 1/3	p = 2/5	p = 1/2
2	0	0.8100	0.6400	0.5625	0.4444	0.3600	0.2500
	1	0.1800	0.3200	0.3750	0.4444	0.4800	0.5000
	2	0.0100	0.0400	0.0625	0.1111	0.1600	0.2500
3	0	0.7290	0.5120	0.4219	0.2963	0.2160	0.1250
	1	0.2430	0.3840	0.4219	0.4444	0.4320	0.3750
	2	0.0270	0.0960	0.1406	0.2222	0.2880	0.3750
	3	0.0010	0.0080	0.0156	0.0370	0.0640	0.1250
4	0	0.6561	0.4096	0.3164	0.1975	0.1296	0.0625
	1	0.2916	0.4096	0.4219	0.3951	0.3456	0.2500
	2	0.0486	0.1536	0.2109	0.2963	0.3456	0.3750
	3	0.0036	0.0256	0.0469	0.0988	0.1536	0.2500
	4	0.0001	0.0016	0.0039	0.0123	0.0256	0.0625
5	0	0.5905	0.3277	0.2373	0.1317	0.0778	0.0312
	1	0.3281	0.4096	0.3955	0.3292	0.2592	0.1562
	2	0.0729	0.2048	0.2637	0.3292	0.3456	0.3125
	3	0.0081	0.0512	0.0879	0.1646	0.2304	0.3125
	4	0.0005	0.0064	0.0146	0.0412	0.0768	0.1562
	5	0.0000	0.0003	0.0010	0.0041	0.0102	0.0312
6	0	0.5314	0.2621	0.1780	0.0878	0.0467	0.0156
	1	0.3543	0.3932	0.3560	0.2634	0.1866	0.0938
	2	0.0984	0.2458	0.2966	0.3292	0.3110	0.2344
	3	0.0146	0.0819	0.1318	0.2195	0.2765	0.3125
	4	0.0012	0.0154	0.0330	0.0823	0.1382	0.2344
	5	0.0001	0.0015	0.0044	0.0165	0.0369	0.0938
	6	0.0000	0.0001	0.0002	0.0014	0.0041	0.0156
7	0	0.4783	0.2097	0.1335	0.0585	0.0280	0.0078
	1	0.3720	0.3670	0.3115	0.2048	0.1306	0.0547
	2	0.1240	0.2753	0.3115	0.3073	0.2613	0.1641
	3	0.0230	0.1147	0.1730	0.2561	0.2903	0.2734
	4	0.0026	0.0287	0.0577	0.1280	0.1935	0.2734
	5	0.0002	0.0043	0.0115	0.0384	0.0774	0.1641
	6	0.0000	0.0004	0.0013	0.0064	0.0172	0.0547
	7	0.0000	0.0000	0.0001	0.0005	0.0016	0.0078

Table A.1

Continued

n	k	p = 1/10	p = 1/5	p = 1/4	p = 1/3	p = 2/5	p = 1/2
8	0	0.4305	0.1678	0.1001	0.0390	0.0168	0.0039
	1	0.3826	0.3355	0.2670	0.1561	0.0896	0.0312
	2	0.1488	0.2936	0.3115	0.2731	0.2090	0.1094
	3	0.0331	0.1468	0.2076	0.2731	0.2787	0.2188
	4	0.0046	0.0459	0.0865	0.1707	0.2322	0.2734
	5	0.0004	0.0092	0.0231	0.0683	0.1239	0.2188
	6	0.0000	0.0011	0.0038	0.0171	0.0413	0.1094
	7	0.0000	0.0001	0.0004	0.0024	0.0079	0.0312
	8	0.0000	0.0000	0.0000	0.0002	0.0007	0.0039
9	0	0.3874	0.1342	0.0751	0.0260	0.0101	0.0020
	1	0.3874	0.3020	0.2253	0.1171	0.0605	0.0176
	2	0.1722	0.3020	0.3003	0.2341	0.1612	0.0703
	3	0.0446	0.1762	0.2336	0.2731	0.2508	0.1641
	4	0.0074	0.0661	0.1168	0.2048	0.2508	0.2461
	5	0.0008	0.0165	0.0389	0.1024	0.1672	0.2461
	6	0.0001	0.0028	0.0087	0.0341	0.0743	0.1641
	7	0.0000	0.0003	0.0012	0.0073	0.0212	0.0703
	8	0.0000	0.0000	0.0001	0.0009	0.0035	0.0176
	9	0.0000	0.0000	0.0000	0.0001	0.0003	0.0020
10	0	0.3487	0.1074	0.0563	0.0173	0.0060	0.0010
	1	0.3874	0.2684	0.1877	0.0867	0.0403	0.0098
	2	0.1937	0.3020	0.2816	0.1951	0.1209	0.0439
	3	0.0574	0.2013	0.2503	0.2601	0.2150	0.1172
	4	0.0112	0.0881	0.1460	0.2276	0.2508	0.2051
	5	0.0015	0.0264	0.0584	0.1366	0.2007	0.2461
	6	0.0001	0.0055	0.0162	0.0569	0.1115	0.2051
	7	0.0000	0.0008	0.0031	0.0163	0.0425	0.1172
	8	0.0000	0.0001	0.0004	0.0030	0.0106	0.0439
	9	0.0000	0.0000	0.0000	0.0003	0.0016	0.0098
	10	0.0000	0.0000	0.0000	0.0000	0.0001	0.0010

Table A.1

Continued

n	k	$p=1/10$	$p=1/5$	$p=1/4$	$p=1/3$	$p=2/5$	$p=1/2$
11	0	0.3138	0.0859	0.0422	0.0116	0.0036	0.0005
	1	0.3835	0.2362	0.1549	0.0636	0.0266	0.0054
	2	0.2131	0.2953	0.2581	0.1590	0.0887	0.0269
	3	0.0710	0.2215	0.2581	0.2384	0.1774	0.0806
	4	0.0158	0.1107	0.1721	0.2384	0.2365	0.1611
	5	0.0025	0.0388	0.0803	0.1669	0.2207	0.2256
	6	0.0003	0.0097	0.0268	0.0835	0.1471	0.2256
	7	0.0000	0.0017	0.0064	0.0298	0.0701	0.1611
	8	0.0000	0.0002	0.0011	0.0075	0.0234	0.0806
	9	0.0000	0.0000	0.0001	0.0012	0.0052	0.0269
	10	0.0000	0.0000	0.0000	0.0001	0.0007	0.0054
	11	0.0000	0.0000	0.0000	0.0000	0.0000	0.0005
12	0	0.2824	0.0687	0.0317	0.0077	0.0022	0.0002
	1	0.3766	0.2062	0.1267	0.0462	0.0174	0.0029
	2	0.2301	0.2835	0.2323	0.1272	0.0639	0.0161
	3	0.0852	0.2362	0.2581	0.2120	0.1419	0.0537
	4	0.0213	0.1329	0.1936	0.2384	0.2128	0.1208
	5	0.0038	0.0532	0.1032	0.1908	0.2270	0.1934
	6	0.0005	0.0155	0.0401	0.1113	0.1766	0.2256
	7	0.0000	0.0033	0.0115	0.0477	0.1009	0.1934
	8	0.0000	0.0005	0.0024	0.0149	0.0420	0.1208
	9	0.0000	0.0001	0.0004	0.0033	0.0125	0.0537
	10	0.0000	0.0000	0.0000	0.0005	0.0025	0.0161
	11	0.0000	0.0000	0.0000	0.0000	0.0003	0.0029
	12	0.0000	0.0000	0.0000	0.0000	0.0000	0.0002

Table A.2

Poisson Mass Functions

$$q(k) = P[X = k] = e^{-\lambda}\lambda^k/k!$$

k	$\lambda = .5$	$\lambda = 1$	$\lambda = 1.5$	$\lambda = 2$	$\lambda = 2.5$	$\lambda = 3$	$\lambda = 3.5$
0	0.6065	0.3679	0.2231	0.1353	0.0821	0.0498	0.0302
1	0.3033	0.3679	0.3347	0.2707	0.2052	0.1494	0.1057
2	0.0758	0.1839	0.2510	0.2707	0.2565	0.2240	0.1850
3	0.0126	0.0613	0.1255	0.1804	0.2138	0.2240	0.2158
4	0.0016	0.0153	0.0471	0.0902	0.1336	0.1680	0.1888
5	0.0002	0.0031	0.0141	0.0361	0.0668	0.1008	0.1322
6	0.0000	0.0005	0.0035	0.0120	0.0278	0.0504	0.0771
7	0.0000	0.0001	0.0008	0.0034	0.0099	0.0216	0.0385
8	0.0000	0.0000	0.0001	0.0009	0.0031	0.0081	0.0169
9	0.0000	0.0000	0.0000	0.0002	0.0009	0.0027	0.0066
10	0.0000	0.0000	0.0000	0.0000	0.0002	0.0008	0.0023
11	0.0000	0.0000	0.0000	0.0000	0.0000	0.0002	0.0007
12	0.0000	0.0000	0.0000	0.0000	0.0000	0.0001	0.0002
13	0.0000	0.0000	0.0000	0.0000	0.0000	0.0000	0.0001

k	$\lambda = 4$	$\lambda = 5$	$\lambda = 6$	$\lambda = 7$	$\lambda = 8$	$\lambda = 9$	$\lambda = 10$
0	0.0183	0.0067	0.0025	0.0009	0.0003	0.0001	0.0000
1	0.0733	0.0337	0.0149	0.0064	0.0027	0.0011	0.0005
2	0.1465	0.0842	0.0446	0.0223	0.0107	0.0050	0.0023
3	0.1954	0.1404	0.0892	0.0521	0.0286	0.0150	0.0076
4	0.1954	0.1755	0.1339	0.0912	0.0573	0.0337	0.0189
5	0.1563	0.1755	0.1606	0.1277	0.0916	0.0607	0.0378
6	0.1042	0.1462	0.1606	0.1490	0.1221	0.0911	0.0631
7	0.0595	0.1044	0.1377	0.1490	0.1396	0.1171	0.0901
8	0.0298	0.0653	0.1033	0.1304	0.1396	0.1318	0.1126
9	0.0132	0.0363	0.0688	0.1014	0.1241	0.1318	0.1251
10	0.0053	0.0181	0.0413	0.0710	0.0993	0.1186	0.1251
11	0.0019	0.0082	0.0225	0.0452	0.0722	0.0970	0.1137
12	0.0006	0.0034	0.0113	0.0263	0.0481	0.0728	0.0948
13	0.0002	0.0013	0.0052	0.0142	0.0296	0.0504	0.0729
14	0.0001	0.0005	0.0022	0.0071	0.0169	0.0324	0.0521
15	0.0000	0.0002	0.0009	0.0033	0.0090	0.0194	0.0347
16	0.0000	0.0000	0.0003	0.0014	0.0045	0.0109	0.0217
17	0.0000	0.0000	0.0001	0.0006	0.0021	0.0058	0.0128
18	0.0000	0.0000	0.0000	0.0002	0.0009	0.0029	0.0071
19	0.0000	0.0000	0.0000	0.0001	0.0004	0.0014	0.0037
20	0.0000	0.0000	0.0000	0.0000	0.0002	0.0006	0.0019

Table A.3

Standard Normal c.d.f.

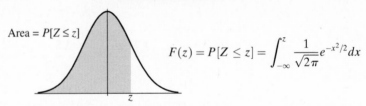

Area = $P[Z \le z]$

$$F(z) = P[Z \le z] = \int_{-\infty}^{z} \frac{1}{\sqrt{2\pi}} e^{-x^2/2} dx$$

z	.00	.01	.02	.03	.04	.05	.06	.07	.08	.09
0.0	.5000	.5040	.5080	.5120	.5160	.5199	.5239	.5279	.5319	.5359
0.1	.5398	.5438	.5478	.5517	.5557	.5596	.5636	.5675	.5714	.5753
0.2	.5793	.5832	.5871	.5910	.5948	.5987	.6026	.6064	.6103	.6141
0.3	.6179	.6217	.6255	.6293	.6331	.6368	.6406	.6443	.6480	.6517
0.4	.6554	.6591	.6628	.6664	.6700	.6736	.6772	.6808	.6844	.6879
0.5	.6915	.6950	.6985	.7019	.7054	.7088	.7123	.7157	.7190	.7224
0.6	.7257	.7291	.7324	.7357	.7389	.7422	.7454	.7486	.7517	.7549
0.7	.7580	.7611	.7642	.7673	.7704	.7734	.7764	.7794	.7823	.7852
0.8	.7881	.7910	.7939	.7967	.7995	.8023	.8051	.8078	.8106	.8133
0.9	.8159	.8186	.8212	.8238	.8264	.8289	.8315	.8340	.8365	.8389
1.0	.8413	.8438	.8461	.8485	.8508	.8531	.8554	.8577	.8599	.8621
1.1	.8643	.8665	.8686	.8708	.8729	.8749	.8770	.8790	.8810	.8830
1.2	.8849	.8869	.8888	.8907	.8925	.8944	.8962	.8980	.8997	.9015
1.3	.9032	.9049	.9066	.9082	.9099	.9115	.9131	.9147	.9162	.9177
1.4	.9192	.9207	.9222	.9236	.9251	.9265	.9279	.9292	.9306	.9319
1.5	.9332	.9345	.9357	.9370	.9382	.9394	.9406	.9418	.9429	.9441
1.6	.9452	.9463	.9474	.9484	.9495	.9505	.9515	.9525	.9535	.9545
1.7	.9554	.9564	.9573	.9582	.9591	.9599	.9608	.9616	.9625	.9633
1.8	.9641	.9649	.9656	.9664	.9671	.9678	.9686	.9693	.9699	.9706
1.9	.9713	.9719	.9726	.9732	.9738	.9744	.9750	.9756	.9761	.9767
2.0	.9772	.9778	.9783	.9788	.9793	.9798	.9803	.9808	.9812	.9817
2.1	.9821	.9826	.9830	.9834	.9838	.9842	.9846	.9850	.9854	.9857
2.2	.9861	.9864	.9868	.9871	.9875	.9878	.9881	.9884	.9887	.9890
2.3	.9893	.9896	.9898	.9901	.9904	.9906	.9909	.9911	.9913	.9916
2.4	.9918	.9920	.9922	.9925	.9927	.9929	.9931	.9932	.9934	.9936
2.5	.9938	.9940	.9941	.9943	.9945	.9946	.9948	.9949	.9951	.9952
2.6	.9953	.9955	.9956	.9957	.9959	.9960	.9961	.9962	.9963	.9964
2.7	.9965	.9966	.9967	.9968	.9969	.9970	.9971	.9972	.9973	.9974
2.8	.9974	.9975	.9976	.9977	.9977	.9978	.9979	.9979	.9980	.9981
2.9	.9981	.9982	.9982	.9983	.9984	.9984	.9985	.9985	.9986	.9986
3.0	.9987	.9987	.9987	.9988	.9988	.9989	.9989	.9989	.9990	.9990

Table A.4

Chi-Square Percentiles

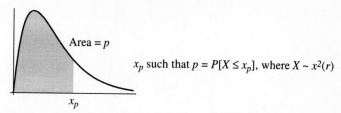

Area = p

x_p such that $p = P[X \leq x_p]$, where $X \sim x^2(r)$

r	$p = .01$	$p = .025$	$p = .05$	$p = .10$	$p = .90$	$p = .95$	$p = .975$	$p = .99$
1	0.00	0.00	0.00	0.02	2.71	3.84	5.02	6.63
2	0.02	0.05	0.10	0.21	4.61	5.99	7.38	9.21
3	0.11	0.22	0.35	0.58	6.25	7.81	9.35	11.34
4	0.30	0.48	0.71	1.06	7.78	9.49	11.14	13.28
5	0.55	0.83	1.15	1.61	9.24	11.07	12.83	15.09
6	0.87	1.24	1.64	2.20	10.64	12.59	14.45	16.81
7	1.24	1.69	2.17	2.83	12.02	14.07	16.01	18.48
8	1.65	2.18	2.73	3.49	13.36	15.51	17.53	20.09
9	2.09	2.70	3.33	4.17	14.68	16.92	19.02	21.67
10	2.56	3.25	3.94	4.87	15.99	18.31	20.48	23.21
11	3.05	3.82	4.57	5.58	17.28	19.68	21.92	24.72
12	3.57	4.40	5.23	6.30	18.55	21.03	23.34	26.22
13	4.11	5.01	5.89	7.04	19.81	22.36	24.74	27.69
14	4.66	5.63	6.57	7.79	21.06	23.68	26.12	29.14
15	5.23	6.26	7.26	8.55	22.31	25.00	27.49	30.58
16	5.81	6.91	7.96	9.31	23.54	26.30	28.85	32.00
17	6.41	7.56	8.67	10.09	24.77	27.59	30.19	33.41
18	7.01	8.23	9.39	10.86	25.99	28.87	31.53	34.81
19	7.63	8.91	10.12	11.65	27.20	30.14	32.85	36.19
20	8.26	9.59	10.85	12.44	28.41	31.41	34.17	37.57
21	8.90	10.28	11.59	13.24	29.62	32.67	35.48	38.93
22	9.54	10.98	12.34	14.04	30.81	33.92	36.78	40.29
23	10.20	11.69	13.09	14.85	32.01	35.17	38.08	41.64
24	10.86	12.40	13.85	15.66	33.20	36.42	39.36	42.98
25	11.52	13.12	14.61	16.47	34.38	37.65	40.65	44.31
26	12.20	13.84	15.38	17.29	35.56	38.89	41.92	45.64
27	12.88	14.57	16.15	18.11	36.74	40.11	43.19	46.96
28	13.56	15.31	16.93	18.94	37.92	41.34	44.46	48.28
29	14.26	16.05	17.71	19.77	39.09	42.56	45.72	49.59
30	14.95	16.79	18.49	20.60	40.26	43.77	46.98	50.89

Table A.5

t-Percentiles

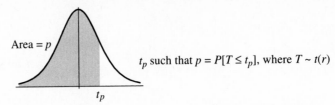

Area = p

t_p such that $p = P[T \le t_p]$, where $T \sim t(r)$

r	$p = .80$	$p = .85$	$p = .90$	$p = .95$	$p = .975$	$p = .99$
1	1.38	1.96	3.08	6.31	12.71	31.82
2	1.06	1.39	1.89	2.92	4.30	6.96
3	0.98	1.25	1.64	2.35	3.18	4.54
4	0.94	1.19	1.53	2.13	2.78	3.75
5	0.92	1.16	1.48	2.02	2.57	3.36
6	0.91	1.13	1.44	1.94	2.45	3.14
7	0.90	1.12	1.41	1.89	2.36	3.00
8	0.89	1.11	1.40	1.86	2.31	2.90
9	0.88	1.10	1.38	1.83	2.26	2.82
10	0.88	1.09	1.37	1.81	2.23	2.76
11	0.88	1.09	1.36	1.80	2.20	2.72
12	0.87	1.08	1.36	1.78	2.18	2.68
13	0.87	1.08	1.35	1.77	2.16	2.65
14	0.87	1.08	1.35	1.76	2.14	2.62
15	0.87	1.07	1.34	1.75	2.13	2.60
16	0.86	1.07	1.34	1.75	2.12	2.58
17	0.86	1.07	1.33	1.74	2.11	2.57
18	0.86	1.07	1.33	1.73	2.10	2.55
19	0.86	1.07	1.33	1.73	2.09	2.54
20	0.86	1.06	1.33	1.72	2.09	2.53
21	0.86	1.06	1.32	1.72	2.08	2.52
22	0.86	1.06	1.32	1.72	2.07	2.51
23	0.86	1.06	1.32	1.71	2.07	2.50
24	0.86	1.06	1.32	1.71	2.06	2.49
25	0.86	1.06	1.32	1.71	2.06	2.49
26	0.86	1.06	1.31	1.71	2.06	2.48
27	0.86	1.06	1.31	1.70	2.05	2.47
28	0.85	1.06	1.31	1.70	2.05	2.47
29	0.85	1.06	1.31	1.70	2.05	2.46
30	0.85	1.05	1.31	1.70	2.04	2.46

Table A.6

F-percentiles

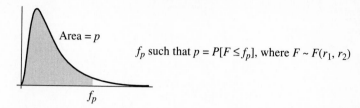

Area = p

f_p such that $p = P[F \le f_p]$, where $F \sim F(r_1, r_2)$

f_p

						r_2					
r_1	p	1	2	3	4	5	6	7	8	9	10
1	.95	161.45	18.51	10.13	7.71	6.61	5.99	5.59	5.32	5.12	4.96
	.975	647.79	38.51	17.44	12.22	10.01	8.81	8.07	7.57	7.21	6.94
	.99	4052.20	98.50	34.12	21.20	16.26	13.75	12.25	11.26	10.56	10.04
2	.95	199.50	19.00	9.55	6.94	5.79	5.14	4.74	4.46	4.26	4.10
	.975	799.50	39.00	16.04	10.65	8.43	7.26	6.54	6.06	5.71	5.46
	.99	4999.50	99.00	30.82	18.00	13.27	10.92	9.55	8.65	8.02	7.56
3	.95	215.71	19.16	9.28	6.59	5.41	4.76	4.35	4.07	3.86	3.71
	.975	864.16	39.17	15.44	9.98	7.76	6.60	5.89	5.42	5.08	4.83
	.99	5403.40	99.17	29.46	16.69	12.06	9.78	8.45	7.59	6.99	6.55
4	.95	224.58	19.25	9.12	6.39	5.19	4.53	4.12	3.84	3.63	3.48
	.975	899.58	39.25	15.10	9.60	7.39	6.23	5.52	5.05	4.72	4.47
	.99	5624.60	99.25	28.71	15.98	11.39	9.15	7.85	7.01	6.42	5.99
5	.95	230.16	19.30	9.01	6.26	5.05	4.39	3.97	3.69	3.48	3.33
	.975	921.85	39.30	14.88	9.36	7.15	5.99	5.29	4.82	4.48	4.24
	.99	5763.60	99.30	28.24	15.52	10.97	8.75	7.46	6.63	6.06	5.64
6	.95	233.99	19.33	8.94	6.16	4.95	4.28	3.87	3.58	3.37	3.22
	.975	937.11	39.33	14.73	9.20	6.98	5.82	5.12	4.65	4.32	4.07
	.99	5859.00	99.33	27.91	15.21	10.67	8.47	7.19	6.37	5.80	5.39
7	.95	236.77	19.35	8.89	6.09	4.88	4.21	3.79	3.50	3.29	3.14
	.975	948.22	39.36	14.62	9.07	6.85	5.70	4.99	4.53	4.20	3.95
	.99	5928.40	99.36	27.67	14.98	10.46	8.26	6.99	6.18	5.61	5.20
8	.95	238.88	19.37	8.85	6.04	4.82	4.15	3.73	3.44	3.23	3.07
	.975	956.66	39.37	14.54	8.98	6.76	5.60	4.90	4.43	4.10	3.85
	.99	5981.10	99.37	27.49	14.80	10.29	8.10	6.84	6.03	5.47	5.06
9	.95	240.54	19.38	8.81	6.00	4.77	4.10	3.68	3.39	3.18	3.02
	.975	963.28	39.39	14.47	8.90	6.68	5.52	4.82	4.36	4.03	3.78
	.99	6022.50	99.39	27.35	14.66	10.16	7.98	6.72	5.91	5.35	4.94
10	.95	241.88	19.40	8.79	5.96	4.74	4.06	3.64	3.35	3.14	2.98
	.975	968.63	39.40	14.42	8.84	6.62	5.46	4.76	4.30	3.96	3.72
	.99	6055.80	99.40	27.23	14.55	10.05	7.87	6.62	5.81	5.26	4.85

Table A.6

Continued

r_1	p	11	12	13	14	15	16	17	18	19	20
1	.95	4.84	4.75	4.67	4.60	4.54	4.49	4.45	4.41	4.38	4.35
	.975	6.72	6.55	6.41	6.30	6.20	6.12	6.04	5.98	5.92	5.87
	.99	9.65	9.33	9.07	8.86	8.68	8.53	8.40	8.29	8.18	8.10
2	.95	3.98	3.89	3.81	3.74	3.68	3.63	3.59	3.55	3.52	3.49
	.975	5.26	5.10	4.97	4.86	4.77	4.69	4.62	4.56	4.51	4.46
	.99	7.21	6.93	6.70	6.51	6.36	6.23	6.11	6.01	5.93	5.85
3	.95	3.59	3.49	3.41	3.34	3.29	3.24	3.20	3.16	3.13	3.10
	.975	4.63	4.47	4.35	4.24	4.15	4.08	4.01	3.95	3.90	3.86
	.99	6.22	5.95	5.74	5.56	5.42	5.29	5.18	5.09	5.01	4.94
4	.95	3.36	3.26	3.18	3.11	3.06	3.01	2.96	2.93	2.90	2.87
	.975	4.28	4.12	4.00	3.89	3.80	3.73	3.66	3.61	3.56	3.51
	.99	5.67	5.41	5.21	5.04	4.89	4.77	4.67	4.58	4.50	4.43
5	.95	3.20	3.11	3.03	2.96	2.90	2.85	2.81	2.77	2.74	2.71
	.975	4.04	3.89	3.77	3.66	3.58	3.50	3.44	3.38	3.33	3.29
	.99	5.32	5.06	4.86	4.69	4.56	4.44	4.34	4.25	4.17	4.10
6	.95	3.09	3.00	2.92	2.85	2.79	2.74	2.70	2.66	2.63	2.60
	.975	3.88	3.73	3.60	3.50	3.41	3.34	3.28	3.22	3.17	3.13
	.99	5.07	4.82	4.62	4.46	4.32	4.20	4.10	4.01	3.94	3.87
7	.95	3.01	2.91	2.83	2.76	2.71	2.66	2.61	2.58	2.54	2.51
	.975	3.76	3.61	3.48	3.38	3.29	3.22	3.16	3.10	3.05	3.01
	.99	4.89	4.64	4.44	4.28	4.14	4.03	3.93	3.84	3.77	3.70
8	.95	2.95	2.85	2.77	2.70	2.64	2.59	2.55	2.51	2.48	2.45
	.975	3.66	3.51	3.39	3.29	3.20	3.12	3.06	3.01	2.96	2.91
	.99	4.74	4.50	4.30	4.14	4.00	3.89	3.79	3.71	3.63	3.56
9	.95	2.90	2.80	2.71	2.65	2.59	2.54	2.49	2.46	2.42	2.39
	.975	3.59	3.44	3.31	3.21	3.12	3.05	2.98	2.93	2.88	2.84
	.99	4.63	4.39	4.19	4.03	3.89	3.78	3.68	3.60	3.52	3.46
10	.95	2.85	2.75	2.67	2.60	2.54	2.49	2.45	2.41	2.38	2.35
	.975	3.53	3.37	3.25	3.15	3.06	2.99	2.92	2.87	2.82	2.77
	.99	4.54	4.30	4.10	3.94	3.80	3.69	3.59	3.51	3.43	3.37

Table A.6

Continued

r_1	p	22	24	26	28	30	32	34	36	38	40
						r_2					
1	.95	4.30	4.26	4.23	4.20	4.17	4.15	4.13	4.11	4.10	4.08
	.975	5.79	5.72	5.66	5.61	5.57	5.53	5.50	5.47	5.45	5.42
	.99	7.95	7.82	7.72	7.64	7.56	7.50	7.44	7.40	7.35	7.31
2	.95	3.44	3.40	3.37	3.34	3.32	3.29	3.28	3.26	3.24	3.23
	.975	4.38	4.32	4.27	4.22	4.18	4.15	4.12	4.09	4.07	4.05
	.99	5.72	5.61	5.53	5.45	5.39	5.34	5.29	5.25	5.21	5.18
3	.95	3.05	3.01	2.98	2.95	2.92	2.90	2.88	2.87	2.85	2.84
	.975	3.78	3.72	3.67	3.63	3.59	3.56	3.53	3.50	3.48	3.46
	.99	4.82	4.72	4.64	4.57	4.51	4.46	4.42	4.38	4.34	4.31
4	.95	2.82	2.78	2.74	2.71	2.69	2.67	2.65	2.63	2.62	2.61
	.975	3.44	3.38	3.33	3.29	3.25	3.22	3.19	3.17	3.15	3.13
	.99	4.31	4.22	4.14	4.07	4.02	3.97	3.93	3.89	3.86	3.83
5	.95	2.66	2.62	2.59	2.56	2.53	2.51	2.49	2.48	2.46	2.45
	.975	3.22	3.15	3.10	3.06	3.03	3.00	2.97	2.94	2.92	2.90
	.99	3.99	3.90	3.82	3.75	3.70	3.65	3.61	3.57	3.54	3.51
6	.95	2.55	2.51	2.47	2.45	2.42	2.40	2.38	2.36	2.35	2.34
	.975	3.05	2.99	2.94	2.90	2.87	2.84	2.81	2.78	2.76	2.74
	.99	3.76	3.67	3.59	3.53	3.47	3.43	3.39	3.35	3.32	3.29
7	.95	2.46	2.42	2.39	2.36	2.33	2.31	2.29	2.28	2.26	2.25
	.975	2.93	2.87	2.82	2.78	2.75	2.71	2.69	2.66	2.64	2.62
	.99	3.59	3.50	3.42	3.36	3.30	3.26	3.22	3.18	3.15	3.12
8	.95	2.40	2.36	2.32	2.29	2.27	2.24	2.23	2.21	2.19	2.18
	.975	2.84	2.78	2.73	2.69	2.65	2.62	2.59	2.57	2.55	2.53
	.99	3.45	3.36	3.29	3.23	3.17	3.13	3.09	3.05	3.02	2.99
9	.95	2.34	2.30	2.27	2.24	2.21	2.19	2.17	2.15	2.14	2.12
	.975	2.76	2.70	2.65	2.61	2.57	2.54	2.52	2.49	2.47	2.45
	.99	3.35	3.26	3.18	3.12	3.07	3.02	2.98	2.95	2.92	2.89
10	.95	2.30	2.25	2.22	2.19	2.16	2.14	2.12	2.11	2.09	2.08
	.975	2.70	2.64	2.59	2.55	2.51	2.48	2.45	2.43	2.41	2.39
	.99	3.26	3.17	3.09	3.03	2.98	2.93	2.89	2.86	2.83	2.80

Table A.6

Continued

r_1	p	1	2	3	4	5	6	7	8	9	10
11	.95	242.98	19.40	8.76	5.94	4.70	4.03	3.60	3.31	3.10	2.94
	.975	973.03	39.41	14.37	8.79	6.57	5.41	4.71	4.24	3.91	3.66
	.99	6083.30	99.41	27.13	14.45	9.96	7.79	6.54	5.73	5.18	4.77
12	.95	243.91	19.41	8.74	5.91	4.68	4.00	3.57	3.28	3.07	2.91
	.975	976.71	39.41	14.34	8.75	6.52	5.37	4.67	4.20	3.87	3.62
	.99	6106.30	99.42	27.05	14.37	9.89	7.72	6.47	5.67	5.11	4.71
13	.95	244.69	19.42	8.73	5.89	4.66	3.98	3.55	3.26	3.05	2.89
	.975	979.84	39.42	14.30	8.71	6.49	5.33	4.63	4.16	3.83	3.58
	.99	6125.90	99.42	26.98	14.31	9.82	7.66	6.41	5.61	5.05	4.65
14	.95	245.36	19.42	8.71	5.87	4.64	3.96	3.53	3.24	3.03	2.86
	.975	982.53	39.43	14.28	8.68	6.46	5.30	4.60	4.13	3.80	3.55
	.99	6142.70	99.43	26.92	14.25	9.77	7.60	6.36	5.56	5.01	4.60
15	.95	245.95	19.43	8.70	5.86	4.62	3.94	3.51	3.22	3.01	2.85
	.975	984.87	39.43	14.25	8.66	6.43	5.27	4.57	4.10	3.77	3.52
	.99	6157.30	99.43	26.87	14.20	9.72	7.56	6.31	5.52	4.96	4.56
16	.95	246.46	19.43	8.69	5.84	4.60	3.92	3.49	3.20	2.99	2.83
	.975	986.92	39.44	14.23	8.63	6.40	5.24	4.54	4.08	3.74	3.50
	.99	6170.10	99.44	26.83	14.15	9.68	7.52	6.28	5.48	4.92	4.52
17	.95	246.92	19.44	8.68	5.83	4.59	3.91	3.48	3.19	2.97	2.81
	.975	988.73	39.44	14.21	8.61	6.38	5.22	4.52	4.05	3.72	3.47
	.99	6181.40	99.44	26.79	14.11	9.64	7.48	6.24	5.44	4.89	4.49
18	.95	247.32	19.44	8.67	5.82	4.58	3.90	3.47	3.17	2.96	2.80
	.975	990.35	39.44	14.20	8.59	6.36	5.20	4.50	4.03	3.70	3.45
	.99	6191.50	99.44	26.75	14.08	9.61	7.45	6.21	5.41	4.86	4.46
19	.95	247.69	19.44	8.67	5.81	4.57	3.88	3.46	3.16	2.95	2.79
	.975	991.80	39.45	14.18	8.58	6.34	5.18	4.48	4.02	3.68	3.44
	.99	6200.60	99.45	26.72	14.05	9.58	7.42	6.18	5.38	4.83	4.43
20	.95	248.01	19.45	8.66	5.80	4.56	3.87	3.44	3.15	2.94	2.77
	.975	993.10	39.45	14.17	8.56	6.33	5.17	4.47	4.00	3.67	3.42
	.99	6208.70	99.45	26.69	14.02	9.55	7.40	6.16	5.36	4.81	4.41

Table A.6

Continued

| | | \multicolumn{10}{c}{r_2} | | | | | | | | |
r_1	p	11	12	13	14	15	16	17	18	19	20
11	.95	2.82	2.72	2.63	2.57	2.51	2.46	2.41	2.37	2.34	2.31
	.975	3.47	3.32	3.20	3.09	3.01	2.93	2.87	2.81	2.76	2.72
	.99	4.46	4.22	4.02	3.86	3.73	3.62	3.52	3.43	3.36	3.29
12	.95	2.79	2.69	2.60	2.53	2.48	2.42	2.38	2.34	2.31	2.28
	.975	3.43	3.28	3.15	3.05	2.96	2.89	2.82	2.77	2.72	2.68
	.99	4.40	4.16	3.96	3.80	3.67	3.55	3.46	3.37	3.30	3.23
13	.95	2.76	2.66	2.58	2.51	2.45	2.40	2.35	2.31	2.28	2.25
	.975	3.39	3.24	3.12	3.01	2.92	2.85	2.79	2.73	2.68	2.64
	.99	4.34	4.10	3.91	3.75	3.61	3.50	3.40	3.32	3.24	3.18
14	.95	2.74	2.64	2.55	2.48	2.42	2.37	2.33	2.29	2.26	2.22
	.975	3.36	3.21	3.08	2.98	2.89	2.82	2.75	2.70	2.65	2.60
	.99	4.29	4.05	3.86	3.70	3.56	3.45	3.35	3.27	3.19	3.13
15	.95	2.72	2.62	2.53	2.46	2.40	2.35	2.31	2.27	2.23	2.20
	.975	3.33	3.18	3.05	2.95	2.86	2.79	2.72	2.67	2.62	2.57
	.99	4.25	4.01	3.82	3.66	3.52	3.41	3.31	3.23	3.15	3.09
16	.95	2.70	2.60	2.51	2.44	2.38	2.33	2.29	2.25	2.21	2.18
	.975	3.30	3.15	3.03	2.92	2.84	2.76	2.70	2.64	2.59	2.55
	.99	4.21	3.97	3.78	3.62	3.49	3.37	3.27	3.19	3.12	3.05
17	.95	2.69	2.58	2.50	2.43	2.37	2.32	2.27	2.23	2.20	2.17
	.975	3.28	3.13	3.00	2.90	2.81	2.74	2.67	2.62	2.57	2.52
	.99	4.18	3.94	3.75	3.59	3.45	3.34	3.24	3.16	3.08	3.02
18	.95	2.67	2.57	2.48	2.41	2.35	2.30	2.26	2.22	2.18	2.15
	.975	3.26	3.11	2.98	2.88	2.79	2.72	2.65	2.60	2.55	2.50
	.99	4.15	3.91	3.72	3.56	3.42	3.31	3.21	3.13	3.05	2.99
19	.95	2.66	2.56	2.47	2.40	2.34	2.29	2.24	2.20	2.17	2.14
	.975	3.24	3.09	2.96	2.86	2.77	2.70	2.63	2.58	2.53	2.48
	.99	4.12	3.88	3.69	3.53	3.40	3.28	3.19	3.10	3.03	2.96
20	.95	2.65	2.54	2.46	2.39	2.33	2.28	2.23	2.19	2.16	2.12
	.975	3.23	3.07	2.95	2.84	2.76	2.68	2.62	2.56	2.51	2.46
	.99	4.10	3.86	3.66	3.51	3.37	3.26	3.16	3.08	3.00	2.94

Table A.6

Continued

r_1	p	22	24	26	28	30	32	34	36	38	40
11	.95	2.26	2.22	2.18	2.15	2.13	2.10	2.08	2.07	2.05	2.04
	.975	2.65	2.59	2.54	2.49	2.46	2.43	2.40	2.37	2.35	2.33
	.99	3.18	3.09	3.02	2.96	2.91	2.86	2.82	2.79	2.75	2.73
12	.95	2.23	2.18	2.15	2.12	2.09	2.07	2.05	2.03	2.02	2.00
	.975	2.60	2.54	2.49	2.45	2.41	2.38	2.35	2.33	2.31	2.29
	.99	3.12	3.03	2.96	2.90	2.84	2.80	2.76	2.72	2.69	2.66
13	.95	2.20	2.15	2.12	2.09	2.06	2.04	2.02	2.00	1.99	1.97
	.975	2.56	2.50	2.45	2.41	2.37	2.34	2.31	2.29	2.27	2.25
	.99	3.07	2.98	2.90	2.84	2.79	2.74	2.70	2.67	2.64	2.61
14	.95	2.17	2.13	2.09	2.06	2.04	2.01	1.99	1.98	1.96	1.95
	.975	2.53	2.47	2.42	2.37	2.34	2.31	2.28	2.25	2.23	2.21
	.99	3.02	2.93	2.86	2.79	2.74	2.70	2.66	2.62	2.59	2.56
15	.95	2.15	2.11	2.07	2.04	2.01	1.99	1.97	1.95	1.94	1.92
	.975	2.50	2.44	2.39	2.34	2.31	2.28	2.25	2.22	2.20	2.18
	.99	2.98	2.89	2.81	2.75	2.70	2.65	2.61	2.58	2.55	2.52
16	.95	2.13	2.09	2.05	2.02	1.99	1.97	1.95	1.93	1.92	1.90
	.975	2.47	2.41	2.36	2.32	2.28	2.25	2.22	2.20	2.17	2.15
	.99	2.94	2.85	2.78	2.72	2.66	2.62	2.58	2.54	2.51	2.48
17	.95	2.11	2.07	2.03	2.00	1.98	1.95	1.93	1.92	1.90	1.89
	.975	2.45	2.39	2.34	2.29	2.26	2.22	2.20	2.17	2.15	2.13
	.99	2.91	2.82	2.75	2.68	2.63	2.58	2.54	2.51	2.48	2.45
18	.95	2.10	2.05	2.02	1.99	1.96	1.94	1.92	1.90	1.88	1.87
	.975	2.43	2.36	2.31	2.27	2.23	2.20	2.17	2.15	2.13	2.11
	.99	2.88	2.79	2.72	2.65	2.60	2.55	2.51	2.48	2.45	2.42
19	.95	2.08	2.04	2.00	1.97	1.95	1.92	1.90	1.88	1.87	1.85
	.975	2.41	2.35	2.29	2.25	2.21	2.18	2.15	2.13	2.11	2.09
	.99	2.85	2.76	2.69	2.63	2.57	2.53	2.49	2.45	2.42	2.39
20	.95	2.07	2.03	1.99	1.96	1.93	1.91	1.89	1.87	1.85	1.84
	.975	2.39	2.33	2.28	2.23	2.20	2.16	2.13	2.11	2.09	2.07
	.99	2.83	2.74	2.66	2.60	2.55	2.50	2.46	2.43	2.40	2.37

Table A.6

Continued

		r_2									
r_1	p	1	2	3	4	5	6	7	8	9	10
22	.95	248.58	19.45	8.65	5.79	4.54	3.86	3.43	3.13	2.92	2.75
	.975	995.36	39.45	14.14	8.53	6.30	5.14	4.44	3.97	3.64	3.39
	.99	6222.80	99.45	26.64	13.97	9.51	7.35	6.11	5.32	4.77	4.36
24	.95	249.05	19.45	8.64	5.77	4.53	3.84	3.41	3.12	2.90	2.74
	.975	997.25	39.46	14.12	8.51	6.28	5.12	4.41	3.95	3.61	3.37
	.99	6234.60	99.46	26.60	13.93	9.47	7.31	6.07	5.28	4.73	4.33
26	.95	249.45	19.46	8.63	5.76	4.52	3.83	3.40	3.10	2.89	2.72
	.975	998.85	39.46	14.11	8.49	6.26	5.10	4.39	3.93	3.59	3.34
	.99	6244.60	99.46	26.56	13.89	9.43	7.28	6.04	5.25	4.70	4.30
28	.95	249.80	19.46	8.62	5.75	4.50	3.82	3.39	3.09	2.87	2.71
	.975	1000.20	39.46	14.09	8.48	6.24	5.08	4.38	3.91	3.58	3.33
	.99	6253.20	99.46	26.53	13.86	9.40	7.25	6.02	5.22	4.67	4.27
30	.95	250.10	19.46	8.62	5.75	4.50	3.81	3.38	3.08	2.86	2.70
	.975	1001.40	39.46	14.08	8.46	6.23	5.07	4.36	3.89	3.56	3.31
	.99	6260.60	99.47	26.50	13.84	9.38	7.23	5.99	5.20	4.65	4.25
32	.95	250.36	19.46	8.61	5.74	4.49	3.80	3.37	3.07	2.85	2.69
	.975	1002.50	39.47	14.07	8.45	6.21	5.05	4.35	3.88	3.55	3.30
	.99	6267.20	99.47	26.48	13.81	9.36	7.21	5.97	5.18	4.63	4.23
34	.95	250.59	19.47	8.61	5.73	4.48	3.79	3.36	3.06	2.85	2.68
	.975	1003.40	39.47	14.06	8.44	6.20	5.04	4.34	3.87	3.53	3.29
	.99	6272.90	99.47	26.46	13.79	9.34	7.19	5.95	5.16	4.61	4.21
36	.95	250.79	19.47	8.60	5.73	4.47	3.79	3.35	3.06	2.84	2.67
	.975	1004.20	39.47	14.05	8.43	6.19	5.03	4.33	3.86	3.52	3.27
	.99	6278.10	99.47	26.44	13.78	9.32	7.17	5.94	5.14	4.59	4.19
38	.95	250.98	19.47	8.60	5.72	4.47	3.78	3.35	3.05	2.83	2.67
	.975	1004.90	39.47	14.04	8.42	6.18	5.02	4.32	3.85	3.51	3.26
	.99	6282.60	99.47	26.43	13.76	9.31	7.16	5.92	5.13	4.58	4.18
40	.95	251.14	19.47	8.59	5.72	4.46	3.77	3.34	3.04	2.83	2.66
	.975	1005.60	39.47	14.04	8.41	6.18	5.01	4.31	3.84	3.51	3.26
	.99	6286.80	99.47	26.41	13.75	9.29	7.14	5.91	5.12	4.57	4.17

Table A.6

Continued

r_1	p	r_2									
		11	**12**	**13**	**14**	**15**	**16**	**17**	**18**	**19**	**20**
22	.95	2.63	2.52	2.44	2.37	2.31	2.25	2.21	2.17	2.13	2.10
	.975	3.20	3.04	2.92	2.81	2.73	2.65	2.59	2.53	2.48	2.43
	.99	4.06	3.82	3.62	3.46	3.33	3.22	3.12	3.03	2.96	2.90
24	.95	2.61	2.51	2.42	2.35	2.29	2.24	2.19	2.15	2.11	2.08
	.975	3.17	3.02	2.89	2.79	2.70	2.63	2.56	2.50	2.45	2.41
	.99	4.02	3.78	3.59	3.43	3.29	3.18	3.08	3.00	2.92	2.86
26	.95	2.59	2.49	2.41	2.33	2.27	2.22	2.17	2.13	2.10	2.07
	.975	3.15	3.00	2.87	2.77	2.68	2.60	2.54	2.48	2.43	2.39
	.99	3.99	3.75	3.56	3.40	3.26	3.15	3.05	2.97	2.89	2.83
28	.95	2.58	2.48	2.39	2.32	2.26	2.21	2.16	2.12	2.08	2.05
	.975	3.13	2.98	2.85	2.75	2.66	2.58	2.52	2.46	2.41	2.37
	.99	3.96	3.72	3.53	3.37	3.24	3.12	3.03	2.94	2.87	2.80
30	.95	2.57	2.47	2.38	2.31	2.25	2.19	2.15	2.11	2.07	2.04
	.975	3.12	2.96	2.84	2.73	2.64	2.57	2.50	2.44	2.39	2.35
	.99	3.94	3.70	3.51	3.35	3.21	3.10	3.00	2.92	2.84	2.78
32	.95	2.56	2.46	2.37	2.30	2.24	2.18	2.14	2.10	2.06	2.03
	.975	3.10	2.95	2.82	2.72	2.63	2.55	2.49	2.43	2.38	2.33
	.99	3.92	3.68	3.49	3.33	3.19	3.08	2.98	2.90	2.82	2.76
34	.95	2.55	2.45	2.36	2.29	2.23	2.17	2.13	2.09	2.05	2.02
	.975	3.09	2.94	2.81	2.71	2.62	2.54	2.47	2.42	2.37	2.32
	.99	3.90	3.66	3.47	3.31	3.18	3.06	2.96	2.88	2.81	2.74
36	.95	2.54	2.44	2.35	2.28	2.22	2.17	2.12	2.08	2.04	2.01
	.975	3.08	2.93	2.80	2.69	2.60	2.53	2.46	2.40	2.35	2.31
	.99	3.89	3.65	3.45	3.29	3.16	3.05	2.95	2.86	2.79	2.72
38	.95	2.54	2.43	2.35	2.27	2.21	2.16	2.11	2.07	2.03	2.00
	.975	3.07	2.92	2.79	2.68	2.59	2.52	2.45	2.39	2.34	2.30
	.99	3.87	3.63	3.44	3.28	3.15	3.03	2.93	2.85	2.77	2.71
40	.95	2.53	2.43	2.34	2.27	2.20	2.15	2.10	2.06	2.03	1.99
	.975	3.06	2.91	2.78	2.67	2.59	2.51	2.44	2.38	2.33	2.29
	.99	3.86	3.62	3.43	3.27	3.13	3.02	2.92	2.84	2.76	2.69

Table A.6

Continued

r_1	p	r_2									
		22	24	26	28	30	32	34	36	38	40
22	.95	2.05	2.00	1.97	1.93	1.91	1.88	1.86	1.85	1.83	1.81
	.975	2.36	2.30	2.24	2.20	2.16	2.13	2.10	2.08	2.05	2.03
	.99	2.78	2.70	2.62	2.56	2.51	2.46	2.42	2.38	2.35	2.33
24	.95	2.03	1.98	1.95	1.91	1.89	1.86	1.84	1.82	1.81	1.79
	.975	2.33	2.27	2.22	2.17	2.14	2.10	2.07	2.05	2.03	2.01
	.99	2.75	2.66	2.58	2.52	2.47	2.42	2.38	2.35	2.32	2.29
26	.95	2.01	1.97	1.93	1.90	1.87	1.85	1.82	1.81	1.79	1.77
	.975	2.31	2.25	2.19	2.15	2.11	2.08	2.05	2.03	2.00	1.98
	.99	2.72	2.63	2.55	2.49	2.44	2.39	2.35	2.32	2.28	2.26
28	.95	2.00	1.95	1.91	1.88	1.85	1.83	1.81	1.79	1.77	1.76
	.975	2.29	2.23	2.17	2.13	2.09	2.06	2.03	2.00	1.98	1.96
	.99	2.69	2.60	2.53	2.46	2.41	2.36	2.32	2.29	2.26	2.23
30	.95	1.98	1.94	1.90	1.87	1.84	1.82	1.80	1.78	1.76	1.74
	.975	2.27	2.21	2.16	2.11	2.07	2.04	2.01	1.99	1.96	1.94
	.99	2.67	2.58	2.50	2.44	2.39	2.34	2.30	2.26	2.23	2.20
32	.95	1.97	1.93	1.89	1.86	1.83	1.80	1.78	1.76	1.75	1.73
	.975	2.26	2.19	2.14	2.10	2.06	2.02	2.00	1.97	1.95	1.93
	.99	2.65	2.56	2.48	2.42	2.36	2.32	2.28	2.24	2.21	2.18
34	.95	1.96	1.92	1.88	1.85	1.82	1.79	1.77	1.75	1.74	1.72
	.975	2.24	2.18	2.13	2.08	2.04	2.01	1.98	1.96	1.93	1.91
	.99	2.63	2.54	2.46	2.40	2.35	2.30	2.26	2.22	2.19	2.16
36	.95	1.95	1.91	1.87	1.84	1.81	1.78	1.76	1.74	1.73	1.71
	.975	2.23	2.17	2.11	2.07	2.03	2.00	1.97	1.94	1.92	1.90
	.99	2.61	2.52	2.45	2.38	2.33	2.28	2.24	2.21	2.17	2.14
38	.95	1.95	1.90	1.86	1.83	1.80	1.78	1.75	1.73	1.72	1.70
	.975	2.22	2.16	2.10	2.06	2.02	1.99	1.96	1.93	1.91	1.89
	.99	2.60	2.51	2.43	2.37	2.31	2.27	2.23	2.19	2.16	2.13
40	.95	1.94	1.89	1.85	1.82	1.79	1.77	1.75	1.73	1.71	1.69
	.975	2.21	2.15	2.09	2.05	2.01	1.98	1.95	1.92	1.90	1.88
	.99	2.58	2.49	2.42	2.35	2.30	2.25	2.21	2.18	2.14	2.11

APPENDIX B

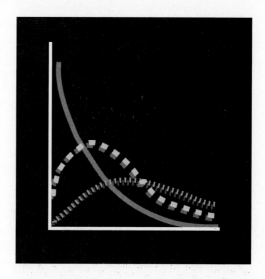

MINITAB PRIMER

Minitab is a sophisticated software package for statistical analysis. It is available for many hardware platforms, although here we will assume that you are using an IBM-compatible microcomputer. The Macintosh version of *Minitab* is little different, and the mainframe versions are only slightly different, so if you are using one of these versions you should not have much difficulty adapting. Since a version of *Minitab* for Windows will be widely available in microcomputer labs at publication of this text, you will see how to do things using the Windows interface and will also encounter *Minitab* command syntax that can be applied in DOS-based or other systems.

Rather than thoroughly exploring all of *Minitab's* many capabilities, the goal of this appendix is to introduce you to the way it handles the few, simple statistical procedures that we cover in this text. For more information, you can check out the documentation that accompanies the program. Also, it is assumed that you have the appropriate statistical knowledge as you read or consult this appendix, so it will concentrate only on *Minitab* details.

B.1 | Basic *Minitab*

You can launch *Minitab* by moving into the directory where it is located and typing "Minitab" at the system prompt or by using the mouse to double-click on the *Minitab* icon in a windowing environment.

While you are interacting with *Minitab* two or more windows will be open. The most important are the Session window, in which your *Minitab* commands are typed (after a prompt displayed by *Minitab* as MTB>), and the Data window, which holds your current data. Examples of each are displayed in Fig. B.1.

When you call for tasks to be done by *Minitab* for Windows using the menus at the top of the screen, you will notice *Minitab* typing the appropriate command syntax in the Session window. You can also enter those commands yourself. If you do so, be aware that *Minitab* is not case-sensitive and that it allows you to freely insert commenting words to aid in documentation. For instance, in the command "REGRESS c3 on 2 predictors c1,c2" the words *on* and *predictors* are unnecessary inclusions that simply make the command more readable. *Minitab* understands the terser expression of the command: "REGRESS c3 2 c1,c2". The double quote marks are for the purpose of demarcating *Minitab* commands in this appendix and are not to be typed as part of the command syntax. Key words that must be included will be displayed in all uppercase format, and optional comments in lower case. *Minitab* itself does not care.

The Data window is a worksheet of columns similar to the worksheet provided by a spreadsheet program. One usually thinks of a sample of observations from some one-dimensional population as residing in a column. Columns are denoted C1, C2, etc., and can hold nonnumeric as well as numeric data. Although they are not displayed, *Minitab* also has constants denoted K1, K2, ... and matrices denoted M1, M2, ... into which you can store values for use in computations.

```
MTB > info

Column    Name    Count
C1        x        3
C2        y        3

MTB > mean c1
   MEAN  =   3.5000
```

	C1	C2
	x	y
1	2.0	4.1
2	3.1	2.5
3	5.4	5.7

(a) Session window   (b) Data window

Figure B.1 *Minitab* Windows

B.2 | File Handling

Minitab begins by supplying you with a new, blank data worksheet and a blank Session window. You can click on the Data Window, or select it from the Window menu, then enter a new data set. If you have an old data set in a *Minitab* worksheet file that you want to process, do the following: Select Open Worksheet off of the File menu, move to the appropriate directory by clicking Select File in the Open dialog box, then select the worksheet and click OK. The data will be loaded in the same format in which it was previously saved. To get an old worksheet using a *Minitab* command, type

$$MTB > RETRIEVE \text{ `FILENAME'}$$

in the session window. The filename is the DOS name in the current directory, or the full pathname if the worksheet is located somewhere other than the current directory. The single quotes around file names are indeed an essential part of the syntax. For information on loading data files that are in ASCII or some format other than *Minitab* worksheet format, see the *Minitab Reference Manual*, included with the software.

You can save a worksheet that is currently in memory by selecting Save Worksheet off the File menu. If you want to move the worksheet to a different location, or if you are naming the worksheet for the first time, use SaveWorksheet As instead. Click on the appropriate folders to reach the directory you want, then type the new filename in the box and click OK. In the Session window you can type

$$MTB > SAVE \text{ `FILENAME'}$$

The name of the file is typed as described in the section on the RETRIEVE command. You will note that *Minitab* gives worksheets a default .MTW extension.

B.3 | Getting Help

You can get helpful information in Windows by opening the Help menu at the top and following instructions. You can also type "HELP" for a list of topics, or "HELP" followed by the topic name of your choice at the MTB> prompt in the Session window to get help on a specific topic. For instance, "HELP RANDOM" would give you information about the "RANDOM" command.

B.4 | Ending a Session

After completing your work and perhaps saving the worksheet you may quit *Minitab* by selecting Exit off the File menu. Alternatively, on the command line in the session window you can type "STOP".

B.5 | Worksheet Information

You can get information about which worksheet columns are in use by opening up the *Minitab* Info Window. Select Info off the Window menu at the top of the screen. The command equivalent is "INFO" at the MTB> prompt.

B.6 | Data Entry and Editing

To enter new data or to edit or view old data, switch to the Data window by selecting it from the Window menu at the top of the screen. Using the mouse and the scroll bars or the keyboard arrows, you can move to whatever row and column you desire. Then just type in the new data. For non-Windows systems, check your documentation to see how to switch to the Data window (the ESC key may toggle back and forth). You may need to learn the SET and READ commands if you are using an early version of *Minitab* without on-screen editing capability; refer to the *Reference Manual* for these commands. For short corrections, the command equivalent of typing over, say, the 5th row in the 3rd column is

MTB > LET C3(5) = new data

B.7 | Reformatting the Worksheet in Windows

The Windows version of *Minitab* simplifies inserting, deleting, and moving existing columns or rows. To insert a blank row above the current cursor position, select Insert Row off the Edit menu. To delete a row, move the cursor to the row header in the left margin for the row you want to delete, and select Delete Cells from the Edit menu. To insert one or more new columns, manually cut and paste existing columns that are to be moved. Specifically, move the cursor to the column header where you want the new blank column(s), and drag to the right to shade all columns that must be moved. Select Cut Cells from the Edit menu to move these data temporarily to the clipboard. There will now be blank columns. Move the cursor to the column header of the column where you want the old data to be placed, and select Paste Cells from the Edit menu. A similar procedure can move whole rows. If you want to delete a whole column, click on its header and select Delete Cells off the Edit menu.

B.8 | Computed Data

In Windows, the Calc menu provides the means for computing new data and setting the results into columns. The Mathematical Expressions selection on that menu has a dialog box in which you specify which column is to contain the result of the calculation, and what the expression for the new column is as a function of the old columns. Multiplication, division, addition, and subtraction use the notation $*$, $/$, $+$, and $-$, respectively, and exponentiation is done by $**$ (e.g., $C1^3$ would be C1**3). Many standard mathematical functions are provided for use in expressions: SQRT ( ) for square root; LOGE ( ) and LOGTEN ( ) for base e and base 10 logs; SIN ( ), COS ( ), etc., for trig functions; ROUND ( ) for rounding; and ABSOLUTE ( ) for absolute value. Special statistical functions such as MEAN ( ), STDEV ( ), SUM ( ), SORT ( ), and RANK ( ) are also provided to give summary information about or transform a column of data. The command equivalent uses LET. To illustrate:

$$\text{MTB} > \text{LET C2} = \text{LOGE(C1)} - \text{SQRT(C3)}$$
$$\text{MTB} > \text{LET K1} = \text{MEAN(C1)}$$
$$\text{MTB} > \text{LET K3} = \text{MEAN(C3)}/(2*\text{STDEV(C3)}).$$

B.9 | Patterned Data

The selection Set Patterned Data from the Calc menu allows you to put data that follow a pattern into a column. You are asked for the pattern and how many times that pattern is to be repeated in the appropriate dialog box, which is relatively self-explanatory. This is convenient for ANOVA problems (see later) in which a column of data needs to be tagged with numbers in another column indicating factor levels for the particular data elements. The command equivalent uses "SET". Type "SET Ck" at the MTB> prompt to begin the process of setting the patterned data into column k. On the next line type

$$n \text{ (low:high) } m$$

to put n copies of the range of integers from low to high, each repeated m times, into column k. End the command by typing the word "END". For example, "SET C1 (1:4)3 END" would enter the pattern 111 222 333 444 into C1. The command "SET C3 2(1:2)" would enter 12 12 into column 3, and the command "SET C4 2(0:1)3" would put 000 111 000 111 into column 4.

B.10 | Data Sampled from Distributions

To place a random sample of observations from a distribution into a column, use the Random Data selection in the Calc menu. You will be asked how many observations you want, which column they should go into, and which distribution they are to be sampled from. *Minitab* names of important distributions include the following:

BINOMIAL n p

POISSON mu

NORMAL mu sigma (the second parameter is the standard deviation)

UNIFORM a b

CHISQUARE r

EXPONENTIAL b (the parameter is the mean of the distribution)

GAMMA a b (the second parameter is $1/\lambda$ in our notation)

WEIBULL a b

Using the Session window to enter the command yourself, you would type something like the following:

MTB > RANDOM 15 C3;

SUBC > uniform 2 12.

This command will place 15 observations from the uniform [2,12] distribution into column 3. The semicolon indicates that a subcommand is to follow. *Minitab* then prints the SUBC> prompt. More than one subcommand is available for some commands, and you can continue to enter them on lines terminated by semicolons. The last subcommand line is terminated by a period.

B.11 | Naming Columns

Usually columns correspond to variables in a data set. It is helpful to give these columns mnemonic names, which will then be used in certain tables and graphs by *Minitab*. To give a column a name, switch to the Data window and click on the column header, then type the name, which must be 8 characters or less. The equivalent command is

$$\text{MTB} > \text{NAME Cn 'name'}$$

The quotes around the name are necessary, both in this command and in any others that use the column by its name (e.g., LET 'heights' = 'old' + 1). Once named, the column name will appear in the INFO table.

B.12 | Producing Plots

Discussion here will be limited to producing simple plots using default options. *Minitab* for Windows offers many ways of customizing graphs using the Annotation, Frame, and Regions options in the Plot dialog box, so in the interest of space I will encourage you to discover them by trial and error or refer to the *Reference Manual*.

The most frequently needed plot is probably the scattergram of values of one variable against another variable with which it is associated. To produce a scattergram in *Minitab* for Windows, open the Graph menu at the top of the screen and select Plot. A dialog box will appear and ask you to choose a column variable for the *y*-axis (by clicking on one of a list of variables) and another column variable for the *x*-axis. Multiple graphs can be specified at once. Click the OK button to produce the plot. For non-Windows *Minitab* versions, the command is

$$\text{MTB} > \text{PLOT Ycolumn vs Xcolumn}$$

For example, "PLOT c3 vs c2" will plot the data points (x, y) found in the rows of columns 2 and 3, respectively.

Probably second in importance is the histogram. Select Histogram off the Graph menu, and choose an X variable column that contains the data to be plotted. The command equivalent is

$$\text{MTB} > \text{HISTOGRAM Xcolumn}$$

A time series plot can be produced by choosing the Time Series Plot entry from the Graph menu. You will have to specify a Y column variable that contains the series data. The command to do the plot is "TSPLOT Ycolumn".

Minitab can use a column of data to produce a box-and-whisker plot using the Boxplot command from the Graph menu. The column that holds the data must be entered in the Y variable list. To get a box-and-whisker plot by issuing a command in the Session window, type "BOXPLOT Ck" at the MTB>prompt.

Finally, dot plots of one or more data sets can be obtained from the Dotplot subcommand of the Character Graphs selection in the Graph menu. More than one plot can be produced at once. The corresponding command syntax is, for example,

$$\text{MTB} > \text{DOTPLOT C2, C3, C5}$$

to give dotplots of three variables in columns 2, 3, and 5. They may be plotted using the same horizontal scale by including the subcommand SAME.

B.13 | Elementary Statistics and Hypothesis Tests

The classical statistical estimation and testing procedures depend on the calculation of the elementary statistics, especially the mean and variance, of a random sample. To obtain these statistics from *Minitab* for Windows, click open the Stat menu at the top of the screen, then open the Basic Statistics submenu, and select Descriptive Statistics. Enter the column or columns for which you want the statistics. The equivalent command to use in the Session window is

$$\text{MTB} > \text{DESCRIBE C}, \ldots, \text{C}$$

The Basic Statistics submenu also includes various normal theory tests and confidence intervals. This feature is a convenience rather than a necessity in light of the Descriptive Statistics command. The procedures for normal mean with known variance are under 1-Sample Z Confidence Interval and 1-Sample Z Test Mean. For the confidence interval you must enter the column the data is in, the level of confidence, and the assumed standard deviation. The hypothesis test also requires the data column and σ, the hypothesized value μ_0 of the mean, and the designation of a one-sided or two-sided test. A p-value is reported. In command form, these routines are invoked by the commands "ZINTERVAL LEVEL SIGMA Ck" and "ZTEST MU0 SIGMA Ck". For example, to obtain a confidence interval of level 90% assuming $\sigma = 4$ for the data in column 2, you would type "ZINTERVAL 90 4 C2". To do a two-sided test of $H_0 : \mu = 1.5$ for the same data, the command would be "ZTEST 1.5 4 C2".

The single sample t-procedures for the unknown variance case are very similar and are located under 1-Sample t in the Basic Statistics submenu of the Stat menu. The confidence interval is requested in the same way as in the known variance case, omitting the sigma argument, and similarly for the hypothesis test. In command language, in place of ZINTERVAL and ZTEST are TINTERVAL and TTEST. In either the z- or t-hypothesis test, you can supply *Minitab* with a subcommand in order to get a p-value appropriate for a one-sided test. Recall that to issue a subcommand of a command, you end the command line with a semicolon, and the subcommand prompts SUBC> appears. At that point type ALTERNATIVE = 1 or ALTERNATIVE = −1, respectively, for the alternative hypotheses $\mu > \mu_0$ and $\mu < \mu_0$. End the subcommand with a period.

A two sample t-test is provided by clicking on the Stat menu, then Basic Statistics, and 2-Sample t. There is an option Samples in Different Columns, which you will probably need frequently. You can specify the two columns that the sample is in, whether to use the pooled estimate of the standard deviation, the confidence level to use in a confidence interval for the difference of means, and the type of alternative to use in the hypothesis test. The command syntax is very simple: "TWOSAMPLE level C1 C2", with possible subcommands ALTERNATIVE = 1 or −1 and POOLED to use S_p as the denominator in the t-statistic as opposed to

$$S = \sqrt{\frac{S_1^2}{n_1} + \frac{S_2^2}{n_2}}.$$

Consult the *Minitab Reference Manual* for information about the form of the 2-Sample t command in which all data are in a single column and a second column indicates which sample each data item belongs to.

You can also calculate sample covariances and correlations by choosing either Covariance or Correlation from the Basic Statistics command menu. You must specify the columns of data for which you want these measures of association. If more than two columns are specified, then a covariance or correlation matrix is output, which can be saved into one of *Minitab*'s matrix variables. To calculate covariances and correlations in the Session window, the Commands "CORRELATION columns" and "COVARIANCE columns" can be used. Optionally you can specify a matrix into which to save the information, as in

MTB > COVARIANCE C1, C2, C3 M1

B.14 | Regression

Under the Stat menu, click on the Regression entry, then again on the word Regression. You will get a dialog box in which you can, among other things, specify a column where the dependent variable data are located and other columns which correspond to the predictor variables. You can also check boxes that save the regression residuals and fitted values into unused columns. The corresponding command is

MTB > REGRESS Ycolumn on K predictors XColumnlist

For example,

$$MTB > REGRESS\ C2\ 3\ C4\text{-}C6$$

does a multiple regression with 3 predictor variables in columns 4 through 6, and a response variable in column 2. Among the many available sub-commands are: "FITS column", which puts the predicted values into the designated column, "RESIDUALS column", which does the same for the residuals; and "PREDICT columns", which compute predicted values of the response variable for the new predictor variable values contained in the specified columns. To learn much more, consult your documentation.

B.15 | Analysis of Variance

For analysis of variance problems, *Minitab* provides several alternatives, but here we will introduce only the most general method. Before discussing it, we must first look at how *Minitab* lays out data in the worksheet and how you write a linear model in *Minitab*.

ANOVA problems sometimes deal with great masses of data, organized by the levels of the treatments (factors) applied to the data. So, in addition to a column for responses, there must be a column corresponding to each factor. The entry in row i of a factor column is an integer code for the level of that factor used for the response in that row. For example, in Fig. B.2 level 0 of factor A and level 0 of factor B were used to produce the response observation 25.6. Entering the data in a patterned way will save time and allow you to use *Minitab*'s patterned data entry facility to quickly set up the correct levels in the factor columns.

A model contains one or more factors, and perhaps interaction terms between pairs, triples, and so on, of factors. *Minitab* uses the asterisk to represent an interaction and provides the vertical bar as a device to include all interactions. For example,

"C1 C2 C1*C2", equivalently "C1|C2"

	Response	Factor A	Factor B
1	25.6	0	0
2	22.1	0	1
3	23.0	0	0
4	27.2	1	1
5	29.5	1	0
6	28.4	1	1

Figure B.2   Data Layout for ANOVA

describes the two-factor model with interaction term where the factor levels are in C1 and C2. The model "C1 C2 C3 C1*C3" is a three-way ANOVA model with factor levels in C1, C2, and C3, and all interactions except the C1-C3 interaction assumed to be 0. In the text we would write this model as

$$Y_{ijkl} = \mu + \alpha_i + \beta_j + \gamma_k + (\alpha\gamma)_{ik} + \epsilon_{ijkl}.$$

To perform the analysis and obtain an ANOVA table, select ANOVA and then General Linear Model from the Stat menu. You will need to fill in the response column and then the model using the syntax described earlier. The equivalent command syntax is, for example,

MTB > GLM Y = C1 C2 C1*C2

Note that this time the full equation is written after the command descriptor GLM.

See the manual for the Balanced ANOVA, OneWay (unstacked and stacked) and TwoWay commands, which are special cases of GLM or which assume that the layout of the data in the worksheet is different.

B.16 | Other Commands

Minitab has a few commands for nonparametric statistical methods, including the sign test, Wilcoxon test, and runs test discussed in this book. These commands are located in the Stat menu under Nonparametrics. A procedure for χ^2-testing of independence can be found in the Tables option under the Stat menu. To use this option, the contingency table entries are assumed to be in columns that you report to *Minitab*. Finally, *Minitab* gives you what amounts to comprehensive tables of cumulative distribution functions for the most common distributions. Under the Calc menu is an option Probability Distribution. If you click on it, you can choose either cumulative or inverse cumulative probabilities for the distribution you enter. In command terms, for example, the command

MTB > CDF 3;
SUBC > NORMAL 0 1.

would give you the c.d.f. value $F(3)$ for the standard normal distribution, and the command

MTB > INVCDF .5;
SUBC > POISSON 3.

would return the 50th percentile of the Poisson (3) distribution. The syntax for distribution names is as described in Section B.10.

APPENDIX C

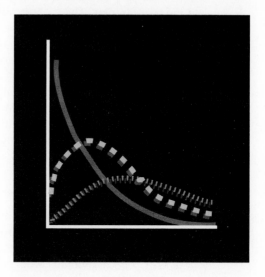

COMPUTER PROJECTS
USING *MATHEMATICA*

Ten sample projects using *Mathematica* will be discussed, and section numbers to which the problems apply most directly will follow the title head of the project. The statements of the problems to be investigated are general, and the lab may be carried out using any available symbolic-graphical package, although we assume that you will be using *Mathematica*. (*Maple* and *Math-CAD* are other possibilities, and some of the work requested can be done in *Minitab* and other statistical packages as well.) Some helpful hints for doing the problem using *Mathematica* follow the problem statement. To avoid repetition, these hints are cumulative, so for a given project you may need to refer to the hints from earlier projects. This appendix is not intended to be a comprehensive introduction to *Mathematica*; rather, in the interest of space it depends on your prior experience. Little is said about the various special options that exist for customizing *Mathematica* commands. Refer to Wolfram (1991) for more information. For information on the external packages, look at the *Guide to Standard Mathematica Packages* manual accompanying the software.

You are given instructions to do certain tasks. These instructions should be interpreted as guidelines for the minimum, not the maximum, amount of work necessary. In particular many of the labs involve simulating replications of random samples. The more replications, the more assurance you have that the phenomena you observe illustrate some general property.

Follow-up questions and side issues will arise. You may even conceive of your own investigation, because the current technology can be used in many ways to shed light on statistical problems beyond those dealt with here. Areas to look at include approximation of binomial probabilities by Poisson probabilities, validation of the Strong Law of Large Numbers, simulation of confidence intervals, investigation of the effect of outliers on statistical methods such as hypothesis testing and regression, and power of F-tests for analysis of variance.

Project 1: Disease Detection (Section 1.4)

A condition or disease has a prevalence p among the general population. A test does exist to detect the condition, but it is imperfect. For individuals who do have the condition, the conditional probability is $q < 1$ that the test detects it. For individuals who do not have the condition, the conditional probability is $r > 0$ that the test reports a positive result anyway, which is an erroneous test. If the condition is an alarming one such as the presence of the HIV virus, we would like our test to have the characteristic that a positive result means that it is very likely that the patient truly is infected, else we will be striking fear into uninfected patients unnecessarily. The question is, To what factors is the conditional probability of infection given a positive test most sensitive? Would it make sense to devote a lot of effort to increasing the chance of a positive result on an infected person, or should we spend more time trying to reduce the probability that an uninfected individual has a positive test result? Does the decision depend on the magnitude of p; in other words, should we allocate our effort differently when a condition is rampant in the population than if it is rare?

Investigate this question by doing the following steps:

1. Write an analytical formula for the conditional probability that a tested person has the condition given a positive test result.
2. Evaluate the conditional probability for numerous combinations of values for the parameters p, q, and r. You must decide what is a reasonable value for each parameter.
3. Produce appropriate plots of the conditional probability as a function of the parameters, and comment on what they are telling you. You may find three-dimensional surface plots to be valuable tools.

Reach a conclusion: If you had a million dollars in research and development money to use to fine-tune the test, how would you spend it, and how would your decision change as the condition spreads in the population?

Mathematica Details. The general syntax for a function definition in *Math-*

ematica is $f[x\_, y\_, \ldots] := (formula)$, for example,

$$f[x\_] := x\hat{}2 + 1 \quad \text{or} \quad g[x\_, y\_] := Sin[x * y].$$

The underscore symbol and colon are very important elements of the syntax. You can evaluate functional values by typing the function's name and the value of the argument, for example, $f[2]$ or $g[Pi, 0]$. Remember that *Mathematica* names are case-sensitive.

Mathematica will attempt to give exact analytical expressions rather than decimal approximations when it makes computations. For instance, $f[Pi]$ would turn out to be $Pi\hat{}2 + 1$ for the preceding function. If you want a decimal approximation of any arithmetical expression, use the numerical evaluation operator $N[\ ]$—for instance, $N[f[Pi]\ ]$ or $N[Pi\hat{}2 + 1]$.

To sketch the graph of a function over a domain $[a, b]$, you can use the *Plot* command, which takes the function, and the desired plotting domain of the function as its arguments. The domain is in the form of a list such as $\{x, a, b\}$, where x is the functional variable. Several functions can be plotted together if the list of functions to be plotted is enclosed in braces. For example,

$$Plot[f[x], \{x, 0, 10\}] \quad \text{or} \quad Plot[\{f[x], 4x\}, \{x, 0, 3\}].$$

Note the absence of the underscore symbol after the functional variable here.

The *Plot3D* command draws a surface graph of a function of two variables. Its syntax is very similar to the *Plot* command, except that an interval of values for each variable must be supplied. For example,

$$Plot3D[g[x, y], \{x, 0, 2Pi\}, \{y, 0, Pi\}].$$

Project 2: Sampling With and Without Replacement (Section 2.3)

A population of size M is partitioned into two types of objects, success and failure, with N_0 total successes and $M - N_0$ failures. A sample of size n is taken at random from the population. Compare the probabilities of exactly k successes in the sample, $k = 0, 1, 2, \ldots, n$, under two alternative assumptions: (a) the sample is drawn without consideration to order and without replacement and (b) the sample is drawn in sequence and each time a population member is drawn it is replaced before the next draw. How large does the population size have to be so that each of the $n + 1$ probabilities in case (a) is uniformly close to the corresponding probability in case (b)? Which case seems to give the higher variance of number of successes? Is there an explanation for this?

Investigate this question by doing the following steps:

1. Write functions for the probability of k successes in each of cases (a) and (b). Leave the parameters of the problem, M, N_0, and n, general for the time being.
2. Specialize to the case where $M = 20$, $N_0 = 10$, and $n = 5$ first. Produce lists of each set of $n + 1$ probabilities. What is the largest difference among all $k = 0, 1, \ldots, 5$ between the probability of k successes in case (a) and the analogous probability in case (b)?

3. Sketch histograms of the probability distributions for case (a) and case (b). Which seems to have the larger variance?

4. Repeat steps 2 and 3 for larger values of M and N_0, keeping $M = 2N_0$ and $n = 5$. Find values of M and N_0 so that each of the six probabilities for case (a) is within .01 of the corresponding probability for case (b).

Try to discover whether the results change very much if you change the proportion of successes N_0/M in the population, or if you increase the sample size n. Write up your results.

Mathematica Details. Functional definition is mentioned in Project 1. If you are writing the functions yourself, you will want to use the built-in function $Binomial[n, m]$. This gives the binomial coefficient $\binom{n}{m}$.

Mathematica has prewritten packages of functions and commands, sub-divided thematically, that supplement the basic commands and that may be loaded on request. One way of loading a package is to issue a command of the form *Needs*[(package pathname)]. For this project there is a helpful package that contains definitions for the common discrete probability mass functions and their c.d.f.'s, among other things. You can load it with the command

$$Needs["Statistics`DiscreteDistributions`"].$$

Mathematica uses the back quote to separate directories and filenames much as DOS uses the backslash. The pathname of the package is enclosed in double quotes. Load the package before trying to use any of its commands, otherwise *Mathematica* will become confused, even after you remember to load the package later.

Mathematica has distribution objects, whose names and parameters are written similarly to the way that they are written in this book. There are also functions *PDF* and *CDF* which can be used to produce the p.m.f. and c.d.f. of the distribution. To give the name f to a $b(n, p)$ p.m.f., for example, you would issue the command

$$f[k_{-}] := PDF[BinomialDistribution[n, p], k].$$

So, distributions have an argument for each parameter that characterizes them, and the *PDF* function takes a distribution as its first argument, and the functional variable as its second. You can do the same with the *HypergeometricDistribution*$[n, N0, M]$ object, where the meaning of the parameters is the same as in our project description. The syntax for the *CDF* function is similar.

Another useful process is to obtain lists of values of functions. The command

$$list1 = Table[f[k], \{k, a, b\}]$$

places in the variable named *list1* the values of f starting with $k = a$ and continuing up to $k = b$ in increments of one. Its output is therefore the list $\{f[a], f[a+1], \ldots, f[b]\}$. Remember that in *Mathematica*, lists are surrounded by braces.

To plot histograms, or bar charts of any kind, you must first load a package called *Graphics'Graphics'* in the manner already described. The command *GeneralizedBarChart* is in this package. If you supply *GeneralizedBarChart* with a list of triples $\{\{pos1, height1, width1\}, \{pos2, height2, width2\}, \ldots, \{posn, heightn, widthn\}\}$, it plots bars of the given heights centered on the given base positions with widths equal to the given widths. Because of the structure of our data, you will probably find it convenient to form three appropriate lists, in which the first contains the bases $\{0, 1, 2, \ldots n\}$, the second the heights $\{f[0], f[1], f[2], \ldots, f[n]\}$, and the third the widths $\{1, 1, 1, \ldots, 1\}$, by using the *Table* command three times. Putting all of this together, you will want a command of the following form:

$$GeneralizedBarChart[Transpose[\{poslist, htlist, widlist\}]].$$

The role of the *Transpose* function is to convert three lists of $n + 1$ elements to one list of $n + 1$ triples.

And finally, since you will want to execute the second and third steps a number of times, you will want to know how to collect your commands together into a single sequence, all of which can be executed by one instruction. For this *Mathematica* has the *Module* command, in which there is a list of variable names used only in the module, followed by your sequence of commands separated by semicolons. Run your module by issuing a command consisting of its name and the arguments that it takes as input, just as you would ask for the value of a function. Example:

$$DoSomething[x\_, y\_] := Module[\{local1, local2, \ldots\}, statement1;$$
$$statement2; \ldots]$$

The list of local variables may be empty, in which case write it as: $\{\}$

Project 3: Simulation of Continuous Distributions (Section 3.3)

The purpose of this particular project is not so much to answer a question as to give first-hand experience at seeing how closely observed data follow the density of a theoretical distribution from which they are simulated. You will produce superimposed sketches of a histogram of an observed sample of data and the underlying probability density function sampled from, for two distributions: the uniform and the normal. Describe in your report of this lab what happened as the number of items in the sample increased, describe whether there was much variation between replications of the same simulation, and note any differences that there might have been between the two distributions. Are you impressed or disappointed by the fit of the sample data to the densities?

1. Simulate 50 values from the uniform [0,1] distribution, and count the frequencies in each of five equally sized subintervals of [0,1]: [0,.2), [.2,.4), [.4,.6), [.6,.8), [.8,1]. On the same graph, show the histogram whose bar heights are relative frequencies divided by subinterval length, and the uniform[0,1] density. Repeat this step twice.
2. Now simulate 100 values as in step 1.

3. Simulate 200 values from uniform [0,1], break [0,1] into 10 equally sized subintervals, and proceed as in step 1.

4. Next simulate 50 observed values of the standard normal distribution. Use class intervals $(-\infty, -3)$, $[-3, -2)$, $[-2, -1)$, $[-1, 0)$, $[0, 1)$, $[1, 2)$, $[2, 3)$ and $[3, +\infty)$, and count the frequency of values in each interval. On the same graph show the $N(0, 1)$ density and the histogram whose bar heights are the relative frequencies. Repeat this step twice.

5. Now simulate 100 values as in step 4.

6. Finally simulate 200 values from $N(0, 1)$. Use class intervals $(-\infty, -3)$, $[-3, -2.5)$, $[-2.5, -2)$, $[-2, -1.5)$, $[-1.5, -1)$, $[-1, -.5)$, $[-.5, 0)$, $[0, .5)$, $[.5, 1)$, $[1, 1.5)$, $[1.5, 2)$, $[2, 2.5)$, $[2.5, 3)$, and $[3, +\infty)$ and proceed as in step 4. On the same graph show the $N(0, 1)$ density and the histogram whose bar heights are the relative frequencies divided by the subinterval size .5.

Mathematica Details. You must load the *Statistics'ContinuousDistributions'* package (see Project 2 for instructions on loading packages) to have access to *NormalDistribution*[*mu, sigma*] and *UniformDistribution*[*min, max*]. You also must load *Graphics'Graphics'* to get the *GeneralizedBarChart* command (see Project 2) and another package called *Statistics'DataManipulation'* to get commands which count class interval frequencies.

A version of the *PDF* command, with the same syntax as described in Project 2, is contained in the *ContinuousDistributions* package, so that to get a $N(\mu, \sigma^2)$ density you would define

$$f[x\_] := PDF[NormalDistribution[mu, sigma], x].$$

Note here that it is the standard deviation, not the variance, that is the second argument. Similarly you can define a function for the uniform distribution.

To simulate random numbers from a chosen distribution, you would call on another function in the *ContinuousDistributions* package called *Random*, which takes a distribution object as its argument. For example, to get a list of 25 simulated observations from uniform[0,6], you would write

$$simlist = Table[Random[UniformDistribution[0, 6]], \{i, 1, 25\}].$$

Two useful functions in the *DataManipulation* package can be used to count frequencies in different cases. First,

$$BinCounts[datalist, \{min, max, dx\}]$$

produces a list of counts of observations in the given list of data that fall into all subintervals of equal length dx of an interval $[min, max]$, that is, the subintervals $[min, min + dx)$, $[min + dx, min + 2dx), \ldots$. Second, for unequally sized intervals, there is a similar command,

$$RangeCounts[datalist, \{cut1, cut2, cut3, \ldots, cutn\}]$$

that produces a list of counts of observations in subintervals with $cut1, cut2,$ ... as cutoff points: $(-\infty, cut1)$, $[cut1, cut2)$, $[cut2, cut3)$, ... , $[cutn, +\infty)$.

Once the frequencies are in hand, bar charts can be drawn as described in Project 2, and plots of the densities can be made using the *Plot* command described in Project 1. (Take care in the use of *GeneralizedBarChart* that you have the same number of positions as heights as widths.) Two graphics objects like bar charts and plots can be displayed on the same graph using the *Show* command. If you have given a plot a name by saying for example $graph1 = Plot[\ldots]$ and you have named another plot $graph2$, then you can instruct *Mathematica* to

$$Show[graph1, graph2]$$

in order to obtain this effect.

Project 4: Bivariate Normal Data (Section 4.4)

In this project you will simulate bivariate normal data and gain intuition about the bivariate normal distribution by comparing those data to the associated bivariate normal density surface. Your work will also illustrate results about covariance and correlation and will anticipate theorems about transformations of normal random vectors. You will be plotting the observed data, the bivariate normal surface that is the theoretical distribution, and the contours of that function. In your report you should draw conclusions about the relationship between the density and the simulated bivariate data, making use of the contours.

1. Begin by simulating 100 pairs (W_1, W_2) of independent normal random variables each with mean 0 and variance $1/169$. Then create from them 100 pairs (X_1, X_2) by the transformation

$$\begin{bmatrix} X_1 \\ X_2 \end{bmatrix} = \begin{bmatrix} 5 & 12 \\ 12 & 5 \end{bmatrix} \begin{bmatrix} W_1 \\ W_2 \end{bmatrix}.$$

   applied to each of the W pairs. Answer the question: If you knew that the joint distribution of X_1 and X_2 is still bivariate normal, what would be its parameters?

2. Now plot the 100 points in the plane. Plot the density surface from your response to the question in step 1. Plot the contours—that is, the curves $f(x, y) = c$—in the plane for various values of the constant c. What seems to be the relationship between the density surface and the data?

3. Repeat steps 1 and 2 for a new set of 100 data points obtained from the same W's by the transformation

$$\begin{bmatrix} X_1 \\ X_2 \end{bmatrix} = \begin{bmatrix} 5 & 12 \\ 26 & 13 \end{bmatrix} \begin{bmatrix} W_1 \\ W_2 \end{bmatrix}.$$

Mathematica Details.   You will need to load the *ContinuousDistributions* package described in Project 3, which contains the command *Random* and the *NormalDistribution* object.

Recall from Project 1 that a surface graph of a function $f(x, y)$ of two real variables can be plotted with the *Plot3D* command. The contours of a function of two variables can be produced with *ContourPlot*, which has virtually the same syntax as *Plot3D*:

$$ContourPlot[f[x, y], \{x, xlow, xhigh\}, \{y, ylow, yhigh\}]$$

where *xlow* and *xhigh* are the low and high boundaries on the *x* domain and *ylow* and *yhigh* are the boundaries on the *y* domain.

The *Table* command mentioned in Project 2 is very versatile. The following command creates a list of pairs $\{\{f[1], g[1]\}, \{f[2], g[2]\}, \ldots, \{f[n], g[n]\}\}$:

$$list = Table[\{f[i], g[i]\}, \{i, 1, n\}].$$

Use *Table* in conjunction with *Random* to produce the list of (W_1, W_2) pairs.

Matrix multiplication is possible in *Mathematica*. A list of lists is viewed as a matrix with each sublist as one of its rows. For example,

$$\{\{1, 0, 0\}, \{0, 1, 1\}\} = \begin{bmatrix} 1 & 0 & 0 \\ 0 & 1 & 1 \end{bmatrix}.$$

Two matrices *A* and *B* of the appropriate size can be multiplied in *Mathematica* by using the period to indicate the multiplication operator, that is, by writing *A.B*. The *Transpose* function could be useful; *Transpose[A]* returns the transpose of matrix *A*. You can find a way to convert your simulated list of pairs of *W*'s to pairs of *X*'s by matrix multiplication. Also, you can find the parameters of the new (X_1, X_2) distribution by matrix multiplication.

The *ListPlot* command can plot a given list of pairs of points in the plane. If you were to call it on the list obtained from the *Table* command, by typing *ListPlot[list]* the points $(f[i], g[i])$ would be plotted in the plane. In case the *ListPlot* command is set by default to connect successive points with a segment, which you will not want, use the following option.

$$ListPlot[list, PlotJoined -> False]$$

Project 5: Distribution of the Sample Mean (Section 6.3)

The convergence of the distribution of the sample mean to normality is illustrated here. You will observe the histogram of a large number of sample means $\overline{X}_1, \overline{X}_2, \overline{X}_3, \ldots, \overline{X}_n$, each based on random samples of some size *m* from certain distributions. The histogram's shape should convince you of the correctness of the Central Limit Theorem.

It is good experience to replicate each of the following steps several times to check that the phenomenon you observe is a consistent one. Describe carefully what you see and how it illustrates what the Central Limit Theorem says.

1. First consider the Bernoulli distribution with parameter $p = 1/2$. Simulate 25 values from this distribution, find their sample mean, and repeat this 100 times to get 100 independent observations of the sample mean. Plot a histogram of the means, from three standard deviations of $\overline{X}$ below to three standard deviations above the expected value of $\overline{X}$.

2. Repeat step 1 for the continuous uniform distribution on the interval [0,1].

3. Repeat step 1 for the exponential distribution with parameter $\lambda = 2$.

You might like to experiment with larger values of the common sample size *m* and larger numbers of sample means *n* to see whether the fit to normality improves dramatically. Further experimentation involves comparing highly skewed distributions to relatively symmetric ones. You can use the Bernoulli distribution with parameters other than 1/2 to do a sequence of experiments along these lines. You can also try other distributions than those suggested here

Mathematica Details. You will have to load the *DiscreteDistributions*, *ContinuousDistributions*, *Graphics*, and *DataManipulation* packages described in Projects 2 and 3 to gain access to the *BinomialDistribution*[*n, p*], *UniformDistribution*[*a, b*], and *ExponentialDistribution*[*lambda*] objects, the *Random* Command, and the *BinCounts*, *RangeCounts* and *GeneralizedBarChart* commands discussed in Project 3. The exponential distribution object uses the same convention on the parameter as we have in this book, namely, the mean of the distribution is 1/*lambda*.

Since you are really doing repeated simulation of a sample of sample means, it will be very much of a time saver to use the *Module* structure discussed in Project 2 to write a *Mathematica* program to do it. This module can take the target distribution, the number of means desired, and the sample size making up each mean as its arguments.

Simulating a list of sample means takes four steps: simulating a single sample value, repeating that operation to get a sample of size *m*, computing its sample mean, and repeating the whole process *n* times. You can achieve the same result by composing four *Mathematica* functions as follows:

$$Table[Mean[Table[Random[dist], \{j, 1, m\}]], \{i, 1, n\}].$$

Look from the inside out. *Random* samples a number from the distribution named. The next *Table* creates *m* such numbers for one random sample. *Mean* is a new function, which computes the average of the list of numbers it works on. The outermost *Table* repeats the process of computing a new sample mean and adding it to the list a total of *n* times.

Other *Mathematica* notes are as follows. The *Mean* function, called with a distribution instead of a list of data as its argument, will produce the theoretical mean of that distribution. Thus the name *Mean* is overloaded, but *Mathematica* can tell from context what is desired. Similarly, the *StandardDeviation* function gives the standard deviation of the distribution or of the data list according to the argument given to it. In experimenting myself, I noticed a tendency for *BinCounts* to make an extra category beyond what is expected, thereby creating inequality between the number of heights and the numbers of midpoints and widths used in the *GeneralizedBarChart* command. An error message results. The problem seems to come from roundoff errors in the computation of category divisions. I have never had a problem as long as I kept the sample size *m* as a perfect square, to avoid irrational square roots in the category widths, and I suggest you do the same.

Project 6: Distribution of the Sample Variance (Section 8.1)

The estimator $S^2 = \sum_{i=1}^{n}(X_i - \overline{X})^2/(n-1)$ of the population variance σ^2 has several good qualities. In this laboratory you will look at the density function of S^2 in order to acquire a deeper appreciation of its properties as an estimator. If the random sample $X_1, X_2, \ldots, X_n$ from which S^2 is constructed is drawn from a normal population, then the random variable $(n-1)S^2/\sigma^2$ has the $\chi^2(n-1)$ distribution. Prepare for this project by computing the density function of S^2 using this result. Your assignment is to study graphically the central tendency of the distribution and the dependence of the density of S^2 on n and σ^2, by doing the following steps:

1. Enter the formula for the density $f(x)$ of S^2, treating n and σ^2 as arguments to the function as well.
2. For $n = 50$, plot the density as a function of x for $\sigma^2 = 1, 4$, and 9. Then produce a three-dimensional plot of f as a function of x and σ^2 for fixed $n = 50$. Comment on what you see. Make numerical computations which show that S^2 is unbiased for σ^2.
3. For $\sigma^2 = 4$, plot the density as a function of x for $n = 25, 50, 100$. Then produce a three-dimensional plot of f as a function of x and n (which is illustrative even though n can only take on integer values) for fixed $\sigma^2 = 4$. Comment on what you see. What have you learned so far about S^2 as an estimator of σ^2?
4. Finally let's try a different exercise. Use a strategic trial and error approach to find n large enough that the interval $[.75\ S^2, 1.25\ S^2]$ contains a true $\sigma^2 = 9$ with probability at least 90%. For this, you will have to do several numerical integrations of the density of S^2. (Could you have solved this problem analytically?)

Mathematica Details. You will use the *Plot* and *Plot3D* commands introduced in Project 1.

In computing and graphing the density of S^2, you have a couple of options. If you are writing it out longhand, you may need the gamma function, which *Mathematica* knows as *Gamma[a]*. Alternatively, the *ContinuousDistributions* package discussed in Project 3 has the *ChiSquareDistribution[r]* object, which could save you some effort if you do your analysis in step 1 correctly. Recall from Project 3 the *PDF* function, which yields the associated density as a function of a variable.

Finally, you can obtain a numerical estimate of the integral of a function f over an interval $[a, b]$ using the command

$$NIntegrate[f[x], \{x, a, b\}].$$

As long as the function is not too complicated, you can get a closed form exact definite integral by using the *Integrate* command in the same way.

Project 7: Simulation of *T*-Statistics (Section 9.2)

The purpose of this exercise is to let you view the results of many t-tests for the two-sided hypothesis $H_0 : \mu = \mu_0$ vs. $H_1 : \mu \neq \mu_0$. Assuming that the random sample $X_1, X_2, \ldots, X_n$ on which the t-statistic is based comes from a normal distribution, you will see the results of the test when H_0 is true and when it is false. Report on the results of several replications of steps 2 and 3.

1. Write a command to simulate the value of a t-statistic based on a normal random sample of size n. It should take as arguments the mean μ and standard deviation σ of the normal distribution being sampled from, the sample size n, and the hypothesized mean μ_0. Then write another command that uses the first one to simulate a given number m of such t-statistics and plots the list of them on the same graph with horizontal lines located at the two-sided level α t-critical values $\pm t_{\alpha/2}$. (This will let you count how many of the simulated t's lead to rejection and to acceptance of the null hypothesis.) This command should have μ, σ, n, μ_0, m, and the level α as its arguments.

2. Simulate 50 t-statistics for the two-sided normal mean test with $H_0 : \mu = 0$ of level .1 based on $N(0, 1)$ random samples, each of size 20. How many of your t-statistics lead to rejection of H_0? Explain what you are seeing.

3. Simulate 50 t-statistics for the two-sided normal mean test $H_0 : \mu = 0$ of level .1 based on $N(2, 1)$ random samples, each of size 20. How many of your t-statistics lead to acceptance of H_0? Explain what you are seeing. About how close to 0 should the true mean μ be so that you see about 10% type II errors?

You can easily pursue many subsidiary investigations using the commands you developed in step 1. For instance, you can study the dependence of the empirical type II error probability on the standard deviation σ, the sample size n, the significance level α, or the number of replications m.

Mathematica Details. You will need to load the *ContinuousDistributions* package to have access to the *NormalDistribution* object and the *Random* function described in Project 3. Also, this package has in it the functions

$$Mean[listname], \quad StandardDeviation[listname]$$

which give the sample mean and standard deviation of the given list of data.

Step 1 suggests that you write two functions using *Module* (see Project 2). In the first, you simulate the normal random sample using *Random* and *Table* as described in Project 3. The functions above and the $Sqrt[x]$ function will compute a t-value, which should be the last step of the module. This will mean that the output of the command is the simulated t.

In the second module, you use *Table* applied to the first function you wrote to get the list of t's. *ListPlot* described in Project 4 can apply to that list to give a plot of the t-values (as a function of the position in the list $1, 2, 3, \ldots, m$). Name this plot. The *ContinuousDistributions* package also has the *StudentTDistribution[r]* object, where r is the degrees of freedom parameter. The critical values $t_{\alpha/2}$ can be computed using the function

$$Quantile[distribution, q]$$

which returns the $q \times 100$th percentile for the given distribution. Then name a plot of the constant functions $f(x) = \pm t_{\alpha/2}$, and use the *Show* command described in Project 3 to superimpose the two named plots.

Project 8: Empirical Power of the Wilcoxon Test (Section 9.2)

Both the Wilcoxon test and the normal mean test can be applied to the problem of testing $H_0 : \mu = 0$ vs. $H_a : \mu > 0$. A question arises: How does the power of a Wilcoxon test of the same significance level compare to the power of the normal mean test on the home ground of the latter, when the underlying population is normal?

To prepare for this project, find the large sample approximate critical region using Wilcoxon W for a test as above at level .05, based on a random sample of size 20. Do the same for the classical test based on $\overline{X}$, assuming the variance is known to be 1. Find also an expression for the power function of the classical $\overline{X}$ test. You will empirically estimate the power of the W test and compare it to the known exact power of the normal test. To do this comparison, do several replications of the following steps.

1. Simulate 100 samples of size 20 from the $N(.25, 1)$ distribution. For each sample, compute the large sample standardized Wilcoxon test statistic. Count the number of times that the null hypothesis is rejected. Compare the relative frequency of rejections to the normal power at $\mu = .25$ and comment.
2. Repeat step 1 for the $N(.5, 1)$ distribution.
3. Repeat step 1 for the $N(.75, 1)$ distribution.

Mathematica Details. This exercise will need the *ContinuousDistributions* and *DataManipulation* packages discussed in Project 3.

Since the same basic steps are to be repeated many times, you will save much work by writing your own functions and commands using the *Module* structure from Project 2. There are several ways to break down your work into two or more modules, depending on how single-minded you want each to be.

You will have to write a line of code as in Project 3 using *Table* and *Random* to simulate a normal random sample. (This may be a function of its own.) The new feature here is the calculation of W based on a list of simulated values. In this project the hypothesized median m is zero, which simplifies the calculation. Recall that the value of W is the sum of the signed ranks of the absolute values of the data. To compute W will require sorting the data on the basis of their absolute values, suppressing the magnitudes of those data once sorted, and using the signs to weight the ranks $1, 2, \ldots, 20$.

Fortunately, *Mathematica* provides a function that sorts a list according to a specified criterion for comparing two list elements. The syntax is *Sort[listname, criterion]*. The criterion is a true–false valued expression written in a special syntax. What you will need here is

$$sortedlist = Sort[listname, (Abs[\#1] < Abs[\#2])\&].$$

(The ampersand indicates an unnamed function that has what is inside parentheses as its argument.) This will produce a sorted list in which the criterion for saying that one member is less than another is that its absolute value is smaller. At this point, the signs are still attached to the data values in the sorted list.

There is also a function *Sign*[*x*] that yields the sign of its argument: that is, it takes the value +1 if $x > 0$ and -1 if $x < 0$. This will be very useful in computing W, because it can be applied to all members of a list at once to yield another list. So the list *Sign*[*sortedlist*] will contain only the signs of the sorted list.

The Wilcoxon statistic can be viewed as the dot product of the vector of signs of the sorted list with the vector [1 2 3 … 20]. This vector can be formed using the *Table* command. The dot product of two vectors written as lists can be computed by using a period to represent multiplication, for example $v1.v2$. With this information you can define a function that computes the Wilcoxon statistic W and its standardized version Z from a simulated normal random sample.

You will then need to simulate a list of such standardized Wilcoxon Z's. The module that you write to make this list should take the true mean μ, the common sample size n, and the desired number m of simulated Wilcoxon Z's as arguments, so that you can repeat your experiments under different conditions. This module will use the previous commands to simulate a normal sample and to calculate W and Z for each in order to have a list of m such Z's.

Finally, you will need to count the number of elements in the list of Z's satisfying the H_0 rejection criterion, then compute the relative frequency among the 100 replications that the Wilcoxon test correctly rejected the null hypothesis. In Project 3 we talked about the *RangeCounts* function, which you can use here on a list of 100 W values with just two categories separated by one cutoff at the critical value for rejection of the null hypothesis. Note that *RangeCounts* will give you a list of two counts if you use it in this way. You are only interested in the count of the second category, those Z's that exceeded the critical value. In *Mathematica* the notation for the second member of a list is *listname*[[2]]. If you run into the problem that *Mathematica* is trying to compute exact answers and you want decimal values, try using the $N[\,]$ function both on the list of Z's and on the output from *RangeCounts* in order to force numerical evaluation. Including a list plot of the Z's, with a horizontal line indicating the critical value, creates a nice visual tool.

For the normal power function, once you derive an expression for it, you should be able to see how to use the *CDF* function (see Project 2) and the *NormalDistribution* object to compute it easily.

Project 9: Robustness of Regression Estimates (Section 10.3)

Statistical procedures are sometimes adversely affected by violations of the assumptions underlying the model. How are regression estimates and test procedures affected when the error distribution is nonnormal?

In this project you will simulate data sets of points (x, y) that satisfy a linear relationship with random disturbance

$$Y = b_0 + b_1 x + \epsilon$$

and compute the slope estimate $\widehat{b}_1$ for each data set. You will then construct a histogram of the estimated slope parameters and compare it to an appropriate normal distribution. Specifically do the following, and comment on whether the nonnormality of the errors will make a big difference in the hypothesis testing and confidence interval procedures involving b_1. Replicate each step a few times, and feel free to increase the number of random samples from 50 to 100 if your computer runs quickly enough to do this in reasonable time.

1. Simulate 30 points (x_i, y_i) following the regression model $Y = 2 + 3x + \epsilon$ where $\epsilon \sim$ uniform $(-1, 1)$. Use x values simulated from the uniform $(0, 10)$ distribution. Obtain the regression estimate $\widehat{b}_1$ for each of 50 such samples, and plot a histogram of the distribution with a normal density superimposed. This density should have the true b_1 as its mean, and the quantity $\sigma^2 / \sum (x_i - \bar{x})^2$ as its variance, where σ^2 is the variance of the error distribution. Let the histogram have categories divided by the points $b_1 + i \cdot \sqrt{\sigma^2 / \sum (x_i - \bar{x})^2}$ for $i = -2, -1, 0, 1, 2$.
2. Repeat step 1, using instead the uniform $(-8, 8)$ distribution for the errors. Use the same x values as in step 1.
3. Repeat step 2, using instead the exp(2) distribution for the errors. Do your results surprise you in light of the fact that the exponential distribution does not have mean 0? What might b_0 have to do with this last question?
4. Repeat step 2, using instead the exp(.5) distribution for the errors.

There are many ways to enlarge and generalize the investigation. You could see how the fit to normality changes as the number of points (here 30) in each random sample increases. You could use other probability distributions for the errors, and you could study the effect of increasing variance within a family of distributions. Also, though we only ask about $\widehat{b}_1$ here, it would be interesting to look at the empirical distribution of $\widehat{b}_0$ as well.

Mathematica Details. To access the distributions needed for simulation and plotting, you will need to load the *Statistics'ContinuousDistributions'* package. You will again need the *Statistics'DataManipulation'* package for the *RangeCounts* function, and the *Graphics'Graphics'* package to use the *GeneralizedBarChart* command.

The problem requires similar simulations to be done over and over using the same list of x-values. This suggests that the structure of your *Mathematica* notebook be as follows. Do all computations involving the list of x's first, then write a *Module* (see Project 2) to simulate one $\widehat{b}_1$ value, and finally write another module to simulate the whole list of $\widehat{b}_1$ values and to do the plotting.

Use *Table, Random*, and the *UniformDistribution* object (see Project 3) to produce your list of x's once and only once. To compute $\bar{x}$, apply the *Mean* function to your list of x's (see Project 7). To compute the sum of squares $\sum (x_i - \bar{x})^2$, which will be used in a few places, take the dot product (see Project 8) of the list $\{x_i - \bar{x}\}$ with itself. Note that when you tell *Mathematica* to subtract a constant from a list (or similarly to perform addition, multiplication, or division by a constant) that constant will be subtracted from every list element. Hence you can create the list of differences from the mean with one expression.

For maximum generality the function that simulates a single $\widehat{b}_1$ should have as arguments the distribution of the ϵ_i's that you are using, the sample size, and the true values of b_0 and b_1. After simulating the ϵ_i's using *Table* and *Random*, the list of Y_i's should be easy to create by an arithmetical operation. Recall that

$$\widehat{b}_1 = \frac{\sum (x_i - \overline{x})Y_i}{\sum (x_i - \overline{x})^2}.$$

The denominator was already computed, and the numerator can be done by taking the dot product of two lists.

The function that simulates all of the $\widehat{b}_1$ values needs all of the arguments that the previous function had, and also the number of repetitions desired (here 50). After using *Table* and your $\widehat{b}_1$ simulator function to get the list of $\widehat{b}_1$'s, use *RangeCounts* as in Project 3 to produce absolute frequencies in the desired categories. Let the histogram heights be the relative frequencies over the subinterval width $\sqrt{\sigma^2 / \sum (x_i - \overline{x})^2}$. Recall that the *ContinuousDistributions* package has the *StandardDeviation[dist]* function, which gives the standard deviation of the desired distribution. After setting a subinterval width and midpoint lists, use *Show*, *Plot*, and *Generalized BarChart* as in Projects 2 and 3 to obtain the superimposed plot.

For your information, *Mathematica* has a package called *Statistics'Linear-Regression'* that can be loaded in order to use the command

$$Regress \; [data, model, variables].$$

Here *data* is the list of pairs which is the data set. The second component of a pair is assumed by *Mathematica* to be the dependent variable. The *model* is a list of variables in the model, such as $\{x\}$ for a linear model without a constant term, $\{1, x, x\hat{}2\}$ for a complete quadratic model, etc. For single variable regression *model* will just be $\{1, x\}$. For single variable regression, the third parameter above would just be $\{x\}$, the lone variable in the regression model. (See the *Guide to Standard Mathematica* packages book for the use of *Regress* to do multiple regression.) The *Regress* command produces the parameter table and ANOVA regression table in the same way that *Minitab* does.

Project 10: Empirical Distribution of Chi-Square Statistic (Section 13.1)

We know from our work on goodness-of-fit tests that for large sample size n the chi-square test statistic

$$U = \sum_{i=1}^{k} \frac{(Y_i - np_i)^2}{np_i}$$

has an approximate $\chi^2 \, (k-1)$ distribution, where k is the number of categories, p_i is the ith category probability, and Y_i is the number of observations in the ith category. It is sometimes said that the approximation is reliable as long as the category probabilities are well balanced. A frequently stated rule of thumb is that the approximation is good when every product np_i is at least 5. We will check the dependence of the quality of the approximation on the balance of category probabilities in this lab. Do the following steps, and report on how well the empirical histogram fits the appropriate χ^2 density.

1. Simulate 50 multinomial observations taking values in one of four categories, with equal category probabilities. Compute the value of the χ^2-statistic. Replicate this process 50 times and produce a histogram of the values of U with five subintervals and heights equal to the empirical probabilities divided by the subinterval lengths. Superimpose a graph of the approximate χ^2 density of U on the histogram.
2. Replicate step 1 to check whether your result is typical.
3. Repeat steps 1 and 2, this time with category probabilities $p_1 = \frac{1}{8}$, $p_2 = \frac{1}{4}$, $p_3 = \frac{1}{2}$, $p_4 = \frac{1}{8}$.
4. Repeat steps 1 and 2, with category probabilities $p_1 = \frac{1}{20}$, $p_2 = \frac{4}{10}$, $p_3 = \frac{1}{2}$, $p_4 = \frac{1}{20}$.

Mathematica Details. A convenient way to simulate multinomial observations with four categories is to simulate uniform $(0,1)$ observations and to count how many of them are less than p_1, between p_1 and $p_1 + p_2$, between $p_1 + p_2$ and $p_1 + p_2 + p_3$, and greater than $p_1 + p_2 + p_3$. If you load the *ContinuousDistributions* package mentioned earlier, you can gain access to the *UniformDistribution* object and the *Random* function. In the *Statistics'DataManipulation'* package is the *RangeCounts* function, which can take your list of uniform observations and produce the counts Y_i for $i = 1, 2, 3, 4$.

It is probably easiest to just write a function to compute by brute force the value of the χ^2-statistic U. This function would need as its inputs the list of category frequencies $\{Y_1, Y_2, Y_3, Y_4\}$ and also n and the p_i's. Compose this function with the Y_i simulator to create a simulator of single U values, then use *Table* to simulate a desired number of them.

Project 3 talked about forming bar charts and superimposing bar charts and density graphs. The *Graphics'Graphics'* package will be necessary to use *GeneralizedBarChart*. You will need the *ChiSquareDistribution[r]* object and the *PDF* command to define the appropriate density.

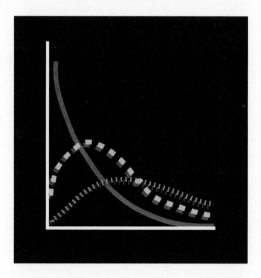

RESEARCH PROJECTS

This appendix provides examples of other applications and topics in probability and statistics and will give you a springboard into expository research. My hope is that you will do some library work, produce a paper in roughly the format of one of the sections of this book, and present that paper in class. It is up to you to prioritize, to think of relevant examples and applications, to discover small extensions to the theory, and in general to have some ownership over a body of mathematics. It is a demanding project, but one that is well worth the effort.

There are eight main topics, the first four having to do predominantly with probability and the last four with statistics. Within each topic are several ideas for specific subtopics on which to concentrate.

D.1 | Markov Chains

Although we have not used the term before, *Markov chains* have already been illustrated a few times, for instance in Exercise 14 of Section 1.5 (the motion of the preoccupied professor). A *Markov chain* is a probabilistic process in which an object moves from one position to another as discrete time instants pass. The assumption is that wherever the object is now, it moves at the next instant of time to a position chosen independently of previous choices, so that the path taken in the past does not matter. The probability distribution of the next position to be visited depends only on the current position. The conditional probabilities of moving from one position to another in a single time step are called *transition probabilities*.

The diagram in Fig. D.1, containing possible positions (generically called *states* of the chain) and transition probabilities from each state to each other state, is called the *transition diagram* of the chain. The game that is afoot in the study of Markov chains is to predict aspects of the future motion of the chain.

To illustrate the idea of a Markov chain, suppose that a soda machine can be in one of three conditions in a particular week: good, damaged, or out of order. From week to week it either maintains its condition or changes to another condition with the transition probabilities shown in the diagram in Fig. D.1. For example, if the machine is in good order one week, it stays that way with probability .9; otherwise it moves to the damaged state. An out-of-order machine will be repaired and restored to good condition next week with probability .6, else it remains in the out-of-order state. We can ask questions like the following: Assuming that the machine begins in good condition in week 0, what is the probability that it will be in good condition in week 3? (Try to answer now.)

A tree is clearly a useful and intuitive way to find short-run probabilities like this. But a tree falls short in the sense that it is unwieldy when the state space is large and when you want information about probabilities for times in the very distant future. An algebraic means of proceeding is essential, and for that the idea of the *transition matrix* of the Markov chain is important. The $i - j$ entry of the transition matrix T is the conditional probability that the next state will be j given that the current state is i. In our soda machine example, the transition matrix can be determined from the transition diagram

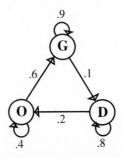

Figure D.1   Transition Diagram of a Markov Chain

in Fig. D.1:

$$T = \text{current} \quad \begin{array}{c} \\ G \\ D \\ O \end{array} \begin{array}{c} \text{Next} \quad \begin{array}{ccc} G & D & O \end{array} \\ \begin{pmatrix} .9 & .1 & 0 \\ 0 & .8 & .2 \\ .6 & 0 & .4 \end{pmatrix} \end{array}$$

Research Topics

Although we have set up some of the ideas, almost everything is left undone. Here are five questions for you to research.

1. **Short-run distributions.** Give a relatively complete treatment of the problem of finding short-run probability distributions of the random variables that make up a Markov chain. Specifically, characterize the probability mass function of the state of the chain at some time m, and the conditional p.m.f. of the state of the chain at time m given the state at the initial time 0. Are there techniques for simplifying the matrix computations that are involved?

2. **First passage times.** Let i be a possible state of a Markov chain. What is the distribution and expectation of the number of steps required for the chain to visit i for the first time? For instance, in the soda machine example, we might be interested in the conditional distribution of the time required for a machine that is in good condition initially to deteriorate to the out-of-order condition.

3. **Long-run distributions.** The first topic concerns the probability distribution of the state for finite time m. Does this distribution converge to some kind of limit as $m \to \infty$? What is the intuitive meaning of such a limiting distribution? Does it depend on the initial state, or does the memory of the initial state wear off with time?

4. **Absorption probabilities.** Suppose that one or more of the possible states of the chain have the special property that once they are entered, the chain never leaves those states. Such states are called *sinks* or *absorbing states*. Is the chain certain to be absorbed into a sink eventually? If there is more than one sink, how do you find the probability that a particular sink absorbs the chain?

5. **Long-run discounted cost problems.** Suppose that a Markov chain incurs a cost of $c(i)$ dollars when it visits state i, and suppose that inflation dictates that a dollar at time n is only worth α^n dollars in present-day terms, where $\alpha \in (0, 1)$. What is the expected total long-run cost incurred by the chain? Can you think of possible applications?

I hope that you will find these to be interesting topics to pursue. There are ample resources to consult, in introductory-level books on finite mathematics and in advanced books on probability, discrete mathematics, stochastic processes and operations research. Hastings (1989), on operations research, has a solid chapter on Markov chains. A classic text on the subject is Kemeny, Snell, and Knapp's *Denumerable Markov Chains* (1966). Erhan Cinlar's fine book (1975) has a fairly deep treatment of Markov chains, as well as other topics in the area of stochastic processes. And the recognized bible of operations research by Hillier and Lieberman has a plethora of interesting applied examples.

D.2 | Reliability

From time to time in this text we have come across examples drawn from the area of reliability. In its broadest form, reliability theory attempts to make predictions about devices or systems that are subject to deterioration and failure at random times. In some problems, such as Example 1.5.5 on redundant bolts, we are predominantly interested in how the probability of failure of a system depends on how the system is built from components. In many of these problems, a simple series or parallel structure is assumed. Other problems involve structures that are not so simple. But all of these problems fall into the category of *structural reliability*.

We have just barely hinted at other categories of problems in the domain of reliability. For example, in the discussion of the Weibull distribution in Section 3.3, we introduced the failure rate function of a distribution, and we said that an increasing failure rate is an indication that a device is aging. There are other ways to characterize aging. An interesting theoretical question involves the relationships among the various aging conditions: Does one imply another?

Still another direction in which to go in this subject involves replacement and inspection policies. One wonders about the influence on long-term reliability of a plan to inspect the device periodically and to repair or replace the device if it is sufficiently deteriorated. An alternative to replacement is a system in which components have backups that step in to perform the function of failed components. This leads naturally to a related class of optimization problems in reliability: How frequently should one do (costly) inspections, and under what conditions should deteriorated components be replaced by (costly) new ones? How many (costly) backup components should be kept on hand, and in what places in the system? The goal is to balance the rewards of improved reliability with the costs of providing for that improvement.

A very different kind of question involves statistical estimation. The way in which data are obtained in reliability problems creates new difficulties seldom seen in other statistical settings. Consider the problem of trying to use empirical data on failure times to estimate the probability distribution of the time of failure of a device. Such a study must be performed in a limited period of time to be of use, which means that some of the devices that we are sampling may not have failed by the time the study is scheduled to end. In other cases, especially in human subject research, our sample values may withdraw from the study before we have had the chance to observe them. So, in reliability problems we encounter the difficulty of incomplete data, which is sometimes systematically rather than randomly incomplete. An area of study within the field of reliability called *lifetime data analysis* addresses these concerns.

In one way or another in these problems, we are concerned with the *reliability function*

$$R(t) = P[T > t] = 1 - F(t),$$

where T is the failure time of the device and F is the c.d.f. of T. We want either to estimate it from incomplete data, to compute it as a function of the structure of the system, or to use it to make decisions.

Research Topics

The preceding paragraphs have attempted to give you a sense of what reliability theory is as a field of inquiry. Besides yielding important results for practitioners, reliability theory has great pedagogical value in illustrating concepts of probability. This discussion and collection of project descriptions have been included for that reason.

Some topics that you might want to write about follow.

1. **Structural reliability.** Give an organized treatment of the problem of structural reliability; that is, show how to find the reliability function of a system in terms of the reliability functions of the components. Investigate the idea of a *module*, which is a subsystem of the main system consisting of one or more components, and how the system reliability depends on module reliabilities.

2. **Notions of aging.** The *increasing failure rate* (IFR) property may be interpreted intuitively as an aging condition as follows: The longer a device survives, the greater is its probability of failure in the next infinitesimally small interval of time. Give a careful definition of the increasing failure rate property, which is one characterization of aging. Define other notions of aging that appear in the literature, such as the *increasing failure rate average* (IFRA) property, the *new better than used* (NBU) property, and the *new better than used in expectation* (NBUE) property. Analyze which is the strongest (the most difficult for a system to satisfy) and which is the weakest (the easiest to satisfy). If a system is known to have one of these aging properties, are there upper or lower bounds on its reliability that do not depend on the specific failure distribution of the system?

3. **Reliability optimization problems.** Formulate and solve a problem in which a controller seeks to improve the reliability of a system at a minimum cost. Here are two examples. (a) A system is made up of parts connected in a known structure. The cost of a part is known and differs from part to part, as does the reliability of a part. How do we purchase backup parts to place in parallel with existing parts so that a threshhold reliability is achieved at minimal cost? (b) A device is subject to shocks, such that the times between shocks are independent and exponentially distributed. Each shock that the

device receives puts it in a higher state of deterioration. The goal is to decide on a level of deterioration at which to repair the device, given that it is more costly and time-consuming to repair it when it is more deteriorated but the device still gives positive benefit when it is partially deteriorated. The criterion for the decision is to maximize the net benefit of the device per unit time.

4. **Lifetime data analysis.** Discuss approaches to estimating the reliability function of a device when the data used for the estimation are incomplete. Include an analysis of the variance of the estimate of the reliability $R(t)$ at a particular time t.

In doing your research, you will find Gertsbakh (1989) to be a useful, well-written introduction to the subject of reliability. In addition to coverage of structural reliability and life distributions, it offers a great deal of interesting material on replacement and inspection schemes to improve system reliability. Another good discussion of structural reliability can be found in Kaufmann, Grouchko, and Cruon (1977). The seminal work on reliability theory is Barlow and Proschan (1975), which is an encyclopedia of relevant information.

D.3 | Portfolio Optimization

Example 4.1.6 introduced you to the idea of combinations of risky investment objects called *portfolios*. We were dealing with the so-called *single-period* portfolio problem, in which there are only two times of interest: now and later. The investor makes one decision on an allocation of wealth among assets now and reaps the benefit of that decision later.

Suppose that we are interested in investing in the kinds of corporations whose stock prices are quoted in the New York Stock Exchange listings in our daily newspaper. We can use many kinds of information to predict the price performance of stocks: general economic conditions, productivity of the company, success of the company's competitors, and so on. However, we would like to consider how to put together an optimal portfolio of assets using only the daily stock quotations to estimate the average rate of return and riskiness of the candidate stocks. The optimal choice of portfolio is quite dependent on our attitude toward risk. A daring risk-taker will typically choose a different balance among the assets in the portfolio than a more cautious individual.

Consider a stock whose price per share now is P, which is known. Let the share price for the same stock later be denoted by Q, which is random. The *rate of return* of the stock—that is, the gain in value per dollar paid—is defined by

$$R = \frac{Q - P}{P}.$$

Then R is also a random variable, and our profit on investing x dollars in the stock is xR.

How do we gain information about the rate of return random variables? If the amount of time that elapses between now and later is one day for instance, then a reasonable approach is to follow the stock over as many days as possible and compute from the price data a corresponding sample of rates of return using the preceding formula.

Typically, assets that have high average rates of return also have high risks. If this were not the case, then investors would all flock to high-reward, low-risk assets, and market imbalances would occur. To increase safety, investors will tend to diversify, to spread their wealth among several assets, hence constructing a collection or *portfolio* of assets. To be precise, a *portfolio* of n investment assets is a vector $\mathbf{w} = (w_1, w_2, \ldots, w_n)$ such that

$$w_1 + w_2 + \cdots + w_n = 1.$$

The component w_i is called the *weight* of the ith asset in the portfolio.

The interpretation of the portfolio vector $\mathbf{w}$ is that its components identify the proportions of the investor's total wealth devoted to each asset, hence the term *portfolio weight* and the summability to 1. We do not a priori force the individual w_i's to be between 0 and 1, because of the real possibility of "selling short" in the market. If the investor sells short, he holds a negative amount of wealth in a stock by acquiring money as if he has just sold the stock, agreeing to buy the stock back (repaying the debt) later.

If the investor begins with total wealth W, then the decision to be made is what fraction w_i of that wealth is to be devoted to asset i, for each asset $i = 1, 2, \ldots, n$. It is reasonable to suppose that the investor chooses portfolio weights to maximize the *rate of return on the portfolio* of assets $\mathbf{w} = (w_1, w_2, \ldots, w_n)$:

$$R = \sum_{i=1}^{n} w_i \cdot R_i.$$

However, the portfolio rate of return R is random, since it depends on the asset rates of return R_i. Therefore it does not make sense for one choice of $\mathbf{w}$ to be optimal for all of the possible values that the R_i's can take on. A more tractable investment goal, suggested in the text, is to try to maximize the expectation of the portfolio rate of return minus a constant times the variance:

$$\max_{\mathbf{w}} \ E[R] - a \cdot \text{Var}(R).$$

The constant a is the *risk aversion* for the investor making the decision; it is large when the investor imposes a large penalty on high variance, and small when the investor cares little about the variance of the return.

Research Topics

As in the previous applications, we have really only set up the study of mathematical portfolio theory without actually deriving anything yet. Following are a few topics that you might choose to research.

1. **Optimization and risk aversion.** Argue that the meaning of the risk aversion constant is as follows: Ask yourself, "If I am given two portfolios with rates of return R and S, and the second portfolio incurs one extra unit of risk, how much extra average return do I want on the second to compensate, so that I am indifferent between the two portfolios?" The answer is your risk aversion. In other words, a is the change in the average return that you demand per unit change in variance that you are willing to accept. Try to decide what your own a value is, then pick four stocks listed on some exchange, estimate the means and variances of their rates of return, and find your optimal portfolio. You may assume independence for now (but see the next project).

2. **Correlated stocks.** Pick at least three stocks that you guess may be correlated. Solve the portfolio problem using some suitable risk aversion constant. For some generic stocks, try some numerical computations with means, variances, and correlations in order to detect diversification effects. In other words, when a pair of stocks is negatively correlated, is it advantageous to have significant portions of both in the portfolio, and why?

3. **Efficient portfolios.** Economists sometimes take a more graphical pedagogical approach to basic single-period portfolio theory. Investigate and report on mean–variance and mean–standard deviation graphs for portfolios, and the so-called *efficient mean–variance frontier*. What shape does that frontier have, and why? What role does risk aversion play?

4. **Utility theory.** We have represented an investor's aversion to risk by a single constant a. This led us to a particular form for the objective of the maximization problem. There are other methods of characterizing attitudes toward risk. In one commonly used formulation, the objective is to maximize some function, called a *utility function*, of final wealth. Study this method, and show that the objective function described here is subsumed as a special case in which the utility is quadratic.

5. **Portfolio separation theorem.** Study, report on, and give a proof of the famous portfolio separation theorem, which is as follows: Suppose that asset 1 is a risk-free savings account or bond, so that $\sigma_1^2 = 0$. Then the ratios that the optimal portfolio weights $w_2, w_3, \ldots, w_n$ bear to one another are independent of the risk aversion a. Consequently, for each risky asset i, the proportion of total wealth in risky assets $2, 3, \ldots, n$ that is devoted to asset i is the same for all investors. For example, if a cautious individual computes that he should hold twice as much wealth in stock 2 as in stock 3, then so will a free-wheeling risk-taker. The role of the risk aversion a is only to determine what proportion of wealth is devoted to the risk-free asset, not the relative mixture of the risky assets.

Sharpe (1970) offers an accessible and broad-ranging treatment of portfolio theory, including some discussion of the impact of correlation on portfolios, the mean–variance frontier, utility theory, and also more coverage than we have given here on investment under constraints. Another general purpose book on finance (Samuelson, 1983) has some material on the kind of portfolio problem that we have introduced. Merton's lengthy survey article reviews and reinforces this material at a higher level, and also discusses the multiple-period investment-consumption problem (not suggested here as a project because it involves a lot of new mathematics), and some very technical continuous-time theory. You may have a more difficult time finding understandable, complete treatments of portfolio theory in book form, however. Your friendly neighborhood economics and business department may be able to help direct you to journal articles and book excerpts.

D.4 | Queueing

The last of our research areas in probability is *queueing theory*. A *queue* is a waiting line, and queueing theory is the study of waiting lines and other characteristics of service systems. In a queueing system one or more servers give service to arriving customers. The customers come at random times, and service of a customer takes a random amount of time. A customer who seeks service from a server who is already occupied must wait in a queue. Some discipline, such as a first-come, first-served rule, is in place to select the next customer to be served from the queue. On completion of service, the customer either proceeds to another service station in the system or departs the system. A schematic representation of a first-come, first-served queueing system with three service stations connected in a circle and one server per station appears in Exercise 10 in Section 1.3.

Our description of queueing systems suggests familiar applications to human service in retail stores, post offices, banks, and the like, but queueing theory is certainly not limited to these applications. Planes (customers) waiting to use runways (servers) also fit the queueing paradigm. Queueing models are frequently used in manufacturing, specifically in modeling flows of goods through assembly stations, warehouses, and shipping areas (see Law and McComas, 1988). Perhaps the most active uses of queueing theory today involve the related areas of telecommunications and computer networks, in which packets of data are switched through a network of stations or processors.

In general, what kind of information would we like to obtain from a queueing model? From the customer's perspective, the most important quantity is time spent waiting in line. If the queueing problem involves production, with customers representing raw materials and service representing the synthesis of those materials into final products, then the manufacturer is concerned with maximizing throughput, that is, number of completed services per unit time. If the server is an expensive piece of equipment whose utilization should be high, then the proportion of time during which the server is utilized would be important. In problems involving managing traffic, we are most concerned with the overall customer congestion in the system, specifically the probability distribution of the number of customers present.

The behavior of a queue is determined by its physical organization and by two basic sequences of times: (1) the sequence $S_1, S_2, S_3, \ldots$ of times between successive arrivals of customers (here S_1 is the first arrival time, S_2 is the elapsed time between the arrivals of the first and second customers, etc.); and (2) the sequence of service times $U_1, U_2, U_3, \ldots$ (U_i is the elapsed time between the beginning and the end of service of customer i).

The following example should convince you of this point. Suppose that the first three interarrival times are 1.2, .8, and 2.5, and the service times of these three customers are 1.0, 2.3, and 1.6, respectively. Let's justify the entries of the following table, which show when the customers arrive, begin service, and depart.

Customer	Interarrival Time	Time of Arrival	Begin Service	Service Time	Departure Time
1	1.2	1.2	1.2	1.0	2.2
2	.8	2.0	2.2	2.3	4.5
3	2.5	4.5	4.5	1.6	6.1

I have recorded the given interarrival times and service times in columns 2 and 5. The time at which a customer arrives is just the total of the interarrival times up through that customer. Thus customer 1 arrives at time 1.2, customer 2 arrives .8 time units later at time 2.0, and customer 3 arrives 2.5 time units after that, which is time 4.5. The Departure Time column is determined by the Begin Service column and the Service Time. Customer 1 enters service immediately on arrival at time 1.2, requires 1.0 time units to be served, and hence departs at time $1.2 + 1.0 = 2.2$. Customer 2 enters service either when customer 1 departs, or immediately if customer 1 has already departed. In other words, the time when customer 2 begins service is the larger of the departure time of customer 1 and the arrival time of customer 2. Here, customer 2 enters service at the departure time of customer 1, which equals 2.2. Customer 2 uses the server for 2.3 time units, hence departs the system at time $2.2 + 2.3 = 4.5$. It happens that customer 3 arrives exactly at that time, enters service immediately, and departs at time $4.5 + 1.6 = 6.1$.

In a real application one cannot expect these interarrival and service times to be known before the actual run of the queueing process. So, we must model the S_i's and U_i's as random variables, whose probability laws are known or can be estimated from historical data.

One frequent assumption is that the interarrival times S_i have the exponential p.d.f. with parameter λ, and the service times U_i have the exponential density with parameter μ. In addition, all of these times are mutually independent. Notice that this means that the sequence of arrivals forms a Poisson process.

By our previous work on the exponential distribution, $\lambda = 1/($average interarrival time$)$; hence λ can be thought of as the average number of arrivals per unit time, that is, the *arrival rate*. Similarly, μ is the *service rate*. The quantity $\rho = \lambda/\mu$ will appear frequently as you do your research. It is a measure of rate of arrivals compared to rate of service, and so is called the *traffic intensity* of the queue.

You should be aware of a shorthand notation used to describe the basic structure of single-station queueing models. It uses a string of symbols of the following form:

$$A/B/n/K,$$

where A is a symbol for the interarrival distribution [M is used to stand for the $\exp(\lambda)$ distribution, E_k for the Erlang(k, λ) distribution, G for an unspecified distribution], B is a symbol for the service time distribution (using similar codes), n is the number of servers present at the station, and K is the size of the waiting area for all customers, including the one in service. In the case that the waiting area is unlimited, the K symbol is omitted from the string. Then, for example, $M/E_3/2$ is a queue that has exponential interarrivals, Erlangian service with k parameter equal to 3, two servers, and unlimited waiting space. The $G/M/1/3$ queue has some unspecified interarrival distribution, exponential service, a single server, and three customer positions.

Research Topics

Here are a few accessible topics to study.

1. **Long-run distribution of *M/M/1/K*.** For the $M/M/1/K$ queue, define

$$X_t = \text{number of customers in system at time } t.$$

This count is meant to include the customer in service, if any, as well as those who might be waiting in queue. Because of the restrictions on available waiting space, the possible values of X_t are $\{0, 1, 2, \ldots, K\}$. Define for each $n = 0, 1, 2, \ldots, K$

$$P_n(t) = P[X_t = n] = P[n \text{ customers in system at time } t].$$

Write a system of differential equations for the functions $P_n(t)$ along the lines of the development of the Poisson process, and find explicitly the limiting probabilities $\lim_{t \to \infty} P_n(t)$.

2. **Poissonian queues with infinite space.** Assume again that inter-arrival and service time distributions are exponential and that queues have unlimited waiting space. Find differential equations for the functions $P_n(t)$ and the limiting system size probabilities. Extend the development to multiple-server queues.

3. **Waiting times.** For the $M/M/1$ queue, find the distribution of the time that a customer waits in the queue, and use it to find the average waiting time. Research *Little's formula*, which relates average queue length and average waiting time, and which holds even for non-Poissonian queues.

4. **M/G/1 and G/M/1 queues.** Discuss how the concept of the *embedded Markov chain* of the queue length process assists in finding the limiting system size probabilities for the $M/G/1$ and $G/M/1$ queues. If you have classmates researching Markov chains, you may want to interact and collaborate with them.

5. **Simulation of queues.** The more complicated a queueing system is, the less amenable it becomes to exact analytical treatment. But we might still be anxious to know about average queue lengths, throughput, waiting times, server utilization, and so on. In this case, software tools are available to simulate, and gather statistical information about the queueing system. One of the better known simulation systems is called GPSS (for General Purpose Simulation System). Report on GPSS modeling, including a simple example GPSS program. If the software is available to you, run the program several times to see whether the simulated results conform to what the analysis predicts. What are some of the strengths and limitations of simulation?

There is a huge body of work on queueing models in the literature, although much of it requires more background than you probably have at this point. A gentle introduction to queueing may be found in Hillier and Lieberman (1980) and in Hastings (1989). Two higher-level expositions of classical queueing theory are Gross and Harris (1985) and Saaty (1961). Solomon (1983) is a good introduction to simulation of queues, as well as some applications that have been done in the real world. For topic 5, the GPSS user's manuals issued by the software vendors are well written and should allow you to pick up the basics of this simulation language without too much effort. A more extensive treatment of simulation in GPSS is given in Schriber (1991). New applications and theory of queueing appear regularly in the journal *Operations Research* and others.

D.5 | Time Series

We have spent most of the second half of this book analyzing random samples $X_1, X_2, \ldots, X_n$. In a random sample the random variables are independent, and the subscripts have little meaning except to distinguish one random variable from another. But there are many statistical settings, particularly in the area of economics, where the subscript indicates the time order in which the data were gathered, and there are good reasons to suspect that each X_i is dependent on previous X_k's for $k < i$. For example, if X_i represents the unemployment rate in an area in month i, one would not expect the ith observation to be independent of $X_{i-1}, X_{i-2}, \ldots$. A high value of X_{i-1} should be associated with a high value of X_i, since employment conditions do not normally change rapidly.

What kinds of models for time series are useful in representing real data and yet are tractable mathematically and susceptible to stastistical analysis of parameters? One class of time series is defined as follows. An *autoregressive time series of index r* satisfies

$$X_{k+r} = \phi_1 X_{k+r-1} + \phi_2 X_{k+r-2} + \cdots + \phi_r X_k + \epsilon_{k+r} \quad \text{for each } k \geq 1,$$

where ϵ_i are i.i.d. normal $(0, \sigma^2)$ random variables. In other words, the value of a member of the sequence is a linear function of the previous r members plus a normally distributed disturbance. A special case is the *random walk* with $r = 1$ and $\phi_1 = 1$ so that

$$X_{k+1} = X_k + \epsilon_{k+1}.$$

We use the notation $AR(r)$ for an autoregressive series with index r; hence the random walk would be an example of an $AR(1)$ process. Notice that since the X's on the right of the defining equation for autoregressive series also depend on previous X's, the dependence on the past extends back indefinitely.

A different type of time series is called the *moving average process*. Here, the dependence on the past is strictly limited. Formally, a time series is called a *moving average process of index r* if it satisfies

$$X_{k+r} = \epsilon_{k+r} - \theta_1 \epsilon_{k+r-1} - \theta_2 \epsilon_{k+r-2} - \cdots - \theta_r \epsilon_k \quad \text{for each } k \geq 1,$$

where ϵ_i are i.i.d. normal $(0, \sigma^2)$ random variables. Thus, an observation in a moving average time series is constructed as a linear combination of normal disturbance terms from the previous r periods, plus a new disturbance. We write $MA(r)$ for such a process. For example, the $MA(1)$ process would satisfy the equation

$$X_{k+1} = \epsilon_{k+1} - \theta_1 \epsilon_k.$$

Research Topics

We have only scratched the surface of time series models, their properties, and their applications. Here are a few particular areas that you might choose to pursue.

1. **Autocorrelation in $AR(r)$ and $MA(r)$ series.** Show that for an $AR(1)$ time series, the covariance between terms dies away geometrically as the separation between the indices of the terms increases, as long as $|\phi_1| < 1$. Contrast this with what happens in the $MA(1)$ series. Formally define and study the properties of the *autocovariance and autocorrelation functions* of the time series. Report on results for series of higher index.

2. **ARMA processes.** Compile a dictionary of information about the properties of combined models called *ARMA* and *ARIMA (autoregressive integrated moving average)*, the latter of which can describe series whose properties change as the time origin changes, that is, nonstationary time series.

3. **Model fitting.** Investigate the fitting of time series data to models using the *sample autocorrelation function*

$$r_l = \frac{\sum_{k=1}^{n-l} (X_k - \overline{X})(X_{k+l} - \overline{X})}{\sum_{k=1}^{n} (X_k - \overline{X})^2}, \qquad l = 0, 1, 2, 3, \ldots.$$

This is the sample correlation between the observed time series values and the series values lagged l times units behind. Gather some data of your own and select an appropriate *ARMA* model.

4. **Parameter estimation.** Study methods for statistical estimation of the parameters in $ARMA$ time series. Does the method of least squares bear fruit?

5. **Seasonal models.** We have not thus far discussed models that describe a very common phenomenon: seasonality. Study some ways in which seasonal behavior can be introduced into a time series.

One very helpful source is Cryer (1986). Another interesting reference (Box and Jenkins, 1976) is viewed as one of the most important texts on this subject. Once you find these on your library shelves, you should also find a wealth of others.

D.6 | Quality Control

The area of *quality control* (called by many *quality improvement*) has been growing in activity in the United States recently in response to strong economic competition from other countries, especially Japan. The ironic fact is that the subject was first developed at Bell Laboratories in the early 1920s, and was espoused by a small corps of statisticians, managers, and engineers who were largely ignored in this country during a long period of prosperity. After the Second World War when Japan was rebuilding itself, however, its industrial leaders had the foresight to recognize that the way in which it could achieve economic success in a world market growing increasingly competitive was to constantly improve the quality of its goods. The phrase Made in Japan, originally disparaging, has undergone an incredible change in meaning in a short time. And one American, W. Edwards Deming, who was brought in by the Japanese as a consultant on quality, is credited by them for a significant part of that transformation. Yet Deming was mostly unknown in this country until fairly recently. Not coincidentally, Deming and others began receiving more attention in the American business community when the cracks in U.S. industry began to show and the country's world economic leadership began to fade.

Much of quality control is far less mathematical than other areas of statistics, since it is concerned with human factors such as the involvement of employees at all levels of the manufacturing process so as to restore pride in excellent workmanship. The idea is to build quality in, not just detect and remove items of low quality. And it is the minds and hands of those people most directly responsible for production who know the product best, and who can therefore provide the most important information about sources of error and ways to improve the product or the process of making it. But the workers must have some stake in the endeavor and must be able to translate their ideas into action, and that requires a radical change in the way management operates and in the way that business is organized. Management must respect and listen to its employees, continual education for productive change must be undergone by all, and employees must give 100% to what they do by their own choice, all in the interest of making better things, not necessarily more things.

There are several more technical aspects to the pursuit of quality. First, the need to detect a process that has gone out of control does not disappear; rather it is just accompanied by the problem of keeping the process from getting out of control in the first place. So, knowledge of how to do acceptance sampling of the kind illustrated by earlier problems in combinatorial probability is important to this area.

Experimental design and analysis is an important tool for quality engineers, especially those interested in proactive, front-end improvement of a system. The application here is to test for significant effects of one or more factors in the manufacturing process (such as line speed, curing temperature, kind of raw material, machine, or operator) on a dependent variable which is a measure of the quality of the manufactured item. Not only would the experimenter be looking for ways of increasing the mean of the quality measure, but also for sources of variablity in quality. These sorts of analyses have paid great benefits to those organizations who have employed them.

Research Topics

Following is a short list of research projects; you may find more as you look into the subject of quality.

1. **Deming and the quality movement.** Research and report on the life and thought of W. Edwards Deming. Discuss his 14 points for quality. Do they seem realistic and attainable?

2. **Control charts.** To detect departures from normal operation, a group of elementary statistical estimators and graphs have been developed that are understandable and usable even by those without much statistical education. Investigate control charts, which come in several varieties, including charts for the mean, and attribute control charts.

3. **Acceptance sampling.** Do a computer investigation of the properties of the function that gives the probability of detecting at least one of N defective items among M items by sampling a total of n items. How does it vary with n, N, and M? Which seems more important in designing a sampling plan: the proportion of the batch sampled or the absolute number sampled?

4. **Factorial design for quality improvement.** Study *factorial design* with many factors, each at two levels. (See also our eighth research area.) Look at designs called *fractional factorials*, which are very economical in terms of the number of replications they use. Comment on the price paid for the economy relative to the ability to test for interaction effects.

The number of references on quality control is growing. Two older books by Burr (1976, 1979) could be good starting points. You also should check out Deming's book (1960) to read firsthand the most influential person in the quality community. The texts by Montgomery (1991) and Box, Hunter, and Hunter (1978) are very comprehensive sources on experimental design applied to quality issues.

D.7 | Bayesian Statistics

One community of statisticians identify themselves as Bayesians and have a special point of view about inference problems. To them, probability distributions depending on parameters, such as $b(n, p)$, Poisson(μ), or $N(\mu, \sigma^2)$, are conditional distributions given particular values of the parameters. These parameters belong to a set of possible parameter values, and there is a probability distribution called a *prior distribution* on that set, which gives some subjective assessment of how likely it is that the true parameter takes on each of its possible values. When data are gathered, how they are distributed provides new information about the distributional parameter. Specifically, you can compute a new *posterior distribution* for the parameter that reflects how the observed data changes the a priori likelihoods making up the prior distribution.

To see how this is done, consider the case where the parameter θ of some discrete p.m.f. $p(x; \theta)$ is given a discrete prior distribution

$$q(\theta_i) = P[\theta = \theta_i], \qquad i = 1, 2, \ldots, m.$$

We take a random sample $X_1, X_2, \ldots, X_n$ from the distribution $p(x; \theta)$, and we let

$$A = \{X_1 = x_1, X_2 = x_2, \ldots, X_n = x_n\},$$

where the x_i's are the observed values. Bayes's formula allows us to compute new probabilities for the events $\{\theta = \theta_i\}$ given A, that is, given the observed data:

$$q_{\theta|x}(\theta_i) = P[\theta = \theta_i | X_1 = x_1, \ldots, X_n = x_n]$$

$$= \frac{P[X_1 = x_1, \ldots, X_n = x_n | \theta = \theta_i] \cdot P[\theta = \theta_i]}{\sum_{j=1}^{m} P[X_1 = x_1, \ldots, X_n = x_n | \theta = \theta_j] \cdot P[\theta = \theta_j]}$$

$$= c \cdot L(\theta_i; x_1, \ldots x_n) \cdot q(\theta_i),$$

where c does not depend on $\theta_1, \theta_2, \ldots, \theta_m$ and L is the likelihood function of the sample data. The fact that c doesn't depend on the θ_j's follows from the law of total probability applied to the denominator of the middle line of the preceding offset computation, where we conditioned and unconditioned on the value of θ. This result can be expressed concisely as: "The posterior is proportional to the likelihood times the prior."

The amount of arbitrariness and subjectivity involved in deciding upon the prior distribution of the parameter actually offends some non-Bayesian statisticians, but the Bayesians respond that it is inefficient not to use whatever expert information is at hand to estimate parameters. Relations between these two camps have at times been uneasy.

Research Topics

With the ideas of prior and posterior distributions in hand, where do we go from here? Is it possible to redo classical statistics from a Bayesian point of view? That is, are there such things as Bayesian point estimators, Bayesian hypothesis tests, or Bayesian regression analysis? The answer is yes, which leads us to our research project descriptions.

1. **Bayesian estimation.** The estimation problem, of course, is to produce a number y from observed data $x_1, x_2, \ldots, x_n$ that is close to a parameter θ of the distribution from which the x's were sampled. The nearer is y to θ, the better is our estimate, but as you have probably seen elsewhere in mathematics, there are numerous ways of measuring the distance between two objects. Study and report on the notions of *loss function*, *risk function*, and estimators that minimize the *Bayes risk*.

2. **Continuous Bayesian statistics.** We have only talked about prior and posterior distributions in the discrete setting. What changes are necessary to adapt the ideas to continuous prior distributions and/or data from continuous distributions? Investigate the idea of *conjugate prior distributions*, especially in the context of normal random samples and the problem of estimating the mean.

3. **Bayesian confidence intervals and tests.** Describe the Bayesian approach to the problems of interval estimation and hypothesis testing. With regard to the latter, explore thoroughly the case of a simple null hypothesis $H_0:\theta = \theta_0$ vs. a simple alternative $H_a:\theta = \theta_1$.

4. **Bayesian regression estimation.** Discuss the idea of Bayesian prior and posterior distributions for the parameters β_0 and β_1 in linear regression.

You will find a number of books on Bayesian statistics in your library. The text by Casella and Berger (1990) has a good in-depth chapter on the loss-risk function idea. An introductory book by Winkler (1972) may help to give you a foothold in the area. Press (1989) discusses Bayesian approaches to many of the statistical problems encountered in this book.

D.8 | More on Experimental Design

The subject of experimental design is a broad one, and the work we did in Chapter 11 is only one step on the road. Consequently, there are a few topics for further study that are within your grasp.

One famous experimental design to which we have not attended has the goal of using data very efficiently: the 2^k-*factorial design*. In a sense the 2^k-factorial design is not new for us. It is nothing more than a completely randomized *k*-factor experiment, with the special stipulation that each factor has exactly two levels. You can therefore use a statistics package like *Minitab* to compute sums of squares and *F*-statistics to test for significance of the factors. Such tests might be pilot studies, in which for some new or poorly understood system there might be numerous potential factors affecting response, some of which must be weeded out with as little data gathering as possible. So, instead of using many levels of a dubious factor, an experimenter might choose two grossly different ones to see whether there is any chance at all that the factor is important. If it is, then it can be kept in the model and tested more carefully at more levels; otherwise it can be disposed of.

An example adapted from Box, Hunter, and Hunter (1978) that illustrates 2^2 experiments is as follows. There is a belief that traces of phosphorus and nitrogen in pond water may have an effect on the algae population of the pond. To check this hypothesis, an experimenter tested two concentrations of phosphorus, .06 mg/liter and .30 mg/liter, and two concentrations of nitrogen, 2.00 mg/liter and 10.00 mg/liter, using eight independent test ponds. Treatment combinations of the low and high concentrations of the two elements were assigned randomly, with each treatment combination used exactly twice. Denoting the low levels of concentration by "−" and the high levels by "+," the following data were observed.

Phosphorus	Nitrogen	Algae Population	Phosphorus	Nitrogen	Algae Population
−	−	.312	−	−	.479
+	−	.391	+	−	.481
−	+	.412	−	+	.465
+	+	.376	+	+	.451

With only eight observations, the experimenter is able to make a rough judgment about whether two factors are necessary to include as predictors of algae population. If you are interested, try to run an analysis of variance on *Minitab* or another statistical package to see whether either of the factors is significant.

Research Topics

We have only begun to work with 2^k-factorial experiments, and the real interest comes when we encounter the problems of parameter estimation, lack of replication, blocking, and so-called *half-fractional experiments* as described below in topic 4. In addition, there are many other experimental designs that remain untouched.

1. **Latin squares.** In Chapter 11 we discussed the randomized block experiment for one-factor models with an additional blocking factor. *Latin square* designs are useful (a) when there is another blocking factor whose influence should be accounted for and (b) when there is an overriding reason for limiting the number of experimental runs. To elaborate upon the latter

point, there may not be enough experimental material to run every possible treatment and block combination. So, the experimenter looks for a sparse design which at least allows every treatment level to appear with every level of blocking factor 1 and with every level of blocking factor 2 (but not with every combination of the two blocking factors). Investigate the appropriate linear model, and discuss the analysis.

2. 2^k **factorial experiments.** Discuss the estimation of treatment effects in 2^k-experiments with $k \geq 2$. How do the estimated effects relate to the sums of squares? How do you compute an error sum of squares if the number n of replications is 1?

3. **Blocking in 2^k factorial experiments.** Study the problem of blocking in 2^k-experiments. Specifically, how do you design the experiment and analyze the data if the experimental material is in two blocks? (This could be the case if it was created in two batches.)

4. **Half-fractional factorial experiments.** The 2^k-experiment is called that because if there are k factors each at two levels, then there are 2^k factor level combinations. One full replication therefore requires at least 2^k data points. Is there a way of gathering and analyzing data to test for factor effects using only 2^{k-1} or half as many data points? What is sacrificed, if anything, for the savings on data collection?

5. **Balanced incomplete block designs.** Another design that involves sparse data is the *balanced incomplete block design*. Research it, describe what it is, and discuss the analysis.

The experimental design books mentioned many times (Box, Hunter, and Hunter, 1978 and Montgomery, 1991) are good references, and you will undoubtedly find some others yourself.

APPENDIX E

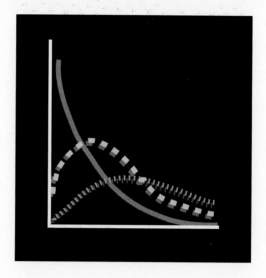

LINEAR ALGEBRA REVIEW

This appendix summarizes some of the key definitions, ideas, and results from linear algebra that are used in this book. It is not meant to be a first introduction to the subject, but rather a refresher and a guide to common language and notation. For more detail, see Anton (1984), Strang (1980), or Schneider (1987).

A *matrix* is a rectangular array of numbers:

$$\begin{bmatrix} a_{11} & a_{12} & \cdots & a_{1n} \\ a_{21} & a_{22} & \cdots & a_{2n} \\ \vdots & \vdots & \ddots & \vdots \\ a_{m1} & a_{m2} & \cdots & a_{mn} \end{bmatrix}.$$

The *dimension*, or *order*, of the preceding matrix is $m \times n$, that is, m rows by n columns. The special case

$$\mathbf{x} = \begin{bmatrix} x_1 \\ x_2 \\ \vdots \\ x_m \end{bmatrix},$$

in which the matrix has just one column, is called a *column vector*, and the special case

$$\mathbf{y} = \begin{bmatrix} y_1 & y_2 & \cdots & y_n \end{bmatrix}$$

is called a *row vector*.

Matrices of the same dimension can be added or subtracted by adding or subtracting all of the corresponding components of the matrices, for example,

$$\begin{bmatrix} 1 & 2 & 3 \\ 2 & -1 & 4 \end{bmatrix} + \begin{bmatrix} 3 & 0 & 0 \\ 0 & 1 & 0 \end{bmatrix} = \begin{bmatrix} 4 & 2 & 3 \\ 2 & 0 & 4 \end{bmatrix},$$

$$\begin{bmatrix} 1 & -1 & 0 \end{bmatrix} - \begin{bmatrix} 0 & 1 & 1 \end{bmatrix} = \begin{bmatrix} 1 & -2 & -1 \end{bmatrix}.$$

A *constant times a matrix* is defined as the matrix whose components are the constant times the components of the original matrix:

$$2 \cdot \begin{bmatrix} 1 & 0 \\ 0 & 1 \end{bmatrix} = \begin{bmatrix} 2 & 0 \\ 0 & 2 \end{bmatrix}.$$

Products of matrices are defined in a more complex way. First, a row vector of n components times a column vector, also of n components, is a number determined as follows:

$$\begin{bmatrix} a_1 & a_2 & \cdots & a_n \end{bmatrix} \cdot \begin{bmatrix} b_1 \\ b_2 \\ \vdots \\ b_n \end{bmatrix} = a_1 b_1 + a_2 b_2 + \cdots + a_n b_n.$$

Then the *product* of matrices A (dimension $m \times p$) and B (dimension $p \times n$) is the $m \times n$ matrix whose i, j component is the product of row i of A with column j of B:

$$\begin{bmatrix} 1 & 0 & -2 \\ 7 & 4 & -3 \end{bmatrix} \begin{bmatrix} 2 & -1 \\ 4 & -1 \\ 6 & 3 \end{bmatrix} = \begin{bmatrix} 1 \cdot 2 + 0 \cdot 4 + (-2) \cdot 6 & 1(-1) + 0(-1) + (-2) \cdot 3 \\ 7 \cdot 2 + 4 \cdot 4 + (-3) \cdot 6 & 7(-1) + 4(-1) + (-3) \cdot 3 \end{bmatrix}$$

$$= \begin{bmatrix} -10 & -7 \\ 12 & -20 \end{bmatrix}.$$

Matrix multiplication is not in general commutative.

Many matrices have enough structure that it is advantageous to look at them as being composed of block submatrices, for example,

$$\begin{bmatrix} 1 & 0 & 0 & 0 \\ 0 & 1 & 0 & 0 \\ 0 & 0 & 2 & 0 \\ 0 & 0 & 0 & 2 \end{bmatrix} = \begin{bmatrix} A & \mathbf{0} \\ \mathbf{0} & B \end{bmatrix},$$

where A consists of the first two rows and columns, B consists of the third and fourth row and column, and $\mathbf{0}$ is a 2×2 matrix of zeros. Matrix arithmetic can be done in blocked form by doing operations to corresponding blocks. For example the second power of the preceding matrix would be

$$\begin{bmatrix} A & \mathbf{0} \\ \mathbf{0} & B \end{bmatrix} \cdot \begin{bmatrix} A & \mathbf{0} \\ \mathbf{0} & B \end{bmatrix} = \begin{bmatrix} A^2 & \mathbf{0} \\ \mathbf{0} & B^2 \end{bmatrix}.$$

The *transpose* of a matrix A, written A' or sometimes A^t, is the matrix obtained by letting the rows of A be the columns of A', for instance,

$$\begin{bmatrix} 2 & 1 & 0 \\ 1 & 2 & 3 \end{bmatrix}' = \begin{bmatrix} 2 & 1 \\ 1 & 2 \\ 0 & 3 \end{bmatrix}, \qquad \begin{bmatrix} 1 & 3 \\ -1 & 2 \end{bmatrix}' = \begin{bmatrix} 1 & -1 \\ 3 & 2 \end{bmatrix}.$$

A square $n \times n$ matrix $A = (a_{ij})$ is *symmetric* if $a_{ij} = a_{ji}$ for all $i \neq j$. This means that a matrix is symmetric if and only if it equals its transpose. So, the following matrices are symmetric:

$$\begin{bmatrix} 1 & -1 \\ -1 & 2 \end{bmatrix}, \qquad \begin{bmatrix} 1 & -7 & 1 \\ -7 & 5 & 3 \\ 1 & 3 & 2 \end{bmatrix}.$$

The usual convention that we follow in this book is that when a vector $\mathbf{x}$ is referred to, it is a column vector rather than a row vector. Then the transpose of $\mathbf{x}$ is a row vector. You will often see special products of three matrices of the following form, called a *quadratic form*, which turn out to be numbers. Usually, Q below is a symmetric matrix:

$$\mathbf{x}'Q\mathbf{x} = \begin{bmatrix} x_1 & x_2 & \cdots & x_n \end{bmatrix} \begin{bmatrix} q_{11} & q_{12} & \cdots & q_{1n} \\ q_{21} & q_{22} & \cdots & q_{2n} \\ \vdots & \vdots & \ddots & \vdots \\ q_{n1} & q_{n2} & \cdots & q_{nn} \end{bmatrix} \begin{bmatrix} x_1 \\ x_2 \\ \vdots \\ x_n \end{bmatrix}$$

$$= \sum_i \sum_j q_{ij} x_i x_j.$$

Vectors $\mathbf{x}$ and $\mathbf{y}$ are called *orthogonal* if $\mathbf{x}'\mathbf{y} = \mathbf{y}'\mathbf{x} = 0$. The intuition is that in their realizations as segments in space, orthogonal vectors are perpendicular.

Other useful properties of transposes are:

$$(A')' = A, \qquad (A \pm B)' = A' \pm B'.$$

The *determinant* is a function from square matrices to numbers that can be computed recursively in the dimension as follows:

$$\det [a_{11}] = a_{11},$$

$$\det \begin{bmatrix} a_{11} & a_{12} \\ a_{21} & a_{22} \end{bmatrix} = a_{11}a_{22} - a_{12}a_{21}$$

$$\det \begin{bmatrix} a_{11} & a_{12} & a_{13} \\ a_{21} & a_{22} & a_{23} \\ a_{31} & a_{32} & a_{33} \end{bmatrix}$$

$$= a_{11} \det \begin{bmatrix} a_{22} & a_{23} \\ a_{32} & a_{33} \end{bmatrix} - a_{12} \det \begin{bmatrix} a_{21} & a_{23} \\ a_{31} & a_{33} \end{bmatrix} + a_{13} \det \begin{bmatrix} a_{21} & a_{22} \\ a_{31} & a_{32} \end{bmatrix},$$

and so on. In general, one can expand the determinant about the first row by taking each first row component, multiplying it by the determinant of the smaller square matrix obtained by deleting the row and column containing the current row component, and taking an alternating sum and difference of the results. (In fact, this means of expansion works using any row, not just the first one, and it also works using any column.) For example, for a 4×4

matrix,

$$\det \begin{bmatrix} 1 & 0 & 1 & 0 \\ -2 & 3 & 3 & 1 \\ 1 & 1 & -1 & 2 \\ 3 & -2 & 1 & 0 \end{bmatrix} = 1 \cdot \det \begin{bmatrix} 3 & 3 & 1 \\ 1 & -1 & 2 \\ -2 & 1 & 0 \end{bmatrix} - 0 \cdot \det \begin{bmatrix} -2 & 3 & 1 \\ 1 & -1 & 2 \\ 3 & 1 & 0 \end{bmatrix}$$

$$+ 1 \cdot \det \begin{bmatrix} -2 & 3 & 1 \\ 1 & 1 & 2 \\ 3 & -2 & 0 \end{bmatrix} - 0 \cdot \det \begin{bmatrix} -2 & 3 & 3 \\ 1 & 1 & -1 \\ 3 & -2 & 1 \end{bmatrix}.$$

For a *diagonal matrix*—one whose components are zero everywhere except on the main upper left to lower right diagonal—it is not difficult to see that

$$\det \begin{bmatrix} a_{11} & 0 & \cdots & 0 \\ 0 & a_{22} & \cdots & 0 \\ \vdots & \vdots & \ddots & \vdots \\ 0 & 0 & \cdots & a_{nn} \end{bmatrix} = a_{11} a_{22} \cdots a_{nn}.$$

Determinants can be computed easily for matrices of a certain block structure. All of the following matrices have determinant equal to $\det(A) \det(B)$:

$$\begin{bmatrix} A & 0 \\ 0 & B \end{bmatrix}, \qquad \begin{bmatrix} A & C \\ 0 & B \end{bmatrix}, \qquad \begin{bmatrix} A & 0 \\ C & B \end{bmatrix}.$$

That is, if A and B are square submatrices as shown, you can express the determinant just the way you would if they were numbers and the overall matrix is just a 2×2 matrix.

The *inverse* of an $n \times n$ matrix A, if it exists, is an $n \times n$ matrix $B = A^{-1}$ such that both $AB = I$ and $BA = I$, where I is the $n \times n$ *identity matrix*:

$$I = \begin{bmatrix} 1 & 0 & \cdots & 0 \\ 0 & 1 & \cdots & 0 \\ \vdots & \vdots & \ddots & \vdots \\ 0 & 0 & \cdots & 1 \end{bmatrix}.$$

The identity matrix acts as a multiplicative identity on the set of matrices that it can legally multiply. Some frequently useful properties relating inverses, transposes, products, and determinants are the following:

$$(AB)^{-1} = B^{-1} A^{-1}$$
$$(A')^{-1} = (A^{-1})'$$
$$\det A^{-1} = 1 / \det A$$
$$\det A' = \det A$$
$$\det A \cdot B = \det A \cdot \det B$$
$$(A \cdot B)' = B' \cdot A'$$
$$A^{-1} \text{ exists } \Leftrightarrow \det A \neq 0$$
$$A = \begin{bmatrix} a & b \\ c & d \end{bmatrix} \text{ and } \det A \neq 0 \Rightarrow A^{-1} = \frac{1}{\det A} \begin{bmatrix} d & -b \\ -c & a \end{bmatrix}$$

For block diagonal matrices, a simple result on inverses follows:

$$\begin{bmatrix} A & \mathbf{0} \\ \mathbf{0} & B \end{bmatrix}^{-1} = \begin{bmatrix} A^{-1} & \mathbf{0} \\ \mathbf{0} & B^{-1} \end{bmatrix},$$

as long as A and B themselves are square matrices that have inverses.

Inverse matrices relate to systems of equations. A linear system

$$a_{11}x_1 + a_{12}x_2 + \cdots + a_{1n}x_n = b_1$$
$$a_{21}x_1 + a_{22}x_2 + \cdots + a_{2n}x_n = b_2$$
$$\vdots$$
$$a_{n1}x_1 + a_{n2}x_2 + \cdots + a_{nn}x_n = b_n$$

can be written in matrix form as $A\mathbf{x} = \mathbf{b}$, where $A = (a_{ij})$, $\mathbf{x} = [x_1 \ x_2 \ \cdots \ x_n]'$, $\mathbf{b} = [b_1 \ b_2 \ \cdots \ b_n]'$. Then if A^{-1} exists, the solution of the system is $\mathbf{x} = A^{-1}\mathbf{b}$, since $AA^{-1}\mathbf{b} = \mathbf{b}$.

Recall that the method of *Gaussian elimination* can be used to solve linear systems directly, and also to compute inverse matrices. We will not discuss it in detail here, but remember that this algorithm begins by augmenting the coefficient matrix A with the right side vector $\mathbf{b}$ (or an identity matrix I when computing the inverse of A). Then, you successively perform one of several legal row operations, such as subtracting one row from another, on the augmented matrix in order to reduce the old A part to the identity matrix. The final augmented part of the matrix is the solution vector (or the inverse matrix in the latter case).

A square matrix A is called *positive definite* if for all nonzero vectors $\mathbf{y}$ of the proper dimension,

$$\mathbf{y}'A\mathbf{y} > 0.$$

If the preceding inequality is nonstrict, then the matrix is called *nonnegative definite*. One reason why positive definite matrices are important is that all their square submatrices, formed by deleting 0 or more rows and the same number of columns, are invertible. In particular, A itself has an inverse. For example, the matrix in the middle is positive definite:

$$\begin{bmatrix} y_1 & y_2 \end{bmatrix} \begin{bmatrix} 4 & 1 \\ 1 & 9 \end{bmatrix} \begin{bmatrix} y_1 \\ y_2 \end{bmatrix} = 4y_1^2 + 2y_1y_2 + 9y_2^2,$$

because, looked at as a quadratic polynomial in y_1, the last expression has a positive leading coefficient and no roots, according to the discriminant test.

Let $\mathbf{x}_1, \mathbf{x}_2, \ldots, \mathbf{x}_n$ be a collection of vectors of the same order. They are called *linearly independent* if whenever a linear combination

$$a_1\mathbf{x}_1 + a_2\mathbf{x}_2 + \cdots + a_n\mathbf{x}_n$$

equals the vector $\mathbf{0}$ of all zeros, then all of the coefficients a_i must be zero. The next two vectors are linearly independent, because the only solution to the system of equations is $a_1 = a_2 = 0$:

$$a_1 \begin{bmatrix} 1 \\ -1 \end{bmatrix} + a_2 \begin{bmatrix} 0 \\ 1 \end{bmatrix} = \begin{bmatrix} 0 \\ 0 \end{bmatrix} \Rightarrow \begin{cases} a_1 + 0 \cdot a_2 = 0 \\ -a_1 + a_2 = 0 \end{cases}.$$

An important fact is that when the columns (or rows) of a square matrix are linearly independent, then the matrix is invertible.

The *linear space spanned by a collection of vectors* $\mathbf{x}_1, \mathbf{x}_2, \ldots, \mathbf{x}_n$ of the same order is the collection of all vectors that can be written as a linear combination of these. The *dimension of the linear space* is the largest number of linearly independent vectors within the spanning set. For instance, the unit coordinate vectors $\mathbf{i} = (1, 0, 0)$, $\mathbf{j} = (0, 1, 0)$, $\mathbf{k} = (0, 0, 1)$ are independent and span all of three-dimensional Euclidean space. The vector $(1, 1, 1)$ could be added to the spanning set without loss, but it is superfluous because it is a linear combination of $\mathbf{i}$, $\mathbf{j}$, and $\mathbf{k}$.

A matrix can be looked at as being made up of a collection of row vectors, or a collection of column vectors. The *rank* of a matrix is the maximum number of linearly independent rows. In the case that the matrix is square, this turns out to be the same as the number of linearly independent columns. A matrix is called *full rank* if its rank equals the number of rows, that is, (if all of the rows are linearly independent). The rows of the matrix

$$\begin{bmatrix} 1 & 0 & 3 \\ 0 & 2 & -1 \\ 0 & 0 & 4 \end{bmatrix}$$

turn out to be independent, as do the columns. This is therefore a full rank matrix of rank 3, and from above, it must also be invertible.

A number λ is an *eigenvalue* of an $n \times n$ matrix A if

$$\det(A - \lambda I) = 0,$$

where again I is an identity matrix. This characterizes the eigenvalues of a matrix nicely as the solutions of a polynomial equation, but a different characterization is very useful, which leads to the idea of diagonalizing a matrix. The number λ is an *eigenvalue* of A with associated *eigenvector* $\mathbf{x}$ if

$$A\mathbf{x} = \lambda \mathbf{x}.$$

For the preceding 3×3 matrix the eigenvalue equation is

$$\det \begin{bmatrix} 1 - \lambda & 0 & 3 \\ 0 & 2 - \lambda & -1 \\ 0 & 0 & 4 - \lambda \end{bmatrix} = (1 - \lambda)(2 - \lambda)(4 - \lambda) - 0 - 0 = 0.$$

Therefore, the eigenvalues are 1, 2, and 4. For eigenvalue $\lambda = 1$ for instance, we can find an eigenvector by solving

$$\begin{bmatrix} 1 & 0 & 3 \\ 0 & 2 & -1 \\ 0 & 0 & 4 \end{bmatrix} \begin{bmatrix} x_1 \\ x_2 \\ x_3 \end{bmatrix} = 1 \cdot \begin{bmatrix} x_1 \\ x_2 \\ x_3 \end{bmatrix}$$

$$\Rightarrow \begin{cases} x_1 + 3x_3 = x_1 \\ 2x_2 - x_3 = x_2 \\ 4x_3 = x_3 \end{cases}.$$

The third equation forces $x_3 = 0$, which in turn forces $x_2 = 0$ in the second equation. But the first equation then allows x_1 to be anything. Thus $[1\ 0\ 0]'$ is one of infinitely many eigenvectors; in fact, from the definition you can see easily that any multiple of an eigenvector is also an eigenvector for the same eigenvalue.

The phenomenon you may have noticed in the last example is true in general: When the matrix is either upper or lower triangular (diagonal matrices being a special case), then the eigenvalues are the diagonal entries of the matrix. Other important results that are true in general about the eigenvalues and vectors are

sum of eigenvalues of A = sum of diagonal entries of $A \equiv trace(A)$

product of eigenvalues of A = det A;

nonzero eigenvectors of different eigenvalues are independent.

A very important theorem comes out of the last of these facts. Suppose that an $n \times n$ matrix A has n distinct eigenvalues $\lambda_1, \lambda_2, \ldots, \lambda_n$, and let $\mathbf{x}_1, \mathbf{x}_2, \ldots \mathbf{x}_n$ be nonzero eigenvectors for them. Then these vectors can be placed as the columns of an $n \times n$ matrix N, which is invertible because of the independence of its columns. Let D be the diagonal matrix containing $\lambda_1, \lambda_2, \ldots, \lambda_n$ on the diagonal (in the same order as the eigenvectors were put into N). Then A can be decomposed as

$$A = NDN^{-1}.$$

This is called the *diagonalization* of A. The distinctness of the eigenvalues is sufficient, not necessary, for the independence of the eigenvectors. Many other matrices can be diagonalized. There are many ways in which diagonalization is useful, among which is an easy method to compute powers of a matrix. The third power, say, of A as above, would be

$$A^3 = NDN^{-1} \cdot NDN^{-1} \cdot NDN^{-1} = ND^3N^{-1}.$$

The powers of the diagonal matrix D are simple: They are again diagonal with the components equal to the powers of the components of D.

A stronger result is true when the matrix A is symmetric. Then eigenvectors corresponding to different eigenvalues are orthogonal to one another and can be normalized such that

$$A = NDN^t,$$

where the inverse of N is the same as the transpose of N. A matrix satisfying the latter property is called *orthogonal*.

We often encounter matrices in statistics with the following property. A symmetric square matrix A is called *idempotent* if $A^2 = A$. Interesting things happen with the eigenvalues of such a matrix, and with its diagonalization. Because of the idempotency, if λ is an eigenvalue of A with eigenvector $\mathbf{x}$, then

$$A\mathbf{x} = A^2\mathbf{x} = A\lambda\mathbf{x} = \lambda A\mathbf{x} = \lambda^2\mathbf{x},$$

hence λ^2 must also be an eigenvalue. Because there can only be finitely many solutions of the determinant equation, the only way for this to happen is for λ to be 0 or 1, which are their own squares. So, all eigenvalues of an idempotent matrix are either 0 or 1, and moreover, it can be shown that the multiplicity of 1 as a solution of the eigenvalue determinant equation is the same as the rank of the matrix A. If r is this rank, then the diagonal matrix D in the decomposition of A has r 1s in the first diagonal positions, followed by zeros. It can be written in blocked form as

$$D = \begin{bmatrix} I_r & \mathbf{0} \\ \mathbf{0} & \mathbf{0} \end{bmatrix},$$

where I_r is an identity matrix.

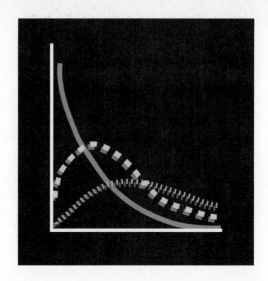

REFERENCES

Able, Kenneth W., Michael P. Fahay, and Gary R. Shepard, "Early Life History of Black Sea Bass," *U.S. Dept. of Commerce Fishery Bulletin*, Vol. 93, No. 3, July 1995, pp. 429–445.

Als, Heidelise, Gretchen Lawhon, Frank H. Duffy, Gloria B. McAnulty, Rita Gibes-Grossman, and Johan G. Blickman, "Individualized Developmental Care for the Very Low Birth Weight Preterm Infant," *JAMA*, Vol. 273, No. 11, March 15, 1995, pp. 853–858.

American Mathematical Society, "AMS-IMS-MAA Annual Survey," *Notices of the American Mathematical Society*, July/August 1994, Volume 41, Number 6.

Anderson, Edwin L., Michael D. Decker, Janet A. Englund, Kathryn M. Edwards, Porter Anderson, Pamela McInnes, and Robert B. Belshe, "Interchangeability of Conjugated Haemophilus Influenzae Type b Vaccines in Infants," *JAMA*, Vol. 273, No. 11, March 15, 1995, pp. 849–853.

Anderson, T. W., *An Introduction to Multivariate Statistical Analysis*, 2nd ed. (New York: John Wiley & Sons, 1984).

Anton, Howard, *Elementary Linear Algebra*, 4th ed. (New York: John Wiley, 1984).

Arnold, Steven F., *Mathematical Statistics* (Englewood Cliffs, NJ: Prentice-Hall, 1990).

Badalamente, Anthony F., Robert J. Langs, and James Robinson, "Lawful Systems Dynamics in How Poets Choose Their Words," *Behavioral Science*, Vol. 39, No. 1, 1994, pp. 46–71.

Barlow, Richard E., and Frank Proschan, *Statistical Theory of Reliability and Life Testing* (New York: Holt, Rinehart and Winston, 1975).

Beckett, James, and Dennis W. Eckes, *The Sport Americana Baseball Card Price Guide,* Number 4 (Lakewood, OH: Edgewater Books, 1982).

Bickel, Peter J., and Kjell A. Doksum, *Mathematical Statistics; Basic Ideas and Selected Topics* (San Francisco: Holden-Day, 1977).

Billingsley, Patrick, *Probability and Measure* (New York: John Wiley & Sons, 1979).

Box, George E. P., William G. Hunter, and J. Stuart Hunter, *Statistics for Experimenters; An Introduction to Design, Data Analysis, and Model Building* (New York: John Wiley & Sons, 1978).

Box, George E. P., and G. M. Jenkins, *Time Series Analysis: Forecasting and Control* (San Francisco: Holden-Day, 1976).

Boyland, Philip, *Guide to Standard Mathematica Packages* (Champaign, IL: Wolfram Research, 1991).

Burr, Irving W., *Elementary Statistical Quality Control* (New York: Marcel Dekker, 1979).

Burr, Irving W., *Statistical Quality Control Methods* (New York: Marcel Dekker, 1976).

"Cardizem" (advertisement), *JAMA*, Vol. 273, No. 11, March 15, 1995.

Casella, George, and Roger L. Berger, *Statistical Inference* (Belmont, CA: Wadsworth, 1990).

Cashen, Valjean M., and Gary C. Ramseyer, "ESP and the Prediction of Test Items in Psychology Examinations," *J. Parapsychology*, Vol. 34, No. 2, June 1970, pp. 117–123.

Centers for Disease Control and Prevention, "Risky Driving Behaviors Among Teenagers—Gwinnett County Georgia, 1993," *JAMA*, Vol. 273, No. 11, March 15, 1995a, pp. 844–845.

Centers for Disease Control and Prevention, "Head Injuries Associated with Motorcycle Use—Wisconsin, 1991," *JAMA*, Vol. 273, No. 11, March 15, 1995b, pp. 845–846.

Chicago Tribune, May 28, 1993.

Chicago Tribune, June 21, 1993.

Chicago Tribune, June 29, 1993.

Chicago Tribune, July 4, 1993.

Chung, K. L., *Elementary Probability Theory with Stochastic Processes* (New York: Springer-Verlag, 1979).

Chung, K. L., *A Course in Probability Theory*, 2nd ed. (New York: Academic Press, 1974).

Cinlar, Erhan, *Introduction to Stochastic Processes* (Englewood Cliffs, NJ: Prentice-Hall, 1975).

Cleveland, William S., *The Elements of Graphing Data*, Bell Telephone Laboratories, Inc., Murray Hill, 1985.

Comrie, Andrew C., "Tracking Ozone: Air-Mass Trajectories and Pollutant Source Regions Influencing Ozone in Pennsylvania Forests," *Annals of the Assn. of Amer. Geographers*, Vol. 84, No. 4, 1994, pp. 635–651.

Cooper, Doug, and Michael Clancy, *Oh! Pascal!*, 2nd ed. (New York: W.W. Norton and Co., 1985).

Craik, F., M. Moscovitch, and J. McDowd, "Surface and Conceptual Information," *J. Exp. Psych: Learning, Memory, and Cognition*, July 1994.

Cryer, Jonathan D., *Time Series Analysis* (Boston: PWS, 1986).

Deming, W. Edwards, *Sample Design in Business Research* (New York: John Wiley & Sons, 1960).

Devore, Jay L., *Probability and Statistics for Engineering and the Sciences*, 3rd ed. (Belmont, CA: Wadsworth, 1991).

Dortch, Shannon, "Coffee at Home," *American Demographics*, Vol. 17, No. 8, August 1995, pp. 4–6.

Douthat, Strat, "Clergy Divorce Rate Not As High As Thought: Survey," *Associated Press*, in Peoria Journal Star, June 24, 1995.

Fama, Eugene, "Multiperiod Consumption—Investment Decision," *American Economic Review*, Vol. 60, pp. 163–174, 1970.

Feller, William, *An Introduction to Probability Theory and Its Applications*, 3rd ed. (New York: John Wiley & Sons, 1968).

Fischbach, Frances, *A Manual of Laboratory Diagnostic Tests* (Philadelphia: J.B. Lippincott, 1980).

Fisher, Lawrence, and James H. Lorie, *A Half-Century of Returns on Stocks and Bonds* (Chicago: University of Chicago Graduate School of Business, 1977).

Forsyth, Randall W., "Tax-Exempts Still Deliver," *Barron's*, August 7, 1995, p. 35.

Franklin, Barry A., Patrick Hogan, Kim Bonzheim, Donovan Bakalyar, Edward Terrien, Seymour Gordon, and Gerald C. Timmis, "Cardiac Demands of Heavy Snow Shoveling," *JAMA*, Vol. 273, No. 11, March 15, 1995, pp. 880–882.

Freund, John E., and Walpole, Ronald E., *Mathematical Statistics*, 4th ed. (Englewood Cliffs, NJ: Prentice-Hall, 1987).

Friedman, Dorian, "A Nation of Significant Change," *U.S. News and World Report*, August 14, 1995, p. 9.

Fulkerson, Jennifer, "The Secret Life of Donors," *American Demographics*, Vol. 17, No. 8, August 1995, pp. 20–21.

Galesburg Register Mail, February 3, 1994.

Gertsbakh, I. B., *Statistical Reliability Theory* (New York: Marcel Dekker, 1989).

Gibbons, Jean Dickinson, *Nonparametric Statistical Inference* (New York: McGraw-Hill, 1971).

Government of Pakistan, *Ten Years of Pakistan in Statistics: 1972:1982*.

Green Book, U.S. Army, October. 1994, pp. 210–231.

Gross, Donald, and Carl M. Harris, *Fundamentals of Queueing Theory*, 2nd ed. (New York: John Wiley & Sons, 1985).

Hamilton, Lawrence C., *Modern Data Analysis; A First Course in Applied Statistics* (Belmont, CA: Wadsworth, 1990).

Hastings, Kevin J., *Introduction to the Mathematics of Operations Research* (New York: Marcel Dekker, 1989).

Hillier, Frederick S., and Gerald J. Lieberman, *Introduction to Operations Research*, 3rd ed. (San Francisco: Holden-Day, 1980).

Hogg, Robert V., and Elliot A. Tanis, *Probability and Statistical Inference*, 3rd ed. (New York: Macmillan, 1988).

Hogg, Robert V., and Allen T. Craig, *Introduction to Mathematical Statistics*, 4th ed. (New York: Macmillan, 1978).

Illinois Academe, AAUP of Illinois, Vol. 12, No. 1, 1991, p. 10.

Institute for Electrical and Electronics Engineers, "Women in Technical/Scientific Professions: Results of Two National Surveys," IEEE, New York, June 1995.

Johnson, Richard A., and Dean W. Wichern, *Applied Multivariate Statistical Analysis*, 2nd ed. (Englewood Cliffs, NJ: Prentice-Hall, 1988).

Johnson, Timothy E., *Investment Principles* (Englewood Cliffs, NJ: Prentice-Hall, 1978).

Johnston, Lloyd D., Patrick M. O'Malley, and Jerald G. Bachman, *National Survey Results on Drug Use from: The Monitoring The Future Study 1975–1992, Vol. II*, National Institute on Drug Abuse, NIH Pub. No. 93-3598, 1993.

Kasser, Tim, Richard M. Ryan, and Melvin Zax, "The Relations of Maternal and Social Environments to Late Adolescents' Materialistic and Prosocial Values," *Developmental Psychology*, Vol. 31, No. 6, 1995, pp. 907–914.

Kaufmann, Arnold, D. Grouchko, and R. Cruon, *Mathematical Models for the Study of the Reliability of Systems* (New York: Academic Press, 1977).

Kemeny, John G., J. Laurie Snell, and A. Knapp, *Denumerable Markov Chains* (New York: Van Nostrand, 1966).

Kemph, Brad T., and Tim Kasser, "Effects of Sexual Orientation of Interviewer on Expressed Attitudes Toward Male Homosexuality," *J. Social Psych.* (to appear)

Kendall, Maurice G., *The Advanced Theory of Statistics* (London: Charles Griffin & Co., 1946).

Kisiel, Ralph, "Holiday Rambler Grows," *Automotive News*, August 7, 1995, p. 18.

Larsen, Richard J., and Morris L. Marx, *An Introduction to Mathematical Statistics and Its Applications* (Englewood Cliffs, NJ: Prentice-Hall, 1986).

Law, Averill M., and M. G. McComas, "How Simulation Pays Off," *Manufacturing Engineering*, February 1988.

Loeve, M., *Probability Theory I* (New York: Springer-Verlag, 1977).

"Markets and Economics," *Chemical Week*, Vol. 157, No. 5, August 9, 1995, p. 30.

McClave, James T., and Frank H. Dietrich, II, *Statistics*, 6th ed. (New York: Macmillan, 1994).

Merton, Robert C., "On the Microeconomic Theory of Investment Under Uncertainty," in *Handbook of Mathematical Economics*, Vol. 11, North-Holland, 1982.

Miller, Irwin R., John E. Freund, and Richard Johnson, *Probability and Statistics for Engineers,* 4th ed. (Englewood Cliffs, NJ: Prentice-Hall, 1990).

Milton, J. S., and Jesse C. Arnold, *Probability and Statistics in the Engineering and Computing Sciences* (New York: McGraw-Hill, 1986).

Minitab Reference Manual Release 9 for Windows, Minitab Inc., 1993.

Montgomery, Douglas C., *Design and Analysis of Experiments*, 3rd ed. (New York: John Wiley & Sons, 1991).

Naor, P., "The Regulation of Queue Size by Levying Tolls," *Econometrica,* Vol. 37, No. 1, 1969.

Nelligan, Chris, "Equality of Opportunity in Physical Education at Fitzharry's School," *Bull. of Phys. Ed.*, Vol. 30, No. 2, 1994, pp. 40–59.

Palm, Risa, and Michael Hodgson, "Earthquake Insurance: Mandated Disclosure and Homeowner Response in California," *Annals of the Assn. of Amer. Geographers*, Vol. 82, No. 2, 1992, pp. 207–272.

Parzen, Emanuel, *Modern Probability Theory and Its Applications* (New York: John Wiley & Sons, 1960).

Peoria Journal Star, February 26, 1993.

Peoria Journal Star, May 28, 1993.

Peoria Journal Star, May 24, 1994.

Peoria Journal Star, November 27, 1994.

Peoria Journal Star, August 1, 1995.

Pockney, B. P., *Soviet Statistics Since 1950* (New York: St. Martin's Press, 1991).

Press, S. James, *Bayesian Statistics: Principles, Models, and Applications* (New York: John Wiley & Sons, 1989).

Ramer, Arthur, "A Note on Defining Conditional Probability," *American Mathematical Monthly*, Vol. 97, No. 4, April 1990, pp. 336–337.

Redfield, Robert R., and Donald S. Burke, "HIV Infection: The Clinical Picture," *Scientific American*, October 1988, pp. 90–98.

Reisinger, William M., Arthur H. Miller, Vicki L. Hesli, and Kristen Hill Maher, "Political Values in Russia, Ukraine, and Lithuania: Sources and Implications for Democracy," *Brit. Jour. Poly. Sci.*, Vol. 24, Part 2, April 1994, pp. 183–223.

Rice, John A., *Mathematical Statistics and Data Analysis,* (Belmont, CA: Wadsworth, 1988).

Rinaman, William C., *Foundations of Probability and Statistics* (Philadelphia: Saunders College Publishing, 1993).

Romana, Elpidio, and Ricardo Jose, "Never Imagine Yourself to Be Otherwise: Filipino Image of Japan over the Centuries," *Asian Studies*, Vol. XXIX, 1991.

Rudin, Walter, *Principles of Mathematical Analysis* (New York: McGraw-Hill, 1964).

Ryan, Thomas P., *Statistical Methods for Quality Improvement* (New York: John Wiley & Sons, 1989).

Saaty, Thomas L., *Elements of Queueing Theory with Application* (New York: Dover, 1961).

Samuelson, Paul A., *Foundations of Economic Analysis* (Cambridge, MA: Harvard University Press, 1983).

Schick, Frank L., ed., *Statistical Handbook on Aging Americans* (Phoenix, AZ: Oryx Press, 1986).

Schneider, Dennis M., *Linear Algebra: A Concrete Introduction*, 2nd ed. (New York: Macmillan, 1987).

Schriber, Thomas J., *An Introduction to Simulation Using GPSS/H* (New York: John Wiley & Sons, 1991).

Sharpe, William F., *Portfolio Theory and Capital Markets* (New York: McGraw-Hill, 1970).

Solomon, Susan L, *Simulation of Waiting Line Systems* (Englewood Cliffs, NJ: Prentice-Hall, 1983).

The Sporting News, "Major League Baseball Final Statistics," October 12, 1992.

Strang, Gilbert, *Linear Algebra and Its Applications*, 2nd ed. (New York: Academic Press, 1980).

Terasaki, Paul I., J. Michael Cecka, David W. Gjertson, and Steven Takemoto, "High Survival Rates of Kidney Transplants from Spousal Donors," *New England J. Medicine*, Vol. 333, No. 6, August 10, 1995.

Thomas, G. Scott, *The Rating Guide to Life in America's Small Cities* (Buffalo, NY: Prometheus Books, 1992).

Tukey, John W., *Exploratory Data Analysis* (Reading, MA: Addison-Wesley, 1977).

United Nations Dept. for Economic and Social Information and Policy Analysis, *Statistical Yearbook*, 39th ed., 1994.

United States Department of Agriculture, *Agricultural Statistics*, 1991, 1976.

United States Department of Health and Human Services, *Vital Statistics of the U.S., 1991, Vol. 1—Natality*, Pub. No. 95-1100, 1995.

Van Buren, Harry J., "The Exploitation of Mexican Workers," *Business and Society Review*, No. 92, Winter 1995, pp. 29–33.

Walpole, Ronald E., and Raymond H. Myers, *Probability and Statistics for Engineers and Scientists*, 4th ed. (New York: Macmillan, 1989).

Wilkie, Maxine, "Scent of a Market," *American Demographics*, Vol. 17, No. 8, August 1995, pp. 40–49.

Williamson, Debra Aho, "Digital Media Future Looks Rosy," *Advertising Age*, August 7, 1995, p. 13.

Winkler, Robert L., *Introduction to Bayesian Inference and Decision* (New York: Holt, Rinehart and Winston, 1972).

Wolfram, Stephen, *Mathematica: A System for Doing Mathematics by Computer*, 2nd ed. (Reading, MA: Addison-Wesley, 1991).

Wood, William B., "Forced Migration: Local Conflicts and International Dilemmas," *Annals of the Assn. of Amer. Geographers*, Vol. 84, No. 4, 1994, pp. 607–634.

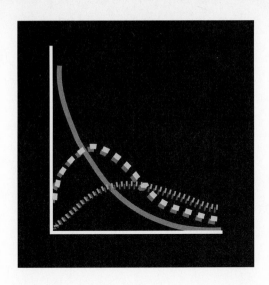

ANSWERS TO
SELECTED EXERCISES

Chapter 1

Section 1.2

1. $P[10 \text{ or more}] = 1/6$
2. $P[\text{even sum}] = 1/2$
4. $50/1000; 850/1000$
6. (a) $7/80$ (b) $22/80$ (c) $8/80$
7. $8/221$
8. $.8; .2; .4; .1$
10. $.9$
11. $95/132$
15. $7/8$
17. $4/9; 4/9$
18. $b = 1/2; 1/8$

Section 1.3

1. $Q(0) = 1/20$, $Q(1) = 9/20$, $Q(2) = 9/20$, $Q(3) = 1/20$

2. $F(x) = (x - 108)/4$ and $f(x) = 1/4$ for $x \in (108, 112)$
3. $F(x) = 1 - (1 - x)^2$
4. $F(s) = s^2$ for $s \in [0, 1]$; $Q((s, t]) = t^2 - s^2$
5. $x_1 : 1/4; x_2 : 3/4$
6. $F(x) = P[X \leq x] = (x - 1)/2$ and $f(x) = 1/2$ for $x \in (1, 3)$
7. $p_X(-1) = 20/38; p_X(1) = 18/38; p_Y(-1) = 37/38;$ $p_Y(34) = 1/38$, assuming you don't get your dollar back
8. (c) $2/3$
9. (a) $1/4$ (b) $15/16$ (c) $y^2/16$ for $y \in [0, 4]$ (d) $3/16$

Section 1.4

1. (a) $1/6$ (b) $1/3$
2. $.433; .449; .277$
3. (a) $1/3$ (b) $2/3$ (c) 1

5. (a) 1/2 (b) 3/4
7. (a) 3/5 (b) 4/5
8. 1/4
9. 3/4
12. 17/60
14. .62
15. 8/9
16. 0–20: .0975; 20–30: .512; 30–40: .244; over 40: .146
17. 787/2162 = .351
18. For stages 1–6, respectively: .2025, .396, .1764, .0791, .0805, .0655

Section 1.5

1. No
4. $8 \cdot \left(\frac{1}{4}\right)^7 \cdot \frac{3}{4} + \left(\frac{1}{4}\right)^8$
5. 13/16
10. Yes; no
11. Distribution of sum: 2: .0278; 3: .0417; 4: .0573; 5: .0729; 6: .0885; 7: .1736; 8: .1302; 9: .1146; 10: .0989; 11: .0833; 12: .1111.
12. $(1 - (1 - p_1)^2) \cdot (1 - (1 - p_2)^3) \cdot (1 - (1 - p_3)^2)$
14. 1:1/9; 3:11/36; 5:1/9; 6:7/36; 8:7/36; 9:1/12

Chapter 2

Section 2.1

1. 26^3; 26^4; $(1/26)^3$
2. (a) $P_{30,28}$ (b) 435 (c) 328,860
3. .271
4. 4/13
7. (a) .001 (b) .027 (c) .014
8. 18 routes
9. (a) 1/16 (b) 1/15 (c) 1/11 (d) 1/5
10. 86/220
12. (a) 3,732,480,000 (b) 663,552
15. (a) .00198 (b) .00144 (c) .04754
16. .8559
19. .443
20. 630,630
21. $\binom{13}{6}\binom{13}{4}\binom{13}{2}\binom{13}{1}/\binom{52}{13}$
24. 1.864×10^{28} games; 5.107×10^{25} years
26. 2/5

Section 2.2

1. $F(x) = 0$ for $x < -1$; $F(x) = 1/4$ for $-1 \le x < 0$; $F(x) = 3/8$ for $0 \le x < 2$; $F(x) = 5/8$ for $2 \le x < 4$; $F(x) = 3/4$ for $4 \le x < 5$; $F(x) = 1$ for $x \ge 5$
2. $P[3 < X < 6] = 3/8$
3. $F(b^-) - F(a)$; $F(b) - F(a^-)$
6. $(d - c)/(b - a + 1)$
7. At least 4 days
8. −$5.26
9. $q(i) = 2i/n(n + 1)$; $6(n - 1)/n(n + 1)$
10. $F(n) = 1 - (8/9)^n$; $q(n) = (8/9)^{n-1}(1/9)$
13. .085
15. (a) $q(k) = \binom{20}{k}\binom{180}{5-k}/\binom{200}{5}$
    (b) $q(k) = \binom{5}{k}(.1)^k(.9)^{5-k}$
17. $q_1(1) = 13/27$, $q_1(2) = 5/27$, $q_1(3) = 1/3$; $q_2(1) = 2/9$, $q_2(2) = 5/9$, $q_2(3) = 2/9$;14/27;2/27;1/3
18. X_1 uniform on $\{0,1\}$; X_2 uniform on $\{0,2\}$; $q_3(0) = 2/3$, $q_3(3) = 1/3$
19. (a) 1/12 (b) 1/4 (c) 29/60

Section 2.3

1. .428
2. .609
3. .805
4. At least 4 books
7. (a) 7 (b) 14 (c) 22
10. $(.507)^n$
12. .156
14. (a) $Np^{N-1}(1 - p) + Np(1 - p)^{N-1}$; $N(1/2)^{N-1}$;
    (b) 1/8
16. .687
18. .5256
19. (a) .91 (b) .0039 (c) .0178

Section 2.4

2. 6.93×10^{-12}
3. .32
4. .04; .655
5. .547
6. $q(n) = \frac{\lambda}{n} q(n - 1)$
10. .05
11. .432
13. $e^{-5}5^m/m!$

14. $.14; 1 - e^{-t}[1 + t + t^2/2]; .44$

16. (a) $.0013$ (b) $.0255$ (c) $.0255$ (d) $.21$
    (e) $\binom{n}{k}(2/5)^k(3/5)^{n-k}$

17. Poisson(6.4)

Section 2.5

1. 7

3. $11

4. $333

5. 0

7. $E[K] = .385, \ \text{Var}(K) = .327$

9. 21.1 °Celsius

11. At least 40 balls; $-$2

12. $300 in asset 1, and $700 in asset 2

14. (a) $E[T] = 10; \ \text{Var}(T) = 90;$ (b) 30

18. $1719.36; $54,980.64

19. Mean 9; StdDev 3

20. (a) 0 (b) 1152/4096 (c) $-1152/4096$

22. (a) 2 (b) 5/3 (c) 10/3 (d) 11/3

23. $c_1 + c_2 + \cdots + c_n = 1$

Chapter 3

Section 3.1

1. $1 - e^{2t}(2t + 1); 5e^{-4} - 7e^{-6}$

2. $3e^{-3s}; \ e^{-3} - e^{-15}$

4. (a) $.875$ (b) $.296$. These are in fairly close
    agreement with the observed proportions of .81
    and .26.

5. (a) e^{-1} (b) e^{-1}

6. $3x^2 - 2x^3; 1/2$

10. (a) $\sqrt{2}/2$ (b) 0

11. (a) $\sqrt[3]{1/4}; \ \sqrt[3]{3/4}$ (b) $1/(1 - p)$

12. $\frac{1}{2}x_2(1 - e^{-x_1}), \ x_1 \in [0, \infty), x_2 \in [0, 2], \ 1 - e^{-x_1},$
    $x_1 \in [0, \infty), x_2 > 2; \ e^{-x_1}, \ x_1 > 0; \ 1 - e^{-x_1},$
    $x_1 > 0; \ \frac{1}{2}, \ x_2 \in [0, 2]; \ \frac{1}{2}x_2, \ x_2 \in [0, 2]$

13. $c = 1/64; \ \frac{1}{16}x + \frac{1}{8}, \ x \in [0, 4]; \ \frac{1}{16}y + \frac{1}{8},$
    $y \in [0, 4]; \ 191/192$

14. 7/16

15. $(1 + .02t)e^{-.02t}$

16. $.5$

17. $475\pi/36$

Section 3.2

2. Mean $= 1$, variance $= 1/2$, standard deviation $=$
   $\sqrt{1/2}$

3. No mean or second moment

4. $409.09

5. $1/2; 1/20$

6. 100

8. Invest $107.14 in the stock, the rest in the bond

12. 2

13. 5

14. 2

16. $[1 \ \ -1 \ \ 21]'$

Section 3.3

1. (a) e^{-3} (b) $2e^{-1} - 4e^{-3}$ (c) 4

4. $2/\lambda^3$

5. $-\frac{1}{\lambda}\ln(1 - p)$

8. $(3.5m + 1)e^{-3.5m} = .5; \ m = .4795$

9. $(\alpha - 1)/\lambda$

12. (a) 15.51 (b) $.8$ (c) 2.73 and 15.51

13. (a) $.1587$ (b) $.1056$ (c) $.383$ (d) 18

14. (a) $.27$ (b) $.715$

16. 82nd percentile

17. (a) $.9544$ (b) $.0026$ (c) $.1587$

18. $.306$

20. (a) $.977$ (b) 1.018 (c) $.0701$

21. $.0591$

Section 3.4

2. $\frac{1}{(2\pi)^{3/2}\sqrt{15}} \exp[-\frac{1}{2}(\frac{4}{15}x_1^2 + \frac{2}{15}x_1x_2 + \frac{4}{15}x_2^2 + (x_3 - 1)^2)]$

5. (a) $.195$ (b) $.345$ (c) $.78$

6. (a) $.5$ (b) $.37$ (c) $.16$ (d) $.0296$ (e) 16
   (f) 160

7. (a) $-.65$ (b) $.023$ (c) $.023$

12. $\mu_x\mu_y + \rho\sigma_x\sigma_y; \ \rho$

Chapter 4

Section 4.1

1. (a) Nearly independent (b) Almost certainly
   dependent

2. $.5795$

3. $f(t_1, t_2, t_3) = (.35)^3(.65)^{t_1-1}(.65)^{t_2-1}(.65)^{t_3-1}$, $t_1, t_2, t_3 \in \{1, 2, 3, \ldots\}$; .107

4. They are very nearly independent.

6. (a) c.d.f. $(1 - e^{-\lambda s_1})(1 - e^{-\lambda s_2})$; p.d.f. $\lambda^2 e^{-\lambda(s_1+s_2)}$, $s_i \geq 0$ (b) $2e^{-6}$

10. 58.3% to asset 1, 30.2% to asset 2, and 11.5% to asset 3

11. $[1.813, 3.787] \times [62.55, 95.45]$

12. c.d.f. $(1 - e^{-t^3})^4$; p.d.f. $12t^2 e^{-t^3}(1 - e^{-t^3})^3$, $t \geq 0$

13. (a) mean $1/\lambda$, variance $1/(3\lambda^2)$ (b) Mean $1/\lambda$, variance $3/(8\lambda^2)$

14. 680

Section 4.2

1. Given Santa Clara: $222/555, 26/555, 307/555$; given L.A.: $133/336, 10/336, 193/336$; given have insurance: $117/581, 222/581, 133/581, 109/581$

2. $q(0|1) = 1/6, q(1|1) = 1/2, q(2|1) = 1/3$

3. (a) $2e^{-2(t_2-t)}, t_2 \geq t$ (b) e^{-1} (c) 1

5. Expected responses: Russia, 2.65; Ukraine, 2.52; Lithuania, 2.75

8. $f(y, z|x) = (1/6)^2$, $x + 1 \leq y \leq x + 6$, $y + 1 \leq z \leq y + 6$

9. (a) $\frac{1}{2}(x + \frac{3}{2})$, $x \in (0, 1)$
   (b) $\frac{1}{2}x + y + \frac{1}{4}$, $x, y \in (0, 1)$ (c) 13/64 (d) 3/8

10. $(4/5)^{s-t-1}(1/5)$, $s = t + 1, t + 2, \ldots$

11. $b(n - m, 1/3)$

12. $480

13. $f(y_2|y_1) = 1/(1 - y_1)$, $y_2 \in (y_1, 1)$

17. $f(\lambda|1) = \lambda e^{-\lambda}/(3e^{-2} - 5e^{-4})$, $\lambda \in [2, 4]$

18. $\beta \lambda^\beta s^{\beta-1} e^{-(\lambda s)^\beta} / e^{-(\lambda t)^\beta}$, $s > t$

19. Conditional p.m.f.: \$0, 1/2; \$5000, 1/2. Conditional expectation, \$2500

Section 4.3

2. Covariance $= .00958$, correlation $= .678$

3. Covariance $= .443$, correlation $= .678$

5. 1/4

7. Covariance $= -2/75$, correlation $= -2/3$, $E[Y \mid X = x] = \frac{2}{3}(1 - x)$

8. 3.9; 2.78

9. Empirical covariance $= \frac{1}{n}\sum_{i=1}^{n}(x_i - \bar{x})(y_i - \bar{y})$, empirical correlation $= \sum_{i=1}^{n}(x_i - \bar{x}) \cdot$
   $(y_i - \bar{y})/\left(\sqrt{\sum_{i=1}^{n}(x_i - \bar{x})^2} \cdot \sqrt{\sum_{i=1}^{n}(y_i - \bar{y})^2}\right)$

11. Invest 126.6% in asset 1, and -26.6% in asset 2;

in economic terms, borrow on asset 2 in order to invest more in asset 1

12. $m\sigma_X^2$

13. mean $= 1$, variance $= 1$

16. 4.2

Section 4.4

2. (a) Mean 473.33, variance 1200 (b) Mean 406.66, variance 1200

3. .002

4. (a) 126.25; (b) .02; (c) 172.56

5. $-.264$; $-.675$; .29

6. Mean 46.8, variance 266.96, interval $[19.9, 73.7]$

7. Bivariate normal, mean vector $[1.156 \; .313]'$, variances .909, .688, covariance $-.056$

8. (a) Bivariate normal, mean vector $[9.5, 9.6]'$, variances .1, .12, covariance .0036 (b) $N(9.8, .08)$
   (c) 9.615 (d) Bivariate normal, mean vector $[9.792, 9.497]'$, variances .0799, .0999, covariance .0013

10. Under the first scenario, the increase is about 11.8, and under the second only 9.9, so the first plan is expected to be more effective.

12. Bivariate normal, mean vector $[52.4 + .376(x_1 - 52.8), 55.3 + .332(x_1 - 52.8)]'$, conditional variances 244.85, 195.22, conditional covariance 98.97

Chapter 5

Section 5.1

2. $f_Z(z) = z/8$, $z \in (0, 4)$

3. (a) $f_Y(y) = \frac{1}{4}y^{-1/2}$, $y \in [1, 9]$
   (b) $f_Y(y) = \frac{1}{2}y^{-1/2}$, $y \in (0, 1)$

5. $f_X(x) = \frac{1}{\pi(1+x^2)}$, $x \in (-\infty, +\infty)$

6. 24λ and β

10. 0; 1.645; 10; 13.29

11. $g(y) = \lambda^2 y e^{-\lambda y}$, $y > 0$

14. $N(0, 2)$

15. $g(y) = e^{-10}10^{y/.3}/(y/.3)!$, $y = 0, .3, .6, \ldots$

Section 5.2

3. Joint density
   $1/4\pi \cdot \exp((-1/2)((5/4)w_1^2 - (1/2)w_1 w_2 + (1/4)w_2^2))$;
   $\rho = \sqrt{1/5}$

4. $(n_1 + n_2 - 1)!/i \, ((n_1 - 1)!(n_2 - 1)!) \cdot y_1^{n_1-1} \cdot (1 - y_1)^{n_2-1}$,
   $0 < u < 1$

5. Joint density $9w^2/z$, $0 < w < z < 1$; marginal density $-9w^2 \ln(w)$, $w \in (0,1)$

7. (a) $f(u/c) \cdot (1/c)$ (b) $f(u/(1-u))/(1-u)^2$

9. 3/4

10. $1/3\pi^2 \cdot (1/(1 + (y_1 + y_2)^2/9)) \cdot (1/(1 + (2y_2 - y_1)^2/9))$, $y_1, y_2 \in (-\infty, +\infty)$

13. Since the chance that the sample median is at least .55 is just .275 if the true median was .5, we have weak evidence that the true median is more than .5.

16. $F_1(t) \cdot F_2(t) \cdots F_n(t)$

18. $105(e^{-5y_3} - 2e^{-6y_3} + e^{-7y_3})$, $y_3 > 0$

20. 57/64

22. 3/5

23. $f(k,l) = \frac{\mu_1^k e^{-\mu_1}}{k!} \cdot \frac{\mu_2^{l-k} e^{-\mu_2}}{(l-k)!}$, $k = 0,1,2,\ldots$,
    $l = k, k+1, k+2, \ldots$

Section 5.3

1. $(e^{tb} - e^{ta})/(tb - ta)$, $t \in \mathbb{R}$

3. $3e^t/t - 6e^t/t^2 + 6e^t/t^3 - 6/t^3$, $t \in \mathbb{R}$

6. $M_Y(t) = e^{td} M_X(ct)$

9. (a) $(1/2)e^{-2t} + (1/6)e^{-t} + 1/4 + (1/4)e^t + (1/6)e^{2t} + (1/12)e^{3t}$ (b) $(1/6)e^t + (1/6)e^{2t} + (1/6)e^{3t} + (1/3)e^{4t} + (1/12)e^{5t} + (1/12)e^{6t}$

10. (a) $pe^t/(1 - (1-p)e^t)$, $(1-p)e^t < 1$;
    (b) $p^2 e^{2t}/(1 - (1-p)e^t)^2$, $(1-p)e^t < 1$

17. $\mu_Y^3 + 3\sigma_Y^2 \mu_Y$

18. $e^{t'b} M_X(At)$

19. $E[X]$; $\text{Var}(X)$

21. (a) $e^{\mu(t-1)}$, $t \in \mathbb{R}$

22. (a) $n! q(n)$ (b) $E[X(X-1)(X-2) \cdots (X - (n-1))]$

Section 5.4

2. .10

4. The sample mean takes the value 288.1, and the chance is near 0 that it could be as large as that if the population mean were 256 or less.

5. $c = .653$

6. $c = .545$; $d = 2.39$

9. $k = .1$: .62; $k = .2$: .32; $k = .3$: .14; $k = .4$: .05

10. $g(y) = (n-1)/\sigma^2 \cdot (1/(2^{(n-1)/2} \Gamma((n-1)/2)) \cdot ((n-1)y/\sigma^2)^{(n-1)/2-1} e^{-(n-1)y/2\sigma^2}$, $y > 0$

11. (a) No   (b) No

13. Multivariate normal with
    $\mu_1 = -5, \mu_2 = -1, \sigma_1^2 = 7, \sigma_2^2 = 3, \rho = 3/\sqrt{21}$

14. Eigenvalues $\lambda_1 = \sigma^2(1 + \rho), \lambda_2 = \sigma^2(1 - \rho)$,
    eigenvectors
    $\mathbf{x}_1 = [1/\sqrt{2}, 1/\sqrt{2}]', \mathbf{x}_2 = [1/\sqrt{2}, -1/\sqrt{2}]'$; $\mathbf{Y}$ is
    bivariate normal with mean $\mathbf{0}$, variances
    $\sigma^2(1 + \rho), \sigma^2(1 - \rho)$, and covariance 0

19. Y_2 is $\chi^2(1)$

Section 5.5

1. .075

2. The sample value of t is 3.47, and the probability is less than .01 that this could happen if the true mean weight loss was 0.

3. $t = 5.03$, which is very strong evidence against $\mu = 256$

4. $t = 2.0$, which is strong evidence against $\mu = 48$

5. (a) $c = 2.11$ (b) $\bar{X} \pm 2.11 \cdot S/\sqrt{n}$

6. (a) $c = 1.70$ for both   (b) for January, $(29.3, 38.1)$, and for February, $(37.3, 47.8)$

9. $F(1, r)$

10. .975

11. About .05

12. Yes, since the ratio of sample variances is .25

13. Observed $f = .736$; no significant difference

14. Observed $f = .796$; no significant difference

15. Observed $f = .97$; no significant difference

17. $r_2/(r_2 - 2)$

18. (a) $c = .26$, $d = 4.30$ (b) $[.23 \cdot S_1^2/S_2^2, 3.85 \cdot S_1^2/S_2^2]$

Chapter 6

Section 6.1

1. .75

2. 500; 375

3. $f(h) = 1 - h/\sqrt{3}$, $0 \le h \le \sqrt{3}$

5. $f(h) = 8/9$ if $h \in [0, \sqrt{3/20})$; $f(h) = 6/9$ if
   $h \in [\sqrt{3/20}, 2\sqrt{3/20})$; $f(h) = 4/9$ if
   $h \in [2\sqrt{3/20}, 3\sqrt{3/20})$; $f(h) = 2/9$ if
   $h \in [3\sqrt{3/20}, 4\sqrt{3/20})$; $f(h) = 0$ if $h \ge 4\sqrt{3/20}$

6. 222

7. .85

9. $1 - (2\sigma^4/((n-1)\epsilon^2))$

11. 32; 63

12. 25,000

16. Yes, because given $\epsilon > 0$, for large enough n it is certain for $|Y_n|$ to be less than ϵ.

Section 6.2

1. $M_n(t) = (1/2 + 1/n) + (1/2 - 1/n)e^{t(1+1/n)}$; $F_n(x) = 0$ if $x < 0$, $F_n(x) = 1/2 + 1/n$ if $0 \le x < 1 + 1/n$, $F_n(x) = 1$ if $x \ge 1 + 1/n$

2. $M(t) = \frac{1}{8}e^{-t} + \frac{3}{4} + \frac{1}{8}e^t$; $F(x) = 1/8$ if $-1 \le x < 0$; $= 7/8$ if $0 \le x < 1$

4. .995

5. .68

6. .56

7. .0045

8. .5; 602

9. 12

10. .02

Chapter 7

Section 7.1

5. $\frac{40}{130}\widehat{p}_f + \frac{35}{130}\widehat{p}_{so} + \frac{30}{130}\widehat{p}_j + \frac{25}{130}\widehat{p}_{se}$

6. $n_i = N_i\sigma_i n / \sum_{j=1}^k N_j\sigma_j$

7. .000, .284, .251, .130, .659, .987, .043, .413, .049, .362; 0, 2, 2, 1, 6, 9, 0, 4, 0, 3

8. Primes: mean 2, variance 18/19; both odds and evens: mean 5/2, variance 75/76

11. Percentiles: 3.3, 5.04, 9.8, 12.1, 17.96; range: 20.5; nearly symmetric, with a little left skew

12. $\overline{Y} = a\overline{X} + b$; $S_y^2 = a^2 S_x^2$; $M_y = aM_x + b$; $IQ_y = aIQ_x$

14. Median: 3.3; mean: 4.88; variance: 29.6; interquartile range: 4.8 million. Skewed positively

15. Correlation matrix: $\begin{bmatrix} 1 & .908 & .929 \\ .908 & 1 & .926 \\ .929 & .926 & 1 \end{bmatrix}$

17. Covariance: 219.3; correlation: .601

18. Mean vector: $\begin{bmatrix} 2.24 \\ 2.36 \\ 2.44 \end{bmatrix}$; covariance matrix: $\begin{bmatrix} .087 & .062 & .085 \\ .062 & .128 & .141 \\ .085 & .141 & .206 \end{bmatrix}$

19. The expected value of a sample covariance matrix is the theoretical covariance matrix.

Section 7.2

2. The state populations are highly right skewed, appearing to be almost exponentially distributed.

3. The unemployment distributions for Europe and other countries are quite similar except for the three extremely large values in the European data.

4. Test 3 is clearly the best on average, and it is also not too variable, except for the lowest value, 62. Tests 1 and 2 are about the same on average, but test 1 is more spread out.

7. A multiple time series graph shows best the rise in NYSE share, the fall in ASE share, and the relative constancy of the other exchanges, which is the main story. A grouped column graph does fairly well at this too, but the distances that the eye must travel are problematic. The stacked bar is fine to show the NYSE progression, but the lack of a common baseline makes it harder to look at the ASE time series.

8. Average salary clearly increases as rank increases, but we also observe that the spread of salary increases with rank. There are also two rather large values, 60,900 and 63,700, for full professors. One could surmise that full professors are stars whose services are bid for, sometimes more successfully by larger and more prestigious universities, whereas there is more consistency to the market worth of professors of the lower ranks.

9. All of the media grow; the dollar amount of growth is similar, with newspapers highest and radio lowest. But the digital media more than doubles, and so is the fastest growing.

11. The main story is that although calculus II and calculus III are closely related, calculus I is not very well related to either of the others.

14. The normal quantile plot appears quite linear, with the exception of the data point corresponding to the largest number of runs

scored, which has a smaller value than would be expected under normality.

15. It is not easy to pick up nonlinearities with the eye for this data, but we would expect to see departures from linearity of some sort, since the underlying distribution is not normal. Some thought suggests that because of the nature of uniform quantiles, we should see for larger sample sizes a plot that is flat on both ends, and rapidly increasing in the middle.

16. A normal quantile plot is very nearly linear. With an estimated mean of 6.87 and standard deviation of .13, the probability that a bond rate of return exceeds 7% is about .16.

Section 7.3

2. There are 17 northern states (including Delaware and West Virginia) and 13 southern states (including Washington, D.C. and Maryland). Group the incomes into classes, such as $17,000 and below, $17,000–$20,000, etc., and plot superposed frequency distributions for both North and South on the same graph. The relative richness of the northern states shows up.

4. A sequence of 6 two-wedge pies, one for each month, can be drawn, but the viewer must compare angles in different pies. Superposed time series show the time trend much better without losing the comparison between sources of cost. Such a graph for all of the data shows remarkable seasonal regularity.

5. Very strikingly, the relative price changes come out nearly identical for the three conditions.

7. Sales volumes of all types except truck campers were down, with the biggest loser being 5th wheel trailers.

8. A direct time series graph of profit = revenue − cost shows clearly the progression of profit with time, and the main story is that there is much more variability in profit than would be expected from the given graph. Steep angles are the problem there. The largest profit is 8 in month 8.

10. Since pentagonal houses are two-dimensional plot symbols, they may exaggerate differences.

11. The rate of change is $\sec^2(\theta)d\theta$. This is very unstable near an angle of $\pi/2$.

13. Though the child figures are eye-catching, the recatangles surrounding them lead one to think more of comparing areas than heights. The visual

impression is that the change is more drastic than it is.

14. The residuals $Y_i - \widehat{Y}_i$ are widely spread.

15. Candidate A might draw a graph of B's support with 0 included; candidate B might draw a graph of A's support with 0 excluded.

Chapter 8

Section 8.1

1. $\overline{X}_n = \frac{1}{n} \sum_{i=1}^{n} X_i$ and $\overline{X}_2 = \frac{1}{2} \sum_{i=1}^{2} X_i$ are unbiased and have standard errors $\sqrt{\mu/n}$ and $\sqrt{\mu/2}$, respectively.

2. $2\overline{X}$ is unbiased for α and has standard error $\sqrt{\alpha/n}$; $\overline{X}/2$ is unbiased for β and has standard error $\beta/\sqrt{2n}$.

6. $Y_2 = .31$

9. $c = \frac{n-1}{n+1}$

10. $(n+1)X_{(1)}$. Its variance is $n\theta^2/(n+2)$, which is larger than that of $2\overline{X}$.

11. $c = \frac{\sqrt{n-1}}{\sqrt{2}} \cdot \frac{\Gamma(\frac{n-1}{2})}{\Gamma(\frac{n}{2})}$

12. μ/n; $\overline{X}$

13. The variance of S^2 is $2\sigma^4/(n-1)$, which does not achieve the R–C bound of $2\sigma^4/n$.

Section 8.2

1. 3

3. 5/8

4. $1/\overline{X}$; yes

5. (a) $1/\overline{X}$ (b) $2/\overline{X}$

7. $\sum_{i=1}^{n}(X_i - \mu)^2/n$

8. $-n/\sum_{i=1}^{n} \ln(X_i)$

9. $X_{(1)}$; no, it is not unbiased.

10. 58

12. $\left(\dfrac{n}{\sum_{i=1}^{n} t_i^{1.5}}\right)^{2/3}$

14. $\widehat{p}_1 = (\text{total \# yes's})/n$; $\widehat{p}_2 = (\text{total \# no's})/n$

17. (a) $\overline{X}$ (b) $\overline{X}$ and $\sum_{i=1}^{n} X_i^2/n - \overline{X}^2$

Section 8.3

5. $Z = \sum_{i=1}^{n} X_i$ is sufficient for λ.

7. $Z = \ln(\prod_{i=1}^{n} X_i)$ is sufficient for θ; the MLE is $-n/\ln(\prod_{i=1}^{n} X_i)$.

8. $Z = \sum_{i=1}^{n} \ln(1 + X_i)$ is sufficient for θ.

10. $Z = \sum_{i=1}^{n} T_i^{\beta}$ is sufficient for λ when β is known.

Section 8.4

1. [331.0, 336.6]

2. 3.6%

3. [4.63, 5.41]

4. [123, 606]; [82, 402]; [50, 247]. Since the intervals all overlap, there is no clear difference.

5. For accident group: [.335, .539]; for nonaccident group: [.237,.336]

8. (b) [242.7, 253.1]   (c) [369.9, 723.3]

10. [3.10, 4.72]

12. $[(\overline{X} + z_{\alpha/2}\sqrt{\overline{X}(\overline{X} - 1)/n})^{-1},$
$(\overline{X} - z_{\alpha/2}\sqrt{\overline{X}(\overline{X} - 1)/n})^{-1}]$

16. (b) $[\exp(-\chi_{.05}^2(2n)/(2\sum_{i=1}^{n} X_i),$
$\exp(-\chi_{.95}^2(2n)/(2\sum_{i=1}^{n} X_i)]$

17. 807

18. 271

Section 8.5

2. An approximate 95% confidence interval for the difference in proportions is [.001, .093]. Since 0 is not in the interval, we conclude that 8 is significantly more likely to occur on the 2nd than the 3rd draw.

3. [−.128, .008]

4. [−14.52, 5.82]

5. At the 95% level of confidence there is no significant difference.

7. For headaches, a 90% confidence interval is [−.021, .029], so there is no significant difference; for edema, a 90% confidence interval is [−.002, .028], so there is no significant difference, but since 0 is close to the edge, there is some indication that Cardizem increases frequency of edema.

8. [.72, 2.6]; yes

9. (a) [−21.2, −14.0]   (b) $x^* \approx 22$

11. (a) The confidence interval is [−12.2, 5.4]; there is no significant difference. (b) The confidence interval is [−14.8, −2.8]; test 3 seems higher than test 2.

12. A 90% confidence interval is [−12.26, 36.01], hence there is no significant difference.

13. [.63, 2.37]

14. An approximate 95% confidence interval for the difference in mean absences is [−2.2, −1.2]. Since 0 is not in the interval, we conclude that store 1 has significantly fewer absences on average than store 2.

Chapter 9

Section 9.1

2. Reject H_0; Type II error probability $\approx .14$

3. (a) Accept H_0; p-value $\approx .13$ (b) Critical region $Y =$ number of households in sample of two or fewer members > 15; exact power $= .242$

4. Accept H_0 at level .01; p-value $= .014$

5. Reject H_0 at level .05; type II error probability $= .20$

7. Do not reject H_0; i.e. there is insufficient evidence that the area will be annexed. Power function: $1 - \Phi((.616 - p)/\sqrt{p(1 - p)/50})$

10. Reject H_0 if $\overline{X} > .932$. Power function: $1 - \Phi(2.5(.932 - \mu))$

11. $\alpha = .216$; $\beta = .58$

12. $\sum X_i > 8$, level .068; .068; .41; .67

14. Reject H_0 at level .091. p-value .024

15. At exact level .02 with critical value 5, accept H_0; p-value $= .41$

16. Observed $z = 1.58$; accept H_0 at level .05

Section 9.2

1. (a) Observed $z = 3.91$; reject H_0 at level .01
   (b) Observed $t = 5.72$; reject H_0 at level .01.

2. .0002

4. $1 - \Phi(z_\alpha + \frac{\mu_0 - \mu}{\sigma/\sqrt{n}})$

5. At least 34

7. Observed $t = 5.89$; reject H_0 at level .025

8. (a) $[\overline{x} - z_\alpha \frac{\sigma}{\sqrt{n}}, \infty)$

9. $H_0: \mu = 26$ vs. $H_2: \mu > 26$; data $t = 2.49$, reject at level .01

10. Observed $t = .69$; accept H_0 at level .05; p-value roughly .3, in view of the trend of percentiles in the $t(20)$ table

11. Data $t = .56$; accept H_0

12. $t_\alpha(n - 1)S/\sqrt{n}$

14. Reject H_0 if $W \leq -39$; observed $w = -39$; hence reject H_0

15. $f(w) = 1/32$ if $w = \pm 15, \pm 13, \pm 11$; $f(w) = 2/32$ if $w = \pm 9, \pm 7$; $f(w) = 3/32$ if $w = \pm 5, \pm 3, \pm 1$. Reject H_0 if $W \leq -11$ or $W \geq 11$, level .1875; accept H_0 for the given data

16. Observed $w = -131$, $z = -2.64$; hence reject H_0

17. Observed $w = 145$, $z = 2.91$; hence reject H_0 for the Wilcoxon test; observed $t = 3.62$, so also reject H_0 for the t-test

Section 9.3

1. For the mean test, $t = .50$, so accept H_0. For the variance test, $U = 31.64$, so H_0 can be rejected even at level .01. The power is at least .99.

2. The observed value of the test statistic is 76.55, which is significant at a level even lower than .01.

4. $K(\sigma^2) = 1 - F(\sigma_0^2 \cdot \chi_\alpha^2(n-1)/\sigma^2)$, where F is the c.d.f. of the $\chi^2(n-1)$ distribution.

5. For the ventilation variable, $f = 9.79$, for feeding, $f = 11.1$, and for charges, $f = 22.1$. Since the critical value for level .05 is 2.20, reject H_0 in all three cases.

6. Since the observed f value is .416, and the two critical values are .42 and 2.29, reject H_0.

7. Since the observed f value is .18, and the one-sided critical value is .42, reject H_0.

9. Type II error probability: $F(f_\alpha(m-1, n-1)/r)$, where F is the c.d.f. of the $F(m-1, n-1)$ distribution

12. $f(x) = 4/10$ if $x = 5$; $f(x) = 2/10$ if $x = 6, 9$; $f(x) = 1/10$ if $x = 2, 8$

13. Rejection region $M \leq 5$ has p-value 3/35

14. Value of M: 1533; value of Z: -2.38; reject H_0 at level .05

15. The value of M is 378.75, and the associated Z value is -2.27, so reject H_0 at level .05 in favor of the hypothesis that the National League data are less disperse.

Section 9.4

1. Using the normal test, $z = -3.67$, reject H_0; using the t-test, $t = -3.66$, reject H_0.

2. Sample $z = -1.46$, so accept H_0

4. Equality of variance test passes; sample $t = .44$ so accept H_0.

6. Using the equal, unknown variance form of the t-test, $t = 4.01$, so reject H_0 at level .01.

7. $\Phi(-z_{\alpha/2} - \frac{d}{\sqrt{\sigma^2/m + \sigma^2/n}}) + 1 - \Phi(z_{\alpha/2} - \frac{d}{\sqrt{\sigma^2/m + \sigma^2/n}})$

8. $z = -.43$, accept H_0

9. The value of the paired t-statistic is $t = 1.23$, so accept H_0 at level .05.

11. 428

12. Sample $t = (\bar{x} - \bar{y} - (-15))/s_d/\sqrt{n} = -2.046$ so accept H_0 at level .01.

15. Perform signed rank test on differences $X_i - Y_i + 15$; $z = -1.58$; p-value: .057

16. (b) Critical region $W \geq 21$ has exact level 2/35.
    (c) Critical region $W \leq 11$ has exact level 2/35, accept H_0 since the observed $w = 12$.

17. Critical region $W \leq 17$ or $W \geq 38$, where W is the sum of the ranks of the New England observations. Observed $w = 37$. Cannot reject H_0 at exact level .032.

20. Sample $w = 324$; $z = -4.01$; reject H_0

22. Sample $w = 426.5$; $z = -1.15$; accept H_0

Section 9.5

3. Reject H_0 if $\sum(X_i - \mu)^2/\sigma_0^2 < c$

5. Reject H_0 if $\overline{X} < d$

6. Reject H_0 if $\overline{X} < d_1$ or $\overline{X} > d_2$

Chapter 10

Section 10.1

1. $\widehat{b}_0 = 1.12$, $\widehat{b}_1 = .192$, estimated tourism in 1993 $= 2.85$ million, over 3 million in 1994

3. $y = 43375.2 - 2.86981x$; maximum production at about 7557 thousand cows

4. Runs $= 495.6 + 1.466$ hr, runs $= -647 + 5142$ avg; prefer the 10 pts. in average

5. Game 2 = 79.19 + .456 game 1; estimate of $\sigma^2 = 125.075$; predicted score 125

6. Before removing the point, $y = -9.13 + 1.03x$, and $R^2 = .94$; after removing it, $y = 2.20 + .046x$, and $R^2 = .006$.

9. Test 3 = 62.1 + .319 test 1, $R^2 = 25\%$, predicted test 3 = 84.4

10. $\widehat{b}_1 = \sum_{i=1}^{n} x_i Y_i / \sum_{i=1}^{n} x_i^2$

15. .22

16. 2.75

Section 10.2

1. Runs = −516 +1.15 hr + 4103 avg; Runs = 364 +1.91 hr + .631 sb; Runs = −674 +5325 avg −.163 sb; Runs = −425 +1.45 hr + 3436 avg + .352 sb; Predicted runs 659.3.

3. (a) $R^2 = .03$ for linear model, $R^2 = .14$ for quadratic model
(b) $R^2 = .04$ for linear model, $R^2 = .05$ for quadratic model
(c) All variables: $glucose = 132.65 + .015 \cdot bb + .405 \cdot ab + .318 \cdot bl - 1.026 \cdot al$ with an R^2 of 58%; using after-meal variables only, $glucose = 158.28 + .541 \cdot ab - 1.174 \cdot al$ with an R^2 of 54%.

4. (a) No variable is very good, but years of experience is best; predicted score: 289. (b) Again, no variable is very good, but years of experience is best; predicted score: 277.

6. (a) $Activity = 53.98 - 1.02 \cdot age$, $R^2 = .09$; $activity = 39.15 + .149 \cdot twin1$, $R^2 = .04$
(b) $Activity = 46.465 - 1.03 \cdot age + .15 \cdot twin1$, $R^2 = .13$
(c) $Activity = 32.04 - .657 \cdot age + 43 \cdot twin1$, $R^2 = .24$

9. R^2 for linear model: .73; R^2 for quadratic model: .89. Regression equation $workforce = 23.29 + .126 \cdot time - .014 \cdot time^2$ applied at time 20 gives about 20.2

10. Linear model: $R^2 = 82\%$; quadratic: $R^2 = 88\%$; $cost = -23.8 + 1.24$ items $-.00162$ items^2

11. (a) $Revenue = -.342 + .401 \cdot time - .011 \cdot time^2$; predicted revenue at time 16: 3.14
(b) R^2 for quadratic: 99.2%; R^2 for cubic: 99.3%

Section 10.3

1. All coefficients in the cubic model are highly significant; confidence interval: [.232, .316].

2. *p*-value is .011, so there is significant evidence of a relationship; [15.6, 342].

3. Accept hypothesis that constant term is 0 at level .05; reject hypothesis that slope term is zero at level .05; [0, 33.3].

6. .225

7. Prediction interval: [2.41, 3.28]; confidence interval: [.139, .245]

8. (a) HR and avg are significant at very low levels; sb is not significant at level .05.
(b) Predicition interval for 100 home runs: [572.3, 712.3]
(c) Prediction interval for .275 avg: [692.4, 841.2]
(d) Confidence interval: [425.71, 565.53]

9. For the quadratic model, R^2 increases by only about 5 pts., and no coefficient is significant at a level less than .1. Confidence interval: [34563, 52188]

10. (a) No; [72.1, 96.8] (b) Yes; [73.2, 96.3]

11. (a) For reading, no predictors are significant at a level less than .2; for math, only years of experience is significant at level .167. (b) [251, 324]

Section 10.4

1. (103, 132) is an outlying data point. Normal scores plot is curved and histogram is asymmetrical.

2. The main problem is that two observations, one with very low average and one with high average, exert a lot of influence on the regression coefficients and make themselves felt in the shape of residual vs. fit graph.

3. All of the diagnostics turn out fairly well.

4. A linear regression of thefts on population omitting the Los Angeles point gives $R^2 = 98.8\%$ with reasonable graphs.

5. (a) $Price = -1286 + 12.5 \cdot age + 83.1 \cdot numbid$, all coefficients very significant, $R^2 = 87.8\%$
(b) There is an inverted U-shaped pattern.
(c) $Price = -2060 + 12.0 \cdot age + 268 \cdot numbid - 9.29 \cdot numbid^2$, all coefficients very significant, $R^2 = 91.6\%$
(d) Price$^{1/2}$ regressed on age, numbid, and numbid^2 gives an R^2 value of 93.2% and good graphs.

6. A transformation $y^{7/8}$ helps, although it appears that the original data are basically linear, but with a nonremovable heteroscedasticity.

7. $r = 1/2$

8. Although the linear model has an R^2 of .895 and very significant slope coefficient, the residual vs. fits graph is very parabolic and the residual histogram very left-skewed. A quadratic model improves the former, but not the latter. However it is a small data set.

9. All diagnostics seem good.

10. The histogram is not good, and the time series plot is somewhat suspicious. The transformed data have an R^2 of over 95%, and except for one strange data point, the diagnostics check out well.

12. Log of amount vs. time should be linear.

13. Transform all variables by logarithms.

14. $Y = 5 + \epsilon$. Lagged residuals seem to correlate with residuals.

15. No strong evidence, high variability

Section 10.5

1. Calc I and Calc II ($r = .592$), Calc I and Calc III ($r = .651$), and Calc II and Calc III ($r = .874$) are significantly correlated at level .05.

2. $r = .257$, not significant at level .05 (even for a one-sided test)

3. Test 1 vs. test 2: $t = 1.57$, $z = 1.45$; test 1 vs. test 3: $t = 2.0$, $z = 1.82$; test 2 vs. test 3: $t = 2.52$, $z = 2.24$

6. The observed correlation is $r = .202$, which produces a t-value of .714. This is not significant at level .1.

8. $r = .611$, $z = 3.56$, and for a two-sided test the p-value is about .0004.

9. $r = .360$, $t = 1.34$, $z = 1.25$. Neither the t-test nor the z-test shows significant correlation at a level less than .1.

10. Aggression and activity: $r = .171$; not significant for a two-sided level .1 test. Aggression and sleep: $r = .250$; not significant for a two-sided level .1 test. Activity and sleep: $r = .193$; not significant for a two-sided level .1 test

11. .41

12. [.98, 1]

14. (b) .786   (c) Values: 1, .6, −.6, −1 all have probability 1/24; values: .2, 0, −.2 all have probability 2/24; values: .8, −.8 each have probability 3/24; values .4, −.4 each have probability 4/24.

Chapter 11

Section 11.1

4. Each field should be divided into four plots. Assign a brand randomly to those plots.

7. $\frac{n!}{n_1! n_2! \cdots n_k!}$

9. $N(\mu + \alpha_i, \sigma^2/n)$

10. $\alpha_i - \alpha_k$; $2\sigma^2/n$

11. $\mu + \beta_j$; σ^2/an

13. $N(\mu, \sigma_a^2 + \sigma^2/n)$

Section 11.2

1. Observed $f = 14.55$; hence the mean delay times are in fact different at levels even lower than .001. The *AMS Bulletin* stands out as having the smallest delays.

2. Observed $f = 1.18$; hence accept H_0 at level .05.

3. Observed $f = 15.5$; hence reject H_0 at levels lower than .001. The South seems much lower.

6. Observed $f = 1.13$; hence accept H_0.

7. Observed $f = 216.1$; hence reject H_0 at level .01.

9. Observed $f = .09$; hence accept H_0 at any reasonable level.

10. For completely randomized model, $f = 7.52$, p-value .003; for randomized block model, $f = 12.35$, p-value .001.

11. Observed $f = 47.4$; hence reject H_0 at level .05.

12. Observed $f = 2.10$; hence (barely) accept H_0 at level .05.

16. Observed $f = .46$; hence accept H_0 at any reasonable level.

17. Randomized block, $f = 9.6$, p-value .001; completely randomized, $f = .54$, p-value .59.

25. Observed $f = 14.69$; hence still reject H_0 at level .001.

27. $(MSF - MSE)/n$

Section 11.3

1. Observed f for gas grade: 3.61; hence reject H_0 at level .05. Observed f for speed: 35.47; hence reject H_0 at very low levels. Observed interaction $f = .36$; accept H_0.

2. Observed f for variety: 0.83; hence accept H_0 at reasonable levels. Observed f for herbicide: 32.43; hence reject H_0 at level .05. Observed interaction $f = 1.02$ has p-value .388.

4. Observed f for interviewer: 18.6; hence reject H_0 at a level even lower than .01. Observed f for gender: 9.9; hence reject H_0 at level .01. Observed interaction $f = 7.4$; hence reject H_0 at level .025.

6. Observed f for drugs: .80; hence accept H_0 at reasonable levels. Observed f for pairing: 34.42; hence reject H_0 at very low levels. Observed interaction $f = .06$ has p-value .943.

8. Observed f-values for grade, drug type, and interaction are 30.57, 300.53, and 25.0, respectively. Reject H_0 for all at levels below .001.

10. Observed f for county: .25; hence accept H_0 at any resonable level. Observed f for percentage poor: 3.93 with p-value .083. Observed interaction $f = .11$; hence accept H_0.

16. Observed f for school: 43.9, highly significant; observed f for course: 1.39, not significant; observed f for interaction: 3.82, barely significant at level .05.

17. Observed f for speed: 68.1; hence reject H_0 at very low levels. Observed f for machine: 2.21; hence accept H_0 at level .05. Observed interaction $f = 1.71$ has p-value .187.

18. Observed f for region: 1.68; not significant at level .05. Observed f for worker type: .007; not significant. Observed f for interaction: 9.41; significant at level .01.

19. Observed f-values for temperature, diet, and interaction are 2.73, .608, and 1.41, respectively. None are significant at level .05. If temperature is considered fixed, the diet f value changes to .856, which is still not significant at level .05.

20. 0

21. σ^2

Chapter 12

Section 12.1

1. The chi-square statistic Q takes the value 11.2. Since $\chi^2_{.025}(3) = 9.348$, reject H_0 at level .025.

2. The chi-square statistic Q takes the value 283; hence the differences are highly significant. They arise mostly in the applied sciences and engineering.

3. The value of Q is 16.5, and since $\chi^2_{.01}(5) = 15.09$, H_0 is rejected at this level.

4. The chi-square statistic Q takes the value 16.49. Since $\chi^2_{.01}(3) = 11.3$, reject H_0 at level .01.

5. The chi-square statistic Q takes the value 54816, so reject H_0 at levels much lower than .001. The biggest contributions to Q are made by the weekend categories.

7. The observed $q = 56.2$, which is highly significant.

8. Combine categories 0 and 1, and categories 4 and 5. The chi-square statistic Q takes the value 11.5. Since $\chi^2_{.05}(3) = 7.815$, reject H_0 at level .05.

9. Since $q = 20.53$ and $\chi^2_{.01}(3) = 11.3$, reject H_0 at level .01.

11. The observed $q = 1.19$, which is not significant at any reasonable level.

12. The observed $q = 178$, which is highly significant.

13. The observed $q = 450$, which is highly significant.

14. The chi-square statistic Q takes the value 16.3. Since $\chi^2_{.01}(5) = 15.09$, reject H_0 at level .01 as well as at level .05.

Section 12.2

1. The chi-square statistic Q takes the value 16.98. Since $\chi^2_{.01}(4) = 13.28$, reject H_0 at level .01.

2. Principally because of the very low expected category frequency in the 2 or more category as well as an imbalance in the first two categories, the chi-square statistic Q takes the value 117; hence reject H_0 at very low levels.

3. The observed value $q = .62$. Accept H_0 at any reasonable level.

4. The chi-square statistic Q roughly takes the value 3.4. Since $\chi^2_{.05}(1) = 3.84$, accept H_0 at level .05.

5. The observed value $q = 4.39$. Since $\chi^2_{.05}(2) = 5.99$, accept H_0 at level .05.

6. The MLE $\hat{r} = 7/12$. The chi-square statistic Q takes the value 11.9. Since $\chi^2_{.01}(1) = 6.635$, reject H_0 at level .01.

7. Hypothesize the $b(4, p)$ distribution. The chi-square statistic Q takes the value .282; hence accept H_0 at any reasonable level.

8. The chi-square statistic Q takes the value 4.0. Since $\chi^2_{.05}(1) = 3.841$, reject H_0 at level .05.

13. The K–S statistic takes the value .2549 and the table gives $d_{.05}(18) = .31$; hence accept H_0 at level .05.

14. The K–S statistic takes the value .117 and the table gives $d_{.05}(15) = .34$; hence accept H_0 at level .05.

15. The K–S statistic takes the value .27 and the table gives $d_{.05}(10) = .41$; hence accept H_0 at level .05.

16. The K–S statistic takes the value .145 and the table gives $d_{.1}(20) = .264$; hence accept H_0 at level .1.

Section 12.3

1. $q = 6.65$. Since $\chi^2_{.05}(9) = 16.92$, accept H_0 at level .05.

2. $q = 14.9$. Since $\chi^2_{.01}(2) = 9.21$, reject H_0 at level .01.

3. Using ACT categories of at least 26 and less than 26, and SAT categories at least 560 and less than 560, the chi-square statistic Q takes the value .04. So, accept H_0 at any reasonable level.

4. $q = 50$. Since $\chi^2_{.05}(6) = 12.59$, reject H_0 at level .05.

6. $q = 24.6$. Since $\chi^2_{.05}(5) = 11.07$, reject H_0 at level .05.

7. $q = 4.38$. Since $\chi^2_{.1}(2) = 4.61$, accept H_0 at level .1.

9. $q = 80.5$; hence reject H_0 at very low levels. There seem to be fewer mathematicians hired in industry, and more statisticians and operations researchers, than expected.

11. Using the helmet status and whether or not the person died as the two factors, a test for independence gives observed $q = 1.8$. Since $\chi^2_{.1}(1) = 2.71$, accept H_0 at level .1.

12. The chi-square statistic Q takes the value 7. Since $\chi^2_{.05}(2) = 5.991$, reject H_0 at level .05.

13. The observed number of runs is $r = 5$. The rejection region $R \leq 3$ or $R \geq 11$ has level about .026, and leads to acceptance of the null hypothesis.

14. The observed number of runs is $r = 6$. The rejection region $R \leq 4$ or $R \geq 12$ has level about .044, and leads to acceptance of the null hypothesis.

15. For the first sequence $R = 6$, and since the critical region $R \leq 3$ has level about .07, we accept H_0 at that level and higher levels. For the second sequence, $R = 8$, and the critical region $R \geq 8$ has level .024; hence reject H_0 at that level

16. The number of runs is 17, the value of the approximate Z-statistic is 2.44, and the p-value is about .014.

17. The observed value of R is 6, and the critical region $R \leq 4$ has level .109, so accept H_0 at this level.

INDEX